T0185695

Lecture Notes in Computer Science 13672

More information about this series at https://link.springer.com/bookseries/558

Shai Avidan · Gabriel Brostow ·
Moustapha Cissé · Giovanni Maria Farinella ·
Tal Hassner (Eds.)

Computer Vision – ECCV 2022

17th European Conference
Tel Aviv, Israel, October 23–27, 2022
Proceedings, Part XII

 Springer

Editors
Shai Avidan
Tel Aviv University
Tel Aviv, Israel

Gabriel Brostow 🆔
University College London
London, UK

Moustapha Cissé
Google AI
Accra, Ghana

Giovanni Maria Farinella 🆔
University of Catania
Catania, Italy

Tal Hassner 🆔
Facebook (United States)
Menlo Park, CA, USA

ISSN 0302-9743   ISSN 1611-3349 (electronic)
Lecture Notes in Computer Science
ISBN 978-3-031-19774-1   ISBN 978-3-031-19775-8 (eBook)
https://doi.org/10.1007/978-3-031-19775-8

This Springer imprint is published by the registered company Springer Nature Switzerland AG
The registered company address is: Gewerbestrasse 11, 6330 Cham, Switzerland

Foreword

Organizing the European Conference on Computer Vision (ECCV 2022) in Tel-Aviv during a global pandemic was no easy feat. The uncertainty level was extremely high, and decisions had to be postponed to the last minute. Still, we managed to plan things just in time for ECCV 2022 to be held in person. Participation in physical events is crucial to stimulating collaborations and nurturing the culture of the Computer Vision community.

There were many people who worked hard to ensure attendees enjoyed the best science at the 16th edition of ECCV. We are grateful to the Program Chairs Gabriel Brostow and Tal Hassner, who went above and beyond to ensure the ECCV reviewing process ran smoothly. The scientific program includes dozens of workshops and tutorials in addition to the main conference and we would like to thank Leonid Karlinsky and Tomer Michaeli for their hard work. Finally, special thanks to the web chairs Lorenzo Baraldi and Kosta Derpanis, who put in extra hours to transfer information fast and efficiently to the ECCV community.

We would like to express gratitude to our generous sponsors and the Industry Chairs, Dimosthenis Karatzas and Chen Sagiv, who oversaw industry relations and proposed new ways for academia-industry collaboration and technology transfer. It's great to see so much industrial interest in what we're doing!

Authors' draft versions of the papers appeared online with open access on both the Computer Vision Foundation (CVF) and the European Computer Vision Association (ECVA) websites as with previous ECCVs. Springer, the publisher of the proceedings, has arranged for archival publication. The final version of the papers is hosted by SpringerLink, with active references and supplementary materials. It benefits all potential readers that we offer both a free and citeable version for all researchers, as well as an authoritative, citeable version for SpringerLink readers. Our thanks go to Ronan Nugent from Springer, who helped us negotiate this agreement. Last but not least, we wish to thank Eric Mortensen, our publication chair, whose expertise made the process smooth.

October 2022

Rita Cucchiara
Jiří Matas
Amnon Shashua
Lihi Zelnik-Manor

Preface

Welcome to the proceedings of the European Conference on Computer Vision (ECCV 2022). This was a hybrid edition of ECCV as we made our way out of the COVID-19 pandemic. The conference received 5804 valid paper submissions, compared to 5150 submissions to ECCV 2020 (a 12.7% increase) and 2439 in ECCV 2018. 1645 submissions were accepted for publication (28%) and, of those, 157 (2.7% overall) as orals.

846 of the submissions were desk-rejected for various reasons. Many of them because they revealed author identity, thus violating the double-blind policy. This violation came in many forms: some had author names with the title, others added acknowledgments to specific grants, yet others had links to their github account where their name was visible. Tampering with the LaTeX template was another reason for automatic desk rejection.

ECCV 2022 used the traditional CMT system to manage the entire double-blind reviewing process. Authors did not know the names of the reviewers and vice versa. Each paper received at least 3 reviews (except 6 papers that received only 2 reviews), totalling more than 15,000 reviews.

Handling the review process at this scale was a significant challenge. To ensure that each submission received as fair and high-quality reviews as possible, we recruited more than 4719 reviewers (in the end, 4719 reviewers did at least one review). Similarly we recruited more than 276 area chairs (eventually, only 276 area chairs handled a batch of papers). The area chairs were selected based on their technical expertise and reputation, largely among people who served as area chairs in previous top computer vision and machine learning conferences (ECCV, ICCV, CVPR, NeurIPS, etc.).

Reviewers were similarly invited from previous conferences, and also from the pool of authors. We also encouraged experienced area chairs to suggest additional chairs and reviewers in the initial phase of recruiting. The median reviewer load was five papers per reviewer, while the average load was about four papers, because of the emergency reviewers. The area chair load was 35 papers, on average.

Conflicts of interest between authors, area chairs, and reviewers were handled largely automatically by the CMT platform, with some manual help from the Program Chairs. Reviewers were allowed to describe themselves as senior reviewer (load of 8 papers to review) or junior reviewers (load of 4 papers). Papers were matched to area chairs based on a subject-area affinity score computed in CMT and an affinity score computed by the Toronto Paper Matching System (TPMS). TPMS is based on the paper's full text. An area chair handling each submission would bid for preferred expert reviewers, and we balanced load and prevented conflicts.

The assignment of submissions to area chairs was relatively smooth, as was the assignment of submissions to reviewers. A small percentage of reviewers were not happy with their assignments in terms of subjects and self-reported expertise. This is an area for improvement, although it's interesting that many of these cases were reviewers hand-picked by AC's. We made a later round of reviewer recruiting, targeted at the list of authors of papers submitted to the conference, and had an excellent response which helped provide enough emergency reviewers. In the end, all but six papers received at least 3 reviews.

The challenges of the reviewing process are in line with past experiences at ECCV 2020. As the community grows, and the number of submissions increases, it becomes ever more challenging to recruit enough reviewers and ensure a high enough quality of reviews. Enlisting authors by default as reviewers might be one step to address this challenge.

Authors were given a week to rebut the initial reviews, and address reviewers' concerns. Each rebuttal was limited to a single pdf page with a fixed template.

The Area Chairs then led discussions with the reviewers on the merits of each submission. The goal was to reach consensus, but, ultimately, it was up to the Area Chair to make a decision. The decision was then discussed with a buddy Area Chair to make sure decisions were fair and informative. The entire process was conducted virtually with no in-person meetings taking place.

The Program Chairs were informed in cases where the Area Chairs overturned a decisive consensus reached by the reviewers, and pushed for the meta-reviews to contain details that explained the reasoning for such decisions. Obviously these were the most contentious cases, where reviewer inexperience was the most common reported factor.

Once the list of accepted papers was finalized and released, we went through the laborious process of plagiarism (including self-plagiarism) detection. A total of 4 accepted papers were rejected because of that.

Finally, we would like to thank our Technical Program Chair, Pavel Lifshits, who did tremendous work behind the scenes, and we thank the tireless CMT team.

October 2022

Gabriel Brostow
Giovanni Maria Farinella
Moustapha Cissé
Shai Avidan
Tal Hassner

Organization

General Chairs

Rita Cucchiara   University of Modena and Reggio Emilia, Italy
Jiří Matas   Czech Technical University in Prague, Czech Republic
Amnon Shashua   Hebrew University of Jerusalem, Israel
Lihi Zelnik-Manor   Technion – Israel Institute of Technology, Israel

Program Chairs

Shai Avidan   Tel-Aviv University, Israel
Gabriel Brostow   University College London, UK
Moustapha Cissé   Google AI, Ghana
Giovanni Maria Farinella   University of Catania, Italy
Tal Hassner   Facebook AI, USA

Program Technical Chair

Pavel Lifshits   Technion – Israel Institute of Technology, Israel

Workshops Chairs

Leonid Karlinsky   IBM Research, Israel
Tomer Michaeli   Technion – Israel Institute of Technology, Israel
Ko Nishino   Kyoto University, Japan

Tutorial Chairs

Thomas Pock   Graz University of Technology, Austria
Natalia Neverova   Facebook AI Research, UK

Demo Chair

Bohyung Han   Seoul National University, Korea

Social and Student Activities Chairs

Tatiana Tommasi　　　　　　　Italian Institute of Technology, Italy
Sagie Benaim　　　　　　　　University of Copenhagen, Denmark

Diversity and Inclusion Chairs

Xi Yin　　　　　　　　　　　Facebook AI Research, USA
Bryan Russell　　　　　　　　Adobe, USA

Communications Chairs

Lorenzo Baraldi　　　　　　　University of Modena and Reggio Emilia, Italy
Kosta Derpanis　　　　　　　York University & Samsung AI Centre Toronto,
　　　　　　　　　　　　　　　Canada

Industrial Liaison Chairs

Dimosthenis Karatzas　　　　　Universitat Autònoma de Barcelona, Spain
Chen Sagiv　　　　　　　　　SagivTech, Israel

Finance Chair

Gerard Medioni　　　　　　　University of Southern California & Amazon,
　　　　　　　　　　　　　　　USA

Publication Chair

Eric Mortensen　　　　　　　MiCROTEC, USA

Area Chairs

Lourdes Agapito　　　　　　　University College London, UK
Zeynep Akata　　　　　　　　University of Tübingen, Germany
Naveed Akhtar　　　　　　　　University of Western Australia, Australia
Karteek Alahari　　　　　　　Inria Grenoble Rhône-Alpes, France
Alexandre Alahi　　　　　　　École polytechnique fédérale de Lausanne,
　　　　　　　　　　　　　　　Switzerland
Pablo Arbelaez　　　　　　　Universidad de Los Andes, Columbia
Antonis A. Argyros　　　　　　University of Crete & Foundation for Research
　　　　　　　　　　　　　　　and Technology-Hellas, Crete
Yuki M. Asano　　　　　　　　University of Amsterdam, The Netherlands
Kalle Åström　　　　　　　　Lund University, Sweden
Hadar Averbuch-Elor　　　　　Cornell University, USA

Hossein Azizpour	KTH Royal Institute of Technology, Sweden
Vineeth N. Balasubramanian	Indian Institute of Technology, Hyderabad, India
Lamberto Ballan	University of Padova, Italy
Adrien Bartoli	Université Clermont Auvergne, France
Horst Bischof	Graz University of Technology, Austria
Matthew B. Blaschko	KU Leuven, Belgium
Federica Bogo	Meta Reality Labs Research, Switzerland
Katherine Bouman	California Institute of Technology, USA
Edmond Boyer	Inria Grenoble Rhône-Alpes, France
Michael S. Brown	York University, Canada
Vittorio Caggiano	Meta AI Research, USA
Neill Campbell	University of Bath, UK
Octavia Camps	Northeastern University, USA
Duygu Ceylan	Adobe Research, USA
Ayan Chakrabarti	Google Research, USA
Tat-Jen Cham	Nanyang Technological University, Singapore
Antoni Chan	City University of Hong Kong, Hong Kong, China
Manmohan Chandraker	NEC Labs America, USA
Xinlei Chen	Facebook AI Research, USA
Xilin Chen	Institute of Computing Technology, Chinese Academy of Sciences, China
Dongdong Chen	Microsoft Cloud AI, USA
Chen Chen	University of Central Florida, USA
Ondrej Chum	Vision Recognition Group, Czech Technical University in Prague, Czech Republic
John Collomosse	Adobe Research & University of Surrey, UK
Camille Couprie	Facebook, France
David Crandall	Indiana University, USA
Daniel Cremers	Technical University of Munich, Germany
Marco Cristani	University of Verona, Italy
Canton Cristian	Facebook AI Research, USA
Dengxin Dai	ETH Zurich, Switzerland
Dima Damen	University of Bristol, UK
Kostas Daniilidis	University of Pennsylvania, USA
Trevor Darrell	University of California, Berkeley, USA
Andrew Davison	Imperial College London, UK
Tali Dekel	Weizmann Institute of Science, Israel
Alessio Del Bue	Istituto Italiano di Tecnologia, Italy
Weihong Deng	Beijing University of Posts and Telecommunications, China
Konstantinos Derpanis	Ryerson University, Canada
Carl Doersch	DeepMind, UK

Matthijs Douze | Facebook AI Research, USA
Mohamed Elhoseiny | King Abdullah University of Science and Technology, Saudi Arabia
Sergio Escalera | University of Barcelona, Spain
Yi Fang | New York University, USA
Ryan Farrell | Brigham Young University, USA
Alireza Fathi | Google, USA
Christoph Feichtenhofer | Facebook AI Research, USA
Basura Fernando | Agency for Science, Technology and Research (A*STAR), Singapore
Vittorio Ferrari | Google Research, Switzerland
Andrew W. Fitzgibbon | Graphcore, UK
David J. Fleet | University of Toronto, Canada
David Forsyth | University of Illinois at Urbana-Champaign, USA
David Fouhey | University of Michigan, USA
Katerina Fragkiadaki | Carnegie Mellon University, USA
Friedrich Fraundorfer | Graz University of Technology, Austria
Oren Freifeld | Ben-Gurion University, Israel
Thomas Funkhouser | Google Research & Princeton University, USA
Yasutaka Furukawa | Simon Fraser University, Canada
Fabio Galasso | Sapienza University of Rome, Italy
Jürgen Gall | University of Bonn, Germany
Chuang Gan | Massachusetts Institute of Technology, USA
Zhe Gan | Microsoft, USA
Animesh Garg | University of Toronto, Vector Institute, Nvidia, Canada
Efstratios Gavves | University of Amsterdam, The Netherlands
Peter Gehler | Amazon, Germany
Theo Gevers | University of Amsterdam, The Netherlands
Bernard Ghanem | King Abdullah University of Science and Technology, Saudi Arabia
Ross B. Girshick | Facebook AI Research, USA
Georgia Gkioxari | Facebook AI Research, USA
Albert Gordo | Facebook, USA
Stephen Gould | Australian National University, Australia
Venu Madhav Govindu | Indian Institute of Science, India
Kristen Grauman | Facebook AI Research & UT Austin, USA
Abhinav Gupta | Carnegie Mellon University & Facebook AI Research, USA
Mohit Gupta | University of Wisconsin-Madison, USA
Hu Han | Institute of Computing Technology, Chinese Academy of Sciences, China

Bohyung Han	Seoul National University, Korea
Tian Han	Stevens Institute of Technology, USA
Emily Hand	University of Nevada, Reno, USA
Bharath Hariharan	Cornell University, USA
Ran He	Institute of Automation, Chinese Academy of Sciences, China
Otmar Hilliges	ETH Zurich, Switzerland
Adrian Hilton	University of Surrey, UK
Minh Hoai	Stony Brook University, USA
Yedid Hoshen	Hebrew University of Jerusalem, Israel
Timothy Hospedales	University of Edinburgh, UK
Gang Hua	Wormpex AI Research, USA
Di Huang	Beihang University, China
Jing Huang	Facebook, USA
Jia-Bin Huang	Facebook, USA
Nathan Jacobs	Washington University in St. Louis, USA
C.V. Jawahar	International Institute of Information Technology, Hyderabad, India
Herve Jegou	Facebook AI Research, France
Neel Joshi	Microsoft Research, USA
Armand Joulin	Facebook AI Research, France
Frederic Jurie	University of Caen Normandie, France
Fredrik Kahl	Chalmers University of Technology, Sweden
Yannis Kalantidis	NAVER LABS Europe, France
Evangelos Kalogerakis	University of Massachusetts, Amherst, USA
Sing Bing Kang	Zillow Group, USA
Yosi Keller	Bar Ilan University, Israel
Margret Keuper	University of Mannheim, Germany
Tae-Kyun Kim	Imperial College London, UK
Benjamin Kimia	Brown University, USA
Alexander Kirillov	Facebook AI Research, USA
Kris Kitani	Carnegie Mellon University, USA
Iasonas Kokkinos	Snap Inc. & University College London, UK
Vladlen Koltun	Apple, USA
Nikos Komodakis	University of Crete, Crete
Piotr Koniusz	Australian National University, Australia
Philipp Kraehenbuehl	University of Texas at Austin, USA
Dilip Krishnan	Google, USA
Ajay Kumar	Hong Kong Polytechnic University, Hong Kong, China
Junseok Kwon	Chung-Ang University, Korea
Jean-François Lalonde	Université Laval, Canada

Ivan Laptev	Inria Paris, France
Laura Leal-Taixé	Technical University of Munich, Germany
Erik Learned-Miller	University of Massachusetts, Amherst, USA
Gim Hee Lee	National University of Singapore, Singapore
Seungyong Lee	Pohang University of Science and Technology, Korea
Zhen Lei	Institute of Automation, Chinese Academy of Sciences, China
Bastian Leibe	RWTH Aachen University, Germany
Hongdong Li	Australian National University, Australia
Fuxin Li	Oregon State University, USA
Bo Li	University of Illinois at Urbana-Champaign, USA
Yin Li	University of Wisconsin-Madison, USA
Ser-Nam Lim	Meta AI Research, USA
Joseph Lim	University of Southern California, USA
Stephen Lin	Microsoft Research Asia, China
Dahua Lin	The Chinese University of Hong Kong, Hong Kong, China
Si Liu	Beihang University, China
Xiaoming Liu	Michigan State University, USA
Ce Liu	Microsoft, USA
Zicheng Liu	Microsoft, USA
Yanxi Liu	Pennsylvania State University, USA
Feng Liu	Portland State University, USA
Yebin Liu	Tsinghua University, China
Chen Change Loy	Nanyang Technological University, Singapore
Huchuan Lu	Dalian University of Technology, China
Cewu Lu	Shanghai Jiao Tong University, China
Oisin Mac Aodha	University of Edinburgh, UK
Dhruv Mahajan	Facebook, USA
Subhransu Maji	University of Massachusetts, Amherst, USA
Atsuto Maki	KTH Royal Institute of Technology, Sweden
Arun Mallya	NVIDIA, USA
R. Manmatha	Amazon, USA
Iacopo Masi	Sapienza University of Rome, Italy
Dimitris N. Metaxas	Rutgers University, USA
Ajmal Mian	University of Western Australia, Australia
Christian Micheloni	University of Udine, Italy
Krystian Mikolajczyk	Imperial College London, UK
Anurag Mittal	Indian Institute of Technology, Madras, India
Philippos Mordohai	Stevens Institute of Technology, USA
Greg Mori	Simon Fraser University & Borealis AI, Canada

Vittorio Murino	Istituto Italiano di Tecnologia, Italy
P. J. Narayanan	International Institute of Information Technology, Hyderabad, India
Ram Nevatia	University of Southern California, USA
Natalia Neverova	Facebook AI Research, UK
Richard Newcombe	Facebook, USA
Cuong V. Nguyen	Florida International University, USA
Bingbing Ni	Shanghai Jiao Tong University, China
Juan Carlos Niebles	Salesforce & Stanford University, USA
Ko Nishino	Kyoto University, Japan
Jean-Marc Odobez	Idiap Research Institute, École polytechnique fédérale de Lausanne, Switzerland
Francesca Odone	University of Genova, Italy
Takayuki Okatani	Tohoku University & RIKEN Center for Advanced Intelligence Project, Japan
Manohar Paluri	Facebook, USA
Guan Pang	Facebook, USA
Maja Pantic	Imperial College London, UK
Sylvain Paris	Adobe Research, USA
Jaesik Park	Pohang University of Science and Technology, Korea
Hyun Soo Park	The University of Minnesota, USA
Omkar M. Parkhi	Facebook, USA
Deepak Pathak	Carnegie Mellon University, USA
Georgios Pavlakos	University of California, Berkeley, USA
Marcello Pelillo	University of Venice, Italy
Marc Pollefeys	ETH Zurich & Microsoft, Switzerland
Jean Ponce	Inria, France
Gerard Pons-Moll	University of Tübingen, Germany
Fatih Porikli	Qualcomm, USA
Victor Adrian Prisacariu	University of Oxford, UK
Petia Radeva	University of Barcelona, Spain
Ravi Ramamoorthi	University of California, San Diego, USA
Deva Ramanan	Carnegie Mellon University, USA
Vignesh Ramanathan	Facebook, USA
Nalini Ratha	State University of New York at Buffalo, USA
Tammy Riklin Raviv	Ben-Gurion University, Israel
Tobias Ritschel	University College London, UK
Emanuele Rodola	Sapienza University of Rome, Italy
Amit K. Roy-Chowdhury	University of California, Riverside, USA
Michael Rubinstein	Google, USA
Olga Russakovsky	Princeton University, USA

Mathieu Salzmann	École polytechnique fédérale de Lausanne, Switzerland
Dimitris Samaras	Stony Brook University, USA
Aswin Sankaranarayanan	Carnegie Mellon University, USA
Imari Sato	National Institute of Informatics, Japan
Yoichi Sato	University of Tokyo, Japan
Shin'ichi Satoh	National Institute of Informatics, Japan
Walter Scheirer	University of Notre Dame, USA
Bernt Schiele	Max Planck Institute for Informatics, Germany
Konrad Schindler	ETH Zurich, Switzerland
Cordelia Schmid	Inria & Google, France
Alexander Schwing	University of Illinois at Urbana-Champaign, USA
Nicu Sebe	University of Trento, Italy
Greg Shakhnarovich	Toyota Technological Institute at Chicago, USA
Eli Shechtman	Adobe Research, USA
Humphrey Shi	University of Oregon & University of Illinois at Urbana-Champaign & Picsart AI Research, USA
Jianbo Shi	University of Pennsylvania, USA
Roy Shilkrot	Massachusetts Institute of Technology, USA
Mike Zheng Shou	National University of Singapore, Singapore
Kaleem Siddiqi	McGill University, Canada
Richa Singh	Indian Institute of Technology Jodhpur, India
Greg Slabaugh	Queen Mary University of London, UK
Cees Snoek	University of Amsterdam, The Netherlands
Yale Song	Facebook AI Research, USA
Yi-Zhe Song	University of Surrey, UK
Bjorn Stenger	Rakuten Institute of Technology
Abby Stylianou	Saint Louis University, USA
Akihiro Sugimoto	National Institute of Informatics, Japan
Chen Sun	Brown University, USA
Deqing Sun	Google, USA
Kalyan Sunkavalli	Adobe Research, USA
Ying Tai	Tencent YouTu Lab, China
Ayellet Tal	Technion – Israel Institute of Technology, Israel
Ping Tan	Simon Fraser University, Canada
Siyu Tang	ETH Zurich, Switzerland
Chi-Keung Tang	Hong Kong University of Science and Technology, Hong Kong, China
Radu Timofte	University of Würzburg, Germany & ETH Zurich, Switzerland
Federico Tombari	Google, Switzerland & Technical University of Munich, Germany

James Tompkin	Brown University, USA
Lorenzo Torresani	Dartmouth College, USA
Alexander Toshev	Apple, USA
Du Tran	Facebook AI Research, USA
Anh T. Tran	VinAI, Vietnam
Zhuowen Tu	University of California, San Diego, USA
Georgios Tzimiropoulos	Queen Mary University of London, UK
Jasper Uijlings	Google Research, Switzerland
Jan C. van Gemert	Delft University of Technology, The Netherlands
Gul Varol	Ecole des Ponts ParisTech, France
Nuno Vasconcelos	University of California, San Diego, USA
Mayank Vatsa	Indian Institute of Technology Jodhpur, India
Ashok Veeraraghavan	Rice University, USA
Jakob Verbeek	Facebook AI Research, France
Carl Vondrick	Columbia University, USA
Ruiping Wang	Institute of Computing Technology, Chinese Academy of Sciences, China
Xinchao Wang	National University of Singapore, Singapore
Liwei Wang	The Chinese University of Hong Kong, Hong Kong, China
Chaohui Wang	Université Paris-Est, France
Xiaolong Wang	University of California, San Diego, USA
Christian Wolf	NAVER LABS Europe, France
Tao Xiang	University of Surrey, UK
Saining Xie	Facebook AI Research, USA
Cihang Xie	University of California, Santa Cruz, USA
Zeki Yalniz	Facebook, USA
Ming-Hsuan Yang	University of California, Merced, USA
Angela Yao	National University of Singapore, Singapore
Shaodi You	University of Amsterdam, The Netherlands
Stella X. Yu	University of California, Berkeley, USA
Junsong Yuan	State University of New York at Buffalo, USA
Stefanos Zafeiriou	Imperial College London, UK
Amir Zamir	École polytechnique fédérale de Lausanne, Switzerland
Lei Zhang	Alibaba & Hong Kong Polytechnic University, Hong Kong, China
Lei Zhang	International Digital Economy Academy (IDEA), China
Pengchuan Zhang	Meta AI, USA
Bolei Zhou	University of California, Los Angeles, USA
Yuke Zhu	University of Texas at Austin, USA

Todd Zickler   Harvard University, USA
Wangmeng Zuo   Harbin Institute of Technology, China

Technical Program Committee

Davide Abati
Soroush Abbasi
    Koohpayegani
Amos L. Abbott
Rameen Abdal
Rabab Abdelfattah
Sahar Abdelnabi
Hassan Abu Alhaija
Abulikemu Abuduweili
Ron Abutbul
Hanno Ackermann
Aikaterini Adam
Kamil Adamczewski
Ehsan Adeli
Vida Adeli
Donald Adjeroh
Arman Afrasiyabi
Akshay Agarwal
Sameer Agarwal
Abhinav Agarwalla
Vaibhav Aggarwal
Sara Aghajanzadeh
Susmit Agrawal
Antonio Agudo
Touqeer Ahmad
Sk Miraj Ahmed
Chaitanya Ahuja
Nilesh A. Ahuja
Abhishek Aich
Shubhra Aich
Noam Aigerman
Arash Akbarinia
Peri Akiva
Derya Akkaynak
Emre Aksan
Arjun R. Akula
Yuval Alaluf
Stephan Alaniz
Paul Albert
Cenek Albl

Filippo Aleotti
Konstantinos P.
    Alexandridis
Motasem Alfarra
Mohsen Ali
Thiemo Alldieck
Hadi Alzayer
Liang An
Shan An
Yi An
Zhulin An
Dongsheng An
Jie An
Xiang An
Saket Anand
Cosmin Ancuti
Juan Andrade-Cetto
Alexander Andreopoulos
Bjoern Andres
Jerone T. A. Andrews
Shivangi Aneja
Anelia Angelova
Dragomir Anguelov
Rushil Anirudh
Oron Anschel
Rao Muhammad Anwer
Djamila Aouada
Evlampios Apostolidis
Srikar Appalaraju
Nikita Araslanov
Andre Araujo
Eric Arazo
Dawit Mureja Argaw
Anurag Arnab
Aditya Arora
Chetan Arora
Sunpreet S. Arora
Alexey Artemov
Muhammad Asad
Kumar Ashutosh

Sinem Aslan
Vishal Asnani
Mahmoud Assran
Amir Atapour-Abarghouei
Nikos Athanasiou
Ali Athar
ShahRukh Athar
Sara Atito
Souhaib Attaiki
Matan Atzmon
Mathieu Aubry
Nicolas Audebert
Tristan T.
    Aumentado-Armstrong
Melinos Averkiou
Yannis Avrithis
Stephane Ayache
Mehmet Aygün
Seyed Mehdi
    Ayyoubzadeh
Hossein Azizpour
George Azzopardi
Mallikarjun B. R.
Yunhao Ba
Abhishek Badki
Seung-Hwan Bae
Seung-Hwan Baek
Seungryul Baek
Piyush Nitin Bagad
Shai Bagon
Gaetan Bahl
Shikhar Bahl
Sherwin Bahmani
Haoran Bai
Lei Bai
Jiawang Bai
Haoyue Bai
Jinbin Bai
Xiang Bai
Xuyang Bai

Yang Bai
Yuanchao Bai
Ziqian Bai
Sungyong Baik
Kevin Bailly
Max Bain
Federico Baldassarre
Wele Gedara Chaminda
    Bandara
Biplab Banerjee
Pratyay Banerjee
Sandipan Banerjee
Jihwan Bang
Antyanta Bangunharcana
Aayush Bansal
Ankan Bansal
Siddhant Bansal
Wentao Bao
Zhipeng Bao
Amir Bar
Manel Baradad Jurjo
Lorenzo Baraldi
Danny Barash
Daniel Barath
Connelly Barnes
Ioan Andrei Bârsan
Steven Basart
Dina Bashkirova
Chaim Baskin
Peyman Bateni
Anil Batra
Sebastiano Battiato
Ardhendu Behera
Harkirat Behl
Jens Behley
Vasileios Belagiannis
Boulbaba Ben Amor
Emanuel Ben Baruch
Abdessamad Ben Hamza
Gil Ben-Artzi
Assia Benbihi
Fabian Benitez-Quiroz
Guy Ben-Yosef
Philipp Benz
Alexander W. Bergman

Urs Bergmann
Jesus Bermudez-Cameo
Stefano Berretti
Gedas Bertasius
Zachary Bessinger
Petra Bevandić
Matthew Beveridge
Lucas Beyer
Yash Bhalgat
Suvaansh Bhambri
Samarth Bharadwaj
Gaurav Bharaj
Aparna Bharati
Bharat Lal Bhatnagar
Uttaran Bhattacharya
Apratim Bhattacharyya
Brojeshwar Bhowmick
Ankan Kumar Bhunia
Ayan Kumar Bhunia
Qi Bi
Sai Bi
Michael Bi Mi
Gui-Bin Bian
Jia-Wang Bian
Shaojun Bian
Pia Bideau
Mario Bijelic
Hakan Bilen
Guillaume-Alexandre
    Bilodeau
Alexander Binder
Tolga Birdal
Vighnesh N. Birodkar
Sandika Biswas
Andreas Blattmann
Janusz Bobulski
Giuseppe Boccignone
Vishnu Boddeti
Navaneeth Bodla
Moritz Böhle
Aleksei Bokhovkin
Sam Bond-Taylor
Vivek Boominathan
Shubhankar Borse
Mark Boss

Andrea Bottino
Adnane Boukhayma
Fadi Boutros
Nicolas C. Boutry
Richard S. Bowen
Ivaylo Boyadzhiev
Aidan Boyd
Yuri Boykov
Aljaz Bozic
Behzad Bozorgtabar
Eric Brachmann
Samarth Brahmbhatt
Gustav Bredell
Francois Bremond
Joel Brogan
Andrew Brown
Thomas Brox
Marcus A. Brubaker
Robert-Jan Bruintjes
Yuqi Bu
Anders G. Buch
Himanshu Buckchash
Mateusz Buda
Ignas Budvytis
José M. Buenaposada
Marcel C. Bühler
Tu Bui
Adrian Bulat
Hannah Bull
Evgeny Burnaev
Andrei Bursuc
Benjamin Busam
Sergey N. Buzykanov
Wonmin Byeon
Fabian Caba
Martin Cadik
Guanyu Cai
Minjie Cai
Qing Cai
Zhongang Cai
Qi Cai
Yancheng Cai
Shen Cai
Han Cai
Jiarui Cai

Bowen Cai
Mu Cai
Qin Cai
Ruojin Cai
Weidong Cai
Weiwei Cai
Yi Cai
Yujun Cai
Zhiping Cai
Akin Caliskan
Lilian Calvet
Baris Can Cam
Necati Cihan Camgoz
Tommaso Campari
Dylan Campbell
Ziang Cao
Ang Cao
Xu Cao
Zhiwen Cao
Shengcao Cao
Song Cao
Weipeng Cao
Xiangyong Cao
Xiaochun Cao
Yue Cao
Yunhao Cao
Zhangjie Cao
Jiale Cao
Yang Cao
Jiajiong Cao
Jie Cao
Jinkun Cao
Lele Cao
Yulong Cao
Zhiguo Cao
Chen Cao
Razvan Caramalau
Marlène Careil
Gustavo Carneiro
Joao Carreira
Dan Casas
Paola Cascante-Bonilla
Angela Castillo
Francisco M. Castro
Pedro Castro

Luca Cavalli
George J. Cazenavette
Oya Celiktutan
Hakan Cevikalp
Sri Harsha C. H.
Sungmin Cha
Geonho Cha
Menglei Chai
Lucy Chai
Yuning Chai
Zenghao Chai
Anirban Chakraborty
Deep Chakraborty
Rudrasis Chakraborty
Souradeep Chakraborty
Kelvin C. K. Chan
Chee Seng Chan
Paramanand Chandramouli
Arjun Chandrasekaran
Kenneth Chaney
Dongliang Chang
Huiwen Chang
Peng Chang
Xiaojun Chang
Jia-Ren Chang
Hyung Jin Chang
Hyun Sung Chang
Ju Yong Chang
Li-Jen Chang
Qi Chang
Wei-Yi Chang
Yi Chang
Nadine Chang
Hanqing Chao
Pradyumna Chari
Dibyadip Chatterjee
Chiranjoy Chattopadhyay
Siddhartha Chaudhuri
Zhengping Che
Gal Chechik
Lianggangxu Chen
Qi Alfred Chen
Brian Chen
Bor-Chun Chen
Bo-Hao Chen

Bohong Chen
Bin Chen
Ziliang Chen
Cheng Chen
Chen Chen
Chaofeng Chen
Xi Chen
Haoyu Chen
Xuanhong Chen
Wei Chen
Qiang Chen
Shi Chen
Xianyu Chen
Chang Chen
Changhuai Chen
Hao Chen
Jie Chen
Jianbo Chen
Jingjing Chen
Jun Chen
Kejiang Chen
Mingcai Chen
Nenglun Chen
Qifeng Chen
Ruoyu Chen
Shu-Yu Chen
Weidong Chen
Weijie Chen
Weikai Chen
Xiang Chen
Xiuyi Chen
Xingyu Chen
Yaofo Chen
Yueting Chen
Yu Chen
Yunjin Chen
Yuntao Chen
Yun Chen
Zhenfang Chen
Zhuangzhuang Chen
Chu-Song Chen
Xiangyu Chen
Zhuo Chen
Chaoqi Chen
Shizhe Chen

Xiaotong Chen

Xiaozhi Chen

Dian Chen

Defang Chen

Dingfan Chen

Ding-Jie Chen

Ee Heng Chen

Tao Chen

Yixin Chen

Wei-Ting Chen

Lin Chen

Guang Chen

Guangyi Chen

Guanying Chen

Guangyao Chen

Hwann-Tzong Chen

Junwen Chen

Jiacheng Chen

Jianxu Chen

Hui Chen

Kai Chen

Kan Chen

Kevin Chen

Kuan-Wen Chen

Weihua Chen

Zhang Chen

Liang-Chieh Chen

Lele Chen

Liang Chen

Fanglin Chen

Zehui Chen

Minghui Chen

Minghao Chen

Xiaokang Chen

Qian Chen

Jun-Cheng Chen

Qi Chen

Qingcai Chen

Richard J. Chen

Runnan Chen

Rui Chen

Shuo Chen

Sentao Chen

Shaoyu Chen

Shixing Chen

Shuai Chen

Shuya Chen

Sizhe Chen

Simin Chen

Shaoxiang Chen

Zitian Chen

Tianlong Chen

Tianshui Chen

Min-Hung Chen

Xiangning Chen

Xin Chen

Xinghao Chen

Xuejin Chen

Xu Chen

Xuxi Chen

Yunlu Chen

Yanbei Chen

Yuxiao Chen

Yun-Chun Chen

Yi-Ting Chen

Yi-Wen Chen

Yinbo Chen

Yiran Chen

Yuanhong Chen

Yubei Chen

Yuefeng Chen

Yuhua Chen

Yukang Chen

Zerui Chen

Zhaoyu Chen

Zhen Chen

Zhenyu Chen

Zhi Chen

Zhiwei Chen

Zhixiang Chen

Long Chen

Bowen Cheng

Jun Cheng

Yi Cheng

Jingchun Cheng

Lechao Cheng

Xi Cheng

Yuan Cheng

Ho Kei Cheng

Kevin Ho Man Cheng

Jiacheng Cheng

Kelvin B. Cheng

Li Cheng

Mengjun Cheng

Zhen Cheng

Qingrong Cheng

Tianheng Cheng

Harry Cheng

Yihua Cheng

Yu Cheng

Ziheng Cheng

Soon Yau Cheong

Anoop Cherian

Manuela Chessa

Zhixiang Chi

Naoki Chiba

Julian Chibane

Kashyap Chitta

Tai-Yin Chiu

Hsu-kuang Chiu

Wei-Chen Chiu

Sungmin Cho

Donghyeon Cho

Hyeon Cho

Yooshin Cho

Gyusang Cho

Jang Hyun Cho

Seungju Cho

Nam Ik Cho

Sunghyun Cho

Hanbyel Cho

Jaesung Choe

Jooyoung Choi

Chiho Choi

Changwoon Choi

Jongwon Choi

Myungsub Choi

Dooseop Choi

Jonghyun Choi

Jinwoo Choi

Jun Won Choi

Min-Kook Choi

Hongsuk Choi

Janghoon Choi

Yoon-Ho Choi

Yukyung Choi
Jaegul Choo
Ayush Chopra
Siddharth Choudhary
Subhabrata Choudhury
Vasileios Choutas
Ka-Ho Chow
Pinaki Nath Chowdhury
Sammy Christen
Anders Christensen
Grigorios Chrysos
Hang Chu
Wen-Hsuan Chu
Peng Chu
Qi Chu
Ruihang Chu
Wei-Ta Chu
Yung-Yu Chuang
Sanghyuk Chun
Se Young Chun
Antonio Cinà
Ramazan Gokberk Cinbis
Javier Civera
Albert Clapés
Ronald Clark
Brian S. Clipp
Felipe Codevilla
Daniel Coelho de Castro
Niv Cohen
Forrester Cole
Maxwell D. Collins
Robert T. Collins
Marc Comino Trinidad
Runmin Cong
Wenyan Cong
Maxime Cordy
Marcella Cornia
Enric Corona
Huseyin Coskun
Luca Cosmo
Dragos Costea
Davide Cozzolino
Arun C. S. Kumar
Aiyu Cui
Qiongjie Cui

Quan Cui
Shuhao Cui
Yiming Cui
Ying Cui
Zijun Cui
Jiali Cui
Jiequan Cui
Yawen Cui
Zhen Cui
Zhaopeng Cui
Jack Culpepper
Xiaodong Cun
Ross Cutler
Adam Czajka
Ali Dabouei
Konstantinos M. Dafnis
Manuel Dahnert
Tao Dai
Yuchao Dai
Bo Dai
Mengyu Dai
Hang Dai
Haixing Dai
Peng Dai
Pingyang Dai
Qi Dai
Qiyu Dai
Yutong Dai
Naser Damer
Zhiyuan Dang
Mohamed Daoudi
Ayan Das
Abir Das
Debasmit Das
Deepayan Das
Partha Das
Sagnik Das
Soumi Das
Srijan Das
Swagatam Das
Avijit Dasgupta
Jim Davis
Adrian K. Davison
Homa Davoudi
Laura Daza

Matthias De Lange
Shalini De Mello
Marco De Nadai
Christophe De
    Vleeschouwer
Alp Dener
Boyang Deng
Congyue Deng
Bailin Deng
Yong Deng
Ye Deng
Zhuo Deng
Zhijie Deng
Xiaoming Deng
Jiankang Deng
Jinhong Deng
Jingjing Deng
Liang-Jian Deng
Siqi Deng
Xiang Deng
Xueqing Deng
Zhongying Deng
Karan Desai
Jean-Emmanuel Deschaud
Aniket Anand Deshmukh
Neel Dey
Helisa Dhamo
Prithviraj Dhar
Amaya Dharmasiri
Yan Di
Xing Di
Ousmane A. Dia
Haiwen Diao
Xiaolei Diao
Gonçalo José Dias Pais
Abdallah Dib
Anastasios Dimou
Changxing Ding
Henghui Ding
Guodong Ding
Yaqing Ding
Shuangrui Ding
Yuhang Ding
Yikang Ding
Shouhong Ding

Haisong Ding
Hui Ding
Jiahao Ding
Jian Ding
Jian-Jiun Ding
Shuxiao Ding
Tianyu Ding
Wenhao Ding
Yuqi Ding
Yi Ding
Yuzhen Ding
Zhengming Ding
Tan Minh Dinh
Vu Dinh
Christos Diou
Mandar Dixit
Bao Gia Doan
Khoa D. Doan
Dzung Anh Doan
Debi Prosad Dogra
Nehal Doiphode
Chengdong Dong
Bowen Dong
Zhenxing Dong
Hang Dong
Xiaoyi Dong
Haoye Dong
Jiangxin Dong
Shichao Dong
Xuan Dong
Zhen Dong
Shuting Dong
Jing Dong
Li Dong
Ming Dong
Nanqing Dong
Qiulei Dong
Runpei Dong
Siyan Dong
Tian Dong
Wei Dong
Xiaomeng Dong
Xin Dong
Xingbo Dong
Yuan Dong

Samuel Dooley
Gianfranco Doretto
Michael Dorkenwald
Keval Doshi
Zhaopeng Dou
Xiaotian Dou
Hazel Doughty
Ahmad Droby
Iddo Drori
Jie Du
Yong Du
Dawei Du
Dong Du
Ruoyi Du
Yuntao Du
Xuefeng Du
Yilun Du
Yuming Du
Radhika Dua
Haodong Duan
Jiafei Duan
Kaiwen Duan
Peiqi Duan
Ye Duan
Haoran Duan
Jiali Duan
Amanda Duarte
Abhimanyu Dubey
Shiv Ram Dubey
Florian Dubost
Lukasz Dudziak
Shivam Duggal
Justin M. Dulay
Matteo Dunnhofer
Chi Nhan Duong
Thibaut Durand
Mihai Dusmanu
Ujjal Kr Dutta
Debidatta Dwibedi
Isht Dwivedi
Sai Kumar Dwivedi
Takeharu Eda
Mark Edmonds
Alexei A. Efros
Thibaud Ehret

Max Ehrlich
Mahsa Ehsanpour
Iván Eichhardt
Farshad Einabadi
Marvin Eisenberger
Hazim Kemal Ekenel
Mohamed El Banani
Ismail Elezi
Moshe Eliasof
Alaa El-Nouby
Ian Endres
Francis Engelmann
Deniz Engin
Chanho Eom
Dave Epstein
Maria C. Escobar
Victor A. Escorcia
Carlos Esteves
Sungmin Eum
Bernard J. E. Evans
Ivan Evtimov
Fevziye Irem Eyiokur
  Yaman
Matteo Fabbri
Sébastien Fabbro
Gabriele Facciolo
Masud Fahim
Bin Fan
Hehe Fan
Deng-Ping Fan
Aoxiang Fan
Chen-Chen Fan
Qi Fan
Zhaoxin Fan
Haoqi Fan
Heng Fan
Hongyi Fan
Linxi Fan
Baojie Fan
Jiayuan Fan
Lei Fan
Quanfu Fan
Yonghui Fan
Yingruo Fan
Zhiwen Fan

Zicong Fan
Sean Fanello
Jiansheng Fang
Chaowei Fang
Yuming Fang
Jianwu Fang
Jin Fang
Qi Fang
Shancheng Fang
Tian Fang
Xianyong Fang
Gongfan Fang
Zhen Fang
Hui Fang
Jiemin Fang
Le Fang
Pengfei Fang
Xiaolin Fang
Yuxin Fang
Zhaoyuan Fang
Ammarah Farooq
Azade Farshad
Zhengcong Fei
Michael Felsberg
Wei Feng
Chen Feng
Fan Feng
Andrew Feng
Xin Feng
Zheyun Feng
Ruicheng Feng
Mingtao Feng
Qianyu Feng
Shangbin Feng
Chun-Mei Feng
Zunlei Feng
Zhiyong Feng
Martin Fergie
Mustansar Fiaz
Marco Fiorucci
Michael Firman
Hamed Firooz
Volker Fischer
Corneliu O. Florea
Georgios Floros

Wolfgang Foerstner
Gianni Franchi
Jean-Sebastien Franco
Simone Frintrop
Anna Fruehstueck
Changhong Fu
Chaoyou Fu
Cheng-Yang Fu
Chi-Wing Fu
Deqing Fu
Huan Fu
Jun Fu
Kexue Fu
Ying Fu
Jianlong Fu
Jingjing Fu
Qichen Fu
Tsu-Jui Fu
Xueyang Fu
Yang Fu
Yanwei Fu
Yonggan Fu
Wolfgang Fuhl
Yasuhisa Fujii
Kent Fujiwara
Marco Fumero
Takuya Funatomi
Isabel Funke
Dario Fuoli
Antonino Furnari
Matheus A. Gadelha
Akshay Gadi Patil
Adrian Galdran
Guillermo Gallego
Silvano Galliani
Orazio Gallo
Leonardo Galteri
Matteo Gamba
Yiming Gan
Sujoy Ganguly
Harald Ganster
Boyan Gao
Changxin Gao
Daiheng Gao
Difei Gao

Chen Gao
Fei Gao
Lin Gao
Wei Gao
Yiming Gao
Junyu Gao
Guangyu Ryan Gao
Haichang Gao
Hongchang Gao
Jialin Gao
Jin Gao
Jun Gao
Katelyn Gao
Mingchen Gao
Mingfei Gao
Pan Gao
Shangqian Gao
Shanghua Gao
Xitong Gao
Yunhe Gao
Zhanning Gao
Elena Garces
Nuno Cruz Garcia
Noa Garcia
Guillermo
    Garcia-Hernando
Isha Garg
Rahul Garg
Sourav Garg
Quentin Garrido
Stefano Gasperini
Kent Gauen
Chandan Gautam
Shivam Gautam
Paul Gay
Chunjiang Ge
Shiming Ge
Wenhang Ge
Yanhao Ge
Zheng Ge
Songwei Ge
Weifeng Ge
Yixiao Ge
Yuying Ge
Shijie Geng

Zhengyang Geng
Kyle A. Genova
Georgios Georgakis
Markos Georgopoulos
Marcel Geppert
Shabnam Ghadar
Mina Ghadimi Atigh
Deepti Ghadiyaram
Maani Ghaffari Jadidi
Sedigh Ghamari
Zahra Gharaee
Michaël Gharbi
Golnaz Ghiasi
Reza Ghoddoosian
Soumya Suvra Ghosal
Adhiraj Ghosh
Arthita Ghosh
Pallabi Ghosh
Soumyadeep Ghosh
Andrew Gilbert
Igor Gilitschenski
Jhony H. Giraldo
Andreu Girbau Xalabarder
Rohit Girdhar
Sharath Girish
Xavier Giro-i-Nieto
Raja Giryes
Thomas Gittings
Nikolaos Gkanatsios
Ioannis Gkioulekas
Abhiram
   Gnanasambandam
Aurele T. Gnanha
Clement L. J. C. Godard
Arushi Goel
Vidit Goel
Shubham Goel
Zan Gojcic
Aaron K. Gokaslan
Tejas Gokhale
S. Alireza Golestaneh
Thiago L. Gomes
Nuno Goncalves
Boqing Gong
Chen Gong

Yuanhao Gong
Guoqiang Gong
Jingyu Gong
Rui Gong
Yu Gong
Mingming Gong
Neil Zhenqiang Gong
Xun Gong
Yunye Gong
Yihong Gong
Cristina I. González
Nithin Gopalakrishnan
   Nair
Gaurav Goswami
Jianping Gou
Shreyank N. Gowda
Ankit Goyal
Helmut Grabner
Patrick L. Grady
Ben Graham
Eric Granger
Douglas R. Gray
Matej Grcić
David Griffiths
Jinjin Gu
Yun Gu
Shuyang Gu
Jianyang Gu
Fuqiang Gu
Jiatao Gu
Jindong Gu
Jiaqi Gu
Jinwei Gu
Jiaxin Gu
Geonmo Gu
Xiao Gu
Xinqian Gu
Xiuye Gu
Yuming Gu
Zhangxuan Gu
Dayan Guan
Junfeng Guan
Qingji Guan
Tianrui Guan
Shanyan Guan

Denis A. Gudovskiy
Ricardo Guerrero
Pierre-Louis Guhur
Jie Gui
Liangyan Gui
Liangke Gui
Benoit Guillard
Erhan Gundogdu
Manuel Günther
Jingcai Guo
Yuanfang Guo
Junfeng Guo
Chenqi Guo
Dan Guo
Hongji Guo
Jia Guo
Jie Guo
Minghao Guo
Shi Guo
Yanhui Guo
Yangyang Guo
Yuan-Chen Guo
Yilu Guo
Yiluan Guo
Yong Guo
Guangyu Guo
Haiyun Guo
Jinyang Guo
Jianyuan Guo
Pengsheng Guo
Pengfei Guo
Shuxuan Guo
Song Guo
Tianyu Guo
Qing Guo
Qiushan Guo
Wen Guo
Xiefan Guo
Xiaohu Guo
Xiaoqing Guo
Yufei Guo
Yuhui Guo
Yuliang Guo
Yunhui Guo
Yanwen Guo

Akshita Gupta
Ankush Gupta
Kamal Gupta
Kartik Gupta
Ritwik Gupta
Rohit Gupta
Siddharth Gururani
Fredrik K. Gustafsson
Abner Guzman Rivera
Vladimir Guzov
Matthew A. Gwilliam
Jung-Woo Ha
Marc Habermann
Isma Hadji
Christian Haene
Martin Hahner
Levente Hajder
Alexandros Haliassos
Emanuela Haller
Bumsub Ham
Abdullah J. Hamdi
Shreyas Hampali
Dongyoon Han
Chunrui Han
Dong-Jun Han
Dong-Sig Han
Guangxing Han
Zhizhong Han
Ruize Han
Jiaming Han
Jin Han
Ligong Han
Xian-Hua Han
Xiaoguang Han
Yizeng Han
Zhi Han
Zhenjun Han
Zhongyi Han
Jungong Han
Junlin Han
Kai Han
Kun Han
Sungwon Han
Songfang Han
Wei Han

Xiao Han
Xintong Han
Xinzhe Han
Yahong Han
Yan Han
Zongbo Han
Nicolai Hani
Rana Hanocka
Niklas Hanselmann
Nicklas A. Hansen
Hong Hanyu
Fusheng Hao
Yanbin Hao
Shijie Hao
Udith Haputhanthri
Mehrtash Harandi
Josh Harguess
Adam Harley
David M. Hart
Atsushi Hashimoto
Ali Hassani
Mohammed Hassanin
Yana Hasson
Joakim Bruslund Haurum
Bo He
Kun He
Chen He
Xin He
Fazhi He
Gaoqi He
Hao He
Haoyu He
Jiangpeng He
Hongliang He
Qian He
Xiangteng He
Xuming He
Yannan He
Yuhang He
Yang He
Xiangyu He
Nanjun He
Pan He
Sen He
Shengfeng He

Songtao He
Tao He
Tong He
Wei He
Xuehai He
Xiaoxiao He
Ying He
Yisheng He
Ziwen He
Peter Hedman
Felix Heide
Yacov Hel-Or
Paul Henderson
Philipp Henzler
Byeongho Heo
Jae-Pil Heo
Miran Heo
Sachini A. Herath
Stephane Herbin
Pedro Hermosilla Casajus
Monica Hernandez
Charles Herrmann
Roei Herzig
Mauricio Hess-Flores
Carlos Hinojosa
Tobias Hinz
Tsubasa Hirakawa
Chih-Hui Ho
Lam Si Tung Ho
Jennifer Hobbs
Derek Hoiem
Yannick Hold-Geoffroy
Aleksander Holynski
Cheeun Hong
Fa-Ting Hong
Hanbin Hong
Guan Zhe Hong
Danfeng Hong
Lanqing Hong
Xiaopeng Hong
Xin Hong
Jie Hong
Seungbum Hong
Cheng-Yao Hong
Seunghoon Hong

Yi Hong
Yuan Hong
Yuchen Hong
Anthony Hoogs
Maxwell C. Horton
Kazuhiro Hotta
Qibin Hou
Tingbo Hou
Junhui Hou
Ji Hou
Qiqi Hou
Rui Hou
Ruibing Hou
Zhi Hou
Henry Howard-Jenkins
Lukas Hoyer
Wei-Lin Hsiao
Chiou-Ting Hsu
Anthony Hu
Brian Hu
Yusong Hu
Hexiang Hu
Haoji Hu
Di Hu
Hengtong Hu
Haigen Hu
Lianyu Hu
Hanzhe Hu
Jie Hu
Junlin Hu
Shizhe Hu
Jian Hu
Zhiming Hu
Juhua Hu
Peng Hu
Ping Hu
Ronghang Hu
MengShun Hu
Tao Hu
Vincent Tao Hu
Xiaoling Hu
Xinting Hu
Xiaolin Hu
Xuefeng Hu
Xiaowei Hu

Yang Hu
Yueyu Hu
Zeyu Hu
Zhongyun Hu
Binh-Son Hua
Guoliang Hua
Yi Hua
Linzhi Huang
Qiusheng Huang
Bo Huang
Chen Huang
Hsin-Ping Huang
Ye Huang
Shuangping Huang
Zeng Huang
Buzhen Huang
Cong Huang
Heng Huang
Hao Huang
Qidong Huang
Huaibo Huang
Chaoqin Huang
Feihu Huang
Jiahui Huang
Jingjia Huang
Kun Huang
Lei Huang
Sheng Huang
Shuaiyi Huang
Siyu Huang
Xiaoshui Huang
Xiaoyang Huang
Yan Huang
Yihao Huang
Ying Huang
Ziling Huang
Xiaoke Huang
Yifei Huang
Haiyang Huang
Zhewei Huang
Jin Huang
Haibin Huang
Jiaxing Huang
Junjie Huang
Keli Huang

Lang Huang
Lin Huang
Luojie Huang
Mingzhen Huang
Shijia Huang
Shengyu Huang
Siyuan Huang
He Huang
Xiuyu Huang
Lianghua Huang
Yue Huang
Yaping Huang
Yuge Huang
Zehao Huang
Zeyi Huang
Zhiqi Huang
Zhongzhan Huang
Zilong Huang
Ziyuan Huang
Tianrui Hui
Zhuo Hui
Le Hui
Jing Huo
Junhwa Hur
Shehzeen S. Hussain
Chuong Minh Huynh
Seunghyun Hwang
Jaehui Hwang
Jyh-Jing Hwang
Sukjun Hwang
Soonmin Hwang
Wonjun Hwang
Rakib Hyder
Sangeek Hyun
Sarah Ibrahimi
Tomoki Ichikawa
Yerlan Idelbayev
A. S. M. Iftekhar
Masaaki Iiyama
Satoshi Ikehata
Sunghoon Im
Atul N. Ingle
Eldar Insafutdinov
Yani A. Ioannou
Radu Tudor Ionescu

Umar Iqbal
Go Irie
Muhammad Zubair Irshad
Ahmet Iscen
Berivan Isik
Ashraful Islam
Md Amirul Islam
Syed Islam
Mariko Isogawa
Vamsi Krishna K. Ithapu
Boris Ivanovic
Darshan Iyer
Sarah Jabbour
Ayush Jain
Nishant Jain
Samyak Jain
Vidit Jain
Vineet Jain
Priyank Jaini
Tomas Jakab
Mohammad A. A. K.
  Jalwana
Muhammad Abdullah
  Jamal
Hadi Jamali-Rad
Stuart James
Varun Jampani
Young Kyun Jang
YeongJun Jang
Yunseok Jang
Ronnachai Jaroensri
Bhavan Jasani
Krishna Murthy
  Jatavallabhula
Mojan Javaheripi
Syed A. Javed
Guillaume Jeanneret
Pranav Jeevan
Herve Jegou
Rohit Jena
Tomas Jenicek
Porter Jenkins
Simon Jenni
Hae-Gon Jeon
Sangryul Jeon

Boseung Jeong
Yoonwoo Jeong
Seong-Gyun Jeong
Jisoo Jeong
Allan D. Jepson
Ankit Jha
Sumit K. Jha
I-Hong Jhuo
Ge-Peng Ji
Chaonan Ji
Deyi Ji
Jingwei Ji
Wei Ji
Zhong Ji
Jiayi Ji
Pengliang Ji
Hui Ji
Mingi Ji
Xiaopeng Ji
Yuzhu Ji
Baoxiong Jia
Songhao Jia
Dan Jia
Shan Jia
Xiaojun Jia
Xiuyi Jia
Xu Jia
Menglin Jia
Wenqi Jia
Boyuan Jiang
Wenhao Jiang
Huaizu Jiang
Hanwen Jiang
Haiyong Jiang
Hao Jiang
Huajie Jiang
Huiqin Jiang
Haojun Jiang
Haobo Jiang
Junjun Jiang
Xingyu Jiang
Yangbangyan Jiang
Yu Jiang
Jianmin Jiang
Jiaxi Jiang

Jing Jiang
Kui Jiang
Li Jiang
Liming Jiang
Chiyu Jiang
Meirui Jiang
Chen Jiang
Peng Jiang
Tai-Xiang Jiang
Wen Jiang
Xinyang Jiang
Yifan Jiang
Yuming Jiang
Yingying Jiang
Zeren Jiang
ZhengKai Jiang
Zhenyu Jiang
Shuming Jiao
Jianbo Jiao
Licheng Jiao
Dongkwon Jin
Yeying Jin
Cheng Jin
Linyi Jin
Qing Jin
Taisong Jin
Xiao Jin
Xin Jin
Sheng Jin
Kyong Hwan Jin
Ruibing Jin
SouYoung Jin
Yueming Jin
Chenchen Jing
Longlong Jing
Taotao Jing
Yongcheng Jing
Younghyun Jo
Joakim Johnander
Jeff Johnson
Michael J. Jones
R. Kenny Jones
Rico Jonschkowski
Ameya Joshi
Sunghun Joung

Felix Juefei-Xu
Claudio R. Jung
Steffen Jung
Hari Chandana K.
Rahul Vigneswaran K.
Prajwal K. R.
Abhishek Kadian
Jhony Kaesemodel Pontes
Kumara Kahatapitiya
Anmol Kalia
Sinan Kalkan
Tarun Kalluri
Jaewon Kam
Sandesh Kamath
Meina Kan
Menelaos Kanakis
Takuhiro Kaneko
Di Kang
Guoliang Kang
Hao Kang
Jaeyeon Kang
Kyoungkook Kang
Li-Wei Kang
MinGuk Kang
Suk-Ju Kang
Zhao Kang
Yash Mukund Kant
Yueying Kao
Aupendu Kar
Konstantinos Karantzalos
Sezer Karaoglu
Navid Kardan
Sanjay Kariyappa
Leonid Karlinsky
Animesh Karnewar
Shyamgopal Karthik
Hirak J. Kashyap
Marc A. Kastner
Hirokatsu Kataoka
Angelos Katharopoulos
Hiroharu Kato
Kai Katsumata
Manuel Kaufmann
Chaitanya Kaul
Prakhar Kaushik

Yuki Kawana
Lei Ke
Lipeng Ke
Tsung-Wei Ke
Wei Ke
Petr Kellnhofer
Aniruddha Kembhavi
John Kender
Corentin Kervadec
Leonid Keselman
Daniel Keysers
Nima Khademi Kalantari
Taras Khakhulin
Samir Khaki
Muhammad Haris Khan
Qadeer Khan
Salman Khan
Subash Khanal
Vaishnavi M. Khindkar
Rawal Khirodkar
Saeed Khorram
Pirazh Khorramshahi
Kourosh Khoshelham
Ansh Khurana
Benjamin Kiefer
Jae Myung Kim
Junho Kim
Boah Kim
Hyeonseong Kim
Dong-Jin Kim
Dongwan Kim
Donghyun Kim
Doyeon Kim
Yonghyun Kim
Hyung-Il Kim
Hyunwoo Kim
Hyeongwoo Kim
Hyo Jin Kim
Hyunwoo J. Kim
Taehoon Kim
Jaeha Kim
Jiwon Kim
Jung Uk Kim
Kangyeol Kim
Eunji Kim

Daeha Kim
Dongwon Kim
Kunhee Kim
Kyungmin Kim
Junsik Kim
Min H. Kim
Namil Kim
Kookhoi Kim
Sanghyun Kim
Seongyeop Kim
Seungryong Kim
Saehoon Kim
Euyoung Kim
Guisik Kim
Sungyeon Kim
Sunnie S. Y. Kim
Taehun Kim
Tae Oh Kim
Won Hwa Kim
Seungwook Kim
YoungBin Kim
Youngeun Kim
Akisato Kimura
Furkan Osman Kınlı
Zsolt Kira
Hedvig Kjellström
Florian Kleber
Jan P. Klopp
Florian Kluger
Laurent Kneip
Byungsoo Ko
Muhammed Kocabas
A. Sophia Koepke
Kevin Koeser
Nick Kolkin
Nikos Kolotouros
Wai-Kin Adams Kong
Deying Kong
Caihua Kong
Youyong Kong
Shuyu Kong
Shu Kong
Tao Kong
Yajing Kong
Yu Kong

Zishang Kong
Theodora Kontogianni
Anton S. Konushin
Julian F. P. Kooij
Bruno Korbar
Giorgos Kordopatis-Zilos
Jari Korhonen
Adam Kortylewski
Denis Korzhenkov
Divya Kothandaraman
Suraj Kothawade
Iuliia Kotseruba
Satwik Kottur
Shashank Kotyan
Alexandros Kouris
Petros Koutras
Anna Kreshuk
Ranjay Krishna
Dilip Krishnan
Andrey Kuehlkamp
Hilde Kuehne
Jason Kuen
David Kügler
Arjan Kuijper
Anna Kukleva
Sumith Kulal
Viveka Kulharia
Akshay R. Kulkarni
Nilesh Kulkarni
Dominik Kulon
Abhinav Kumar
Akash Kumar
Suryansh Kumar
B. V. K. Vijaya Kumar
Pulkit Kumar
Ratnesh Kumar
Sateesh Kumar
Satish Kumar
Vijay Kumar B. G.
Nupur Kumari
Sudhakar Kumawat
Jogendra Nath Kundu
Hsien-Kai Kuo
Meng-Yu Jennifer Kuo
Vinod Kumar Kurmi

Yusuke Kurose
Keerthy Kusumam
Alina Kuznetsova
Henry Kvinge
Ho Man Kwan
Hyeokjun Kweon
Heeseung Kwon
Gihyun Kwon
Myung-Joon Kwon
Taesung Kwon
YoungJoong Kwon
Christos Kyrkou
Jorma Laaksonen
Yann Labbe
Zorah Laehner
Florent Lafarge
Hamid Laga
Manuel Lagunas
Shenqi Lai
Jian-Huang Lai
Zihang Lai
Mohamed I. Lakhal
Mohit Lamba
Meng Lan
Loic Landrieu
Zhiqiang Lang
Natalie Lang
Dong Lao
Yizhen Lao
Yingjie Lao
Issam Hadj Laradji
Gustav Larsson
Viktor Larsson
Zakaria Laskar
Stéphane Lathuilière
Chun Pong Lau
Rynson W. H. Lau
Hei Law
Justin Lazarow
Verica Lazova
Eric-Tuan Le
Hieu Le
Trung-Nghia Le
Mathias Lechner
Byeong-Uk Lee

Chen-Yu Lee
Che-Rung Lee
Chul Lee
Hong Joo Lee
Dongsoo Lee
Jiyoung Lee
Eugene Eu Tzuan Lee
Daeun Lee
Saehyung Lee
Jewook Lee
Hyungtae Lee
Hyunmin Lee
Jungbeom Lee
Joon-Young Lee
Jong-Seok Lee
Joonseok Lee
Junha Lee
Kibok Lee
Byung-Kwan Lee
Jangwon Lee
Jinho Lee
Jongmin Lee
Seunghyun Lee
Sohyun Lee
Minsik Lee
Dogyoon Lee
Seungmin Lee
Min Jun Lee
Sangho Lee
Sangmin Lee
Seungeun Lee
Seon-Ho Lee
Sungmin Lee
Sungho Lee
Sangyoun Lee
Vincent C. S. S. Lee
Jaeseong Lee
Yong Jae Lee
Chenyang Lei
Chenyi Lei
Jiahui Lei
Xinyu Lei
Yinjie Lei
Jiaxu Leng
Luziwei Leng

Jan E. Lenssen
Vincent Lepetit
Thomas Leung
María Leyva-Vallina
Xin Li
Yikang Li
Baoxin Li
Bin Li
Bing Li
Bowen Li
Changlin Li
Chao Li
Chongyi Li
Guanyue Li
Shuai Li
Jin Li
Dingquan Li
Dongxu Li
Yiting Li
Gang Li
Dian Li
Guohao Li
Haoang Li
Haoliang Li
Haoran Li
Hengduo Li
Huafeng Li
Xiaoming Li
Hanao Li
Hongwei Li
Ziqiang Li
Jisheng Li
Jiacheng Li
Jia Li
Jiachen Li
Jiahao Li
Jianwei Li
Jiazhi Li
Jie Li
Jing Li
Jingjing Li
Jingtao Li
Jun Li
Junxuan Li
Kai Li

Kailin Li
Kenneth Li
Kun Li
Kunpeng Li
Aoxue Li
Chenglong Li
Chenglin Li
Changsheng Li
Zhichao Li
Qiang Li
Yanyu Li
Zuoyue Li
Xiang Li
Xuelong Li
Fangda Li
Ailin Li
Liang Li
Chun-Guang Li
Daiqing Li
Dong Li
Guanbin Li
Guorong Li
Haifeng Li
Jianan Li
Jianing Li
Jiaxin Li
Ke Li
Lei Li
Lincheng Li
Liulei Li
Lujun Li
Linjie Li
Lin Li
Pengyu Li
Ping Li
Qiufu Li
Qingyong Li
Rui Li
Siyuan Li
Wei Li
Wenbin Li
Xiangyang Li
Xinyu Li
Xiujun Li
Xiu Li

Xu Li
Ya-Li Li
Yao Li
Yongjie Li
Yijun Li
Yiming Li
Yuezun Li
Yu Li
Yunheng Li
Yuqi Li
Zhe Li
Zeming Li
Zhen Li
Zhengqin Li
Zhimin Li
Jiefeng Li
Jinpeng Li
Chengze Li
Jianwu Li
Lerenhan Li
Shan Li
Suichan Li
Xiangtai Li
Yanjie Li
Yandong Li
Zhuoling Li
Zhenqiang Li
Manyi Li
Maosen Li
Ji Li
Minjun Li
Mingrui Li
Mengtian Li
Junyi Li
Nianyi Li
Bo Li
Xiao Li
Peihua Li
Peike Li
Peizhao Li
Peiliang Li
Qi Li
Ren Li
Runze Li
Shile Li

Sheng Li
Shigang Li
Shiyu Li
Shuang Li
Shasha Li
Shichao Li
Tianye Li
Yuexiang Li
Wei-Hong Li
Wanhua Li
Weihao Li
Weiming Li
Weixin Li
Wenbo Li
Wenshuo Li
Weijian Li
Yunan Li
Xirong Li
Xianhang Li
Xiaoyu Li
Xueqian Li
Xuanlin Li
Xianzhi Li
Yunqiang Li
Yanjing Li
Yansheng Li
Yawei Li
Yi Li
Yong Li
Yong-Lu Li
Yuhang Li
Yu-Jhe Li
Yuxi Li
Yunsheng Li
Yanwei Li
Zechao Li
Zejian Li
Zeju Li
Zekun Li
Zhaowen Li
Zheng Li
Zhenyu Li
Zhiheng Li
Zhi Li
Zhong Li

Zhuowei Li
Zhuowan Li
Zhuohang Li
Zizhang Li
Chen Li
Yuan-Fang Li
Dongze Lian
Xiaochen Lian
Zhouhui Lian
Long Lian
Qing Lian
Jin Lianbao
Jinxiu S. Liang
Dingkang Liang
Jiahao Liang
Jianming Liang
Jingyun Liang
Kevin J. Liang
Kaizhao Liang
Chen Liang
Jie Liang
Senwei Liang
Ding Liang
Jiajun Liang
Jian Liang
Kongming Liang
Siyuan Liang
Yuanzhi Liang
Zhengfa Liang
Mingfu Liang
Xiaodan Liang
Xuefeng Liang
Yuxuan Liang
Kang Liao
Liang Liao
Hong-Yuan Mark Liao
Wentong Liao
Haofu Liao
Yue Liao
Minghui Liao
Shengcai Liao
Ting-Hsuan Liao
Xin Liao
Yinghong Liao
Teck Yian Lim

Che-Tsung Lin
Chung-Ching Lin
Chen-Hsuan Lin
Cheng Lin
Chuming Lin
Chunyu Lin
Dahua Lin
Wei Lin
Zheng Lin
Huaijia Lin
Jason Lin
Jierui Lin
Jiaying Lin
Jie Lin
Kai-En Lin
Kevin Lin
Guangfeng Lin
Jiehong Lin
Feng Lin
Hang Lin
Kwan-Yee Lin
Ke Lin
Luojun Lin
Qinghong Lin
Xiangbo Lin
Yi Lin
Zudi Lin
Shijie Lin
Yiqun Lin
Tzu-Heng Lin
Ming Lin
Shaohui Lin
SongNan Lin
Ji Lin
Tsung-Yu Lin
Xudong Lin
Yancong Lin
Yen-Chen Lin
Yiming Lin
Yuewei Lin
Zhiqiu Lin
Zinan Lin
Zhe Lin
David B. Lindell
Zhixin Ling

Zhan Ling
Alexander Liniger
Venice Erin B. Liong
Joey Litalien
Or Litany
Roee Litman
Ron Litman
Jim Little
Dor Litvak
Shaoteng Liu
Shuaicheng Liu
Andrew Liu
Xian Liu
Shaohui Liu
Bei Liu
Bo Liu
Yong Liu
Ming Liu
Yanbin Liu
Chenxi Liu
Daqi Liu
Di Liu
Difan Liu
Dong Liu
Dongfang Liu
Daizong Liu
Xiao Liu
Fangyi Liu
Fengbei Liu
Fenglin Liu
Bin Liu
Yuang Liu
Ao Liu
Hong Liu
Hongfu Liu
Huidong Liu
Ziyi Liu
Feng Liu
Hao Liu
Jie Liu
Jialun Liu
Jiang Liu
Jing Liu
Jingya Liu
Jiaming Liu

Jun Liu
Juncheng Liu
Jiawei Liu
Hongyu Liu
Chuanbin Liu
Haotian Liu
Lingqiao Liu
Chang Liu
Han Liu
Liu Liu
Min Liu
Yingqi Liu
Aishan Liu
Bingyu Liu
Benlin Liu
Boxiao Liu
Chenchen Liu
Chuanjian Liu
Daqing Liu
Huan Liu
Haozhe Liu
Jiaheng Liu
Wei Liu
Jingzhou Liu
Jiyuan Liu
Lingbo Liu
Nian Liu
Peiye Liu
Qiankun Liu
Shenglan Liu
Shilong Liu
Wen Liu
Wenyu Liu
Weifeng Liu
Wu Liu
Xiaolong Liu
Yang Liu
Yanwei Liu
Yingcheng Liu
Yongfei Liu
Yihao Liu
Yu Liu
Yunze Liu
Ze Liu
Zhenhua Liu

Zhenguang Liu
Lin Liu
Lihao Liu
Pengju Liu
Xinhai Liu
Yunfei Liu
Meng Liu
Minghua Liu
Mingyuan Liu
Miao Liu
Peirong Liu
Ping Liu
Qingjie Liu
Ruoshi Liu
Risheng Liu
Songtao Liu
Xing Liu
Shikun Liu
Shuming Liu
Sheng Liu
Songhua Liu
Tongliang Liu
Weibo Liu
Weide Liu
Weizhe Liu
Wenxi Liu
Weiyang Liu
Xin Liu
Xiaobin Liu
Xudong Liu
Xiaoyi Liu
Xihui Liu
Xinchen Liu
Xingtong Liu
Xinpeng Liu
Xinyu Liu
Xianpeng Liu
Xu Liu
Xingyu Liu
Yongtuo Liu
Yahui Liu
Yangxin Liu
Yaoyao Liu
Yaojie Liu
Yuliang Liu

Yongcheng Liu
Yuan Liu
Yufan Liu
Yu-Lun Liu
Yun Liu
Yunfan Liu
Yuanzhong Liu
Zhuoran Liu
Zhen Liu
Zheng Liu
Zhijian Liu
Zhisong Liu
Ziquan Liu
Ziyu Liu
Zhihua Liu
Zechun Liu
Zhaoyang Liu
Zhengzhe Liu
Stephan Liwicki
Shao-Yuan Lo
Sylvain Lobry
Suhas Lohit
Vishnu Suresh Lokhande
Vincenzo Lomonaco
Chengjiang Long
Guodong Long
Fuchen Long
Shangbang Long
Yang Long
Zijun Long
Vasco Lopes
Antonio M. Lopez
Roberto Javier
    Lopez-Sastre
Tobias Lorenz
Javier Lorenzo-Navarro
Yujing Lou
Qian Lou
Xiankai Lu
Changsheng Lu
Huimin Lu
Yongxi Lu
Hao Lu
Hong Lu
Jiasen Lu

Juwei Lu
Fan Lu
Guangming Lu
Jiwen Lu
Shun Lu
Tao Lu
Xiaonan Lu
Yang Lu
Yao Lu
Yongchun Lu
Zhiwu Lu
Cheng Lu
Liying Lu
Guo Lu
Xuequan Lu
Yanye Lu
Yantao Lu
Yuhang Lu
Fujun Luan
Jonathon Luiten
Jovita Lukasik
Alan Lukezic
Jonathan Samuel Lumentut
Mayank Lunayach
Ao Luo
Canjie Luo
Chong Luo
Xu Luo
Grace Luo
Jun Luo
Katie Z. Luo
Tao Luo
Cheng Luo
Fangzhou Luo
Gen Luo
Lei Luo
Sihui Luo
Weixin Luo
Yan Luo
Xiaoyan Luo
Yong Luo
Yadan Luo
Hao Luo
Ruotian Luo
Mi Luo

Tiange Luo
Wenjie Luo
Wenhan Luo
Xiao Luo
Zhiming Luo
Zhipeng Luo
Zhengyi Luo
Diogo C. Luvizon
Zhaoyang Lv
Gengyu Lyu
Lingjuan Lyu
Jun Lyu
Yuanyuan Lyu
Youwei Lyu
Yueming Lyu
Bingpeng Ma
Chao Ma
Chongyang Ma
Congbo Ma
Chih-Yao Ma
Fan Ma
Lin Ma
Haoyu Ma
Hengbo Ma
Jianqi Ma
Jiawei Ma
Jiayi Ma
Kede Ma
Kai Ma
Lingni Ma
Lei Ma
Xu Ma
Ning Ma
Benteng Ma
Cheng Ma
Andy J. Ma
Long Ma
Zhanyu Ma
Zhiheng Ma
Qianli Ma
Shiqiang Ma
Sizhuo Ma
Shiqing Ma
Xiaolong Ma
Xinzhu Ma

Gautam B. Machiraju
Spandan Madan
Mathew Magimai-Doss
Luca Magri
Behrooz Mahasseni
Upal Mahbub
Siddharth Mahendran
Paridhi Maheshwari
Rishabh Maheshwary
Mohammed Mahmoud
Shishira R. R. Maiya
Sylwia Majchrowska
Arjun Majumdar
Puspita Majumdar
Orchid Majumder
Sagnik Majumder
Ilya Makarov
Farkhod F.
    Makhmudkhujaev
Yasushi Makihara
Ankur Mali
Mateusz Malinowski
Utkarsh Mall
Srikanth Malla
Clement Mallet
Dimitrios Mallis
Yunze Man
Dipu Manandhar
Massimiliano Mancini
Murari Mandal
Raunak Manekar
Karttikeya Mangalam
Puneet Mangla
Fabian Manhardt
Sivabalan Manivasagam
Fahim Mannan
Chengzhi Mao
Hanzi Mao
Jiayuan Mao
Junhua Mao
Zhiyuan Mao
Jiageng Mao
Yunyao Mao
Zhendong Mao
Alberto Marchisio

Diego Marcos
Riccardo Marin
Aram Markosyan
Renaud Marlet
Ricardo Marques
Miquel Martí i Rabadán
Diego Martin Arroyo
Niki Martinel
Brais Martinez
Julieta Martinez
Marc Masana
Tomohiro Mashita
Timothée Masquelier
Minesh Mathew
Tetsu Matsukawa
Marwan Mattar
Bruce A. Maxwell
Christoph Mayer
Mantas Mazeika
Pratik Mazumder
Scott McCloskey
Steven McDonagh
Ishit Mehta
Jie Mei
Kangfu Mei
Jieru Mei
Xiaoguang Mei
Givi Meishvili
Luke Melas-Kyriazi
Iaroslav Melekhov
Andres Mendez-Vazquez
Heydi Mendez-Vazquez
Matias Mendieta
Ricardo A. Mendoza-León
Chenlin Meng
Depu Meng
Rang Meng
Zibo Meng
Qingjie Meng
Qier Meng
Yanda Meng
Zihang Meng
Thomas Mensink
Fabian Mentzer
Christopher Metzler

Gregory P. Meyer
Vasileios Mezaris
Liang Mi
Lu Mi
Bo Miao
Changtao Miao
Zichen Miao
Qiguang Miao
Xin Miao
Zhongqi Miao
Frank Michel
Simone Milani
Ben Mildenhall
Roy V. Miles
Juhong Min
Kyle Min
Hyun-Seok Min
Weiqing Min
Yuecong Min
Zhixiang Min
Qi Ming
David Minnen
Aymen Mir
Deepak Mishra
Anand Mishra
Shlok K. Mishra
Niluthpol Mithun
Gaurav Mittal
Trisha Mittal
Daisuke Miyazaki
Kaichun Mo
Hong Mo
Zhipeng Mo
Davide Modolo
Abduallah A. Mohamed
Mohamed Afham
    Mohamed Aflal
Ron Mokady
Pavlo Molchanov
Davide Moltisanti
Liliane Momeni
Gianluca Monaci
Pascal Monasse
Ajoy Mondal
Tom Monnier

Aron Monszpart
Gyeongsik Moon
Suhong Moon
Taesup Moon
Sean Moran
Daniel Moreira
Pietro Morerio
Alexandre Morgand
Lia Morra
Ali Mosleh
Inbar Mosseri
Sayed Mohammad
    Mostafavi Isfahani
Saman Motamed
Ramy A. Mounir
Fangzhou Mu
Jiteng Mu
Norman Mu
Yasuhiro Mukaigawa
Ryan Mukherjee
Tanmoy Mukherjee
Yusuke Mukuta
Ravi Teja Mullapudi
Lea Müller
Matthias Müller
Martin Mundt
Nils Murrugarra-Llerena
Damien Muselet
Armin Mustafa
Muhammad Ferjad Naeem
Sauradip Nag
Hajime Nagahara
Pravin Nagar
Rajendra Nagar
Naveen Shankar Nagaraja
Varun Nagaraja
Tushar Nagarajan
Seungjun Nah
Gaku Nakano
Yuta Nakashima
Giljoo Nam
Seonghyeon Nam
Liangliang Nan
Yuesong Nan
Yeshwanth Napolean

Dinesh Reddy
    Narapureddy
Medhini Narasimhan
Supreeth
    Narasimhaswamy
Sriram Narayanan
Erickson R. Nascimento
Varun Nasery
K. L. Navaneet
Pablo Navarrete Michelini
Shant Navasardyan
Shah Nawaz
Nihal Nayak
Farhood Negin
Lukáš Neumann
Alejandro Newell
Evonne Ng
Kam Woh Ng
Tony Ng
Anh Nguyen
Tuan Anh Nguyen
Cuong Cao Nguyen
Ngoc Cuong Nguyen
Thanh Nguyen
Khoi Nguyen
Phi Le Nguyen
Phong Ha Nguyen
Tam Nguyen
Truong Nguyen
Anh Tuan Nguyen
Rang Nguyen
Thao Thi Phuong Nguyen
Van Nguyen Nguyen
Zhen-Liang Ni
Yao Ni
Shijie Nie
Xuecheng Nie
Yongwei Nie
Weizhi Nie
Ying Nie
Yinyu Nie
Kshitij N. Nikhal
Simon Niklaus
Xuefei Ning
Jifeng Ning

Yotam Nitzan
Di Niu
Shuaicheng Niu
Li Niu
Wei Niu
Yulei Niu
Zhenxing Niu
Albert No
Shohei Nobuhara
Nicoletta Noceti
Junhyug Noh
Sotiris Nousias
Slawomir Nowaczyk
Ewa M. Nowara
Valsamis Ntouskos
Gilberto Ochoa-Ruiz
Ferda Ofli
Jihyong Oh
Sangyun Oh
Youngtaek Oh
Hiroki Ohashi
Takahiro Okabe
Kemal Oksuz
Fumio Okura
Daniel Olmeda Reino
Matthew Olson
Carl Olsson
Roy Or-El
Alessandro Ortis
Guillermo Ortiz-Jimenez
Magnus Oskarsson
Ahmed A. A. Osman
Martin R. Oswald
Mayu Otani
Naima Otberdout
Cheng Ouyang
Jiahong Ouyang
Wanli Ouyang
Andrew Owens
Poojan B. Oza
Mete Ozay
A. Cengiz Oztireli
Gautam Pai
Tomas Pajdla
Umapada Pal

Simone Palazzo
Luca Palmieri
Bowen Pan
Hao Pan
Lili Pan
Tai-Yu Pan
Liang Pan
Chengwei Pan
Yingwei Pan
Xuran Pan
Jinshan Pan
Xinyu Pan
Liyuan Pan
Xingang Pan
Xingjia Pan
Zhihong Pan
Zizheng Pan
Priyadarshini Panda
Rameswar Panda
Rohit Pandey
Kaiyue Pang
Bo Pang
Guansong Pang
Jiangmiao Pang
Meng Pang
Tianyu Pang
Ziqi Pang
Omiros Pantazis
Andreas Panteli
Maja Pantic
Marina Paolanti
Joao P. Papa
Samuele Papa
Mike Papadakis
Dim P. Papadopoulos
George Papandreou
Constantin Pape
Toufiq Parag
Chethan Parameshwara
Shaifali Parashar
Alejandro Pardo
Rishubh Parihar
Sarah Parisot
JaeYoo Park
Gyeong-Moon Park

Hyojin Park
Hyoungseob Park
Jongchan Park
Jae Sung Park
Kiru Park
Chunghyun Park
Kwanyong Park
Sunghyun Park
Sungrae Park
Seongsik Park
Sanghyun Park
Sungjune Park
Taesung Park
Gaurav Parmar
Paritosh Parmar
Alvaro Parra
Despoina Paschalidou
Or Patashnik
Shivansh Patel
Pushpak Pati
Prashant W. Patil
Vaishakh Patil
Suvam Patra
Jay Patravali
Badri Narayana Patro
Angshuman Paul
Sudipta Paul
Rémi Pautrat
Nick E. Pears
Adithya Pediredla
Wenjie Pei
Shmuel Peleg
Latha Pemula
Bo Peng
Houwen Peng
Yue Peng
Liangzu Peng
Baoyun Peng
Jun Peng
Pai Peng
Sida Peng
Xi Peng
Yuxin Peng
Songyou Peng
Wei Peng

Weiqi Peng
Wen-Hsiao Peng
Pramuditha Perera
Juan C. Perez
Eduardo Pérez Pellitero
Juan-Manuel Perez-Rua
Federico Pernici
Marco Pesavento
Stavros Petridis
Ilya A. Petrov
Vladan Petrovic
Mathis Petrovich
Suzanne Petryk
Hieu Pham
Quang Pham
Khoi Pham
Tung Pham
Huy Phan
Stephen Phillips
Cheng Perng Phoo
David Picard
Marco Piccirilli
Georg Pichler
A. J. Piergiovanni
Vipin Pillai
Silvia L. Pintea
Giovanni Pintore
Robinson Piramuthu
Fiora Pirri
Theodoros Pissas
Fabio Pizzati
Benjamin Planche
Bryan Plummer
Matteo Poggi
Ashwini Pokle
Georgy E. Ponimatkin
Adrian Popescu
Stefan Popov
Nikola Popović
Ronald Poppe
Angelo Porrello
Michael Potter
Charalambos Poullis
Hadi Pouransari
Omid Poursaeed

Shraman Pramanick
Mantini Pranav
Dilip K. Prasad
Meghshyam Prasad
B. H. Pawan Prasad
Shitala Prasad
Prateek Prasanna
Ekta Prashnani
Derek S. Prijatelj
Luke Y. Prince
Véronique Prinet
Victor Adrian Prisacariu
James Pritts
Thomas Probst
Sergey Prokudin
Rita Pucci
Chi-Man Pun
Matthew Purri
Haozhi Qi
Lu Qi
Lei Qi
Xianbiao Qi
Yonggang Qi
Yuankai Qi
Siyuan Qi
Guocheng Qian
Hangwei Qian
Qi Qian
Deheng Qian
Shengsheng Qian
Wen Qian
Rui Qian
Yiming Qian
Shengju Qian
Shengyi Qian
Xuelin Qian
Zhenxing Qian
Nan Qiao
Xiaotian Qiao
Jing Qin
Can Qin
Siyang Qin
Hongwei Qin
Jie Qin
Minghai Qin

Yipeng Qin
Yongqiang Qin
Wenda Qin
Xuebin Qin
Yuzhe Qin
Yao Qin
Zhenyue Qin
Zhiwu Qing
Heqian Qiu
Jiayan Qiu
Jielin Qiu
Yue Qiu
Jiaxiong Qiu
Zhongxi Qiu
Shi Qiu
Zhaofan Qiu
Zhongnan Qu
Yanyun Qu
Kha Gia Quach
Yuhui Quan
Ruijie Quan
Mike Rabbat
Rahul Shekhar Rade
Filip Radenovic
Gorjan Radevski
Bogdan Raducanu
Francesco Ragusa
Shafin Rahman
Md Mahfuzur Rahman
  Siddiquee
Hossein Rahmani
Kiran Raja
Sivaramakrishnan
  Rajaraman
Jathushan Rajasegaran
Adnan Siraj Rakin
Michaël Ramamonjisoa
Chirag A. Raman
Shanmuganathan Raman
Vignesh Ramanathan
Vasili Ramanishka
Vikram V. Ramaswamy
Merey Ramazanova
Jason Rambach
Sai Saketh Rambhatla

Clément Rambour
Ashwin Ramesh Babu
Adín Ramírez Rivera
Arianna Rampini
Haoxi Ran
Aakanksha Rana
Aayush Jung Bahadur
  Rana
Kanchana N. Ranasinghe
Aneesh Rangnekar
Samrudhdhi B. Rangrej
Harsh Rangwani
Viresh Ranjan
Anyi Rao
Yongming Rao
Carolina Raposo
Michalis Raptis
Amir Rasouli
Vivek Rathod
Adepu Ravi Sankar
Avinash Ravichandran
Bharadwaj Ravichandran
Dripta S. Raychaudhuri
Adria Recasens
Simon Reiß
Davis Rempe
Daxuan Ren
Jiawei Ren
Jimmy Ren
Sucheng Ren
Dayong Ren
Zhile Ren
Dongwei Ren
Qibing Ren
Pengfei Ren
Zhenwen Ren
Xuqian Ren
Yixuan Ren
Zhongzheng Ren
Ambareesh Revanur
Hamed Rezazadegan
  Tavakoli
Rafael S. Rezende
Wonjong Rhee
Alexander Richard

Christian Richardt
Stephan R. Richter
Benjamin Riggan
Dominik Rivoir
Mamshad Nayeem Rizve
Joshua D. Robinson
Joseph Robinson
Chris Rockwell
Ranga Rodrigo
Andres C. Rodriguez
Carlos Rodriguez-Pardo
Marcus Rohrbach
Gemma Roig
Yu Rong
David A. Ross
Mohammad Rostami
Edward Rosten
Karsten Roth
Anirban Roy
Debaditya Roy
Shuvendu Roy
Ahana Roy Choudhury
Aruni Roy Chowdhury
Denys Rozumnyi
Shulan Ruan
Wenjie Ruan
Patrick Ruhkamp
Danila Rukhovich
Anian Ruoss
Chris Russell
Dan Ruta
Dawid Damian Rymarczyk
DongHun Ryu
Hyeonggon Ryu
Kwonyoung Ryu
Balasubramanian S.
Alexandre Sablayrolles
Mohammad Sabokrou
Arka Sadhu
Aniruddha Saha
Oindrila Saha
Pritish Sahu
Aneeshan Sain
Nirat Saini
Saurabh Saini

Takeshi Saitoh
Christos Sakaridis
Fumihiko Sakaue
Dimitrios Sakkos
Ken Sakurada
Parikshit V. Sakurikar
Rohit Saluja
Nermin Samet
Leo Sampaio Ferraz
    Ribeiro
Jorge Sanchez
Enrique Sanchez
Shengtian Sang
Anush Sankaran
Soubhik Sanyal
Nikolaos Sarafianos
Vishwanath Saragadam
István Sárándi
Saquib Sarfraz
Mert Bulent Sariyildiz
Anindya Sarkar
Pritam Sarkar
Paul-Edouard Sarlin
Hiroshi Sasaki
Takami Sato
Torsten Sattler
Ravi Kumar Satzoda
Axel Sauer
Stefano Savian
Artem Savkin
Manolis Savva
Gerald Schaefer
Simone Schaub-Meyer
Yoni Schirris
Samuel Schulter
Katja Schwarz
Jesse Scott
Sinisa Segvic
Constantin Marc Seibold
Lorenzo Seidenari
Matan Sela
Fadime Sener
Paul Hongsuck Seo
Kwanggyoon Seo
Hongje Seong

Dario Serez
Francesco Setti
Bryan Seybold
Mohamad Shahbazi
Shima Shahfar
Xinxin Shan
Caifeng Shan
Dandan Shan
Shawn Shan
Wei Shang
Jinghuan Shang
Jiaxiang Shang
Lei Shang
Sukrit Shankar
Ken Shao
Rui Shao
Jie Shao
Mingwen Shao
Aashish Sharma
Gaurav Sharma
Vivek Sharma
Abhishek Sharma
Yoli Shavit
Shashank Shekhar
Sumit Shekhar
Zhijie Shen
Fengyi Shen
Furao Shen
Jialie Shen
Jingjing Shen
Ziyi Shen
Linlin Shen
Guangyu Shen
Biluo Shen
Falong Shen
Jiajun Shen
Qiu Shen
Qiuhong Shen
Shuai Shen
Wang Shen
Yiqing Shen
Yunhang Shen
Siqi Shen
Bin Shen
Tianwei Shen

Xi Shen
Yilin Shen
Yuming Shen
Yucong Shen
Zhiqiang Shen
Lu Sheng
Yichen Sheng
Shivanand Venkanna
    Sheshappanavar
Shelly Sheynin
Baifeng Shi
Ruoxi Shi
Botian Shi
Hailin Shi
Jia Shi
Jing Shi
Shaoshuai Shi
Baoguang Shi
Boxin Shi
Hengcan Shi
Tianyang Shi
Xiaodan Shi
Yongjie Shi
Zhensheng Shi
Yinghuan Shi
Weiqi Shi
Wu Shi
Xuepeng Shi
Xiaoshuang Shi
Yujiao Shi
Zenglin Shi
Zhenmei Shi
Takashi Shibata
Meng-Li Shih
Yichang Shih
Hyunjung Shim
Dongseok Shim
Soshi Shimada
Inkyu Shin
Jinwoo Shin
Seungjoo Shin
Seungjae Shin
Koichi Shinoda
Suprosanna Shit

Palaiahnakote
    Shivakumara
Eli Shlizerman
Gaurav Shrivastava
Xiao Shu
Xiangbo Shu
Xiujun Shu
Yang Shu
Tianmin Shu
Jun Shu
Zhixin Shu
Bing Shuai
Maria Shugrina
Ivan Shugurov
Satya Narayan Shukla
Pranjay Shyam
Jianlou Si
Yawar Siddiqui
Alberto Signoroni
Pedro Silva
Jae-Young Sim
Oriane Siméoni
Martin Simon
Andrea Simonelli
Abhishek Singh
Ashish Singh
Dinesh Singh
Gurkirt Singh
Krishna Kumar Singh
Mannat Singh
Pravendra Singh
Rajat Vikram Singh
Utkarsh Singhal
Dipika Singhania
Vasu Singla
Harsh Sinha
Sudipta Sinha
Josef Sivic
Elena Sizikova
Geri Skenderi
Ivan Skorokhodov
Dmitriy Smirnov
Cameron Y. Smith
James S. Smith
Patrick Snape

Mattia Soldan
Hyeongseok Son
Sanghyun Son
Chuanbiao Song
Chen Song
Chunfeng Song
Dan Song
Dongjin Song
Hwanjun Song
Guoxian Song
Jiaming Song
Jie Song
Liangchen Song
Ran Song
Luchuan Song
Xibin Song
Li Song
Fenglong Song
Guoli Song
Guanglu Song
Zhenbo Song
Lin Song
Xinhang Song
Yang Song
Yibing Song
Rajiv Soundararajan
Hossein Souri
Cristovao Sousa
Riccardo Spezialetti
Leonidas Spinoulas
Michael W. Spratling
Deepak Sridhar
Srinath Sridhar
Gaurang Sriramanan
Vinkle Kumar Srivastav
Themos Stafylakis
Serban Stan
Anastasis Stathopoulos
Markus Steinberger
Jan Steinbrener
Sinisa Stekovic
Alexandros Stergiou
Gleb Sterkin
Rainer Stiefelhagen
Pierre Stock

Ombretta Strafforello
Julian Straub
Yannick Strümpler
Joerg Stueckler
Hang Su
Weijie Su
Jong-Chyi Su
Bing Su
Haisheng Su
Jinming Su
Yiyang Su
Yukun Su
Yuxin Su
Zhuo Su
Zhaoqi Su
Xiu Su
Yu-Chuan Su
Zhixun Su
Arulkumar Subramaniam
Akshayvarun Subramanya
A. Subramanyam
Swathikiran Sudhakaran
Yusuke Sugano
Masanori Suganuma
Yumin Suh
Yang Sui
Baochen Sun
Cheng Sun
Long Sun
Guolei Sun
Haoliang Sun
Haomiao Sun
He Sun
Hanqing Sun
Hao Sun
Lichao Sun
Jiachen Sun
Jiaming Sun
Jian Sun
Jin Sun
Jennifer J. Sun
Tiancheng Sun
Libo Sun
Peize Sun
Qianru Sun

Shanlin Sun
Yu Sun
Zhun Sun
Che Sun
Lin Sun
Tao Sun
Yiyou Sun
Chunyi Sun
Chong Sun
Weiwei Sun
Weixuan Sun
Xiuyu Sun
Yanan Sun
Zeren Sun
Zhaodong Sun
Zhiqing Sun
Minhyuk Sung
Jinli Suo
Simon Suo
Abhijit Suprem
Anshuman Suri
Saksham Suri
Joshua M. Susskind
Roman Suvorov
Gurumurthy Swaminathan
Robin Swanson
Paul Swoboda
Tabish A. Syed
Richard Szeliski
Fariborz Taherkhani
Yu-Wing Tai
Keita Takahashi
Walter Talbott
Gary Tam
Masato Tamura
Feitong Tan
Fuwen Tan
Shuhan Tan
Andong Tan
Bin Tan
Cheng Tan
Jianchao Tan
Lei Tan
Mingxing Tan
Xin Tan

Zichang Tan
Zhentao Tan
Kenichiro Tanaka
Masayuki Tanaka
Yushun Tang
Hao Tang
Jingqun Tang
Jinhui Tang
Kaihua Tang
Luming Tang
Lv Tang
Sheyang Tang
Shitao Tang
Siliang Tang
Shixiang Tang
Yansong Tang
Keke Tang
Chang Tang
Chenwei Tang
Jie Tang
Junshu Tang
Ming Tang
Peng Tang
Xu Tang
Yao Tang
Chen Tang
Fan Tang
Haoran Tang
Shengeng Tang
Yehui Tang
Zhipeng Tang
Ugo Tanielian
Chaofan Tao
Jiale Tao
Junli Tao
Renshuai Tao
An Tao
Guanhong Tao
Zhiqiang Tao
Makarand Tapaswi
Jean-Philippe G. Tarel
Juan J. Tarrio
Enzo Tartaglione
Keisuke Tateno
Zachary Teed

Ajinkya B. Tejankar
Bugra Tekin
Purva Tendulkar
Damien Teney
Minggui Teng
Chris Tensmeyer
Andrew Beng Jin Teoh
Philipp Terhörst
Kartik Thakral
Nupur Thakur
Kevin Thandiackal
Spyridon Thermos
Diego Thomas
William Thong
Yuesong Tian
Guanzhong Tian
Lin Tian
Shiqi Tian
Kai Tian
Meng Tian
Tai-Peng Tian
Zhuotao Tian
Shangxuan Tian
Tian Tian
Yapeng Tian
Yu Tian
Yuxin Tian
Leslie Ching Ow Tiong
Praveen Tirupattur
Garvita Tiwari
George Toderici
Antoine Toisoul
Aysim Toker
Tatiana Tommasi
Zhan Tong
Alessio Tonioni
Alessandro Torcinovich
Fabio Tosi
Matteo Toso
Hugo Touvron
Quan Hung Tran
Son Tran
Hung Tran
Ngoc-Trung Tran
Vinh Tran

Phong Tran
Giovanni Trappolini
Edith Tretschk
Subarna Tripathi
Shubhendu Trivedi
Eduard Trulls
Prune Truong
Thanh-Dat Truong
Tomasz Trzcinski
Sam Tsai
Yi-Hsuan Tsai
Ethan Tseng
Yu-Chee Tseng
Shahar Tsiper
Stavros Tsogkas
Shikui Tu
Zhigang Tu
Zhengzhong Tu
Richard Tucker
Sergey Tulyakov
Cigdem Turan
Daniyar Turmukhambetov
Victor G. Turrisi da Costa
Bartlomiej Twardowski
Christopher D. Twigg
Radim Tylecek
Mostofa Rafid Uddin
Md. Zasim Uddin
Kohei Uehara
Nicolas Ugrinovic
Youngjung Uh
Norimichi Ukita
Anwaar Ulhaq
Devesh Upadhyay
Paul Upchurch
Yoshitaka Ushiku
Yuzuko Utsumi
Mikaela Angelina Uy
Mohit Vaishnav
Pratik Vaishnavi
Jeya Maria Jose Valanarasu
Matias A. Valdenegro Toro
Diego Valsesia
Wouter Van Gansbeke
Nanne van Noord

Simon Vandenhende
Farshid Varno
Cristina Vasconcelos
Francisco Vasconcelos
Alex Vasilescu
Subeesh Vasu
Arun Balajee Vasudevan
Kanav Vats
Vaibhav S. Vavilala
Sagar Vaze
Javier Vazquez-Corral
Andrea Vedaldi
Olga Veksler
Andreas Velten
Sai H. Vemprala
Raviteja Vemulapalli
Shashanka
    Venkataramanan
Dor Verbin
Luisa Verdoliva
Manisha Verma
Yashaswi Verma
Constantin Vertan
Eli Verwimp
Deepak Vijaykeerthy
Pablo Villanueva
Ruben Villegas
Markus Vincze
Vibhav Vineet
Minh P. Vo
Huy V. Vo
Duc Minh Vo
Tomas Vojir
Igor Vozniak
Nicholas Vretos
Vibashan VS
Tuan-Anh Vu
Thang Vu
Mårten Wadenbäck
Neal Wadhwa
Aaron T. Walsman
Steven Walton
Jin Wan
Alvin Wan
Jia Wan

Jun Wan
Xiaoyue Wan
Fang Wan
Guowei Wan
Renjie Wan
Zhiqiang Wan
Ziyu Wan
Bastian Wandt
Dongdong Wang
Limin Wang
Haiyang Wang
Xiaobing Wang
Angtian Wang
Angelina Wang
Bing Wang
Bo Wang
Boyu Wang
Binghui Wang
Chen Wang
Chien-Yi Wang
Congli Wang
Qi Wang
Chengrui Wang
Rui Wang
Yiqun Wang
Cong Wang
Wenjing Wang
Dongkai Wang
Di Wang
Xiaogang Wang
Kai Wang
Zhizhong Wang
Fangjinhua Wang
Feng Wang
Hang Wang
Gaoang Wang
Guoqing Wang
Guangcong Wang
Guangzhi Wang
Hanqing Wang
Hao Wang
Haohan Wang
Haoran Wang
Hong Wang
Haotao Wang

Hu Wang
Huan Wang
Hua Wang
Hui-Po Wang
Hengli Wang
Hanyu Wang
Hongxing Wang
Jingwen Wang
Jialiang Wang
Jian Wang
Jianyi Wang
Jiashun Wang
Jiahao Wang
Tsun-Hsuan Wang
Xiaoqian Wang
Jinqiao Wang
Jun Wang
Jianzong Wang
Kaihong Wang
Ke Wang
Lei Wang
Lingjing Wang
Linnan Wang
Lin Wang
Liansheng Wang
Mengjiao Wang
Manning Wang
Nannan Wang
Peihao Wang
Jiayun Wang
Pu Wang
Qiang Wang
Qiufeng Wang
Qilong Wang
Qiangchang Wang
Qin Wang
Qing Wang
Ruocheng Wang
Ruibin Wang
Ruisheng Wang
Ruizhe Wang
Runqi Wang
Runzhong Wang
Wenxuan Wang
Sen Wang

Shangfei Wang
Shaofei Wang
Shijie Wang
Shiqi Wang
Zhibo Wang
Song Wang
Xinjiang Wang
Tai Wang
Tao Wang
Teng Wang
Xiang Wang
Tianren Wang
Tiantian Wang
Tianyi Wang
Fengjiao Wang
Wei Wang
Miaohui Wang
Suchen Wang
Siyue Wang
Yaoming Wang
Xiao Wang
Ze Wang
Biao Wang
Chaofei Wang
Dong Wang
Gu Wang
Guangrun Wang
Guangming Wang
Guo-Hua Wang
Haoqing Wang
Hesheng Wang
Huafeng Wang
Jinghua Wang
Jingdong Wang
Jingjing Wang
Jingya Wang
Jingkang Wang
Jiakai Wang
Junke Wang
Kuo Wang
Lichen Wang
Lizhi Wang
Longguang Wang
Mang Wang
Mei Wang

Min Wang
Peng-Shuai Wang
Run Wang
Shaoru Wang
Shuhui Wang
Tan Wang
Tiancai Wang
Tianqi Wang
Wenhai Wang
Wenzhe Wang
Xiaobo Wang
Xiudong Wang
Xu Wang
Yajie Wang
Yan Wang
Yuan-Gen Wang
Yingqian Wang
Yizhi Wang
Yulin Wang
Yu Wang
Yujie Wang
Yunhe Wang
Yuxi Wang
Yaowei Wang
Yiwei Wang
Zezheng Wang
Hongzhi Wang
Zhiqiang Wang
Ziteng Wang
Ziwei Wang
Zheng Wang
Zhenyu Wang
Binglu Wang
Zhongdao Wang
Ce Wang
Weining Wang
Weiyao Wang
Wenbin Wang
Wenguan Wang
Guangting Wang
Haolin Wang
Haiyan Wang
Huiyu Wang
Naiyan Wang
Jingbo Wang

Jinpeng Wang
Jiaqi Wang
Liyuan Wang
Lizhen Wang
Ning Wang
Wenqian Wang
Sheng-Yu Wang
Weimin Wang
Xiaohan Wang
Yifan Wang
Yi Wang
Yongtao Wang
Yizhou Wang
Zhuo Wang
Zhe Wang
Xudong Wang
Xiaofang Wang
Xinggang Wang
Xiaosen Wang
Xiaosong Wang
Xiaoyang Wang
Lijun Wang
Xinlong Wang
Xuan Wang
Xue Wang
Yangang Wang
Yaohui Wang
Yu-Chiang Frank Wang
Yida Wang
Yilin Wang
Yi Ru Wang
Yali Wang
Yinglong Wang
Yufu Wang
Yujiang Wang
Yuwang Wang
Yuting Wang
Yang Wang
Yu-Xiong Wang
Yixu Wang
Ziqi Wang
Zhicheng Wang
Zeyu Wang
Zhaowen Wang
Zhenyi Wang

Zhenzhi Wang
Zhijie Wang
Zhiyong Wang
Zhongling Wang
Zhuowei Wang
Zian Wang
Zifu Wang
Zihao Wang
Zirui Wang
Ziyan Wang
Wenxiao Wang
Zhen Wang
Zhepeng Wang
Zi Wang
Zihao W. Wang
Steven L. Waslander
Olivia Watkins
Daniel Watson
Silvan Weder
Dongyoon Wee
Dongming Wei
Tianyi Wei
Jia Wei
Dong Wei
Fangyun Wei
Longhui Wei
Mingqiang Wei
Xinyue Wei
Chen Wei
Donglai Wei
Pengxu Wei
Xing Wei
Xiu-Shen Wei
Wenqi Wei
Guoqiang Wei
Wei Wei
XingKui Wei
Xian Wei
Xingxing Wei
Yake Wei
Yuxiang Wei
Yi Wei
Luca Weihs
Michael Weinmann
Martin Weinmann

Congcong Wen
Chuan Wen
Jie Wen
Sijia Wen
Song Wen
Chao Wen
Xiang Wen
Zeyi Wen
Xin Wen
Yilin Wen
Yijia Weng
Shuchen Weng
Junwu Weng
Wenming Weng
Renliang Weng
Zhenyu Weng
Xinshuo Weng
Nicholas J. Westlake
Gordon Wetzstein
Lena M. Widin Klasén
Rick Wildes
Bryan M. Williams
Williem Williem
Ole Winther
Scott Wisdom
Alex Wong
Chau-Wai Wong
Kwan-Yee K. Wong
Yongkang Wong
Scott Workman
Marcel Worring
Michael Wray
Safwan Wshah
Xiang Wu
Aming Wu
Chongruo Wu
Cho-Ying Wu
Chunpeng Wu
Chenyan Wu
Ziyi Wu
Fuxiang Wu
Gang Wu
Haiping Wu
Huisi Wu
Jane Wu

Jialian Wu
Jing Wu
Jinjian Wu
Jianlong Wu
Xian Wu
Lifang Wu
Lifan Wu
Minye Wu
Qianyi Wu
Rongliang Wu
Rui Wu
Shiqian Wu
Shuzhe Wu
Shangzhe Wu
Tsung-Han Wu
Tz-Ying Wu
Ting-Wei Wu
Jiannan Wu
Zhiliang Wu
Yu Wu
Chenyun Wu
Dayan Wu
Dongxian Wu
Fei Wu
Hefeng Wu
Jianxin Wu
Weibin Wu
Wenxuan Wu
Wenhao Wu
Xiao Wu
Yicheng Wu
Yuanwei Wu
Yu-Huan Wu
Zhenxin Wu
Zhenyu Wu
Wei Wu
Peng Wu
Xiaohe Wu
Xindi Wu
Xinxing Wu
Xinyi Wu
Xingjiao Wu
Xiongwei Wu
Yangzheng Wu
Yanzhao Wu

Yawen Wu
Yong Wu
Yi Wu
Ying Nian Wu
Zhenyao Wu
Zhonghua Wu
Zongze Wu
Zuxuan Wu
Stefanie Wuhrer
Teng Xi
Jianing Xi
Fei Xia
Haifeng Xia
Menghan Xia
Yuanqing Xia
Zhihua Xia
Xiaobo Xia
Weihao Xia
Shihong Xia
Yan Xia
Yong Xia
Zhaoyang Xia
Zhihao Xia
Chuhua Xian
Yongqin Xian
Wangmeng Xiang
Fanbo Xiang
Tiange Xiang
Tao Xiang
Liuyu Xiang
Xiaoyu Xiang
Zhiyu Xiang
Aoran Xiao
Chunxia Xiao
Fanyi Xiao
Jimin Xiao
Jun Xiao
Taihong Xiao
Anqi Xiao
Junfei Xiao
Jing Xiao
Liang Xiao
Yang Xiao
Yuting Xiao
Yijun Xiao

Yao Xiao
Zeyu Xiao
Zhisheng Xiao
Zihao Xiao
Binhui Xie
Christopher Xie
Haozhe Xie
Jin Xie
Guo-Sen Xie
Hongtao Xie
Ming-Kun Xie
Tingting Xie
Chaohao Xie
Weicheng Xie
Xudong Xie
Jiyang Xie
Xiaohua Xie
Yuan Xie
Zhenyu Xie
Ning Xie
Xianghui Xie
Xiufeng Xie
You Xie
Yutong Xie
Fuyong Xing
Yifan Xing
Zhen Xing
Yuanjun Xiong
Jinhui Xiong
Weihua Xiong
Hongkai Xiong
Zhitong Xiong
Yuanhao Xiong
Yunyang Xiong
Yuwen Xiong
Zhiwei Xiong
Yuliang Xiu
An Xu
Chang Xu
Chenliang Xu
Chengming Xu
Chenshu Xu
Xiang Xu
Huijuan Xu
Zhe Xu

Jie Xu
Jingyi Xu
Jiarui Xu
Yinghao Xu
Kele Xu
Ke Xu
Li Xu
Linchuan Xu
Linning Xu
Mengde Xu
Mengmeng Frost Xu
Min Xu
Mingye Xu
Jun Xu
Ning Xu
Peng Xu
Runsheng Xu
Sheng Xu
Wenqiang Xu
Xiaogang Xu
Renzhe Xu
Kaidi Xu
Yi Xu
Chi Xu
Qiuling Xu
Baobei Xu
Feng Xu
Haohang Xu
Haofei Xu
Lan Xu
Mingze Xu
Songcen Xu
Weipeng Xu
Wenjia Xu
Wenju Xu
Xiangyu Xu
Xin Xu
Yinshuang Xu
Yixing Xu
Yuting Xu
Yanyu Xu
Zhenbo Xu
Zhiliang Xu
Zhiyuan Xu
Xiaohao Xu

Yanwu Xu
Yan Xu
Yiran Xu
Yifan Xu
Yufei Xu
Yong Xu
Zichuan Xu
Zenglin Xu
Zexiang Xu
Zhan Xu
Zheng Xu
Zhiwei Xu
Ziyue Xu
Shiyu Xuan
Hanyu Xuan
Fei Xue
Jianru Xue
Mingfu Xue
Qinghan Xue
Tianfan Xue
Chao Xue
Chuhui Xue
Nan Xue
Zhou Xue
Xiangyang Xue
Yuan Xue
Abhay Yadav
Ravindra Yadav
Kota Yamaguchi
Toshihiko Yamasaki
Kohei Yamashita
Chaochao Yan
Feng Yan
Kun Yan
Qingsen Yan
Qixin Yan
Rui Yan
Siming Yan
Xinchen Yan
Yaping Yan
Bin Yan
Qingan Yan
Shen Yan
Shipeng Yan
Xu Yan

Yan Yan
Yichao Yan
Zhaoyi Yan
Zike Yan
Zhiqiang Yan
Hongliang Yan
Zizheng Yan
Jiewen Yang
Anqi Joyce Yang
Shan Yang
Anqi Yang
Antoine Yang
Bo Yang
Baoyao Yang
Chenhongyi Yang
Dingkang Yang
De-Nian Yang
Dong Yang
David Yang
Fan Yang
Fengyu Yang
Fengting Yang
Fei Yang
Gengshan Yang
Heng Yang
Han Yang
Huan Yang
Yibo Yang
Jiancheng Yang
Jihan Yang
Jiawei Yang
Jiayu Yang
Jie Yang
Jinfa Yang
Jingkang Yang
Jinyu Yang
Cheng-Fu Yang
Ji Yang
Jianyu Yang
Kailun Yang
Tian Yang
Luyu Yang
Liang Yang
Li Yang
Michael Ying Yang

Yang Yang
Muli Yang
Le Yang
Qiushi Yang
Ren Yang
Ruihan Yang
Shuang Yang
Siyuan Yang
Su Yang
Shiqi Yang
Taojiannan Yang
Tianyu Yang
Lei Yang
Wanzhao Yang
Shuai Yang
William Yang
Wei Yang
Xiaofeng Yang
Xiaoshan Yang
Xin Yang
Xuan Yang
Xu Yang
Xingyi Yang
Xitong Yang
Jing Yang
Yanchao Yang
Wenming Yang
Yujiu Yang
Herb Yang
Jianfei Yang
Jinhui Yang
Chuanguang Yang
Guanglei Yang
Haitao Yang
Kewei Yang
Linlin Yang
Lijin Yang
Longrong Yang
Meng Yang
MingKun Yang
Sibei Yang
Shicai Yang
Tong Yang
Wen Yang
Xi Yang

Xiaolong Yang
Xue Yang
Yubin Yang
Ze Yang
Ziyi Yang
Yi Yang
Linjie Yang
Yuzhe Yang
Yiding Yang
Zhenpei Yang
Zhaohui Yang
Zhengyuan Yang
Zhibo Yang
Zongxin Yang
Hantao Yao
Mingde Yao
Rui Yao
Taiping Yao
Ting Yao
Cong Yao
Qingsong Yao
Quanming Yao
Xu Yao
Yuan Yao
Yao Yao
Yazhou Yao
Jiawen Yao
Shunyu Yao
Pew-Thian Yap
Sudhir Yarram
Rajeev Yasarla
Peng Ye
Botao Ye
Mao Ye
Fei Ye
Hanrong Ye
Jingwen Ye
Jinwei Ye
Jiarong Ye
Mang Ye
Meng Ye
Qi Ye
Qian Ye
Qixiang Ye
Junjie Ye

Sheng Ye
Nanyang Ye
Yufei Ye
Xiaoqing Ye
Ruolin Ye
Yousef Yeganeh
Chun-Hsiao Yeh
Raymond A. Yeh
Yu-Ying Yeh
Kai Yi
Chang Yi
Renjiao Yi
Xinping Yi
Peng Yi
Alper Yilmaz
Junho Yim
Hui Yin
Bangjie Yin
Jia-Li Yin
Miao Yin
Wenzhe Yin
Xuwang Yin
Ming Yin
Yu Yin
Aoxiong Yin
Kangxue Yin
Tianwei Yin
Wei Yin
Xianghua Ying
Rio Yokota
Tatsuya Yokota
Naoto Yokoya
Ryo Yonetani
Ki Yoon Yoo
Jinsu Yoo
Sunjae Yoon
Jae Shin Yoon
Jihun Yoon
Sung-Hoon Yoon
Ryota Yoshihashi
Yusuke Yoshiyasu
Chenyu You
Haoran You
Haoxuan You
Yang You

Quanzeng You
Tackgeun You
Kaichao You
Shan You
Xinge You
Yurong You
Baosheng Yu
Bei Yu
Haichao Yu
Hao Yu
Chaohui Yu
Fisher Yu
Jin-Gang Yu
Jiyang Yu
Jason J. Yu
Jiashuo Yu
Hong-Xing Yu
Lei Yu
Mulin Yu
Ning Yu
Peilin Yu
Qi Yu
Qian Yu
Rui Yu
Shuzhi Yu
Gang Yu
Tan Yu
Weijiang Yu
Xin Yu
Bingyao Yu
Ye Yu
Hanchao Yu
Yingchen Yu
Tao Yu
Xiaotian Yu
Qing Yu
Houjian Yu
Changqian Yu
Jing Yu
Jun Yu
Shujian Yu
Xiang Yu
Zhaofei Yu
Zhenbo Yu
Yinfeng Yu

Zhuoran Yu
Zitong Yu
Bo Yuan
Jiangbo Yuan
Liangzhe Yuan
Weihao Yuan
Jianbo Yuan
Xiaoyun Yuan
Ye Yuan
Li Yuan
Geng Yuan
Jialin Yuan
Maoxun Yuan
Peng Yuan
Xin Yuan
Yuan Yuan
Yuhui Yuan
Yixuan Yuan
Zheng Yuan
Mehmet Kerim Yücel
Kaiyu Yue
Haixiao Yue
Heeseung Yun
Sangdoo Yun
Tian Yun
Mahmut Yurt
Ekim Yurtsever
Ahmet Yüzügüler
Edouard Yvinec
Eloi Zablocki
Christopher Zach
Muhammad Zaigham
    Zaheer
Pierluigi Zama Ramirez
Yuhang Zang
Pietro Zanuttigh
Alexey Zaytsev
Bernhard Zeisl
Haitian Zeng
Pengpeng Zeng
Jiabei Zeng
Runhao Zeng
Wei Zeng
Yawen Zeng
Yi Zeng

Yiming Zeng
Tieyong Zeng
Huanqiang Zeng
Dan Zeng
Yu Zeng
Wei Zhai
Yuanhao Zhai
Fangneng Zhan
Kun Zhan
Xiong Zhang
Jingdong Zhang
Jiangning Zhang
Zhilu Zhang
Gengwei Zhang
Dongsu Zhang
Hui Zhang
Binjie Zhang
Bo Zhang
Tianhao Zhang
Cecilia Zhang
Jing Zhang
Chaoning Zhang
Chenxu Zhang
Chi Zhang
Chris Zhang
Yabin Zhang
Zhao Zhang
Rufeng Zhang
Chaoyi Zhang
Zheng Zhang
Da Zhang
Yi Zhang
Edward Zhang
Xin Zhang
Feifei Zhang
Feilong Zhang
Yuqi Zhang
GuiXuan Zhang
Hanlin Zhang
Hanwang Zhang
Hanzhen Zhang
Haotian Zhang
He Zhang
Haokui Zhang
Hongyuan Zhang

Hengrui Zhang
Hongming Zhang
Mingfang Zhang
Jianpeng Zhang
Jiaming Zhang
Jichao Zhang
Jie Zhang
Jingfeng Zhang
Jingyi Zhang
Jinnian Zhang
David Junhao Zhang
Junjie Zhang
Junzhe Zhang
Jiawan Zhang
Jingyang Zhang
Kai Zhang
Lei Zhang
Lihua Zhang
Lu Zhang
Miao Zhang
Minjia Zhang
Mingjin Zhang
Qi Zhang
Qian Zhang
Qilong Zhang
Qiming Zhang
Qiang Zhang
Richard Zhang
Ruimao Zhang
Ruisi Zhang
Ruixin Zhang
Runze Zhang
Qilin Zhang
Shan Zhang
Shanshan Zhang
Xi Sheryl Zhang
Song-Hai Zhang
Chongyang Zhang
Kaihao Zhang
Songyang Zhang
Shu Zhang
Siwei Zhang
Shujian Zhang
Tianyun Zhang
Tong Zhang

Tao Zhang
Wenwei Zhang
Wenqiang Zhang
Wen Zhang
Xiaolin Zhang
Xingchen Zhang
Xingxuan Zhang
Xiuming Zhang
Xiaoshuai Zhang
Xuanmeng Zhang
Xuanyang Zhang
Xucong Zhang
Xingxing Zhang
Xikun Zhang
Xiaohan Zhang
Yahui Zhang
Yunhua Zhang
Yan Zhang
Yanghao Zhang
Yifei Zhang
Yifan Zhang
Yi-Fan Zhang
Yihao Zhang
Yingliang Zhang
Youshan Zhang
Yulun Zhang
Yushu Zhang
Yixiao Zhang
Yide Zhang
Zhongwen Zhang
Bowen Zhang
Chen-Lin Zhang
Zehua Zhang
Zekun Zhang
Zeyu Zhang
Xiaowei Zhang
Yifeng Zhang
Cheng Zhang
Hongguang Zhang
Yuexi Zhang
Fa Zhang
Guofeng Zhang
Hao Zhang
Haofeng Zhang
Hongwen Zhang

Hua Zhang	Zhizhong Zhang	Bowen Zhao
Jiaxin Zhang	Qilong Zhangli	Pu Zhao
Zhenyu Zhang	Bingyin Zhao	Bingchen Zhao
Jian Zhang	Bin Zhao	Borui Zhao
Jianfeng Zhang	Chenglong Zhao	Fuqiang Zhao
Jiao Zhang	Lei Zhao	Hanbin Zhao
Jiakai Zhang	Feng Zhao	Jian Zhao
Lefei Zhang	Gangming Zhao	Mingyang Zhao
Le Zhang	Haiyan Zhao	Na Zhao
Mi Zhang	Hao Zhao	Rongchang Zhao
Min Zhang	Handong Zhao	Ruiqi Zhao
Ning Zhang	Hengshuang Zhao	Shuai Zhao
Pan Zhang	Yinan Zhao	Wenda Zhao
Pu Zhang	Jiaojiao Zhao	Wenliang Zhao
Qing Zhang	Jiaqi Zhao	Xiangyun Zhao
Renrui Zhang	Jing Zhao	Yifan Zhao
Shifeng Zhang	Kaili Zhao	Yaping Zhao
Shuo Zhang	Haojie Zhao	Zhou Zhao
Shaoxiong Zhang	Yucheng Zhao	He Zhao
Weizhong Zhang	Longjiao Zhao	Jie Zhao
Xi Zhang	Long Zhao	Xibin Zhao
Xiaomei Zhang	Qingsong Zhao	Xiaoqi Zhao
Xinyu Zhang	Qingyu Zhao	Zhengyu Zhao
Yin Zhang	Rui Zhao	Jin Zhe
Zicheng Zhang	Rui-Wei Zhao	Chuanxia Zheng
Zihao Zhang	Sicheng Zhao	Huan Zheng
Ziqi Zhang	Shuang Zhao	Hao Zheng
Zhaoxiang Zhang	Siyan Zhao	Jia Zheng
Zhen Zhang	Zelin Zhao	Jian-Qing Zheng
Zhipeng Zhang	Shiyu Zhao	Shuai Zheng
Zhixing Zhang	Wang Zhao	Meng Zheng
Zhizheng Zhang	Tiesong Zhao	Mingkai Zheng
Jiawei Zhang	Qian Zhao	Qian Zheng
Zhong Zhang	Wangbo Zhao	Qi Zheng
Pingping Zhang	Xi-Le Zhao	Wu Zheng
Yixin Zhang	Xu Zhao	Yinqiang Zheng
Kui Zhang	Yajie Zhao	Yufeng Zheng
Lingzhi Zhang	Yang Zhao	Yutong Zheng
Huaiwen Zhang	Ying Zhao	Yalin Zheng
Quanshi Zhang	Yin Zhao	Yu Zheng
Zhoutong Zhang	Yizhou Zhao	Feng Zheng
Yuhang Zhang	Yunhan Zhao	Zhaoheng Zheng
Yuting Zhang	Yuyang Zhao	Haitian Zheng
Zhang Zhang	Yue Zhao	Kang Zheng
Ziming Zhang	Yuzhi Zhao	Bolun Zheng

Haiyong Zheng
Mingwu Zheng
Sipeng Zheng
Tu Zheng
Wenzhao Zheng
Xiawu Zheng
Yinglin Zheng
Zhuo Zheng
Zilong Zheng
Kecheng Zheng
Zerong Zheng
Shuaifeng Zhi
Tiancheng Zhi
Jia-Xing Zhong
Yiwu Zhong
Fangwei Zhong
Zhihang Zhong
Yaoyao Zhong
Yiran Zhong
Zhun Zhong
Zichun Zhong
Bo Zhou
Boyao Zhou
Brady Zhou
Mo Zhou
Chunluan Zhou
Dingfu Zhou
Fan Zhou
Jingkai Zhou
Honglu Zhou
Jiaming Zhou
Jiahuan Zhou
Jun Zhou
Kaiyang Zhou
Keyang Zhou
Kuangqi Zhou
Lei Zhou
Lihua Zhou
Man Zhou
Mingyi Zhou
Mingyuan Zhou
Ning Zhou
Peng Zhou
Penghao Zhou
Qianyi Zhou

Shuigeng Zhou
Shangchen Zhou
Huayi Zhou
Zhize Zhou
Sanping Zhou
Qin Zhou
Tao Zhou
Wenbo Zhou
Xiangdong Zhou
Xiao-Yun Zhou
Xiao Zhou
Yang Zhou
Yipin Zhou
Zhenyu Zhou
Hao Zhou
Chu Zhou
Daquan Zhou
Da-Wei Zhou
Hang Zhou
Kang Zhou
Qianyu Zhou
Sheng Zhou
Wenhui Zhou
Xingyi Zhou
Yan-Jie Zhou
Yiyi Zhou
Yu Zhou
Yuan Zhou
Yuqian Zhou
Yuxuan Zhou
Zixiang Zhou
Wengang Zhou
Shuchang Zhou
Tianfei Zhou
Yichao Zhou
Alex Zhu
Chenchen Zhu
Deyao Zhu
Xiatian Zhu
Guibo Zhu
Haidong Zhu
Hao Zhu
Hongzi Zhu
Rui Zhu
Jing Zhu

Jianke Zhu
Junchen Zhu
Lei Zhu
Lingyu Zhu
Luyang Zhu
Menglong Zhu
Peihao Zhu
Hui Zhu
Xiaofeng Zhu
Tyler (Lixuan) Zhu
Wentao Zhu
Xiangyu Zhu
Xinqi Zhu
Xinxin Zhu
Xinliang Zhu
Yangguang Zhu
Yichen Zhu
Yixin Zhu
Yanjun Zhu
Yousong Zhu
Yuhao Zhu
Ye Zhu
Feng Zhu
Zhen Zhu
Fangrui Zhu
Jinjing Zhu
Linchao Zhu
Pengfei Zhu
Sijie Zhu
Xiaobin Zhu
Xiaoguang Zhu
Zezhou Zhu
Zhenyao Zhu
Kai Zhu
Pengkai Zhu
Bingbing Zhuang
Chengyuan Zhuang
Liansheng Zhuang
Peiye Zhuang
Yixin Zhuang
Yihong Zhuang
Junbao Zhuo
Andrea Ziani
Bartosz Zieliński
Primo Zingaretti

Nikolaos Zioulis
Andrew Zisserman
Yael Ziv
Liu Ziyin
Xingxing Zou
Danping Zou
Qi Zou

Shihao Zou
Xueyan Zou
Yang Zou
Yuliang Zou
Zihang Zou
Chuhang Zou
Dongqing Zou

Xu Zou
Zhiming Zou
Maria A. Zuluaga
Xinxin Zuo
Zhiwen Zuo
Reyer Zwiggelaar

Contents – Part XII

Explicit Model Size Control and Relaxation via Smooth Regularization for Mixed-Precision Quantization

Vladimir Chikin[1]([⊠]), Kirill Solodskikh[1], and Irina Zhelavskaya[2]📵

[1] Huawei Noah's Ark Lab, Moscow, Russia
{vladimir.chikin,solodskikh.kirill1}@huawei.com
[2] Skolkovo Institute of Science and Technology (Skoltech), Moscow, Russia
irina.zhelavskaya@skolkovotech.ru

Abstract. While Deep Neural Networks (DNNs) quantization leads to a significant reduction in computational and storage costs, it reduces model capacity and therefore, usually leads to an accuracy drop. One of the possible ways to overcome this issue is to use different quantization bit-widths for different layers. The main challenge of the mixed-precision approach is to define the bit-widths for each layer, while staying under memory and latency requirements. Motivated by this challenge, we introduce a novel technique for explicit complexity control of DNNs quantized to mixed-precision, which uses smooth optimization on the surface containing neural networks of constant size. Furthermore, we introduce a family of smooth quantization regularizers, which can be used jointly with our complexity control method for both post-training mixed-precision quantization and quantization-aware training. Our approach can be applied to any neural network architecture. Experiments show that the proposed techniques reach state-of-the-art results.

Keywords: Neural network quantization · Mixed-precision quantization · Regularization for quantization

1 Introduction

Modern DNNs allow solving a variety of practical problems with accuracy comparable to human perception. The commonly used DNN architectures, however, do not take into account the deployment stage. One popular way to optimize neural networks for that stage is quantization. It significantly reduces memory, time and power consumption due to the usage of integer arithmetic that speeds up the addition and multiplication operations.

V. Chikin and K. Solodskikh—These authors contributed equally to this work.

Supplementary Information The online version contains supplementary material available at https://doi.org/10.1007/978-3-031-19775-8_1.

© The Author(s), under exclusive license to Springer Nature Switzerland AG 2022
S. Avidan et al. (Eds.): ECCV 2022, LNCS 13672, pp. 1–16, 2022.
https://doi.org/10.1007/978-3-031-19775-8_1

There are several types of quantization that are commonly used. The fastest in terms of application time and implementation is post-training quantization [1,3,7,17], in which the weights and activations of the full-precision (FP) network are approximated by fixed-point numbers. While being the quickest approach, it usually leads to the decrease in accuracy of a quantized network. To reduce the accuracy drop, quantization-aware training (QAT) algorithms can be used [5,9,15,26]. QAT employs stochastic gradient descent with quantized weights and activations on the forward pass and full-precision weights on the backward pass of training. It usually leads to better results but requires more time and computational resources than PTQ.

One technique to make QAT converge faster to the desired quantization levels is regularization. Periodic functions, such as mean squared quantization error (MSQE) or the sine function, are typically used as regularizers [6,11,18]. The high-precision weights are then naturally pushed towards the desired quantization values corresponding to the minima of those periodic functions. The sine function is smooth and therefore has advantages over MSQE, which is not smooth. The sine function has an infinite number of minima, however, which may lead to a high clipping error in quantization, and therefore, can have a negative impact on the model accuracy.

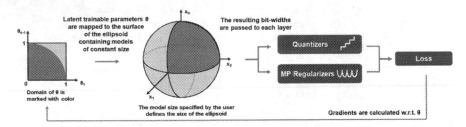

Fig. 1. Illustration of the proposed pipeline. Parameters θ are mapped to the ellipsoid of constant-size DNNs. Each point of this ellipsoid defines the bit-width distribution across different layers in order to obtain the model size pre-defined by the user. Next, the bit-widths are passed to a DNN for the forward pass. Our special smooth mixed-precision (MP) regularizers are computed jointly with the task loss and help reduce the gradient mismatch problem. The DNN parameters, quantization parameters, and θ are updated according to the calculated loss on the backward pass.

If a trained quantized network still does not reach an acceptable quality, mixed-precision quantization may be applied. Such algorithms allocate different bit-widths to different layers and typically perform better than the fixed bit-width counterparts. At the same time, it is also important to keep the memory and latency constraints in mind, so that a quantized model meets specific hardware requirements. Too much compression may lead to a significant loss in accuracy, while not enough compression may not meet the given memory budget. Many existing studies use gradient descent to tune the bit-width distribution, but this approach inherently does not have the ability to explicitly set the required compression ratio of the model. One way to constrain the model size is to use additional regularizers on the size of weights and activations [22]. However, regularizers also do not allow setting the model size, and hence the compression ratio, explicitly, and multiple experiments may still be needed to obtain a desired model size. Other state-of-the-art methods are based on reinforcement learning [12,23], or general neural architecture search (NAS) algorithms [4,24,27], where the search space is defined by the set of possible bit-widths for each layer. However, such methods imply training of multiple instances of a neural network, which usually require large computational and memory resources.

In this paper, we address the problems mentioned above. First, we show that models of the same size quantized to different mixed-precision lie on the surface of a multi-dimensional ellipsoid. We suggest a parametrization of this ellipsoid by using latent continuous trainable variables, using which the discrete problem of quantized training can be reformulated to a smooth one. This technique imposes almost no computational overhead compared to conventional QAT algorithms and certainly requires much less computational time and resources compared to reinforcement or NAS algorithms. Furthermore, we suggest a universal family of smooth quantization regularizers, which are bounded, and therefore reduce the clipping error and lead to better performance. Our main contributions are the following:

1. We propose a novel method of mixed-precision quantization that strictly controls the model size. It can be applied to both weights and activations. To the best of our knowledge, this is the first method of mixed-precision quantization that allows *explicit* model size control.
2. We construct a family of bounded quantization regularizers for smooth optimization of bit-width parameters. It allows for faster convergence to the discrete values and avoids high clipping error. It also avoids difficulties of the discrete optimization.
3. We validate our approach on image classification and image super-resolution tasks and compare it to the recent methods of mixed-precision quantization. We show that the proposed approach reaches state-of-the-art results.

Notation. We use $x, \mathbf{x}, \mathbf{X}$ to denote a scalar, a vector and a matrix (or a tensor, as is clear from the context). $\lfloor . \rfloor, \lceil . \rceil$ and $\lfloor . \rceil$ are the floor, ceiling, and round operators. $\mathbb{E}[\cdot]$ denotes the expectation operator. We call the total model size the following value: $\sum_{i=1}^{n} k_i \cdot b_i + 32 \cdot k_0$, where $\{k_i\}_{i=1}^{n}$ and $\{b_i\}_{i=1}^{n}$ are the sizes of weight tensors of quantized layers and the corresponding bit-widths, and k_0 is the number of non-quantized parameters.

2   Related Work

Quantization-Aware Training. Many existing quantization techniques, such as [5,9,15,26] train quantized DNNs using gradient descent-based algorithms. QAT is based on the usage of quantized weights and activations on the forward pass and full-precision weights on the backward pass of training. It is difficult to train the quantized DNNs, however, as the derivative of the round function is zero almost everywhere. To overcome this limitation, straight-through estimators (STE) were proposed [2]. They approximate the derivative of the round function and allow backpropagating the gradients through it. In that way, the network can be trained using standard gradient descent [15].

Quantization Through Regularization. These methods involve additional regularization for quantized training. [6] used mean squared quantization error (MSQE) regularizer to push the high-precision weights and activations towards the desired quantized values. [11,18] employed the trigonometric sine function for regularization. Sinusoidal functions are differentiable everywhere and have periodic minima which can be utilized to drive the weights towards the quantized values. However, the sine function has an infinite number of minima points, which may lead to a high clipping error.

Mixed-Precision Quantization. In some cases, quantization of the whole network to low bit-width could produce unacceptable accuracy drop. In such cases, some parts of a network could be quantized to a higher precision. Mixed-precision quantization algorithms are used to obtain a trade-off between quality and acceleration. This task involves searching in a large configuration space. The state-of-the-art methods are based on reinforcement learning [12,23], or general neural architecture search algorithms [4,24,27], where the search space is defined by the set of possible bit-widths for each layer. Such methods imply training of multiple instances of a neural network, which may require large memory resources. Differently in [22], the bit-width is learned from the round function parameterized by quantization scale and range. The derivative of this parametrization is obtained using STE. In all these studies, an additional complexity regularizer is used to control the compression ratio of the model. This technique does not allow to specify the desired compression ratio explicitly, and multiple experiments may be needed to obtain a desired model size.

3   Motivation and Preliminaries

In this section, we briefly describe the quantization process of neural networks and the motivation for the proposed methods. Consider a neural network with n layers parameterized by weights $\hat{\mathbf{W}} = \{\mathbf{W}_i\}_{i=1}^n$. Let each layer correspond to some function $\mathcal{F}_i(\mathbf{W}_i, \mathbf{A}_i)$, where $\mathbf{W}_i$ are the weights of layer i, and $\mathbf{A}_i$ are the input activations to a layer.

Quantization. Quantization of a neural network implies obtaining a model, whose parameters belong to some finite set. As a rule, this finite set is specified by a function that maps the model parameters to this set. We can define the following *uniform quantization function*:

$$Q_U^b(x) = clip(\lfloor x \rceil, -2^{b-1}, 2^{b-1} - 1) \tag{1}$$

where $[-2^{b-1}, 2^{b-1} - 1]$ is the quantization range, b is the quantization bit-width. Under the assumption that layer $\mathcal{F}_i(\mathbf{W}_i, \mathbf{A}_i)$ commutes with scalar multiplication, quantization is reduced to optimizing the quality of a quantized model relative to its parameters $\hat{\mathbf{W}}$ and *quantization scale parameters* $\mathbf{s}_w = \{s_{w_i}\}_{i=1}^n$ and $\mathbf{s}_a = \{s_{a_i}\}_{i=1}^n$. The scale parameters determine the quantization range; if all FP values of weights are covered, then $s_i = \max |\mathbf{W}_i|/(2^b - 1)$ for the symmetric quantization scheme. Layers $\mathcal{F}_i(\mathbf{W}_i, \mathbf{A}_i)$ are then replaced with *quantized layers*:

$$\mathcal{F}_i^q = s_{w_i} s_{a_i} \mathcal{F}_i \left(\mathcal{Q}_U^{b_w} \left(\frac{\mathbf{W}_i}{s_{w_i}} \right), \mathcal{Q}_U^{b_a} \left(\frac{\mathbf{A}_i}{s_{a_i}} \right) \right), \tag{2}$$

where b_w is the quantization bit-width of the weight tensor and b_a is the quantization bit-width of the activation tensor. The most commonly used layers, such as convolutional and fully-connected layers, satisfy this assumption.

Backpropagation Through Quantization. To train such a quantized network, we need to pass the gradients through the quantization function on the backward pass. Since the derivative of the round function is zero almost everywhere, straight-through estimators (STE) are used to approximate it. The following STE was proposed in [2] and is commonly used (we will use it in this paper as well):

$$\frac{d\mathcal{Q}_U(x)}{dx} = \begin{cases} 1, & \text{if } -2^{b-1} \leq x \leq 2^{b-1} - 1, \\ 0, & \text{otherwise.} \end{cases} \tag{3}$$

Model Complexity Control. The problem of training a quantized model using STE is reduced to optimization of the loss function $\mathcal{L}_Q$:

$$\mathcal{L}_Q = \mathbb{E}[\mathcal{L}(\mathcal{F}^q(\hat{\mathbf{W}}, \mathbf{X}))]. \tag{4}$$

Some modern quantization methods optimize $\mathcal{L}_Q$ relative to variables $(\hat{\mathbf{W}}, \mathbf{s}_w, \mathbf{s}_a)$, that is, the scale parameters and weights are tuned as trainable variables. In the case of traditional quantization, the bit-width values are fixed during training. In mixed precision quantization, bit-widths of different layers can change during training. A number of modern mixed-precision quantization methods use gradient descent to tune the bit-width distribution, but they do not have the ability to explicitly set the required compression ratio of the model. In this paper, we propose a method for training the mixed-precision quantized models with the explicit, user-defined total model size. In the proposed framework, bit-widths are tuned as trainable variables in addition to the scale parameters and weights. We describe it in detail in the next section.

Soft Regularization. In training of (4), high-precision weights are used in the backward pass of gradient descent, and low-precision weights and activations are used in the forward pass. To reduce the discrepancy in the backward and forward passes, regularization can be utilized. A possible choice for the regularizer is MSQE [6]. However, it has a non-stable behavior of gradients in the neighborhood of the transition points. It can be seen from Fig. 2 that the derivative of MSQE changes drastically in the transition points and pushes the points in the opposite direction due to the large derivative there. In this paper, we propose to use smooth quantization regularizers instead of MSQE. Figure 2 also shows an example of the sinusoidal regularizer, which has a more stable gradient behavior in the neighborhood of the transition points. In particular, the gradient of the main loss (4) has a larger impact than the regularization loss there. It is worth noting that some previous works [11,18] also employed smooth quantization regularizers. However, they were unbounded (see Fig. 2), which led to accumulation of a clipping error due to the points outside the quantization range. We propose to use bounded smooth regularizers, which we describe in detail next.

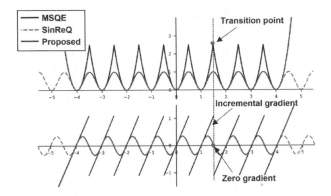

Fig. 2. Comparison of MSQE and smooth bounded (proposed in this work) and unbounded (e.g., SinReQ [11]) regularizers. Smooth quantization regularizers have similar properties to MSQE but different behavior close to the transition points.

4   Methodology

In this section, we first describe the proposed method of mixed-precision QAT with explicit model size control. Then, we describe the proposed family of smooth regularizers that allow faster convergence to quantized values. Finally, we describe an additional technique for bit-widths stabilization during training. We provide a description for the case of symmetric mixed-precision quantization of weights into *int* scheme. The proposed method can be easily generalized to the case of mixed-precision quantization of activations without any additional constraints and other quantization schemes.

4.1   Explicit Model Size Control Using Surfaces of Constant-Size Neural Networks

We propose to use a special parametrization of the bit-width parameters of the model layers. This parametrization imposes restrictions on the size of the quantized model. To do this, we build a surface of neural networks of the same size in the bit-width space. We parameterize this surface using latent independent variables, which are tuned during training. The proposed technique is applicable to any layer of a DNN.

Suppose that we want to quantize n layers of a DNN. For simplicity, we assume that activations of those layers are quantized to fixed bit-width b_a. We denote by $\{k_i\}_{i=1}^n$ and $\{b_i\}_{i=1}^n$ the sizes of weight tensors of those layers and their corresponding quantization bit-widths. The size of the quantized part of the network is $\sum_{i=1}^n k_i b_i$. Our goal is to preserve it during training.

Consider the n-dimensional *ellipsoid equation*:

$$\sum_{i=1}^n k_i x_i^2 = C. \tag{5}$$

We can parameterize the surface of an ellipsoid in the first orthant ($x_i > 0$) using $n-1$ independent variables $\theta = \{\theta_i\}_{i=1}^{n-1}$, each from segment $[0,1]$:

$$\begin{cases} x_1 = \sqrt{\frac{C}{k_1}}\theta_1, \\ \dots, \\ x_{n-1} = \sqrt{\frac{C}{k_{n-1}}}\theta_{n-1}, \\ x_n = \sqrt{\frac{C-\sum_{i=1}^{n-1} k_i x_i^2}{k_n}}. \end{cases} \tag{6}$$

Figure 1 illustrates the process of mapping trainable parameters θ_i to the ellipsoid of models with the same size. We do not consider other orthants for parametrization as they are redundant and increase the search space during the bit-width tuning. It is important to note that such parametrization does not satisfy the ellipsoid equation for all possible θ in $[0,1]^{n-1}$ (see Fig. 1). In particular, during training we clamp the outliers. From our empirical evaluation, such outliers appear with almost zero probability.

Additionally, let us define variables $t_i = 2^{x_i^2-1}$. We can use x_i and t_i for weight quantization of the i-th layer by replacing the bit-width parameter b_w by x_i^2 and the maximum absolute value of the quantization range 2^{b_w-1} by t_i in Eq. (2):

$$\mathcal{F}_i^q = \frac{\tilde{s}_{w_i}}{t_i}\frac{\tilde{s}_{a_i}}{2^{b_a-1}}\mathcal{F}_i\left(\mathcal{Q}_U^{x_i^2}\left(t_i \cdot \frac{\mathbf{W}_i}{\tilde{s}_{w_i}}\right), \mathcal{Q}_U^{b_a}\left(2^{b_a-1} \cdot \frac{\mathbf{A}_i}{\tilde{s}_{a_i}}\right)\right), \tag{7}$$

where $\tilde{s}$ corresponds to absolute maximum values of FP weights/activations (see (3)).

As a result, $\mathbf{W}_i$ is quantized using a grid of integers from the range $[-\lfloor t_i \rceil, \lfloor t_i \rceil - 1]$ consisting of $2\lfloor t_i \rceil$ integers. If x_i^2 is an integer, the value of $2\lfloor t_i \rceil$ equals to $2^{x_i^2}$ – an integer power of two, which corresponds to quantization to x_i^2 bits. For example, if $x_i^2 = 8$, then $t_i = 128$ and the corresponding tensor $\mathbf{W}_i$ is quantized using a grid of integers consisting of 256 elements, or to 8 bits. Thus, x_i^2 serves as a continuous equivalent of the bit-width of the weights of the i-th layer. To estimate the real bit-width of the i-th layer during the training, we use the smallest sufficient integer value, namely:

$$b_i = \lceil \log_2(2 \cdot \lfloor 2^{x_i^2-1} \rceil) \rceil \approx x_i^2. \tag{8}$$

We propose to train the quantized model by minimizing loss $\mathcal{L}_Q$ (4) relative to variables $(\mathbf{W}, \mathbf{s}_w, \mathbf{s}_a, \theta)$. Thus, we can train quantized models, whose mixed-precision bit-widths are tuned during model training in a continuous space. Despite the fact that parameters θ, and hence the bit-widths of the layers, change during training, it follows from the definition of the ellipsoid parametrization that (5) is always satisfied, which means that the size of a model is preserved during training. Due to the error of rounding the bit-widths to integer values (8), the size of the quantized model may differ from C, but not significantly.

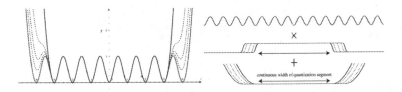

Fig. 3. The proposed family of smooth quantization regularizers. It can be implemented as a multiplication of the periodic function by the hat function, and then adding a function, which is square outside and zero inside the quantization range.

4.2 Smooth Bounded Regularization as a Booster for Quantized Training

Additional regularization is often used as a special technique for training quantized models, see (2) and (3). We propose to use special regularizers that allow improving the quality of bit-width tuning and training of mixed-precision quantized models. We consider a family of smooth bounded regularizers $\phi(x, t)$ that smoothly depend on parameter t determining the width of the quantization range. We require the following from those regularizers:

- For fixed t, integers from the range $[-\lfloor t \rfloor, \lfloor t \rfloor - 1]$ are the roots and minima of function $\phi(\cdot, t)$, in particular, functions ϕ and ϕ' are zero in these points.
- If $t \in \mathbb{Z}$, then there are no other minima.
- If $t \notin \mathbb{Z}$, then a maximum of 2 more local minima outside $[-\lfloor t \rfloor, \lfloor t \rfloor - 1]$ are possible.

There are various types of regularizers satisfying the above conditions. In our quantization experiments, we use an easy-to-implement function (shown in Fig. 3):

$$\phi(x, t) = \sin^2(\pi x) \cdot \sigma(x + t) \cdot \sigma(t - 1 - x) +$$
$$\begin{cases} \pi^2(x + t)^2, & x \leq -t, \\ 0, & x \in [-t, t - 1], \\ \pi^2(x - t + 1)^2, & x \geq t - 1, \end{cases} \tag{9}$$

where σ is the sigmoid function. For quantization of weights of the i-th layer, scaled tensor $\mathbf{W}_i$ must be passed to regularizer ϕ as argument x, and parameter $t_i = t_i(\theta)$ as argument t. For quantization of activations of the i-th layer, scaled tensor A_i must be passed to the regularizer as argument x, and the maximum absolute value of the activation quantization grid 2^{b_a-1} as argument t.

Analysis of the Taylor series expansion of function $\frac{s^2}{t^2}\phi(t \cdot \frac{x}{s}, t)$ shows that this function is a good estimate of the mean squared quantization error in the neighborhood of integer points of the range $[-\lfloor t \rfloor, \lfloor t \rfloor - 1]$ (see Appendix A for a detailed proof). Therefore, we propose to use the following functions as regularizers:

$$\mathcal{L}_w = \sum_i \frac{s_{w_i}^2}{t_i^2(\theta)} \phi\left(t_i(\theta) \cdot \frac{\mathbf{W}_i}{s_{w_i}}, t_i(\theta)\right) \tag{10}$$

$$\mathcal{L}_a = \sum_i \mathbb{E}\left[\frac{s_{a_i}^2}{2^{b_a}} \phi\left(2^{b_a-1} \cdot \frac{\mathbf{A}_i}{s_{a_i}}, 2^{b_a-1}\right)\right] \tag{11}$$

As a result, training of a mixed-precision quantized model can be done by optimizing the following loss function relative to parameters (W, s_w, s_a, θ):

$$\mathcal{L}_Q = \mathbb{E}\left[\mathcal{L}\left(\mathcal{F}^q(\mathbf{W}, \mathbf{X})\right)\right] + \lambda_w \mathcal{L}_w + \lambda_a \mathcal{L}_a. \tag{12}$$

The proposed family of smooth regularizers also allows tuning the bit-widths as a post-training algorithm without involving other parameters in the training procedure (see Sect. 5.1 for an example).

4.3 Regularizers for Bit-Width Stabilization

During training, parameters t_i can converge to values that are not degree of 2, corresponding to non-integer bit-widths. For example, if $x_i^2 = 3.2$, then the scaled tensor is quantized to a grid of integers consisting of $2 \cdot \lceil 2^{3.2-1} \rceil = 10$ elements. In this case, 3 bits are not sufficient to perform the calculations (for this, there should be not more than 8 elements), and we do not fully use 4 bits (only 10 from 16 possible elements are used). In order for the bit-widths to converge to integer values, we add a sinusoidal regularizer for bit-width values $\mathcal{L}_b = \sum_{i=1}^n \sin^2(\pi x_i^2)$ to loss $\mathcal{L}_Q$ (12).

In some tasks, we may need the ability to use a specific set of bit-width values for mixed-precision quantization, for example, 4, 8 or 16 bits. In this case, we add a special regularizer for bit-widths values to the loss $\mathcal{L}_Q$ (12), which is aimed at contracting them to the required set. We propose smooth regularizers that have local minima at the points from the required set of bit-widths and have no other local minima. Using such regularizers during training makes the bit-width parameters converge to a given set of values. We normalize these regularizers using regularization parameter λ_b.

4.4 Algorithm Overview

Here, we describe the overall algorithm combining the proposed techniques. A pseudo code for the algorithm and more specifics are given in Appendix B. We minimize loss function $\mathcal{L}_Q$ (12) relative to trainable variables $(\mathbf{W}, \mathbf{s}_w, \mathbf{s}_a, \theta)$ using stochastic gradient descent. We quantize weights and activations of a model on the forward pass of training. We use a straight-through estimator (3) to propagate through round function $\mathcal{Q}_U$ on the backward pass.

We initialize the scale parameters of the weight tensors with their maximum absolute values. To initialize the scale parameters of the activation tensors, we pass several data samples through the model and estimate the average maximum absolute values of inputs to all quantized layers. To set the required model size, a user must specify parameter b_{init} – an initial value of the continuous bit-widths x_i^2 of the quantized layers. Using Eqs. (6), we initialize parameters θ_i so that continuous bit-widths x_i^2 of all quantized layers are equal to b_{init}. Parameter b_{init} determines the total size of the quantized part of the model using Eq. (5), which means that it determines the ellipsoid that is used for training.

Table 1. Influence of the proposed techniques. Weights only quantization of ResNet-20 on CIFAR-10. MC – model compression ratio (times), Top 1 – Top-1 quantized accuracy in %, MP – mixed precision.

Quantization	# bits W/A	No regularizers		With regularizers	
		Top-1	MC	Top-1	MC
Full precision	32/32	91.73	–	91.73	–
Post-training	2/32	19.39	14.33	–	–
BN tuning	2/32	64.34	14.33	–	–
Bit-widths tuning	MP/32	65.94	14.55	77.05	14.55
Bit-widths and scales tuning	MP/32	79.20	15.02	86.85	14.55
All model parameters tuning	MP/32	**91.27**	14.55	**91.57**	14.55
Post-training	3/32	75.57	9.92	–	–
BN tuning	3/32	89.58	9.92	–	–
Bit-widths tuning	MP/32	89.70	10.11	90.13	10.05
Bit-widths and scales tuning	MP/32	90.65	10.19	90.70	10.19
All model parameters tuning	MP/32	**91.69**	10.16	**91.89**	10.25

5 Experiments

We evaluate the performance of the proposed techniques on several computer vision tasks and models. In Sect. 5.1, we study how each of the proposed techniques influences the quality of mixed precision quantization. In Sect. 5.2, we compare the proposed algorithm to other QAT methods for image classification.

Experimental Setup. In our experiments, we quantize weights and activations only, and some of the parameters of the models remain non-quantized (for example, biases and batch normalization parameters). The bit-width of the non-quantized parameters is 32. All quantized models use pre-trained float32 networks for initialization. As a model compression metric, we use the compression ratio of the total model size (not just weights), and as an activation compression metric, we use the mean input compression ratio over a set of model layers (same metric as in [22]). We set a specific value for the model compression by setting the corresponding value for the parameter b_{init} in each of the tasks. We normalize the quantization regularizers to have the same order as the main loss $\mathcal{L}_Q$ (4) by using coefficients λ_w, λ_a, which are chosen as powers of 10. After some number of epochs, we adjust λ_w by multiplying it by 10 to reduce the weights quantization error; λ_a does not change during training. The corresponding training strategy for each experiment is described in Appendix C.

5.1   Ablation Study

Impact of the Proposed Techniques. The proposed method is based on several techniques described above and involves tuning bit-widths, scale parameters and model parameters (i.e., weights). We investigate the impact of the proposed techniques by applying them separately for quantization of ResNet-20 [14] on the CIFAR-10 dataset [16]. We also compare the obtained results with the results of post-training quantization to fixed bit-width with and without batch-normalization (BN) tuning. Additionally, we explore the influence of smooth regularizers suggested in Sect. 4.2. The results are presented in Table 1.

The proposed technique of the bit-width tuning used as a PTQ without training the model parameters and scale parameters obtains a quantized model with a better quality and the same compression ratio as post-training with BN tuning for both 2-bit and 3-bit quantization. The subsequent addition of the proposed methods leads to an increasing improvement in the quality of the quantized model, and the joint training of the bit-widths, scales and model parameters leads to the best accuracy. One can also note that the use of smooth regularizers further improves the quality of the resulting quantized models. Thus, all of the proposed techniques contribute to the increase of quantized model accuracy.

Optimality of the Determined Mixed Precision. We further investigate whether the bit-widths of different layers found with our approach are optimal. We show that on the task of $3\times$ image super-resolution for *Efficient Sub-Pixel CNN* [21] with global residual connection. We choose this task to demonstrate the effectiveness of our approach visually as well.

The full precision Efficient Sub-Pixel CNN consists of 6 convolutions. We quantize all layers of the model except for the first layer. The bit-widths of activations are equal to the bit-widths of the corresponding weights of each layer. The last 4 convolutions of the model are almost the same size. We use the proposed method to quantize all layers to 8-bit, except for one of those 4 convolutions, the size of which we set to 4 bits. Our algorithm selects the

Table 2. Validation PSNR, dB, for ESPCN quantized to mixed-precision. MC – model compression ratio (times).

Dataset	Our	Bit #1	Bit #2	Bit #3
Vimeo-90K	**31.10**	30.93	30.88	30.74
Set5	**30.55**	30.46	30.47	30.02
Set14	**26.93**	26.88	26.88	26.63
MC	**3.37**	3.37	3.37	3.34

Table 3. Validation PSNR, dB, for ESPCN quantized to mixed-precision (for both weights and activations).

Network	Vimeo-90K	Set5	Set14
FP	31.28	30.74	27.06
8 bit	**31.19**	**30.74**	**27.05**
MixPr1	31.10	30.55	30.55
MixPr2	30.95	30.50	26.89
4 bit	30.62	29.84	26.53
Bicubic	29.65	28.92	25.91

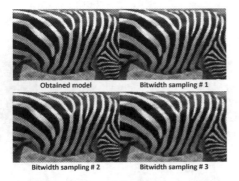

Fig. 4. Ablation study for mixed precision quantization of ESPCN.

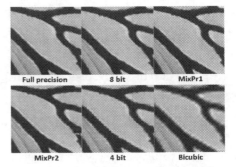

Fig. 5. Quantization of ESPCN: comparison of different bit-widths.

second of these four convolutions to be quantized to 4 bits. To prove that our algorithms converges to an optimal configuration, we train other configurations with fixed bit-widths, in which one of the last four layers is quantized to 4 bit. The three possible configurations are denoted by Bit #1, Bit #2, Bit #3, and their performance is shown in Table 2 and Fig. 4. We can see that the model obtained with our method has the best perceptual quality out of other possible model choices. The PSNR of the model obtained with our method is also larger than PSNR of other models. This means that our algorithm determines the best layer for 4-bit quantization while preserving the model compression rate.

Influence of Model Compression. We test the effect of different proportion of 4-bit and 8-bit quantization of a model on its accuracy for the same task as in the previous experiment. To investigate that, we train several models with different compression ratios. First, we train two mixed precision quantized models: *MixPr1*, in which 15.6% of model is quantized to 4 bit and 75.6% to 8 bit, and *MixPr2*, in which 46.8% of model is quantized to 4 bit and 44.4% to 8 bit. We compare these models to the ones quantized to 4 bit and 8 bit only (see

Table 3). We train these mixed-precision quantized models using a regularizer for bit-width stabilization for 4 and 8 bits (see Sect. 4.3).

We observe that as the proportion of 8 bit increases, the perceptual quality of the resulting images improves (Fig. 5). The perceptual quality of the 8-bit model produced by our method matches the perceptual quality of the full precision model. More examples are provided in Appendix D.

5.2 Comparison with Existing Studies

CIFAR-10. We compare our method to several methods for mixed-precision quantization of ResNet-20 in Table 4. In these experiments, we quantize all layers of the models. The first and last layers are quantized to a fixed bit-width, and the rest of the layers are used for mixed-precision quantization. We test cases when activations are quantized to 4 bits and when they are not quantized.

Our method leads to the best compression ratio when activations are quantized to 4 bits and wins over other methods in terms of accuracy except for HAWQ, even though we have used a weaker baseline FP model. Regarding comparison with HAWQ, one can note that the relative differences between the baseline and the resulting quantized accuracies are similar: 0.16% for HAWQ and 0.12% for our method, but the compression ratio for our method is much higher: 15.13 vs. 13.11. The proposed method outperforms other methods when activations are not quantized. The reason for only slight difference of models with and without our regularizers in Table 4 may lie in that both models are very close to the FP accuracy and therefore, are close to saturation in accuracy.

ImageNet. We also test our method for quantization of ResNet-18, ResNet-50 [14], and MobileNet-v2 [20] to 4 and 8 bits on ImageNet [19], and compare it with other methods in Table 5. The accuracies of the baseline full precision (FP) models used by all the methods are noted in the table. We used the baseline with the highest accuracy for comparison with other methods. The obtained bit-width distributions are given in the Appendix E. The proposed approach performs significantly better then the other methods both in terms of accuracy and compression ratio. It reaches accuracy larger than the FP model for ResNet-50 and MobileNet-V2, while all other methods do not. It is worth noting that these results were obtained in less than 7 epochs for all the models.

Table 4. Quantization of ResNet-20 on CIFAR-10. MC – model compression ratio (times), AC – activation compression ratio (times), FP Top-1 – the baseline FP model accuracy in %, Quant. Top-1 – Top-1 quantized accuracy in %, Difference – difference between quantized and FP Top-1 accuracy in %.

Method	MC	AC	FP Top-1	Quant. Top-1	Difference
MP DNNs [22]	14.97	8	92.71	91.40	−1.31
HAWQ [10]	13.11	8	92.37	**92.22**	−0.15
PDB [8]	11.94	8	91.60	90.54	−1.06
Ours (no regularizers)	**16.17**	8	91.73	91.55	−0.18
Ours (with regularizers)	15.13	8	91.73	91.62	−0.11
MP DNNs [22]	14.97	1	92.71	91.41	−1.3
DoReFa + SinReQ [11]	10.67	1	93.50	88.70	−4.8
Ours (no regularizers)	**16.19**	1	91.73	91.75	+0.02
Ours (with regularizers)	14.73	1	91.73	**91.97**	+0.24

Table 5. Quantization of ResNet-18, ResNet-50 and MobileNet-v2 on ImageNet. MC – model compression ratio (times), AC – activation compression ratio (times), FP Top-1 – the baseline FP model accuracy in %, Quant. Top-1 – Top-1 quantized accuracy in %, Difference – difference between quantized and FP Top-1 accuracy in %.

Method	MC	AC	FP Top-1	Quant. Top-1	Difference
ResNet-18					
MP DNNs [22]	4.24	2.9	70.28	70.66	+0.38
LSQ [13]	4.00	4	70.50	71.10	+0.6
HAWQ-V3 [25]	4.02	4	71.47	71.56	+0.09
Ours	**4.40**	4	71.47	**71.81**	+0.34
DoReFa + WaveQ	7.98	8	70.10	70.00	−0.1
FracBits-SAT	7.61	8	70.20	70.60	+0.4
MP DNNs	8.25	8	70.28	70.08	−0.2
DoReFa + SinReQ	7.61	8	70.50	64.63	−5.87
HAWQ-V3	7.68	8	71.47	68.45	−3.02
Ours	**8.57**	8	71.47	**70.64**	−0.83
ResNet-50					
LSQ [13]	4.00	4	76.90	76.80	−0.1
HAWQ-V3 [25]	3.99	4	77.72	77.58	−0.14
Ours	**4.33**	4	79.23	**79.45**	+0.22
HAWQ-V3	7.47	8	77.72	74.24	−3.48
Ours	**8.85**	8	79.23	**76.38**	−2.85
MobileNet-V2					
MP DNNs [22]	**4.21**	2.9	70.18	70.59	+0.41
HAQ [23]	4.00	4	71.87	71.81	−0.06
Ours	4.13	4	71.87	**71.90**	+0.03

6   Conclusions

In this paper, we propose a novel technique for mixed-precision quantization with explicit model size control, that is, the final model size can be specified by a user unlike in any other mixed-precision quantization method. In particular, we define the mixed-precision quantization problem as a constrained optimization problem and solve it together with soft regularizers, as well as a bit-width regularizer to constrain the quantization bit-widths to a pre-defined set. We validate the effectiveness of the proposed methods by conducting experiments on CIFAR10, ImageNet, and an image super resolution task, and show that the method reaches state-of-the-art results with no significant overhead compared to conventional QAT methods.

References

1. Banner, R., Nahshan, Y., Soudry, D.: Post training 4-bit quantization of convolutional networks for rapid-deployment. In: Wallach, H., Larochelle, H., Beygelzimer, A., d'Alché-Buc, F., Fox, E., Garnett, R. (eds.) Advances in Neural Information Processing Systems, vol. 32. Curran Associates, Inc. (2019). https://proceedings. neurips.cc/paper/2019/file/c0a62e133894cdce435bcb4a5df1db2d-Paper.pdf
2. Bengio, Y., Léonard, N., Courville, A.C.: Estimating or propagating gradients through stochastic neurons for conditional computation. CoRR abs/1308.3432 (2013). https://arxiv.org/abs/1308.3432
3. Cai, Y., Yao, Z., Dong, Z., Gholami, A., Mahoney, M.W., Keutzer, K.: ZeroQ: a novel zero shot quantization framework. CoRR abs/2001.00281 (2020). https:// arxiv.org/abs/2001.00281
4. Cai, Z., Vasconcelos, N.: Rethinking differentiable search for mixed-precision neural networks (2020)
5. Choi, J., Wang, Z., Venkataramani, S., Chuang, P.I., Srinivasan, V., Gopalakrishnan, K.: PACT: parameterized clipping activation for quantized neural networks. CoRR abs/1805.06085 (2018)
6. Choi, Y., El-Khamy, M., Lee, J.: Learning low precision deep neural networks through regularization (2018)
7. Choukroun, Y., Kravchik, E., Kisilev, P.: Low-bit quantization of neural networks for efficient inference. CoRR abs/1902.06822 (2019). https://arxiv.org/abs/1902. 06822
8. Chu, T., Luo, Q., Yang, J., Huang, X.: Mixed-precision quantized neural network with progressively decreasing bitwidth for image classification and object detection. CoRR abs/1912.12656 (2019). https://arxiv.org/abs/1912.12656
9. Courbariaux, M., Hubara, I., Soudry, D., El-Yaniv, R., Bengio, Y.: Binarized neural networks: training deep neural networks with weights and activations constrained to +1 or -1 (2016)
10. Dong, Z., Yao, Z., Gholami, A., Mahoney, M.W., Keutzer, K.: HAWQ: Hessian aware quantization of neural networks with mixed-precision. CoRR abs/1905.03696 (2019). https://arxiv.org/abs/1905.03696
11. Elthakeb, A.T., Pilligundla, P., Esmaeilzadeh, H.: SinReQ: generalized sinusoidal regularization for low-bitwidth deep quantized training (2019)

12. Elthakeb, A.T., Pilligundla, P., Mireshghallah, F., Yazdanbakhsh, A., Esmaeilzadeh, H.: ReLeQ: a reinforcement learning approach for deep quantization of neural networks (2020)
13. Esser, S.K., McKinstry, J.L., Bablani, D., Appuswamy, R., Modha, D.S.: Learned step size quantization. CoRR abs/1902.08153 (2019). https://arxiv.org/abs/1902.08153
14. He, K., Zhang, X., Ren, S., Sun, J.: Deep residual learning for image recognition (2015)
15. Hubara, I., Courbariaux, M., Soudry, D., El-Yaniv, R., Bengio, Y.: Quantized neural networks: training neural networks with low precision weights and activations (2016)
16. Krizhevsky, A.: Learning multiple layers of features from tiny images. Technical report (2009)
17. Nagel, M., van Baalen, M., Blankevoort, T., Welling, M.: Data-free quantization through weight equalization and bias correction. CoRR abs/1906.04721 (2019). https://arxiv.org/abs/1906.04721
18. Naumov, M., Diril, U., Park, J., Ray, B., Jablonski, J., Tulloch, A.: On periodic functions as regularizers for quantization of neural networks (2018)
19. Russakovsky, O., et al.: ImageNet large scale visual recognition challenge. CoRR abs/1409.0575 (2014). https://arxiv.org/abs/1409.0575
20. Sandler, M., Howard, A., Zhu, M., Zhmoginov, A., Chen, L.C.: MobileNetV 2: inverted residuals and linear bottlenecks (2018)
21. Shi, W., et al.: Real-time single image and video super-resolution using an efficient sub-pixel convolutional neural network. CoRR abs/1609.05158 (2016). https://arxiv.org/abs/1609.05158
22. Uhlich, S., et al.: Differentiable quantization of deep neural networks. CoRR abs/1905.11452 (2019). https://arxiv.org/abs/1905.11452
23. Wang, K., Liu, Z., Lin, Y., Lin, J., Han, S.: HAQ: hardware-aware automated quantization with mixed precision (2019)
24. Wu, B., Wang, Y., Zhang, P., Tian, Y., Vajda, P., Keutzer, K.: Mixed precision quantization of convnets via differentiable neural architecture search. CoRR abs/1812.00090 (2018)
25. Yao, Z., et al.: HAWQV3: dyadic neural network quantization. CoRR abs/2011.10680 (2020). https://arxiv.org/abs/2011.10680
26. Zhou, S., Wu, Y., Ni, Z., Zhou, X., Wen, H., Zou, Y.: DoReFa-Net: training low bitwidth convolutional neural networks with low bitwidth gradients (2016)
27. Zur, Y., et al.: Towards learning of filter-level heterogeneous compression of convolutional neural networks (2019). https://doi.org/10.48550/ARXIV.1904.09872. https://arxiv.org/abs/1904.09872

BASQ: Branch-wise Activation-clipping Search Quantization for Sub-4-bit Neural Networks

Han-Byul Kim[1,2], Eunhyeok Park[3,4], and Sungjoo Yoo[1,2]

[1] Department of Computer Science and Engineering,
Seoul National University, Seoul, Korea
sungjoo.yoo@gmail.com
[2] Neural Processing Research Center (NPRC),
Seoul National University, Seoul, Korea
shinestarhb@gmail.com
[3] Department of Computer Science and Engineering, POSTECH, Pohang, Korea
canusglow@gmail.com
[4] Graduate School of Artificial Intelligence, POSTECH, Pohang, Korea

Abstract. In this paper, we propose Branch-wise Activation-clipping Search Quantization (BASQ), which is a novel quantization method for low-bit activation. BASQ optimizes clip value in continuous search space while simultaneously searching L2 decay weight factor for updating clip value in discrete search space. We also propose a novel block structure for low precision that works properly on both MobileNet and ResNet structures with branch-wise searching. We evaluate the proposed methods by quantizing both weights and activations to 4-bit or lower. Contrary to the existing methods which are effective only for redundant networks, e.g., ResNet-18, or highly optimized networks, e.g., MobileNet-v2, our proposed method offers constant competitiveness on both types of networks across low precisions from 2 to 4-bits. Specifically, our 2-bit MobileNet-v2 offers top-1 accuracy of 64.71% on ImageNet, outperforming the existing method by a large margin (2.8%), and our 4-bit MobileNet-v2 gives 71.98% which is comparable to the full-precision accuracy 71.88% while our uniform quantization method offers comparable accuracy of 2-bit ResNet-18 to the state-of-the-art non-uniform quantization method. Source code is on https://github.com/HanByulKim/BASQ.

Keywords: Mobile network · Quantization · Neural architecture search

1 Introduction

Neural network optimization is becoming more and more important with the increasing demand for efficient computation for both mobile and server applications. When we reduce the data bit-width via quantization, the memory footprint

Supplementary Information The online version contains supplementary material available at https://doi.org/10.1007/978-3-031-19775-8_2.

S. Avidan et al. (Eds.): ECCV 2022, LNCS 13672, pp. 17–33, 2022.
https://doi.org/10.1007/978-3-031-19775-8_2

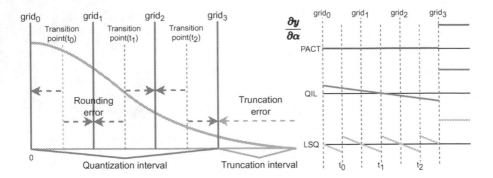

Fig. 1. Error components of quantization for activation distribution (left) and gradient of clipping threshold (right)

is reduced significantly, and the computation performance could be improved when the hardware acceleration is available. However, one major drawback of quantization is the output quality degradation due to the limited number of available values. In particular, when we apply sub-4-bit quantization to optimized networks, e.g., MobileNet-v2 [36], because the backbone structure is already highly optimized and has limited capacity, the accuracy drop is significant compared to the sub-4-bit quantization of the conventional, well-known redundant network, e.g., ResNet-18 [16]. However, because the computation efficiency of the optimized network is much higher than that of the redundant network, it is highly desirable to be able to quantize the advanced networks.

To preserve the quality of output or minimize the accuracy loss as much as possible, diverse quantization schemes have been proposed. Differentiable quantization is the representative method that determines the hyper-parameter of the quantization operator via back-propagation. For instance, in the case of PACT [8], the range of activation quantization is determined by an activation clipping parameter α, which denotes the value of the largest quantization level. In the case of QIL [21], the interval of quantization is parameterized by two learnable parameters for the center and size of the quantization interval. On the contrary, LSQ [11] learns the step size, which denotes the difference of quantization levels. PACT and QIL consider the single component of quantization error, truncation error (truncation and clipping are interchangeably used in this paper), and rounding error, respectively, while LSQ tries to optimize both components at the same time, as shown in Fig. 1. Due to this characteristic, LSQ often outperforms other quantization methods and exhibits state-of-the-art results.

However, according to our observation, LSQ becomes unstable and converges to sub-optimal points, especially when applying quantization to an optimized structure, e.g., a network with depthwise-separable convolution. In such an optimized network, the activation distribution gets skewed at every iteration depending on the weight quantization, as pointed out in the PROFIT study [33]. Under this circumstance, we empirically observe that PACT with a judicious tuning of weights on L2 decay of activation clipping parameter (called L2 decay weight parameter throughout in this paper) in its loss function shows a potential of lowering the final loss, especially for the quantization of optimized networks.

Since the value range of activation exhibits distinct per-layer characteristics, the L2 decay weight parameter, which determines the clip value, i.e., the value range of activation quantization, should be tuned judiciously in a layerwise manner. In this paper, we propose a novel idea, branch-wise activation-clipping search quantization (BASQ), that automates the L2 decay weight tuning process via the design space exploration technique motivated by neural architecture search studies. BASQ is designed to assume that the quantization operators with different L2 decay weights are selection candidates and decide, in a discrete space, the candidate that minimizes the accuracy degradation under quantization. According to our extensive studies, this scheme stabilizes the overall quantization process in the optimized network, offering state-of-the-art accuracy.

To further improve ultra-low-bit quantization, we adopt the recent advancement in the network structure for binary neural networks (BNN). In these studies [29,47], newly designed block design schemes are proposed to stabilize network training in low precision and accelerate the convergence via a batch normalization arrangement and new activation function. In this paper, motivated by these studies, we propose a novel extended block design for sub-4-bit quantization with search algorithm, contributing to accuracy improvement in low precision.

2   Related Works

2.1   Low-bit Quantization

Recently, various quantization algorithms for low precision, such as 4-bit or less, have been proposed. In uniform quantization, mechanisms such as clip value training [8], quantization interval learning [21], PACT with statistics-aware weight binning [7], and differentiable soft quantization [12] show good results in 2-bit to 4-bit precision. LSQ [11], which applies step size learning, shows good 2-bit to 4-bit accuracy in ResNet [16] networks.

Non-uniform quantizations [26,43] offer outstanding results than uniform quantizations thanks to their capability of fitting with various distributions. However, they have a critical limitation: they need special compute functions and data manipulations to support quantization values mapped to non-uniform levels.

Mixed precision quantizations [1,31,37–39] show cost reduction by varying bit precision, e.g., in a layer-wise manner. For instance, they can use high-bits in important layers while using low-bits in non-sensitive layers. But they also have a limitation in that the hardware accelerators must be capable of supporting various bit precisions in their compute units. Also, neural architecture search (NAS) solutions [13,14,40] apply low-bit quantization by constructing search space with operations of diverse quantization bit candidates.

Thus far, existing quantization methods have struggled with low-bit quantization of networks targeting low cost and high performance, such as MobileNet-v1 [19], v2 [36], and v3 [18]. In PROFIT [33], the progressive training schedule is applied to achieve promising 4-bit accuracy from the MobileNet series.

Recently, binary neural networks (BNN) [3,28,29,32,34,47] are also being actively studied. Unlike the low-bit network of 2-bit to 4-bit, there is a difference in BNN by using the sign function as a quantization function. Various works based on XNOR operation [34], improvements in training schedule [32], NAS solution [3], block structure and activation function [28,29,47] contribute to good results in 1-bit precision.

2.2   Neural Architecture Search

While hand-crafted architectures [16,18–20,30,36,41,46] have evolved with good accuracy and low computation, neural architecture search (NAS) has introduced an automated solution to explore the optimal structure, showing better accuracy and lower computation cost than most hand-crafted architectures. Amoebanet [35] and NAS-Net [49] show the potential of architecture search in the early days but require high training costs. Differential NAS and one-shot NAS have emerged with the concept of supernet for efficient NAS. Differential NAS [4,6,27,42] uses shared parameters so that weights and architecture parameters are jointly trained on supernet training, and the architecture decisions are made in the final design stage. One-shot NAS [9,13,44] trains a supernet by continuously sampling subnets on the supernet. Space exploration algorithm such as evolutionary algorithm (EA) is often used to determine the best subnet architecture from the trained supernet.

In this paper, we adopt the one-shot NAS method to conduct our search-based quantization. Specifically, we propose to find the L2 decay weight of the clip value by considering the multiple L2 decay weights as selection candidates.

3   Preliminary

Considering that, under quantization, the number of available levels is highly restricted to 2^{bit}, to maintain the quality of output after quantization, hyper-parameters for the quantization should be selected carefully. In this work, our quantization method for activation is based on the well-known differentiable quantization method, PACT [8]. Before explaining the details of our proposed methods, BASQ and novel structure design, we will briefly introduce PACT.

As a conventional quantization method, PACT is also implemented based on the straight-through estimator [2]. The rounding operation in the quantization interval is ignored through back-propagation and bypasses the gradient. The output function of the PACT activation quantizer is designed as in Eq. 1,

$$y = \frac{1}{2}(|x| - |x - \alpha| + \alpha), \quad \frac{\partial y}{\partial \alpha} = \begin{cases} 0, & \text{if } x < \alpha \\ 1, & \text{if } x \geq \alpha, \end{cases} \quad (1)$$

where α is the learnable parameter for the activation clipping threshold. The input values larger than the truncation interval are clamped to the clipping threshold, and the input values smaller than the threshold are linearly quantized. According to Eq. 1, the gradient of the clipping threshold comes from the

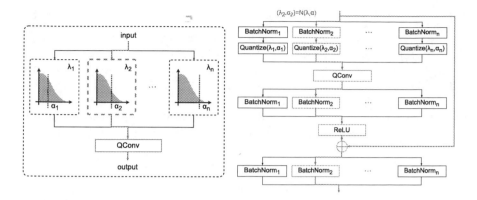

(a) BASQ on activation quantization   (b) BASQ search block

Fig. 2. BASQ: branch-wise activation-clipping search quantization (execution path by selected branch from uniform sampling is highlighted.)

values in the truncation interval; thereby, the clipping threshold tends to become larger in order to minimize the truncation error. In order to prevent the clipping threshold from exploding, L2 decay weight is introduced to guide the convergence of the clipping threshold, balancing the rounding error and truncation error. The updating process of the clipping threshold works as follows:

$$\alpha_{new} = \alpha - \eta(\sum_i^N \frac{\partial \mathcal{L}}{\partial y_{q,i}} \frac{\partial y_i}{\partial \alpha} + \lambda|\alpha|). \tag{2}$$

4   Branch-wise Activation-clipping Search Quantization

Branch-wise activation-clipping search quantization (BASQ) is designed to search the optimal L2 decay weight for the activation clipping threshold in the PACT algorithm. When the optimal value of the L2 decay weight hyperparameter is determined in a layer-wise manner, an exhaustive search of candidates is impractical. For instance, in case of ResNet-18 having eight available L2 decay weights per block, the number of candidates is as much as 17 Million. Searching for such a large space is prohibitively expensive; thus, we adopt an alternative approach motivated by NAS. In this algorithm, we introduce multiple branches having different L2 decay weights (λ) as selection candidates, as shown in Fig. 2. There are multiple branches of quantization operators with different L2 decay weights for the activation clipping. One of the branches could be selected to determine the L2 decay weight used to learn the clipping threshold for quantizing the input activation. To minimize quality degradation after quantization, BASQ searches the best configurations of L2 decay weights among candidates.

4.1   Search Space Design

The search candidates of BASQ for each activation quantization layer are defined by a pair of two parameters, L2 decay weight (λ) and clip value (α), (λ, α). Let the super-set of λ be Λ, and the super-set of α be A. $\Lambda \times A$ is the search space of BASQ with element pair (λ, α). As in Eq. 3, Λ is a finite set by restricting it with L2 decay weight candidates. A is a subset of the real numbers, and each element (α) exists as a learnable parameter. As a result, BASQ searches discrete space with Λ and simultaneously searches continuous space with A.

$$\Lambda = \{\lambda_1, \lambda_2, \cdots, \lambda_n\} \, (\lambda_i \in \mathbb{R}), \quad A \subset \mathbb{R} \tag{3}$$

The network model of BASQ consists of multiple quantization branches, as shown in Fig. 2a. Each branch has a (λ, α) pair and can be seen as an activation quantization operator following the PACT algorithm. When one of the branches is selected, the forward operation is performed as the left-hand side of Eq. 1 and backward operations as the right-hand side of Eq. 1 and Eq. 2.

In case of BASQ, the training can be considered as the joint training of the continuous learnable parameter α and discrete parameter for the architecture structure λ. The difference from the search structure of existing NAS methods is that only the activation quantization operators are designed to be explored while the computation, e.g., convolution, is shared across different branches. In short, the computation operation is identical across branches while the quantization strategy (for λ) is searched. According to our observation, the accuracy is rather inferior when taking a private computation on each branch. In addition, note that batch normalization layers are private on each quantization branch to solve statistics diversity [5] that may occur from the updated clip value.

The branch structure of BASQ looks similar to that of [45]. Our difference is that we adopt the activation quantization branches to obtain optimal quantization policies for the activation having the same bit-width across branches while the branches in [45] have different bit-widths. In addition, we adopt the space exploration algorithm to determine the best configuration reducing the training loss, while the previous study exploits each path selectively for different bit-width configurations.

4.2   Search Strategy

BASQ is designed to exploit the supernet-based one-shot training as [13], based on the parameter sharing technique. The search process of BASQ is composed of three steps; First, the base network with quantization operators is jointly trained as a supernet covering all possible configurations. Then, we explore the search space to determine the optimal selection among candidates. Finally, we stabilize batch statistics of the optimal selection. The details are provided in the following.

Supernet Training. At every iteration, one branch of activation quantization is selected via uniform random sampling. The parameters in the sampled path, or the selected L2 decay weight (λ) and clip value (α), are used to quantize the input activation. The activation quantization process follows the conventional PACT algorithm, as shown in Eqs. 1 and 2. In this stage, the weight of the network and the quantization parameters are jointly optimized. When the network is trained for long enough, the weights of the supernet are optimized for the possible configurations of multiple branches, which enables us to evaluate the quality of quantization configuration without additional training.

Architecture Search. After training the super-network, we obtain the best configuration of truncation parameter trained with L2 decay weight (α^*), based on the evolutionary algorithm. The problem definition is as follows (ACC_{val} is network accuracy of the validation set):

$$\alpha^* = \underset{\alpha \in A}{\mathrm{argmax}} \ ACC_{val}(w, \alpha). \tag{4}$$

Finetuning. After selecting the best architecture, we additionally stabilize the batch statistics by performing additional finetuning for the subset of the network, similar to [33]. We freeze all layers except the normalization layers and clip values in the network, and have a finetuning step of 3 epochs.

5   Block Structure for Low-bit Quantization

5.1   New Building Block

Many BNN studies focus on designing an advanced network architecture that allows stable and fast convergence with binary operators. The representative study [29] proposes a ReAct block that allows skipping connection end-to-end across all layers in the network. [47] provides better accuracy by modifying the arrangement of the activation function and batch normalization in the ReAct block. In this paper, we present an innovative building block that can have a good property in low precision. As shown in Fig. 3a and 3b, the new building block for BASQ additionally places a batch normalization layer in front of activation quantization in the block structure of [47] including the skip connection. Originally, [47] places batch normalization after the shortcut to obtain the effect of balanced activation distribution through the affine transformation of the batch normalization layer. However, the gradient through the skip connection in the following block affects the batch normalization at the end of the current block. This has adverse effects on scale & shift of the quantization operator on the next block. Our new building block places a dedicated batch normalization in front of the quantization layer right after the starting point of the skip connection to stabilize activation distribution entering the quantization layer. In addition, since this study targets 2 to 4-bit low precision, the sign and RPReLU functions are replaced by multi-bit quantization and ReLU functions, respectively.

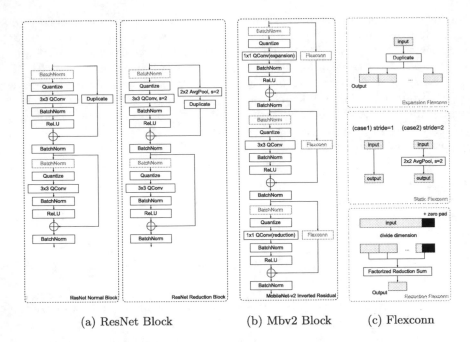

(a) ResNet Block   (b) Mbv2 Block   (c) Flexconn

Fig. 3. (a), (b) New building block structures and (c) block skip connections (Flexconn) for low-bit quantization (differences from [29] & [47] are highlighted and Mbv2 represents MobileNet-v2.)

5.2 Flexconn: A Flexible Block Skip Connection for Fully Skip-Connected Layers

In order to bring stable effect to search algorithm with low-bit quantization through the new building block in Sect. 5.1, it is important to build fully skip-connected layers by connecting skip connections over all layers as in [29] and [47]. This fully skip-connected layer refers to a skip connection made of full precision without any quantized operation, unlike skip connection using 1×1 convolution in ResNets. However, it is non-trivial to build a fully skip-connected layer, especially if the number of output channels is not an integer multiple of that of input channels, e.g., a reduction layer in the inverted residual block of MobileNet-v2. Therefore, in this paper, we propose Flexconn, which enables the formation of a fully skip-connected layer, even in the dynamic change (expansion and reduction at any ratio) of channel dimension. Flexconn forms a fully skip-connected layer for the following three cases, as shown in Fig. 3c.

- **Expansion Flexconn on 1×1 channel expansion convolution**
  In channel expansion, Expansion Flexconn copies the input tensor up to the integer multiple of channel expansion and concatenates the copied tensors in the channel dimension. It can be used for the 1×1 channel expansion convolution of the inverted residual block in MobileNet-v2.

- **Static Flexconn on 3×3 same channel convolution**
  In the case of the same channel size, Static Flexconn is used. It handles the skip connection in the same way as in [29]. If the stride is 1, we add the skip connection as is. If the stride is 2, we use 2×2 average pooling to adjust the resolution. Static Flexconn can be used for 3×3 convolution in the middle of the inverted residual block in MobileNet-v2.
- **Reduction Flexconn on 1×1 channel reduction convolution**
  Reduction Flexconn is used to make a connection in case of channel reduction. It performs a factorized reduction sum operation after dividing the input channels by the integer multiple of the reduced channels. In MobileNet-v2, it is used for the last 1×1 channel reduction convolution of the inverted residual block. However, when the number of input channels is not an integer multiple of that of reduced channels, Flexconn uses zero tensors and concatenates them to the input channels on the last channel dimension to match the integer multiple reductions. Flexconn solves the corner case of non-integer multiple reductions in MobileNet-v2, thereby enabling the end-to-end identity connection.

6   Experiments

To evaluate the effectiveness of the proposed methods, we perform the BASQ experiment in ImageNet dataset [10]. The networks are trained along with the learning schedule of 2-phase training used in [32] for stable clipping parameter training. Therefore, training is performed with only activations quantized in phase-1 and both activations and weights quantized in phase-2. In MobileNet-v2, we use a batch size of 768 with an initial learning rate of 0.000375 for 466,000 iterations per phase. For selection candidates, seven L2 decay weights are used in MobileNet-v2. In MobileNet-v1 and ResNet-18, we use a batch size of 1024 with an initial learning rate of 0.0005 for 400,000 iterations per phase. For selection candidates, eight L2 decay weights are used in both networks. For all experiments, we use Adam optimizer [23] with cosine learning rate decay without restart. The weight decay is set to 5×10^{-6}. In addition, knowledge distillation [17] is applied with ResNet-50 as a teacher. The KL-divergence between teacher and student's softmax output is used as loss [29].

Note that we apply LSQ as weight quantization in all experiments. Also, in MobileNet-v2, we apply BASQ only on activation quantization in front of the 3×3 convolution in the middle of the inverted residual block. In the evolutionary algorithm, we use 45 populations with 15 mutations and 15 crossovers in 20 iterations. We evaluate 2 to 4-bit quantizations for MobileNet-v2 and ResNet-18 and 4-bit for MobileNet-v1.

To evaluate BASQ, we apply the k-fold evaluation ($k = 10$) method in the ImageNet validation set by constructing a test set with exception to the dataset used in architecture selection.[1] The method is as follows.

---

[1] In [13], the accuracy is evaluated in a different way. [13] use about 16% of the validation set for architecture selection. The test set is constructed and evaluated using the entire validation set. In order to avoid such a duplicate use of the same data in architecture selection and evaluation, we adopt k-fold evaluation.

1. We divide the validation set into 10 subsets.
2. One subset is used to select the architecture while the other subsets are used for the test set to evaluate the accuracy of the selected architecture.
3. We perform step 2 for each of the remaining nine subsets that are not used for architecture selection. Each one is used for selecting the architecture and the rest are used as a test set. We average 10 evaluation results to calculate the final accuracy.

6.1 Evaluation with MobileNet-v2 and MobileNet-v1

Table 1. Top-1 ImageNet accuracy (%) of MobileNet-v2 and v1 models

Method	A2/W2	A3/W3	A4/W4
PACT [8]			61.4
DSQ [12]			64.8
QKD [22]	45.7	62.6	67.4
LSQ [11]	46.7	65.3	69.5
LSQ + BR [15]	50.6	67.4	70.4
LCQ [43]			70.8
PROFIT [33]	61.9	69.6	71.56
BASQ (Ours)	**64.71**	**70.25**	**71.98**

(a) MobileNet-v2

Method	A4/W4
PACT [8]	62.44
LSQ [11]	63.60
PROFIT [33]	69.06
BASQ (Ours)	**72.05**

(b) MobileNet-v1

Table 1a shows the accuracy of 2 to 4-bit quantized MobileNet-v2. Our 2-bit model (under ResNet-50 teacher) outperforms the state-of-the-art model, PROFIT [33] (with ResNet-101 teacher) by a large margin (64.71% vs 61.9%). Our 3-bit and 4-bit models also exhibit slightly better accuracy than PROFIT, while our 4-bit model gives 71.98% which is slightly better than the full-precision accuracy of 71.88%. Table 1b shows accuracy of MobileNet-v1. BASQ gives significantly better results on 4-bit (72.05%) compared to the state-of-the-art method, PROFIT (69.06%).

6.2 Evaluation with ResNet-18

Table 2 shows the accuracy of 2 to 4-bit quantized ResNet-18. Our models offer competitive results with the state-of-the-art one, LCQ [43]. Specifically, our 3-bit and 4-bit models give better results than LCQ by 0.8% and 1.1%, respectively, while our 2-bit model offers comparable (within 0.3%) results to LCQ. Note that LCQ is a non-uniform quantization method. Thus, it is meaningful that BASQ, as a uniform quantization method, shows comparable results to the state-of-the-art non-uniform method, LCQ.

Table 2. Top-1 ImageNet accuracy (%) of ResNet-18 models

Method	A2/W2	A3/W3	A4/W4
Dorefa [48]	62.6	67.5	68.1
PACT [8]	64.4	68.1	69.2
DSQ [12]	65.17	68.66	69.56
QIL [21]	65.7	69.2	70.1
PACT + SAWB + fpsc [7]	67.0		
APOT [26]	67.3	69.9	70.7
QKD [22]	67.4	70.2	71.4
LSQ [11]	67.6	70.2	71.1
LCQ [43]	**68.9**	70.6	71.5
BASQ (Ours)	68.60	**71.40**	**72.56**

Table 1 and 2 demonstrate that MobileNets are more difficult to quantize than ResNet-18. Thus, the existing methods, e.g., [8,11,12,22], which perform well on ResNet-18 in Table 2 give poor results on MobileNets in Table 1. It is also challenging to offer constant competitiveness across low-bit precisions. Some existing works, e.g., [11,22] show comparable results on 4-bit models while giving poor results in 2-bit cases. Our proposed BASQ is unique in that it constantly offers competitive results on both ResNet-18 and MobileNets across bit precisions from 2 to 4-bits. As will be explained in the next section, such benefits result from our proposed joint training of continuous and discrete parameters and the proposed block structures.

6.3   Ablation Study

Effects of Components. We use 2-bit ResNet-20 and MobileNet-v2 models on CIFAR10 [24] to evaluate the effect of each component in our proposed method. Table 3 shows the effect of adopting BASQ in activation quantization in the 2-phase training. The table shows that BASQ, adopted for activation quantization, can contribute to accuracy improvement in both ResNet-20 and MobileNet-v2.

Table 3. BASQ results of 2-bit models on CIFAR10

Model	Activation quantization	Weight quantization	Accuracy (%)
ResNet-20	LSQ	LSQ	89.40
	BASQ	LSQ	90.21
MobileNet-v2	LSQ	LSQ	89.62
	BASQ	LSQ	90.47

We also evaluate the effect of new building block (in Sect. 5.1), shown in Table 4. Note that, in the case of MobileNet-v2, we apply the new building block structure in Fig. 3b (with new batch normalization layer and without layer skip connections as original MobileNet-v2 block). The results show that the new building block improves the accuracy in both ResNet-20 and MobileNet-v2. MobileNet-v2 shows lower advances than ResNet-20 due to the absence of layer skip connections that obstructs the intention of the new batch normalization layer. We also evaluate the block structure of [47] in both networks and obtain accuracy degradation. This shows that the dedicated batch normalization layer in the new building block has the potential of stabilizing the convergence of low precision networks.

Table 4. Effect of new building block in 2-bit models on CIFAR10

Model	Block	Accuracy (%)
ResNet-20	preactivation ResNet	89.40
	Fracbnn [47]	89.39
	new building block	89.72
MobileNet-v2	basic inverted residual block	89.62
	Fracbnn [47]	89.32
	new building block	89.74

Table 5 shows the effect of Flexconn (in Sect. 5.2). Basically, we use the residual block connection originally used in MobileNet-v2 while using Flexconns. The table shows that we obtain the highest accuracy when all the three cases of Flexconn are adopted in the inverted residual block. This shows that constructing a fully skip-connected layer, making all layers connected in high precision, is critical in low precision networks. (For detailed analysis of our proposed block structure with BASQ, see Sect. 3 of supplementary.)

Table 5. Effect of Flexconn in 2-bit MobileNet-v2 on CIFAR10

Methods on Inverted Residual Block				Accuracy (%)
New Building Block	Expansion Flexconn	Static Flexconn	Reduction Flexconn	
				89.62
✓				89.74
✓	✓			90.02
✓		✓		88.95
✓			✓	89.61
✓	✓	✓		90.35
✓	✓		✓	89.86
✓		✓	✓	90.08
✓	✓	✓	✓	90.94

Importance of Searching L2 Decay Weight in Discrete Search Space. BASQ jointly optimizes the weight and clip value in continuous search space while searching L2 decay weight for updating clip value in discrete search space. To investigate the importance of searching for L2 decay weight in discrete search space, we obtain the accuracy of 100 architectures by randomly selecting each block's clip value within the supernet. 2-bit ResNet-18 and MobileNet-v2 are evaluated on ImageNet. The results are shown in Fig. 4.

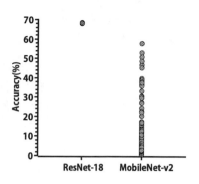

Fig. 4. Accuracies for 100 random architectures from 2-bit ResNet-18 and MobileNet-v2

In the case of ResNet-18, the overall accuracy deviation between architectures is as small as [67.98%, 68.53%]. However, in the case of MobileNet-v2, the difference of accuracy between architectures is quite large in [0.08%, 57.76%], much greater than that of ResNet-18. This demonstrates that a simple selection of L2 decay weight (or just one learning solution) adopted in the previous works [8, 11, 12, 22] worked well for ResNets while such a strategy does not work in MobileNets as shown in our experiments, which advocates the necessity of selection methods for L2 decay weights like ours.

Loss Landscape Comparison. Figure 5 illustrates the complexity of loss surface between LSQ and BASQ. We apply the loss-landscape [25] method to 2-bit ResNet-20 and MobileNet-v2 on CIFAR10 with LSQ and BASQ. We do not apply the new building block and Flexconn to evaluate pure quantization effects. As shown in Fig. 5a, 5b of ResNet-20, BASQ shows improved loss surface with clear convexity while LSQ has several local minima. As shown in Fig. 5c, 5d, in case of MobileNet-v2, both LSQ and BASQ do not show a clear loss surface, possibly due to the fact that MobileNet-v2 is an optimized network. Nevertheless, BASQ shows improved convexity while LSQ has a local minima near the center. The comparison shows that BASQ has the potential of better convergence in training low-bit networks, including optimized ones like MobileNet-v2.

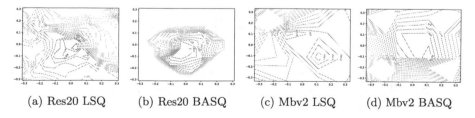

(a) Res20 LSQ   (b) Res20 BASQ   (c) Mbv2 LSQ   (d) Mbv2 BASQ

Fig. 5. Loss landscape of 2-bit quantized models on CIFAR10 (Res20 and Mbv2 represents ResNet-20 and MobileNet-v2, respectively.)

Table 6. Effect of training schedule in MobileNet-v2 on ImageNet

Method	A2/W2	A3/W3	A4/W4
LSQ [11]	50.0	66.5	70.2
BASQ	64.7	70.3	72.0

Table 7. BASQ results with shared and private convolution in search block

BASQ Method	Accuracy (%)
Shared conv	90.21
Private conv	89.90

Training Schedule. Quantization algorithms usually start with a full-precision pre-trained model and proceed with finetuning while quantizing activation and weight. However, BASQ trains itself without a full-precision pre-trained model. It starts training from scratch and proceeds with a 2-phase schedule from [32]. Progressive training is performed with only activation quantized (phase-1) and then with both activation and weight quantized (phase-2). This training schedule can affect MobileNet-v2 results. For example, [33] proposes good MobileNet results by special training schedule without changing the network. To evaluate the effect of the training schedule, we train LSQ [11] with the same training schedule as BASQ. The results are shown in Table 6. Although the accuracy of LSQ gets improved, there is still a large accuracy gap in 2-bit and 3-bit models.

Search Block. Unlike typical NAS methods, BASQ consists of search blocks where branches are made up of quantization operators so that the computation is shared across selections as shown in Fig. 2b. To evaluate the effect of shared computation, we measure the accuracy of private computation, where each branch is equipped with private convolution as in typical NAS approaches. Table 7, on 2-bit ResNet-20 with CIFAR10, shows that the shared one offers better accuracy.

7   Conclusion

In this paper, we propose a novel quantization method BASQ and block structures (new building block and Flexconn). BASQ judiciously exploits NAS to search for hyper-parameters, namely, L2 decay weights of clipping threshold. Our proposed block structure helps form a fully skip-connected layer with a dedicated normalization layer, which contributes to the quantization for low-bit precisions. The experiments show that our proposed method offers constant competitiveness for both ResNet and MobileNets across low-bit precisions from 2 to 4-bits. Specifically, our 2-bit MobileNet-v2 model (though trained under a weaker teacher) outperforms the state-of-the-art 2-bit model by a large margin. In addition, our uniformly quantized 2-bit ResNet-18 model offers comparable accuracy to the non-uniformly quantized model of the state-of-the-art method.

Acknowledgment. This work was supported by IITP and NRF grants funded by the Korea government (MSIT, 2021-0-00105, NRF-2021M3F3A2A02037893) and Samsung Electronics (Memory Division, SAIT, and SRFC-TC1603-04).

References

1. Bai, H., Cao, M., Huang, P., Shan, J.: BatchQuant: quantized-for-all architecture search with robust quantizer. Advances in Neural Information Processing Systems 34 (2021)
2. Bengio, Y., Léonard, N., Courville, A.: Estimating or propagating gradients through stochastic neurons for conditional computation. arXiv preprint arXiv:1308.3432 (2013)
3. Bulat, A., Martinez, B., Tzimiropoulos, G.: Bats: Binary architecture search. In: European Conference on Computer Vision (ECCV) (2020)
4. Cai, H., Zhu, L., Han, S.: ProxylessNAS: Direct neural architecture search on target task and hardware. In: International Conference on Learning Representations (ICLR) (2019)
5. Chen, P., Liu, J., Zhuang, B., Tan, M., Shen, C.: Towards accurate quantized object detection. In: Computer Vision and Pattern Recognition (CVPR) (2021)
6. Chen, X., Xie, L., Wu, J., Tian, Q.: Progressive differentiable architecture search: bridging the depth gap between search and evaluation. In: International Conference on Computer Vision (ICCV) (2019)
7. Choi, J., Venkataramani, S., Srinivasan, V., Gopalakrishnan, K., Wang, Z., Chuang, P.: Accurate and efficient 2-bit quantized neural networks. Proc. Mach. Learn. Syst. 1, 348–359 (2019)
8. Choi, J., Wang, Z., Venkataramani, S., Chuang, P., Srinivasan, V., Gopalakrishnan, K.: Pact: parameterized clipping activation for quantized neural networks. arXiv:1805.06085 (2018)
9. Chu, X., Zhang, B., Xu, R.: Fairnas: rethinking evaluation fairness of weight sharing neural architecture search. In: International Conference on Computer Vision (ICCV) (2021)
10. Deng, J., Dong, W., Socher, R., Li, L., Li, K., Fei-Fei, L.: ImageNet: a largescale hierarchical image database. In: Computer Vision and Pattern Recognition (CVPR) (2009)
11. Esser, S., McKinstry, J., Bablani, D., Appuswamy, R., Modha, D.: Learned step size quantization. In: International Conference on Learning Representations (ICLR) (2020)
12. Gong, R., et al.: Differentiable soft quantization: Bridging full-precision and low-bit neural networks. In: International Conference on Computer Vision (ICCV), vol. 1, pp. 348–359 (2019)
13. Guo, Z., et al.: Single path one-shot neural architecture search with uniform sampling. In: European Conference on Computer Vision (ECCV) (2020)
14. Habi, H., Jennings, R., Netzer, A.: HMQ: Hardware friendly mixed precision quantization block for CNNs. In: European Conference on Computer Vision (ECCV) (2020)
15. Han, T., Li, D., Liu, J., Tian, L., Shan, Y.: Improving low-precision network quantization via bin regularization. In: Proceedings of the IEEE/CVF International Conference on Computer Vision, pp. 5261–5270 (2021)
16. He, K., Zhang, X., Ren, S., Sun, J.: Deep residual learning for image recognition. In: Computer Vision and Pattern Recognition (CVPR) (2016)
17. Hinton, G., Vinyals, O., Dean, J.: Distilling the knowledge in a neural network. arXiv:1503.02531 (2015)
18. Howard, A., et al.: Searching for MobileNetV3. In: International Conference on Computer Vision (ICCV) (2019)

19. Howard, A., et al.: MobileNets: efficient convolutional neural networks for mobile vision applications. arXiv:1704.04861 (2017)
20. Hu, J., Shen, L., Sun, G.: Squeeze-and-excitation networks. In: Computer Vision and Pattern Recognition (CVPR) (2018)
21. Jung, S., et al.: Learning to quantize deep networks by optimizing quantization intervals with task loss. In: Computer Vision and Pattern Recognition (CVPR) (2018)
22. Kim, J., Bhalgat, Y., Lee, J., Patel, C., Kwak, N.: QKD: quantization-aware knowledge distillation. arXiv:1911.12491 (2019)
23. Kingma, D., Ba, J.: Adam: a method for stochastic optimization. arXiv:1412.6980 (2014)
24. Krizhevsky, A.: Learning multiple layers of features from tiny images. Technical report (2009)
25. Li, H., Xu, Z., Taylor, G., Studer, C., Goldstein, T.: Visualizing the loss landscape of neural nets. In: Advances in neural information processing systems (NIPS) 31 (2018)
26. Li, Y., Dong, X., Wang, W.: Additive powers-of-two quantization: an efficient non-uniform discretization for neural networks. In: International Conference on Learning Representations (ICLR) (2020)
27. Liu, H., Simonyan, K., Yang, Y.: Darts: differentiable architecture search. In: International Conference on Learning Representations (ICLR) (2019)
28. Liu, Z., Shen, Z., Li, S., Helwegen, K., Huang, D., Cheng, K.: How do Adam and training strategies help BNNs optimization. In: International Conference on Machine Learning (ICML) (2021)
29. Liu, Z., Shen, Z., Savvides, M., Cheng, K.: ReActNet: Towards precise binary neural network with generalized activation functions. In: European Conference on Computer Vision (ECCV) (2020)
30. Ma, N., Zhang, X., Zheng, H., Sun, J.: ShuffleNet V2: Practical guidelines for efficient cnn architecture design. In: European Conference on Computer Vision (ECCV) (2018)
31. Ma, Y., et al.: OMPQ: Orthogonal mixed precision quantization. arXiv:2109.07865 (2021)
32. Martinez, B., Yang, J., Bulat, A., Tzimiropoulos, G.: Training binary neural networks with real-to-binary convolutions. In: International Conference on Learning Representations (ICLR) (2020)
33. Park, E., Yoo, S.: Profit: a novel training method for sub-4-bit MobileNet models. In: European Conference on Computer Vision (ECCV) (2020)
34. Rastegari, M., Ordonez, V., Redmon, J., Farhadi, A.: XNOR-Net: ImageNet classification using binary convolutional neural networks. In: European Conference on Computer Vision (ECCV) (2016)
35. Real, E., Aggarwal, A., Huang, Y., Le, Q.: Regularized evolution for image classifier architecture search. In: AAAI Conference on Artificial Intelligence 33, 4780–4789 (2019)
36. Sandler, M., Howard, A., Zhu, M., Zhmoginov, A., Chen, L.: MobileNet V2: inverted residuals and linear bottlenecks. In: Computer Vision and Pattern Recognition (CVPR) (2018)
37. Uhlich, S., et al.: Mixed precision DNNs: All you need is a good parametrization. In: International Conference on Learning Representations (ICLR) (2020)
38. Wang, K., Liu, Z., Lin, Y., Lin, J., Han, S.: HAQ: hardware-aware automated quantization with mixed precision. In: Computer Vision and Pattern Recognition (CVPR) (2019)

39. Wang, T., et al.: APQ: Joint search for network architecture, pruning and quantization policy. In: Computer Vision and Pattern Recognition (CVPR) (2020)
40. Wu, B., Wang, Y., Zhang, P., Tian, Y., Vajda, P., Keutzer, K.: Mixed precision quantization of convnets via differentiable neural architecture search. arXiv:1812.00090 (2018)
41. Xie, S., Girshick, R., Dollár, P., Tu, Z., He, K.: Aggregated residual transformations for deep neural networks. In: Computer Vision and Pattern Recognition (CVPR) (2017)
42. Xie, S., Zheng, H., Liu, C., Lin, L.: SNAS: stochastic neural architecture search. International Conference on Learning Representations (ICLR) (2018)
43. Yamamoto, K.: Learnable companding quantization for accurate low-bit neural networks. In: Computer Vision and Pattern Recognition (CVPR) (2021)
44. You, S., Huang, T., Yang, M., Wang, F., Qian, C., Zhang, C.: GreedyNAS: Towards fast one-shot NAS with greedy supernet. In: Computer Vision and Pattern Recognition (CVPR) (2020)
45. Yu, H., Li, H., Shi, H., Huang, T., Hua, G.: Any-precision deep neural networks. arXiv:1911.07346 (2019)
46. Zhang, X., Zhou, X., Lin, M., Sun, J.: ShuffleNet: an extremely efficient convolutional neural network for mobile devices. In: Computer Vision and Pattern Recognition (CVPR) (2018)
47. Zhang, Y., Pan, J., Liu, X., Chen, H., Chen, D., Zhang, Z.: FracBNN: accurate and FPGA-efficient binary neural networks with fractional activations. ACM/SIGDA International Symposium on Field-Programmable Gate Arrays, pp. 171–182 (2021)
48. Zhou, S., Ni, Z., Zhou, X., Wen, H., Wu, Y., Zou, Y.: DoReFa-net: Training low bitwidth convolutional neural networks with low bitwidth gradients. arXiv:1606.06160 (2016)
49. Zoph, B., Vasudevan, V., Shlens, J., Le, Q.: Learning transferable architectures for scalable image recognition. In: Computer Vision and Pattern Recognition (CVPR) (2018)

You Already Have It: A Generator-Free Low-Precision DNN Training Framework Using Stochastic Rounding

Geng Yuan[1], Sung-En Chang[1], Qing Jin[1], Alec Lu[2], Yanyu Li[1],
Yushu Wu[1], Zhenglun Kong[1], Yanyue Xie[1], Peiyan Dong[1],
Minghai Qin[1], Xiaolong Ma[3], Xulong Tang[4], Zhenman Fang[2],
and Yanzhi Wang[1(✉)]

[1] Northeastern University, Boston, MA 02115, USA
yanz.wang@northeastern.edu
[2] Simon Fraser University, Burnaby, BC V5A 1S6, Canada
[3] Clemson University, Clemson, SC 29634, USA
[4] University of Pittsburgh, Pittsburgh, PA 15260, USA

Abstract. Stochastic rounding is a critical technique used in low-precision deep neural networks (DNNs) training to ensure good model accuracy. However, it requires a large number of random numbers generated on the fly. This is not a trivial task on the hardware platforms such as FPGA and ASIC. The widely used solution is to introduce random number generators with extra hardware costs. In this paper, we innovatively propose to employ the stochastic property of DNN training process itself and directly extract random numbers from DNNs in a self-sufficient manner. We propose different methods to obtain random numbers from different sources in neural networks and a generator-free framework is proposed for low-precision DNN training on a variety of deep learning tasks. Moreover, we evaluate the quality of the extracted random numbers and find that high-quality random numbers widely exist in DNNs, while their quality can even pass the NIST test suite.

Keywords: Efficient training · Quantization · Stochastic rounding

1 Introduction

To fully unleash the full power of the deep neural networks (DNNs) on various resource-constrained edge computing devices, DNN model compression [5,39,40] [17,18,22,26,28,29,31,34,50] has become the fundamental element and core enabler to bridge the gap between algorithm innovation and hardware implementation [6,11–13,21,25,27,30,35,47–49]. Recently, a surge of research efforts has

---

G. Yuan and S.E. Chang—These authors contributed equally.

Supplementary Information The online version contains supplementary material available at https://doi.org/10.1007/978-3-031-19775-8_3.

been devoted to low-precision DNN training to better satisfy the limitation of the computation and storage resources on edge devices [52,53]. However, the commonly used rounding schemes, such as round-up (ceiling), round-down (floor), or rounding-to-nearest, usually lead to severe accuracy degradation in low-precision training scenarios [14]. The reason is that the small gradients below the minimum representation precision are always rounded to zero, hence an information loss for weight updates [16]. Therefore, prior works propose the stochastic rounding scheme to help preserve information and enable low-precision training achieving similar accuracy as the full-precision floating-point training [14,32].

Stochastic rounding is to round up or round down a number (e.g., the gradient) in a probabilistic way. And the probability is proportional to the proximity of the number and its nearby low-precision representation level. In specific, for each time of stochastic rounding and each number to be rounded, a random number needs to be generated to indicate the probabilistic rounding decision.

However, generating a large number of high-quality random numbers in parallel on hardware-based deep learning platforms such as FPGA and ASIC is not a trivial task [2,24]. The commonly used solution is to incorporate a large number of random number generators (RNGs) [19] to generate random numbers on the fly [32,38]. This will inevitably introduce extra hardware costs and complexity. Many prior works just assume the stochastic rounding can be appropriately incorporated in their design by default but barely care about its actual implementation [43,52,53]. For example, considering a representative FPGA-based DNN training design [23], incorporating stochastic rounding will increase 23% hardware costs (i.e., LUTs) [33] to fulfill its computation parallelism.

With the trend of stochastic rounding becoming a "must-do" step in low-precision DNN training [43,46,52,53], we may raise the following question. *Is there a more efficient way to obtain the random numbers without using extra hardware?* Fortunately, the answer is positive. One important thing that is neglected by the prior works is that the neural network training process is based on a certain degree of randomness, for example, the randomness introduced by the mini-batch training with stochastic gradient descent and the randomly shuffled training samples. This indicates that the neural network itself is supposed to be a potential source of random numbers.

In this paper, we innovatively propose to employ the stochastic property of the neural network training process to directly extract random numbers from neural networks. We consider the dynamically changed trainable parameters, training data, and intermediate results during the training process can be regarded as source data with randomness, and random numbers with arbitrary bits can be extracted from them. We propose two methods of random number extraction. One is to extract the corresponding number of bits directly from a source data, but the random numbers obtained in this way are heavily affected by the distribution of the source data. Therefore, we further propose the method of extracting the least significant bit (LSB) from multiple source data and synthesizing a random number with multiple bits (e.g., 8 bits). This method greatly improves the randomness of the obtained random numbers.

Based on that, we argue that when a random number is needed during DNN training, it is not necessary to use an additional random number generator to generate one because you already have it. We can directly extract random numbers from the network by leveraging the stochastic property of the DNN training process. Therefore, we propose a generator-free framework as a more flexible and hardware-economic solution for low-precision DNN training. And we utilize the extracted random numbers for stochastic rounding in a self-sufficient manner to achieve high model accuracy.

We investigate the quality of random numbers extracted from different types of source data (e.g., trainable parameters, intermediate results). And we find that high-quality random numbers can be widely extracted from certain types of source data (e.g., the gradient of weights) without delicate selections. Most impressively, we find that the extracted random numbers can pass the entire NIST test suite SP800-22 [37], which is one of the most widely used testing suits for random numbers. Note that even the most widely used linear-feedback shift register (LFSR)-based random number generator fails some of the tests.

Moreover, besides obtaining random numbers with uniform distribution needed for stochastic rounding, we further propose a method that can obtain random numbers with arbitrary distributions (e.g., Gaussian and Beta distribution) using the pixels of training data. This further enhances the flexibility of our framework, and it is much more difficult to be achieved when using the conventional hardware-based random number generators.

To validate the effectiveness of our proposed methods, we conduct comprehensive experiments on a variety of deep learning tasks, including image classification task on CIFAR-10/100 dataset using ResNet-20/ResNet-32 and ImageNet dataset using ResNet-18, image super-resolution task on Set-5 and DIV2K using WDSR, and various natural language processing tasks using BERT. Compared to conventional methods that use random number generators, our *generator-free* methods can achieve similar accuracy with a 9% reduction in hardware costs.

The contributions of our paper are summarized as follows:

- Unlike the conventional methods that require many random number generators, we innovatively propose a generator-free framework, which directly extracts random numbers from the neural network by employing the randomness of the training process.
- We explore the validity of different sources for random number extraction. Then, we propose different random number extraction methods and analyze the quality of the extracted random numbers.
- Our methods can widely extract high-quality random numbers that can pass the entire NIST test suite SP800-22, while the widely used LFSR-based random number generator cannot.
- Besides successfully extracting the uniformly distributed random numbers for stochastic rounding, we further propose an image pixel-based method that can obtain random numbers with arbitrary distribution, which is hard to achieve using hardware random number generators.
- Finally, we validate the effectiveness of our generator-free framework on various tasks, including image classification, image super-resolution, and natural

language processing. Our framework successfully achieves the same accuracy as the convention methods while eliminating the hardware costs of random number generators.

2 Background

2.1 Rounding Schemes

Rounding technique has been widely used in a range of scientific fields. It usually occurs when compressing the representation precision of a number. The rounding technique can be generally formulated as follows:

$$Round(x) = \begin{cases} \lfloor * \rfloor x, & \text{with probability } p(x), \\ \lfloor * \rfloor x + \epsilon, & \text{with probability } 1 - p(x), \end{cases} \tag{1}$$

where $Round(x)$ denotes the rounding scheme applied to a given value x. The $\lfloor * \rfloor x$ represents to floor the x to the its nearest representation level. And ϵ is the representation precision. The probability $p(x) \in [0, 1]$ and the different rounding schemes can be distinguished by using different constraints on $p(x)$.

For example, the round-up (ceiling) or round-down (floor) scheme sets $p(x) = 0$ or $p(x) = 1$ consistently. On the other hand, the probability constraint can also relate to x, such as in the round-to-nearest scheme [1] and the stochastic rounding scheme [14, 16]. In the round-to-nearest scheme, the probability is set to $p(x) = 1$ for $x \in [0, \frac{\epsilon}{2})$ and $p(x) = 0$ for $x \in [\frac{\epsilon}{2}, \epsilon)$. Instead of the deterministic rounding schemes above, the stochastic rounding scheme lets $p(x) = 1 - \frac{x - \lfloor * \rfloor x}{\epsilon}$, making the expected rounding error to be zero, i.e., $\mathbb{E}(Round(x)) = x$. Therefore, stochastic rounding is considered an unbiased rounding scheme [45].

2.2 Stochastic Rounding in Low-Precision DNN Training

Due to the challenges of the intensive computation and storage in DNN training, quantization techniques are commonly used to save hardware resources, which is especially critical for resource-limited devices such as FPGAs and ASICs. In a low-precision DNN training process, data from several sources are mainly to be quantized, including weights, activations, gradients, and errors [44]. The later works [8, 46] further propose to quantize the

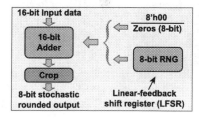

Fig. 1. Conventional stochastic rounding unit design on hardware.

batch normalization (BN) and the optimizer. Among those sources, the model accuracy is largely sensitive to gradient quantization.

The prior work [14] shows that the DNN training is hard to converge under 16-bit precision gradients when using the conventional rounding-to-nearest scheme. And the 16-bit precision is not even a considerably low-precision compared to 8-bit, 4-bit, even the binary precision. The reason is that when using

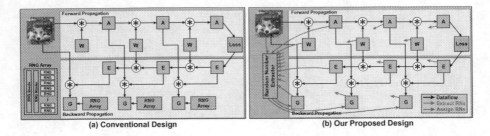

Fig. 2. Overall dataflow in DNN training and the comparison of (a) the conventional design that uses random number generators for stochastic rounding and (b) our proposed generator-free design. The A, W, E, and G stand for Activations, Weights, Errors, and Gradients, respectively. And the ⊛ represents the convolution operation.

the rounding-to-nearest scheme on low-precision gradients the small gradients below the minimum representation precision are always rounded to zero. This will incur the information loss for weight updates [16]. And by using stochastic rounding scheme, this issue can be mitigated [14]. With the help of stochastic rounding, recent works further quantize the gradients to 8-bit precision while still maintaining a comparable accuracy as the full-precision (i.e., floating-point 32 bits) training.

Many literatures that focus on algorithm optimization for low-precision training assume the stochastic rounding can be appropriately incorporated in their design by default [43,52,53], and they barely care about the actual implementation of stochastic rounding on hardware. However, implementing stochastic rounding for low-precision DNN training on hardware is not a trivial task. The stochastic rounding units (SRUs) are commonly used in prior designs [32,38]. Figure 1 shows the general SRU design on hardware. Assume 8-bit precision is used for DNN training. The 16-bit input data of a SRU is obtained from the convolution result of 8-bit activations and 8-bit errors. A 8-bit random number generator is needed in each SRU to generate random numbers on the fly. The linear-feedback shift register (LFSR) is usually used as the RNG [33]. The generated random number will concatenate with eight zeros as its higher bits and add to the input data. Then the stochastic rounded 8-bit gradient can be obtained by cropping the lower 8 bits away on the 16-bit output data of the adder.

3   A Generator-Free Framework for Low-Precision DNN Training

3.1   Framework Overview

As we mentioned in Sect. 2.2, in order to mitigate the information loss and achieve high accuracy, the stochastic rounding technique is indispensable in low-precision DNN training. Since each gradient requires independent stochastic rounding and a RNG can only generate one random number at each time, in the conventional design, the RNG array modules are used to fulfill the computation parallelism, as shown in Fig. 2 (a). In each RNG array module, there are a large number of RNGs needed. This will introduce considerable hardware costs, especially for high throughput designs. And to make RNGs work independently (i.e., generate independent random numbers), each RNG also needs to have its corresponding seed and mask. This also introduces extra storage overhead.

In this work, we argue that it is unnecessary to use RNGs to generate random numbers during the DNN training process. The reason is that the DNN training process is based on a certain degree of randomness, such as the mini-batch training with stochastic gradient descent and the randomly shuffled training samples. We are supposed to find random numbers directly from the neural network. Therefore, we propose a generator-free framework for low-precision DNN training. As shown in Fig. 2 (b), instead of using RNG arrays, we propose to use a random number extractor module to extract random numbers from different sources in the neural network (in Sect. 3.2). And we propose two methods to extract random numbers from the source data (in Sect. 3.3). Besides obtaining the uniformly distributed random numbers used for stochastic rounding, we also find a method that can obtain random numbers with arbitrary distribution (in Sect. 3.4).

3.2   Source of Random Numbers

We consider the accessible data during the neural network training as the source data. The source that can be potentially used for random number extracting should satisfy certain characteristics. The first characteristic is that the source data should be dynamically changing and have stochasticity over time during the training process. The second characteristic is that the source should have a large amount of source data that can fulfill the computation parallelism.

By considering the above characteristics, several candidate sources can be potentially used for random number extraction, including the trainable parameters, intermediate computation results, and input data from training samples, as shown in Fig. 2 (b). However, not all the candidate sources are suitable for random number extraction. For example, the intermediate results (activations) after the ReLU layers will contain a large number of zeros, which significantly biased data distribution. Extracting random numbers from such sources cannot obtain high-quality random numbers with uniform distribution, which is desired for stochastic rounding. Therefore, the model accuracy will degrade considerably.

In this work, we explore the quality of random numbers extracted from different sources and evaluate the performance of low-precision training using the extracted random numbers. We find that the weights and the gradients of weights can generally be used as good sources to extract high-quality random numbers. On the contrary, the sources such as activations and the errors (the gradient of the Conv layers' outputs) are bad sources for random number extraction, and hence a bad low-precision training accuracy. Note that this bad accuracy can be improved using our proposed number-mapping strategy (will be explained in Sect. 3.3), but it is still lower than the accuracy achieved by using random

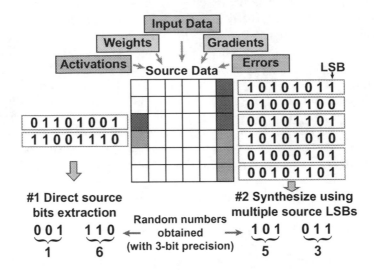

Fig. 3. The two proposed methods of extracting random numbers from source data in DNN training process.

numbers extracted from good sources. More details and comparison results are presented in Sect. 4.3.

3.3   Methods of Random Number Extraction

After determining the potential sources for random number extraction, we also need to have an appropriate method to extract the actual random numbers.

Method #1: direct source bits extraction. The first method is to directly extract random number with certain bits from the source data. As shown in the left-hand side of Fig. 3, assuming the source data matrix/tensor could be one layer's weights, gradients, activations, errors, or input pixels of training samples, from each source data in the matrix/tensor we can extract a random number. The number of bits to extract depends on the required representation precision of the random number. We prefer to use the n lowest bits (e.g., 3 bits) since they are usually the fraction bits that will change frequently. The random numbers obtained in this method do have a certain degree of randomness; however, they are heavily affected by the distribution of the source data. If the distribution of source data is far from uniform, the accuracy of low-precision training will be compromised. This is because non-uniformly distributed random numbers introduce rounding bias during the training process, while the ideal stochastic rounding is unbiased.

Method #2: synthesize using multiple source LSBs. To overcome the non-ideal distribution of the source data, we further propose to synthesize a random number using the least significant bit (LSB) extracted from different source data. As shown in the right-hand side of Fig. 3, a 3-bit random number

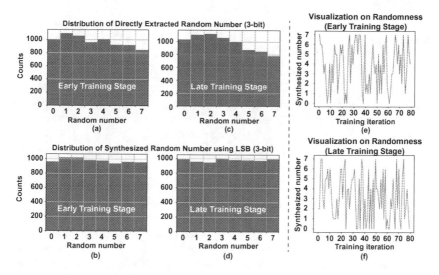

Fig. 4. The performance of the proposed two random number extraction methods in low-precision DNN training using ResNet-20 on CIFAR-10. (a), (b), (c), and (d) show the distribution of the extracted random numbers collected over 20 training epochs at the early training stage and late training stage, respectively. (e) and (f) show the changing trace (randomness) of the extracted (synthesized) random numbers at the early and late training stages, respectively.

is synthesized by using three LSBs from three different source data. The LSB of source data is the most frequently changed bit and is independent to other source data. Compared to method #1, the synthesized random numbers have much better quality that is closer to the ideal uniform distribution, making them valid for low-precision DNN training. Though synthesizing random numbers consumes more source data than directly extracting random numbers with certain bits, the synthesized random numbers are still enough to fulfill the computation parallelism due to the adequate sources for the extraction.

During DNN training, the extracted random numbers will be directly assigned to the corresponding stochastic rounding unit to replace the RNGs. In this way, the hardware costs of RNGs can be eliminated. Another advantage of our proposed methods is that the extracted/synthesized random numbers are free from periodic repetition, which is a common issue for the RNGs [36].

Figure 4 shows the quality of extracted random numbers obtained using our methods during a real low-precision DNN training process using ResNet-20 on CIFAR-10 dataset. We pick a location from the second last layer's gradients and observed the random numbers extracted from the same source data location (three locations are picked for the synthesizing method). We collect the extracted random numbers over 20 training epochs. From Fig. 4 (a), (b), (c), and (d) we can see that the synthesizing method provides random numbers with higher quality than the directly extracting method in both the early training stage and late training stage. And Fig. 4 (e) and (f) shows the synthesized random numbers obtained in 80 consecutive training iterations. The high randomness can be observed in both the early training stage and the late training stage.

Extraction with Number-Mapping Strategy. As we mentioned in Sect. 3.2, some source data such as activations (has a large number of zeros) are significantly biased. None of the above two extraction methods can obtain high-quality random numbers. When using such bad random numbers for stochastic rounding in low-precision training, the network even cannot converge. However, if we extract random numbers (e.g., 3-bits) and use a simple mapping strategy which maps the extracted even numbers to the middle numbers within the range (e.g., $\{0, 1, 2, 3, 4, 5, 6, 7\} \rightarrow \{3, 1, 2, 7, 4, 6, 5, 0\}$), then the extracted random numbers will form a bell-like distribution. This can effectively improve the trained model accuracy. Though this method cannot outperform the model accuracy obtained by extracting random numbers from good sources, it still has better accuracy than using nearest rounding. The reason is that the bell-like distribution makes the rounding becomes a hybrid scheme between the ideal stochastic rounding (with uniform random number distribution) and nearest rounding. It mitigates the biases introduced by the bad data distribution while preserving a certain degree of randomness. But we still suggest directly extracting high-quality random numbers from good sources. More details are discussed in Appendix.

3.4   Obtaining Random Numbers with Arbitrary Distribution

In our generator-free framework, besides extracting the uniformly distributed random numbers for stochastic rounding, we further propose a novel image-pixel-based random number extraction method that can extract random numbers with an arbitrary distribution. And this is hard to be achieved in conventional methods using fixed hardware random number generators.

As shown in Fig. 5, in DNN training, the training samples in a training dataset will be divided into several mini-batches. If we look at the value of the same pixel location over a training epoch (i.e., n mini-batches), we will find the values of a pixel location scattered within a range following a certain distribution. In each training epoch, every training sample is guaranteed to be used once, but the order of the training samples used is different due to the dataset shuffling for each training epoch. This indicates two facts: ① the value distribution of a pixel location will remain the same over the entire training process; ② the order of values presented in a pixel location is varied between different epochs. These two properties give us a unique opportunity for random number extraction. In our method, we divide the value range of a pixel location into 2^n intervals (e.g., 8 intervals for 3-bit random number, as shown in Fig. 5) using $2^n - 1$ threshold levels. Then, depending on the interval in which the pixel value is located, the corresponding random number can be obtained. Since the value distributions of different pixel locations are varied, each pixel requires its own threshold levels. The threshold levels are unique for a certain dataset and can be easily obtained offline. For each training mini-batch, each pixel of input image can create a random number.

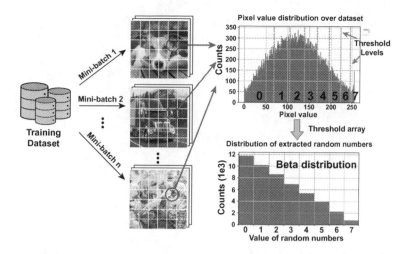

Fig. 5. Extract random numbers with arbitrary distribution using the image-pixel-based extraction method.

It is worth noting that the content in the edge pixels of images are often the background. As a result, the values of those edge pixels are more likely to be concentrated in very large or small, making it possible that some intervals are indistinguishable and fail to obtain the well-distributed random numbers. Therefore, we exclude the edge pixels for random number extraction.

In Fig. 5, we show a real example that uses our image-pixel-based method obtaining random numbers with the beta distribution ($\alpha = 1, \beta = 2$). It can be observed that a high-quality beta distribution is obtained. Besides, we can also obtain other desired distributions by simply modifying the threshold levels. And the threshold arrays can be easily generated offline by going over the training dataset once. More details are shown in Appendix.

4   Results

4.1   Experiments Setup

In this Section, we evaluate our proposed framework and methods on various practical deep learning tasks, including image classification task, image super-resolution task , and natural language processing tasks. All of our models including the full-precision (FP) models and the low-precision (LP) models are trained on a GPU server with 4× NVIDIA 2080Ti GPUs. The PyTorch framework is used for model training. For the low-precision training, we quantize all weights, activations, gradients, and errors using fewer bits (detailed numbers are given in corresponding result tables). The "Ours" method in the result tables indicates the results obtained using our generator-free framework, which extracts random numbers from the network as we proposed. To make fair comparisons, in the "LP" method, we adopt the same training hyperparameters as our generator-free method, but with the random numbers generated from the simulated RNGs.

Table 1. Accuracy comparison of ResNet-20 and ResNet-32 on CIFAR-10 and CIFAR-100 dataset. W: weight, A: activation, G: gradient.

Model	Method	Precision (W/A/G)	Generator Free	Accuracy	Model	Method	Precision (W/A/G)	Generator Free	Accuracy
CIFAR10					CIFAR100				
ResNet-20	FP	32-bit	-	92.17	ResNet-32	FP	32-bit	-	74.53
	Zhu et al. [53]	8-bit	×	92.12		$LP_{stochastic}$	8-bit	×	74.47
	Ours	8-bit	✓	92.15		**Ours**	8-bit	✓	74.41
	$LP_{stochastic}$	6-bit	×	91.83		$LP_{stochastic}$	6-bit	×	74.03
	Ours	6-bit	✓	91.88		**Ours**	6-bit	✓	74.06

We simulate the behavior of LFSR-based RNGs to generate random numbers for the "LP" method. For all our results, if not specified, they are extracted from the gradients of the model's second layer by synthesizing using multiple source LSBs method (i.e., Method#2) and without using number-mapping strategy. We mainly evaluate four different extraction sources including the activations (i.e., Conv outputs, before BN and ReLU), errors (the gradients of activations), weights, and gradient (of weights).

4.2   Accuracy of Low-precision DNN Training

Accuracy for Image Classification. We evaluate our framework on CIFAR-10/100 dataset [20] using ResNet-20/32 [15]. The results are shown in Table 1. The 8-bit precision and 6-bit precision are used, respectively. For the 8-bit precision results on CIFAR-10, we compare our method with Zhu et al. [53]. Both of methods achieve similar accuracy as the FP result, thanks to the superiority of stochastic rounding. However, 16-bit RNGs are required in Zhu et al. [53], while we are generator-free. With the 6-bit precision, though both the LP method and our method have around 0.3% accuracy degradation, our method still achieves similar accuracy as the LP method. For the results on CIFAR-100 using ResNet-32, since the CIFAR-100 task is harder than CIFAR-10, all results with 8-bit precision start to have a minor accuracy degradation compared to the FP result. And for both 8-bit precision and 6-bit precision, our generator-free method can achieve similar accuracy as the LP method that requires RNGs.

Table 2 shows the results on ImageNet dataset [7] using ResNet-18. Our generator-free method achieve the same accuracy as the floating point training result. We have a clear advantage compared to nearest rounding method. More importantly, our method also outperforms the hardware random number generator-based method ($LP_{stochastic}$) with a 0.4% margin. The reason is that our extracted random numbers have higher quality, which has a more critical impact on the harder dataset (ImageNet v.s. CIFAR).

Accuracy for Other Tasks. Table 3 shows the comparison results for image super-resolution task using WDSR-B network [10]. The model is trained on DIV2K [41] dataset and tested on Set-5 [3] and DIV2K dataset. We evaluate our

Table 2. Comparison with existing works using ResNet-18 on ImageNet dataset.

Model	Method	Precision (W/A/G)	Generator Free	Accuracy
ResNet-18	WAGEUBN [46]	8-bit	×	67.40
	FP8 [43]	8-bit	×	67.34
	Uint8 [53]	8-bit	×	69.67
	ADint8 [52]	8-bit	×	70.21
	FP	32-bit	-	71.12
	$LP_{nearest}$	8-bit	-	68.13
	$LP_{stochastic}$	8-bit	×	70.72
	Ours	8-bit	✓	71.10

framework using input size of 640×360 and 320×180 for 2× and 4× resolution up scaling, respectively. And Table 4 shows the evaluation results for natural language processing (NLP) tasks using BERT [9] on a variety of datasets from the General Language Understanding Evaluation (GLUE) [42] benchmark. From the results, we can observe that the image super-resolution task is more sensitive to low-precision training compared to classification and NLP tasks. Our generator-free method consistently achieves similar accuracy as the conventional method, which validates the effectiveness of our proposed methods and framework.

4.3   Comparison of Different Extraction Sources

We first investigate the low-precision training model accuracy using extracted random numbers from different sources including activations, errors, weights, and gradients. And for each type of source, we also cover different layers from the first, middle, and last part of the model. From Table 5 we can find consistent phenomenons on both datasets. The errors are the worst source for random number extraction, and its later layers are relatively better than the front layers. This is because the errors generally contain a dominant number of zeros in low-precision training, which will significantly bias the distribution of random numbers. The weights and gradients are the two of the good sources which can

Table 3. Accuracy comparison for image super-resolution using WDSR-B network on SET-5 and DIV2K dataset.

Upscale rate	Method	bits	PSNR	PSNR_Y	SSIM	PSNR	PSNR_Y	SSIM
			SET-5			DIV2K		
2x	FP	32	34.9591	37.0072	0.9564	33.3994	34.9098	0.9356
	$LP_{stochastic}$	8	34.1163	36.4235	0.9516	32.8305	34.5763	0.9317
	Ours	8	34.1971	36.5799	0.9535	32.8124	34.5243	0.9308
4x	FP	32	28.8708	30.6974	0.8691	27.9794	29.4714	0.8133
	$LP_{stochastic}$	8	28.3235	30.2768	0.8647	27.6667	29.2359	0.8045
	Ours	8	28.3351	30.2786	0.8662	27.6182	29.2295	0.8039

Table 4. Accuracy comparison for natural language processing tasks using BERT.

Method	bits	MRPC	STS-B	RTE	COLA	MNLI	QQP	SST2	QNLI	Avg
FP	32	89.66	89.19	66.43	57.27	84.37	91.18	92.66	91.40	82.77
$LP_{stochastic}$	8	89.58	88.86	66.43	56.92	84.25	91.13	92.61	91.22	82.61
Ours	8	89.62	88.82	66.37	57.04	84.35	90.92	92.65	91.14	82.62

Table 5. Comparison of low-precision training model accuracy using extracted random numbers from different sources. We use 6-bit quantization on weights, activations, errors, and gradeints. 6-bit random numbers are extracted for stochastic rounding.

Model	Source	layer-2	layer-10	layer-19	Model	Source	layer-2	layer-18	layer-31
CIFAR10					CIFAR100				
ResNet-20	Activation	90.36	89.97	87.40	ResNet-32	Activation	70.19	70.41	68.66
	Error	71.11	72.87	74.91		Error	8.34	11.59	54.22
	Weight	91.76	91.69	91.62		Weight	73.32	73.30	73.39
	Gradient	91.88	91.47	91.68		Gradient	74.06	73.91	73.87

achieve high training accuracy because they generally tend to have a normal distribution which will make the LSB has relatively balanced zeros and ones. However, based on our observations, we found the weights have a much lower flipping rate on their LSB compared to the gradients. We conjecture that this is caused by the low-precision representation in training, which limits the small perturbation on weights to a certain degree compared to the floating-point 32-bit training. This makes the gradient a better source for random number extraction. The evaluation of information entropy will be shown in Sect. 4.4 and more discussions can be found in Appendix.

4.4 Randomness Tests and Hardware Saving

To quantitatively evaluate the quality of the extracted random numbers, we test the information entropy of the random numbers throughout the training process. The results are shown in Table 6. We can find that the random numbers extracted from weight and gradient have highest information entropy and they are close to 6-bit. Combining the accuracy results from the Table 5, we can find a positive relationship between the information entropy and the final low-precision training accuracy. We also test the quality of extracted random numbers using the widely used NIST test suite SP800-22 [37]. And we surprisingly find that the random numbers extracted from gradients can generally pass the entire tests, which indicates the random numbers has very high quality. Note that even the widely used LFSR-based random number generators can fail the NIST test. More details are discussed in Appendix.

From the hardware perspective, compared to hardware-based random number generators, our generator-free approach does not require any arithmetic or logical expressions (do no use LUT), and thus is more hardware-friendly and enables resource-efficient design implementations. We estimate the LUT saving

Table 6. The information entropy of extracted random numbers. The results are tested on CIFAR-100 dataset ResNet-32 with 6-bit quantization and 6-bit random numbers.

Source	Ideal	Activation	Error	Weight	Gradient
Entropy (bits)	6	4.9738	2.8017	5.8429	5.8774

using Xilinx Vitis high-level synthesis (HLS) tool and based on the representative FPGA-based DNN training design methodology [23] with LFSR-based stochastic rounding unit (SRU) design [33]. We can successfully save 38.5% and 9% LUTs costs of the SRU and the overall design, respectively. Note that the LUTs are considered a tight resource in FPGA-based DNN training design. Generally, a design prefers not to reach a high LUTs utilization rate since it will lead to routing problems on the FPGA that can significantly slow down the working frequency [51]. Even in the FPGA vendor's (Xilinx) design [4], the LUT utilization does not pass 70% on edge FPGA (PYNQ-Z1) or 50% on cloud FPGAs (AWS F1). Therefore, the LUTs reduction achieved by our method is considerable.

4.5 Discussion and Future Works

In this paper, we pave a new way to obtain random numbers for low-precision training. We explore the performance of different types of sources heuristically, and a systematic exploration can be done for future works. By finding a large number of high-quality random numbers that can be easily extracted from DNNs, our work may also inspire more research in the security field. Moreover, besides the random numbers used by stochastic rounding, our image-pixel-based method can also extract random numbers with arbitrary distribution, which can be potentially used for a broader range of tasks and is worth being further investigated.

5 Conclusion

In this paper, we argue that when random numbers are needed during the DNN training, it is unnecessary to pay extra hardware costs for random number generators because we already have them. Therefore, we explore the validity of different sources and methods for high-quality random number extraction. We propose a generator-free framework to extract and use the random numbers during low-precision DNN training. Moreover, we propose an image-pixel-based method that can extract random numbers with arbitrary distribution, which is hard to achieve using hardware random number generators. Our framework successfully achieves the same accuracy as the convention methods while eliminating the hardware costs of random number generators.

Acknowledgement. This work was partly supported by NSF CCF-1919117 and CCF-1937500; NSERC Discovery Grant RGPIN-2019-04613, DGECR-2019-00120, Alliance Grant ALLRP-552042-2020; CFI John R. Evans Leaders Fund.

References

1. IEEE standard for floating-point arithmetic. IEEE Std 754–2019 (Revision of IEEE 754–2008), pp. 1–84 (2019). https://doi.org/10.1109/IEEESTD.2019.8766229
2. Best, S., Xu, X.: An all-digital true random number generator based on chaotic cellular automata topology. In: 2019 IEEE/ACM International Conference on Computer-Aided Design (ICCAD), pp. 1–8. IEEE (2019)
3. Bevilacqua, M., Roumy, A., Guillemot, C., Alberi-Morel, M.L.: Low-complexity single-image super-resolution based on nonnegative neighbor embedding. In: Proceedings of the British Machine Vision Conference, pp. 135.1–135.10. BMVA Press (2012)
4. Blott, M., et al.: FINN-R: an end-to-end deep-learning framework for fast exploration of quantized neural networks. ACM Trans. Reconfigurable Technol. Syst. (TRETS) **11**(3), 1–23 (2018)
5. Chang, S.E., et al.: Mix and match: a novel FPGA-centric deep neural network quantization framework. In: 2021 IEEE International Symposium on High-Performance Computer Architecture (HPCA), pp. 208–220. IEEE (2021)
6. Chu, C., Wang, Y., Zhao, Y., Ma, X., Ye, S., Hong, Y., Liang, X., Han, Y., Jiang, L.: PIM-prune: fine-grain DCNN pruning for crossbar-based process-in-memory architecture. In: 2020 57th ACM/IEEE Design Automation Conference (DAC), pp. 1–6. IEEE (2020)
7. Deng, J., Dong, W., Socher, R., Li, L.J., Li, K., Fei-Fei, L.: ImageNet: a large-scale hierarchical image database. In: 2009 IEEE conference on computer vision and pattern recognition, pp. 248–255. IEEE (2009)
8. Dettmers, T., Lewis, M., Shleifer, S., Zettlemoyer, L.: 8-bit optimizers via block-wise quantization (2021)
9. Devlin, J., Chang, M.W., Lee, K., Toutanova, K.: BERT: pre-training of deep bidirectional transformers for language understanding. arXiv preprint arXiv:1810.04805 (2018)
10. Fan, Y., Yu, J., Huang, T.S.: Wide-activated deep residual networks based restoration for BPG-compressed images. In: Proceedings of the IEEE Conference on Computer Vision and Pattern Recognition Workshops, pp. 2621–2624 (2018)
11. Fang, H., Mei, Z., Shrestha, A., Zhao, Z., Li, Y., Qiu, Q.: Encoding, model, and architecture: Systematic optimization for spiking neural network in FPGAs. In: 2020 IEEE/ACM International Conference On Computer Aided Design (ICCAD), pp. 1–9. IEEE (2020)
12. Fang, H., Shrestha, A., Zhao, Z., Qiu, Q.: Exploiting neuron and synapse filter dynamics in spatial temporal learning of deep spiking neural network. In: Proceedings of the Twenty-Ninth International Joint Conference on Artificial Intelligence. IJCAI'20 (2021)
13. Fang, H., Taylor, B., Li, Z., Mei, Z., Li, H.H., Qiu, Q.: Neuromorphic algorithm-hardware codesign for temporal pattern learning. In: 2021 58th ACM/IEEE Design Automation Conference (DAC), pp. 361–366. IEEE (2021)
14. Gupta, S., Agrawal, A., Gopalakrishnan, K., Narayanan, P.: Deep learning with limited numerical precision. In: International conference on machine learning, pp. 1737–1746. PMLR (2015)
15. He, K., Zhang, X., Ren, S., Sun, J.: Deep residual learning for image recognition. In: Proceedings of the IEEE conference on computer vision and pattern recognition (CVPR), pp. 770–778 (2016)

16. Höhfeld, M., Fahlman, S.E.: Probabilistic rounding in neural network learning with limited precision. Neurocomputing **4**(6), 291–299 (1992)
17. Hou, Z., et al.: Chex: channel exploration for CNN model compression. In: Proceedings of the IEEE/CVF Conference on Computer Vision and Pattern Recognition (CVPR), pp. 12287–12298 (2022)
18. Kong, Z., et al.: SPViT: Enabling faster vision transformers via soft token pruning. arXiv preprint arXiv:2112.13890 (2021)
19. Krawczyk, H.: LFSR-based hashing and authentication. In: Annual International Cryptology Conference, pp. 129–139. Springer, Heidelberg (1994). https://doi.org/10.1007/3-540-48658-5_15
20. Krizhevsky, A., Hinton, G., et al.: Learning multiple layers of features from tiny images (2009)
21. Li, Y., Fang, H., Li, M., Ma, Y., Qiu, Q.: Neural network pruning and fast training for DRL-based UAV trajectory planning. In: 2022 27th Asia and South Pacific Design Automation Conference (ASP-DAC), pp. 574–579. IEEE (2022)
22. Liu, N., et al.: Lottery ticket preserves weight correlation: is it desirable or not? In: International Conference on Machine Learning (ICML), pp. 7011–7020. PMLR (2021)
23. Luo, C., Sit, M.K., Fan, H., Liu, S., Luk, W., Guo, C.: Towards efficient deep neural network training by FPGA-based batch-level parallelism. J. Semiconduct. **41**(2), 022403 (2020)
24. Luo, Y., Wang, W., Best, S., Wang, Y., Xu, X.: A high-performance and secure TRNG based on chaotic cellular automata topology. IEEE Trans. Circuit Syst. I: Regul. Pap. **67**(12), 4970–4983 (2020)
25. Ma, X., et al.: PCONV: the missing but desirable sparsity in DNN weight pruning for real-time execution on mobile devices. In: Proceedings of the AAAI Conference on Artificial Intelligence (AAAI), vol. 34, pp. 5117–5124 (2020)
26. Ma, X., et al.: Non-structured DNN weight pruning-is it beneficial in any platform? IEEE Transactions on Neural Networks and Learning Systems (TNNLS) (2021)
27. Ma, X., et al.: An image enhancing pattern-based sparsity for real-time inference on mobile devices. In: Vedaldi, A., Bischof, H., Brox, T., Frahm, J.-M. (eds.) ECCV 2020. LNCS, vol. 12358, pp. 629–645. Springer, Cham (2020). https://doi.org/10.1007/978-3-030-58601-0_37
28. Ma, X., et al.: Effective model sparsification by scheduled grow-and-prune methods. In: Proceedings of the International Conference on Learning Representations (ICLR) (2021)
29. Ma, X., et al.: BLCR: Towards real-time DNN execution with block-based reweighted pruning. In: International Symposium on Quality Electronic Design (ISQED), pp. 1–8. IEEE (2022)
30. Ma, X., et al.: Tiny but accurate: a pruned, quantized and optimized memristor crossbar framework for ultra efficient DNN implementation. In: 2020 25th Asia and South Pacific Design Automation Conference (ASP-DAC), pp. 301–306. IEEE (2020)
31. Ma, X., et al.: Sanity checks for lottery tickets: does your winning ticket really win the jackpot? In: Advances in Neural Information Processing Systems (NeurIPS), vol. 34 (2021)
32. Mikaitis, M.: Stochastic rounding: algorithms and hardware accelerator. In: 2021 International Joint Conference on Neural Networks (IJCNN), pp. 1–6. IEEE (2021)
33. Na, T., Ko, J.H., Kung, J., Mukhopadhyay, S.: On-chip training of recurrent neural networks with limited numerical precision. In: 2017 International Joint Conference on Neural Networks (IJCNN), pp. 3716–3723. IEEE (2017)

34. Niu, W., et al.: GRIM: a general, real-time deep learning inference framework for mobile devices based on fine-grained structured weight sparsity. IEEE Transactions on Pattern Analysis and Machine Intelligence (TPAMI) (2021)

35. Niu, W., et al.: PatDNN: achieving real-time DNN execution on mobile devices with pattern-based weight pruning. In: Proceedings of the Twenty-Fifth International Conference on Architectural Support for Programming Languages and Operating Systems (ASPLOS), pp. 907–922 (2020)

36. Roth Jr, C.H., John, L.K.: Digital systems design using VHDL. Cengage Learning (2016)

37. Rukhin, A., Soto, J., Nechvatal, J., Smid, M., Barker, E.: A statistical test suite for random and pseudorandom number generators for cryptographic applications. Technical report, Booz-allen and hamilton inc mclean va (2001)

38. Su, C., Zhou, S., Feng, L., Zhang, W.: Towards high performance low bitwidth training for deep neural networks. J. Semiconduct. **41**(2), 022404 (2020)

39. Sun, M., et al.: FILM-QNN: Efficient FPGA acceleration of deep neural networks with intra-layer, mixed-precision quantization. In: Proceedings of the 2022 ACM/SIGDA International Symposium on Field-Programmable Gate Arrays, pp. 134–145 (2022)

40. Sun, M., et al.: VAQF: fully automatic software-hardware co-design framework for low-bit vision transformer. arXiv preprint arXiv:2201.06618 (2022)

41. Timofte, R., Gu, S., Wu, J., Van Gool, L.: NTIRE 2018 challenge on single image super-resolution: methods and results. In: Proceedings of the IEEE conference on computer vision and pattern recognition workshops, pp. 852–863 (2018)

42. Wang, A., Singh, A., Michael, J., Hill, F., Levy, O., Bowman, S.R.: Glue: A multi-task benchmark and analysis platform for natural language understanding. In: Proceedings of the 2018 EMNLP Workshop BlackboxNLP: Analyzing and Interpreting Neural Networks for NLP, pp. 353–355 (2018)

43. Wang, N., Choi, J., Brand, D., Chen, C.Y., Gopalakrishnan, K.: Training deep neural networks with 8-bit floating point numbers. In: Advances in neural information processing systems, vol. 31 (2018)

44. Wu, S., Li, G., Chen, F., Shi, L.: Training and inference with integers in deep neural networks. arXiv preprint arXiv:1802.04680 (2018)

45. Xia, L., Anthonissen, M., Hochstenbach, M., Koren, B.: A simple and efficient stochastic rounding method for training neural networks in low precision. arXiv preprint arXiv:2103.13445 (2021)

46. Yang, Y., Deng, L., Wu, S., Yan, T., Xie, Y., Li, G.: Training high-performance and large-scale deep neural networks with full 8-bit integers. Neural Netw. **125**, 70–82 (2020)

47. Yuan, G., et al.: TinyADC: Peripheral circuit-aware weight pruning framework for mixed-signal DNN accelerators. In: 2021 Design, Automation & Test in Europe Conference & Exhibition (DATE), pp. 926–931. IEEE (2021)

48. Yuan, G., et al.: Improving DNN fault tolerance using weight pruning and differential crossbar mapping for ReRAM-based edge AI. In: 2021 22nd International Symposium on Quality Electronic Design (ISQED), pp. 135–141. IEEE (2021)

49. Yuan, G., et al.: An ultra-efficient memristor-based DNN framework with structured weight pruning and quantization using ADMM. In: 2019 IEEE/ACM International Symposium on Low Power Electronics and Design (ISLPED), pp. 1–6. IEEE (2019)

50. Yuan, G., et al.: MEST: Accurate and fast memory-economic sparse training framework on the edge. In: Advances in Neural Information Processing Systems (NeurIPS), vol. 34 (2021)

51. Zhang, C., Sun, G., Fang, Z., Zhou, P., Pan, P., Cong, J.: Caffeine: toward uniformed representation and acceleration for deep convolutional neural networks. IEEE Trans. Comput. Aid. Design Integr. Circ. Syst. **38**(11), 2072–2085 (2018)
52. Zhao, K., et al.: Distribution adaptive INT8 quantization for training CNNs. In: Proceedings of the Thirty-Fifth AAAI Conference on Artificial Intelligence (2021)
53. Zhu, F., et al.: Towards unified INT8 training for convolutional neural network. In: Proceedings of the IEEE/CVF Conference on Computer Vision and Pattern Recognition, pp. 1969–1979 (2020)

Real Spike: Learning Real-Valued Spikes for Spiking Neural Networks

Yufei Guo, Liwen Zhang, Yuanpei Chen, Xinyi Tong, Xiaode Liu,
YingLei Wang, Xuhui Huang[✉], and Zhe Ma[✉]

Intelligent Science and Technology Academy of CASIC, Beijing 100854, China
yfguo@pku.edu.cn, starhxh@126.com, mazhe_thu@163.com

Abstract. Brain-inspired spiking neural networks (SNNs) have recently drawn more and more attention due to their event-driven and energy-efficient characteristics. The integration of storage and computation paradigm on neuromorphic hardwares makes SNNs much different from Deep Neural Networks (DNNs). In this paper, we argue that SNNs may not benefit from the weight-sharing mechanism, which can effectively reduce parameters and improve inference efficiency in DNNs, in some hardwares, and assume that an SNN with unshared convolution kernels could perform better. Motivated by this assumption, a training-inference decoupling method for SNNs named as **Real Spike** is proposed, which not only enjoys both unshared convolution kernels and binary spikes in inference-time but also maintains both shared convolution kernels and **Real**-valued **Spike**s during training. This decoupling mechanism of SNN is realized by a re-parameterization technique. Furthermore, based on the training-inference-decoupled idea, a series of different forms for implementing **Real Spike** on different levels are presented, which also enjoy shared convolutions in the inference and are friendly to both neuromorphic and non-neuromorphic hardware platforms. A theoretical proof is given to clarify that the Real Spike-based SNN network is superior to its vanilla counterpart. Experimental results show that all different **Real Spike** versions can consistently improve the SNN performance. Moreover, the proposed method outperforms the state-of-the-art models on both non-spiking static and neuromorphic datasets.

Keywords: Spiking neural network · Real spike · Binary spike · Training-inference-decoupled · Re-parameterization

1 Introduction

Spiking Neural Networks (SNNs) have received increasing attention as a novel brain-inspired computing model that adopts binary spike signals to communicate between units. Different from the Deep Neural Networks (DNNs), SNNs

---

Y. Guo and L. Zhang—Equal contribution.

Supplementary Information The online version contains supplementary material available at https://doi.org/10.1007/978-3-031-19775-8_4.

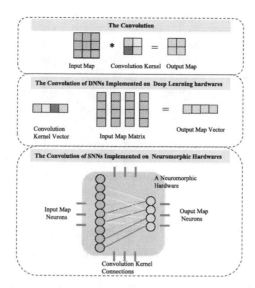

Fig. 1. The difference of convolution computing process between DNNs and SNNs. For DNNs, the calculation is conducted in a highly-paralleled way on conventional hardwares. However, for SNNs, each connection between neurons will be mapped into a synapse on some neuromorphic hardwares, which cannot benefit from the advantages of the weight-shared convolution kernel, *e.g.*, inference acceleration and parameter reduction.

transmit information by spike events, and the computation dominated by the addition operation occurs only when the unit receives spike events. Benefitting from this characteristic, SNNs can greatly save energy and run efficiently when implementing on neuromorphic hardwares, *e.g.*, SpiNNaker [17], TrueNorth [1], Darwin [28], Tianjic [32], and Loihi [5].

The success of DNNs inspires the SNNs in many ways. Nonetheless, the rich spatio-temporal dynamics, event-driven paradigm, and friendly to neuromorphic hardwares make SNNs much different from DNNs, and directly applying the successful experience of DNNs to SNNs may limit the performance of SNNs. As one of the most widely used techniques in DNNs, the weight-shared convolution kernel shows great advantages. It can reduce the parameters of the network and accelerate the inference. However, SNNs show great advantages in the condition of being implemented on neuromorphic hardwares which is very different from DNNs being implemented on deep learning hardwares. As shown in Fig. 1, in DNNs, the calculation is carried out in a highly-paralleled way on deep learning hardwares, thus sharing convolution kernel can improve the computing efficiency by reducing the data transferring between separated memory and processing units. However, for an ideal situation, to take full advantage of the storage-computation-integrated paradigm of neuromorphic hardwares, each unit and connection of the SNNs in the inference phase should be mapped into a neuron and synapse in neuromorphic hardware, respectively. Though these hardwares could be multiplexed, it also increases the complexity of deploymention and extra cost of data transfer. As far as we know, at least Darwin [28],

Tianjic [32], and other memristor-enabled neuromorphic computing systems [41] adopt this one-to-one mapping form at present. Hence, all the components of an SNN will be deployed as a fixed configuration on these hardwares, no matter they share the same convolution kernel or not. Unlike the DNNs, the shared convolution kernels will not bring SNNs the advantages of parameter reduction and inference acceleration in this situation. Hence we argue that it would be better to learn unshared convolution kernels for each output feature map in SNNs.

Unfortunately, whether in theory or technology, it is not feasible to directly train an unshared convolution kernels-based SNN. First, there is no obvious proof that learning different convolution kernels directly will surely benefit the network performance. Second, due to the lack of mature development platforms for SNNs, many efforts are focusing on training SNNs with DNN-oriented programming frameworks, which usually do not support the learning of unshared convolution kernels for each feature map directly. Considering these limitations, we focus on training SNNs with unshared convolution kernels based on the modern DNN-oriented frameworks indirectly.

Driven by the above reasons, a training-time and inference-time decoupled SNN is proposed, where a neuron can emit *real-valued spikes* during training but binary spikes during inference, dubbed **Real Spike**. The training-time real-valued spikes can be converted to inference-time binary spikes via convolution kernel re-parameterization and a shared convolution kernel, which can be derived into multiples then (see details in Sect. 3.3). In this way, an SNN with different convolution kernels for every output feature map can be obtained as we expected. Specifically, in the training phase, the SNN will learn real-valued spikes and a shared convolution kernel for every output feature map. While in the inference phase, every real-valued spike will be transformed into a binary spike by folding a part of the value to its corresponding kernel weight. Due to the diversity of the real-valued spikes, by absorbing part of the value from each real spike, the original convolution kernel shared by each output map can be converted into multiple forms. Thus different convolution kernels for each feature map of SNNs can be obtained indirectly. It can be guaranteed theoretically that the **Real Spike** method can improve the performance due to the richer representation capability of real-valued spikes than binary spikes (see details in Sect. 3.4). Besides, **Real Spike** is well compatible with present DNN-oriented programming frameworks, and it still retains the advantages of DNN-oriented frameworks in terms of the convolution kernel sharing mechanism in the training. Furthermore, we extract the essential idea of training-inference-decoupled and extend **Real Spike** to a more generalized form, which is friendly to both neuromorphic and non-neuromorphic hardwares (see details in Sect. 3.5). The overall workflow of the proposed method is illustrated in Fig. 2.

Our main contributions are summarized as follows:

– We propose the **Real Spike**, a simple yet effective method to obtain SNNs with unshared convolution kernels. The Real Spike-SNN can be trained in DNN-oriented frameworks directly. It can effectively enhance the information representation capability of the SNN without introducing training difficulty.

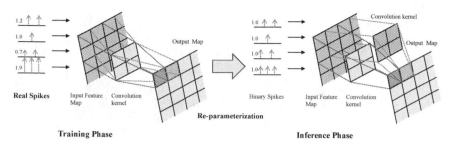

Fig. 2. The overall workflow of **Real Spike**. In the training phase, the SNN learns real-valued spikes and the shared convolution kernel for each output feature map. In the inference phase, the real spikes can be converted to binary spikes via convolution kernel re-parameterization. Then an SNN with different convolution kernels for every feature map is obtained.

– The convolution kernel re-parameterization is introduced to decouple a training-time SNN with real-valued spikes and shared convolution kernels, and an inference-time SNN with binary spikes and unshared convolution kernels.
– We extend the **Real Spike** to other different granularities (layer-wise and channel-wise). These extensions can keep shared convolution kernels in the inference and show advantages independent of specific hardwares.
– The effectiveness of **Real Spike** is verified on both static and neuromorphic datasets. Extensive experimental results show that our method performs remarkably.

2   Related Work

This work starts from training more accurate SNNs with unshared convolution kernels for each feature map. Considering the lack of specialized suitable platforms that can support the training of deep SNNs, powerful DNN-oriented programming frameworks are adopted. To this end, we briefly overview recent works of SNNs in three aspects: (i) learning algorithms of SNNs; (ii) SNN programming frameworks; (iii) convolutions.

2.1   Learning Algorithms of SNNs

The learning algorithms of SNNs can be divided into three categories: converting ANN to SNN (ANN2SNN) [2,13,25,36], unsupervised learning [6,14], and supervised learning [12,16,26,29,37]. ANN2SNN converts a pre-trained ANN to an SNN by transforming the real-valued output of the activation function to binary spikes. Due to the success of ANNs, ANN2SNN can generate an SNN in a short time with competitive performance. However, the converted ANN inherits the limitation of ignoring rich temporal dynamic behaviors from DNNs, it cannot handle neuromorphic datasets well. On the other hand, ANN2SNN usually requires hundreds of timesteps to approach the accuracy of pre-trained DNNs. Unsupervised learning is considered a more biologically plausible method. Unfortunately, the lack of sufficient understanding of the biological mechanism prevents the network from going deep, thus it is usually limited to small datasets and non-ideal performance. Supervised learning trains SNNs with error backpropagation. Supervised learning-based methods can not only achieve high accuracy within very few timesteps but also handle neuromorphic datasets. Our work falls under this category.

2.2   SNN Programming Frameworks

There exist several specialized programming frameworks for SNN modeling. However, most of them cannot support the direct training of deep SNNs. NEURON [3], a simulation environment for modeling computational models of neurons and networks of neurons, mainly focuses on simulating neuron issues with complex anatomical and physiological characteristics, which is more suitable for neuroscience research. BRIAN2 [11] and NEST [10], simulators for SNNs, aim at making the writing of simulation code as quick and easy as possible for the user, but they are not designed for the supervised learning for deep SNNs. SpikingJelly [7], an open-source deep learning framework for SNNs based on PyTorch, provides a solution to establish SNNs by mature DNN-oriented programming frameworks. However, so far most functions of the SpikingJelly are still under development. Instead of developing a new framework, we attempt to investigate an ingenious solution to develop the SNNs that have unshared convolutional kernels for each output feature map in the DNN-oriented frameworks, which have demonstrated an easy-to-use interface.

2.3   Convolutions

Convolutional Neural Networks have recently harvested a huge success in large-scale computer vision tasks [15,20,22,35,38,39]. Due to the abilities of input scale adaptation, translation invariance, and parameter reduction, the stacked convolution layers help train a deep and well-performed neural network with fewer parameters compared to dense-connected fully connected layers. For a convolution layer, feature maps of the previous layer are convolved with some learnable kernels and presented through the activation function to form the output feature maps as the current layer. Each learnable kernel is corresponding to an output map. In general, we have that

$$\mathbf{X}_{i,j} = f(\mathbf{I}_j * \mathbf{K}_i), \tag{1}$$

where $\mathbf{I}_j \in \mathbb{R}^{k \times k \times C}$ denotes j-th block of the input maps with total C channels, $\mathbf{K}_i \in \mathbb{R}^{k \times k \times C}$ is the i-th convolution kernel with $i = 1, \ldots, C$, then $\mathbf{X}_{i,j} \in \mathbb{R}$ is the j-th element in i-th output feature map. It can be seen that all the outputs in each channel share a same convolution kernel. However, when implementing a convolution-based SNN on above mentioned neuromorphic hardwares, all the kernels will be mapped as synapses no matter they are shared or not. Hence, keeping shared convolution kernel cannot show the same advantages for SNNs as DNNs. We argue that learning different convolution kernels for each output map may improve the performance of the SNN further. In this case, a convolution layer for SNNs can be written as

$$\mathbf{X}_{i,j} = f(\mathbf{I}_j * \mathbf{K}_{i,j}), \tag{2}$$

where $\mathbf{K}_{i,j} \in \mathbb{R}^{k \times k \times C}$ is the j-th convolution kernel for i-th output map. Unfortunately, it is not easy to directly implement the learning of unshared convolution kernel for SNNs in the DNN-oriented frameworks. And dealing with this issue is one of the important works in this paper.

3   Materials and Methodology

Aiming at training more accurate SNNs, we adopt the explicitly iterative leaky Integrate-and-Fire (LIF) model, which can be implemented on mature DNN-oriented frameworks easily to simulate the fundamental computing unit of SNNs. Considering that it is not easy to realize unshared convolution kernels for each output feature map with existing DNN-oriented frameworks, a modified LIF model that can emit the real-valued spike is proposed first. By using this modified LIF model, our SNNs will learn real spikes along with the shared convolution kernel for every output channel as DNNs during training. While for the inference phase, real spikes will be transformed into binary spikes and each convolution kernel will be re-parameterized as multiple different convolution kernels, so the advantages of SNNs can be recovered and unshared convolution kernel for each output map can be obtained. Then, to make the proposed design more general, we also propose layer-wise and channel-wise **Real Spike**, which can keep shared convolution kernels in both training phase and inference phase and will introduce no more parameters than its vanilla counterpart. In this section, we will introduce explicitly iterative LIF model, **Real Spike**, re-parameterization, and extensions of **Real Spike** successively.

3.1   Explicitly Iterative Leaky Integrate-and-Fire Model

The spiking neuron is the fundamental computing unit of SNNs. The LIF neuron is most commonly used in supervised learning-based methods. Mathematically, a LIF neuron can be described as

$$\tau \frac{\partial u}{\partial t} = -u + I, \quad u < V_{th} \tag{3}$$

$$u = u_{rest} \quad \& \quad fire \ a \ spike, \quad u \geq V_{th} \tag{4}$$

where u, u_{rest}, τ, I, and V_{th} represent the membrane potential, membrane resting potential, membrane time constant that controls the decaying rate of u, pre-synaptic input, and the given firing threshold, respectively. When u is below

V_{th}, it acts as a leaky integrator of I. On the contrary, when u exceeds V_{th}, it will fire a spike and propagate the spike to the next layer, then it will be reset to the resting potential, u_{rest}, which is usually set as 0.

The LIF model has a complex dynamics structure that is incompatible with nowadays DNN-oriented framework. By discretizing and transforming the LIF model to an explicitly iterative LIF model, SNNs can be implemented in these mature frameworks. The hardware friendly iterative LIF model can be described as

$$u(t) = \tau u(t-1) + I(t), \quad u(t) < V_{th} \tag{5}$$

$$o(t) = \begin{cases} 1, & if\ u(t) \geq V_{th}, \\ 0, & otherwise. \end{cases} \tag{6}$$

where V_{th} is set to 0.5 in this work. Up to now, there is still an obstacle for training SNNs in a direct way, i.e., Eq. (6) is non-differentiable. As in other work [4,33,40], we appoint a rectangular function as the particular pseudo derivative of spike firing as follows,

$$\frac{do}{du} = \begin{cases} 1, & if\ 0 \leq u \leq 1, \\ 0, & otherwise. \end{cases} \tag{7}$$

With all these settings, now we can train an SNN on DNN-oriented frameworks.

3.2   Real Spike

As aforementioned, driven by the suppressed advantages of the shared convolution kernel and the expectation of enhancing the capacity of information representation for SNNs, we turn the problem of learning unshared convolution kernels to learning real spikes. To be more specific, the output of our modified LIF model in Eq. (6) is further rewritten as

$$\tilde{o}(t) = a \cdot o(t) \tag{8}$$

where a is a learnable coefficient. With this modification, our LIF model can emit a *real-valued spike*, dubbed **Real Spike**. Then we train SNNs with this modified LIF model along with the shared convolution kernel for each output map, which can be easily implemented in DNN-oriented frameworks. Obviously, unlike the binary spike, the real-valued spike will lose the advantage of computation efficiency of SNNs, since the corresponding multiplication cannot be substituted to addition. And another problem is that the learned convolution kernels for each output map are still shared at this time. Therefore, to jointly deal with these problems, we propose a training-inference decoupled framework, which can transform real spikes into binary spikes and convert the shared convolution kernel as different kernels by using re-parameterization, which will be introduced in the next subsection.

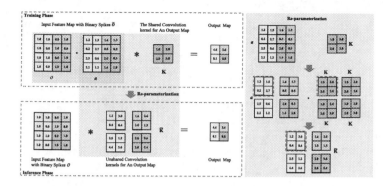

Fig. 3. The diagram of re-parameterization by a simple example.

3.3 Re-parameterization

Consider a convolution layer, which takes $\mathbf{F} \in \mathbb{R}^{D_F \times D_F \times M}$ as input feature maps, and generates the output feature maps, $\mathbf{G} \in \mathbb{R}^{D_G \times D_G \times N}$, where D_F and M denote the size of the square feature maps and the number of channels (depths) for the input, respectively; D_G and N denote the size of the square feature maps and the number of channels (depths) for the output, respectively. The convolution layer is actually parameterized by a group of convolution kernels, which can be denoted as a tensor, $\mathbf{K} \in \mathbb{R}^{D_K \times D_K \times M \times N}$ with a spatial dimension of D_K. Then, each element in $\mathbf{G}$ is computed as

$$\mathbf{G}_{k,l,n} = \sum_{i,j,m} \mathbf{K}_{i,j,m,n} \cdot \mathbf{F}_{k+i-1,l+j-1,m} \qquad (9)$$

For standard SNNs, the elements of input maps are binary spikes, while in this work, the SNN is trained with real-valued spikes for the purpose of enhancing the network representation capacity. In this case, we can further denote the input feature map, $\mathbf{F}$, according to Eq. (8) as follows

$$\mathbf{F} = \mathbf{a} \odot \mathbf{B} \qquad (10)$$

where $\mathbf{B}$ and $\mathbf{a}$ denote a binary tensor and a learnable coefficient tensor, respectively. With this element-wise multiplication in Eq. (10), we can extract a part of the value from each element in $\mathbf{F}$, and fold it into the shared convolution kernel one-by-one according to the corresponding position during inference. Then the single shared convolution kernel can be turned into multiples without changing the values of the output maps. Through this decoupling process, a new SNN that can emit binary spikes and enjoy different convolution kernels will be obtained. This process can be illustrated from Eq. (11)–Eq. (13) as follows:

$$\mathbf{G}_{k,l,n} = \sum_{i,j,m} \mathbf{K}_{i,j,m,n} \cdot \left(\mathbf{a}_{k+i-1,l+j-1,m} \cdot \mathbf{B}_{k+i-1,l+j-1,m} \right) \qquad (11)$$

$$\mathbf{G}_{k,l,n} = \sum_{i,j,m} (\mathbf{a}_{k+i-1,l+j-1,m} \cdot \mathbf{K}_{i,j,m,n}) \cdot \mathbf{B}_{k+i-1,l+j-1,m} \qquad (12)$$

$$\mathbf{G}_{k,l,n} = \sum_{i,j,m} \tilde{\mathbf{K}}_{k,l,i,j,m,n} \cdot \mathbf{B}_{k+i-1,l+j-1,m} \qquad (13)$$

where $\tilde{\mathbf{K}}$ is the unshared convolution kernel tensor.

The whole process described above is called re-parameterization, which allows us to convert a real-valued-spike-based SNN into an output-invariant binary-spike-based SNN with unshared convolution kernels for each output map. That is, re-parameterization provides a solution to obtain an SNN with unshared convolution kernels under DNN-oriented frameworks by decoupling the training-time SNN and inference-time SNN. Figure 3 illustrates the details of re-parameterization by a simple example.

3.4 Analysis and Discussions

In this work, we assume that firing real-valued spikes during training can help increase the representation capacity of the SNNs. To verify our assumption, a series of analyses and discussions are conducted by using the information entropy concept. Given a scalar, vector, matrix, or tensor, $\mathbf{X}$, its representation capability is denoted as $\mathcal{R}(\mathbf{X})$, which can be measured by the information entropy of $\mathbf{X}$, as follows

$$\mathcal{R}(\mathbf{X}) = \max \mathcal{H}(\mathbf{X}) = \max(-\sum_{x \in \mathbf{X}} p_{\mathbf{X}}(x) log p_{\mathbf{X}}(x)) \qquad (14)$$

where $p_{\mathbf{X}}(x)$ is the probability of a sample from $\mathbf{X}$. When $p_{\mathbf{X}}(x_1) = \cdots = p_{\mathbf{X}}(x_n)$, $\mathcal{H}(\mathbf{X})$ reaches its maximum (see Appendix A.1 for detailed proofs). For a binary spike o, it can be expressed with 1 bit, and the number of samples from o is 2. While the real-valued spike $\tilde{o}$ needs 32 bits, which consists of 2^{32} samples. Hence, $\mathcal{R}(o) = 1$ and $\mathcal{R}(\tilde{o}) = 32$ according to Eq. (14). Obviously, the representation capability of real spikes far exceeds that of binary spikes. This indicates that real spikes will enhance the information expressiveness of SNNs, which accordingly benefit the performance improvement. To further show the difference between real spikes and binary spikes intuitively, the visualizations of some channels expressed by real spikes and binary spikes are given respectively in the appendix.

Another intuitive conjecture to explain why the SNN with real-valued spikes performs better than its counterpart with binary spikes is that, for the former one, the information loss can be restrained to some extent by changing the fixed spike value to an appropriate value with a scalable coefficient, a; while for the later one, the firing function would inevitably induce the quantization error.

It can be concluded from the above analysis that, learning real-valued spikes instead of the binary spikes in the training phase, enables us to train a more accurate SNN. And by performing re-parameterization in the inference phase, real spikes can be converted to binary spikes and unshared convolution kernels can be obtained. In another word, learning **Real Spike** is actually used to generate a better information encoder, while the re-parameterization will transfer the

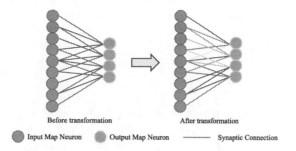

Before transformation   After transformation

◉ Input Map Neuron   ◉ Output Map Neuron   ------- Synaptic Connection

Fig. 4. The difference of adjacent layers in SNNs implemented on neuromorphic hardwares before re-parameterization and after re-parameterization. The convolution kernel re-parameterization will only change the values of the connections between neurons
without introducing any additional computation and storage resource since the entire
architecture topology of an SNN must be completely mapped into the hardwares.

rich information encoding capacity from **Real Spike** into information decoding. Moreover, the transformed model after re-parameterization can also retain
the advantage coming from the binary spike information processing paradigm
in standard SNNs. As shown in Fig. 4, when deploying an SNN into some neuromorphic hardwares, all the units and their connections of the network will be
mapped as neurons and synaptic connections one by one. Hence, the SNN with
multiple different convolution kernels will not introduce any computation and
storage resource.

3.5 Extensions of Real Spike

The key observation made in the **Real Spike** is that re-scaling the binary spike
of the performance neuron with a real-valued coefficient, a, can increase the
representation capability and accuracy of SNNs. As shown in Eq. (10), the default
Real Spike is performed in the element-wise way. In a similar fashion, we argue
that introducing scaling coefficient by layer-wise or channel-wise manners will
also retain the benefit to some extent. Then, we further propose to re-formulate
Eq. (8) for one layer as follows

$$\tilde{\mathbf{o}}(t) = \mathbf{a} \cdot \mathbf{o}(t) \tag{15}$$

With this new formulation we can explore various ways of introducing $\mathbf{a}$
during training. Specifically, we propose to introduce $\mathbf{a}$ for each layer in the
following 3 ways:
Layer-wise:

$$\mathbf{a} \in \mathbb{R}^{1 \times 1 \times 1} \tag{16}$$

Channel-wise:

$$\mathbf{a} \in \mathbb{R}^{C \times 1 \times 1} \tag{17}$$

Element-wise:

$$\mathbf{a} \in \mathbb{R}^{C \times H \times H} \tag{18}$$

Table 1. Ablation study for **Real Spike**.

Dataset	Architecture	Timestep	Accuracy
CIFAR10	ResNet20 w/ BS	2	88.91%
		4	91.73%
		6	92.98%
	ResNet20 w/ RS	2	90.47%
		4	92.53%
		6	93.44%
CIFAR100	ResNet20 w/ BS	2	62.59%
		4	63.27%
		6	67.12%
	ResNet20 w/ RS	2	63.40%
		4	64.87%
		6	68.60%

RS represents real spikes and BS represents binary spikes.

where C is the number of channels and H is the spatial dimension of a square feature map. Obviously, for layer-wise and channel-wise Real Spike, reparameterization will only re-scale the shared convolution kernels without transferring them to different ones. In these two forms, our **Real Spike** act in the same way as conventional SNNs on any hardwares. That is to say, starting from the motivation of how to take full advantage of the integration of memory and computation, we propose element-wise **Real Spike**. Then we extract the essential idea of training-inference-decoupled and extend **Real Spike** to a more generalized form, which is friendly to both neuromorphic and non-neuromorphic hardware platforms.

4 Experiment

The performance of the **Real Spike**-based SNNs were evaluated on several traditional static datasets (CIFAR-10 [19], CIFAR-100 [19], ImageNet [20]) and one neuromorphic dataset (CIFAR10-DVS [24]). And multiple widely-used spiking archetectures including ResNet20 [33,36], VGG16 [33], ResNet18 [8], and ResNet34 [8] were used to verify the effectiveness of our **Real Spike**. Detailed introduction of the datasets and experimental settings are provided in the appendix. Extensive ablation studies were conducted first to compare the SNNs with real-valued spikes and their binary spike counterpart. Then, we comprehensively compared our SNNs with the existing state-of-the-art SNN methods.

4.1 Ablation Study for Real Spike

The ablation study of **Real Spike** was conducted on CIFAR-10/100 datasets based on ResNet20. The SNNs with real-valued spikes and binary spikes were trained with the same configuration, with timestep varying from 2 to 6. Results in

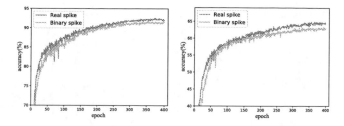

Fig. 5. The accuracy curves of ResNet20 with real-valued spikes and binary spikes with a timestep = 4 on CIFAR-10 (left) and CIFAR-100 (right). The real-valued spikes-based SNNs obviously enjoy higher accuracy and easier convergence.

Table 2. Performance comparison for different **Real Spike** versions.

Dataset	Architecture	Version	Accuracy
CIFAR10	ResNet20	Vanilla	91.73%
		Layer-wise	92.12%
		Channel-wise	92.25%
		Element-wise	92.53%
CIFAR100	ResNet20	Vanilla	63.27%
		Layer-wise	64.28%
		Channel-wise	64.71%
		Element-wise	64.87%

Table 1 show that the test accuracy of the SNNs with real-valued spikes is always higher than that with binary spikes. This is due to the richer representation capacity from Real Spike, which benefits the performance improvement of SNNs. Figure 5 illustrates the test accuracy curves of ResNet20 with real-valued spikes and its counterpart with binary spikes on CIFAR-10/100 during training. It can be seen that the SNNs with real-valued spikes obviously perform better on the accuracy and convergence speed.

4.2 Ablation Study for Extensions of Real Spike

The performance of SNNs with different extensions of Real Spike was evaluated on CIFAR-10(100). Results in Table 2 show that element-wise **Real Spike** always outperforms the other versions. But all the Real Spike-based SNNs conformably outperform the vanilla (standard) SNN. Another observation is that the accuracy difference between layer-wise **Real Spike** and the binary spike is greater than that between layer-wise **Real Spike** and element-wise **Real Spike**. This shows that our method touches on the essence of improving the SNN accuracy and is very effective.

4.3 Comparison with the State-of-the-Art

In this section, we compared the **Real Spike**-based SNNs with other state-of-the-art SNNs on several static datasets and a neuromorphic dataset. For each

Table 3. Performance comparison with SOTA methods.

Dataset	Method	Type	Architecture	Timestep	Accuracy
CIFAR-10	SpikeNorm [36]	ANN2SNN	VGG16	2500	91.55%
	Hybrid-Train [34]	Hybrid	VGG16	200	92.02%
	SBBP [23]	SNN training	ResNet11	100	90.95%
	STBP [40]	SNN training	CIFARNet	12	90.53%
	TSSL-BP [42]	SNN training	CIFARNet	5	91.41%
	PLIF [9]	SNN training	PLIFNet	8	93.50%
	Diet-SNN [33]	SNN training	VGG16	5	92.70%
				10	93.44%
			ResNet20	5	91.78%
				10	92.54%
	STBP-tdBN [43]	SNN training	ResNet19	2	92.34%
				4	92.92%
				6	93.16%
	Real Spike	SNN training	ResNet19	2	**94.01%** $\pm$ 0.10
				4	**95.60%** $\pm$ 0.08
				6	**95.71%** $\pm$ 0.07
			ResNet20	5	**93.01%** $\pm$ 0.07
				10	**93.65%** $\pm$ 0.05
			VGG16	5	**92.90%** $\pm$ 0.09
				10	**93.58%** $\pm$ 0.06
CIFAR-100	BinarySNN [27]	ANN2SNN	VGG15	62	63.20%
	Hybrid-Train [34]	Hybrid	VGG11	125	67.90%
	T2FSNN [31]	ANN2SNN	VGG16	680	68.80%
	Burst-coding [30]	ANN2SNN	VGG16	3100	68.77%
	Phase-coding [18]	ANN2SNN	VGG16	8950	68.60%
	Diet-SNN [33]	SNN training	ResNet20	5	64.07%
			VGG16	5	69.67%
	Real Spike	SNN training	ResNet20	5	**66.60%** $\pm$ 0.11
			VGG16	5	**70.62%** $\pm$ 0.08
			VGG16	10	**71.17%** $\pm$ 0.07
ImageNet	Hybrid-Train [34]	Hybrid	ResNet34	250	61.48%
	SpikeNorm [36]	ANN2SNN	ResNet34	2500	69.96%
	STBP-tdBN [43]	SNN training	ResNet34	6	63.72%
	SEW ResNet [8]	SNN training	ResNet18	4	63.18%
			ResNet34	4	67.04%
	Real Spike	SNN training	ResNet18	4	**63.68%** $\pm$ 0.08
			ResNet34	4	**67.69%** $\pm$ 0.07
CIFAR10-DVS	Rollout [21]	Streaming	DenseNet	10	66.80%
	STBP-tdBN [43]	SNN training	ResNet19	10	67.80%
	Real Spike	SNN training	ResNet19	10	**72.85%** $\pm$ 0.12
			ResNet20	10	**78.00%** $\pm$ 0.10

run, we report the mean accuracy as well as the standard deviation with 3 trials. Experimental results are shown in Table 3.

CIFAR-10(100). On CIFAR-10, our SNNs achieve higher accuracy than the other state-of-the-art methods, and the best result of 95.71% top-1 accuracy is achieved by ResNet19 with 6 timesteps. And even trained with much fewer timesteps, *i.e.*, 2, our ResNet19 still outperforms the STBP-tdBN under 6 timesteps with 0.85% higher accuracy. This comparison shows that our method also enjoys the advantage of latency reduction for SNNs. On CIFAR-100, Real Spike also performs better and achieves a 2.53% increment on ResNet20 and a 0.95% increment on VGG16.

ImageNet. The ResNet18 and ResNet34 were selected as the backbones. Considering that the image size of the samples is much larger, the channel-wise Real Spike was used. For a fair comparison, we made our architectures consistent with SEW-ResNets, which are not typical SNNs, where the IF model and modified residual structure are adopted. Results show that, even with 120 fewer epochs of training (200 for ours, 320 for SEW-ResNets), the channel-wise Real Spike-based SNNs can still outperform SEW-ResNets. In particular, our method achieves a 0.5% increment on ResNet18 and a 0.65% increment on ResNet34. Moreover, Real Spike-based ResNet34 with 4 timesteps still outperforms STBP-tdBN-based RersNet34 with 6 timesteps by 3.97% higher accuracy.

CIFAR10-DVS. To further verify the generalization of the Real Spike, we also conducted experiments on the neuromorphic dataset CIFAR10-DVS. Using ResNet20 as the backbone, Real Spike achieves the best performance with 78.00% accuracy in 10 timesteps. For ResNet19, Real Spike obtains 5.05% improvement compared with STBP-tdBN. It's worth noting that, as a more complex model, ResNet19 performs worse than ResNet20. This is because that neuromorphic datasets usually suffer much more noise than static ones, thus more complex models are easier to overfit on these noisier datasets.

5   Conclusions

In this work, we focused on the difference between SNNs and DNNs and speculated that the unshared convolution kernel-based SNNs would enjoy more advantages than those with shared convolution kernels. Motivated by this idea, we proposed **Real Spike**, which aims at enhancing the representation capacity for an SNN by learning real-valued spikes during training and transferring the rich representation capacity into inference-time SNN by re-parameterizing the shared convolution kernel to different ones. Furthermore, a series of **Real Spike**s in different granularities were explored, which are enjoy shared convolution kernels in both training and inference phases and friendly to both neuromorphic and non-neuromorphic hardware platforms. Proof of why **Real Spike** has a better performance than vanilla SNNs was provided. Extensive experiments verified that our proposed method consistently achieves better performance than the existed state-of-the-art SNNs.

References

1. Akopyan, F., et al.: TrueNorth: design and tool flow of a 65 mW 1 million neuron programmable neurosynaptic chip. IEEE Trans. Compute. Aided Des. Integr. Circuits Syst. **34**(10), 1537–1557 (2015)
2. Cao, Y., Chen, Y., Khosla, D.: Spiking deep convolutional neural networks for energy-efficient object recognition. Int. J. Comput. Vis. **113**(1), 54–66 (2015)
3. Carnevale, N.T., Hines, M.L.: The NEURON Book. Cambridge University Press, Cambridge (2006)
4. Cheng, X., Hao, Y., Xu, J., Xu, B.: LISNN: improving spiking neural networks with lateral interactions for robust object recognition. In: IJCAI, pp. 1519–1525 (2020)
5. Davies, M., et al.: Loihi: a neuromorphic manycore processor with on-chip learning. IEEE Micro **38**(1), 82–99 (2018)
6. Diehl, P.U., Cook, M.: Unsupervised learning of digit recognition using spike-timing-dependent plasticity. Front. Comput. Neurosci. **9**, 99 (2015)
7. Fang, W., et al.: Spikingjelly (2020). http://github.com/fangwei123456/spikingjelly
8. Fang, W., Yu, Z., Chen, Y., Huang, T., Masquelier, T., Tian, Y.: Deep residual learning in spiking neural networks. In: Advances in Neural Information Processing Systems 34, pp. 21056–21069 (2021)
9. Fang, W., Yu, Z., Chen, Y., Masquelier, T., Huang, T., Tian, Y.: Incorporating learnable membrane time constant to enhance learning of spiking neural networks. In: Proceedings of the IEEE/CVF International Conference on Computer Vision, pp. 2661–2671 (2021)
10. Gewaltig, M.O., Diesmann, M.: Nest (neural simulation tool). Scholarpedia **2**(4), 1430 (2007)
11. Goodman, D.F., Brette, R.: The Brian simulator. Front. Neurosci. **3**, 26 (2009)
12. Guo, Y., et al.: RecDis-SNN: rectifying membrane potential distribution for directly training spiking neural networks. In: Proceedings of the IEEE/CVF Conference on Computer Vision and Pattern Recognition (CVPR), pp. 326–335, June 2022
13. Han, B., Roy, K.: Deep spiking neural network: energy efficiency through time based coding. In: Vedaldi, A., Bischof, H., Brox, T., Frahm, J.-M. (eds.) ECCV 2020. LNCS, vol. 12355, pp. 388–404. Springer, Cham (2020). https://doi.org/10.1007/978-3-030-58607-2_23
14. Hao, Y., Huang, X., Dong, M., Xu, B.: A biologically plausible supervised learning method for spiking neural networks using the symmetric STDP rule. Neural Netw. **121**, 387–395 (2020)
15. He, K., Zhang, X., Ren, S., Sun, J.: Deep residual learning for image recognition. In: Proceedings of the IEEE Conference on Computer Vision and Pattern Recognition, pp. 770–778 (2016)
16. Huh, D., Sejnowski, T.J.: Gradient descent for spiking neural networks. In: Advances in Neural Information Processing Systems 31 (2018)
17. Khan, M.M., et al.: SpiNNaker: mapping neural networks onto a massively-parallel chip multiprocessor. In: 2008 IEEE International Joint Conference on Neural Networks (IEEE World Congress on Computational Intelligence), pp. 2849–2856. IEEE (2008)
18. Kim, J., Kim, H., Huh, S., Lee, J., Choi, K.: Deep neural networks with weighted spikes. Neurocomputing **311**, 373–386 (2018)

19. Krizhevsky, A., Nair, V., Hinton, G.: CIFAR-10 (Canadian Institute for Advanced Research) **5**(4), 1 (2010). http://wwwcs.toronto.edu/kriz/cifar.html
20. Krizhevsky, A., Sutskever, I., Hinton, G.E.: ImageNet classification with deep convolutional neural networks. In: Advances in Neural Information Processing Systems 25 (2012)
21. Kugele, A., Pfeil, T., Pfeiffer, M., Chicca, E.: Efficient processing of spatio-temporal data streams with spiking neural networks. Front. Neurosci. **14**, 439 (2020)
22. LeCun, Y., Bottou, L., Bengio, Y., Haffner, P.: Gradient-based learning applied to document recognition. Proc. IEEE **86**(11), 2278–2324 (1998)
23. Lee, C., Sarwar, S.S., Panda, P., Srinivasan, G., Roy, K.: Enabling spike-based backpropagation for training deep neural network architectures. Front. Neurosci. **14**, 119 (2020)
24. Li, H., Liu, H., Ji, X., Li, G., Shi, L.: CIFAR10-DVS: an event-stream dataset for object classification. Front. Neurosci. **11**, 309 (2017)
25. Li, Y., Deng, S., Dong, X., Gong, R., Gu, S.: A free lunch from ANN: towards efficient, accurate spiking neural networks calibration. In: International Conference on Machine Learning, pp. 6316–6325. PMLR (2021)
26. Li, Y., Guo, Y., Zhang, S., Deng, S., Hai, Y., Gu, S.: Differentiable spike: rethinking gradient-descent for training spiking neural networks. In: Advances in Neural Information Processing Systems 34, pp. 23426–23439 (2021)
27. Lu, S., Sengupta, A.: Exploring the connection between binary and spiking neural networks. Front. Neurosci. **14**, 535 (2020)
28. Ma, D., et al.: Darwin: a neuromorphic hardware co-processor based on spiking neural networks. J. Syst. Archit. **77**, 43–51 (2017)
29. Neftci, E.O., Mostafa, H., Zenke, F.: Surrogate gradient learning in spiking neural networks: bringing the power of gradient-based optimization to spiking neural networks. IEEE Signal Process. Mag. **36**(6), 51–63 (2019)
30. Park, S., Kim, S., Choe, H., Yoon, S.: Fast and efficient information transmission with burst spikes in deep spiking neural networks. In: 2019 56th ACM/IEEE Design Automation Conference (DAC), pp. 1–6. IEEE (2019)
31. Park, S., Kim, S., Na, B., Yoon, S.: T2FSNN: deep spiking neural networks with time-to-first-spike coding. In: 2020 57th ACM/IEEE Design Automation Conference (DAC), pp. 1–6. IEEE (2020)
32. Pei, J., et al.: Towards artificial general intelligence with hybrid Tianjic chip architecture. Nature **572**(7767), 106–111 (2019)
33. Rathi, N., Roy, K.: Diet-SNN: direct input encoding with leakage and threshold optimization in deep spiking neural networks. arXiv preprint arXiv:2008.03658 (2020)
34. Rathi, N., Srinivasan, G., Panda, P., Roy, K.: Enabling deep spiking neural networks with hybrid conversion and spike timing dependent backpropagation. arXiv preprint arXiv:2005.01807 (2020)
35. Ren, S., He, K., Girshick, R., Sun, J.: Faster R-CNN: towards real-time object detection with region proposal networks. In: Advances in Neural Information Processing Systems 28 (2015)
36. Sengupta, A., Ye, Y., Wang, R., Liu, C., Roy, K.: Going deeper in spiking neural networks: VGG and residual architectures. Front. Neurosci. **13**, 95 (2019)
37. Shrestha, S.B., Orchard, G.: SLAYER: Spike layer error reassignment in time. In: Bengio, S., Wallach, H., Larochelle, H., Grauman, K., Cesa-Bianchi, N., Garnett, R. (eds.) Advances in Neural Information Processing Systems 31, pp. 1419–1428. Curran Associates, Inc. (2018), http://papers.nips.cc/paper/7415-slayer-spike-layer-error-reassignment-in-time.pdf

38. Simonyan, K., Zisserman, A.: Very deep convolutional networks for large-scale image recognition. arXiv preprint arXiv:1409.1556 (2014)
39. Szegedy, C., et al.: Going deeper with convolutions. In: Proceedings of the IEEE Conference on Computer Vision and Pattern Recognition, pp. 1–9 (2015)
40. Wu, Y., Deng, L., Li, G., Zhu, J., Xie, Y., Shi, L.: Direct training for spiking neural networks: faster, larger, better. In: Proceedings of the AAAI Conference on Artificial Intelligence, vol. 33, pp. 1311–1318 (2019)
41. Yao, P., et al.: Fully hardware-implemented memristor convolutional neural network. Nature **577**(7792), 641–646 (2020)
42. Zhang, W., Li, P.: Temporal spike sequence learning via backpropagation for deep spiking neural networks. In: Advances in Neural Information Processing Systems 33, pp. 12022–12033 (2020)
43. Zheng, H., Wu, Y., Deng, L., Hu, Y., Li, G.: Going deeper with directly-trained larger spiking neural networks. In: Proceedings of the AAAI Conference on Artificial Intelligence, vol. 35, pp. 11062–11070 (2021)

FedLTN: Federated Learning for Sparse and Personalized Lottery Ticket Networks

Vaikkunth Mugunthan[1,2]([✉]), Eric Lin[1,3], Vignesh Gokul[4], Christian Lau[1], Lalana Kagal[2], and Steve Pieper[3]

[1] DynamoFL, San Francisco, USA
{vaik,eric,christian}@dynamofl.com
[2] CSAIL, Massachusetts Institute of Technology, Cambridge, USA
lkagal@csail.mit.edu
[3] Harvard University, Cambridge, USA
pieper@bwh.harvard.edu
[4] University of California San Diego, San Diego, USA
vgokul@eng.ucsd.edu

Abstract. Federated learning (FL) enables clients to collaboratively train a model, while keeping their local training data decentralized. However, high communication costs, data heterogeneity across clients, and lack of personalization techniques hinder the development of FL. In this paper, we propose FedLTN, a novel approach motivated by the well-known Lottery Ticket Hypothesis to learn sparse and personalized lottery ticket networks (LTNs) for communication-efficient and personalized FL under non-identically and independently distributed (non-IID) data settings. Preserving batch-norm statistics of local clients, postpruning without rewinding, and aggregation of LTNs using server momentum ensures that our approach significantly outperforms existing state-of-the-art solutions. Experiments on CIFAR-10 and Tiny ImageNet datasets show the efficacy of our approach in learning personalized models while significantly reducing communication costs.

Keywords: Federated learning · Lottery ticket hypothesis · Statistical heterogeneity · Personalization · Sparse networks

1 Introduction

Federated Learning (FL) allows decentralized clients to collaboratively learn without sharing private data. Clients exchange model parameters with a central server to train high-quality and robust models while keeping local data private. However, FL faces a myriad of performance and training-related challenges:

V. Mugunthan and E. Lin—Equal contributions.

Supplementary Information The online version contains supplementary material available at https://doi.org/10.1007/978-3-031-19775-8_5.

S. Avidan et al. (Eds.): ECCV 2022, LNCS 13672, pp. 69–85, 2022.
https://doi.org/10.1007/978-3-031-19775-8_5

1. *Personalization*: Vanilla FL constructs a server model for all clients by averaging their local models, while postulating that all clients share a single common task. However, this scheme does not adapt the model to each client. For example, platforms like Youtube and Netflix require a unique personalized model for each of their clients. Most FL algorithms focus on improving the average performance across clients, aiming to achieve high accuracy for the global server model. The global model may perform well for some clients but perform extremely poorly for others. This is not the ideal scenario for a fair and optimal FL algorithm. When deployed on edge devices in the real world, local test accuracy is instead a more important metric for success.
2. *Statistical heterogeneity/Non-identically independently distributed (non-IID) data*: When different clients have different data distributions, the performance of FL degrades and results in slower model convergence.
3. *Communication Cost*: Sending and receiving model parameters is a huge bottleneck in FL protocols as it could be expensive for resource-constrained clients. It is important to reduce the total number of communication rounds and the size of the packets that are transmitted during every round. Unfortunately, there is usually a tradeoff between model accuracy and communication cost accrued during the federation process. For instance, techniques that speed up accuracy convergence or decrease the model size may result in a small decrease in accuracy.
4. *Partial Participation*: Real-world FL scenarios involve hundreds of clients. Hence, it is important for an FL algorithm to be robust even with low participation rates and if clients drop out and rejoin during the FL process.

While there have been numerous papers that address each of these challenges individually (See Sect. 2 for related work), finding one approach that provides solutions for all of them has proven to be difficult. LotteryFL [18] provided the first attempt at addressing the above-mentioned issues in one protocol. It is motivated by the Lottery Ticket hypothesis (LTH), which states that there exist subnetworks in an initialized model that provides the same performance as an unpruned model. Finding these high-performing subnetworks are referred to as finding winning lottery tickets. LotteryFL obtains personalized models by averaging the lottery tickets at every federated round. This also improves the communication costs as only the lottery tickets are communicated across the client and server instead of the whole model. However, LotteryFL fails to achieve the same performance in terms of pruning as obtained in Lottery Ticket Hypothesis (LTH). LotteryFL models are pruned only up to 50% compared to 90% or more in the non-federated LTH setting. This is due to the fact that pruning in LotteryFL takes a lot of time – the authors claim that it takes around 2000 federated rounds to prune around 50% of the model.

The slow pruning process of LotteryFL presents major drawbacks. In many experimental settings, local clients have difficulty reaching the accuracy threshold to prune for most rounds. On more difficult tasks, some clients may never reach the accuracy threshold. Consequently, they fail to find winning lottery tickets and do not reach personalized and more cost-efficient models. To avoid this in the LotteryFL approach, the accuracy threshold must be lowered, inhibiting the efficacy of finding the right lottery ticket networks for each client.

Moreover, LotteryFL uses evaluation measures such as average test accuracy to compare their work with baselines. We argue that these measures can easily misrepresent the performance of a federated training paradigm. In FL, it is also essential that each client achieves a fair performance, an aspect that is not captured by the average test accuracy. We introduce an evaluation metric based on the minimum client test accuracy to solve this problem. We find that LotteryFL sometimes achieves a lower minimum client accuracy than FedAvg.

To address the above-mentioned challenges, we present FedLTN, a novel approach motivated by the Lottery Ticket Hypothesis to learn sparse and personalized Lottery Ticket Networks (LTNs) for communication-efficient and personalized FL under non-IID data settings. Our contributions are as follows:

- We propose postpruning without rewinding to achieve faster and greater model sparsity. Although our pruning method is contrary to non-federated pruning practices, we find that it can be leveraged in federated learning due to the special properties of averaging across multiple devices that mitigate overfitting.
- We introduce Jump-Start and aggregation of LTNs using server momentum to significantly shorten the rounds needed to reach convergence. In particular, FedLTN with Jump-Start reduces communication costs by 4.7X and 6X compared to LotteryFL and FedAvg respectively.
- We learn more personalized FL models by preserving batch normalization layers. FedLTN achieves 6.8% and 9.8% higher local client test accuracy than LotteryFL and FedAvg respectively on CIFAR-10 [15].
- We demonstrate the efficacy of our techniques on the CIFAR-10 and Tiny ImageNet [17] datasets. Moreover, our optimizations are effective in both high-client (100) and low-client (10) experimental setups.

2 Related Work

FedAvg is a simple and commonly used algorithm in federated learning proposed in the seminal work of [21]. In FedAvg, the server sends a model to participating clients. Each client trains using its local dataset and sends the updated model to the server. The server aggregates via simple averaging of the received models and the process continues for a fixed number of rounds or until convergence is achieved. Most existing FL algorithms are derived from FedAvg where the goal is to train a server model that tries to perform well on most FL clients. However, FL still faces numerous challenges, among which convergence on heterogeneous data, personalization, and communication cost are the most pressing problems.

2.1 Performance on Heterogeneous (Non-IID) Data

FedAvg is successful when clients have independent and identically distributed data. However, in many real-world settings, data is distributed in a non-IID manner to edge clients and hence leads to client drift [11]. That is, gradients computed by different clients are skewed and consequently local models move away from globally optimal models. This substantially affects the performance of FedAvg as simple averaging leads to conflicting updates that hinder learning, especially in scenarios where clients have a large number of local training steps and a high degree of variance in their data distributions. Under such scenarios, introducing adaptive learning rates [24] and momentum [9,12,23,30,31] to the server aggregation process are beneficial as they incorporate knowledge from prior iterations. Gradient masking [28] and weighted averaging [25,32] of models have also been shown to be useful.

2.2 Personalization

Under the traditional FL setting, a "one-fit-for-all" single server model is trained and updated by all clients. However, this model may underperform for specific clients if their local distribution differs drastically from the global distribution. For example, if there is extreme non-IID data distribution skew amongst clients, the server model trained through FL may only reach a mediocre performance for each local test set. Another failure scenario occurs when there are uneven data distributions amongst clients. In these cases, the federated server model may learn to perform well only on a subset of data. Clients with data distributions that are different from this subset would then have subpar performance. Consequently, it is important to evaluate FL frameworks not only for their global performance but also for their average and worst-case (minimum) local performance.

A variety of papers have aimed to introduce personalized FL, where each client can learn a model more properly finetuned based on their data distribution. In [5], the authors apply the use of a model-agnostic meta-learning framework in the federated setting. In this framework, clients first find an initial shared model and then update it in a decentralized manner via gradient descent with different loss functions specific to their data. Other papers have proposed similar finetuning approaches based on the use of transfer learning [20,29].

Another category of personalization techniques relies on user clustering and multi-task learning [14,26]. These techniques cluster together clients with similar data distributions and train personalized models for each cluster. Then, they use multi-task learning techniques to arrive at one model that may perform well for all clients.

Lastly, preserving local batch normalization while averaging models has been used to address the domain and feature shift non-IID problems [3,4,10,19]. Since these batch normalization layers are unique to each client, they help personalize each model to the underlying local data distribution.

However, [16] mentions drawbacks to these various personalization approaches in the FL setting. Namely, most of these approaches incur increased communication costs, such as a greater number of federation rounds – both transfer learning and user clustering techniques require clients to learn a shared base model first. Furthermore, many personalization approaches result in greater model sizes, such as the addition of batch normalization layers.

In our work, clients can learn personalized lottery ticket networks (LTNs) similar to user clustering techniques without the overhead of communication costs. We show in our paper that by preserving local batch norm properties while learning these LTNs, clients can improve their accuracy (i.e. achieve better personalization) while compressing model sizes.

2.3 Communication Cost

Communication cost is a huge problem for FL as clients frequently communicate with the server. There are three major components of cost during federation – model size, gradient compression (limiting the data needed to be transmitted between each local edge device and the server), and an overall number of rounds of the federation. Model compression techniques [1,2,8] like quantization, pruning, and sparsification are usually taken from the classic single centralized setting and applied to FL. In particular, sparse models not only lead to lower memory footprints but also result in faster training times.

For gradient compression, [1,13,27] have proposed various update methods to reduce uplink communication costs. These include structured updates, which restrict the parameter space used to learn an update, and sketched updates, which compress model updates through a combination of quantization, rotations, subsampling, and other techniques.

2.4 Lottery Ticket Hypothesis

The Lottery Ticket Hypothesis (LTH) [6] states that there exists a subnetwork in a randomly initialized network, such that the subnetwork when trained in isolation can equal the performance of the original network in atmost the same number of iterations. The steps involved in LTH usually include the following: First, the randomly initialized model is trained for a few iterations. Second, the model is pruned based on the magnitude of its weights. Then, the unpruned weights are reinitialized to the weights at the start of the training (rewinding to round 0). This process continues iteratively until the target pruning is achieved.

Though LTH initially showed promising results only for small datasets and network architectures, recent works have expanded LTH to more complex networks and datasets. [7] notably demonstrates better results by rewinding weights to a previous iteration rather than that of the initial iteration (round 0).

3   FedLTN: Federated Learning for Sparse and Personalized Lottery Ticket Networks

In this section, we present the motivation and reasoning behind various components of our FedLTN framework that achieve higher degrees of personalization, pruning, and communication efficiency in finding lottery ticket networks (LTNs).

These components focus on improving the averaged server LTN to inform better localized performance in terms of accuracy and memory footprint. Along with FedLTN, we propose Jump-Start, a technique for drastically reducing the number of federated communication rounds needed to learn LTNs:

Algorithm 1 (FedLTN): To learn complex image classification tasks with greater non-IID feature skew, FedLTN utilizes batch-norm preserved LTN, postpruning, no rewinding, and accelerated global momentum. This combination uses batch-norm preserved averaging to find a server LTN that helps individual clients learn without imposing on their personalized accuracy. Postpruning without rewinding parameters significantly increases the rate of pruning and also helps networks find more personalized LTNs. Finally, our accelerated aggregation of LTNs helps speed up convergence.

Algorithm 2 (FedLTN with Jump-Start): Jump-Start can be utilized to skip communication costs of the first few rounds of training by replacing them with local training, without causing a loss in local test accuracy. This technique is especially useful for scenarios where there are resource constraints on local client devices. One client can be first trained and pruned, then all other clients transfer-learn off this smaller model.

3.1   Personalization

Batch Normalization-Preserved LTN: In order to personalize LTNs, we introduce a batch normalization-preserved protocol. During the federated aggregation process, batch normalization (BN) layers are not averaged while computing the server model nor uploaded/downloaded by local clients. Since BN layers have been commonly used for domain shift adaptation in single client settings, these layers help personalize each client to its individual data distribution. Preserving BN layers during aggregation leads to higher personalized accuracy by avoiding conflicting updates to clients' individualized BN. It also decreases communication costs as batch normalization layers are not transmitted to the server.

3.2   Smaller Memory Footprint/Faster Pruning

Postpruning: As reported in Sect. 1, one problem with the LotteryFL's naive approach in applying LTH to FL is that the server model sent back to clients each round suffers drastic losses in accuracy before any local training. Moreover, each client decides to prune the model immediately after receiving it from the server, **prior** to any local training. If the server model reaches a certain threshold for the client's local test accuracy, they prune $r_p\%$ of the parameters that have the least L1-norm magnitude. For clients with relatively low levels of pruning, this means that most of the parameters pruned will be the same amongst all the clients that decide to prune that federated round. This approach hopes that only clients with the same archetype (who have the same data distribution) will prune on the same round. However, due to the above-stated challenges in slow prune rates, LotteryFL sets the accuracy threshold to be 50% for clients

---

Algorithm 1 FedLTN

---

function SERVEREXECUTE(θ_0):

$\quad \theta_g \leftarrow \theta_0$ $\qquad\qquad\qquad$ ▷ random init model unless Algorithm 2 is used

$\quad k \leftarrow max(N \cdot K, 1)$ $\qquad\qquad$ ▷ N available clients, participation rate K

$\quad S_t \leftarrow \{C_1, \ldots, C_k\}$ $\qquad\qquad$ ▷ k randomly sampled clients

$\quad$**for** each $k = 1, 2, \ldots, N$ **do**

$\qquad \theta_k^t \leftarrow \theta_g$ $\qquad\qquad\qquad$ ▷ each client starts with same init model

$\quad$**end for**

$\quad$**for** each round $t = 1, 2, \ldots, T$ **do**

$\qquad$**for** each client $k \in S_t$ **in parallel do**

$\qquad\quad \theta_k^t = \theta_g^t \odot m_k^t$ $\qquad$ ▷ m_k is the mask of client k and indicates its LTN

$\qquad\quad \theta_k^{t+1}, m_k^{t+1} \leftarrow$ ClientUpdate($C_k, \theta_k^t, \theta_0$)

$\qquad$**end for**

$\qquad$// BN-preserved aggregation: simple average of non-BN layers of LTNs

$\qquad \theta_g^{t+1} \leftarrow$ aggregate(θ_k^{t+1}, ignore_batchnorm=True)

$\qquad$// Aggregation of LTNs using server momentum

$\qquad \theta_g^{t+1} \leftarrow \tau \theta_g^{t+1} + (1 - \tau)(\theta_g^t - \lambda \Delta^t)$ $\qquad$ ▷ τ is a hyperparameter

$\qquad \Delta^{t+1} \leftarrow -(\theta_g^{t+1} - \theta_g^t)$

$\quad$**end for**

end function

function CLIENTUPDATE($C_k, \theta_k^t, \theta_0$):

$\quad$// BN-preserved LTN

$\quad C_k^t, \theta_k^t \leftarrow$ copy_ignore_batchnorm($C_k, \theta_k^t, C_k^{t-1}, \theta_k^{t-1}$)

$\qquad\qquad\qquad\qquad\qquad$ ▷ Retain previous round's batch normalization layers

$\quad \mathcal{B} \leftarrow$ split (local data D_k^{train} into batches)

$\quad$// Regularization when using server momentum

$\quad \theta_{init} \leftarrow \theta_k^t$

$\quad$**for** each local epoch i from 1 to E **do**

$\qquad$**for** batch b $\in \mathcal{B}$ **do**

$\qquad\quad \theta_k^{t+1} \leftarrow \theta_k^t - \eta \nabla_{\theta_k^t} \ell(\theta_k^t; b) + \|\theta_{init} - \theta_k^t\|$ $\qquad$ ▷ η is learning rate, $\ell(\cdot)$ loss function

$\qquad$**end for**

$\quad$**end for**

$\quad$// Postpruning without rewinding

$\quad acc \leftarrow$ eval(θ_k^t, local val_set D_k^{val})

$\quad$// r_{target} is the target pruning rate, r_k^t is the current pruning rate for client k at round t

$\quad$**if** $acc > acc_{threshold}$ and $r_k^t < r_{target}$ **then**

$\qquad m_k^{t+1} \leftarrow$ prune(θ_k^t, pruning step r_p) $\qquad\qquad$ ▷ new mask for LTN

$\qquad \theta_k^{t+1} \leftarrow \theta_k^{t+1} \odot m_k^{t+1}$ $\qquad\qquad$ ▷ do not rewind parameters

$\qquad \mathcal{B} \leftarrow$ split (local data D_k^{train} into batches) $\qquad$ ▷ train again after pruning

$\qquad$**for** each local epoch i from 1 to E **do**

$\qquad\quad$**for** batch b $\in \mathcal{B}$ **do**

$\qquad\qquad \theta_k^{t+1} \leftarrow \theta_k^t - \eta \nabla_{\theta_k^t} \ell(\theta_k^t; b)$ $\qquad$ ▷ η is learning rate, $\ell(\cdot)$ loss function

$\qquad\quad$**end for**

$\qquad$**end for**

$\quad$**end if**

$\quad$return $\theta_k^{t+1}, m_k^{t+1}$

end function

---

trained on a binary classification problem. Consequently, models merely need to be slightly better than random chance – which means that clients of different archetypes pruning on the same round (on mostly the same parameters) are a common occurrence. This process hinders the degree of personalization achieved via LotteryFL.

Here, we present an alternative pruning protocol. Instead of pruning *before* any local training, we prune the client model based on the magnitude of the weights *after* n local_epochs of training. This speeds up the pruning process since it is much more likely for the validation accuracy threshold to be reached. Furthermore, postpruning actively encourages diversity in pruned parameters, as each client's parameters will be different after they locally train. Since a freshly pruned model needs to be retrained, we stipulate that if a client prunes it retrains for another n epochs.

Rewinding: Conventionally, pruned models rewind weights and learning rate to a previous round T. LotteryFL rewinds to $T = 0$ (global initial model). Although in the Lottery Ticket Hypothesis rewinding is needed to avoid convergence to a local minima in the loss function, we hypothesize that this isn't needed during federation, since averaging across multiple models helps mitigate overfitting. That is, instead of resetting all parameters back to the global_init model after pruning, the non-pruned weights stay the same.

Aggregation of LTNs Using Server Momentum: One of our main contributions is to fasten the convergence of each client model. This is crucial to obtain a performance greater than the threshold so that the client model can prune at a faster rate. This, in turn, reduces the number of communication rounds and hence the overall communication cost, as fewer parameters have to be sent each round. In our problem setting, our server 'model' is an aggregate of all the clients' winning ticket networks. To improve the convergence speed, the server sends an anticipatory update of the ticket networks. We outline the steps Algorithm 1 follows to achieve faster convergence.

– At each round, the server sends an accelerated update (θ_t) to each of the clients. This means that the clients receive an accelerated winning lottery ticket update at every round. This initialization helps each client to train faster. More formally, we have in the server, for round t

$$\theta_g^{t+1} = \tau \sum_k \text{LTNs}(\theta_k^t) + (1 - \tau)(\theta_g^t - \lambda \Delta^t)$$

$$\Delta^{t+1} = \theta_g^{t+1} - \theta_g^t$$

where τ is a hyperparameter.
– The server sends the corresponding parameters to each of the clients based on the client mask. We have for client k at iteration 0:

$$\theta_{k0}^{t+1} = m_k . \theta_g^{t+1}$$

– The initial weights of the client are used in the regularization term while training the client to align the local gradients with the accelerated global updates. β is a hyperparameter that weights the regularization term. For client k at iteration i, we have

$$L(\theta_{ki}^{t+1}) = l(\theta_{ki}^{t+1}) + \beta\|\theta_{ki}^{t+1} - \theta_{k0}^{t+1}\|$$

FedLTN with Jump-Start. Communication cost is an important factor while implementing FL in the real world. Furthermore, clients with resource-constrained devices may only be able to locally store and train a model after achieving a certain degree of sparsification. We present FedLTN with Jump-Start in Algorithm 2 as an extension of FedLTN to address these challenges.

Before federated training, k clients (usually, $k = N$ all clients) locally train for T_{jump} rounds without communicating with the server nor with each other. During local training, they prune to a small degree (e.g. 30%). Then, we choose the model with the highest validation accuracy from the local training Jump-Start and send it to all clients as a model for transfer learning. Then, FedLTN begins with much fewer communication rounds required. Jump-Start is motivated by the work of [22], which found that an LTN trained on one image classification dataset can be finetuned to other datasets without much loss in accuracy.

Algorithm 2 FedLTN with Jump-Start
___

  function JUMPSTART(T_{jump}):

    $\theta_g \leftarrow \theta_0$ ▷ init random global model

    $k \leftarrow max(N \cdot K, 1)$ ▷ N available clients, participation rate K

    $S_t \leftarrow \{C_1, \ldots, C_k\}$ ▷ k randomly sampled clients

    for each round $t = 1, 2, \ldots, T_{jump}$ **do**

        $\theta_k^t = \theta_k^{t-1} \odot m_k^t$ ▷ m_k is the mask of client k and indicates its LTN

        $\theta_k^{t+1}, m_k^{t+1} \leftarrow$ ClientUpdate$(C_k, \theta_k^t, \theta_0)$ ▷ local update only

    end for

    $\theta_g \leftarrow \arg\max_{\theta_k^t} \frac{\Sigma_n \text{eval}(\theta_k^t, D_n^{val})}{N}$ ▷ choose client with highest val acc

    ServerExecute(θ_g) ▷ pass θ_g to FedLTN for clients to transfer learn

  end function
___

4 Experiments

We evaluate the performance of FedLTN against different baselines in several different types of experiments, where we vary the task (image classification), environment setting (large vs. small number of clients), heterogeneity settings (non-IID distributions), and client participation rates.

4.1   Experiment Setup

Datasets. We use the CIFAR-10 [15] and Tiny ImageNet datasets for our experiments. To simulate the non-IID scenario, each client consists of 2 classes and 25 datapoints per class for training, 25 datapoints for validation, and 200 datapoints as the test set. For Tiny ImageNet, we randomly sample 10 classes from the 200 classes as our dataset. To simulate a more challenging scenario, we also consider the Dirichlet non-IID data skew. Each client consists of all 10 classes, with the proportions of data volume per class based on the Dirichlet distribution.

Compared Methods

1. **FedAvg**: This baseline indicates the performance of an unpruned federated model in our settings. FedAvg computes the federated model by averaging the weights (including BN layers) of all the participating clients' models at every round.
2. **FedBN**: FedBN proposes an improvement over FedAvg to obtain better personalization for clients. The server model in FedBN does not aggregate the batch-norm parameters of the participating clients. We use this as a baseline to compare our personalization performance.
3. **LotteryFL**: LotteryFL presented the first attempt at using the Lottery Ticket Hypothesis to achieve personalized submodels for each client. We use this as a baseline to analyze the performance of a pruned federated model.

Model Architecture. We utilize ResNet18 as the standard architecture for our experiments. Since LotteryFL does not account for batch normalization (BN) layers, we also conduct experiments on a custom CNN model without BN layers on CIFAR-10 for a baseline comparison. Results for the high-client setting are shown below. Custom CNN results for the low-client setting and other implementation details can be found in our supplementary material (Appendix C).

Hyperparameters. We set the hyperparameters local epochs $(E) = 3$, batch size $(B) = 8$, $\tau = 0.5$, accuracy threshold $(acc_{threshold}) = 0.6$, prune step $(r_p) = 0.1$ and $\beta = 0.01$ for FedLTN. For LotteryFL, we use the same hyperparameters the authors mentioned in [18]. We fix the number of communication rounds to 50 for the low-client (10) setting and 2000 rounds for the high-client (100) setting. We denote the models with the target pruning rate within paranthesis. For example, FedLTN (0.9) refers to FedLTN paradigm with 90% as the target pruning percent. For experiments using FedLTN with Jump-Start, we use 25 Jump-Start and 25 FedLTN rounds with $K = 50\%$ participation and 10% r_p prune step. We set a max prune of 30% for Jump-Start and 90% for FedLTN.

Evaluation Metrics. To evaluate the personalization achieved by our models, we compute the average classification accuracy achieved by each client model on its corresponding test set. Although average test accuracy indicates overall test performance across all clients, it is also important to measure the minimum client test accuracy. This is to ensure that all clients participating in the federated training paradigm learn a personalized LTN that performs well on their local data distribution. Hence, we also use the minimum client test accuracy as an evaluative measure. To compute communication costs, we sum the data volume communicated at every round during the training process.

4.2   Evaluation

We demonstrate the success of FedLTN in learning personalized, sparse models in this section. We analyze the test accuracy performance, maximum pruning achieved, communication costs, and the convergence speed of FedLTN with baselines. We report the results of all the baselines and our method in Table 1.

Table 1. Comparison of performance of FedLTN with all the baselines on the CIFAR-10 and Tiny ImageNet datasets in the low-client setting with ResNet18. FedLTN (0.9; jumpstart) refers to 90% target pruning with 25 rounds of Jump-Start and 25 rounds of FedLTN. Rewinding resets model parameters to randomly initialized model in round 0. **Bolded** numbers represent best performance and <u>underlined</u> numbers represent the second best.

Dataset	Algorithm	Avg Test Acc (%)	Min Test Acc (%)	Comm. Cost (MB)
CIFAR-10	FedAvg	72.8	<u>64.9</u>	11,150.0
	FedBN	78.72	**65.75**	11,137.5
	LotteryFL (0.1)	72.0	56.3	10,613.95
	LotteryFL (0.5)	75.8	52.5	8,675.7
	FedLTN (0.1)	74.5	59.8	10,134.6
	FedLTN (0.5)	81.1	61.5	6,563.3
	FedLTN (0.9)	**82.6**	63.0	<u>3,753.9</u>
	FedLTN (0.9; jumpstart)	<u>82.2</u>	64.8	**1,846.0**
	FedLTN (0.9; rewind)	71.5	53.0	4,940.7
Tiny ImageNet	FedAvg	68.9	51.8	11,150.0
	FedBN	73.0	55.5	11,137.5
	LotteryFL (0.1)	72.6	41.3	10,370.7
	LotteryFL (0.5)	71.3	50.8	6,885.5
	FedLTN (0.1)	68.4	38.3	10,169.0
	FedLTN (0.5)	73.3	50.0	6,885.5
	FedLTN (0.9)	<u>74.5</u>	<u>59.8</u>	<u>4,650.9</u>
	FedLTN (0.9; jumpstart)	**83.1**	**61.8**	**1,778.4**
	FedLTN (0.9; rewind)	71.8	53.8	5,144.0

Personalization: We analyze the level of personalization achieved by each client participating in FL. We consider the average test accuracy across all clients, i.e. the average performance for each client model on the test datasets. We find that our method achieves better accuracy than an unpruned FedAvg baseline and 50% pruned LotteryFL baseline models. For example, in CIFAR-10, FedLTN (0.9) achieves average test accuracy of 82.6% when compared to LotteryFL (0.5), which achieves 75.8%, and FedBN which achieves 78.72%. For both CIFAR-10 and Tiny ImageNet, FedLTN (0.9) with and without Jump-Start performs better than all the baselines, even when pruning 40% more parameters.

We find that our method achieves the highest minimum test accuracy compared with all the other baselines. In Tiny ImageNet, FedLTN (0.9) with 25 Jump-Start rounds achieves 59.8% minimum test accuracy compared to 50.8% in the same setting using LotteryFL (0.5). We observe that while increasing the target pruning rate of the experiment, the overall test accuracy increases. Our highest pruned models give the best overall test accuracy. This is due to clients learning more personalized models as the pruning rate increases, which leads to better test performance.

Pruning: In our experiments, we evaluate FedLTN's performance while using different target pruning rates ranging from 10 to 90%. As shown in Table 1, we can see that our method improves average test accuracy despite pruning 40% more parameters compared to baseline LotteryFL. Our pruning achieves substantial reductions in memory footprint. When models are pruned to 90% sparsity, the parameters take up a mere 0.031 MB for our custom CNN architecture and 4.46 MB for ResNet18. This degree of sparsity is important as it makes it feasible to train on even resource-constrained edge devices. We also compare the rate of pruning between our method and the baselines. As seen in Fig. 1, we observe that our method prunes around 70% in the first 20 rounds, while LotteryFL (0.9) prunes around 10%. Our postpruning method can achieve larger pruning rates quickly in a few rounds, even though we set a higher accuracy threshold than LotteryFL.

Convergence: One of our objectives is to achieve faster convergence to facilitate faster pruning. Figure 1 shows the performance obtained by the clients on their validation set in each training round. Our method converges faster than FedAvg and LotteryFL. This is due to the server broadcasting accelerated aggregated LTNs during each round. This anticipatory update and the client initialization, along with the modified regularization term align the client gradients to avoid any local minima. This boosts the convergence speed of our method compared to the baselines.

Communication Costs: Since our method prunes more than LotteryFL, the communication costs are 2.3x and 3x times lower than that of LotteryFL and FedAvg. As our method prunes faster than LotteryFL, we send lower parameters

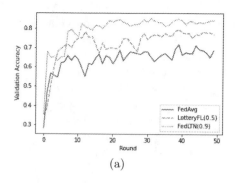
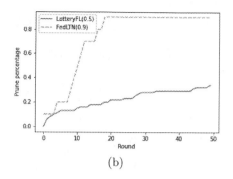

(a)   (b)

Fig. 1. Left (a): Comparison of validation accuracies at each round. We observe that our method converges faster than other baselines. Right (b): Comparison of pruning rate at each round. our method prunes around 70% in 20 rounds while baseline LotteryFL prunes around 10%.

during each communication round, reducing the overall communication cost. We observe that the communication costs reduce even more when using jumpstart. For example, with jumpstart, we obtain 4.7x and 6x lower communication costs than LotteryFL(0.5) and FedAvg.

Impact of Number of Clients on Performance: We run experiments to compare the performance of FedLTN in a high-client setting (100 clients) with low client participation of 5%. Table 2 shows the performance of FedLTN compared with other baselines. We observe that FedLTN achieves the best average and minimum local test accuracy. FedAvg performs the worst among all baselines. We also note that FedLTN (0.9) achieves better performance than LotteryFL (0.5) even if FedLTN is pruning 40% more parameters than the latter.

Table 2. Performance of FedLTN and other baselines on CIFAR-10 in the high-client setting with 100 clients over 2000 rounds.

Setup	Algorithm	Avg Test Acc (%)	Min Test Acc (%)	Comm. Cost (MB)
	FedAvg	50.6	42.0	2398.3
CIFAR-10	LotteryFL (0.1)	75.1	47.5	2173.1
Custom CNN	LotteryFL (0.5)	74.9	51.3	1304.4
No BN Layers	FedLTN (0.1)	**77.9**	50.5	2166.1
100 clients	FedLTN (0.5)	76.3	**54.3**	1254.7
	FedLTN (0.9)	75.4	50.5	**472.5**

Impact of Number of Classes on Performance: We run experiments with each client consisting of all 10 classes in CIFAR-10. The data volume of these classes is given by the Dirichlet distribution. The Dirichlet distribution is controlled by the parameter α. For low α (close to 0) the data volume is heavily skewed towards one particular class, while as α increases (close to 1), the data volume is distributed in an iid manner across all classes. Since our goal is to learn more personalized models, higher values of α pose a challenge. Table 3 shows the performance of FedLTN (0.9) and FedLTN (0.5) compared with LotteryFL (0.5) for α values of 0.5 and 0.7. As we can see, FedLTN (0.9) learns better-personalized models for high values of α. For example, FedLTN (0.5) achieves 6% more test accuracy than LotteryF L (0.5). We also observe that FedLTN (0.9) despite pruning more parameters, can achieve better overall performance than LotteryFL (0.5). This means that our method is capable of learning more personalized models for high values of α when we have more number of classes.

Table 3. Performance on the CIFAR-10 dataset with dirichlet $\alpha = \{0.5, 0.7\}$

Algorithm	$\alpha = 0.5$		$\alpha = 0.7$	
	Avg Test	Min	Avg Test	Min
LotteryFL (0.5)	61.19	24.00	54.00	36.00
FedLTN (0.5)	**67.20**	**40.00**	**56.00**	**40.00**
FedLTN (0.9)	64.80	**40.00**	55.60	36.00

Impact of Rewinding on Performance: Table 1 shows that rewinding to round 0 after pruning leads to significant drops in FedLTN's accuracy to similar levels as LotteryFL. Moreover, rewinding leads to slower pruning and thus convergence, as seen in our supplementary material (Appendix B).

Impact of Architecture on Performance: When training on the custom CNN, our BN-preserved LTN aggregation cannot be applied and leads to lower performance on sparser models in Table 2. Comparing the results from Table 1 and the equivalent CIFAR-10 low-client experiment in the supplementary material (Appendix C), BN-preserved LTN provides a major boost in our test accuracy (especially for sparser models). Despite this, we see that FedLTN still performs better than LotteryFL in Table 2. On the other hand, we find that LotteryFL does not benefit from the BN layers and greater model depth of ResNet18, which points to the efficacy of our BN-preserved LTN aggregation.

Impact of Jump-Start on Performance and Communication Cost: We observe that using Jump-Start drastically reduces the communication cost without any compromise on the performance. For example, we see from Table 1 and the supplementary material (Appendix D) that FedLTN (0.9) achieves 74.5% on Tiny ImageNet, while FedLTN (0.9; jumpstart) achieves 83.1% with up to 60% lower communication costs. This is due to the efficacy of transfer learning from the best-performing model after local training. Consequently, Jump-Start allows clients to skip the communication cost of the initial FL rounds.

5 Conclusion

The Lottery Ticket Hypothesis has shown promising results in reducing the model size without loss of accuracy for models trained on a single client. In this work, we address the most pressing challenges of applying LTH to the FL setting – slow model pruning and convergence. We propose a new framework, FedLTN, for learning Lottery Ticket Networks via postpruning without rewinding, preserving batch normalization layers, and aggregation using server momentum. We extend FedLTN with Jump-Start, which uses local pre-training to reduce communication costs. FedLTN and FedLTN with Jump-Start achieve higher local test accuracies, significantly accelerate model pruning, and reduce communication cost by 4.7x compared to existing FL approaches.

References

1. Alistarh, D., Grubic, D., Li, J., Tomioka, R., Vojnovic, M.: QSGD: communication-efficient SGD via gradient quantization and encoding. In: Advances in Neural Information Processing Systems 30 (2017)
2. Barnes, L.P., Inan, H.A., Isik, B., Özgür, A.: rTop-k: a statistical estimation approach to distributed SGD. IEEE J. Sel. Areas Inf. Theory **1**(3), 897–907 (2020)
3. Chen, Y., Lu, W., Wang, J., Qin, X.: FedHealth 2: weighted federated transfer learning via batch normalization for personalized healthcare. arXiv preprint arXiv:2106.01009 (2021)
4. Chen, Y., Lu, W., Wang, J., Qin, X., Qin, T.: Federated learning with adaptive batchnorm for personalized healthcare. arXiv preprint arXiv:2112.00734 (2021)
5. Fallah, A., Mokhtari, A., Ozdaglar, A.: Personalized federated learning: a meta-learning approach. arXiv preprint arXiv:2002.07948 (2020)
6. Frankle, J., Carbin, M.: The lottery ticket hypothesis: finding sparse, trainable neural networks. arXiv preprint arXiv:1803.03635 (2018)
7. Frankle, J., Dziugaite, G.K., Roy, D.M., Carbin, M.: Stabilizing the lottery ticket hypothesis. arXiv preprint arXiv:1903.01611 (2019)
8. Hamer, J., Mohri, M., Suresh, A.T.: FedBoost: a communication-efficient algorithm for federated learning. In: International Conference on Machine Learning, pp. 3973–3983. PMLR (2020)
9. Hsu, T.M.H., Qi, H., Brown, M.: Measuring the effects of non-identical data distribution for federated visual classification. arXiv preprint arXiv:1909.06335 (2019)
10. Idrissi, M.J., Berrada, I., Noubir, G.: FEDBS: learning on Non-IID data in federated learning using batch normalization. In: 2021 IEEE 33rd International Conference on Tools with Artificial Intelligence (ICTAI), pp. 861–867. IEEE (2021)

11. Karimireddy, S.P., Kale, S., Mohri, M., Reddi, S., Stich, S., Suresh, A.T.: Scaffold: stochastic controlled averaging for federated learning. In: International Conference on Machine Learning, pp. 5132–5143. PMLR (2020)
12. Kim, G., Kim, J., Han, B.: Communication-efficient federated learning with acceleration of global momentum. arXiv preprint arXiv:2201.03172 (2022)
13. Konečnỳ, J., McMahan, H.B., Yu, F.X., Richtárik, P., Suresh, A.T., Bacon, D.: Federated learning: strategies for improving communication efficiency. arXiv preprint arXiv:1610.05492 (2016)
14. Kopparapu, K., Lin, E.: FedFMC: sequential efficient federated learning on Non-IID data. arXiv preprint arXiv:2006.10937 (2020)
15. Krizhevsky, A., Hinton, G., et al.: Learning multiple layers of features from tiny images (2009)
16. Kulkarni, V., Kulkarni, M., Pant, A.: Survey of personalization techniques for federated learning. In: 2020 Fourth World Conference on Smart Trends in Systems, Security and Sustainability (WorldS4), pp. 794–797. IEEE (2020)
17. Le, Y., Yang, X.: Tiny ImageNet visual recognition challenge. CS 231N **7**(7), 3 (2015)
18. Li, A., Sun, J., Wang, B., Duan, L., Li, S., Chen, Y., Li, H.: LotteryFL: personalized and communication-efficient federated learning with lottery ticket hypothesis on Non-IID datasets. arXiv preprint arXiv:2008.03371 (2020)
19. Li, X., Jiang, M., Zhang, X., Kamp, M., Dou, Q.: FedBN: federated learning on Non-IID features via local batch normalization. arXiv preprint arXiv:2102.07623 (2021)
20. Mansour, Y., Mohri, M., Ro, J., Suresh, A.T.: Three approaches for personalization with applications to federated learning. arXiv preprint arXiv:2002.10619 (2020)
21. McMahan, B., Moore, E., Ramage, D., Hampson, S., y Arcas, B.A.: Communication-efficient learning of deep networks from decentralized data. In: Artificial Intelligence and Statistics, pp. 1273–1282. PMLR (2017)
22. Morcos, A., Yu, H., Paganini, M., Tian, Y.: One ticket to win them all: generalizing lottery ticket initializations across datasets and optimizers. In: Advances in Neural Information Processing Systems 32 (2019)
23. Ozfatura, E., Ozfatura, K., Gündüz, D.: FedADC: accelerated federated learning with drift control. In: 2021 IEEE International Symposium on Information Theory (ISIT), pp. 467–472. IEEE (2021)
24. Reddi, S., et al.: Adaptive federated optimization. arXiv preprint arXiv:2003.00295 (2020)
25. Reyes, J., Di Jorio, L., Low-Kam, C., Kersten-Oertel, M.: Precision-weighted federated learning. arXiv preprint arXiv:2107.09627 (2021)
26. Smith, V., Chiang, C.K., Sanjabi, M., Talwalkar, A.S.: Federated multi-task learning. In: Advances in Neural Information Processing Systems 30 (2017)
27. Suresh, A.T., Felix, X.Y., Kumar, S., McMahan, H.B.: Distributed mean estimation with limited communication. In: International Conference on Machine Learning, pp. 3329–3337. PMLR (2017)
28. Tenison, I., Sreeramadas, S.A., Mugunthan, V., Oyallon, E., Belilovsky, E., Rish, I.: Gradient masked averaging for federated learning. arXiv preprint arXiv:2201.11986 (2022)
29. Wang, K., Mathews, R., Kiddon, C., Eichner, H., Beaufays, F., Ramage, D.: Federated evaluation of on-device personalization. arXiv preprint arXiv:1910.10252 (2019)
30. Xu, A., Huang, H.: Double momentum SGD for federated learning. arXiv preprint arXiv:2102.03970 (2021)

31. Xu, J., Wang, S., Wang, L., Yao, A.C.C.: FedCM: federated learning with client-level momentum. arXiv preprint arXiv:2106.10874 (2021)
32. Yeganeh, Y., Farshad, A., Navab, N., Albarqouni, S.: Inverse distance aggregation for federated learning with Non-IID data. In: Albarqouni, S., et al. (eds.) DART/DCL -2020. LNCS, vol. 12444, pp. 150–159. Springer, Cham (2020). https://doi.org/10.1007/978-3-030-60548-3_15

Theoretical Understanding of the Information Flow on Continual Learning Performance

Joshua Andle and Salimeh Yasaei Sekeh$^{(\boxtimes)}$

University of Maine, Orono, ME 04469, USA
salimeh.yasaei@maine.edu

Abstract. Continual learning (CL) requires a model to continually learn new tasks with incremental available information while retaining previous knowledge. Despite the numerous previous approaches to CL, most of them still suffer forgetting, expensive memory cost, or lack sufficient theoretical understanding. While different CL training regimes have been extensively studied empirically, insufficient attention has been paid to the underlying theory. In this paper, we establish a probabilistic framework to analyze information flow through layers in networks for sequential tasks and its impact on learning performance. Our objective is to optimize the information preservation between layers while learning new tasks. This manages task-specific knowledge passing throughout the layers while maintaining model performance on previous tasks. Our analysis provides novel insights into information adaptation within the layers during incremental task learning. We provide empirical evidence and practically highlight the performance improvement across multiple tasks. Code is available at https://github.com/Sekeh-Lab/InformationFlow-CL.

Keywords: Continual learning · Information flow · Forgetting

1 Introduction

Humans are continual learning systems that have been very successful at adapting to new situations while not forgetting about their past experiences. Similar to the human brain, continual learning (CL) tackles the setting of learning new tasks sequentially without forgetting information learned from the previous tasks [3,15,20]. A wide variety of CL methods mainly either minimize a loss function which is a combination of forgetting and generalization loss to reduce catastrophic forgetting [11,13,18,30,31] or improve quick generalization [7,27]. While these approaches have demonstrated state-of-the-art performance and achieve

Supplementary Information The online version contains supplementary material available at https://doi.org/10.1007/978-3-031-19775-8_6.

some degree of continual learning in deep neural networks, there has been limited prior work extensively and analytically investigating the impact that different training regimes can have on learning a sequence of tasks. Although major advances have been made in the field, one recurring problem that still remains not completely solved is that of catastrophic forgetting (CF). An approach to address this goal is to gradually extend acquired knowledge learned within layers in the network and use it for future learning. While the CF issue has been extensively studied empirically, little attention has been paid from a theoretical angle [10,18,21]. To the best of our knowledge, there are no works which explain what occurs when certain portions of a network are more important than others for passing information of a given task downstream to the end of the network. In this paper, we explore the CL performance and CF problem from a probabilistic perspective. We seek to understand the connection of the passing of information downstream through layers in the network and learning performance at a more in-depth and fundamental theoretical level. We integrate these studies into two central questions:

(1) *Given a sequence of joint random variables and tasks, how much does information flow between layers affect learning performance and alleviate CF?*
(2) *Given a sequence of tasks, how much does the sparsity level of layers on task-specific training influence the forgetting?*

The answers to these questions are theoretically and practically important for continual learning research because: (1) despite the tangible improvements in task learning, the core problem of deep network efficiency on performance assists selective knowledge sharing through downstream information within layers; (2) a systematic understanding of learning tasks provides schemes to accommodate more tasks to learn; and (3) monitoring information flow in the network for each task alleviates forgetting.

Toward our analysis, we measure the information flow between layers given a task by using dependency measures between filters in consecutive layers conditioned on tasks. Given a sequence of joint random variables and tasks, we compute the forgetting by the correlation between task and trained model's loss on the tasks in the sequence. To summarize, our contributions in this paper are,

– Introducing the new concept of *task-sensitivity*, which targets task-specific knowledge passing through layers in the network.
– Providing a theoretical and in-depth analysis of information flow through layers in networks for task sequences and its impact on learning performance.
– Optimizing the information preservation between layers while learning new tasks by freezing task-specific important filters.
– Developing a new bound on expected forgetting using optimal freezing mask.
– Providing experimental evidence and practical observations of layer connectivities in the network and their impact on accuracy.

Organization: The paper is organized as follows. In Sect. 2 we briefly review the continual learning problem formulation and fundamental definitions of the performance. In addition, a set of new concepts including task sensitivity and task usefulness of layers in the network is introduced. In Sect. 3 we establish a series of foundational and theoretical findings that focus on performance and forgetting analysis in terms of the sensitivity property of layers. A new bound on expected forgetting based on the optimal freezing mask is given in this section. Finally, in Sect. 5 we provide experimental evidence of our analysis using the CIFAR-10/100 and Permuted MNIST datasets. The main proofs of the theorems in the paper are given in the supplementary materials, although in Sect. 3.4 we provide the key components and techniques that we use for the proofs.

2   Problem Formulation

In supervised continual learning, we are given a sequence of joint random variables $(\mathbf{X}_t, T_t)$, with realization space $\mathcal{X}_t \times \mathcal{T}_t$ where $(\mathbf{x}_t, y_t)$ is an instance of the $\mathcal{X}_t \times \mathcal{T}_t$ space. We use $\|.\|$ to denote the Euclidean norm for vectors and $\|.\|_F$ to denote the Frobenius norm for matrices. In this section, we begin by presenting a brief list of notations and then provide the key definitions.

Notations: We assume that a given DNN has a total of L layers where,

- $F^{(L)}$: A function mapping the input space $\mathcal{X}$ to a set of classes $\mathcal{T}$, i.e. $F^{(L)} : \mathcal{X} \mapsto \mathcal{T}$.
- $f^{(l)}$: The l-th layer of $F^{(L)}$ with M_l as number of filters in layer l.
- $f_i^{(l)}$: i-th filter in layer l.
- $F^{(i,j)} := f^{(j)} \circ \ldots \circ f^{(i)}$: A subnetwork which is a group of consecutive layers.
- $F^{(j)} := F^{(1,j)} = f^{(j)} \circ \ldots \circ f^{(1)}$: First part of the network up to layer j.
- $\sigma^{(l)}$: The activation function in layer l.
- $\widetilde{f}_t^{(l)}$: Sensitive layer for task t.
- $\widetilde{F}_t^{(L)} := F_t^{(L)} / \widetilde{f}_t^{(l)}$: The network with L layers when l-th sensitive layer $\widetilde{f}^{(l)}$ is frozen while training on task t.
- $\pi(T_t)$: The prior probability of class label $T_t \in \mathcal{T}_t$.
- η_{tl}, γ_{tl}: Thresholds for sensitivity and usefulness of l-th layer $f^{(l)}$ for task t.

In this section, we revisit the standard definition of training performance and forgetting and define the new concepts *task-sensitive layer* and *task-useful layer*.

Definition 1. (Task-Sensitive Layer) *The l-th layer, $f^{(l)}$, is called a t-task-sensitive layer if the average information flow between filters in consecutive layers l and $l+1$ is high i.e.*

$$\Delta_t(f^{(l)}, f^{(l+1)}) := \frac{1}{M_l \, M_{l+1}} \sum_{i=1}^{M_l} \sum_{j=1}^{M_{l+1}} \rho\left(f_i^{(l)}, f_j^{(l+1)} | T_t\right) \geq \eta_{lt}, \qquad (1)$$

where ρ is a connectivity measure given task T_t such as conditional Pearson correlation or conditional Mutual Information [4,5]. In this work we focus on only Pearson correlation as the connectivity measure between layers l and $l+1$.

Without loss of generality, in this work we assume that filters $f_i^{(l)}$, $i = 1, \ldots, M_l$, are normalized such that

$$\mathbb{E}_{(\mathbf{X}_t, T_t) \sim D_t} \left[f_i^{(l)}(\mathbf{X}_t) | T_t \right] = 0 \text{ and } \mathbb{V} \left[f_i^{(l)}(\mathbf{X}_t) | T_t \right] = 1, \quad l = 1, \ldots, L,$$

Therefore the Pearson correlation between the i-th filter in layer l and the j-th filter in layer $l + 1$ becomes

$$\rho(f_i^{(l)}, f_j^{(l+1)} | T_t) := \mathbb{E}_{(\mathbf{X}_t, T_t) \sim D_t} \left[f_i^{(l)}(\mathbf{X}_t) f_j^{(l+1)}(\mathbf{X}_t) | T_t \right]. \tag{2}$$

Note that in this paper we consider the absolute value of ρ in the range $[0, 1]$.

Definition 2. (Task-Useful Layer) *Suppose input $\mathbf{X}_t$ and task T_t have joint distribution $\mathcal{D}_t$. For a given distribution $\mathcal{D}_t$, the l-layer $f^{(l)}$ is called t-task-useful if there exist two mapping functions $G_l : \mathcal{L}_l \mapsto T_t$ and $K_l : \mathcal{X}_t \mapsto \mathcal{L}_l$ such that*

$$\mathbb{E}_{(\mathbf{X}_t, T_t) \sim \mathcal{D}_t} \left[T_t \cdot G_l \circ f^{(l)}(K_{l-1} \circ \mathbf{X}_t) \right] \geq \gamma_{tl}. \tag{3}$$

Note that here $f^{(l)}$ is a map function $f^{(l)} : \mathcal{L}_{l-1} \mapsto \mathcal{L}_l$.

Within this formulation, two parameters determine the contributions of the l-th layer of network $F^{(l)}$ on task T_t: η_{tl} the contribution of passing forward the information flow to the next consecutive layer, and γ_{tl}, the contribution of the l-th layer in learning task T_t. Training a neural network $F_t^{(L)} \in \mathcal{F}$ is performed by minimizing a loss function (empirical risk) that decreases with the correlation between the weighted combination of the networks and the label:

$$\mathbb{E}_{(\mathbf{X}_t, T_t) \sim D_t} \left\{ L_t(F_t^{(L)}(\mathbf{X}_t), T_t) \right\} = -\mathbb{E}_{(\mathbf{X}_t, T_t) \sim D_t} \left\{ T_t \cdot \left(b + \sum_{F_t \in \mathcal{F}} w_{F_t} \cdot F_t^{(L)}(\mathbf{X}_t) \right) \right\}. \tag{4}$$

We remove offset b without loss of generality. Define

$$\ell_t(\omega) := - \sum_{F_t \in \mathcal{F}} w_{F_t} \cdot F_t^{(L)}(\mathbf{X}_t), \tag{5}$$

therefore the loss function in (4) becomes $\mathbb{E}_{(\mathbf{X}_t, T_t) \sim D_t} \{T_t \cdot \ell_t(\omega)\}$. Let ω_t^* be the set of parameters when the network is trained on task T_t that minimizes (4):

$$\omega_t^* := argmin_{\omega_t} \mathbb{E}_{(\mathbf{X}_t, T_t) \sim D_t} \left\{ T_t \cdot (\ell_t(\omega_t)) \right\}, \tag{6}$$

where ℓ_t is defined in (5). The total risk of all seen tasks $t < \tau$ is given by

$$\sum_{t=1}^{\tau} \mathbb{E}_{(\mathbf{X}_t, T_t) \sim D_t} \left\{ T_t \cdot \ell_t(\omega_\tau) \right\}. \tag{7}$$

The set of parameters when the network $F^{(l)}$ is trained after seeing all tasks is the solution of minimizing the risk in (7) and is denoted by ω_τ^*.

Definition 3. (Performance Difference) *Suppose input* $\mathbf{X}_t$ *and task* T_t *have joint distribution* $\mathcal{D}_t$. *Let* $\widetilde{F}_t^{(L)} := F_t^{(L)}/\widetilde{f}_t^{(l)} \in \mathcal{F}$ *be the network with* L *layers when l-layer* $f^{(l)}$ *is frozen while training on task t. The performance difference between training* $F_t^{(L)}$ *and* $\widetilde{F}_t^{(L)}$ *is defined as*

$$d(F_t^{(L)}, \widetilde{F}_t^{(L)}) := \mathbb{E}_{(\mathbf{X}_t, T_t) \sim \mathcal{D}_t} \left\{ L_t(F_t^{(L)}(\mathbf{X}_t), T_t) - L_t(\widetilde{F}_t^{(L)}(\mathbf{X}_t), T_t) \right\}. \quad (8)$$

Let ω_t^* *and* $\widetilde{\omega}_t^*$ *be the convergent parameters after training* $F_t^{(L)}$ *and* $\widetilde{F}_t^{(L)}$ *has been finished for task* T_t, *respectively. Define the training deviation for* T_t *as:*

$$\delta_t(\omega_t^*|\widetilde{\omega}_t^*) := \ell_t(\omega_t^*) - \ell_t(\widetilde{\omega}_t^*). \quad (9)$$

The optimal performance difference in Definition 3 is the average of δ_t *in (9):*

$$d(F_t^{(L)}, \widetilde{F}_t^{(L)}) = \mathbb{E}_{(\mathbf{X}_t, T_t) \sim \mathcal{D}_t} [T_t \cdot \delta_t(\omega_t^*|\widetilde{\omega}_t^*)] = \mathbb{E}_{(\mathbf{X}_t, T_t) \sim \mathcal{D}_t} [T_t \cdot (\ell_t(\omega_t^*) - \ell_t(\widetilde{\omega}_t^*))].$$

3 Continual Learning Performance Study

Our goal is to decide which filters trained for intermediate task T_t to prune/freeze when training the network on task T_{t+1}, given the sensitivity scores of layers introduced in (1), so that the predictive power of the network is maximally retained and not only does forgetting not degrade performance but we also gain a performance improvement. In this section, we first take an in-depth look at the layers and show the relationship between task sensitive and task useful layers. Second we provide an analysis in which we show that sensitive layers affect performance if they get frozen while training the network on the new task.

3.1 Performance Analysis

The motivation of our objective in this section is that the difference between the loss functions produced by the original network $F^{(L)}$ and the frozen network $\widetilde{F}_t^{(L)}$ should be maximized with respect to sensitive and important filters. We begin by showing that sensitive layers are useful in improving network performance.

Theorem 1. *For a given sequence of joint random variables* $(\mathbf{X}_t, T_t) \sim \mathcal{D}_t$ *and network* $F^{(L)}$, *if the l-th layer,* $f^{(l)}$ *is t-task-sensitive then it is t-task-useful.*

Theorem 2. *Suppose input* $\mathbf{x}_t$ *and label* y_t *are samples from* $(\mathbf{X}_t, T_t)$ *with joint distribution* $\mathcal{D}_t$. *For a given distribution* $\mathcal{D}_t$, *if the layer l is a t-task-useful layer,*

$$\mathbb{E}_{(\mathbf{X}_t, T_t) \sim \mathcal{D}_t} [T_t \cdot G_l \circ f^{(l)}(K_{l-1} \circ \mathbf{X}_t)] \geq \gamma_{tl}, \quad (10)$$

where $G_l : \mathcal{L}_l \mapsto T_t$ *and* $K_l : \mathcal{X}_t \mapsto \mathcal{L}_l$ *are map functions. Then removing layer l decreases the performance i.e.*

$$d(F_t^{(L)}, \widetilde{F}_t^{(L)}) := \mathbb{E}_{(\mathbf{X}_t, T_t) \sim \mathcal{D}_t} \left\{ L_t(F_t^{(L)}(\mathbf{X}_t), T_t) - L_t(\widetilde{F}_t^{(L)}(\mathbf{X}_t), T_t) \right\} > K(\gamma_{tl}). \quad (11)$$

Here $\widetilde{F}_t^{(L)} := F_t^{(L)}/\widetilde{f}_t^{(l)} \in \mathcal{F}$ *is the network with* L *layers when layer l is frozen while training on task t. The function* $K(\gamma_{tl})$ *is increasing in* γ_{tl}.

An immediate result from the combination of Theorems 1 and 2 is stated below:

Theorem 3. *Suppose input $\mathbf{x}_t$ and label y_t are samples from joint random variables $(\mathbf{X}_t, T_t)$ with distribution $\mathcal{D}_t$. For a given distribution $\mathcal{D}_t$, if the layer l is a t-task-sensitive layer i.e. $\Delta_t(f^{(l)}, f^{(l+1)}) \geq \eta_{tl}$, then the performance difference between $d(F_t^{(L)}, \widetilde{F}_t^{(L)})$ is bounded as*

$$d(F_t^{(L)}, \widetilde{F}_t^{(L)}) := \mathbb{E}_{(\mathbf{X}_t, T_t) \sim \mathcal{D}_t} \left\{ L_t(F_t^{(L)}(\mathbf{X}_t), T_t) - L_t(\widetilde{F}_t^{(L)}(\mathbf{X}_t), T_t) \right\} \geq g(\eta_{tl}), \tag{12}$$

where g is an increasing function of η_{tl}. Here $\widetilde{F}_t^{(L)} := F_t^{(L)} / \widetilde{f}_{t-1}^{(l)} \in \mathcal{F}$ is the network with L layers when layer l is frozen while training on task t.

One important takeaway from this theorem is that as sensitivity between layers η_{tl} increases the performance gap between the original and frozen network's loss functions increases. An important property of filter importance is that it is a probabilistic measure and can be computed empirically along the network. The total loss (empirical risk) on the training set for task T_t is approximated by $\frac{1}{|T_t|} \sum_{(\mathbf{x}_t, y_t)} y_t \ell_t(\omega_t; \mathbf{x}_t, y_t)$, where ℓ_t is a differentiable loss function (5) associated with data point $(\mathbf{x}_t, y_t)$ for task T_t or we use cross entropy loss.

3.2   Forgetting Analysis

When sequentially learning new tasks, due to restrictions on access to examples of previously seen tasks, managing the forgetting becomes a prominent challenge. In this section we focus on measuring the forgetting in CL with two tasks. It is potentially possible to extend these findings to more tasks.

Let ω_t^* and ω_{t+1}^* be the convergent parameters after training has been finished for the tasks T_t and T_{t+1} sequentially. Forgetting of the t task is defined as

$$O_t := \ell_t(\omega_{t+1}^*) - \ell_t(\omega_t^*) \tag{13}$$

In this work, we propose the expected forgetting measure based on correlation between task T_t and forgetting (13) given distribution $\mathcal{D}_t$:

Definition 4. (Expected Forgetting) *Let ω_t^* and ω_{t+1}^* be the convergent or optimum parameters after training has been finished for the t and $t+1$ task sequentially. The expected forgetting denoted by EO_t is defined as*

$$EO_t := \mathbb{E}_{(\mathbf{X}_t, T_t) \sim \mathcal{D}_t} \left[T_t \cdot \left| \left(\ell_t(\omega_{t+1}^*) - \ell_t(\omega_t^*) \right) \right| \right]. \tag{14}$$

Theorem 4. *Suppose input $\mathbf{x}_t$ and label y_t are samples from joint distribution $\mathcal{D}_t$. For a given distribution $\mathcal{D}_t$, if the layer l is a t-task-useful layer,*

$$\mathbb{E}_{(\mathbf{X}_t, T_t) \sim \mathcal{D}_t} \left[T_t \cdot G_l \circ f^{(l)}(K_{l-1} \circ \mathbf{X}_t) \right] \geq \gamma_{tl}, \tag{15}$$

then expected forgetting EO_t defined in (14) is bounded by $\epsilon(\gamma_{tl})$, a decreasing function of γ_{tl} i.e.

$$\widetilde{EO}_t := \mathbb{E}_{(\mathbf{X}_t, T_t) \sim D_t} \left\{ L_t(\widetilde{F}_{t+1}^{(L)}(\mathbf{X}_t), T_t) - L_t(F_t^{(L)}(\mathbf{X}_t), T_t) \right\} < \epsilon(\gamma_{tl}), \quad (16)$$

where $\widetilde{F}_{t+1}^{(L)} := F_{t+1}^{(L)} / \widetilde{f}_{t+1}^{(l)} \in \mathcal{F}$ is the network with L layers when layer l is frozen while training on task $t + 1$.

A few notes on this bound: (i) based on our finding in (16), we analytically show that under the assumption that the l-th layer is highly t-task-useful i.e. when the hyperparameter γ_{tl} is increasing then average forgetting is decreasing if we freeze the layer l during training the network on new task T_{t+1}. This is achieved because $\epsilon(\gamma_{tl})$ is a decreasing function with respect to γ_{tl}; (ii) by a combination of Theorems 1 and 2 we achieve an immediate result that if layer l is t-task-sensitive then forgetting is bounded by a decreasing function of threshold η_{tl}, $\epsilon(\eta_{tl})$; (iii) We prove that the amount of forgetting that a network exhibits from learning the tasks sequentially correlates with the connectivity properties of the filters in consecutive layers. In particular, the larger these connections are, the less forgetting happens. We empirically verify the relationship between expected forgetting and average connectivity in Sect. 5.

3.3   A Bound on EO_t Using Optimal Freezing Mask

Let ω_t^* be the set of parameters when the network is trained on task T_t, the optimal sparsity for layer $f^{(l)}$ with optimal mask $m^{*(l)}_{t+1}$ while training on task T_{t+1} is achieved by

$$(\omega^*_{t+1}, m^{*(l)}_{t+1}) := \underset{\omega_{t+1}, m}{\arg \min} \, \mathbb{E}_{(\mathbf{X}_t, T_t) \sim D_t} \left\{ \left| T_t \cdot \left(\ell_t(m^{(l)}_{t+1} \odot \omega_{t+1}) - \ell_t(\omega_t^*) \right) \right| \right\}, \quad (17)$$

where $m^{*(l)}_{t+1}$ is the binary mask matrix created after freezing filters in the l-th layer after training on task T_t (masks are applied to the past weights) and before training on task T_{t+1}. Denote $P^{*(l)}_m = \frac{\|m^{*(l)}_{t+1}\|_0}{|\omega^{*(l)}_{t+1}|}$ the optimal sparsity of frozen filters in layer l in the original network $F^{(L)}$.

Definition 5. (Task-Fully-Sensitive Layer) *The l-th layer, $f^{(l)}$, is called a t-task-fully-sensitive layer if the average information flow between filters in layers l and $l + 1$ is maximum i.e. $\Delta_t(f^{(l)}, f^{(l+1)}) \to 1$ (a.s.). Note that here ρ in (1) is a connectivity measure which varies in $[0, 1]$.*

Theorem 5. *Suppose input $\mathbf{x}_t$ and label y_t in space $\mathcal{X}_t \times \mathcal{T}_t$ are samples from random variables $(\mathbf{X}_t, T_t)$ with joint distribution $\mathcal{D}_t$. For a given distribution $\mathcal{D}_t$, if layer l is t-task-fully-sensitive and $P^{*(l)}_m = \frac{\|m^{*(l)}_{t+1}\|_0}{|\omega^{*(l)}_{t+1}|} \to 1$ (a.s.), this means that the entire layer l is frozen when training on task T_{t+1}. Let $\widetilde{\omega}^{*(l)}_{t+1}$ be the optimal*

weight set for layer l, masked and trained on task T_{t+1}, $\widetilde{\omega}_{t+1}^{*(l)} = m_{t+1}^{*(l)} \odot \omega_{t+1}^{*(l)}$,
Then the expected forgetting $\widetilde{EO}_t$ defined in

$$\widetilde{EO}_t = \mathbb{E}_{(\mathbf{X}_t, T_t) \sim D_t} \left\{ \left| T_t \cdot \left(\ell_t(\widetilde{\omega}_{t+1}^*) - \ell_t(\omega_t^*) \right| \right) \right\}, \text{ is bounded by}$$

$$\widetilde{EO}_t \leq \frac{1}{2} \mathbb{E}_{(\mathbf{X}_t, T_t) \sim D_t} \left\{ T_t \cdot \lambda_t^{max} \left(C + \frac{C_\epsilon}{\lambda_t^{max}} \right)^2 \right\}, \; C \; \& \; C_\epsilon \text{ are constants, } \quad (18)$$

and λ_t^{max} is the maximum eigenvalue of Hessian $\nabla^2 \ell_t(\omega_t^*)$.

Based on the argumentation of this section, we believe the bound found in (18) can provide a supportive study in how freezing rate affects forgetting explicitly. In [18], it has been shown that lower λ_t^{max} or equivalently wider loss function L_t leads to less forgetting however, our bound in (18) is not a monotonic function of maximum eigenvalue of Hessian. Therefore we infer that when a layer has highest connectivity, freezing the entire layer and blocking it for a specific task does not necessarily control the forgetting. Our inference is not only tied to the reduction of λ_t^{max} which describes the width of a local minima [12], but we also need to rely on other hidden factors that is undiscovered for us up to this time. Although we believe that to reduce forgetting, each task should push its learning towards information preservation by protecting sensitive filters and can possibly employ the same techniques used to widen the minima to improve generalization.

3.4   Key Components to Prove Theorems

The main proofs of Theorems 1–5 are provided in supplementary materials, however in this section, we describe a set of widely used key strategies and components that are used to prove findings in Sect. 3.

Theorem 1. To prove that a task-sensitive layer is a task-useful layer, we use key components: **(I)** Set $\overline{\sigma}_j(s) = s.\sigma_j(s)$ where σ_j is activation function:

$$\Delta_t(f^{(l)}, f^{(l+1)}) \propto \sum_{i=1}^{M_l} \sum_{y_t \in T_t} \pi(y_t) \mathbb{E} \left[\sum_{j=1}^{M_{l+1}} \overline{\sigma}_j \left(f_i^{(l)}(\mathbf{X}_t) \right) | T_t = y_t \right]. \quad (19)$$

(II) There exist a constant C_t such that

$$C_t \sum_{i=1}^{M_l} \sum_{y_t \in T_t} y_t \pi(y_t) \mathbb{E}_{\mathbf{X}_t | y_t} \left[f_i^{(l)}(\mathbf{X}_t) | T_t = y_t \right]$$

$$\geq \sum_{i=1}^{M_l} \sum_{y_t \in T_t} \pi(y_t) \mathbb{E} \left[\sum_{j=1}^{M_{l+1}} \overline{\sigma}_j \left(f_i^{(l)}(\mathbf{X}_t) \right) | T_t = y_t \right]. \quad (20)$$

Theorem 2. Let ω_t^* and $\widetilde{\omega}_t^*$ be the convergent or optimum parameters after training $F_t^{(L)}$ and $\widetilde{F}_t^{(L)}$ has been finished for task t, respectively. Here we establish three important components:

(I) Using Taylor approximation of ℓ_t around $\widetilde{\omega}_t^*$:

$$\ell_t(\omega_t^*) - \ell_t(\widetilde{\omega}_t^*) \approx \frac{1}{2}(\omega_t^* - \widetilde{\omega}_t^*)^T \nabla^2 \ell_t(\widetilde{\omega}_t^*)(\omega_t^* - \widetilde{\omega}_t^*). \tag{21}$$

(II) Let $\widetilde{\lambda}_t^{min}$ be the minimum eigenvalue of $\nabla^2 \ell_t(\widetilde{\omega}_t^*)$, we show

$$\frac{1}{2}\mathbb{E}_{(\mathbf{X}_t,T_t) \sim D_t}\left[T_t \cdot \left((\omega_t^* - \widetilde{\omega}_t^*)^T \nabla^2 \ell_t(\widetilde{\omega}_t^*)(\omega_t^* - \widetilde{\omega}_t^*)\right)\right]$$
$$\geq \frac{1}{2}\mathbb{E}_{(\mathbf{X}_t,T_t) \sim D_t}\left[T_t \cdot \left(\widetilde{\lambda}_t^{min}\|\omega_t^* - \widetilde{\omega}_t^*\|^2\right)\right]. \tag{22}$$

(III) There exist a constant $C^{(l)}$ and a map function $G_l : \mathcal{L}_l \mapsto \mathcal{T}_t$ such that

$$\mathbb{E}_{(\mathbf{X}_t,T_t) \sim D_t}\left[T_t \cdot G_l \circ \sigma_t^{(l)}\left((\omega_t^* - \widetilde{\omega}_t^*)\mathbf{X}_t\right)\right]$$
$$\leq C^{(l)} \,\mathbb{E}_{(\mathbf{X}_t,T_t) \sim D_t}\left[T_t \cdot G_l \circ |\sigma_t^{(l)}(\omega_t^*\mathbf{X}_t) - \sigma_t^{(l)}(\widetilde{\omega}_t^*\mathbf{X}_t)|\right]. \tag{23}$$

Theorem 4. Let $\widetilde{\omega}_{t+1}^*$ be the optimal weight after training $\widetilde{F}_{t+1}^{(L)}$ on task $t + 1$. Here are the key components we need to use to prove the theorem: (I) we show

$$\mathbb{E}_{(\mathbf{X}_t,T_t) \sim D_t}\left\{T_t \cdot \left(\ell_t(\widetilde{\omega}_{t+1}^*) - \ell_t(\widetilde{\omega}_t^*)\right)\right\} \leq \frac{1}{2}\mathbb{E}_{(\mathbf{X}_t,T_t) \sim D_t}\left\{T_t \cdot \widetilde{\lambda}_t^{max}\|\widetilde{\omega}_{t+1}^* - \widetilde{\omega}_t^*\|^2\right\}, \tag{24}$$

(II) Let $\widetilde{w}_t'$ be the convergent or (near-) optimum parameters after training $\widetilde{F}_t^{(L)}$ and $\widetilde{\lambda}_t^{max}$ be the maximum eigenvalue of $\nabla^2 \ell_t(\widetilde{\omega}_t^*)$:

$$\nabla \ell_t(\widetilde{\omega}_t') - \nabla \ell_t(\widetilde{\omega}_t^*) \approx \nabla^2 \ell_t(\widetilde{\omega}_t^*)(\widetilde{\omega}_t' - \widetilde{\omega}_t^*) \leq \widetilde{\lambda}_t^{max}\|\widetilde{\omega}_t' - \widetilde{\omega}_t^*\|, \tag{25}$$

(III) If the convergence criterion is satisfied in the ϵ-neighborhood of $\widetilde{\omega}_t^*$, then

$$\|\widetilde{\omega}_{t+1}^* - \widetilde{\omega}_t^*\| \leq \frac{C_\epsilon}{\widetilde{\lambda}_t^{max}}, \qquad C_\epsilon = \max\{\epsilon, 2\sqrt{\epsilon}\}.$$

Theorem 5. Denote $\widetilde{\omega}_{t+1}^{*(l)} = m_{t+1}^{*(l)} \odot \omega_{t+1}^{*(l)}$ where $m_{t+1}^{*(l)}$ is the binary freezing mask for layer l. For the optimal weight matrix $\widetilde{\omega}_{t+1}^*$ with mask m_{t+1}^*, define

$$\widetilde{EO}_t = \mathbb{E}_{(\mathbf{X}_t,T_t) \sim D_t}\left\{|T_t \cdot \left(\ell_t(\widetilde{\omega}_{t+1}^*) - \ell_t(\omega_t^*)\right)|\right\}.$$

(I) Once we assume that only one connection is frozen in the training process, we can use the following upper bound of the model [14]:

$$|\ell_t(\widetilde{\omega}_{t+1}^*) - \ell_t(\omega_{t+1}^*)| \leq \frac{\|\omega_{t+1}^{*(l)} - \widetilde{\omega}_{t+1}^{*(l)}\|_F}{\|\omega_{t+1}^{*(l)}\|_F} \prod_{j=1}^L \|\omega_{t+1}^{*(l)}\|_F, \tag{26}$$

(II) Under the assumption $P_m^{*(l)} = \frac{\|m_{t+1}^{*(l)}\|_0}{|\omega_{t+1}^{*(l)}|} \to 1$, we show

$$\widetilde{EO}_t \leq \mathbb{E}_{(\mathbf{X}_t,T_t) \sim D_t}\left\{T_t \cdot |\left(\ell_t(\omega_{t+1}^*) - \ell_t(\omega_t^*)\right)|\right\}. \tag{27}$$

4   Related Work

In recent years significant interest has been given to methods for the sequential training of a single neural network on multiple tasks. One of the primary obstacles to achieving this is catastrophic forgetting (CF), the decrease in performance observed on previously trained tasks after learning a new task. As such, overcoming CF is a primary desiderata of CL methods. Several approaches have been taken to address this problem, including various algorithms which mitigate forgetting, as well as investigation into the properties of CF itself.

Catastrophic Forgetting: The issue of catastrophic forgetting isn't new [1,17], however the popularity of deep learning methods has brought it renewed attention. Catastrophic forgetting occurs in neural networks due to the alterations of weights during the training of new tasks. This changes the network's parameters from the optimized state achieved by training on the previous task. Recent works have aimed to better understand the causes and behavior of forgetting [6,22], as well as to learn how the specific tasks being trained influence it and to empirically study its effects [9,19]. Such theoretical research into CF provides solutions to mitigate catastrophic forgetting beyond the design of the algorithm. Similarly, our investigation into the relationship between information flow and CF provides a useful tool for reducing forgetting independent of a specific algorithm.

Continual Learning: Several methods have been applied to the problem of CL. These generally fall into four categories: Regularization [13,32], Pruning-Based [16,26,28], Replay [25,29], and Dynamic Architecture approaches [23,31]. Regularization approaches attempt to reduce the amount of forgetting by implementing a regularization term on previously optimized weights based on their importance for performance. Replay methods instead store or generate samples of past tasks in order to limit forgetting when training for a new task. Dynamic architectures expand the network to accommodate new tasks. Lastly, Pruning-based methods aim to freeze the most important partition of weights in the network for a given task before pruning any unfrozen weights.

While pruning-based methods are able to remove forgetting by freezing and masking weights, they are often implemented to make simple pruning decisions, either using fixed pruning percents for the full network or relying on magnitude-based pruning instead of approaches which utilize available structural information of the network. Other recent works have demonstrated the importance of structured pruning [2,8], suggesting that pruning-based CL methods would benefit from taking advantage of measures of information such as connectivity. While these methods commonly use fixed pruning percentages across the full network, some works outside of the domain of CL investigate different strategies for selecting layer-wise pruning percents, and together they demonstrate the importance of a less homogeneous approach to pruning [14,24].

5   Experimental Evidence

To evaluate the influence of considering knowledge of information flow when training on sequential tasks, we perform multiple experiments demonstrating improved performance when reducing pruning on the task-sensitive layers defined in Definition 1. The experimental results section is divided into two main parts aligning with the overall goal of analyzing downstream information across layers. The first part discusses the performance of CL in the context of protecting highly task-sensitive layers during pruning when adding multiple tasks in a single neural network as in [16]. The second part focuses on the connectivity across layers given tasks and how connectivity varies across the layers and between tasks.

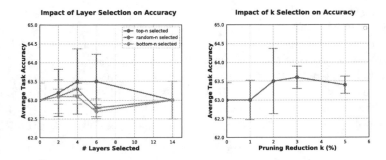

Fig. 1. The average accuracy across tasks is reported for varying values of n when $k = 2\%$(left) and k when $n = 4$(right), where n is the number of layers selected for reduced pruning and k is the hyper-parameter dictating how much the pruning on selected layers is reduced by. We compare the performance when the n layers are selected as the most (top-n), least (bottom-n), or randomly chosen (random-n) connected layers

Setting: We carry out training with a VGG16 model on a split CIFAR-10/100 dataset, where task 1 is CIFAR-10 and tasks 2–6 are each 20 consecutive classes of CIFAR-100. We perform experiments on the Permuted MNIST dataset to determine how the characteristics of information flow differ between datasets (supporting experiments on MNIST are included in the supplementary materials). Three trials were run per experiment. After training on a given task T_t, and prior to pruning, we calculate $\Delta_t(f^{(l)}, f^{(l+1)})$ between each adjacent pair of convolutional or linear layers as in 1. Connectivity figures are plotted by layer index, which includes all VGG16 layers (ReLu, pooling, conv2D, etc.), however only trainable layers are plotted. As a baseline we prune 80% of the lowest-magnitude, unfrozen weights in each layer (freezing the remaining 20%).

5.1   How Do Task Sensitive Layers Affect Performance?

Top-Connectivity Layer Freezing: For this experiment we select the n layers with the highest value of Δ_t and prune $k\%$ fewer weights in those layers for Task T_t,

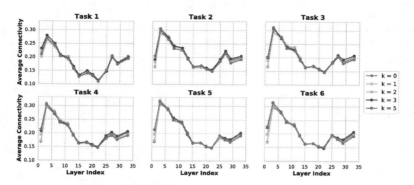

Fig. 2. For each layer the average connectivity value with the subsequent layer is reported. The connectivities are plotted for each task in CIFAR-10/100, for various k when $n = 4$ most connected layers are selected for reduced pruning.

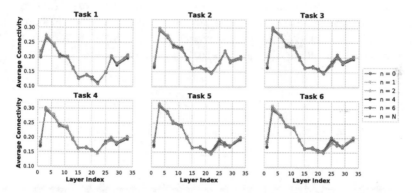

Fig. 3. The average connectivities across layers with the subsequent layer is reported. The scores are plotted for each task in CIFAR-10/100, when the n most-connected layers are selected to have their pruning percent reduced by $k = 2\%$.

where both n and k are hyper-parameters. This reduction is determined individually for each task, and only applies to the given task. By reducing pruning on the most task-sensitive layer, information flow through the network is better maintained, preserving performance on the current task. This is demonstrated in Fig. 1, in which selecting the most connected layers for reduced pruning outperforms selecting the least connected or random layers. Although n and k have the same values in each case, by selecting the top-n layers we better maintain the flow of information by avoiding pruning highly-connected weights. By taking values of $n > 1$, we can account for cases where reducing the pruning on a single layer doesn't sufficiently maintain the flow of information above the baseline. Figure 1 also shows that the performance increase for pruning the top-n connected layers varies depending on the reduction in pruning k.

Connectivity Analysis: To better characterize our measure of information flow and determine which layers are most task-sensitive, we plot the values of Δ_t

for each convolutional or linear layer, as in Figs. 2, 3, and 4. These figures show how connectivity varies over several experimental setups as we change n and k during the freezing of the top-n connected layers. We compare these trends to those seen when performing the baseline $(n, k = 0)$ on Permuted MNIST.

6   Discussion

In-depth Analysis of Bounds: The bound established in Theorem 3 shows that the performance gap between original and adapted networks with task-specific frozen layers grows as the layers contribute more in passing the information to the next layer given tasks. This gap has a direct relationship with the activations' lipschitz property and the minimum eigenvalues of the Hessian at optimal weights for the pruned network. From the forgetting bound in Theorem 4, we infer that as a layer is more useful for a task then freezing it reduces the forgetting more. In addition from Theorem 5, we establish that the average forgetting is a non-linear function of width of a local minima and when the entire filters of a fully sensitive layer is frozen the forgetting tends to a tighter bound.

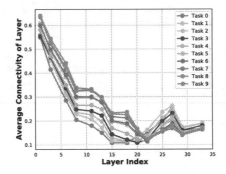

Fig. 4. The average connectivity for each layer is reported for training on Permuted MNIST. Training was done with the baseline setting when $n, k = 0$. Each of the 10 tasks in the Permuted MNIST dataset are plotted.

Information Flow: The connectivities plotted in Figs. 2, 3, and 4 display patterns which remain generally consistent for a given dataset, but have noticeable differences between each dataset. For Figs. 2 and 3 tasks 2–6, which correspond to CIFAR-100, show larger connectivities across most of the network compared to CIFAR-10, particularly in the early and middle layers. Meanwhile, for MNIST we observe connectivities which are much different from those of CIFAR-10 and CIFAR-100. These observations suggest that when applied to different datasets, the task sensitivity of the layers in a network (VGG16 in this case) differ, indicating that the optimal freezing masks and pruning decisions differ as well. Further, Fig. 4 prominently shows that as subsequent tasks are trained, the connectivity of the early layers increases while the later layers' connectivity values decrease. This can also be seen to a lesser extent in Figs. 2 and 3, where the peak in the last four layers decreases, while the first three layers take larger values for later tasks. This indicates that not only is the data important for determining which layers are task-sensitive, but the position of a given task in the training order is as well. For the data shown here, we would suspect the optimal freezing mask to more readily freeze the earlier, more highly connected layers, in the network.

Top-n Layer Freezing: The selection of the most connected layers in Fig. 1 demonstrated an improvement over the baseline, least connected, or randomly chosen layers, showing that the improved performance isn't simply due to freezing more weights. While the performance improves for top-n freezing, the standard deviation also noticeably increases and overlaps the bottom-n results. This was observed for all top-n experiments, and may be linked to the observations in Figs. 2 and 3 that the top-n connected layers are found at the beginning of the network, as perhaps repeated freezing of early layers has a more destabilizing effect. While further work is needed to see if these results can be further improved upon, these observations lend support to the idea that making pruning decisions by utilizing knowledge of information flow in the network is an available tool to retain performance in pruning-based continual learning applications.

7 Conclusion

We've theoretically established a relationship between information flow and catastrophic forgetting and introduced new bounds on the expected forgetting. We've shown empirically how the information flow (measured by the connectivity between layers) varies between the layers of a network, as well as between tasks. Looking ahead these results highlight future possible directions of research in investigating differences in connectivity trends between various datasets, using a probabilistic connectivity measure like mutual information, and investigation on which portions of a network would be most important for passing information.

Finally, we have also empirically demonstrated that utilizing the knowledge of information flow when implementing a pruning-based CL method can improve overall performance. While these core experiments would benefit from further supporting investigations, such as the effects of different networks or tuning hyper-parameters beyond n and k, the reported results nonetheless show promising support for the utility of information flow. Here we limited our investigation to using connectivity when determining the extent of pruning/freezing within a layer, however it would be of significant interest to see possible applications in determining which weights are pruned (as an alternative to magnitude-based pruning), or even the use of information flow in CL methods which don't utilize pruning. These are left as a very interesting future work.

While this paper uses common CL datasets for validation of our theoretical work and focuses on pruning-based methods, applying the methods to a number of larger/more complex datasets will be the focus of more empirical future work, and may help further assess our method's capabilities as well as whether or not the connectivity trends seen here also reflect other, more complex datasets. The core theory and measure of information flow are independent of the scale of the data, so the method is expected to still work with larger datasets.

Acknowledgment. This work has been partially supported by NSF 2053480 and NSF 5409260; the findings are those of the authors only and do not represent any position of these funding bodies.

References

1. Ans, B., Rousset, S.: Neural networks with a self-refreshing memory: knowledge transfer in sequential learning tasks without catastrophic forgetting. Connect. Sci. **12**(1), 1–19 (2000)
2. Chen, T., Zhang, Z., Liu, S., Chang, S., Wang, Z.: Long live the lottery: the existence of winning tickets in lifelong learning. In: International Conference on Learning Representations (2020)
3. Chen, Z., Liu, B.: Lifelong machine learning. Synth. Lect. Artif. Intell. Mach. Learn. **12**(3), 1–207 (2018)
4. Cover, T., Thomas, J.A.: Elements of Information Theory, 1st edn. Wiley, Chichester (1991)
5. Csiszár, I., Shields, P.C.: Information theory and statistics: a tutorial. J. R. Stat. Soc. Ser. B (Methodol.) **1**, 417–528 (2004)
6. Doan, T., Bennani, M.A., Mazoure, B., Rabusseau, G., Alquier, P.: A theoretical analysis of catastrophic forgetting through the NTK overlap matrix. In: International Conference on Artificial Intelligence and Statistics, pp. 1072–1080. PMLR (2021)
7. Finn, C., Abbeel, P., Levine, S.: Model-agnostic meta-learning for fast adaptation of deep networks. In: International Conference on Machine Learning, pp. 1126–1135. PMLR (2017)
8. Golkar, S., Kagan, M., Cho, K.: Continual learning via neural pruning. arXiv preprint arXiv:1903.04476 (2019)
9. Goodfellow, I.J., Mirza, M., Xiao, D., Courville, A., Bengio, Y.: An empirical investigation of catastrophic forgetting in gradient-based neural networks. arXiv preprint arXiv:1312.6211 (2013)
10. Jung, S., Ahn, H., Cha, S., Moon, T.: Continual learning with node-importance based adaptive group sparse regularization. Adv. Neural Inf. Process. Syst. **33**, 3647–3658 (2020)
11. Ke, Z., Liu, B., Huang, X.: Continual learning of a mixed sequence of similar and dissimilar tasks. Adv. Neural Inf. Process. Syst. **33**, 18493–18504 (2020)
12. Keskar, N.S., Mudigere, D., Nocedal, J., Smelyanskiy, M., Tang, P.T.P.: On large-batch training for deep learning: generalization gap and sharp minima. arXiv preprint arXiv:1609.04836 (2016)
13. Kirkpatrick, J., et al.: Overcoming catastrophic forgetting in neural networks. Proc. Natl. Acad. Sci. **114**(13), 3521–3526 (2017)
14. Lee, J., Park, S., Mo, S., Ahn, S., Shin, J.: Layer-adaptive sparsity for the magnitude-based pruning. In: International Conference on Learning Representations (2020)
15. Li, Z., Hoiem, D.: Learning without forgetting. IEEE Trans. Pattern Anal. Mach. Intell. **40**(12), 2935–2947 (2017)

16. Mallya, A., Lazebnik, S.: PackNet: adding multiple tasks to a single network by iterative pruning. In: Proceedings of the IEEE Conference on Computer Vision and Pattern Recognition, pp. 7765–7773 (2018)
17. McCloskey, M., Cohen, N.J.: Catastrophic interference in connectionist networks: the sequential learning problem. In: Psychology of Learning and Motivation, vol. 24, pp. 109–165. Elsevier (1989)
18. Mirzadeh, S.I., Farajtabar, M., Pascanu, R., Ghasemzadeh, H.: Understanding the role of training regimes in continual learning. Adv. Neural Inf. Process. Syst. **33**, 7308–7320 (2020)
19. Nguyen, C.V., Achille, A., Lam, M., Hassner, T., Mahadevan, V., Soatto, S.: Toward understanding catastrophic forgetting in continual learning. arXiv preprint arXiv:1908.01091 (2019)
20. Parisi, G.I., Kemker, R., Part, J.L., Kanan, C., Wermter, S.: Continual lifelong learning with neural networks: a review. Neural Netw. **113**, 54–71 (2019)
21. Raghavan, K., Balaprakash, P.: Formalizing the generalization-forgetting trade-off in continual learning. In: Advances in Neural Information Processing Systems 34 (2021)
22. Ramasesh, V.V., Dyer, E., Raghu, M.: Anatomy of catastrophic forgetting: hidden representations and task semantics. In: International Conference on Learning Representations (2020)
23. Rusu, A.A., et al.: Progressive neural networks. arXiv preprint arXiv:1606.04671 (2016)
24. Saha, G., Garg, I., Ankit, A., Roy, K.: Space: structured compression and sharing of representational space for continual learning. IEEE Access **9**, 150480–150494 (2021)
25. Shin, H., Lee, J.K., Kim, J., Kim, J.: Continual learning with deep generative replay. In: Advances in Neural Information Processing Systems 30 (2017)
26. Sokar, G., Mocanu, D.C., Pechenizkiy, M.: SpaceNet: make free space for continual learning. Neurocomputing **439**, 1–11 (2021)
27. Vinyals, O., Blundell, C., Lillicrap, T., Wierstra, D., et al.: Matching networks for one shot learning. In: Advances in Neural Information Processing Systems 29 (2016)
28. Wang, Z., Jian, T., Chowdhury, K., Wang, Y., Dy, J., Ioannidis, S.: Learn-prune-share for lifelong learning. In: 2020 IEEE International Conference on Data Mining (ICDM), pp. 641–650. IEEE (2020)
29. Wu, Y., et al.: Incremental classifier learning with generative adversarial networks. arXiv preprint arXiv:1802.00853 (2018)
30. Yin, D., Farajtabar, M., Li, A., Levine, N., Mott, A.: Optimization and generalization of regularization-based continual learning: a loss approximation viewpoint. arXiv preprint arXiv:2006.10974 (2020)
31. Yoon, J., Yang, E., Lee, J., Hwang, S.J.: Lifelong learning with dynamically expandable networks. In: International Conference on Learning Representations (2018)
32. Zenke, F., Poole, B., Ganguli, S.: Continual learning through synaptic intelligence. In: International Conference on Machine Learning, pp. 3987–3995. PMLR (2017)

Exploring Lottery Ticket Hypothesis in Spiking Neural Networks

Youngeun Kim$^{(\boxtimes)}$ ⓘD, Yuhang Li ⓘD, Hyoungseob Park ⓘD,
Yeshwanth Venkatesha, Ruokai Yin ⓘD, and Priyadarshini Panda ⓘD

Department of Electrical Engineering, Yale University, New Haven, CT, USA
{youngeun.kim,yuhang.li,hyoungseob.park,yeshwanth.venkatesha,
ruokai.yin,priya.panda}@yale.edu

Abstract. Spiking Neural Networks (SNNs) have recently emerged as a new generation of low-power deep neural networks, which is suitable to be implemented on low-power mobile/edge devices. As such devices have limited memory storage, neural pruning on SNNs has been widely explored in recent years. Most existing SNN pruning works focus on shallow SNNs (2–6 layers), however, deeper SNNs ($\geq$16 layers) are proposed by state-of-the-art SNN works, which is difficult to be compatible with the current SNN pruning work. To scale up a pruning technique towards deep SNNs, we investigate Lottery Ticket Hypothesis (LTH) which states that dense networks contain smaller subnetworks (*i.e.*, winning tickets) that achieve comparable performance to the dense networks. Our studies on LTH reveal that the winning tickets consistently exist in deep SNNs across various datasets and architectures, providing up to 97% sparsity without huge performance degradation. However, the iterative searching process of LTH brings a huge training computational cost when combined with the multiple timesteps of SNNs. To alleviate such heavy searching cost, we propose Early-Time (ET) ticket where we find the important weight connectivity from a smaller number of timesteps. The proposed ET ticket can be seamlessly combined with a common pruning techniques for finding winning tickets, such as Iterative Magnitude Pruning (IMP) and Early-Bird (EB) tickets. Our experiment results show that the proposed ET ticket reduces search time by up to 38% compared to IMP or EB methods. Code is available at Github.

Keywords: Spiking neural networks · Neural network pruning · Lottery ticket hypothesis · Neuromorphic computing

1 Introduction

Spiking Neural Networks (SNNs) [11,20,34,37,63,74–76] have gained significant attention as promising low-power alternative to Artificial Neural Networks

Supplementary Information The online version contains supplementary material available at https://doi.org/10.1007/978-3-031-19775-8_7.

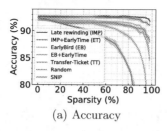

(a) Accuracy

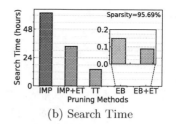

(b) Search Time

Fig. 1. Accuracy and Search time comparison of various pruning methods on SNNs including Iterative Magnitude Pruning (IMP) [21], Early-Bird (EB) ticket [80], Early-Time (ET) ticket (ours), Transferred winning ticket from ANN (TT), Random pruning, and SNIP [41]. We use VGG16 on the CIFAR10 dataset and show mean/standard deviation from 5 random runs

(ANNs). Inspired by the biological neuron, SNNs process visual information through discrete spikes over multiple timesteps. This event-driven behavior of SNNs brings huge energy-efficiency, therefore they are suitable to be implemented on low-power neuromorphic chips [1,13,23,57] which compute spikes in an asynchronous manner. However, as such devices have limited memory storage, neural pruning can be one of the essential techniques by reducing memory usage for weight parameters, thus promoting the practical deployment.

Accordingly, researchers have made certain progress on the pruning technique for SNNs. Neftci et al. [55] and Rathi et al. [60] prune weight connections of SNNs using a predefined threshold value. Guo et al. [25] propose an unsupervised online adaptive weight pruning algorithm that dynamically removes non-critical weights over time. Moreover, Shi et al. [65] present a soft-pruning method where both weight connections and a pruning mask are trained during training. Recently, Deng et al. [14] adapt ADMM optimization tool with sparsity regularization to compress SNNs. Chen et al. [9] propose a gradient-based rewiring method for pruning, where weight values and connections are jointly optimized. However, although existing SNN pruning works significantly increase weight sparsity, they focus on a shallow architecture such as 2-layer MLP [25,55,60] or 6–7 convolutional layers [9,14]. Such pruning techniques are difficult to scale up to the recent state-of-the-art deep SNN architectures where the number of parameters and network depth is scaled up [20,43,44,61,83].

In this paper, we explore sparse deep SNNs based on the recently proposed Lottery Ticket Hypothesis (LTH) [21]. They assert that an over-parameterized neural network contains sparse sub-networks that achieve a similar or even better accuracy than the original dense networks. The discovered sub-networks and their corresponding initialization parameters are referred as *winning tickets*. Based on LTH, a line of works successfully have shown the existence of winning tickets across various tasks such as standard recognition task [22,24,80], reinforcement learning [69,81], natural language processing [4,7,53], and generative model [31]. Along the same line, our primary research objective is to investigate the existence of winning tickets in deep SNNs which has a different type of neuronal dynamics from the common ANNs.

Furthermore, applying LTH on SNNs poses a practical challenge. In general, finding winning tickets requires **Iterative Magnitude Pruning (IMP)** where a network becomes sparse by repetitive initialization-training-pruning operations [21]. Such iterative training process goes much slower with SNNs where multiple feedforward steps (*i.e.*, timesteps) are required. To make LTH with SNNs more practical, we explore several techniques for reducing search costs. We first investigate **Early-Bird (EB) ticket** phenomenon [80] that states the sub-networks can be discovered in the early-training phase. We find that SNNs contain EB tickets across various architectures and datasets. Moreover, SNNs convey information through multiple timesteps, which provides a new dimension for computational cost reduction. Focusing on such temporal property, we propose **Early-Time (ET) ticket** phenomenon: winning tickets can be drawn from the network trained from a smaller number of timesteps. Thus, during the search process, SNNs use a smaller number of timesteps, which can significantly reduce search costs. As ET ticket is a temporal-crafted method, our proposed ET ticket can be combined with IMP [21] and EB tickets [80]. Furthermore, we also explore whether **ANN winning ticket can be transferred to SNN** since search cost at ANN is much cheaper than SNN. Finally, we examine **pruning at initialization** method on SNNs, *i.e.*, SNIP [41], which finds the winning tickets from backward gradients at initialization. In Fig. 1, we compare the accuracy and search time of the above-mentioned pruning methods.

In summary, we explore LTH for SNNs by conducting extensive experiments on two representative deep architectures, *i.e.*, VGG16 [67] and ResNet19 [28], on four public datasets including SVHN [56], Fashion-MNIST [77], CIFAR10 [36] and CIFAR100 [36]. Our key observations are as follows:

- We confirm that Lottery Ticket Hypothesis is valid for SNNs.
- We found that IMP [21] discovers winning tickets with up to 97% sparsity. However, IMP requires over 50 h on GPU to find sparse ($\geq$ 95%) SNNs.
- EB tickets [80] discover sparse SNNs in 1 h on GPUs, which reduces search cost significantly compared to IMP. Unfortunately, they fail to detect winning tickets over 90% sparsity.
- Applying ET tickets to both IMP and EB significantly reduces search time by up to 41% while showing $\leq$1% accuracy drop at less than 95% sparsity.
- Winning ticket obtained from ANN can be transferable at less than 90% sparsity. However, huge accuracy drop is incurred at high sparsity levels ($\geq$ 95%), especially for complex datasets such as CIFAR100.
- Pruning at initialization method [41] fails to discover winning tickets in SNNs owing to the non-differentiability of spiking neurons.

2   Related Work

2.1   Spiking Neural Networks

Spiking Neural Networks (SNNs) transfer binary and asynchronous information through networks in a low-power manner [11,12,43,52,63,64,68,78]. The major difference of SNNs from standard ANNs is using a Leak-Integrate-and-Fire (LIF) neuron [30] as a non-linear activation. The LIF neuron accumulates incoming spikes in membrane potential and generates an output spike when the neuron has a higher membrane potential than a firing threshold. Such integrate-and-fire behavior brings a non-differentiable input-output transfer function where standard backpropagation is difficult to be applied [54]. Recent SNN works circumvent non-differentiable backpropagation problem by defining a surrogate function for LIF neurons when calculating backward gradients [32,39,40,45,54,66,72-74]. Our work is also based on a gradient backpropagation method with a surrogate function (details are provided in Supplementary A). The gradient backpropagation methods enable SNNs to have deeper architectures. For example, adding Batch Normalization (BN) [29] to SNNs [33,38,83] improves the accuracy with deeper architectures such as VGG16 and ResNet19. Also, Fang et al. [20] revisit deep residual connection for SNNs, showing higher performance can be achieved by adding more layers. Although the recent state-of-the-art SNN architecture goes deeper [15,20,83], pruning for such networks has not been explored. We assert that showing the existence of winning tickets in deep SNNs brings a practical advantage to resource-constrained neuromorphic chips and edge devices.

2.2   Lottery Ticket Hypothesis

Pruning has been actively explored in recent decades, which compresses a huge model size of the deep neural networks while maintaining its original performance [26,27,42,47,71]. In the same line of thought, Frankle & Carbin [21] present Lottery Ticket Hypothesis (LTH) which states that an over-parameterized neural network contains sparse sub-networks with similar or even better accuracy than the original dense networks. They search winning tickets by iterative magnitude pruning (IMP). Although IMP methods [2,5,18,46,84] provide higher performance compared to existing pruning methods, such iterative training-pruning-retraining operations require a huge training cost. To address this, a line of work [7,16,50,51] discovers the existence of transferable winning tickets from the source dataset and successfully transfers it to the target dataset, thus eliminating search cost. Further, You et al. [80] introduce early bird ticket hypothesis where they conjecture winning tickets can be achieved in the early training phase, reducing the cost for training till convergence. Recently, Zhang et al. [82] discover winning tickets with a carefully selected subset of training data, called pruning-aware critical set. To completely eliminate training costs, several works [41,70] present searching algorithms from initialized networks, which finds winning tickets without training. Unfortunately, such techniques do not show comparable performance with the original IMP methods, thus mainstream LTH leverages IMP as a pruning scheme [24,51,82]. Based on IMP technique, researchers found the existence of LTH in various applications including visual recognition tasks [24], natural language processing [4,7,53], reinforcement learning [69,81], generative model [31], low-cost neural network ensembling [46], and improving robustness [8]. Although LTH has been actively explored in ANN domain, LTH for SNNs is rarely studied. It is worth mentioning that Martinelli

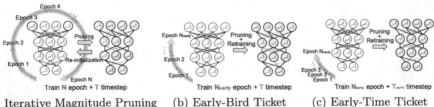

(a) Iterative Magnitude Pruning   (b) Early-Bird Ticket   (c) Early-Time Ticket

Fig. 2. Illustration of the concept of Iterative Magnitude Pruning (IMP), Early-Bird (EB) ticket, and the proposed Early-Time (ET) ticket applied to EB ticket. Our ET ticket reduces search cost for winning tickets by using a smaller number of timesteps during the search process. Note, ET can be applied to both IMP and EB, here we only illustrate ET with EB

et al. [49] apply LTH to two-layer SNN on voice activity detection task. Different from the previous work, our work shows the existence of winning tickets in much deeper networks such as VGG16 and ResNet19, which shows state-of-the-art performance on the image recognition task. We also explore Early-Bird ticket [80], SNIP [41], transferability of winning tickets from ANN, and propose a new concept of winning tickets in the temporal dimension.

3   Drawing Winning Tickets from SNN

In this section, we present the details of pruning methods based on LTH, explored in our experiments. We first introduce the LTH [21] and Early-Bird (EB) Ticket [80]. Then, we propose Early-Time (ET) Tickets where we reduce search cost in the temporal dimension of SNNs. We illustrate the overall search process of each method in Fig. 2.

3.1   Lottery Ticket Hypothesis

In LTH [21], winning tickets are discovered by iterative magnitude pruning (IMP). The whole pruning process of LTH goes through K iterations for the target pruning ratio $p^K\%$. Consider a randomly initialized dense network $f(x; \theta)$ where $\theta \in \mathbb{R}^n$ is the network parameter weights. For the first iteration, initialized network $f(x; \theta)$ is trained till convergence, then mask $m_1 \in \{0,1\}^n$ is generated by removing $p\%$ lowest absolute value weights parameters. Given a pruning mask m_1, we can define subnetworks $f(x; \theta \odot m_1)$ by removing some connections. For the next iteration, we reinitialize the network with $\theta \odot m_1$, and prune p% weights when the network is trained to convergence. This pruning process is repeated for K iterations. In our experiments, we set p and K to 25% and 15, respectively. Also, Frankle *et al.* [22] present *Late Rewinding*, which rewinds the network to the weights at epoch i rather than initialization. This enables IMP to discover the winning ticket with less performance drop in a high sparsity regime by providing a more stable starting point. We found that *Late*

Rewinding shows better performance than the original IMP in deep SNNs (see Supplementary B). Throughout our paper, we apply *Late Rewinding* to IMP for experiments where we rewind the network to epoch 20.

3.2 Early-Bird Tickets

Using IMP for finding lottery ticket incurs huge computational costs. To address this, You *et al.* [80] propose an efficient pruning method, called Early-bird (EB) tickets, where they show winning tickets can be discovered at an early training epoch. Specifically, at searching iteration k, they obtain a mask m_k and measure the mask difference between the current mask m_k and the previous masks within time window q, *i.e.*, $m_{k-1}, m_{k-2}, ..., m_{k-q}$. If the maximum difference is less than hyperparamter τ, they stop training and use m_k as an EB ticket. As searching winning tickets in SNN takes a longer time than ANN, we explore the existence of EB tickets in SNNs. In our experiments, we set q and τ to 5 and 0.02, respectively. Although the original EB tickets use channel pruning, we use unstructured weight pruning for finding EB tickets in order to achieve a similar sparsity level with other pruning methods used in our experiments.

3.3 Early-Time Tickets

Even though EB tickets significantly reduce searching time for winning tickets. For SNNs, one image is passed to a network through multiple timesteps, which provides a new dimension for computational cost reduction. We ask: *can we find the important weight connectivity for the SNN trained with timestep T from the SNN trained with shorter timestep $T' < T$?*

Preliminary Experiments. To answer this question, we conduct experiments on two representative deep architectures (VGG16 and ResNet19) on two datasets (CIFAR10 and CIFAR100). Our experiment protocol is shown in Fig. 3 (left panel). We first train the networks with timestep T_{pre} till convergence. After that, we prune $p\%$ of the low-magnitude weights and re-initialize the networks. Finally, we re-train the pruned networks with longer timestep $T_{post} > T_{pre}$, and measure the test accuracy. Thus, this experiment shows the performance of SNNs where the structure is obtained from the lower timestep. In our experiments, we set $T_{pre} = \{2, 3, 4, 5\}$ and $T_{post} = 5$. Surprisingly, the connections founded from $T_{pre} \geq 3$ can bring similar and even better accuracy compared to the unpruned baseline, as shown in Fig. 3. Note, in the preliminary experiments, we use a common post-training pruning based on the magnitude of weights [27]. Thus, we can extrapolate the existence of early-time winning tickets to a more sophisticated pruning method such as IMP [21], which generally shows better performance than post-training pruning. We call such a winning ticket as *Early-Time tickets*; a winning ticket drawn with a trained network from a smaller number of timesteps T_{early}, which shows matching performance with a winning ticket from the original timesteps T.

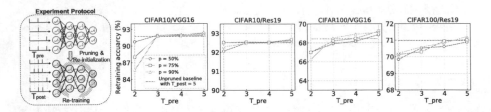

Fig. 3. Preliminary experiments for Early-Time ticket. We conduct experiments on VGG16/ResNet19 on CIFAR10/CIFAR100. We report retraining accuracy ($T_{post} = 5$) with respect to the timestep (T_{pre}) for searching the important connectivity in SNNs

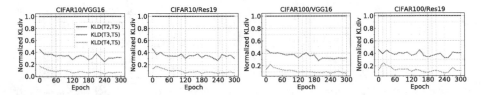

Fig. 4. Kullback-Leibler (KL) divergence between the class prediction distribution from different timesteps. The network is trained with the original timestep $T = 5$. We measure KL divergence between the predicted class probabilities from different timesteps. We use the training set for calculating KL divergence

Proposed Method. Then, *how to practically select a timestep for finding Early-Time tickets?* The main idea is to measure the similarity between class predictions between the original timestep T and a smaller number of timesteps, and select a minimal timestep that shows a similar representation with the target timestep. Specifically, let P_T be the class probability from the last layer of networks by accumulating output values across T timesteps [39]. In this case, our search space can be $S = \{2, 3, ..., T - 1\}$. Note, timestep 1 is not considered since it cannot use the temporal behavior of LIF neurons. To measure the statistical distance between class predictions P_T and $P_{T'}$, we use Kullback-Leibler (KL) divergence:

$$D_{KL}(P_{T'}||P_T) = \sum_x P_{T'}(x)\ln\frac{P_{T'}(x)}{P_T(x)}. \qquad (1)$$

The value of KL divergence goes smaller when the timestep T' is closer to the original timestep T, *i.e.*, $D_{KL}(P_{T-1}||P_T) \leq D_{KL}(P_{T-2}||P_T) \leq ... \leq D_{KL}(P_2||P_T)$. Note, for any $t \in \{1, .., T\}$, we compute P_t by accumulating output layer's activation from 1 to t timesteps. Therefore, due to the accumulation, if the timestep difference between timestep t and t' becomes smaller, the KL divergence $D_{KL}(P_{t'}||P_t)$ becomes lower. After that, we rescale all KL divergence values to $[0, 1]$ by dividing them with $D_{KL}(P_2||P_T)$. In Fig. 4, we illustrate normalized KL divergence of VGG16 and ResNet19 on CIFAR10 and CIFAR100,

---

Algorithm 1. Early-Time (ET) ticket

Input: Training data D; Winning ticket searching method $F(\cdot)$ – IMP or EB ticket; Original timestep T; Threshold λ
Output: Pruned SNN_{pruned}

1: Training SNN with N epochs for stability
2: Memory = [ ]
3: **for** $t \leftarrow 2$ to T **do**
4:     $P_t \leftarrow SNN(t, D)$ ▷ Storing class prediction from each timestep
5:     Memory.append(P_t)
6: **end for**
7: $[D_{KL}(P_{T-1}||P_T), ..., D_{KL}(P_2||P_T)] \leftarrow$ Memory   ▷ Computing KL div.
8: **for** $t \leftarrow 2$ to $T - 1$ **do**
9:     $\hat{D}_{KL}(P_t||P_T) \leftarrow \frac{D_{KL}(P_t||P_T)}{D_{KL}(P_2||P_T)}$ ▷ Normalization
10:     **if** $\hat{D}_{KL}(P_t||P_T) < \lambda$ **then**   ▷ Select timestep when KL div. is less than λ
11:         $T_{early} = t$
12:         **break**
13:     **end if**
14: **end for**
15: $WinningTicket \leftarrow F(SNN, T_{early})$ ▷ Finding winning ticket with T_{early}
16: $SNN_{pruned} \leftarrow F(WinningTicket, T)$ ▷ Train with the original timestep T
17: **return** SNN_{pruned}

---

where we found two observations: (1) The KL divergence with $T = 2$ has a relatively higher value than other timesteps. If the difference in the class probability is large (*i.e.*, larger KL divergence), weight connections are likely to be updated in a different direction. This supports our previous observation that important connectivity founded by $T = 2$ shows huge performance drop at $T = 5$ (Fig. 3). Therefore, we search T_{early} that has KL divergence less than λ while minimizing the number of timesteps. The λ is a hyperparameter for the determination of T_{early} in the winning ticket search process (Algorithm 1, line 15). A higher λ leads to smaller T_{early} and vice versa (we visualize the change of T_{early} versus different λ values in Fig. 7(b)). For example, if we set $\lambda = 0.6$, timestep 3 is used for finding early-time tickets. In our experiments, we found that a similar value of threshold λ can be applied across various datasets. (2) The normalized KL divergence shows fairly consistent values across training epochs. Thus, we can find the suitable timestep T_{early} for obtaining early-time tickets at the very beginning of the training phase.

The early-time ticket approach can be seamlessly applied to both IMP and Early-bird ticket methods. Algorithm 1 illustrates the overall process of Early-Time ticket. For stability with respect to random initialization, we start to search T_{early} after $N = 2$ epoch of training (Line 1). We show the variation of KL divergence results across training epochs in Supplementary C. We first find the T_{early} from KL Divergence of difference timesteps (Line 2–13). After that, we discover the winning ticket using either IMP or EB ticket with T_{early} (Line 14). Finally, the winning ticket is trained with the original timestep T (Line 15).

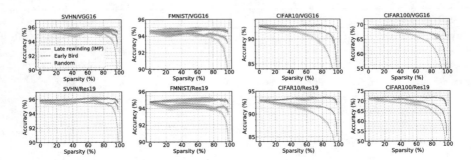

Fig. 5. The accuracy of winning ticket with respect to sparsity level. We report mean and standard deviation from 5 random runs

Table 1. Effect of the proposed Early-Time ticket. We compare the accuracy and search time of Iterative Magnitude Pruning (IMP), Early-Bird (EB) ticket, Early-Time (ET) ticket on four sparsity levels. We show search speed gain and accuracy change from applying ET

Setting	Method	Accuracy				Winning ticket search time (hours)			
		p = 68.30%	p = 89.91%	p = 95.69%	p = 98.13%	p = 68.30%	p = 89.91%	p = 95.69%	p = 98.13%
CIFAR10	IMP	92.66	92.54	92.38	91.81	14.97	29.86	40.84	51.99
VGG16	IMP + ET	92.49	92.09	91.54	91.10	11.19	22.00	30.11	38.26
	Δ Acc./Speed Gain	−0.17	−0.45	−0.84	−0.71	×1.34	×1.35	×1.35	×1.35
	EB	91.74	91.05	89.55	84.64	1.96	0.74	0.11	0.09
	EB + ET	91.27	90.66	88.95	84.86	1.44	0.55	0.07	0.06
	Δ Acc./Speed Gain	−0.47	−0.39	−0.60	+0.22	×1.36	×1.34	×1.18	×1.12
CIFAR10	IMP	93.47	93.49	93.22	92.43	21.01	42.20	58.91	73.54
Res19	IMP + ET	93.10	92.72	92.68	91.36	13.35	26.62	37.27	46.40
	Δ Acc./Speed Gain	−0.37	−0.77	−0.54	−1.07	×1.57	×1.59	×1.58	×1.58
	EB	91.00	90.84	89.90	85.22	2.49	0.87	0.24	0.08
	EB + ET	90.83	91.21	89.65	85.45	1.63	0.58	1.70	0.07
	Δ Acc./Speed Gain	−0.17	+0.37	−0.50	−1.09	×1.52	×1.49	×1.38	×1.16
CIFAR100	IMP	69.08	68.90	68.00	66.02	15.02	29.99	41.03	52.05
VGG16	IMP + ET	68.27	67.99	66.51	64.41	11.24	22.42	30.53	38.32
	Δ Acc./Speed Gain	−0.81	−0.91	−1.49	−1.61	×1.33	×1.34	×1.34	×1.36
	EB	67.35	65.82	61.90	52.11	2.27	0.99	0.32	0.06
	EB + ET	67.26	64.18	61.81	52.77	1.66	0.73	0.24	0.05
	Δ Acc./Speed Gain	−0.09	−1.64	−0.09	+0.66	×1.36	×1.35	×1.31	×1.12
CIFAR100	IMP	71.64	71.38	70.45	67.35	21.21	42.29	59.17	73.52
ResNet19	IMP + ET	71.06	70.45	69.23	65.49	13.56	27.05	37.88	46.65
	Δ Acc./Speed Gain	−0.58	−0.93	−1.22	−1.86	×1.56	×1.56	×1.56	×1.57
	EB	69.41	65.87	62.18	52.92	3.08	1.71	0.43	0.09
	EB + ET	68.98	65.76	62.20	51.50	2.00	1.12	0.29	0.07
	Δ Acc./Speed Gain	−0.43	−0.12	+0.02	−1.42	×1.53	×1.52	×1.45	×1.16

4   Experimental Results

4.1   Implementation Details

We comprehensively evaluate various pruning methods on four public datasets: SVHN [56], Fashion-MNIST [77], CIFAR10 [36] and CIFAR100 [36]. In our work, we focus on pruning deep SNNs, therefore we evaluate two representative architectures; VGG16 [67] and ResNet19 [28]. Our implementation is based on PyTorch [59]. We train the network using SGD optimizer with momentum

0.9 and weight decay 5e−4. Our image augmentation process and loss function follow the previous SNN work [15]. We set the training batch size to 128. The base learning rate is set to 0.3 for all datasets with cosine learning rate scheduling [48]. Here, we set the total number of epochs to 150, 150, 300, 300 for SVHN, F-MNIST, CIFAR10, CIFAR100, respectively. We set the default timesteps T to 5 across all experiments. Also, we use $\lambda = 0.6$ for finding Early-Time tickets. Experiments were conducted on an RTX 2080Ti GPU with PyTorch implementation. We use SpikingJelly [19] package for implementation.

4.2 Winning Tickets in SNNs

Performance of IMP and EB Ticket. In Fig. 5, we show the performance of the winning tickets from IMP and EB. The performance of random pruning is also provided as a reference. Both IMP and EB can successfully find the winning ticket, which shows better performance than random pruning. Especially, IMP finds winning tickets over ∼97% sparsity across all configurations. Also, we observe that the winning ticket sparsity is affected by dataset complexity. EB ticket can find winning tickets (>95% sparsity) for relatively simple datasets such as SVHN and Fashion-MINST. However, they are limited to discovering the winning ticket having ≤95% sparsity on CIFAR10 and CIFAR100. We further provide experiments on ResNet34-TinyImageNet and AlexNet-CIFAR10 in Supplementary G.

Effect of ET Ticket. In Table 1, we report the change in accuracy and search speed gain from applying ET to IMP and EB, on CIFAR10 and CIFAR100 datasets (SVHN and Fashion-MNIST results are provided in Supplementary E). Although IMP achieves highest accuracy across all sparsity levels, they require 26–73 h to search winning tickets with 98.13% sparsity. By applying ET to IMP, the search speed increases up to ×1.59 without a huge accuracy drop (of course, there is an accuracy-computational cost trade-off because the ET winning ticket cannot exactly match with the IMP winning ticket). Compared to IMP, EB ticket provides significantly less search cost for searching winning tickets. Combining ET with EB ticket brings faster search speed, even finding a winning ticket in one hour (on GPU). At high sparsity levels of EB ($p = 98.13\%$), search speed gain from ET is upto ×1.16. Also, applying ET on ResNet19 brings a better speed gain than VGG16 since ResNet19 requires a larger computational graph from multiple timestep operations (details are in Supplementary D). Overall, the results support our hypothesis that important weight connectivity of the SNN can be discovered from shorter timesteps.

Observations from Pruning Techniques. We use *Late Rewinding* [22] (IMP) for obtaining stable performance at high sparsity regime (refer Sect. 3.1). To analyze the effect of the rewinding epoch, we change the rewinding epoch and report the accuracy on four sparsity levels in Fig. 6(a). We observe that the rewinding epoch does not cause a huge accuracy change with sparsity ≤95.69%. However, a high sparsity level (98.13%) shows non-trivial accuracy drop ∼1.5% at epoch 260, which requires careful rewinding epoch selection. Frankle *et al.* [22] also show that using the same pruning percentage for both shallow and deep layers (*i.e.*,

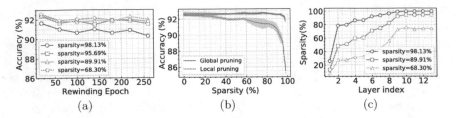

Fig. 6. Observations from Iterative Magnitude Pruning (IMP). (a) The performance change with respect to the rewinding epoch. (b) The performance of global pruning and local pruning. (c) Layer-wise sparsity across different sparsity levels with global pruning. We use VGG16 on the CIFAR10 dataset for experiments

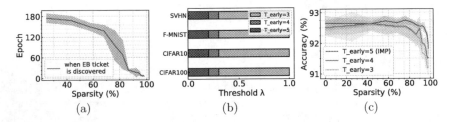

Fig. 7. Observations from Early-Bird (EB) ticket and Early-Time (ET) ticket. (a) Epoch when the EB ticket is discovered. (b) The change of T_{early} with respect to the threshold λ for KL divergence. (c) The performance of winning tickets from different T_{early}. We use VGG16 on the CIFAR10 dataset and show standard deviation from 5 random runs

local pruning) degrades the accuracy. Instead of local pruning, they apply different pruning percentages for each layer (*i.e.*, global pruning). We compare global pruning and local pruning in Fig. 6(b). In SNN, global pruning achieves better performance than local pruning, especially for high sparsity levels. Figure 6(c) illustrates layer-wise sparsity obtained from global pruning. The results show that deep layers have higher sparsity compared to shallow layers.

We also visualize the epoch when an EB ticket is discovered with respect to sparsity levels, in Fig. 7(a). The EB ticket obtains a pruning mask based on the mask difference between the current mask and the masks from the previous epochs. Here, we observe that a highly sparse mask is discovered earlier than a lower sparsity mask according to EB ticket's mask detection algorithm (refer Fig. 3 of [80]). Furthermore, we conduct a hyperparameter analysis of the proposed ET ticket. In Fig. 7(b), we show the change of T_{early} with respect to the threshold λ used for selection (Algorithm 1). We search λ with intervals of 0.1 from 0 to 1. Low λ value indicates that we select T_{early} that is similar to the original timestep, and brings less efficiency gain. Interestingly, the trend is similar across different datasets, which indicates our KL divergence works as a consistent metric. Figure 7(c) shows the accuracy of sparse SNNs from three different T_{early} (Note, $T_{early} = 5$ is the original IMP). The results show that smaller T_{early} also captures important connections in SNNs.

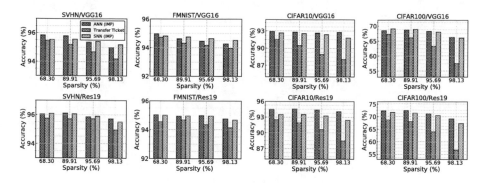

Fig. 8. Transferability study of ANN winning tickets on SNNs

4.3 Transferred Winning Tickets from ANN

The transferability of winning tickets has been actively explored in order to eliminate search costs. A line of work [7, 16, 50, 51] discover the existence of transferable winning tickets from the source dataset and successfully transfers it to the target dataset. With a different perspective from the prior works which focus on cross-dataset configuration, we discover the transferable winning ticket between ANN and SNN where the activation function is different. In Fig. 8, we illustrate the accuracy of IMP on ANN, IMP on SNN, Transferred Ticket, across four sparsity levels (68.30%, 89.91%, 95.69%, 98.13%). Specifically, Transferred Ticket (*i.e.*, initialized weight parameters and pruning mask) is discovered by IMP on ANN, and trained on SNN framework where we change ReLU neuron to LIF neuron. For relatively simple datasets such as SVHN and F-MNIST, Transferred Ticket shows less than 2% accuracy drop even at 98.13% sparsity. However, for CIFAR10 and CIFAR100, Transferred Ticket fails to detect a winning ticket and shows a huge performance drop. The results show that ANN and SNN share common knowledge, but are not exactly the same, which can be supported by the previous SNN works [17, 35, 62] where a pretrained ANN provides better initialization for SNN. Although Transferred Ticket shows limited performance than IMP, searching Transferred Ticket from ANN requires ∼14 h for 98.13% sparsity, which is ∼ 5× faster than IMP on SNN.

4.4 Finding Winning Tickets from Initialization

A line of work [41, 70] effectively reduces search cost for winning tickets by conducting a search process at initialization. This technique should be explored with SNNs where multiple feedforward steps corresponding to multiple timesteps bring expensive search costs. To show this, we conduct experiments on a representative *pruning at initialization* method, SNIP [41]. SNIP computes the importance of each weight connection from the magnitude of backward gradients at initialization. In Fig. 9, we illustrate the accuracy of ANN and SNN with SNIP on VGG16/CIFAR10 configuration.

Table 2. Performance comparison of IMP [21] with the previous works

Pruning method	Architecture	Dataset	Baseline acc. (%)	ΔAcc (%)	Sparsity (%)
Deng *et al.* [14]	7Conv, 2FC	CIFAR10	89.53	−0.38	50.00
				−2.16	75.00
				−3.85	90.00
Bellec *et al.* [3]	6Conv, 2FC	CIFAR10	92.84	−1.98	94.76
				−2.56	98.05
				−3.53	98.96
Chen *et al.* [9]	6Conv, 2FC	CIFAR10	92.84	−0.30	71.59
				−0.81	94.92
				−1.47	97.65
Chen *et al.* [9]*	ResNet19	CIFAR10	93.22	−0.54	76.90
				−1.31	94.25
				−2.10	97.56
IMP on SNN (ours)	**ResNet19**	**CIFAR10**	**93.22**	**+0.28**	**76.20**
				+0.24	**94.29**
				−0.04	**97.54**
Chen *et al.* [9]*	ResNet19	CIFAR100	71.34	−1.98	77.03
				−3.87	94.92
				−4.03	97.65
IMP on SNN (ours)	**ResNet19**	**CIFAR100**	**71.34**	**+0.11**	**76.20**
				−0.34	**94.29**
				−2.29	**97.54**

* We reimplement ResNet19 experiments.

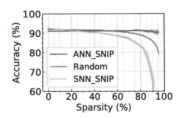

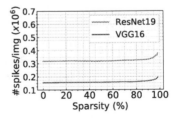

Fig. 9. Performance comparison across pruned ANN and pruned SNN with SNIP

Fig. 10. Number of spikes with respect to sparsity on CIFAR10

Surprisingly, SNN shows huge performance degradation at high sparsity regime (>80%), even worse than random pruning. The results imply that the previous *pruning at initialization* based on the backward gradient (tailored for ANNs) is not compatible with SNNs where the backward gradient is approximated because of the non-differentiability of LIF neuron [54, 74].

4.5   Performance Comparison with Previous Works

In Table 2, we compare the LTH method, especially IMP, with state-of-the-art SNN pruning works [3, 9, 14] in terms of accuracy and achieved sparsity. We show the baseline accuracy of the unpruned model and report the accuracy drop at three different sparsity levels. Note that the previous works use a shallow model with 6–7 conv and 2 FC layers. The ResNet19 architecture achieves higher baseline accuracy compared to the previous shallow architectures. To compare the results on a ResNet19 model, we prune ResNet19 with method proposed by Chen *et al.* [9] using the official code[1] provided by authors. We observe that the previous SNN pruning works fail to achieve matching performance at a high sparsity on deeper SNN architectures. On the other hand, IMP shows less performance drop (in some cases even performance improvement) compared to the previous pruning techniques. Especially, at 97.54% sparsity, [9] shows 2.1% accuracy drop whereas IMP degrades the accuracy by 0.04%. Further, we compare the accuracy on the CIFAR100 dataset to explore the effectiveness of the pruning method with respect to complex datasets. Chen *et al.* 's method fails to discover the winning ticket at sparsity 94.29%, and IMP shows better performance across all sparsity levels. The results imply that the LTH-based method might bring a huge advantage as SNNs are scaled up in the future.

4.6   Observation on the Number of Spikes

In general, the energy consumption of SNNs is proportional to the number of spikes [1, 13, 79] and weight sparsity. In Fig. 10, we measure the number of spikes per image across various sparsity levels of VGG16 and ResNet19 architectures. We use IMP for pruning SNNs. We observe that the sparse SNN maintains a similar number of spikes across all sparsity levels. The results indicate that sparse SNNs obtained with LTH do not bring an additional MAC energy-efficiency gain from spike sparsity. Nonetheless, weight sparsity brings less memory occupation and can alleviate the memory movement overheads [10, 58]. As discussed in [6], with over 98% of the sparse weights, SNNs can reduce the memory movement energy by up to 15×.

5   Conclusion

In this work, we aim to explore lottery ticket hypothesis based pruning for implementing sparse deep SNNs at lower search costs. Such an objective is important as SNNs are a promising candidate for deployment on resource-constrained edge devices. To this end, we apply various techniques including Iterative Magnitude Pruning (IMP), Early-Bird ticket (EB), and the proposed Early-Time ticket (ET). Our key observations are summarized as follows: (1) IMP can achieve higher sparsity levels of deep SNN models compared to previous works on SNN pruning. (2) EB ticket reduces search time significantly, but it cannot achieve a

---

[1] https://github.com/Yanqi-Chen/Gradient-Rewiring

winning ticket over 90% sparsity. (3) Adding ET ticket accelerates search speed for both IMP and EB, by up to ×1.66. We can find a winning ticket in one hour with EB+ET, which can enable practical pruning on edge devices.

Acknowledgment. We would like to thank Anna Hambitzer for her helpful comments. This work was supported in part by C-BRIC, a JUMP center sponsored by DARPA and SRC, Google Research Scholar Award, the National Science Foundation (Grant#1947826), TII (Abu Dhabi) and the DARPA AI Exploration (AIE) program.

References

1. Akopyan, F., et al.: TrueNorth: design and tool flow of a 65 mw 1 million neuron programmable neurosynaptic chip. IEEE Trans. Comput. Aided Des. Integr. Circuits Syst. **34**(10), 1537–1557 (2015)
2. Bai, Y., Wang, H., Tao, Z., Li, K., Fu, Y.: Dual lottery ticket hypothesis. In: International Conference on Learning Representations (2022). https://openreview.net/forum?id=fOsN52jn25l
3. Bellec, G., Salaj, D., Subramoney, A., Legenstein, R., Maass, W.: Long short-term memory and learning-to-learn in networks of spiking neurons. In: Advances in Neural Information Processing Systems 31 (2018)
4. Brix, C., Bahar, P., Ney, H.: Successfully applying the stabilized lottery ticket hypothesis to the transformer architecture. arXiv preprint arXiv:2005.03454 (2020)
5. Burkholz, R., Laha, N., Mukherjee, R., Gotovos, A.: On the existence of universal lottery tickets. arXiv preprint arXiv:2111.11146 (2021)
6. Chen, G.K., Kumar, R., Sumbul, H.E., Knag, P.C., Krishnamurthy, R.K.: A 4096-neuron 1M-synapse 3.8-pJ/SOP spiking neural network with on-chip STDP learning and sparse weights in 10-nm finfet CMOS. IEEE J. Solid-State Circuits **54**(4), 992–1002 (2018)
7. Chen, T., et al.: The lottery ticket hypothesis for pre-trained BERT networks. In: Advances in Neural Information Processing Systems, vol. 33, pp. 15834–15846 (2020)
8. Chen, T., et al.: Sparsity winning twice: better robust generalization from more efficient training. In: International Conference on Learning Representations (2022). https://openreview.net/forum?id=SYuJXrXq8tw
9. Chen, Y., Yu, Z., Fang, W., Huang, T., Tian, Y.: Pruning of deep spiking neural networks through gradient rewiring. arXiv preprint arXiv:2105.04916 (2021)
10. Chen, Y.H., Emer, J., Sze, V.: Eyeriss: a spatial architecture for energy-efficient dataflow for convolutional neural networks. ACM SIGARCH Comput. Archit. News **44**(3), 367–379 (2016)
11. Christensen, D.V., et al.: 2022 roadmap on neuromorphic computing and engineering. Neuromorphic Computing and Engineering (2022)
12. Comsa, I.M., Fischbacher, T., Potempa, K., Gesmundo, A., Versari, L., Alakuijala, J.: Temporal coding in spiking neural networks with alpha synaptic function. In: ICASSP 2020–2020 IEEE International Conference on Acoustics, Speech and Signal Processing (ICASSP), pp. 8529–8533. IEEE (2020)
13. Davies, M., et al.: Loihi: a neuromorphic manycore processor with on-chip learning. IEEE Micro **38**(1), 82–99 (2018)
14. Deng, L., et al.: Comprehensive SNN compression using ADMM optimization and activity regularization. IEEE Trans. Neural Networks and Learn. Syst. (2021)

15. Deng, S., Li, Y., Zhang, S., Gu, S.: Temporal efficient training of spiking neural network via gradient re-weighting. In: International Conference on Learning Representations (2022., https://openreview.net/forum?id=_XNtisL32jv
16. Desai, S., Zhan, H., Aly, A.: Evaluating lottery tickets under distributional shifts. arXiv preprint arXiv:1910.12708 (2019)
17. Ding, J., Yu, Z., Tian, Y., Huang, T.: Optimal ann-snn conversion for fast and accurate inference in deep spiking neural networks. arXiv preprint arXiv:2105.11654 (2021)
18. Ding, S., Chen, T., Wang, Z.: Audio lottery: speech recognition made ultra-lightweight, noise-robust, and transferable. In: International Conference on Learning Representations (2022). https://openreview.net/forum?id=9Nk6AJkVYB
19. Fang, W., et al.: Spikingjelly (2020). https://github.com/fangwei123456/spikingjelly
20. Fang, W., Yu, Z., Chen, Y., Huang, T., Masquelier, T., Tian, Y.: Deep residual learning in spiking neural networks. Advances in Neural Information Processing Systems 34 (2021)
21. Frankle, J., Carbin, M.: The lottery ticket hypothesis: finding sparse, trainable neural networks. arXiv preprint arXiv:1803.03635 (2018)
22. Frankle, J., Dziugaite, G.K., Roy, D.M., Carbin, M.: Stabilizing the lottery ticket hypothesis. arXiv preprint arXiv:1903.01611 (2019)
23. Furber, S.B., Galluppi, F., Temple, S., Plana, L.A.: The spinnaker project. Proc. IEEE **102**(5), 652–665 (2014)
24. Girish, S., Maiya, S.R., Gupta, K., Chen, H., Davis, L.S., Shrivastava, A.: The lottery ticket hypothesis for object recognition. In: Proceedings of the IEEE/CVF Conference on Computer Vision and Pattern Recognition, pp. 762–771 (2021)
25. Guo, W., Fouda, M.E., Yantir, H.E., Eltawil, A.M., Salama, K.N.: Unsupervised adaptive weight pruning for energy-efficient neuromorphic systems. Frontiers in Neuroscience p. 1189 (2020)
26. Han, S., et al.: Dsd: Dense-sparse-dense training for deep neural networks. arXiv preprint arXiv:1607.04381 (2016)
27. Han, S., Pool, J., Tran, J., Dally, W.: Learning both weights and connections for efficient neural network. Advances in neural information processing systems 28 (2015)
28. He, K., Zhang, X., Ren, S., Sun, J.: Deep residual learning for image recognition. In: CVPR, pp. 770–778 (2016)
29. Ioffe, S., Szegedy, C.: Batch normalization: Accelerating deep network training by reducing internal covariate shift. arXiv preprint arXiv:1502.03167 (2015)
30. Izhikevich, E.M.: Simple model of spiking neurons. IEEE Trans. Neural Networks **14**(6), 1569–1572 (2003)
31. Kalibhat, N.M., Balaji, Y., Feizi, S.: Winning lottery tickets in deep generative models. arXiv preprint arXiv:2010.02350 (2020)
32. Kim, Y., Li, Y., Park, H., Venkatesha, Y., Panda, P.: Neural architecture search for spiking neural networks. arXiv preprint arXiv:2201.10355 (2022)
33. Kim, Y., Panda, P.: Revisiting batch normalization for training low-latency deep spiking neural networks from scratch. Frontiers in neuroscience, p. 1638 (2020)
34. Kim, Y., Panda, P.: Visual explanations from spiking neural networks using inter-spike intervals. Sci. Rep. **11**, 19037 (2021). https://doi.org/10.1038/s41598-021-98448-0 (2021)
35. Kim, Y., Venkatesha, Y., Panda, P.: Privatesnn: privacy-preserving spiking neural networks. In: Proceedings of the AAAI Conference on Artificial Intelligence, vol. 36, pp. 1192–1200 (2022)

36. Krizhevsky, A., Hinton, G., et al.: Learning multiple layers of features from tiny images (2009)
37. Kundu, S., Pedram, M., Beerel, P.A.: Hire-snn: harnessing the inherent robustness of energy-efficient deep spiking neural networks by training with crafted input noise. In: Proceedings of the IEEE/CVF International Conference on Computer Vision, pp. 5209–5218 (2021)
38. Ledinauskas, E., Ruseckas, J., Juršėnas, A., Buračas, G.: Training deep spiking neural networks. arXiv preprint arXiv:2006.04436 (2020)
39. Lee, C., Sarwar, S.S., Panda, P., Srinivasan, G., Roy, K.: Enabling spike-based backpropagation for training deep neural network architectures. Front. Neurosci. **14** (2020)
40. Lee, J.H., Delbruck, T., Pfeiffer, M.: Training deep spiking neural networks using backpropagation. Front. Neurosci. **10**, 508 (2016)
41. Lee, N., Ajanthan, T., Torr, P.H.: Snip: Single-shot network pruning based on connection sensitivity. arXiv preprint arXiv:1810.02340 (2018)
42. Li, H., Kadav, A., Durdanovic, I., Samet, H., Graf, H.P.: Pruning filters for efficient convnets. arXiv preprint arXiv:1608.08710 (2016)
43. Li, Y., Deng, S., Dong, X., Gong, R., Gu, S.: A free lunch from ann: towards efficient, accurate spiking neural networks calibration. arXiv preprint arXiv:2106.06984 (2021)
44. Li, Y., Deng, S., Dong, X., Gu, S.: Converting artificial neural networks to spiking neural networks via parameter calibration. arXiv preprint arXiv:2205.10121 (2022)
45. Li, Y., Guo, Y., Zhang, S., Deng, S., Hai, Y., Gu, S.: Differentiable spike: Rethinking gradient-descent for training spiking neural networks. Advances in Neural Information Processing Systems 34 (2021)
46. Liu, S., et al.: Deep ensembling with no overhead for either training or testing: the all-round blessings of dynamic sparsity. arXiv preprint arXiv:2106.14568 (2021)
47. Liu, Z., Sun, M., Zhou, T., Huang, G., Darrell, T.: Rethinking the value of network pruning. arXiv preprint arXiv:1810.05270 (2018)
48. Loshchilov, I., Hutter, F.: Sgdr: Stochastic gradient descent with warm restarts. arXiv preprint arXiv:1608.03983 (2016)
49. Martinelli, F., Dellaferrera, G., Mainar, P., Cernak, M.: Spiking neural networks trained with backpropagation for low power neuromorphic implementation of voice activity detection. In: ICASSP 2020–2020 IEEE International Conference on Acoustics, Speech and Signal Processing (ICASSP), pp. 8544–8548. IEEE (2020)
50. Mehta, R.: Sparse transfer learning via winning lottery tickets. arXiv preprint arXiv:1905.07785 (2019)
51. Morcos, A., Yu, H., Paganini, M., Tian, Y.: One ticket to win them all: generalizing lottery ticket initializations across datasets and optimizers. In: Advances in Neural Information Processing Systems 32 (2019)
52. Mostafa, H.: Supervised learning based on temporal coding in spiking neural networks. IEEE Trans. Neural Networks Learn. Syst. **29**(7), 3227–3235 (2017)
53. Movva, R., Zhao, J.Y.: Dissecting lottery ticket transformers: structural and behavioral study of sparse neural machine translation. arXiv preprint arXiv:2009.13270 (2020)
54. Neftci, E.O., Mostafa, H., Zenke, F.: Surrogate gradient learning in spiking neural networks. IEEE Signal Process. Mag. **36**, 61–63 (2019)
55. Neftci, E.O., Pedroni, B.U., Joshi, S., Al-Shedivat, M., Cauwenberghs, G.: Stochastic synapses enable efficient brain-inspired learning machines. Front. Neurosci. **10**, 241 (2016)

56. Netzer, Y., Wang, T., Coates, A., Bissacco, A., Wu, B., Ng, A.Y.: Reading digits in natural images with unsupervised feature learning (2011)
57. Orchard, G., et al.: Efficient neuromorphic signal processing with loihi 2. In: 2021 IEEE Workshop on Signal Processing Systems (SiPS), pp. 254–259. IEEE (2021)
58. Parashar, A., Rhu, M., Mukkara, A., Puglielli, A., Venkatesan, R., Khailany, B., Emer, J., Keckler, S.W., Dally, W.J.: Scnn: an accelerator for compressed-sparse convolutional neural networks. ACM SIGARCH Computer Architecture News 45(2), 27–40 (2017)
59. Paszke, A., et al.: Automatic differentiation in pytorch. In: NIPS-W (2017)
60. Rathi, N., Panda, P., Roy, K.: Stdp-based pruning of connections and weight quantization in spiking neural networks for energy-efficient recognition. IEEE Trans. Comput. Aided Des. Integr. Circuits Syst. 38(4), 668–677 (2018)
61. Rathi, N., Roy, K.: Diet-SNN: a low-latency spiking neural network with direct input encoding and leakage and threshold optimization. IEEE Trans. Neural Networks Learn Syst. (2021)
62. Rathi, N., Srinivasan, G., Panda, P., Roy, K.: Enabling deep spiking neural networks with hybrid conversion and spike timing dependent backpropagation. arXiv preprint arXiv:2005.01807 (2020)
63. Roy, K., Jaiswal, A., Panda, P.: Towards spike-based machine intelligence with neuromorphic computing. Nature 575(7784), 607–617 (2019)
64. Schuman, C.D., Kulkarni, S.R., Parsa, M., Mitchell, J.P., Kay, B., et al.: Opportunities for neuromorphic computing algorithms and applications. Nature Comput. Sci. 2(1), 10–19 (2022)
65. Shi, Y., Nguyen, L., Oh, S., Liu, X., Kuzum, D.: A soft-pruning method applied during training of spiking neural networks for in-memory computing applications. Front. Neurosci. 13, 405 (2019)
66. Shrestha, S.B., Orchard, G.: Slayer: Spike layer error reassignment in time. arXiv preprint arXiv:1810.08646 (2018)
67. Simonyan, K., Zisserman, A.: Very deep convolutional networks for large-scale image recognition. ICLR (2015)
68. Venkatesha, Y., Kim, Y., Tassiulas, L., Panda, P.: Federated learning with spiking neural networks. arXiv preprint arXiv:2106.06579 (2021)
69. Vischer, M.A., Lange, R.T., Sprekeler, H.: On lottery tickets and minimal task representations in deep reinforcement learning. arXiv preprint arXiv:2105.01648 (2021)
70. Wang, C., Zhang, G., Grosse, R.: Picking winning tickets before training by preserving gradient flow. arXiv preprint arXiv:2002.07376 (2020)
71. Wen, W., Wu, C., Wang, Y., Chen, Y., Li, H.: Learning structured sparsity in deep neural networks. In: Advances in Neural Information Processing Systems 29 (2016)
72. Wu, H., et al.: Training spiking neural networks with accumulated spiking flow. ijo 1(1) (2021)
73. Wu, J., Xu, C., Zhou, D., Li, H., Tan, K.C.: Progressive tandem learning for pattern recognition with deep spiking neural networks. arXiv preprint arXiv:2007.01204 (2020)
74. Wu, Y., Deng, L., Li, G., Zhu, J., Shi, L.: Spatio-temporal backpropagation for training high-performance spiking neural networks. Front. Neurosci. 12, 331 (2018)
75. Wu, Y., Deng, L., Li, G., Zhu, J., Xie, Y., Shi, L.: Direct training for spiking neural networks: Faster, larger, better. In: Proceedings of the AAAI Conference on Artificial Intelligence, vol. 33, pp. 1311–1318 (2019)

76. Wu, Y., Zhao, R., Zhu, J., Chen, F., Xu, M., Li, G., Song, S., Deng, L., Wang, G., Zheng, H., et al.: Brain-inspired global-local learning incorporated with neuromorphic computing. Nat. Commun. **13**(1), 1–14 (2022)
77. Xiao, H., Rasul, K., Vollgraf, R.: Fashion-mnist: a novel image dataset for benchmarking machine learning algorithms. arXiv preprint arXiv:1708.07747 (2017)
78. Yao, M., Gao, H., Zhao, G., Wang, D., Lin, Y., Yang, Z., Li, G.: Temporal-wise attention spiking neural networks for event streams classification. In: Proceedings of the IEEE/CVF International Conference on Computer Vision, pp. 10221–10230 (2021)
79. Yin, R., Moitra, A., Bhattacharjee, A., Kim, Y., Panda, P.: Sata: Sparsity-aware training accelerator for spiking neural networks. arXiv preprint arXiv:2204.05422 (2022)
80. You, H., et al.: Drawing early-bird tickets: Towards more efficient training of deep networks. arXiv preprint arXiv:1909.11957 (2019)
81. Yu, H., Edunov, S., Tian, Y., Morcos, A.S.: Playing the lottery with rewards and multiple languages: lottery tickets in RL and NLP. arXiv preprint arXiv:1906.02768 (2019)
82. Zhang, Z., Chen, X., Chen, T., Wang, Z.: Efficient lottery ticket finding: less data is more. In: International Conference on Machine Learning, pp. 12380–12390. PMLR (2021)
83. Zheng, H., Wu, Y., Deng, L., Hu, Y., Li, G.: Going deeper with directly-trained larger spiking neural networks. arXiv preprint arXiv:2011.05280 (2020)
84. Zhou, H., Lan, J., Liu, R., Yosinski, J.: Deconstructing lottery tickets: Zeros, signs, and the supermask. In: Advances in Neural Information Processing Systems 32 (2019)

On the Angular Update
and Hyperparameter Tuning
of a Scale-Invariant Network

Juseung Yun[1], Janghyeon Lee[2], Hyounguk Shon[1], Eojindl Yi[1],
Seung Hwan Kim[2], and Junmo Kim[1(✉)]

[1] Korea Advanced Institute of Science and Technology, Daejeon, South Korea
{juseung_yun,hyounguk.shon,djwld93,junmo.kim}@kaist.ac.kr
[2] LG AI Research, Seoul, South Korea
{janghyeon.lee,sh.kim}@lgresearch.ai

Abstract. Modern deep neural networks are equipped with normalization layers such as batch normalization or layer normalization to enhance and stabilize training dynamics. If a network contains such normalization layers, the optimization objective is invariant to the scale of the neural network parameters. The scale-invariance induces the neural network's output to be only affected by the weights' direction and not the weights' scale. We first find a common feature of good hyperparameter combinations on such a scale-invariant network, including learning rate, weight decay, number of data samples, and batch size. Then we observe that hyperparameter setups that lead to good performance show similar degrees of angular update during one epoch. Using a stochastic differential equation, we analyze the angular update and show how each hyperparameter affects it. With this relationship, we can derive a simple hyperparameter tuning method and apply it to the efficient hyperparameter search.

Keywords: Scale-invariant network · normalization · effective learning rate · angular update · hyperparameter tuning

1 Introduction

Many recent deep neural network architectures are equipped with normalization layers such as batch normalization (BN) [12], layer normalization (LN) [1], group normalization (GN) [31], or instance normalization (IN) [27]. Such a normalization layer stabilizes deep neural networks' training and boosts generalization performance. These normalization layers make the optimization objective scale-invariant to its parameter, *i.e.*, the weight magnitude does not affect the output of the neural network; only direction does. In stochastic gradient descent (SGD) of a weight x and learning rate η, the direction update of x is proportional to the

---

J. Yun—Work done during an internship at LG AI Research.

Supplementary Information The online version contains supplementary material available at https://doi.org/10.1007/978-3-031-19775-8_8.

effective learning rate $\eta/\|\boldsymbol{x}\|^2$ [6,28,35]. Similarly, Wan *et al.* [29] also proposed a concept of angular update, which is the angle between the current and previous weights. Intriguingly, unlike the traditional concept of weight decay (WD), which regularizes the neural network's capacity by preventing the growth of weight norm, scale-invariant networks have the same expressive power regardless of the weight norm. However, weight decay still implicitly regulates the training dynamics by controlling the effective learning rate [6,28,35].

Recent studies have examined the relationship between hyperparameters and SGD dynamics on a scale-invariant network [3,7,15,18,19,24,25,29]. Li *et al.* [18] showed that SGD with fixed learning rate and weight decay is equivalent to the exponentially increasing learning rate schedule without weight decay. However, as the learning rate exponentially increases, the weight norms gradually overflow in calculations, making practical use difficult. Some studies have investigated the relationship between batch size and learning rate [3,24,25]. Goyal *et al.* [3] proposed a simple linear scaling rule, that is, when the minibatch size is multiplied by α, the learning rate is multiplied by α. Several studies [19,23–25] adopted stochastic differential equation (SDE), which captures the stochastic gradient noise to derive this linear scaling rule. However, this rule only considers learning rate, momentum, and batch size, disregarding factors such as weight decay and the number of training data samples. Yun *et al.* [33] proposed a weight decay scheduling method according to the number of data samples but without analyzing other hyperparameters.

Contribution. This study investigates the relationship between hyperparameters including learning rate η, momentum μ, weight decay λ, batch size B, and the number of data samples N of SGD, especially on a scale-invariant network with a fixed epoch budget.

Thus far, we have attempted to find a common feature of hyperparameter combinations with good performance in Sect. 3. Specifically, we tune the learning rate and weight decay to find the optimal combination that yields the best performance. Then, we observe three factors: effective learning rate, effective learning rate per epoch, and angular update per epoch. Although N, B, η, and λ are different, well-tuned hyperparameter combinations show similar angular updates during one epoch (See Fig. 2a).

Based on the novel findings, we propose a simple hyperparameter tuning rule for SGD. We aim to find the relationship between the angular update per epoch and hyperparameters. Then, conversely, if we keep the relation between hyperparameters, the neural network would show similar angular updates and also show good performance. Here, we adopt SDE to derive the angular update formula. The result shows that we should keep the tuning factor $\frac{N\eta\lambda}{B(1-\mu)}$ for optimum performance. For instance, when B is multiplied by α, multiply λ by α. Or, when N is multiplied by α, we divide λ by α. We apply the tuning factor for efficient hyperparameter search and show that we can find near optimum hyperparameters even with a small portion of training samples. Although the hyperparameter tuning rule might not be the precise optimal policy and is overly simplified, it is expected to provide valuable insight on how to tune the hyperparameters efficiently.

Scope of This Work. Our goal is not to prove why the angular update per epoch is important for tuning but to provide empirical evidence for the interesting observation on the angular update, and simultaneously highlight some insight into our observation. This work only considers training with SGD and a fixed epoch budget. We also assume that SDE can approximate SGD. The theoretical justification for this approximation is provided in [16,20].

2   Preliminaries

SGD and SDE Approximation. In SGD, the update of the neural network parameter $\boldsymbol{x}_k \in \mathbb{R}^d$ for a randomly sampled mini-batch $\mathcal{B}_k$ at k-th step is

$$\boldsymbol{x}_{k+1} \leftarrow \boldsymbol{x}_k - \eta \nabla \mathcal{L}(\boldsymbol{x}_k; \mathcal{B}_k), \tag{1}$$

where η is the learning rate and $\mathcal{L}(\boldsymbol{x}_k; \mathcal{B}_k)$ is the averaged mini-batch loss. Previous studies [2,16,17,19,25,30] used SDE as a surrogate for SGD with a continuous time limit. Assume that the gradient of each sample is independent and follows a short-tailed distribution. Then, the gradient noise, which is the difference between expected gradient $\mathcal{L}(\boldsymbol{x}) = \mathbb{E}_{\mathcal{B}}[\mathcal{L}(\boldsymbol{x}; \mathcal{B})]$ and mini-batch gradient $\mathcal{L}(\boldsymbol{x}; \mathcal{B})$, can be modeled by Gaussian noise with zero mean and covariance $\boldsymbol{\Sigma}(\boldsymbol{x}) := \mathbb{E}_{\mathcal{B}}[(\nabla \mathcal{L}(\boldsymbol{x}; \mathcal{B}) - \nabla \mathcal{L}(\boldsymbol{x}))(\nabla \mathcal{L}(\boldsymbol{x}; \mathcal{B}) - \nabla \mathcal{L}(\boldsymbol{x}))^\top]$. The corresponding SDE for the Eq. 1 is

$$d\boldsymbol{X}_t = -\nabla \mathcal{L}(\boldsymbol{X}_t)dt + (\eta \boldsymbol{\Sigma}(\boldsymbol{X}_t))^{\frac{1}{2}} d\boldsymbol{W}_t, \tag{2}$$

where the time correspondence is $t = k\eta$, i.e., $\boldsymbol{X}_t \approx \boldsymbol{x}_k$, and $\boldsymbol{W}_t \in \mathbb{R}^d$ is the standard Wiener process on interval $t \in [0, T]$, which satisfies $\boldsymbol{W}_t - \boldsymbol{W}_s \sim \mathcal{N}(\boldsymbol{0}, (t-s)\boldsymbol{I}_d)$. Here, $\mathcal{N}(\boldsymbol{0}, \boldsymbol{I}_d)$ is a normal distribution with zero mean and unit variance. Li *et al.* [17] rigorously proved this continuous time limit approximation holds for infinitesimal learning rate. Li *et al.* [20] showed that the approximation is also valid for finite learning rate and theoretically analyzed the conditions for the approximation to hold. Smith *et al.* [23,25] used the SDE to derive the linear scaling rule [3].

From now on, with an abuse of notation, we will omit the mini-batch $\mathcal{B}$ in the loss function $\mathcal{L}(\boldsymbol{x}; \mathcal{B})$ for brevity, if the mini-batch dependency is clear from the context. The stochastic gradient descent with momentum (SGDM) can be written as the following updates

$$\begin{aligned} \boldsymbol{v}_{k+1} &= \mu \boldsymbol{v}_k - \eta \nabla \mathcal{L}(\boldsymbol{x}_k) \\ \boldsymbol{x}_{k+1} &= \boldsymbol{x}_k + \boldsymbol{v}_{k+1}, \end{aligned} \tag{3}$$

where μ is the momentum parameter, and $\boldsymbol{v}$ is the velocity. Similar to the SGD case, the following SDE is an *order-1 weak approximation* of SGDM [16,17]:

$$\begin{aligned} d\boldsymbol{V}_t &= \left(-\eta^{-1}(1-\mu)\boldsymbol{V}_t - \nabla \mathcal{L}(\boldsymbol{X}_t)\right) dt + (\eta \boldsymbol{\Sigma}(\boldsymbol{X}_t))^{\frac{1}{2}} d\boldsymbol{W}_t \\ d\boldsymbol{X}_t &= \eta^{-1} \boldsymbol{V}_t dt. \end{aligned} \tag{4}$$

Scale-Invariance. Let $\mathcal{L}(\boldsymbol{x}) : \mathbb{R}^d \rightarrow \mathbb{R}$ be the loss function of a neural network parameterized by $\boldsymbol{x}$. $\mathcal{L}(\boldsymbol{x})$ is said to be a *scale-invariant function* if it satisfies $\mathcal{L}(\alpha\boldsymbol{x}) = \mathcal{L}(\boldsymbol{x})$ for any $\boldsymbol{x} \in \mathbb{R}^d$ and any $\alpha > 0$, and the neural network is said to be a *scale-invariant neural network*. For a scale-invariant function $\mathcal{L}(\boldsymbol{x})$, the following properties hold:

$$1. \quad \nabla\mathcal{L}(\alpha\boldsymbol{x}) = \frac{1}{\alpha}\nabla\mathcal{L}(\boldsymbol{x}) \tag{5}$$

$$2. \quad \langle \boldsymbol{x}, \nabla\mathcal{L}(\boldsymbol{x}) \rangle = 0 \tag{6}$$

$$3. \quad \|\boldsymbol{\Sigma}(\boldsymbol{x})^{1/2}\boldsymbol{x}\| = \sqrt{\boldsymbol{x}^\top \boldsymbol{\Sigma}(\boldsymbol{x})\boldsymbol{x}} = 0, \tag{7}$$

where $\| \cdot \|$ denots the L_2 norm of a vector. These properties are previously discussed [6,18,19,28,35], but we also derive them in the Appendix.

Effective Learning Rate. For a scale-invariant network parameterized by weight $\boldsymbol{x}$, the magnitude of $\boldsymbol{x}$ does not affect the output, and only the direction does. Defining $\boldsymbol{y} := \frac{\boldsymbol{x}}{\|\boldsymbol{x}\|}$, then SGD update of unit vector $\boldsymbol{y}$ can be approximated as follows:

$$\boldsymbol{y}_{k+1} \approx \boldsymbol{y}_k - \frac{\eta}{\|\boldsymbol{x}_k\|^2}\nabla\mathcal{L}(\boldsymbol{y}_k). \tag{8}$$

The result shows that the update of $\boldsymbol{y}$ is proportional to $\eta/\|\boldsymbol{x}\|^2$, or the *effective learning rate* [6,28,35].

3   Common Feature of Good Hyperparameter Combinations

This section finds a common property of hyperparameter combinations that performs well, and demonstrates novel observations on an update of weight direction. These observations serve as a key intuition for our simple hyperparameter tuning rule. We first describe the architectural modification to make a convolutional neural network scale-invariant, and define novel indices – *effective learning rate per epoch* and *angular update per epoch* – in Sect. 3.1. Then, we provide experimental setups and observation results in Sect. 3.2.

3.1   Measuring Updates of a Scale-Invariant Network

Network Modification. We first describe how to convert a neural network scale-invariant. Many convolutional neural networks (CNNs) are equipped with normalization layers and consist of several *convolution → normalization → activation function* sequences. Before output there is a global average pooling and a fully connected layer [4,5,9,11,32,34]. Such a network is invariant to all convolutional parameters followed by a normalization layer. However, a normalization layer (*e.g.*, BN) also has an affine parameter and the last fully connected layer

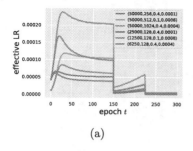
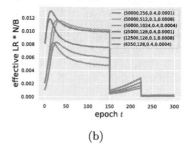

(a)   (b)

Fig. 1. (a) Effective learning rate and (b) effective learning rate per epoch of well-tuned hyperparameter combinations.

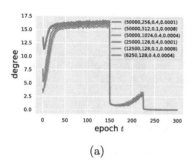
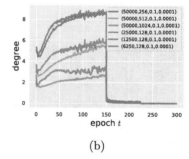

(a)   (b)

Fig. 2. (a) Angular update per epoch of well-tuned hyperparameter combinations and (b) poorly tuned hyperparameter combinations.

has no followed normalization layer. To increase the reliability of the observation in scale-invariant networks, we modify the ResNet [5] to make the training objective scale-invariant to all trainable parameters. We add normalization without affine transform after global average pooling and fix the parameters of the FC layer as done in [8,18,19]. We use ghost batch normalization (GBN) [7] with a ghost batch size of 64 as the normalization layers. GBN is often used in large batch training [23,24] to ensure the mean gradient is independent of batch size [25]. We name this modified architecture as *scale-invariant* ResNet. However, we also observe the result on unmodified ResNet in Sect. 5.2.

Measure for Directional Updates. As described in Sect. 2, the *effective learning rate* is defined as

$$\eta_{eff} := \eta/\|\boldsymbol{x}\|^2. \tag{9}$$

We define a novel index, *effective learning rate per epoch* η_{epoch} as

$$\eta_{epoch} := \eta_{eff}\frac{N}{B} = \frac{\eta N}{\|\boldsymbol{x}\|^2 B}. \tag{10}$$

The number of iterations during one epoch is N/B, and thus η_{epoch} means η_{eff} multiplied by the number of updates per epoch. We also propose another new index, *angular update per epoch* $\Delta\theta$ for measuring directional change of a vector. At e-th epoch, $\Delta\theta$ is defined as

$$\Delta\theta := \angle\left(\boldsymbol{x}_{eN/B}, \boldsymbol{x}_{(e-1)N/B}\right) = \arccos\left(\frac{\langle\boldsymbol{x}_{eN/B}, \boldsymbol{x}_{(e-1)N/B}\rangle}{\|\boldsymbol{x}_{eN/B}\| \cdot \|\boldsymbol{x}_{(e-1)N/B}\|}\right). \tag{11}$$

3.2   Observation

Experimental Setup. This experiment trains a scale-invariant ResNet-18 [5] on the CIFAR-10 dataset [13]. We train for 300 epochs regardless of the number of training samples using SGD with a momentum coefficient of 0.9. For all setups, the learning rate is divided by 10 at epochs 150 and 225. We use standard augmentation settings such as resizing, cropping, and flipping. We tune the learning rate and weight decay for six different hyperparameter combinations (N, B, η, λ):

1. $(50000, 256, \eta, 0.0001)$
2. $(50000, 256, 0.1, \lambda)$
3. $(50000, 1024, \eta, \lambda)$

4. $(25000, 128, \eta, 0.0001)$
5. $(12500, 128, 0.1, \lambda)$
6. $(6250, 128, \eta, \lambda)$.

Values written in Arabic numerals are fixed, and hyperparameters written in symbols are searched (although N is typically not regarded as a hyperparameter, for convenience, we include it in the combination notation). When tuning only η, we search $\eta \in \{0.1, 0.2, 0.4, 0.8, 1.6, 3.2\}$. When tuning only λ, we search $\lambda \in \{0.0001, 0.0002, 0.0004, 0.00008, 0.0016, 0.0032\}$. When tuning both η and λ, we search from $(\eta, \lambda) = (0.1, 0.0001)$ to $(0.8, 0.0008)$ while multiplying both with a factor of $\sqrt{2}$. We evaluate the average performance of three runs for tuning.

Result. The searched values exhibiting the optimal test accuracy are shown in legends of Fig. 1 and Fig. 2a. Figure 1 represents the effective learning rate η_{eff} of the found hyperparameter combinations. In all cases, η_{eff} initially increases and converges to a certain value until the first learning rate decays. It implies $\|\boldsymbol{x}\|$ also converges to a certain value. However, we cannot find a common feature of the converging value of η_{eff}. Figure 1b represents the effective learning rate per epoch η_{epoch} of the found hyperparameter combinations. For cases with the same number of data, η_{epoch} showed similar values as training proceeded (blue, orange, and green lines), although they differed at initial transient phases. If the number of iterations is α times higher, a well-tuned hyperparameter combination has α times smaller η_{eff}. However, for the cases with a different number of data (red, purple, and brown lines), even η_{epoch} showed different values. Figure 2a shows the angular update per epoch $\Delta\theta$ of well-tuned hyperparameter combinations and Fig. 2b shows the result of the poorly tuned combinations. Although N, B, η, and

λ were different, hyperparameter combinations resulting in good performance showed similar angular updates during one epoch. In contrast, poorly tuned hyperparameter combinations showed different values of η_{epoch}.

4   Angular Update

This section adopts SDE to describe the angular update according to the hyperparameters. Then we propose a simple hyperparameter tuning rule.

Angular Update of SGD. Similar to Eq. 2, we obtain the following SDE by considering weight decay:

$$dX_t = -\nabla\mathcal{L}(X_t)dt - \lambda X_t dt + (\eta\Sigma(X_t))^{\frac{1}{2}}\,dW_t, \tag{12}$$

where the time correspondence is $t = k\eta$, i.e., $X_t \approx x_k$. In such a SDE, however, the time scale does not include information on how many iterations the neural network is trained. Assuming that the network is trained for the same number of epochs, the number of entire training iterations is proportional to the ratio of the number of data to batch size N/B. To make the time correspondence inversely proportional to N/B, we rescale the time by $\frac{B}{N}$. We rescale the above SDE by considering $\tilde{X}_t = X_{\frac{N}{B}t}$; then, we have

$$d\tilde{X}_t = dX_{\frac{N}{B}t} \tag{13}$$

$$= -\nabla\mathcal{L}(X_{\frac{N}{B}t})d\left(\frac{N}{B}t\right) - \lambda X_{\frac{N}{B}t}d\left(\frac{N}{B}t\right) + \left(\eta\Sigma(X_{\frac{N}{B}t})\right)^{\frac{1}{2}}dW_{\frac{N}{B}t} \tag{14}$$

$$= -\frac{N}{B}\nabla\mathcal{L}(\tilde{X}_t)dt - \frac{N\lambda}{B}\tilde{X}_t dt + \left(\eta\Sigma(\tilde{X}_t)\right)^{\frac{1}{2}}dW_{\frac{N}{B}t}, \tag{15}$$

where the time correspondence is $t = \frac{B}{N}k\eta$, i.e., evolving for η time with the above SDE approximates $\Omega(N/B)$ steps of SGD. Similar to deriving the linear scaling rule [3,23–25], we also assume that during a small time interval η, the parameters do not move far enough for the gradients to significantly change, i.e., $\|\tilde{X}_t\| \approx \|\tilde{X}_{t+\eta}\|$ and $\nabla\mathcal{L}(\tilde{X})$ are nearly constant. Then, we obtain the update as follows:

$$\tilde{X}_{t+\eta} \approx \tilde{X}_t - \frac{N\eta}{B}\nabla\mathcal{L}(\tilde{X}_t) - \frac{N\eta\lambda}{B}\tilde{X}_t + \left(\eta\Sigma(\tilde{X}_t)\right)^{\frac{1}{2}}W_{\frac{N}{B}\eta}. \tag{16}$$

Define $Y = \frac{\tilde{X}}{\|\tilde{X}\|}$, then we get

$$Y_{t+\eta} \approx \left(1 - \frac{N\eta\lambda}{B}\right)Y_t - \frac{N\eta}{B\|\tilde{X}\|^2}\nabla\mathcal{L}(Y_t) + \frac{1}{\|\tilde{X}\|}\left(\eta\Sigma(Y_t)\right)^{\frac{1}{2}}W_{\frac{N}{B}\eta}. \tag{17}$$

The angle between $\tilde{X}_t$ and $\tilde{X}_{t+\eta}$ is $\arccos\langle Y_t, Y_{t+\eta}\rangle$. By Eq. 6 and Eq. 7, $\langle Y_t, \nabla\mathcal{L}(Y_t)\rangle = 0, Y_t^\top \Sigma(Y_t)^{\frac{1}{2}} = 0$. Then, we have

$$\langle Y_t, Y_{t+\eta}\rangle \approx 1 - \frac{N\eta\lambda}{B}. \tag{18}$$

The result suggests that for $\Theta(N/B)$ steps, the angular update is proportional to $1 - \frac{N\eta\lambda}{B}$. Although we assumed that the parameters do not move far during the update, we observed that such a relationship still holds for one epoch interval (Sect. 5). For instance, when the number of data is multiplied by α, multiplying η by $1/\alpha$ can maintain the angular update per epoch. The result also coincides with that of the linear scaling rule [3,24].

Angular Update of SGDM. Similar to Eq. 3, by considering weight decay, we obtain the following combined SDE for SGDM

$$dV_t = \left(-\eta^{-1}(1-\mu)V_t - \nabla\mathcal{L}(X_t) - \lambda X_t\right) dt + (\eta\Sigma(X_t))^{\frac{1}{2}} dW_t, \quad (19)$$

$$dX_t = \eta^{-1}V_t dt. \quad (20)$$

To make time correspondence inversely proportional to N/B, we rescale the above SDE by considering $\tilde{X}_t = X_{\frac{N}{B}t}$ and $\tilde{V}_t = V_{\frac{N}{B}t}$, which yields

$$d\tilde{V}_t = -\frac{N}{B}\left(\frac{1-\mu}{\eta}\tilde{V}_t + \nabla\mathcal{L}(\tilde{X}_t) + \lambda\tilde{X}_t\right) dt + \left(\eta\Sigma(\tilde{X}_t)\right)^{\frac{1}{2}} dW_{\frac{N}{B}t}, \quad (21)$$

$$d\tilde{X}_t = \frac{N}{B\eta}\tilde{V}_t dt. \quad (22)$$

Using Eq. 21, we can rewrite the right hand side of Eq. 22 as

$$\frac{N}{B\eta}\tilde{V}_t dt = -\frac{1}{1-\mu}d\tilde{V}_t - \frac{N}{B(1-\mu)}\left(\nabla\mathcal{L}(\tilde{X}_t) + \lambda\tilde{X}_t\right) dt \\ + \frac{1}{1-\mu}\left(\eta\Sigma(\tilde{X}_t)\right)^{\frac{1}{2}} dW_{\frac{N}{B}t}. \quad (23)$$

We also assume that during a short time interval η, the parameters do not move far enough for the gradients and velocity to change significantly. Then, by integrating Eq. 22 from t to $t+\eta$, we get

$$\tilde{X}_{t+\eta} \approx \tilde{X}_t - \frac{\tilde{V}_{t+\eta} - \tilde{V}_t}{1-\mu} - \frac{N\eta}{B(1-\mu)}\left(\nabla\mathcal{L}(\tilde{X}_t) + \lambda\tilde{X}_t\right) \\ + \frac{1}{1-\mu}\left(\eta\Sigma(\tilde{X}_t)\right)^{\frac{1}{2}} W_{\frac{N}{B}\eta}. \quad (24)$$

We define $Y = \frac{\tilde{X}}{\|\tilde{X}\|}$; futher, the direction update is

$$Y_{t+\eta} \approx Y_t - \frac{\tilde{V}_{t+\eta} - \tilde{V}_t}{(1-\mu)\|\tilde{X}_t\|} - \frac{N\eta}{B(1-\mu)\|\tilde{X}_t\|}\nabla\mathcal{L}(Y_t) - \frac{N\eta\lambda}{B(1-\mu)}Y_t \\ - \frac{1}{(1-\mu)\|\tilde{X}_t\|}\left(\eta\Sigma(Y_t)\right)^{\frac{1}{2}} W_{\frac{N}{B}\eta}. \quad (25)$$

Thus, the angle between $\boldsymbol{Y}_t$ and $\boldsymbol{Y}_{t+\eta}$ is $\arccos\langle\boldsymbol{Y}_t,\boldsymbol{Y}_{t+\eta}\rangle$ and we get

$$\langle\boldsymbol{Y}_t,\boldsymbol{Y}_{t+\eta}\rangle \approx 1 - \frac{N\eta\lambda}{B(1-\mu)} - \frac{\langle\boldsymbol{Y}_t,\tilde{\boldsymbol{V}}_{t+\eta}-\tilde{\boldsymbol{V}}_t\rangle}{(1-\mu)\|\tilde{\boldsymbol{X}}_t\|} \tag{26}$$

$$\approx 1 - \frac{N\eta\lambda}{B(1-\mu)}. \tag{27}$$

We refer to $\frac{N\eta\lambda}{B(1-\mu)}$ as the *tuning factor*, which determines how much the angular update occurs during a single epoch of training. Thus, based on the observations in Sect. 3 and Eq. 27, we argue that rather than searching for a hyperparameter individually, one should search for the tuning factor $\frac{N\eta\lambda}{B(1-\mu)}$ for an efficient tuning. Finding the tuning factor is sufficient for obtaining near optimum performance, *i.e.*, even if any hyperparameter is changed, a good performance can still be obtained by maintaining the tuning factor. With an intuition for the range of appropriate learning rate and momentum coefficient (*e.g.*, 0.1 for initial learning rate and 0.9 for momentum coefficient is used in many architectures) and by choosing a moderate batch size considering GPU memory or computation budget, we only need to search for a weight decay.

5   Experiments

In this section, we demonstrate that hyperparameter combinations with good performance show similar angular updates per epoch. With the same tuning factor, we can make these angular updates per epoch similar. We first show the experimental results on a modified scale-invariant architecture in Sect. 5.1. Next, we show the results on an unmodified architecture that is equipped with BN but has scale-variant parameters in Sect. 5.2.

5.1   Scale-Invariant Network

Experimental Setup. For clarity, in the main text, we only report experiments using ResNet-18 [5] (or scale-invariant ResNet-18) on CIFAR-10 [13]; however, we provide additional experiments using DenseNet-100 [11] on CIFAR-100 [13], and ResNet-18 on Tiny ImageNet [14] in the appendix. We train for 300 epochs regardless of the number of training samples, and the learning rate is divided by 10 at epochs 150 and 225. For SGDM, we set the base values as LR=0.1 and WD=0.0001, search them by multiplying the factor of 2 or $\sqrt{2}$, and set the momentum coefficient as 0.9. For SGD, we set the base value as LR=1, which makes the value $\frac{N\eta\lambda}{B(1-\mu)}$ the same as LR=0.1 for the SGDM setting. We use the standard augmentation setting such as resizing, cropping, and flipping. Because the tuning factor comprises five components $(N, B, \eta, \lambda, \mu)$, there are many possible combinations. To validate the tuning factor, we categorize the combinations into three: fixed B, fixed N, and fixed $\eta\lambda$. We report the average performance of three runs. However, if there is divergence among the three runs, we exclude the run when calculating the average; if all runs diverge, we leave the result table blank.

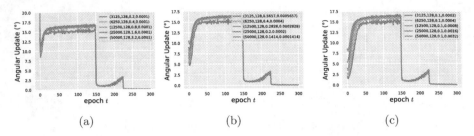

Fig. 3. Angular update per epoch of scale-invariant network with the batch size 128: (a) tuning LR, (b) tuning WD, and (c) tuning both LR and WD.

Table 1. Scale-invariant ResNet-18 on CIFAR-10 with SGDM and $B = 128$.

LR	WD	3125	6250	12500	25000	50000	LR	WD	3125	6250	12500	25000	50000	LR	WD	3125	6250	12500	25000	50000
6.4	0.0001	76.99	81.71	83.07	81.77	80.28	0.1	0.0064	77.31	81.71	82.94	83.79	80.18	0.8	0.0008	76.93	81.78	83.19	82.95	79.39
3.2	0.0001	**78.28**	83.70	86.93	89.25	89.92	0.1	0.0032	**78.59**	83.83	87.44	89.43	90.55	0.5657	0.0005657	**78.48**	84.00	87.18	89.21	90.63
1.6	0.0001	77.87	84.36	88.75	91.60	93.38	0.1	0.0016	77.57	**85.22**	89.06	91.94	93.39	0.4	0.0004	78.39	**84.71**	88.92	92.07	93.16
0.8	0.0001	76.73	**84.69**	89.06	92.75	94.79	0.1	0.0008	76.68	85.11	**89.47**	**93.10**	94.89	0.2828	0.0002828	76.67	84.80	**89.63**	92.86	94.95
0.4	0.0001	74.78	83.90	**89.06**	**92.81**	**95.13**	0.1	0.0004	74.14	84.06	89.31	92.90	95.13	0.2	0.0002	74.64	83.85	89.12	**92.96**	**95.26**
0.2	0.0001	73.17	83.34	88.70	92.76	95.08	0.1	0.0002	73.14	82.66	88.56	92.74	95.00	0.1414	0.0001414	73.66	82.93	88.62	92.44	95.09
0.1	0.0001	71.62	81.61	87.79	92.07	94.86	0.1	0.0001	71.62	81.61	87.79	92.07	94.86	0.1	0.0001	71.62	81.61	87.79	92.07	94.86

Table 2. Scale-invariant ResNet-18 on CIFAR-10 with SGD and $B = 128$.

LR	WD	3125	6250	12500	25000	50000	LR	WD	3125	6250	12500	25000	50000	LR	WD	3125	6250	12500	25000	50000
64	0.0001	76.75	80.58	79.50	78.98		1	0.0064	77.28	81.53	78.34	78.76	78.95	8	0.0008	77.48	81.22	79.81	79.40	78.84
32	0.0001	**78.85**	84.36	88.07	90.27	91.43	1	0.0032	**79.42**	84.67	88.40	91.11	91.60	5.657	0.0005657	**79.62**	84.30	88.03	90.50	91.72
16	0.0001	78.24	84.34	89.10	92.15	93.58	1	0.0016	79.06	**85.14**	89.63	92.52	94.02	4	0.0004	78.75	84.68	89.19	92.32	94.03
8	0.0001	77.08	**84.72**	89.19	92.53	94.80	1	0.0008	77.71	84.94	**89.67**	**93.15**	95.02	2.828	0.0002828	77.19	**85.08**	**89.25**	92.70	94.96
4	0.0001	75.90	83.96	89.05	**92.79**	**95.23**	1	0.0004	75.86	84.22	89.34	92.96	**95.10**	2	0.0002	75.97	84.28	89.18	**92.98**	**95.24**
2	0.0001	74.90	83.46	88.49	92.65	95.14	1	0.0002	74.40	83.53	88.67	92.53	95.05	1.414	0.0001414	74.17	83.17	88.87	92.66	95.12
1	0.0001	73.24	82.18	88.12	92.26	94.83	1	0.0001	73.24	82.18	88.12	92.26	94.83	1	0.0001	73.24	82.18	88.12	92.26	94.83

Experiment on Fixed B. In these experiments, we fixed B as 128 and tuned LR or (and) WD for each number of data samples. Table 1 and Table 2 show the results on SGDM and SGD, respectively. Yellow-colored cells satisfy the tuning factor $\frac{N\eta\lambda}{B(1-\mu)} = \frac{10}{128}$, and bold values represent the highest accuracy for each column. For all columns, the best performance or the second best accuracy was in the yellow cell. Figure 3 shows the angular update per epoch of the yellow cells of Table 1. They show distinct aspects in an initial transient phase where the weight changes rapidly. However, after 50 epochs, they show very similar angular updates, indicating the tuning factor's validity.

Experiment on Fixed N. In these experiments, we fixed N at 50k and tuned LR or (and) WD for each B. Table 3 and Table 4 show the results on SGDM and SGD, respectively. Yellow color indicates the cells that satisfy the tuning factor $\frac{N\eta\lambda}{B(1-\mu)} = \frac{10}{128}$, and bold values represent the highest accuracy for each column. For all columns of Table 3, the best performance or the second-best accuracy

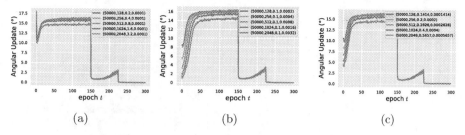

Fig. 4. Angular update per epoch of scale-invariant network with the number of data being 50k. (a) Tuning the learning rate, (b) tuning the weight decay, and (c) tuning both the learning rate and the weight decay.

Table 3. Scale-invariant ResNet-18 on CIFAR-10 with SGDM and 50k data samples.

LR	WD	2048	1024	512	256	128	LR	WD	2048	1024	512	256	128	LR	WD	2048	1024	512	256	128
6.4	0.0001	94.52	93.89	91.90	88.55	80.28	0.1	0.0064	**94.87**	94.23	92.41	89.27	80.18	0.8	0.0008	94.69	94.22	92.44	89.43	79.39
3.2	0.0001	**94.87**	**94.92**	94.31	92.80	89.92	0.1	0.0032	94.82	95.01	94.37	93.17	90.55	0.5657	0.0005657	**94.98**	95.02	94.39	93.12	90.63
1.6	0.0001	94.74	94.91	95.03	94.66	93.38	0.1	0.0016	94.74	**95.05**	95.02	94.78	93.39	0.4	0.0004	94.76	**95.05**	**95.13**	94.79	93.16
0.8	0.0001	94.21	94.83	**95.04**	94.99	94.79	0.1	0.0008	94.16	94.71	**95.16**	**95.30**	94.89	0.2828	0.0002828	94.19	94.70	95.01	**95.24**	94.95
0.4	0.0001	93.71	94.44	94.87	**95.13**	**95.13**	0.1	0.0004	93.46	94.17	94.76	**95.19**	**95.13**	0.2	0.0002	93.75	94.49	94.76	95.11	**95.26**
0.2	0.0001	92.96	93.93	94.45	94.85	95.08	0.1	0.0002	92.69	93.75	94.52	94.95	95.00	0.1414	0.0001414	93.09	93.70	94.40	94.91	95.09
0.1	0.0001	92.36	93.31	93.85	94.50	94.86	0.1	0.0001	92.36	93.31	93.85	94.50	94.86	0.1	0.0001	92.36	93.31	93.85	94.50	94.86

Table 4. Scale-invariant ResNet-18 on CIFAR-10 with SGD and 50k data samples.

LR	WD	2048	1024	512	256	128	LR	WD	2048	1024	512	256	128	LR	WD	2048	1024	512	256	128
64	0.0001	86.40	83.72	82.38	82.70		1	0.0064	86.48	82.41	82.81	81.44	78.95	8	0.0008	86.08	84.91	83.29	81.19	78.84
32	0.0001	91.25	93.62	94.31	93.17	91.43	1	0.0032	92.63	93.66	94.46	93.95	91.60	5.657	0.00056569	92.95	93.94	94.68	93.31	91.72
16	0.0001	93.13	94.63	95.03	94.70	93.58	1	0.0016	93.66	94.60	**95.17**	94.79	94.02	4	0.0004	93.28	**94.75**	**95.16**	94.92	94.03
8	0.0001	**93.50**	**94.69**	**95.15**	95.06	94.80	1	0.0008	**93.79**	**94.66**	95.00	95.22	95.02	2.828	0.00028284	**93.72**	94.68	95.10	**95.21**	94.96
4	0.0001	93.18	94.54	95.04	95.01	**95.23**	1	0.0004	93.47	94.36	94.90	**95.22**	**95.10**	2	0.0002	93.28	94.32	94.96	95.17	**95.24**
2	0.0001	93.06	93.76	94.33	94.76	95.14	1	0.0002	92.63	93.62	94.31	94.85	95.05	1.414	0.00014142	92.77	93.69	94.56	94.91	95.12
1	0.0001	92.30	93.20	93.97	94.57	94.83	1	0.0001	92.30	93.20	93.97	94.57	94.83	1	0.0001	92.30	93.20	93.97	94.57	94.83

was in the yellow cell. An interesting point here is not only the LR linear scaling rule but also a WD linear scaling rule is possible as batch size changes. For SGD (Table 4), there exists a case in which even the second-best value is not in yellows cells for B bigger than 512. This may because the excessively large LR has caused unstable training. Figure 4 shows the angular update per epoch of the yellow cells of Table 3. At the initial transient phase, where the weight changes rapidly, they showed different aspects. However, after 50 epoch, they showed very similar angular update, and it shows the validity of tuning factor.

Experiment on Fixed LR×WD. In these experiments, we fixed the LR as 0.1 and WD as 0.0016, and then tuned B for each number of data samples. Table 5 shows the results on SGDM. We also used yellow to indicate cells that satisfy the tuning factor $\frac{N\eta\lambda}{B(1-\mu)} = \frac{10}{128}$; the bold values represent the highest accuracy for each column. For all columns of Table 5, the best performance or the second-best

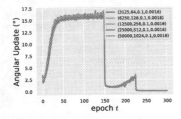

Fig. 5. Angular update per epoch of scale-invariant network with $\eta = 0.1$ and $\lambda = 0.0016$.

Table 5. Scale-invariant ResNet-18 on CIFAR-10 with SGDM, $\eta = 0.1$, and $\lambda = 0.0016$.

B	LR	WD	3125	6250	12500	25000	50000
32	0.1	0.0016	78.64	83.10	85.30	87.08	87.44
64	0.1	0.0016	**79.25**	84.30	87.83	89.52	91.01
128	0.1	0.0016	77.57	**85.22**	89.06	91.94	93.39
256	0.1	0.0016	75.65	84.60	**89.31**	**92.84**	94.78
512	0.1	0.0016	72.02	82.90	89.10	92.78	95.02
1024	0.1	0.0016	68.19	79.62	87.90	92.58	**95.05**
2048	0.1	0.0016	51.51	74.71	85.90	91.41	94.74

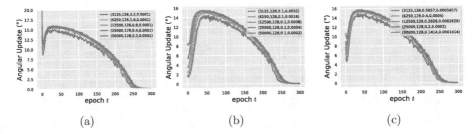

(a)   (b)   (c)

Fig. 6. Angular update per epoch of scale-invariant network with the number of data = 50k and using cosine LR scheduling. (a) Tuning LR, (b) tuning WD, and (c) tuning both LR rate and WD.

accuracy was in the yellow cell. Figure 5 shows the angular update per epoch of the yellow cells of Table 5. They showed remarkably similar angular updates, indicating the validity of the tuning factor.

Experiment on Cosine LR Scheduling. In these experiments, we fixed N as 50k and tuned LR or (and) WD for each B, but used cosine LR scheduling. Table 6 shows the results on SGDM. We also used yellow color to indicate cells that satisfy the tuning factor $\frac{N\eta\lambda}{B(1-\mu)} = \frac{10}{128}$; bold values represent the highest accuracy for each column. Figure 6 shows the angular update per epoch of the yellow cells of Table 6. The result shows that the tuning factor remains valid for cosine LR scheduling.

5.2   Unmodified Network

Previously, we demonstrated the validity of the tuning factor on a scale-invariant network. We now demonstrate that the tuning factor is still valid on a ResNet-18 which is not modified, as described in Sect. 3.1. In the main text, we only report experiments with fixed batch size $B = 128$. We report other results in the Appendix. Table 7 shows the results. Figure 7 shows the angular update per

Table 6. Scale-invariant ResNet-18 on CIFAR-10 with SGDM, $B = 128$, and cosine learning rate scheduling.

LR	WD	3125	6250	12500	25000	50000	LR	WD	3125	6250	12500	25000	50000	LR	WD	3125	6250	12500	25000	50000
6.4	0.0001	78.36	83.08	85.96	87.98	87.36	0.1	0.0064	78.69	83.37	85.96	87.70	87.26	0.8	0.0008	78.92	83.06	85.98	87.92	87.39
3.2	0.0001	**79.02**	84.51	88.21	90.89	92.78	0.1	0.0032	**79.42**	84.74	88.40	91.17	92.95	0.5657	0.0005657	**79.33**	84.61	88.21	90.94	92.81
1.6	0.0001	78.35	**84.97**	89.32	92.52	94.30	0.1	0.0016	78.88	**85.06**	**89.67**	92.52	94.87	0.4	0.0004	78.53	**85.20**	89.57	92.32	94.69
0.8	0.0001	76.80	84.88	**89.48**	**93.00**	95.08	0.1	0.0008	76.13	84.94	89.64	**93.18**	95.39	0.2828	0.0002828	76.53	85.00	**89.77**	**93.16**	95.21
0.4	0.0001	75.18	84.09	88.99	92.89	**95.32**	0.1	0.0004	74.02	84.16	89.17	92.95	**95.51**	0.2	0.0002	74.61	84.21	89.29	92.83	**95.37**
0.2	0.0001	72.60	83.01	88.79	92.81	95.14	0.1	0.0002	71.53	82.05	88.82	92.74	95.25	0.1414	0.0001414	72.68	82.76	88.67	92.60	95.11
0.1	0.0001	71.12	82.15	87.89	92.12	94.92	0.1	0.0001	71.12	82.15	87.89	92.12	94.92	0.1	0.0001	71.12	82.15	87.89	92.12	94.92

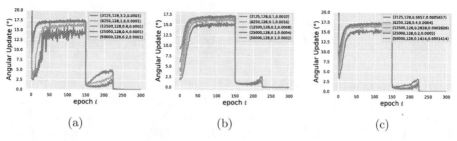

(a) (b) (c)

Fig. 7. Angular update per epoch of ResNet-18 with SGDM, and $B = 128$. (a) Tuning LR, (b) tuning WD, and (c) tuning both LR and WD.

Table 7. ResNet-18 on CIFAR-10 with SGDM, $B=128$.

LR	WD	3125	6250	12500	25000	50000	LR	WD	3125	6250	12500	25000	50000	LR	WD	3125	6250	12500	25000	50000
6.4	0.0001						0.1	0.0064	77.43	80.63	81.95	80.19	81.90	0.8	0.0008	76.39	80.29	80.94	78.81	51.82
3.2	0.0001	73.58	77.20	82.51	83.25	79.93	0.1	0.0032	79.06	83.74	87.23	88.80	89.85	0.5657	0.0005657	77.98	83.49	87.12	89.10	89.71
1.6	0.0001	77.23	83.23	87.45	91.16	92.86	0.1	0.0016	**79.20**	**84.88**	89.38	92.38	93.90	0.4	0.0004	**78.05**	**84.81**	88.70	91.68	93.63
0.8	0.0001	**77.35**	**84.26**	**89.01**	92.62	94.69	0.1	0.0008	77.78	84.83	**89.80**	**93.33**	95.14	0.2828	0.0002828	75.89	84.67	**89.48**	92.99	94.86
0.4	0.0001	73.75	83.22	88.66	**92.69**	**95.18**	0.1	0.0004	75.23	83.75	89.43	93.06	**95.37**	0.2	0.0002	73.19	84.03	89.19	**93.04**	**95.42**
0.2	0.0001	71.86	82.69	88.48	92.58	95.09	0.1	0.0002	72.83	83.31	88.80	92.67	95.18	0.1414	0.0001414	72.98	83.07	88.68	92.64	95.06
0.1	0.0001	72.32	81.71	88.02	91.96	94.87	0.1	0.0001	72.32	81.71	88.02	91.96	94.87	0.1	0.0001	72.32	81.71	88.02	91.96	94.87

epoch of the yellow cells of Table 7. Here, the angular update per epoch was obtained with only scale-invariant weights.

6 Efficient Hyperparameter Search

In this section, we discuss a practical application for efficient hyperparameter search. Finding the tuning factor is sufficient for obtaining near optimum performance, *i.e.*, even if any hyperparameter is changed, a good performance can be obtained if the tuning factor is maintained. This motivates the question: can we find a near optimal hyperparameter with a small portion of data samples? Here, we show that it is not necessary to use all the training data to find the hyperparameter. We search the hyperparameter of training EfficientNet-B0 [26] on ImageNet dataset [22] which comprisis 1.28M samples. EfficientNet has scale-variant parts, *e.g.*, squeeze and excitation module [10] and SiLU activation function [21]. We show that even in the presence of such scale-variant parts, the tuning factor still works well as long as the network has a normalization layer.

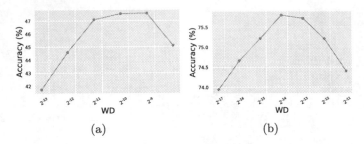

(a)   (b)

Fig. 8. Classification accuracy of EfficientNet-B0 on ImageNet according to λ: (a) $N = 80k$, and (b) $N = 1.28M$.

We first find the tuning factor with 80k randomly sampled data, which is $1/16$ of the entire sample. We apply data augmentation, including random horizontal flip and resizing ratio between 0.08 and 1 and aspect ratio between $3/4$ and $4/3$. We train for 200 epochs using SGD with batch size 512, LR 0.1, momentum coefficient 0.9; the LR is divided by 10 at 100, 150, 180 epochs (50%, 75%, and 90% of the entire duration) and only the WD is tuned. Such a LR and momentum values are typically used in many architectures [4,5,11] which are trained using SGDM. The original EfficientNet study used an RMSprop optimizer with WD 1e−5 and an initial LR of 0.256 that decays by 0.97 every 2.4 epoch. Figure 8(a) shows the result. The best result is obtained when $\lambda = 0.0016$; we obtain the tuning factor $\frac{80k \cdot 0.1 \cdot 0.0016}{512 \cdot (1 - 0.9)}$. Thus, using the factor, we can expect that the optimum is around $\lambda = 1e−04$ when trained with the entire data sample. Figure 8(b) shows the result for the entire dataset. The best result can be seen when $\lambda = 5e−05$; but the second best result is seen when $\lambda = 1e−04$. This shows that we can find near optimal hyperparameters with significantly fewer iterations.

7 Conclusion

This study observed that if hyperparameters are well tuned, the scale-invariant network shows a similar degree of angular update during an epoch. We derived the relation between hyperparameters and angular update per epoch based on this novel observation by adopting SDE. We proposed the concept of a tuning factor, and performed rigorous hyperparameter tuning to show the validity. We also proposed an efficient hyperparameter search method only using a small portion of training samples.

Acknowledgment. This research was supported by the Engineering Research Center Program through the National Research Foundation of Korea (NRF) funded by the Korean Government MSIT (NRF-2018R1A5A1059921).

References

1. Ba, J.L., Kiros, J.R., Hinton, G.E.: Layer normalization. arXiv preprint arXiv:1607.06450 (2016)
2. Gardiner, C.W., et al.: Handbook of stochastic methods, vol. 3. Springer, Berlin (1985)
3. Goyal, P., et al.: Accurate, large minibatch sgd: Training imagenet in 1 hour. arXiv preprint arXiv:1706.02677 (2017)
4. Han, D., Kim, J., Kim, J.: Deep pyramidal residual networks. In: Proceedings of the IEEE Conference on Computer Vision and Pattern Recognition, pp. 5927–5935 (2017)
5. He, K., Zhang, X., Ren, S., Sun, J.: Deep residual learning for image recognition. In: Proceedings of the IEEE Conference on Computer Vision and Pattern Recognition, pp. 770–778 (2016)
6. Hoffer, E., Banner, R., Golan, I., Soudry, D.: Norm matters: efficient and accurate normalization schemes in deep networks. In: Bengio, S., Wallach, H., Larochelle, H., Grauman, K., Cesa-Bianchi, N., Garnett, R. (eds.) Advances in Neural Information Processing Systems, vol. 31. Curran Associates, Inc. (2018). http://www.proceedings.neurips.cc/paper/2018/file/a0160709701140704575d499c997b6ca-Paper.pdf
7. Hoffer, E., Hubara, I., Soudry, D.: Train longer, generalize better: closing the generalization gap in large batch training of neural networks. arXiv preprint arXiv:1705.08741 (2017)
8. Hoffer, E., Hubara, I., Soudry, D.: Fix your classifier: the marginal value of training the last weight layer (2018)
9. Howard, A., et al.: Searching for mobilenetv3. In: Proceedings of the IEEE/CVF International Conference on Computer Vision, pp. 1314–1324 (2019)
10. Hu, J., Shen, L., Sun, G.: Squeeze-and-excitation networks. In: Proceedings of the IEEE Conference on Computer Vision and Pattern Recognition, pp. 7132–7141 (2018)
11. Huang, G., Liu, Z., Van Der Maaten, L., Weinberger, K.Q.: Densely connected convolutional networks. In: Proceedings of the IEEE Conference on Computer Vision and Pattern Recognition, pp. 4700–4708 (2017)
12. Ioffe, S., Szegedy, C.: Batch normalization: Accelerating deep network training by reducing internal covariate shift. In: International Conference on Machine Learning, pp. 448–456. PMLR (2015)
13. Krizhevsky, A., Hinton, G., et al.: Learning multiple layers of features from tiny images (2009)
14. Le, Y., Yang, X.: Tiny imagenet visual recognition challenge. CS 231N $7(7)$, 3 (2015)
15. Lewkowycz, A., Gur-Ari, G.: On the training dynamics of deep networks with l_2 regularization. arXiv preprint arXiv:2006.08643 (2020)
16. Li, Q., Tai, C., Weinan, E.: Stochastic modified equations and adaptive stochastic gradient algorithms. In: International Conference on Machine Learning, pp. 2101–2110. PMLR (2017)
17. Li, Q., Tai, C., Weinan, E.: Stochastic modified equations and dynamics of stochastic gradient algorithms i: mathematical foundations. J. Mach. Learn. Res. $20(1)$, 1474–1520 (2019)
18. Li, Z., Arora, S.: An exponential learning rate schedule for deep learning. In: International Conference on Learning Representations (2020), http://www.openreview.net/forum?id=rJg8TeSFDH

19. Li, Z., Lyu, K., Arora, S.: Reconciling modern deep learning with traditional optimization analyses: the intrinsic learning rate. In: Advances in Neural Information Processing Systems 33 (2020)
20. Li, Z., Malladi, S., Arora, S.: On the validity of modeling sgd with stochastic differential equations (sdes). arXiv preprint arXiv:2102.12470 (2021)
21. Ramachandran, P., Zoph, B., Le, Q.V.: Searching for activation functions. arXiv preprint arXiv:1710.05941 (2017)
22. Russakovsky, O., et al.: ImageNet large scale visual recognition challenge. Int. J. Comput. Vision **115**(3), 211–252 (2015). https://doi.org/10.1007/s11263-015-0816-y
23. Smith, S., Elsen, E., De, S.: On the generalization benefit of noise in stochastic gradient descent. In: International Conference on Machine Learning, pp. 9058–9067. PMLR (2020)
24. Smith, S.L., Kindermans, P.J., Ying, C., Le, Q.V.: Don't decay the learning rate, increase the batch size. arXiv preprint arXiv:1711.00489 (2017)
25. Smith, S.L., Le, Q.V.: A bayesian perspective on generalization and stochastic gradient descent. arXiv preprint arXiv:1710.06451 (2017)
26. Tan, M., Le, Q.: Efficientnet: rethinking model scaling for convolutional neural networks. In: International Conference on Machine Learning, pp. 6105–6114. PMLR (2019)
27. Ulyanov, D., Vedaldi, A., Lempitsky, V.: Instance normalization: the missing ingredient for fast stylization. arXiv preprint arXiv:1607.08022 (2016)
28. Van Laarhoven, T.: L2 regularization versus batch and weight normalization. arXiv preprint arXiv:1706.05350 (2017)
29. Wan, R., Zhu, Z., Zhang, X., Sun, J.: Spherical motion dynamics: learning dynamics of normalized neural network using sgd and weight decay. In: Advances in Neural Information Processing Systems 34 (2021)
30. Welling, M., Teh, Y.W.: Bayesian learning via stochastic gradient langevin dynamics. In: Proceedings of the 28th International Conference on Machine Learning (ICML-11), pp. 681–688. Citeseer (2011)
31. Wu, Y., He, K.: Group normalization. In: Ferrari, V., Hebert, M., Sminchisescu, C., Weiss, Y. (eds.) ECCV 2018. LNCS, vol. 11217, pp. 3–19. Springer, Cham (2018). https://doi.org/10.1007/978-3-030-01261-8_1
32. Xie, S., Girshick, R., Dollár, P., Tu, Z., He, K.: Aggregated residual transformations for deep neural networks. In: Proceedings of the IEEE Conference on Computer Vision and Pattern Recognition, pp. 1492–1500 (2017)
33. Yun, J., Kim, B., Kim, J.: Weight decay scheduling and knowledge distillation for active learning. In: Vedaldi, A., Bischof, H., Brox, T., Frahm, J.-M. (eds.) ECCV 2020. LNCS, vol. 12371, pp. 431–447. Springer, Cham (2020). https://doi.org/10.1007/978-3-030-58574-7_26
34. Zagoruyko, S., Komodakis, N.: Wide residual networks. In: British Machine Vision Conference 2016. British Machine Vision Association (2016)
35. Zhang, G., Wang, C., Xu, B., Grosse, R.: Three mechanisms of weight decay regularization. In: International Conference on Learning Representations (2019). http://www.openreview.net/forum?id=B1lz-3Rct7

LANA: Latency Aware Network Acceleration

Pavlo Molchanov[1]([✉]), Jimmy Hall[2], Hongxu Yin[1], Jan Kautz[1], Nicolo Fusi[2],
and Arash Vahdat[1]

[1] NVIDIA, Mountain View, USA
pmolchanov@nvidia.com
[2] Microsoft Research, Redmond, USA

Abstract. We introduce latency-aware network acceleration (LANA)-
an approach that builds on neural architecture search technique to accelerate neural networks. LANA consists of two phases: in the first phase,
it trains many alternative operations for every layer of a target network
using layer-wise feature map distillation. In the second phase, it solves the
combinatorial selection of efficient operations using a novel constrained
integer linear optimization (ILP) approach. ILP brings unique properties as it (i) performs NAS within a few seconds to minutes, (ii) easily
satisfies budget constraints, (iii) works on the layer-granularity, (iv) supports a huge search space $O(10^{100})$, surpassing prior search approaches
in efficacy and efficiency. In extensive experiments, we show that LANA
yields efficient and accurate models constrained by a target latency budget, while being significantly faster than other techniques. We analyze
three popular network architectures: EfficientNetV1, EfficientNetV2 and
ResNeST, and achieve accuracy improvement (up to 3.0%) for all models
when compressing larger models. LANA achieves significant speed-ups
(up to 5×) with minor to no accuracy drop on GPU and CPU. Project
page: https://bit.ly/3Oja2IF.

1 Introduction

In many applications, we may have access to a neural network that satisfies desired performance needs in terms of accuracy but is computationally too expensive to deploy. The goal of hardware-aware network acceleration [5,13,28,59,67,101] is to accelerate a given neural network such that it
meets efficiency criteria on a device without sacrificing accuracy dramatically.
Network acceleration plays a key role in reducing the operational cost, power
usage, and environmental impact of deploying deep neural networks in real-world
applications.

Given a trained neural network (teacher, base model), the current network
acceleration techniques can be grouped into: *(i) pruning* that removes inactive

Supplementary Information The online version contains supplementary material
available at https://doi.org/10.1007/978-3-031-19775-8_9.

neurons [4, 8, 18–20, 22, 24–27, 29, 33, 35, 36, 39, 43–45, 47–49, 51, 53, 92, 95, 96, 102], (ii) *compile-time optimization* [62] *or kernel fusion* [14, 15, 80, 98] that combines multiple operations into an equivalent operation, (iii) *quantization* that reduces the precision in which the network operates at [7, 11, 16, 34, 55, 65, 82, 85, 99], and *(iv) knowledge distillation* that distills knowledge from a larger *teacher* network into a smaller *student* network [1, 31, 46, 52, 64, 89, 93]. The approaches within (i) to (iii) are restricted to the underlying network operations and they do not change the architecture. Knowledge distillation changes the network architecture from teacher to student, however, the student design requires domain knowledge and is done usually manually.

In this paper, we propose latency-aware network transformation (LANA), a network acceleration framework that replaces inefficient operations in a given trained network with more efficient counterparts (see Fig. 1). Given a convolutional network as target and base model to accelerate, we formulate the problem as searching in a large pool of candidate operations to find efficient operations for different layers of the base model (teacher). The search problem is combinatorial in nature with a space that grows exponentially with the depth of the network. To solve this problem, we can turn to neural architecture search (NAS) [6, 57, 70, 77, 103, 104], which has been proven successful in discovering novel architectures. However, existing NAS solutions are computationally expensive, and usually handle only a small number of candidate operations (ranging from 5 to 15) in each layer and they often struggle with larger candidate pools.

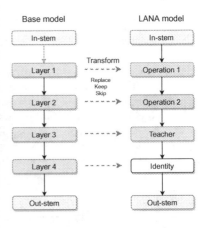

Fig. 1. LANA is a post-training model optimization method that keeps, replaces or skips layers of the trained base model (teacher).

To tackle the search problem with a large number of candidate operations in an efficient and scalable way, we propose a two-phase approach. In the first phase, we define a large candidate pool of operations ranging from classic residual blocks [21] to recent blocks [2, 15, 17, 66], with varying hyperparameters. Candidate operations are pretrained to mimic the teacher's operations via a simple layer-wise optimization. Distillation-based pretraining enables a very quick preparation of all candidate operations, offering a much more competitive starting point for subsequent searching.

In the second phase, we search among the pre-trained operations as well as the teacher's own operations to construct an efficient network. Since our operation selection problem can be considered as searching in the proximity of the teacher network in the architecture space, we assume that the accuracy of a candidate architecture can be approximated by the teacher's accuracy and a simple linear function that measures changes in the accuracy for individual operations. Our approximation allows us to relax the search problem into a constrained integer linear optimization problem that is solved in a few seconds. As we show extensively in our experiments, such relaxation can drastically cut down on the cost of our search and it can be easily applied to a huge pool of operations (197 operations per layer), while offering improvements in model acceleration by a large margin.

In summary, we make the following contributions: **(i)** We propose a simple two-phase approach for accelerating a teacher network using NAS-like search. **(ii)** We propose an effective search algorithm using constrained integer optimization that can find an architecture in seconds tailored to our setting where a fitness measure is available for each operation. **(iii)** We examine a large pool of operations including the recent vision transformers and new variants of convolutional networks. We provide insights into the operations selected by our framework and into final model architectures.

1.1 Related Work

Since our goal is to accelerate a trained model by modifying architecture, in this section we focus on related NAS-based approaches.

Hardware-Aware NAS: The goal of hardware-aware NAS is to design efficient and accurate architectures from scratch while targeting a specific hardware platform. This has been the focus of an increasingly large body of work on multiobjective neural architecture search [6,71,77,78,85,86,90,91]. The goal here is to solve an optimization problem maximizing accuracy while meeting performance constraints specified in terms of latency, memory consumption or number of parameters. Given that the optimization problem is set up from scratch for each target hardware platform, these approaches generally require the search to start from scratch for every new deployment target (*e.g.*, GPU/CPU family) or objective, incurring a search cost that increases linearly as the number of constraints and targets increases. [5] circumvents this issue by training a supernetwork containing every possible architecture in the search space, and then applying a progressive shrinking algorithm to produce multiple high-performing architectures. This approach incurs a high pretraining cost, but once training is complete, new architectures are relatively inexpensive to find. On the other hand, the high computational complexity of pretraining limits the number of operations that can be considered. Adding new operations is also costly, since the supernetwork must be pretrained from scratch every time for a new operation. Recent work [56] also explores prioritized paths within the supernetwork to help guide the learning of weak subnets through distillation, such that all blocks can be trained simultaneously and the strongest path constitutes a final architecture. Despite remarkable insights the searching still imposes GPU days to converge towards a final strong path overseeing a small search space.

Table 1. Related method comparison. Time is mentioned in GPU hours by **h**, or ImageNet epochs by **e**. Our method assumes 197 candidate operations for the full pool, and only 2 (teacher and identity) for single shot mode. L is the number of target architectures.

Method	Knowledge distillation	Diverse operators	Design space size	Pretrain cost	Search cost	Train cost	Total cost
					To Train L architectures		
Once-For-All [5]	None		$>O\left(10^{19}\right)$	1205e	40h	75eL	1205e + 75eL
AKD [42]	Network	✓	$\left(>O\left(10^{13}\right)\right)$	0	50000eL	400eL	50400eL
DNA [38]	Block		$\left(>O\left(10^{15}\right)\right)$	320e	14hL	450eL	320e + 450eL
DONNA [50]	Block	✓	$\left(>O\left(10^{13}\right)\right)$	1920e[a]	1500e + ≤1hL	50eL	3420e + 50eL
Cream [56]	Network	✓	$\left(>O\left(10^{16}\right)\right)$	0	120eL	500eL	620eL
This Work	Layer	✓	$>O\left(10^{100}\right)$	197e	<1hL	100eL	197e + 100eL
This Work - single shot	None		$>O\left(10^{2}\right)$	0	~0	100eL	100eL

[a] can potentially be improved by parallelization

Teacher-Based NAS: Our work is more related to the line of work that focuses on modifying *existing* architectures. Approaches in this area build on teacher-student knowledge distillation, performing multiobjective NAS on the student to mimic the teacher network.

The AKD approach [42] applies knowledge distillation at the network level, training a reinforcement learning agent to construct an efficient student network given a teacher network and a constraint and then training that student from scratch using knowledge distillation. DNA [38] and DONNA [50] take a more fine-grained approach, dividing the network into a small number of blocks, each of which contain several layers. During knowledge distillation, they both attempt to have student blocks mimic the output of teacher blocks, but [38] samples random paths through a mix of operators in each block, whereas [50] trains several candidate blocks with a repeated single operation for each teacher block. They then both search for an optimal set of blocks, with DNA [38] using a novel ranking algorithm to predict the best set of operations within each block, and then applying a traversal search, while [50] trains a linear model that predicts accuracy of a set of blocks and use that to guide an evolutionary search. While both methods deliver impressive results, they differ from our approach in important ways. DNA [38] ranks each path within a block, and then use this ranking to search over the blocks, relying on the low number of blocks to accelerate search. DONNA [50] samples and finetunes 30 models to build a linear accuracy predictor, which incurs a significant startup cost for search. In contrast, we formulate the search problem as an integer linear optimization problems that can be solved very quickly for large networks and large pool of operations.

Table 1 compares our work to these works in detail. We increase the granularity of network acceleration, focusing on each *layer* individually instead of blocks as done in DNA and DONNA. The main advantage of focusing on layers is that it allows us to accelerate the teacher by simply *replacing* inefficient layers whereas blockwise algorithms such as DNA and DONNA require *searching* for an efficient subnetwork that mimics the whole block. The blockwise search introduces additional constraints. For example, both DNA and DONNA enumerate over different depth values (multiplying the search space) while we reduce depth simply using an identity operation. Additionally, DONNA assumes that the same layer in each block is repeated whereas we have more expressivity by assigning differ-

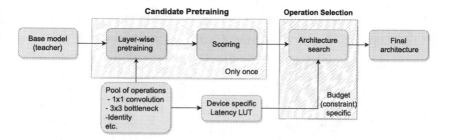

Fig. 2. LANA framework: A set of candidate operations is pretrained to mimic layers of the trained base (teacher) model. Then, operations are scored on their goodness metric to approximate the teacher. These 2 steps are only performed once for a given base model. Finally, an architecture search is performed to select operations per every layer to satisfy a predefined budget constraint.

ent operation to different layers. The expressivity can be seen from the design space size in Table 1 in which our search space is orders of magnitude larger. On the other hand, this extremely large space necessitate the development of a highly efficient search method based on integer linear optimization (presented in Section 2). As we can see from Table 1, even with significantly larger search space, our total cost is lower than prior work. We additionally introduce one-shot formulation when LANA transforms an architecture by simply skipping blocks (cells). This setting imposes no pretraining and search cost is negligibly small.

2   Method

Our goal in this paper is *to accelerate* a given pre-trained teacher/base network by replacing its inefficient operations with more efficient alternatives. Our method, visualized in Fig. 2, is composed of two phases: (i) *Candidate pretraining phase* (Sect. 2.1), in which we use distillation to train a large set of operations to approximate different layers in the original teacher architecture; and (ii) *Operation selection phase* (Sect. 2.2), in which we search for an architecture composed of a combination of the original teacher layers and pretrained efficient operations via linear optimization.

2.1   Candidate Pretraining Phase

We represent the teacher (base) network as the composition of N teacher operations by $\mathcal{T}(x) = t_N \circ t_{N-1} \circ \ldots \circ t_1(x)$, where x is the input tensor, t_i is the ith operation (i.e., layer) in the network. We then define the set of *candidate student operations* $\bigcup_{i=1}^{N} \{s_{ij}\}_{j=1}^{M}$, which will be used to approximate the teacher operations. Here, M denotes the number of candidate operations per layer. The student operations can draw from a wide variety of operations – the only requirement is that all candidate operations for a given layer must have the same input and output tensor dimensions as the teacher operation t_i. We denote all the parameters (e.g., trainable convolutional filters) of the operations as $\mathbf{W} = \{w_{ij}\}_{i,j}^{N,M}$, where w_{ij} denotes the parameters of the student operation s_{ij}. We use a set of binary vectors $\mathbf{Z} = \{\mathbf{z}_i\}_{i=1}^{N}$, where $\mathbf{z}_i = \{0,1\}^M$ is a one-hot vector, to represent operation selection parameters. We denote the candidate network architecture specified by $\mathbf{Z}$ using $\mathcal{S}(x; \mathbf{Z}, \mathbf{W})$.

The problem of optimal selection of operations is often tackled in NAS. This problem is usually formulated as a bi-level optimization that selects operations and optimizes their weights jointly [41,103].

Finding the optimal architecture in hardware-aware NAS reduces to:

$$\min_{\mathbf{Z}} \min_{\mathbf{W}} \underbrace{\sum_{(x,y)\in X_{tr}} \mathcal{L}\big(\mathcal{S}(x; \mathbf{Z}, \mathbf{W}), y\big)}_{\text{objective}}, \tag{1}$$

$$\text{s.t.} \quad \underbrace{\sum_{i=1}^{N} \mathbf{b}_i^T \mathbf{z}_i \leqslant \mathcal{B}}_{\text{budget constraint}} \; ; \quad \underbrace{\mathbf{1}^T \mathbf{z}_i = 1 \; \forall \; i \in [1..N]}_{\text{one op per layer}}$$

where $\mathbf{b}_i \in \mathbb{R}_+^M$ is a vector of corresponding cost of each student operation (latency, number of parameters, FLOPs, etc.) in layer i. The total budget constraint is defined via scalar $\mathcal{B}$. The objective is to minimize the loss function $\mathcal{L}$ that estimates the error with respect to the correct output y while meeting a budget constraint. In general, the optimization problem in Eq. 1 is an NP-hard combinatorial problem with an exponentially large state space (i.e., M^N). The existing NAS approaches often solve this optimization using evolutionary search [60], reinforcement learning [103] or differentiable search [41].

However, the goal of NAS is to find an architecture in the whole search space from scratch, whereas our goal is to improve efficiency of a given teacher network by replacing operations. Thus, our search can be considered as searching in the architecture space in the proximity of the already trained model. That is why we assume that the functionality of each candidate operation is also similar to the teacher's operation, and we train each candidate operation to mimic the teacher operation

using layer-wise feature map distillation with the mean squared error (MSE) loss:

$$\min_{\mathbf{W}} \sum_{x\in X_{\mathrm{tr}}} \sum_{i,j}^{N,M} \|t_i(x_{i-1}) - s_{ij}(x_{i-1}; w_{ij})\|_2^2, \tag{2}$$

where X_{tr} is a set of training samples, and $x_{i-1} = t_{i-1} \circ t_{i-2} \circ \ldots \circ t_1(x)$ is the output of the previous layer of the teacher, fed to both the teacher and student operations.

Our layer-wise pretraining has several advantages. First, the minimization in Eq. 2 can be decomposed into $N \times M$ independent minimization problems as $w_{i,j}$ is specific to one minimization problem per operation and layer. This allows us to train all candidate operations simultaneously in parallel. Second, since each candidate operation is tasked with an easy problem of approximating *one* layer in the teacher network, we can train the student operation quickly in one epoch. In this paper, instead of solving all $N \times M$ problems in separate processes, we train a single operation for each layer in the same forward pass of the teacher to maximize reusing the output features produced in all the teacher layers. This way the pretraining phase roughly takes $O(M)$ epochs of training a full network.

2.2   Operation Selection Phase

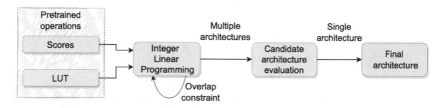

Fig. 3. Architecture search with LANA: Given scores of pretrained operations and their associated cost, LANA formulates the architecture search as an integer linear programming problem. Multiple architectures are found to satisfy the budget constraint via penalizing overlaps. Then, a single architecture is picked during candidate architecture evaluation phase. The cost of the overall search is minor comparing to existing NAS approaches as no training is involved.

Figure 3 shows steps involved to find an accelerated architecture. Since our goal in search is to discover an efficient network in the proximity of the teacher network, we propose a simple linear relaxation of candidate architecture loss using

$$\sum_{X_{tr}} \mathcal{L}\big(\mathcal{S}(x; \mathbf{Z}), y\big) \approx \sum_{X_{tr}} \mathcal{L}\big(\mathcal{T}(x), y\big) + \sum_{i=1}^{N} \mathbf{a}_i^T \mathbf{z}_i, \tag{3}$$

where the first term denotes the training loss of teacher which is constant and $\mathbf{a}_i$ is a vector of change values in the training loss per operation for layer i. We refer to $\mathbf{a}_i$ as a score vector. Our approximation bears similarity to the first-degree Taylor expansion of the student loss with the teacher as the reference point (since the teacher architecture is a member of the search space). To compute $\{\mathbf{a}_i\}_i^N$, after pretraining operations in the first stage, we plug each candidate operation one-by-one in the teacher network and we measure the change on training loss on a small labeled set. Our approximation relaxes the non-linear loss to a linear function. Although this is a weak approximation that ignores how different layers influence the final loss together, we empirically observe that it performs well in practice as a proxy for searching the student.

Approximating the architecture loss with a linear function allows us to formulate the search problem as solving an integer linear program (ILP). This has several main advantages: (i) Although solving integer linear programs is generally NP-hard, there exist many off-the-shelf libraries that can obtain a high-quality solutions in a few seconds. (ii) Since integer linear optimization libraries easily scale up to millions of variables, our search also scales up easily to very large number of candidate operations per layer. (iii) We can easily formulate the search problem such that instead of one architecture, we obtain a set of diverse candidate architectures. Formally, we denote the kth solution with $\left\{\mathbf{Z}^{(k)}\right\}_{k=1}^{K}$, which is obtained by solving:

$$\min_{\mathbf{Z}^{(k)}} \underbrace{\sum_{i=1}^{N} \mathbf{a}_i^T \mathbf{z}_i^{(k)}}_{\text{objective}}, \quad \text{s.t.} \quad \underbrace{\sum_{i=1}^{N} \mathbf{b}_i^T \mathbf{z}_i^{(k)} \leqslant \mathcal{B};}_{\text{budget constraint}} \quad \underbrace{\mathbf{1}^T \mathbf{z}_i^{(k)} = 1 \ \forall \ i;}_{\text{one op per layer}}$$

$$\underbrace{\sum_{i=1}^{N} \mathbf{z}_i^{(k)^T} \mathbf{z}_i^{(k')} \leqslant \mathcal{O}, \forall k' < k}_{\text{overlap constraint}} \tag{4}$$

where we minimize the change in the loss while satisfying the budget and overlap constraint. The scalar $\mathcal{O}$ sets the maximum overlap with any previous solution which is set to $0.7N$ in our case. We obtain K diverse solutions by solving the minimization above K times.

Solving the Integer Linear Program (ILP). We use the off-the-shelf PuLP Python package to find feasible candidate solutions. The cost of finding the first solution is very small, often less than 1 CPU-second. As K increases, so does the difficulty of finding a feasible solution. We limit K to ∼100.

Candidate Architecture Evaluation. Solving Eq. 4 provides us with K architectures. The linear proxy used for candidates loss is calculated in an isolated setting for each operation. To reduce the approximation error, we evaluate all K architectures with pretrained weights from phase one on a small part of the training set (6k images on ImageNet) and select the architecture with the lowest loss. This step assumes that the accuracy of the model before finetuning is positively correlated with the accuracy after finetuning. Batch normalization layers have to use current batch statistics (instead of precompted) to adopted for distribution change with new operations.

Candidate Architecture Fine-Tuning. After selecting the best architecture among the K candidate architectures, we fine-tune it for 100 epochs using the original objective used for training the teacher. Additionally, we add the distillation loss from teacher to student during fine-tuning.

3   Experiments

We apply LANA to the family of EfficientNetV1 [72], EfficientNetV2 [74] and ResNeST50 [100]. When naming our models, we use the latency reduction ratio compared to the original model according to latency look-up table (LUT). For example, $0.25\times$ B6 indicates $4\times$ target speedup for the B6 model. For experiments, ImageNet-1K [61] is used for pretraining (1 epoch), candidate evaluation (6k training images) and finetuning (100 epochs).

We use the NVIDIA V100 GPU and Intel Xeon Silver 4114 CPU as our target hardware. A hardware specific look-up table is precomputed for each candidate operation (vectors $\mathbf{b}_i$ in Eq. 4). We measure latency in 2 settings: (i) in Pytorch framework, and (ii) TensorRT [54]. The latter performs kernel fusion for additional model optimization making it even harder to accelerate models. The exact same setup is used for evaluating latency of **all** competing models, our models, and baselines. Actual latency on target platforms is reported in tables.

Table 2. Comparison to prior art on ImageNet1K. Latency is measured on NVIDIA V100 with various batch size and inference precision.

Method	Accuracy	Latency (ms), bs128/fp16	bs32/fp32
EfficientNetV1-B0	77.7	35.6	
Cream-S [56]	77.6	36.7	
DNA-D [38]	77.1	33.8	
DONNA [50]	78.9		20.0
OFA_flops@482M [5]	79.6	39.3	
LANA(0.45xEFNv1-B2)	79.7	30.2	18.9
LANA(0.4xEFNv2-B3)	**79.9**	39.0	
EfficientNetV1-B1	78.8	59.0	
DNA-D [38]	78.4	61.3	
DONNA [50]	79.5		25.0
OFA_flops@595M [5]	80.0	50.0	
Cream-L [56]	80.0	84.0	
LANA(0.55xEFNv1-B2)	80.1	48.7	24.1
LANA(0.5xEFNv2-B3)	**80.8**	48.1	

Table 3. Models optimized with LANA for latency-accuracy trade-off on ImageNet1K.

Method	Variant	Res px	Accuracy (%)	Latency(ms) TensorRT	PyTorch
EfficientNetV1					
EfficientNetV1-B1		240	78.83	29.3	59.0
LANA	0.25xB4	380	**81.83 (+3.00)**	30.4	64.5
EfficientNetV1-B2		260	80.07	38.2	77.1
LANA	0.3xB4	380	**82.16 (+2.09)**	38.8	81.8
EfficientNetV1-B3		300	81.67	67.2	125.9
LANA	0.5xB4	380	**82.66 (+0.99)**	61.4	148.1
EfficientNetV1-B4		380	83.02	132.0	262.4
LANA	0.25xB6	528	**83.77 (+0.75)**	128.8	282.1
EfficientNetV1-B5		456	83.81	265.7	525.6
LANA	0.5xB6	528	**83.99 (+0.18)**	266.5	561.2
EfficientNetV1-B6		528	84.11	466.7	895.2
EfficientNetV2					
EfficientNetV2-B1		240	79.46	17.9	44.7
LANA	0.45xB3	300	**80.30 (+0.84)**	17.8	**43.0**
EfficientNetV2-B2		260	80.21	24.3	58.9
LANA	0.6xB3	300	**81.14 (+0.93)**	23.8	**56.1**
EfficientNetV2-B3		300	81.97	41.2	91.6
ResNeST50d_1s4x24d					
ResNeST50		224	80.99	32.3	74.0
LANA	0.7x	224	80.85	**22.3(1.45x)**	**52.7**

Candidate Operations. We construct a large of pool of diverse candidate operation including $M = 197$ operations for each layer of teacher. Our operations include:

Teacher operation is used as is in the pretrained (base) model with teacher model accuracy.

Identity is used to skip teacher's operation. It changes the depth of the network.

Inverted residual blocks `efn` [63] and `efnv2` [74] with varying expansion factor $e = \{1, 3, 6\}$, squeeze and excitation ratio $se = \{No, 0.04, 0.025\}$, and kernel size $k = \{1, 3, 5\}$.

Dense convolution blocks inspired by [22] with (i) two stacked convolution (`cb_stack`) with CBRCB structure, C-conv, B-batchnorm, R-Relu; (ii) bottleneck architecture (`cb_bottle`) with CBR-CBR-CB; (ii) CB pair (`cb_res`); (iii) RepVGG block [15]; (iv) CBR pairs with perturbations as `conv_cs`. For all models we vary kernel size $k = \{1, 3, 5, 7\}$ and width $w = \{1/16, 1/10, 1/8, 1/5, 1/4, 1/2, 1, 2, 3, 4\}$.

Transformer variations (i) visual transformer block (`vit`) [17] with depth $d = \{1, 2\}$, dimension $w = \{2^5, 2^6, 2^7, 2^8, 2^9, 2^{10}\}$ and heads $h = \{4, 8, 16\}$; (ii) bottleneck transformers [66] with 4 heads and expansion factor $e = \{1/4, 1/2, 1, 2, 3, 4\}$; (iii) lambda bottleneck layers [2] with expansion $e = \{1/4, 1/2, 1, 2, 3, 4\}$.

With the pool of 197 operations, distilling from an EfficientNet-B6 model with 46 layers yields a design space of the size $197^{46} \approx 10^{100}$.

3.1   EfficientNet and ResNeST Derivatives

Our experimental results on accelerating EfficientNetV1 (B2, B4, B6), Efficient-NetV2 (B3), and ResNeST50 family for GPUs are shown in Tables 2 and 3. Comparison with more models from `timm` is in the Appendix.

At first we compare to prior NAS-like models tailored to EfficientNetV1-B0 and B1 in Table 2. LANA has clear advantages in terms of accuracy and latency.

Next, we demonstrate a capability of LANA for a variety of larger architectures in Table 3. Results show that:

- LANA achieves an accuracy boost of 0.18–3.0% for all models when compressing larger models to the latency level of smaller models (see EfficientNet models and the corresponding LANA models in the same latency group).
- LANA achieves significant real speed-ups with little to no accuracy drop: (i) EfficieintNetV1-B6 accelerated by 3.6x and 1.8x times by trading-off 0.34% and 0.11% accuracy, respectively; (ii) B2 variant is accelerated 2.4x and 1.9x times with 0.36% and no accuracy loss, respectively. (iii) ResNeST50 is accelerated 1.5x with 0.14% accuracy drop.

Table 4. Local pretraining of the teacher model initialized from scratch on ImageNet1K at different granularity.

Model	Original	Distillation	
		Block-wise [38], [50]	Layer-wise (LANA)
EfficientNetV1-B2	82.01	68.21	76.52
EfficientNetV2-B3	81.20	73.53	76.52
ResNeST-50	86.35	77.54	80.61

Table 5. Comparing methods for candidate selection (NAS). Our proposed ILP is better (+0.43%) and 821× faster.

Method	Accuracy	Search cost
ILP, K=100 (ours)	79.28	4.5 CPU/m
Random, found 80 arch	76.44	1.4 CPU/m
SNAS [88]	74.20	16.3 GPU/h
E-NAS [57]	78.85	61.6 GPU/h

Table 6. Impact of the search space on 0.55× EfficientNetV1-B2 compression. Two operations correspond to Single-shot LANA.

Operations	2	5	10	All
Space size	$O(10^7)$	$O(10^{16})$	$O(10^{23})$	$O(10^{46})$
Accuracy	79.40	79.52	79.66	80.00
STD		±0.208	±0.133	

Table 7. Single-shot LANA with only skip connections.

Setup	Top-1 Acc.		Latency (ms)
	Single-shot	All	TensorRT
0.45xEfficientNetV1-B2	78.68	79.71	16.2
0.55xEfficientNetV1-B2	79.40	80.11	20.6

A detailed look on EfficientNets results demonstrate that EfficientNetV1 models are accelerated beyond EfficientNetV2. LANA generates models that have better accuracy-throughput trade-off when comparing models under the same accuracy or the same latency. LANA also allows us to optimize models for different hardware at a little cost. Only a new LUT is required to get optimal model for a new hardware without pretraining the candidate operations again. We present models optimized for CPU in supplementary materials Table 8 which are obtained using a different LUT only.

3.2 Analysis

Here, we provide detailed ablations to analyze our design choices in LANA for both pretraining and search phases, along with observed insights. We use EfficientNetV1-B2 and EfficientNetV2-B3 as our base models for the ablation.

Operator Pretraining. Previous work ([38] and [50]) applies per block distillation for pretraining. Main reason for that is costly search if they operate in per layer setting. With per layer pretraining the search cost increases exponentially. Main advantage of our work is ILP that it is faster by several orders of magnitude. Therefore, we can perform per layer distillation.

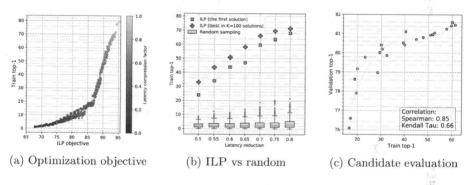

(a) Optimization objective   (b) ILP vs random   (c) Candidate evaluation

Fig. 4. Analyzing ILP performance on EfficientNetV2-B3. ILP results in significantly higher model accuracy before finetuning than 1k randomly sampled architectures in (b). Accuracy monotonically increases with ILP objective (a). Model accuracy before finetuning correctly ranks models after finetuning (c). Train top-1 is measured *before* finetuning, while Validation top-1 is *after*.

To study the benefit of using per layer distillation we perform a teacher mimicking test where all parameters of the teacher are re-initialized. Then, we perform isolated pretraining (with MSE loss) of teacher blocks/layers and report the final top1 accuracy on the training set on the ImageNet. Results in Table 4 clearly demonstrate significant advantage of per layer pretraining. By carefully studying we found that per block supervision provides very little guidance to first layers of the block and therefore lacks performance. We conclude that per layer distillation is more efficient and provides extra advantages.

Linear Relaxation in architecture search assumes that a candidate architecture can be scored by a fitness metric measured independently for all operations. Although this relaxation is not accurate, we observe a strong correlation between our linear objective and the training loss of the full architecture. This assumption is verified by sampling 1000 architectures (different budget constraints), optimizing the ILP objective, and measuring the real loss function. Results are shown in Fig. 4a using the train accuracy as the loss. We observe that ILP objective ranks models with respect to the measured accuracy correctly under different latency budgets. The Kendall Tau correlation is 0.966.

To evaluate the quality of the solution provided with ILP, we compare it with random sampling. The comparison is shown in Fig. 4b, where we sample 1000 random architectures for 7 latency budgets. The box plots indicate the poor performance of the randomly sampled architectures. The first ILP solution has significantly higher accuracy than random architecture. Furthermore, finding multiple diverse solutions is possible with ILP using the overlap constraint. If we increase the number of solutions found by ILP from $K = 1$ to $K = 100$, performance improves further. When plugging pretrained operations (Fig. 4b) into the teacher, the accuracy is high (it is above 30%, even at an acceleration factor of 2×). For EfficientNetV1, this is above 50% for the same compression factor.

Candidate architecture evaluation plays an important role in LANA. This step finds the best architecture quickly out of the diverse candidates generated by the ILP solver, by evaluating them on 6k images from the train data. The procedure is built on the assumption that the accuracy of the model on the training data before finetuning (just by plugging all candidate operations) is a reasonable indicator of the relative model performance after finetuning. We verify this hypothesis in the Fig. 4c and see positive correlation. Same observation was is present in other works like [37].

Comparing with Other NAS Approaches. We compare our search algorithm with other popular approaches to solve Eq. (1), including: (i) *Random* architecture sampling within a latency constrain; (ii) *Differentiable search with Gumbel Softmax* – a popular approach in NAS to relax binary optimization as a continuous variable optimization via learning the sampling distribution [77,86,88]. We follow SNAS [88] in this experiment; (iii) *REINFORCE* is a stochastic optimization framework borrowed from reinforcement learning and adopted for architecture search [57,71,104]. We follow an E-NAS-like [57] architecture search for (iii) and use weight sharing for (ii) and (iii).

Experiments are conducted on EfficientNetV1-B2 accelerated to 0.45× original latency. The final validation top-1 accuracy after finetuning are presented in Table 5. Our proposed ILP achieves higher accuracy (+0.43%) compared to the second best method E-NAS while being 821× faster in search.

Single-Shot LANA. Our method can be applied without pretraining procedure if only teacher cells and *Identity* (skip) operation are used ($M = 2$ operations per layer). Only the vector for the change in loss for the *Identity* operator will be required alongside the LUT for the teacher operations. This allows us to do single-shot network acceleration without any pretraining as reported in Table 7. We observe that LANA efficiently finds residual blocks that can be skipped. This unique property of LANA is enabled because of layer-wise granularity.

Pretraining Insights. To gain more insights into the tradeoff between the accuracy and speed of each operation, we analyze the pretrained candidate operation pool for EffientNetV1B2. A detailed figure is shown in the appendix. Here, we provide general observations.

We observe that no operation outperforms the teacher in terms of accuracy; changing pretraining loss from per-layer MSE to full-student cross-entropy may change this but that comes with an increased costs of pretraining. We also see that it is increasingly difficult to recover the teacher's accuracy as the depth in the network increases. The speedups achievable are roughly comparable across different depths, however, we note that achieving such speedups earlier in the network is particularly effective towards reducing total latency due to the first third of the layers accounting for 54% of the total inference time.

Looking at individual operations, we observe that inverted residual blocks (efn, efnv2) are the most accurate throughout the network, at the expense of increased computational cost (*i.e.,* lower speedups). Dense convolutions (cb_stack, cb_bottle, conv_cs, cb_res) exhibit a good compromise between accuracy and speed, with stacked convolutions being particularly effective earlier in the network.

Visual transformer blocks (ViT) and bottleneck transformer blocks (bot_trans) show neither a speedup advantage nor ability to recover teacher accuracy.

Common Operations. In Appendix 5, we analyze the distribution of the selected operations by solving for 100 architecture candidates with Eq. 4 for $0.5\times$ EfficientNetV1B2. We observe: (i) teacher operations are selected most frequently, it is expected as we only transform the model and no operations can beat teacher in terms of accuracy; (ii) identity operation is selected more often for large architectures with higher compression rates; (iii) dense convolution blocks tend to appear more for larger models like EfficientNet-B6 to speed them up.

Architecture Insights. In the appendices, we visualize the final architectures discovered by LANA. Next, we share the insights observed on these architectures.

Common Across All Models: (i) Teacher's operations are usually selected in the tail of the network as they are relatively fast and approximating them results in the highest error. (ii) Teacher operation is preferred for downsampling layers (e.g. stride is 2). Those cells are hard to approximate as well, we hypothesize this is due to high nonlinearity of these blocks. (iii) Transformer blocks are never selected, probably because they require significantly longer pretraining on larger datasets. (v) LANA automatically adjusts depth of every resolutional block by replacing teacher cells with Identity operations.

EfficientNetV1. Observing final architectures obtained by LANA on the EfficientNetV1 family, particularly the 0.55× B2 version optimized for GPUs, we discover that most of the modifications are done to the first half of the model: (i) squeeze-and-excitation is removed in the early layers; (ii) dense convolutions (like inverted stacked or bottleneck residual blocks) replace depth-wise separable counterparts; (iii) the expansion factor is reduced from 6 to 3.5 on average. *Surprisingly, LANA automatically discovers the same design choices that are introduced in the EfficientNetV2 family when optimized for datacenter inference.*

EfficientNetV2. LANA accelerates EfficientNetV2-B3 by 2×, with the following conclusions: (i) the second conv-bn-act layer is not needed and can be removed; (ii) the second third of the model benefits from reducing the expansion factor from 4 to 1 without squeeze-and-excitation operations. With these simplifications, accelerated model still outperforms EfficientNetV2-B2 and B1.

ResNeST50. LANA discovers that cardinality can be reduced from 4 to 1 or 2 for most blocks without any loss in accuracy, yielding a 1.45× speedup.

Ablations on Finetuning

We select 0.45× EfficientNetV1-B2 with the final accuracy of 79.71%.

Pretrained Weights. We look deeper into the finetuning step. Reinitializing all weights in the model, as opposed to loading them from the pretraining stage, results in 79.42%. The result indicates the importance of pretraining stage (i) to find a strong architecture and (ii) to boost finetuning.

Knowledge distillation plays a key role in student-teacher setups. When it is disabled, we observe an accuracy degradation of 0.65%. This emphasizes the benefit of training a larger model and then self-distilling to a smaller one. We further verify whether we can achieve a similar high accuracy using knowledge distillation from EfficientNetV1-B2 to vanilla EfficientNetV1-B0 in the same setting. The top-1 accuracy of 78.72% is still 1% less than LANA's accuracy. When both models are trained from scratch with the distillation loss, LANA 0.45xB2 achieves 79.42% while EfficientNetV1-B0 achieves 78.01%.

Length of Finetuning. Pretrained operations have already been trained to mimic the teacher layer. Therefore, even before finetuning the student model can be already adept at the task. Next, we evaluate how does the length of finetuning affects the final accuracy in Table 9 (Supplementary). Even with only 5 epochs LANA outperforms the vanilla EfficientNet counterparts.

Search Space Size. ILP enables us to perform NAS in a very large space ($O(10^{100})$). To verify the benefit of large search space, we experiment with a restricted search space. For this, we randomly pick 2, 5 and 10 operations per layer to participate in search and finetuning for 50 epochs. We observe clear improvements from increasing the search space, shown Table 6 (results are averaged over 5 runs). When more than 2 operations are present we select the teacher cell, identity operation and the rest to be random operations with the constrain to have score difference of no more than 5.0% and being faster.

4   Conclusion

In this paper, we proposed LANA, a hardware-aware network transformation framework for accelerating pretrained neural networks. LANA uses a NAS-like search to replace inefficient operations with more efficient alternatives. It tackles this problem in two phases including a candidate pretraining phase and a search phase. The availability of the teacher network allows us to estimate the change in accuracy for each operation at each layer. Using this, we formulate the search problem as solving a linear integer optimization problem, which outperforms the commonly used NAS algorithms while being orders of magnitude faster. We applied our framework to accelerate EfficientNets (V1 and V2) and ResNets with a pool of 197 operations per layer and we observed that LANA accelerates these architectures by several folds with only a small drop in the accuracy.

Limitations. The student performance in LANA is bounded by the base model, and it rarely passes in terms of accuracy. Additionally, the output dimension of layers in the student can not be changed, and must remain the same as in the original base model.

Future Work. We envision that a layer-wise network acceleration framework like LANA can host a wide range of automatically and manually designed network operations, developed in the community. Our design principals in LANA consisting of extremely large operation pool, efficient layer-wise pretraining, and lightening fast search help us realize this vision. Components of LANA can be further improved in the follow-up research: i) pretraining stage to consider error propagation; ii) scoring metric; iii) ILP with finetuning; iv) candidate architecture evaluation.

References

1. Belagiannis, V., Farshad, A., Galasso, F.: Adversarial network compression. In: Leal-Taixé, L., Roth, S. (eds.) ECCV 2018. LNCS, vol. 11132, pp. 431–449. Springer, Cham (2019). https://doi.org/10.1007/978-3-030-11018-5_37
2. Bello, I.: Lambdanetworks: modeling long-range interactions without attention. arXiv preprint arXiv:2102.08602 (2021)
3. Bello, I., et al.: Revisiting ResNets: improved training and scaling strategies (2021)
4. Blalock, D., Ortiz, J.J.G., Frankle, J., Guttag, J.: What is the state of neural network pruning? arXiv preprint arXiv:2003.03033 (2020)
5. Cai, H., Gan, C., Wang, T., Zhang, Z., Han, S.: Once for all: train one network and specialize it for efficient deployment. In: International Conference on Learning Representations (2020). https://arxiv.org/pdf/1908.09791.pdf
6. Cai, H., Zhu, L., Han, S.: ProxylessNAS: direct neural architecture search on target task and hardware. In: International Conference on Learning Representations (2019). https://openreview.net/forum?id=HylVB3AqYm
7. Cai, Y., Yao, Z., Dong, Z., Gholami, A., Mahoney, M.W., Keutzer, K.: ZeroQ: a novel zero shot quantization framework. In: Proceedings of the IEEE/CVF Conference on Computer Vision and Pattern Recognition, pp. 13169–13178 (2020)

8. Chauvin, Y.: A back-propagation algorithm with optimal use of hidden units. In: NIPS (1989)

9. Chen, L.-C., Zhu, Y., Papandreou, G., Schroff, F., Adam, H.: Encoder-decoder with atrous separable convolution for semantic image segmentation. In: Ferrari, V., Hebert, M., Sminchisescu, C., Weiss, Y. (eds.) ECCV 2018. LNCS, vol. 11211, pp. 833–851. Springer, Cham (2018). https://doi.org/10.1007/978-3-030-01234-2_49

10. Chen, Y., Li, J., Xiao, H., Jin, X., Yan, S., Feng, J.: Dual path networks. arXiv preprint arXiv:1707.01629 (2017)

11. Choi, J., Wang, Z., Venkataramani, S., Chuang, P., Srinivasan, V., Gopalakrishnan, K.: Pact: parameterized clipping activation for quantized neural networks. arXiv:abs/1805.06085 (2018)

12. Chollet, F.: Xception: deep learning with depthwise separable convolutions (2017)

13. Dai, X., et al.: ChamNet: towards efficient network design through platform-aware model adaptation. In: CVPR (2019)

14. Ding, X., Guo, Y., Ding, G., Han, J.: ACNet: strengthening the kernel skeletons for powerful CNN via asymmetric convolution blocks. In: Proceedings of the IEEE/CVF International Conference on Computer Vision, pp. 1911–1920 (2019)

15. Ding, X., Zhang, X., Ma, N., Han, J., Ding, G., Sun, J.: RepVGG: making VGG-style convnets great again. arXiv preprint arXiv:2101.03697 (2021)

16. Dong, Z., Yao, Z., Gholami, A., Mahoney, M.W., Keutzer, K.: HAWQ: hessian aware quantization of neural networks with mixed-precision. In: Proceedings of the IEEE/CVF International Conference on Computer Vision, pp. 293–302 (2019)

17. Dosovitskiy, A., et al.: An image is worth 16x16 words: transformers for image recognition at scale. arXiv preprint arXiv:2010.11929 (2020)

18. Frankle, J., Carbin, M.: The lottery ticket hypothesis: finding sparse, trainable neural networks. In: International Conference on Learning Representations (2018)

19. Gordon, A., et al.: MorphNet: fast & simple resource-constrained structure learning of deep networks. In: CVPR (2018)

20. Hanson, S.J., Pratt, L.Y.: Comparing biases for minimal network construction with back-propagation. In: NIPS (1989)

21. He, K., Zhang, X., Ren, S., Sun, J.: Deep residual learning for image recognition. In: CVPR (2016)

22. He, K., Zhang, X., Ren, S., Sun, J.: Identity mappings in deep residual networks. In: Leibe, B., Matas, J., Sebe, N., Welling, M. (eds.) ECCV 2016. LNCS, vol. 9908, pp. 630–645. Springer, Cham (2016). https://doi.org/10.1007/978-3-319-46493-0_38

23. He, T., Zhang, Z., Zhang, H., Zhang, Z., Xie, J., Li, M.: Bag of tricks for image classification with convolutional neural networks (2018)

24. He, Y., Dong, X., Kang, G., Fu, Y., Yang, Y.: Progressive deep neural networks acceleration via soft filter pruning. arXiv preprint arXiv:1808.07471 (2018)

25. He, Y., Kang, G., Dong, X., Fu, Y., Yang, Y.: Soft filter pruning for accelerating deep convolutional neural networks. In: Proceedings of the 27th International Joint Conference on Artificial Intelligence, pp. 2234–2240 (2018)

26. He, Y., Liu, P., Wang, Z., Hu, Z., Yang, Y.: Filter pruning via geometric median for deep convolutional neural networks acceleration. In: Proceedings of the IEEE/CVF Conference on Computer Vision and Pattern Recognition, pp. 4340–4349 (2019)

27. He, Y., Han, S.: ADC: automated deep compression and acceleration with reinforcement learning. arXiv preprint arXiv:1802.03494 (2018)

28. He, Y., et al.: AMC: AutoML for model compression and acceleration on mobile devices. In: Ferrari, V., Hebert, M., Sminchisescu, C., Weiss, Y. (eds.) ECCV 2018. LNCS, vol. 11211, pp. 815–832. Springer, Cham (2018). https://doi.org/10.1007/978-3-030-01234-2_48
29. He, Y., Zhang, X., Sun, J.: Channel pruning for accelerating very deep neural networks. In: Proceedings of the IEEE International Conference on Computer Vision, pp. 1389–1397 (2017)
30. Heo, B., Yun, S., Han, D., Chun, S., Choe, J., Oh, S.J.: Rethinking spatial dimensions of vision transformers (2021)
31. Hinton, G., Vinyals, O., Dean, J.: Distilling the knowledge in a neural network. arXiv preprint arXiv:1503.02531 (2015)
32. Hu, J., Shen, L., Albanie, S., Sun, G., Wu, E.: Squeeze-and-excitation networks (2019)
33. Huang, G., Liu, Z., Van Der Maaten, L., Weinberger, K.Q.: Densely connected convolutional networks. In: CVPR (2017)
34. Krishnamoorthi, R.: Quantizing deep convolutional networks for efficient inference: a whitepaper. arXiv:abs/1806.08342 (2018)
35. Lebedev, V., Lempitsky, V.: Fast convnets using group-wise brain damage. In: CVPR, pp. 2554–2564 (2016)
36. LeCun, Y., Denker, J.S., Solla, S., Howard, R.E., Jackel, L.D.: Optimal brain damage. In: NIPS (1990)
37. Li, B., Wu, B., Su, J., Wang, G.: EagleEye: fast sub-net evaluation for efficient neural network pruning. In: Vedaldi, A., Bischof, H., Brox, T., Frahm, J.-M. (eds.) ECCV 2020. LNCS, vol. 12347, pp. 639–654. Springer, Cham (2020). https://doi.org/10.1007/978-3-030-58536-5_38
38. Li, C., et al.: Block-wisely supervised neural architecture search with knowledge distillation. In: Proceedings of the IEEE/CVF Conference on Computer Vision and Pattern Recognition (CVPR), June 2020
39. Li, H., Kadav, A., Durdanovic, I., Samet, H., Graf, H.P.: Pruning filters for efficient ConvNets. In: ICLR (2017)
40. Li, X., Wang, W., Hu, X., Yang, J.: Selective kernel networks (2019)
41. Liu, H., Simonyan, K., Yang, Y.: Darts: differentiable architecture search. arXiv preprint arXiv:1806.09055 (2018)
42. Liu, Y., et al.: Search to distill: pearls are everywhere but not the eyes. In: Proceedings of the IEEE/CVF Conference on Computer Vision and Pattern Recognition (CVPR), June 2020
43. Liu, Z., Sun, M., Zhou, T., Huang, G., Darrell, T.: Rethinking the value of network pruning. arXiv preprint arXiv:1810.05270 (2018)
44. Louizos, C., Welling, M., Kingma, D.P.: Learning sparse neural networks through l_0 regularization. arXiv preprint arXiv:1712.01312 (2017)
45. Luo, J.H., Wu, J., Lin, W.: ThiNet: a filter level pruning method for deep neural network compression. In: ICCV (2017)
46. Mishra, A., Marr, D.: Apprentice: using knowledge distillation techniques to improve low-precision network accuracy. In: ICLR (2018)
47. Molchanov, D., Ashukha, A., Vetrov, D.: Variational dropout sparsifies deep neural networks. In: International Conference on Machine Learning, pp. 2498–2507. PMLR (2017)
48. Molchanov, P., Mallya, A., Tyree, S., Frosio, I., Kautz, J.: Importance estimation for neural network pruning. In: CVPR (2019)
49. Molchanov, P., Tyree, S., Karras, T., Aila, T., Kautz, J.: Pruning convolutional neural networks for resource efficient transfer learning. In: ICLR (2017)

50. Moons, B., et al.: Distilling optimal neural networks: rapid search in diverse spaces. arXiv preprint arXiv:2012.08859 (2020)
51. Mozer, M.C., Smolensky, P.: Skeletonization: a technique for trimming the fat from a network via relevance assessment. In: NIPS (1989)
52. Nayak, G.K., Mopuri, K.R., Shaj, V., Babu, R.V., Chakraborty, A.: Zero-shot knowledge distillation in deep networks. In: CVPR (2019)
53. Neklyudov, K., Molchanov, D., Ashukha, A., Vetrov, D.P.: Structured Bayesian pruning via log-normal multiplicative noise. In: Advances in Neural Information Processing Systems, pp. 6775–6784 (2017)
54. NVIDIA: TensorRT Library. https://developer.nvidia.com/tensorrt (2021). Accessed 10 May 2021
55. Park, E., Yoo, S., Vajda, P.: Value-aware quantization for training and inference of neural networks. In: Ferrari, V., Hebert, M., Sminchisescu, C., Weiss, Y. (eds.) ECCV 2018. LNCS, vol. 11208, pp. 608–624. Springer, Cham (2018). https://doi.org/10.1007/978-3-030-01225-0_36
56. Peng, H., Du, H., Yu, H., Li, Q., Liao, J., Fu, J.: Cream of the crop: distilling prioritized paths for one-shot neural architecture search. In: Advances in Neural Information Processing Systems, vol. 33, 17955–17964 (2020)
57. Pham, H., Guan, M.Y., Zoph, B., Le, Q.V., Dean, J.: Efficient neural architecture search via parameter sharing. arXiv preprint arXiv:1802.03268 (2018)
58. Radosavovic, I., Kosaraju, R.P., Girshick, R., He, K., Dollár, P.: Designing network design spaces (2020)
59. Rastegari, M., Ordonez, V., Redmon, J., Farhadi, A.: XNOR-net: ImageNet classification using binary convolutional neural networks. In: Leibe, B., Matas, J., Sebe, N., Welling, M. (eds.) ECCV 2016. LNCS, vol. 9908, pp. 525–542. Springer, Cham (2016). https://doi.org/10.1007/978-3-319-46493-0_32
60. Real, E., et al.: Large-scale evolution of image classifiers. In: International Conference on Machine Learning, pp. 2902–2911. PMLR (2017)
61. Russakovsky, O., et al.: ImageNet large scale visual recognition challenge. Int. J. Comput. Vis. **115**(3), 211–252 (2015). https://doi.org/10.1007/s11263-015-0816-y
62. Ryoo, S., Rodrigues, C., Baghsorkhi, S.S., Stone, S.S., Kirk, D., Hwu, W.: Optimization principles and application performance evaluation of a multithreaded GPU using CUDA. In: Proceedings of the 13th ACM SIGPLAN Symposium on Principles and Practice of Parallel Programming (2008)
63. Sandler, M., Howard, A., Zhu, M., Zhmoginov, A., Chen, L.C.: MobileNetV 2: inverted residuals and linear bottlenecks. In: CVPR (2018)
64. Sau, B.B., Balasubramanian, V.N.: Deep model compression: distilling knowledge from noisy teachers. arXiv preprint arXiv:1610.09650 (2016)
65. Shen, S., et al.: Q-BERT: Hessian based ultra low precision quantization of BERT. In: Proceedings of the AAAI Conference on Artificial Intelligence, pp. 8815–8821 (2020)
66. Srinivas, A., Lin, T.Y., Parmar, N., Shlens, J., Abbeel, P., Vaswani, A.: Bottleneck transformers for visual recognition. arXiv preprint arXiv:2101.11605 (2021)
67. Sze, V., Chen, Y.H., Yang, T.J., Emer, J.S.: Efficient processing of deep neural networks: a tutorial and survey. Proc. IEEE **105**(12), 2295–2329 (2017)
68. Szegedy, C., Ioffe, S., Vanhoucke, V., Alemi, A.: Inception-v4, inception-ResNet and the impact of residual connections on learning. In: Proceedings of the AAAI Conference on Artificial Intelligence (2017)

69. Szegedy, C., Vanhoucke, V., Ioffe, S., Shlens, J., Wojna, Z.: Rethinking the inception architecture for computer vision. In: Proceedings of the IEEE Conference on Computer Vision and Pattern Recognition, pp. 2818–2826 (2016)
70. Tan, M., Chen, B., Pang, R., Vasudevan, V., Le, Q.V.: MnasNet: platform-aware neural architecture search for mobile. arXiv preprint arXiv:1807.11626 (2018)
71. Tan, M., et al.: MnasNet: platform-aware neural architecture search for mobile. In: Proceedings of the IEEE/CVF Conference on Computer Vision and Pattern Recognition (CVPR), pp. 2820–2828, June 2019
72. Tan, M., Le, Q.: EfficientNet: rethinking model scaling for convolutional neural networks. In: International Conference on Machine Learning, pp. 6105–6114. PMLR (2019)
73. Tan, M., Le, Q.V.: MixConv: mixed depthwise convolutional kernels. arXiv preprint arXiv:1907.09595 (2019)
74. Tan, M., Le, Q.V.: EfficientNetV2: smaller models and faster training. arXiv preprint arXiv:2104.00298 (2021)
75. Touvron, H., Cord, M., Douze, M., Massa, F., Sablayrolles, A., Jégou, H.: Training data-efficient image transformers & distillation through attention. arXiv preprint arXiv:2012.12877 (2020)
76. Touvron, H., Cord, M., Sablayrolles, A., Synnaeve, G., Jégou, H.: Going deeper with image transformers. arXiv preprint arXiv:2103.17239 (2021)
77. Vahdat, A., Mallya, A., Liu, M.Y., Kautz, J.: UNAS: differentiable architecture search meets reinforcement learning. In: Proceedings of the IEEE/CVF Conference on Computer Vision and Pattern Recognition, pp. 11266–11275 (2020)
78. Veniat, T., Denoyer, L.: Learning time/memory-efficient deep architectures with budgeted super networks. arXiv preprint arXiv:1706.00046 (2017)
79. Wang, C.Y., Liao, H.Y.M., Wu, Y.H., Chen, P.Y., Hsieh, J.W., Yeh, I.H.: CSPNet: a new backbone that can enhance learning capability of CNN. In: Proceedings of the IEEE/CVF Conference on Computer Vision and Pattern Recognition Workshops, pp. 390–391 (2020)
80. Wang, G., Lin, Y., Yi, W.: Kernel fusion: an effective method for better power efficiency on multithreaded GPU. In: 2010 IEEE/ACM International Conference on Green Computing and Communications and International Conference on Cyber, Physical and Social Computing, pp. 344–350 (2010)
81. Wang, J., et al.: Deep high-resolution representation learning for visual recognition. IEEE Trans. Pattern Anal. Mach. Intell. **43**, 3349–3364 (2020)
82. Wang, K., Liu, Z., Lin, Y., Lin, J., Han, S.: HAQ: hardware-aware automated quantization with mixed precision. In: CVPR (2019)
83. Wang, Q., Wu, B., Zhu, P., Li, P., Zuo, W., Hu, Q.: ECA-Net: efficient channel attention for deep convolutional neural networks. In: 2020 IEEE/CVF Conference on Computer Vision and Pattern Recognition (CVPR) (2020)
84. Wightman, R.: PyTorch image models. https://github.com/rwightman/pytorch-image-models (2019). https://doi.org/10.5281/zenodo.4414861
85. Wu, B., Wang, Y., Zhang, P., Tian, Y., Vajda, P., Keutzer, K.: Mixed precision quantization of convnets via differentiable neural architecture search. arXiv:abs/1812.00090 (2018)
86. Wu, B., et al.: FBNet: hardware-aware efficient convnet design via differentiable neural architecture search. In: Proceedings of the IEEE/CVF Conference on Computer Vision and Pattern Recognition (CVPR), June 2019
87. Xie, S., Girshick, R.B., Dollár, P., Tu, Z., He, K.: Aggregated residual transformations for deep neural networks. CoRR abs/1611.05431 (2016), http://arxiv.org/abs/1611.05431

88. Xie, S., Zheng, H., Liu, C., Lin, L.: SNAS: stochastic neural architecture search. In: International Conference on Learning Representations (2019). https://openreview.net/forum?id=rylqooRqK7

89. Xu, Z., Hsu, Y.C., Huang, J.: Training shallow and thin networks for acceleration via knowledge distillation with conditional adversarial networks. In: ICLR Workshop (2018)

90. Yang, T.J., et al.: NetAdapt: platform-aware neural network adaptation for mobile applications. Energy **41**, 46 (2018)

91. Yang, Y., et al.: Synetgy: algorithm-hardware co-design for convnet accelerators on embedded FPGAs. arXiv preprint arXiv:1811.08634 (2018)

92. Ye, J., Lu, X., Lin, Z., Wang, J.Z.: Rethinking the smaller-norm-less-informative assumption in channel pruning of convolution layers. ICLR (2018)

93. Yin, H., et al.: Dreaming to distill: data-free knowledge transfer via deepinversion. In: CVPR (2020)

94. Yu, F., Wang, D., Shelhamer, E., Darrell, T.: Deep layer aggregation. In: Proceedings of the IEEE Conference on Computer Vision and Pattern Recognition, pp. 2403–2412 (2018)

95. Yu, R., et al.: NISP: pruning networks using neuron importance score propagation. In: CVPR (2017)

96. Yu, R., et al.: NISP: pruning networks using neuron importance score propagation. In: CVPR (2018)

97. Zagoruyko, S., Komodakis, N.: Wide residual networks. CoRR abs/1605.07146 (2016). http://arxiv.org/abs/1605.07146

98. Zagoruyko, S., Komodakis, N.: DiracNets: training very deep neural networks without skip-connections. arXiv preprint arXiv:1706.00388 (2017)

99. Zhang, D., Yang, J., Ye, D., Hua, G.: LQ-Nets: learned quantization for highly accurate and compact deep neural networks. arXiv:abs/1807.10029 (2018)

100. Zhang, H., et al.: ResNest: split-attention networks. arXiv preprint arXiv:2004.08955 (2020)

101. Zhang, X., Zhou, X., Lin, M., Sun, J.: ShuffleNet: an extremely efficient convolutional neural network for mobile devices. arXiv preprint arXiv:1707.01083 (2017)

102. Zhu, C., Han, S., Mao, H., Dally, W.J.: Trained ternary quantization. In: ICLR (2017)

103. Zoph, B., Le, Q.V.: Neural architecture search with reinforcement learning. arXiv preprint arXiv:1611.01578 (2016)

104. Zoph, B., Vasudevan, V., Shlens, J., Le, Q.V.: Learning transferable architectures for scalable image recognition. In: CVPR, pp. 8697–8710 (2018)

RDO-Q: Extremely Fine-Grained Channel-Wise Quantization via Rate-Distortion Optimization

Zhe Wang[1], Jie Lin[1(✉)], Xue Geng[1], Mohamed M. Sabry Aly[2], and Vijay Chandrasekhar[1,2]

[1] Institute for Infocomm Research, A*STAR, Singapore 138632, Singapore
{wangz,geng_xue}@i2r.a-star.edu.sg, jie.dellinger@gmail.com
[2] Nanyang Technological University, 50 Nanyang Ave, Singapore 639798, Singapore
msabry@ntu.edu.sg

Abstract. Allocating different bit widths to different channels and quantizing them independently bring higher quantization precision and accuracy. Most of prior works use equal bit width to quantize all layers or channels, which is sub-optimal. On the other hand, it is very challenging to explore the hyperparameter space of channel bit widths, as the search space increases exponentially with the number of channels, which could be tens of thousand in a deep neural network. In this paper, we address the problem of efficiently exploring the hyperparameter space of channel bit widths. We formulate the quantization of deep neural networks as a rate-distortion optimization problem, and present an ultra-fast algorithm to search the bit allocation of channels. Our approach has only linear time complexity and can find the optimal bit allocation within a few minutes on CPU. In addition, we provide an effective way to improve the performance on target hardware platforms. We restrict the bit rate (size) of each layer to allow as many weights and activations as possible to be stored on-chip, and incorporate hardware-aware constraints into our objective function. The hardware-aware constraints do not cause additional overhead to optimization, and have very positive impact on hardware performance. Experimental results show that our approach achieves state-of-the-art results on four deep neural networks, ResNet-18, ResNet-34, ResNet-50, and MobileNet-v2, on ImageNet. Hardware simulation results demonstrate that our approach is able to bring up to 3.5× and 3.0× speedups on two deep-learning accelerators, TPU and Eyeriss, respectively.

Keywords: Deep learning · Quantization · Rate-distortion theory

J. Lin and V. Chandrasekhar—did this work when they were with Institute for Infocomm Research, Singapore.

Supplementary Information The online version contains supplementary material available at https://doi.org/10.1007/978-3-031-19775-8_10.

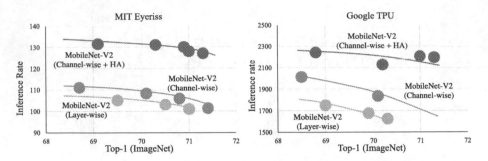

Fig. 1. Channel-wise bit allocation plus hardware-aware constraints (HA) achieves the best performance on Eyeriss and TPU. Channel-wise bit allocation outperforms layer-wise bit allocation because of higher quantization precision. Inference Rate: number of images processed per second.

1   Introduction

Deep Learning [20] has become the de-facto technique in Computer Vision, Natural Language Processing, Speech Recognition, and many other fields. However, the high accuracy of deep neural networks [19] comes at the cost of high computational complexity. Due to the large model size and huge computational cost, deploying deep neural networks on mobile devices is very challenging, especially on tiny devices. It is therefore important to make deep neural networks smaller and faster through model compression [12], to deploy deep neural networks on resource-limited devices.

Quantization [12] is one of the standard techniques for neural network compression. One problem existed in prior works is that they typically use equal bit width to quantize weights and activations of all layers, which is sub-optimal because weights and activations in different layers react differently on quantization. They should be treated independently and quantized with un-equal bit widths. Moreover, most of prior works only consider the model size and accuracy in their methods and do not consider the system-level performance when deploying quantized models on hardware platforms. As illustrated in prior works [30], a well quantized network can not guarantee superior performance on hardware platforms. To address these two issues, the recently proposed mixed-precision quantization methods assign un-equal bit widths across layers, and optimize the hardware metrics directly in the quantization mechanism. For example, HAQ [30] proposed a reinforcement learning method to learn the bit widths of weights and activations across layers and minimized latency and energy in their objective function directly.

Although noticeable improvement has been obtained by mixed-precision quantization, the layer-wise bit allocation scheme is still sub-optimal, since all channels in a CONV layer are quantized with equal bit width. In fact, different channels react very distinctively to quantization. Higher precision can be obtained if allocating un-equal bit widths to channels. However, the challenge is that the hyper-parameter space of channel bit widths increases exponentially with the number of channels. Given N channels and C bit widths, the search complexity is $O(C^N)$, where in a deep neural network, N can be tens of thousand or even more. Such huge search space could make it unaffordable for a heuristic search method, like reinforcement learning [27], to find solution within limited time.

In this paper, we propose a new approach to efficiently explore the hyperparameter space of channel bit widths. We apply the classic coding theories [28], and formulate the quantization of weights and activations as a rate-distortion optimization problem. Since the output distortion is highly related to accuracy, by minimizing the output distortion induced by quantization, our approach is able to well maintain the accuracy at very low bit widths. We then search the optimal bit allocation across channels in a rate-distortion optimized manner. Through utilizing the additivity property of output distortion, we present an ultra-fast algorithm with linear time complexity to find the optimal channel-wise bit allocation, by using Lagrangian formulation. Our algorithm only costs a few minutes on CPU for a deep neural network.

What's more, we present an alternative way to improve the system-level performance when deploying quantized networks on target hardware platforms. Prior works typically optimize the hardware metrics of a whole network directly, and need real-time feedback from simulators in their learning procedure, which could cause additional overhead to optimization. Instead, our approach improves hardware performance by restricting the size of each individual layer, and does not require feedback from simulators. Our key insight is that the volume of weights and activations in some layers is particularly significant, which exceeds the capacity of on-chip memory. As a result, these layers significantly prolong the inference time due to the necessity of slow data access to off-chip memory. We thus constrain the size of these large layers to ensure that all variables can be stored on-chip.

To our best knowledge, only one prior work, AutoQ [22], finds channel-wise bit allocation, and optimizes the performance on hardware platforms simultaneously. AutoQ employs reinforcement learning to solve the bit allocation problem, which is time-consuming, and could fall into a local optimum in their heuristic search method. Our approach adopts Lagrangian formulation for fast optimization, and is able to find global optimal solution in a rate-distortion optimized manner. We summarize the main contributions of our paper as following:

- We formulate the quantization of deep neural networks as a rate-distortion optimization problem, and optimize the channel-wise bit allocation for higher accuracy. We present an ultra-fast algorithm with linear time complexity to efficiently explore the hyperparameter space of channel bit widths.
- We present a simple yet effective way to improve the performance on target hardware platforms, through restricting the size of each individual layer and incorporating hardware-aware constraints into our objective function. The hardware-aware constraints can be integrated seamlessly, without causing additional overhead to optimization.
- Our approach achieves state-of-the-art results on various deep neural networks on ImageNet. Hardware simulation results demonstrate that our approach is able to bring considerable speedups for deep neural networks on two hardware platforms.

Table 1. A comparison with prior mixed-precision quantization works.

Approach	Bit Allocation Scheme	Hardware -Aware	Optimization	Complexity
ReLeQ [7]	Layer-Wise	No	Reinforcement Learning	High
HAQ [30]	Layer-Wise	Yes	Reinforcement Learning	High
DNAS [31]	Layer-Wise	No	Neural Architecture Search	High
DQ [29]	Layer-Wise	No	Training from Scratch	High
HAWQ [6]	Layer-Wise	No	Training from Scratch	High
ALQ [23]	Layer-Wise	No	Training from Scratch	High
AQ [18]	Element-Wise	No	Closed-Form Approximation	Low
PTQ [1]	Channel-Wise	No	Analytic Solution	Low
FracBits [32]	Channel-Wise	No	Training from Scratch	High
DMBQ [34]	Channel-Wise	No	Training from Scratch	High
AutoQ [22]	Channel-Wise	Yes	Reinforcement Learning	High
RDO-Q (Ours)	Channel-Wise	Yes	Lagrangian Formulation	Low

2   Related Works

We discuss prior mixed-precision quantization works related to our work. ReLeQ
[7] proposed an end-to-end deep reinforcement learning (RL) framework to automate the process of discovering quantization bit widths. Alternatively, HAQ
[30] leveraged reinforcement learning to determine quantization bit widths, and
employed a hardware simulator to generate direct feedback signals to the RL
agent. DNAS [31] proposed a differentiable neural architecture search framework
to explore the hyperparameter space of quantization bit widths. Differentiable
Quantization (DQ) [29] learned quantizer parameters, including step size and
range, by training with straight-through gradients, and then inferred quantization bit widths based on the learned step size and range. Hessian AWare Quantization (HAWQ) [6] introduced a second-order quantizatino method to select
the quantization bit width of each layer, based on the layer's Hessian spectrum.
Adaptive Loss-aware Quantization (ALQ) [23] directly minimized network loss
w.r.t. quantized weights, and used network loss to decide quantization bit widths.
Layer-wise bit allocation scheme is employed in all the above methods.

Adaptive Quantization (AQ) [18] found a unique, optimal precision for each
network parameter (element-wise), and provided a closed-form approximation
solution. Post Training Quantization (PTQ) [1] adopted channel-wise bit allocation to improve quantization precision, and provided an analytic solution to
find quantization bit widths, assuming that parameters obey certain distributions. FracBits [32] generalized quantization bit widths to arbitrary real numbers to make them differentiable, and learned channel-wise (or kernel-wise) bit
allocation during training. Distribution-aware Multi-Bit Quantization (DMBQ)
[34] proposed loss-guided bit-width allocation strategy to adjust the bit widths of
weights and activations channel-wisely. AQ, PTQ, FracBits, and DMBQ all did
not take the impact on hardware platforms into account. AutoQ [22] proposed a

hierarchical deep reinforcement learning approach to find quantization bit widths of channels and optimize hardware metrics (e.g., latency and energy) simultaneously. Different with AutoQ, our approach provides an alternative way to quickly explore the hyperparameter space of bit widths with linear time complexity, and is able to find global optimal solution in a rate-distortion optimized manner. Table 1 illustrates the differences between the mixed-precision quantization approaches.

One prior work [9] interpreted neural network compression from a rate-distortion's perspective. The main focus of [9] was giving an upper bound analysis of compression and discussing the limitations. [9] did not give a way to search bit allocation, and there was no practical results provided.

3 Approach

We utilize classic coding theories [28], and formulate the quantization of deep neural networks as a rate-distortion optimization problem [36,37]. The differentiability of input-output relationships for the layers of neural networks allows us to relate output distortion to the bit rate (size) of quantized weights and activations. We add hardware-aware constraints into the objective function to improve the performance on hardware platforms. We will discuss the formulation of our approach and its optimization in this section.

3.1 Formulation

Let $\mathcal{F}$ denote a deep neural network. Given an input $\mathbf{I}$, we denote $\mathbf{Y}$ as the output of $\mathcal{F}$, i.e. $\mathbf{Y} = \mathcal{F}(\mathbf{I})$. When performing quantization on weights and activations, a modified output vector $\widehat{\mathbf{Y}}$ would be received. The output distortion is measured by the distance between $\mathbf{Y}$ and $\widehat{\mathbf{Y}}$, which is defined as

$$\delta = \|\mathbf{Y} - \widehat{\mathbf{Y}}\|_2^2 \tag{1}$$

Here Euclidean distance $\|.\|_2$ is adopted. Our approach allocates different quantization bit widths to weight channels and activation layers. We aim to minimize the output distortion under the constraint of bit rate (size). Given the bit rate constraint r, the rate-distortion optimization problem is formulated as

$$\min \delta = \|\mathbf{Y} - \widehat{\mathbf{Y}}\|_2^2 \quad s.t. \sum_{i=1}^{l} \sum_{j=1}^{n_i} R_{i,j}^w + \sum_{i=1}^{l} R_i^a \leq r, \tag{2}$$

where $R_{i,j}^w$ denotes the bit rate of weight channel j in layer i, R_i^a denotes the bit rate of activations in layer i, n_i denotes the number of channels in layer i, and l denotes the total number of layers. Specifically, $R_{i,j}^w$ equals to the quantization bit width of channel j in layer i, denoted as $B_{i,j}^w$, multiplied by the number of weights in that channel; R_i^a equals to the quantization bit width of activations in layer i, denoted as B_i^a, multiplied by the number of activations in that layer.

We noticed that output distortion is highly related to network accuracy. By minimizing output distortion induced by quantization, our approach is able to maintain the accuracy at very high compression ratio.

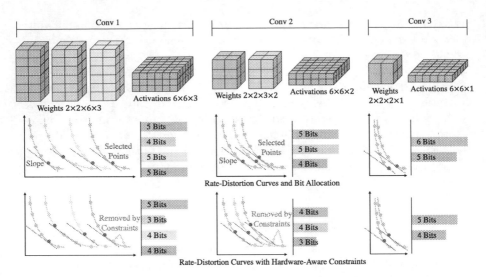

Fig. 2. Examples of finding optimal bit allocation w/ and w/o hardware-aware constraints.

3.2   Optimizing Channel-Wise Bit Allocation

We explored the additivity property of output distortion when performing quantization on weight channels and activation layers, and found that the additivity property holds, similar to the observation made in [37,39]. Utilizing the additivity property, we develop an efficient Lagrangian formulation method to solve the bit allocation problem.

Specifically, let $\delta_{i,j}^w$ and δ_i^a denote the output distortion caused by quantizing an individual weight channel and an individual activation layer, respectively. The output distortion δ, caused by quantizing all weight channels and activation layers, equals the sum of output distortion due to the quantization of each individual item

$$\delta = \sum_{i=1}^{l} \sum_{j=1}^{n_i} \delta_{i,j}^w + \sum_{i=1}^{l} \delta_i^a \tag{3}$$

Equation (3) can be derived mathematically by linearizing the output distortion using Taylor series expansion with the assumption that the neural network is continuously differentiable and quantization errors can be considered as small deviations. The mathematical derivation of Eq. (3) is provided in supplementary material.

We then apply Lagrangian formulation [26] to solve objective function (2). The Lagrangian cost function of (2) is defined as

$$\mathcal{J} = \delta - \lambda \cdot \Big(\sum_{i=1}^{l} \sum_{j=1}^{n_i} R_{i,j}^w + \sum_{i=1}^{l} R_i^a - r \Big), \tag{4}$$

in which λ decides the trade-off between bit rate and output distortion. Setting the partial derivations of $\mathcal{J}$ to zero with respect to each $R_{i,j}^w$ and R_i^a and utilizing the additivity property in (3), we obtain the optimal condition

$$\frac{\partial \tilde{\delta}_{i,1}^w}{\partial r_{i,1}^w} = \ldots = \frac{\partial \tilde{\delta}_{i,n_i}^w}{\partial r_{i,n_i}^w} = \frac{\partial \delta_i^a}{\partial r_i^a} = \lambda, \tag{5}$$

for all $1 \leq i \leq l$. Equation (5) expresses that the slopes of all rate-distortion curves (output distortion versus bit rate functions) should be equal to obtain optimal bit allocation with minimal output distortion. According to (5), we are able to solve objective function (2) efficiently by enumerating slope λ and then choosing the point on each rate-distortion curve with slope equal to λ as solution.

The algorithm works as follows. Before optimization, we quantize each weight channel and activation layer with different bit widths and calculate the output distortion caused by quantization to generate the rate-distortion curve for each weight channel and activation layer. After that, we assign a real value to λ, and select the point with slope equal to λ on each curve. The selected points on all curves correspond to a group of solution for bit allocation. In practice, we explore multiple values for λ until the network bit rate exceeds constraint r. We randomly select 50 images from ImageNet dataset to calculate output distortion caused by quantization. Given the number of λ evaluated, t, and the total number of bit widths, b, the time complexity of optimization is $O((l + \sum_{i=1}^{l} i) \cdot t \cdot b)$. The algorithm has only linear time complexity, which can find the answer in a few minutes on a normal CPU.

3.3 Choice of Quantizer

We adopt uniform quantizer in our approach. The quantization step size Δ is defined as a value of a power of 2, ranging from 2^{-16} to 2^0, where the one with minimal quantization error is selected. We clip all weights by $(-2^{b-1} \cdot \Delta, (2^{b-1} - 1) \cdot \Delta)$ and all activations by $(0, (2^b - 1))$, in which b is the quantization bit width. Note that our approach is compatible with other quantizers, including both uniform quantizer and non-uniform quantizer (e.g., K-Means [10]). Since the focus of this paper is not the design of quantizer, we only evaluate uniform quantizer in our approach. It is worth mentioning that applying a non-uniform quantizer could further improve the accuracy, but non-uniform quantizers are more complicated for computation, and require additional resources (e.g., look-up tables) for implementation. Similar as prior mixed-precision methods [22, 30, 31], our approach employs uniform quantizer, as it is more hardware-friendly and is straightforward for implementation.

3.4 Improving Performance on Hardware

We consider improving inference rate as a guide to the design of our quantization mechanism. Inference rate is defined as the maximum number of images that a neural network can process per unit time. Memory access, especially data

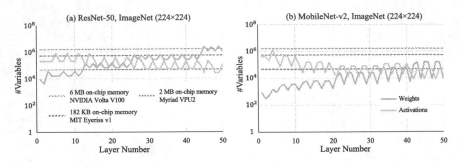

Fig. 3. The number of weights and activations across layers and the on-chip capacity on different hardware platforms.

movement from off-chip memory to on-chip memory, dominates inference time, rather than convolutional operations [12,13]. We thus aim to maintain as many weights and activations as possible stored on-chip, and avoid data movement from off-chip memory to improve inference speed.

Our key insight is that the volume of weights and activations in some layers is particularly significant. As a result, part of weights and activations can not be stored on-chip, which leads to significant memory-access traffic to off-chip DRAM. Figure 3 illustrates the number of parameters across layers and the on-chip memory capacity on different hardware platforms. As we can see, on-chip memory capacity is very limited, and the size of some layers exceeds the capacity. To this end, we restrict the quantization bit widths in these large layers to make sure that the size of these layers is less than on-chip memory capacity. Specifically, for layer i, we have an independent bit rate constraint,

$$\sum_{j=1}^{n_i}\left(K_{i,j}^w \cdot B_{i,j}^w\right)+K_i^a \cdot B_i^a \leq m_{on}, \tag{6}$$

in which $K_{i,j}^w$ denotes the number of weights of channel j in layer i, K_i^a denotes the number of activations in layer i, and m_{on} denotes the on-chip memory capacity. In practice, we relax (6) into two items, and incorporate them to objective function (2),

$$B_{i,j}^w \leq \frac{m_{on}}{\sum_j K_{i,j}^w + \frac{\beta}{1-\beta}K_i^a}, \; B_i^a \leq \frac{\alpha m_{on}}{\frac{1-\beta}{\beta} \sum_j K_{i,j}^w + K_i^a}, \tag{7}$$

for all $1 \leq i \leq l$ and $1 \leq j \leq n_i$, where α and β are two hyperparameters, ranging from 0 to 1. Incorporating constraints (7) into (2), we have the objective function with the bit rate of each weight channel and activation layer constrained to improve hardware performance.

$$\min \delta = \|\mathbf{Y} - \widehat{\mathbf{Y}}\|_2^2 \ \ s.t. \ \sum_{i=1}^{l} \sum_{j=1}^{n_i} R_{i,j}^w + \sum_{i=1}^{l} R_i^a \leq r,$$

$$B_{i,j}^w \leq \frac{m_{on}}{\sum_j K_{i,j}^w + \frac{\beta}{1-\beta} K_i^a}, \ B_i^a \leq \frac{\alpha \cdot m_{on}}{\frac{1-\beta}{\beta} \sum_j K_{i,j}^w + K_i^a}$$

(8)

Note that the optimization of (8) is the same as that of (2). The only difference is that in (8) we have different search range for quantization bit widths. In (2), the range is from 1 to b where $b = 16$ is the maximal bit width, while in (8), it is from 1 to $\frac{m_{on}}{\sum_j K_{i,j}^w + \frac{\beta}{1-\beta} K_i^a}$ for weight channels, and is from 1 to $\frac{\alpha m_{on}}{\frac{1-\beta}{\beta} \sum_j K_{i,j}^w + K_i^a}$ for activation layers. Incorporating bit rate constraints into objective function (2) does not increase the search time. Actually, it even slightly decreases the time as it reduces the search range of bit widths. Figure 2 illustrates examples of the optimization procedure with and without constraints (7).

3.5 Discussion

Prior works [22,30] typically use hardware simulators to guide the design of quantizatino mechanism. Implementing a simulator is complicated and calculating simulation results costs time. Alternatively, we provide a simple yet effective way to improve performance on hardware platforms. We directly restrict the bit rate of each layer to have weights and activations saved on-chip. Our method is easy to implement and does not cause additional overhead to optimization. Another advantage is that, once the rate-distortion curves are generated, our approach is able to find the bit allocation under any network size, by just changing the slope λ. This is better than most prior works which need to re-run the whole method every time searching the bit allocation for a network size.

4 Experiments

We report experimental results in this section. We first show quantization results on four deep neural networks, ResNet-18 [14], ResNet-34, ResNet-50 and MobileNet-v2 [25], on the ImageNet dataset [5]. We then report the results of inference rate on two hardware platforms, Google TPU [16] and MIT Eyeriss [3].

4.1 Parameter Settings

We set hyperparameters α and β as values between 0 and 1, where the combination with best hardware performance is chosen. We enumerate slope λ from -2^{-20} to -2^{20} until network size meets constraint r. Similar as prior works [22,30], we fine-tune the model after quantization up to 100 epochs with learning rate 0.0001. Our approach is able to obtain high accuracy after 2 epochs, and can almost converge after 5 to 10 epochs. We fine-tune 100 epochs to make sure that the quantized network completely converges. Straight-through estimator (STE) [2] is applied to perform back-propagation through non-differentiable

Table 2. Top-1 image classification accuracy at 2 bits on ImageNet.

Method	ResNet-18	ResNet-34	ResNet-50
Original [14]	69.3%	73.0%	75.5%
LQ-Nets [33]	64.9%	68.8%	71.5%
PTG [41]	–	–	70.0%
DSQ [11]	65.2%	70.0%	–
QIL [17]	65.7%	70.6%	–
ALQ [23]	66.4%	71.0%	–
APoT [21]	67.3%	70.9%	73.4%
SAT [15]	65.5%	–	73.3%
LSQ [8]	67.6%	71.6%	73.7%
AUXI [40]	–	–	73.8%
DMBQ [34]	67.8%	72.1%	–
RDO-Q (Ours)	**68.8%**	**72.6%**	**75.0%**

Table 3. Results on MobileNet-v2 at 4 bits on ImageNet.

Method	Top-1 Accuracy	Top-5 Accuracy
Original [25]	71.8%	90.2%
HAQ [30]	67.0% (−4.8)	87.3 (−3.1)%
DQ [29]	69.7% (−2.1)	–
DSQ [11]	64.8% (−7.0)	–
AutoQ [22]	69.0% (−2.8)	89.4% (−0.8)
SAT [15]	71.1% (−0.7)	89.7% (−0.5)
LLSQ [35]	67.4% (−4.4)	88.0% (−2.2)
RDO-Q (Ours)	**71.3% (−0.5)**	**90.0% (−0.2)**

quantization functions in fine-tuning. We randomly select 50 images from ImageNet to generate the rate-distortion curves. We noticed that using more images to generate the curve doesn't affect the final accuracy.

4.2   Quantization Results

Table 2 and Table 3 list the results on four deep neural networks, ResNet-18, ResNet-34, ResNet-50, and MobileNet-v2, when weights and activations are quantized to very low bit widths (i.e., 2 bits or 4 bits). As our approach allocates unequal bit widths to different weight channels and activation layers, we report the results when networks are quantized to the target size on average for fair comparison, same as prior works [22,29]. Our approach, named as Rate-Distortion-Optimized Quantization (RDO-Q), improves state-of-the-arts on the four neural networks. Specifically, our approach outperforms SOTAs by 1.0%, 0.5%, and 1.2%, on ResNet-18, ResNet-34, and ResNet-50, respectively.

Table 4. Inference Rate on Google TPU and MIT Eyeriss. We show the results of our approach with hardware-aware (HA) constrains in this table.

Method	Accuracy top-1	Google TPU	MIT Eyeriss
ResNet-50			
Original [14]	75.5%	361	5.6
DoReFa+PACT [4]	76.5%	646	12.6
DoReFa+PACT [4]	72.2%	920	13.7
RDO-Q+HA (Ours)	76.5%	769	13.9
RDO-Q+HA (Ours)	76.2%	904	15.0
RDO-Q+HA (Ours)	75.0%	1254	17.0
MobileNet-V2			
Original [25]	71.1%	1504	64
DoReFa+PACT [4]	71.2%	1698	104
DoReFa+PACT [4]	70.4%	1764	108
HAQ [30]	71.2%	2067	124
HAQ [30]	68.9%	2197	128
RDO-Q+HA (Ours)	71.3%	2197	127
RDO-Q+HA (Ours)	71.0%	2207	128
RDO-Q+HA (Ours)	70.9%	2256	130

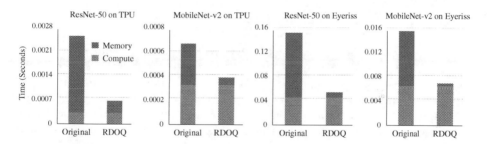

Fig. 4. A breakdown for the compute time and memory time on two hardware platforms, Google TPU and MIT Eyeriss.

4.3 Performance on Hardware Platforms

We examined the inference rate on two hardware platforms, Google TPU [16] and MIT Eyeriss [3], both of which are state-of-the-art architectures, inspired by current embedded and high-performance neural-network-targeted accelerators. We adopt the SCALE-Sim software [24] to simulate the time cycles of ResNet-50 and MobileNet-v2, when mapped into the two considered hardware platforms.

Table 4 illustrates the inference rate on TPU and Eyeriss. Our approach significantly improves the inference rate, compared with originally uncompressed neural networks. We speed up the inference rate by 3.5× and 3.0× for ResNet-50 on TPU and Eyeriss, and by 1.5× and 2.0× for MobileNet-V2 on TPU and Eyeriss, without hurting the accuracy (loss ≤ 0.5%). We also compare our approach with the competitive mixed-precision quantization method, HAQ [30], and the competitive equal bit quantization method, DoReFa+PACT [4,38]. Our approach outperforms both HAQ and DoReFa+PACT. We notice that although equal bit quantization method DoReFa+PACT obtains superior quantization results, the performance on hardware platforms is not high. This is because DoReFa+PACT is not hardware-aware as they do not optimize the hardware performance in their quantization mechanism.

Note that both TPU and Eyeriss do not support computation with mixed precision. The computation on TPU is with 8-bit integers, and that on Eyeriss is with 16-bit integers. We clarify that we pad quantized parameters to 8-bit or 16-bit integers when we do the computation on TPU or Eyeriss, respectively. Although TPU and Eyeriss do not support computation with mixed precision, they still benefit from mixed-precision quantization, because the bottleneck of deep neural networks is memory access and mixed-precision quantization helps to further reduce the network size to reduce memory. Figure 4 illustrates a breakdown of the inference time for memory access and computation on TPU and Eyeriss. We believe that hardware with specialized integer arithmetic units can further improve the performance, since the computation can also be more efficient. As the main focus of our paper is the quantization algorithm, we did not implement a hardware architecture to support mixed-precision computation.

4.4   Time Cost

Table 5 lists the time of our approach to find channel-wise bit allocation on four deep neural networks. Our approach takes about a few minutes on a normal CPU (Intel Core i7 6600U CPU with 2.60 GHZ) to find the solution. We also evaluated HAQ [30]—the competitive mixed-precision quantization method built upon reinforcement learning. As we can see, the reinforcement-learning-based approach requires several days on multiple GPUs to search the bit allocation for one time, which is orders of magnitude slower. Our approach provides an alternative way to quickly explore the hyperparameter space of bit widths, and is particularly suitable for the case without powerful computation resources.

4.5   Distributions of Bit Rate Across Layers

Figure 5 illustrates the distribution of the bit rate under different hardware-aware constraints. Intuitively, by balancing the size between layers, our approach assigns lower bit widths to large layers and higher bit widths to small layers, to meet the constraints. Figure 6 illustrates the number of activations that hardware platforms can accommodate, under different bit widths given to the activation layer. We can see that on both ResNet-50 and MobileNet-v2, some layers have more than 1 million activations, and the bit widths assigned to these layers have to be very small when the on-chip memory capacity is only a few KBs.

4.6   Discussion of Additivity Property

Based on our mathematical analysis, the additivity property holds if quantization errors can be considered as small deviations. In that case, second (or higher)

Table 5. Time cost to find channel-wise bit allocation.

Method	HAQ [30]	RDO-Q
Device	GPU × 4	CPU × 1
ResNet-18	–	2 min
ResNet-34	–	3 min
ResNet-50	117 h	7 min
MobileNet-v2	79 h	6 min

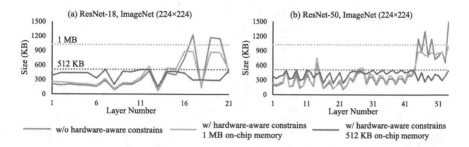

Fig. 5. Distributions of bit rate across layers given different memory constraints.

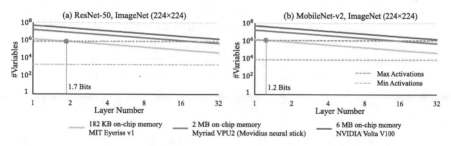

Fig. 6. The number of variables that the on-chip memory on specific hardware platforms can accommodate, give different bit widths per variable. The minimum and maximum numbers of activations that a single layer can have are highlighted.

order items in Taylor series expansion are small values, and we can use the zero and first order items to approximate output distortion. We tested the Mean Square Error (MSE) of quantized parameters in practice, and found that MSEs are really small deviations. Moreover, we evaluated the relation between the output distortion and the sum of individual distortion using real examples. The practical results also consist with our theoretical analysis. The mathematical analysis of additivity property is provided in supplementary material.

4.7 Implementation Details of Optimization

We formulate the quantization of weight channels and activation layers as a rate-distortion optimization problem, as illustrated in Sect. 3. We utilize the additvity of the output distortion and apply the classical Lagrangian formulation to solve (2). The optimal condition in Eq. (4) expresses that the slopes of all output distortion versus bit rate curves should be equal. Based on this equation, the optimization problem can be solved by enumerating λ and selecting the point with slope equal to λ on each rate-distortion curve.

Specifically, we first generate the rate-distortion curve for each channel and activation layer. In our case, the rate-distortion curves are comprised of discrete points. For example, if the range of bit width is from 1-bit to 8-bits, then a rate-distortion curve is a discrete curve with 8 points. We enumerate λ and select the point on each curve with slope equal to λ. We may enumerate different λ to find the best solution with minimal output distortion under the size constraint. Assume that we have N curves and M points on each curve. The total time complexity to generate rate-distortion curves and find optimal bit allocation is $O(I \cdot M \cdot N \cdot C + K \cdot M \cdot N)$, where I denotes the number of images used to generate rate-distortion curves, C is a constant which denotes the cost to perform the inference, and K is the total number of slope λ to be evaluated.

5   Conclusion

Channel-wise bit allocation brings higher quantization precision and superior accuracy. Our approach provides an ultra-fast way to explore the hyperparameter space of channel bit widths with linear time complexity, using Lagrangian Formulation. The quantization of deep neural networks is formulated as a rate-distortion optimization problem, and the fast optimization method is proposed, by utilizing the additivity of output distortion. Moreover, we consider the impact on hardware platforms in the design of our quantization mechanism, and present a simple yet effective method to improve hardware performance. We restrict the bit rate of each layer to allow as many weights and activations as possible saved on-chip, and add hardware-aware constraints in our objective function to improve inference rate on target hardware platforms. The hardware-aware constraints can be incorporated into our objective function seamlessly, without incurring additional overhead for optimization. Extensive experiments show that our approach improves state-of-the-arts on four deep neural networks. Hardware simulation results demonstrate that our approach is able to accelerate deep learning inference considerably on two hardware platforms.

Acknowledgments. This research is supported by the Agency for Science, Technology and Research (A*STAR) under its Funds (Project Number A1892b0026, A19E3b0099, and C211118009). Any opinions, findings and conclusions or recommendations expressed in this material are those of the author(s) and do not reflect the views of the A*STAR.

References

1. Banner, R., Nahshan, Y., Hoffer, E., Soudry, D.: Post-training 4-bit quantization of convolution networks for rapid-deployment. arXiv preprint. arXiv:1810.05723 (2018)

2. Bengio, Y., Leonard, N., Courville, A.: Estimating or propagating gradients through stochastic neurons for conditional computation. In: arXiv:1308.3432 (2013)
3. Chen, Y.H., Krishna, T., Emer, J.S., Sze, V.: Eyeriss: an energy-efficient reconfigurable accelerator for deep convolutional neural networks. IEEE J. Solid-State Circ. **52**(1), 127–138 (2016)
4. Choi, J., Wang, Z., Venkataramani, S., Chuang, P.I.J., Srinivasan, V., Gopalakrishnan, K.: Pact: parameterized clipping activation for quantized neural networks. In: arXiv (2018)
5. Deng, J., Dong, W., Socher, R., Li, L.J., Li, K., Fei-Fei, L.: ImageNet: a large-scale hierarchical image database. In: CVPR (2009)
6. Dong, Z., Yao, Z., Gholami, A., Mahoney, M.W., Keutzer, K.: Hawq: Hessian aware quantization of neural networks with mixed-precision. In: arXiv (2019)
7. Elthakeb, A.T., Pilligundla, P., Mireshghallah, F., Yazdanbakhsh, A., Esmaeilzadeh, H.: Releq: a reinforcement learning approach for deep quantization of neural networks. In: NeurIPS Workshop on ML for Systems (2018)
8. Esser, S.K., McKinstry, J.L., Bablani, D., Appuswamy, R., Modha, D.: Learned step size quantization. In: ICLR (2020)
9. Gao, W., Liu, Y.H., Wang, C., Oh, S.: Rate distortion for model compression: from theory to practice. In: International Conference on Machine Learning, pp. 2102–2111. PMLR (2019)
10. Gersho, A., Gray, R.: Vector Quantization and Signal Compression. Kluwer Academic Publishers, Dordrecht (1991)
11. Gong, R., et al.: Differentiable soft quantization: Bridging full-precision and low-bit neural networks. In: ICCV (2019)
12. Han, S., Mao, H., Dally, W.J.: Deep compression: compressing deep neural networks with pruning, trained quantization and huffman coding. In: ICLR (2016)
13. Han, S., Pool, J., Tran, J., Dally, W.J.: Learning both weights and connections for efficient neural networks. arXiv preprint. arXiv:1506.02626 (2015)
14. He, K., Zhang, X., Ren, S., Sun, J.: Deep residual learning for image recognition. In: CVPR (2016)
15. Jin, Q., Yang, L., Liao, Z., Qian, X.: Neural network quantization with scale-adjusted training. BMVC (2020)
16. Jouppi, N., et al.: In-datacenter performance analysis of a tensor processing unit. In: Proceedings of the 44th Annual International Symposium on Computer Architecture, pp. 1–12. ISCA'17, ACM, New York, NY, USA (2017). https://doi.org/10.1145/3079856.3080246, http://doi.acm.org/10.1145/3079856.3080246
17. Jung, S., et al.: Learning to quantize deep networks by optimizing quantization intervals with task loss. In: CVPR (2019)
18. Khoram, S., Li, J.: Adaptive quantization of neural networks. In: ICLR (2018)
19. Krizhevsky, A., Sutskever, I., Hinton, G.: ImageNet classification with deep convolutional neural networks. In: NIPS (2012)
20. LeCun, Y., Bengio, Y., Hinton, G.: Deep learning. Nature **521**(7553), 436–444 (2015)
21. Li, Y., Dong, X., Wang, W.: Additive powers-of-two quantization: an efficient non-uniform discretization for neural networks. In: ICLR (2020)
22. Lou, Q., Guo, F., Kim, M., Liu, L., Jiang, L.: Autoq: automated kernel-wise neural network quantizations. In: ICLR (2020)
23. Qu, Z., Zhou, Z., Cheng, Y., Thiele, L.: Adaptive loss-aware quantization for multi-bit networks. In: arXiv (2020)

24. Samajdar, A., Zhu, Y., Whatmough, P.N., Mattina, M., Krishna, T.: Scale-sim: systolic CNN accelerator. CoRR abs/1811.02883, http://arxiv.org/abs/1811.02883 (2018)
25. Sandler, M., Howard, A., Zhu, M., Zhmoginov, A., Chen, L.C.: Mobilenetv 2: inverted residuals and linear bottlenecks. In: arXiv (2018)
26. Shoham, Y., Gersho, A.: Efficient bit allocation for an arbitrary set of quantizers (speech coding. In: IEEE Transactions on Acoustics, Speech, and Signal Processing (1988)
27. Sutton, R.S., Barto, A.G.: Reinforcement Learning: An introduction. MIT press, Cambridge (2018)
28. Taubman, D.S., Marcellin, M.W.: JPEG 2000 Image Compression Fundamentals, Standards and Practice. Springer, New York (2002). https://doi.org/10.1007/978-1-4615-0799-4
29. Uhlich, S., et al.: Mixed precision dnns: all you need is a good parametrization. arXiv preprint. arXiv:1905.11452 (2019)
30. Wang, K., Liu, Z., Lin, Y., Lin, J., Han, S.: Haq: hardware-aware automated quantization with mixed precision. In: CVPR (2019)
31. Wu, B., Wang, Y., Zhang, P., Tian, Y., Vajda, P., Keutzer, K.: Mixed precision quantization of convnets via differentiable neural architecture search. In: ICLR (2019)
32. Yang, L., Jin, Q.: Fracbits: mixed precision quantization via fractional bit-widths. arXiv preprint. arXiv:2007.02017 (2020)
33. Zhang, D., Yang, J., Ye, D., Hua, G.: Lq-nets: learned quantization for highly accurate and compact deep neural networks. In: ECCV (2018)
34. Zhao, S., Yue, T., Hu, X.: Distribution-aware adaptive multi-bit quantization. In: CVPR (2021)
35. Zhao, X., Wang, Y., Cai, X., Liu, C., Zhang, L.: Linear symmetric quantization of neural networks for low-precision integer hardware. ICLR (2020)
36. Zhe, W., Lin, J., Aly, M.S., Young, S., Chandrasekhar, V., Girod, B.: Rate-distortion optimized coding for efficient cnn compression. In: DCC (2021)
37. Zhe, W., Lin, J., Chandrasekhar, V., Girod, B.: Optimizing the bit allocation for compression of weights and activations of deep neural networks. In: ICIP (2019)
38. Zhou, S., Wu, Y., Ni, Z., Zhou, X., Wen, H., Zou, Y.: Dorefa-net: training low bitwidth convolutional neural networks with low bitwidth gradients. In: arXiv preprint. arXiv:1606.06160 (2016)
39. Zhou, Y., Moosavi-Dezfooli, S.M., Cheung, N.M., Frossard, P.: Adaptive quantization for deep neural network. In: AAAI (2018)
40. Zhuang, B., Liu, L., Tan, M., Shen, C., Reid, I.: Training quantized neural networks with a full-precision auxiliary module. In: CVPR (2020)
41. Zhuang, B., Shen, C., Tan, M., Liu, L., Reid, I.: Towards effective low-bitwidth convolutional neural networks. In: CVPR (2018)

U-Boost NAS: Utilization-Boosted Differentiable Neural Architecture Search

Ahmet Caner Yüzügüler[(✉)][iD], Nikolaos Dimitriadis[iD], and Pascal Frossard[iD]

EPFL, Lausanne, Switzerland
{ahmet.yuzuguler,nikolaos.dimitriadis,pascal.frossard}@epfl.ch

Abstract. Optimizing resource utilization in target platforms is key to achieving high performance during DNN inference. While optimizations have been proposed for inference latency, memory footprint, and energy consumption, prior hardware-aware neural architecture search (NAS) methods have omitted resource utilization, preventing DNNs to take full advantage of the target inference platforms. Modeling resource utilization efficiently and accurately is challenging, especially for widely-used array-based inference accelerators such as Google TPU. In this work, we propose a novel hardware-aware NAS framework that does not only optimize for task accuracy and inference latency, but also for resource utilization. We also propose and validate a new computational model for resource utilization in inference accelerators. By using the proposed NAS framework and the proposed resource utilization model, we achieve $2.8 - 4\times$ speedup for DNN inference compared to prior hardware-aware NAS methods while attaining similar or improved accuracy in image classification on CIFAR-10 and Imagenet-100 datasets. (Source code is available at https://github.com/yuezuegu/UBoostNAS).

Keywords: Hardware-aware neural architecture search · DNN inference · Hardware accelerator · Resource utilization

1 Introduction

Deep neural networks (DNN) have drastically evolved in recent years to push the limits in numerous computer vision tasks such as image recognition, object detection, and semantic segmentation [14,20]. To reach state-of-the-art performance, today's DNN models contain hundreds of layers to boost their performance. However, this comes at the expense of high computational complexity, which often leads to long inference latency in resource-constraint settings (e.g., mobile devices) [31,34]. It therefore becomes important to co-optimize model accuracy with inference runtime metrics, which is an important area of research in the design of effective DNN architectures [34].

Supplementary Information The online version contains supplementary material available at https://doi.org/10.1007/978-3-031-19775-8_11.

The effective usage of hardware resources (i.e., hardware utilization) in target inference platforms may vary depending on the architecture of a DNN model (e.g., layer types or channel dimensions). For instance, the depthwise convolution operation, which is popularly used in DNNs, has been shown to reduce the hardware utilization down to 1% in inference platforms [13]. Likewise, the channel and filter dimensions of DNN layers also have a significant impact on hardware utilization due to mismatches between DNN dimensions and target inference platforms [9,30]. As a result, unoptimized DNN models unfortunately run on inference platforms with low hardware utilization, hindering their performance (FLOPS/s) and increasing the latency. For example, average FLOPS/s utilization in Google's TPUv4 accelerator is 33% [17], which results in about three times slower inference than what could be achieved with a fully-utilized platform.

Prior works have proposed hardware-aware neural architecture search methods to co-optimize model accuracy and hardware performance metrics [32]. These methods use latency [34,35,37], energy consumption [41], or memory footprint [26] as the hardware performance metrics, which allows to improve the computational efficiency of the DNN architectures. However, no prior work uses hardware utilization as an optimization objective, which leads to DNN models with low efficiency in inference platforms. Moreover, prior hardware-aware NAS methods rely on either "black-box" hardware models, where these metrics are measured in physical devices and stored in look-up tables, or simplistic models such as roofline [13,23] to estimate the hardware performance metrics of the target inference platforms. Unfortunately, these models are impractical, have limited precision, or are non-differentiable, which hinders their effective use in NAS methods.

While prior hardware-aware NAS frameworks mostly focus on inference latency (i.e., execution time in terms of seconds), we argue that this does not necessarily lead to effective usage of hardware resources (i.e., percentage of processing elements actively used during computation) at the inference platforms. Therefore, we propose a NAS method that co-optimizes hardware utilization along with model accuracy and latency. To do so, we develop a hardware utilization model for inference platforms and use it to estimate the hardware utilization while searching for the optimal DNN architecture in image classification tasks. Moreover, we provide a smooth relaxation for the proposed utilization model to allow differentiable NAS, which is orders of magnitude less costly than other NAS methods. To the best of our knowledge, this is the first work that addresses hardware utilization in DNN inference using neural architecture search. We demonstrate through extensive experiments and hardware simulations that DNN models produced by our proposed NAS method run $2.8 - 4\times$ faster in target inference platforms compared to prior hardware-aware NAS methods that are agnostic to resource utilization.

In this paper, we make the following contributions:

- We show that hardware utilization in DNN inference is sensitive to layer types and dimensions of the architecture, and that fine-tuning a DNN architecture may significantly improve hardware utilization while maintaining the model accuracy.

- We propose a computational model for hardware utilization in modern inference platforms that estimates the measured utilization with significantly higher accuracy compared to prior models. We also provide a smooth relaxation of the proposed computational model to enable gradient-based optimization.
- We propose a differential neural architecture search framework that does not only optimize for task accuracy and inference latency, but also resource utilization at target inference platforms.
- We perform image classification experiments on the CIFAR-10 and Imagenet-100 datasets as well as detailed hardware simulations to show that the proposed utilization-aware NAS method significantly improves the hardware utilization and inference latency on typical computer vision tasks.

2   Related Work

Neural architecture search methods aim to automate the design process for DNN architectures that can achieve high accuracy on the given machine learning tasks with low latency and improved efficiency in target inference platforms. In fact, recent work has shown that DNNs produced with hardware-aware NAS methods outperform the hand-crafted DNNs in terms of accuracy and latency [34]. However, NAS methods require vast amounts of computational power, which motivates researchers to study more efficient methods.

Early versions of NAS methods used reinforcement learning [28,34,44,45], evolutionary algorithms [26,29], and Bayesian optimization [3]. However, such methods operate on a discrete search space and require vast amounts of computational resources, as they need to perform many trials while searching for an optimal architecture in an exponentially-increasing hyperparameter space. To mitigate the prohibitive cost of architecture search, many techniques such as weight-sharing [28] and one-shot NAS [2] have been proposed. While these techniques reduce the cost of each trial by allowing to reuse trained parameters, they still require many trials to find the optimal DNN architecture.

Recent works proposed differentiable NAS methods [4,5,24,27,38,40] to optimize DNNs both at microarchitecture [25] and macroarchitecture [37] levels using gradient-based algorithms. In these methods, a continuous relaxation is applied to the categorical decisions using a set of trainable weights (i.e., architectural parameters). Because differentiable NAS methods use the information from gradients with respect to the architectural parameters during training, they achieve faster convergence than their non-differentiable counterparts. Moreover, Wan et al. [35] introduced a differentiable masking technique, which allows to fine-tune channel dimensions and improve the resulting DNN's accuracy.

NAS methods have also been proposed towards optimizing additional performance metrics along with task accuracy, such as hardware related ones. To that end, prior works focused on accelerating inference on resource-constrained target platforms and proposed hardware(platform)-aware neural architecture search [33–35,37,42]. This type of NAS methods typically use a multi-objective loss function that includes terms for the model's predictive accuracy (e.g., cross-entropy) and hardware performance metric (e.g., latency or energy). While the accuracy term is easily calculated based on the given task using a validation dataset, the hardware performance metric depends on multiple variables such as the DNN architecture and the hardware specifications of the target platform, making its accurate estimation complex and leading to various proposed techniques. Early versions of hardware-aware NAS used real-time measurements from inference platforms [34, 41]. However, this approach is not practical because it requires the physical devices to be accessible during architecture search. More recent hardware-aware NAS methods consider the target hardware as a black-box [9,33,35,37], where a look-up table stores hardware measurements for all possible combinations of architectural decisions. This technique is also impractical because the number of required measurements grows combinatorially with the number of hyperparameters in the search space and the resulting models are not differentiable; therefore, they are not eligible to be used in differentiable NAS methods, which are among the most effective NAS methods.

To make the hardware performance metric differentiable, prior work proposed to use surrogate models such as linear regression [39] or neural networks [8]. However, such models require large numbers of samples for training and are hard to interpret. Some prior works also exploit the fact that a DNN's total latency is equal to the sum of individual layers' latency to obtain a differentiable latency model [33,35,37]. While this approach allows making inter-layer optimizations (e.g., which layers to keep or discard), it does not allow for intra-layer optimizations (e.g., operator type and channel dimensions); thus, they do not offer a complete solution. Other prior works proposed analytical hardware models, which estimates the hardware performance metrics using a cycle-accurate model [26] or a roofline model [13,23]. However, those models consider only memory bottlenecks, ignoring the other major sources of underutilization (e.g., dimension mismatches), leading to significant discrepancies between the estimated and actual values of runtime measurements. Unlike previously proposed hardware models, our novel analytical model for hardware utilization offers accurate estimation of the utilization in inference platforms while allowing gradient descent to perform both inter- and intra-layer optimizations in the NAS solution.

3   Modeling Resource Utilization in Inference Platforms

Prior hardware-aware NAS frameworks optimize DNN architectures solely for inference latency, leading to poor resource utilization. For instance, such hardware-aware NAS frameworks can easily reduce the inference latency by limiting the number of layers in DNN architectures but can not improve hardware utilization unless specific characteristics (e.g., operator types, channel dimensions) of the layers are taken into consideration while performing the architecture search. We adopt a different approach and use both latency and utilization as optimization goals along with task accuracy. Modeling hardware utilization is, however, challenging especially for specialized hardware architectures such as systolic arrays [22], which are widely used in DNN inference platforms (e.g.,

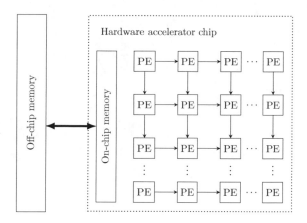

Fig. 1. Illustration of an array-based hardware accelerator.

Google TPU [18] or Tesla FSD chip [1]) due to their unique dataflow patterns. In this section, we first briefly explain these dataflow patterns, and then introduce a novel utilization model for such accelerators.

3.1 Dataflows on Hardware Accelerators

Matrix multiplication operations constitute the vast majority ($\sim$98% [1]) of DNN operations; thus, inference platforms adopt array-based architectures [1,6,18,30]. Figure 1 depicts a typical array-based hardware accelerator, which consists of an array of processing elements (PE), on-chip, and off-chip memory. Unlike general-purpose CPU and GPUs, PEs in such architectures can easily share data between each other through an on-chip interconnection, which allows them to perform matrix multiplication with high efficiency and minimum delay.

While there exist various mapping and dataflow schemas to perform a matrix multiplication on an array-based architecture [6], without loss of generality, we assume one of the most commonly used dataflow in this paper, namely weight stationary [18]. In this dataflow, the accelerator first loads model weights and activations from an off-chip memory, and stores them on the on-chip memory. Then, the weight matrix is first spatially mapped onto the two-dimensional array, the activation matrix is streamed along the PE rows, and partial sums are accumulated along the PE columns [18]. The partial sums that are obtained at the last PE row correspond to the results of the matrix multiplication. The final results are either stored in the on-chip memory to be used in next layers, or written back to the off-chip memory.

While theoretically allowing faster multiplication, array-based accelerators in practice often suffer from low resource utilization due to unoptimized DNN architectures. For instance, the average utilization of Google's TPUv1 and TPUv4 are 20% [18] and 33% [17], where the leading source of underutilization is the mismatches between DNN layer and array dimensions. In such cases, the accelerator can run only at a fraction of its processing capacity (FLOPS/s), resulting in

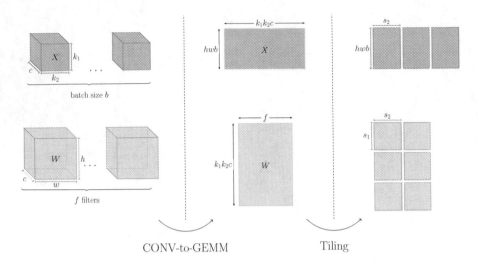

CONV-to-GEMM   Tiling

Fig. 2. Mapping stages for CONV operations onto array-based architectures.

slower execution and longer runtime. Hence, it is crucial to optimize DNN architectures in a way to improve the target platform's resource utilization, which will allow faster DNN inference. To that end, we argue that resource utilization must be addressed while designing DNN architectures with NAS.

3.2   Proposed Utilization Model

To be processed on an array-based accelerator, a DNN layer is first converted into a general matrix multiplication (CONV-to-GEMM) [16] and then tiled to match the dimensions of the array of processing elements. Figure 2 illustrates the CONV-to-GEMM conversion and tiling processes. Let us consider the following convolutional operation:

$$Y_{h \times w \times f \times b} = X_{h \times w \times c \times b} * W_{k_1 \times k_2 \times c \times f} \tag{1}$$

where h and w are the input image sizes, c is the number of input channels, b is the batch size, k_1 and k_2 are kernel sizes, and f is the number of filters, assuming a stride of 1. The matrix multiplication equivalent to the convolution operation is:

$$\hat{Y}_{hwb \times f} = \hat{X}_{hwb \times k_1 k_2 c} \hat{W}_{k_1 k_2 c \times f} \tag{2}$$

where $\hat{X}$, $\hat{W}$, and $\hat{Y}$ are obtained by rearranging the dimensions of X, W, and Y.

Let us consider the mapping of this matrix multiplication operation onto the array of processing elements with s_1 rows and s_2 columns. Since such an array can process a matrix with a maximum size of $s_1 \times s_2$, $\hat{X}$ and $\hat{W}$ must be divided into smaller tiles. The multiplication operation with the tiled operands is:

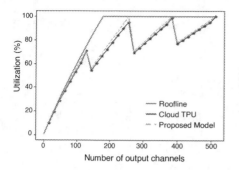

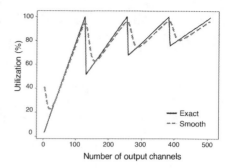

Fig. 3. Measured utilization on Cloud TPUv2 versus predicted utilization with roofline and the proposed model.

Fig. 4. Proposed utilization model with exact ceil function and its smooth approximation using the generalised logistic function.

$$\hat{y}^j_{hwb \times s_2} = \sum_{i=1}^{I} \hat{x}^i_{hwb \times s_1} \hat{w}^{ij}_{s_1 \times s_2} \tag{3}$$

where $\hat{x}^i$, $\hat{w}^{ij}$, and $\hat{y}^j$ are obtained from $\hat{X}$, $\hat{W}$, and $\hat{Y}$ as follows:

$$\hat{Y} = \begin{bmatrix} \hat{y}^1 \dots \hat{y}^J \end{bmatrix}, \quad \hat{X} = \begin{bmatrix} \hat{x}^1 \dots \hat{x}^I \end{bmatrix}, \quad \hat{W} = \begin{bmatrix} \hat{w}^{11} & \dots & \hat{w}^{1J} \\ \vdots & \ddots & \vdots \\ \hat{w}^{I1} & \dots & \hat{w}^{IJ} \end{bmatrix} \tag{4}$$

where I and J represent the number of tiles obtained from first and second dimensions of the matrix $\hat{W}$ and they are equal to $\left\lceil \frac{k_1 k_2 c}{s_1} \right\rceil$[1] and $\left\lceil \frac{f}{s_2} \right\rceil$, respectively. In the computation of the output matrix $\hat{Y}$, the number of tile multiplication operations $(\hat{x}^i \hat{w}^{ij})$ is, therefore, equal to $\left\lceil \frac{k_1 k_2 c}{s_1} \right\rceil \cdot \left\lceil \frac{f}{s_2} \right\rceil$.

As mentioned in Sect. 3.1, we assume a weight stationary dataflow, in which the elements of $\hat{w}^{ij}$ are spatially distributed to the array and $\hat{x}^i$ are loaded onto the array row by row. Processing a tile operation, thus, takes as many cycles as the number of rows in $\hat{x}^i$, namely hwb. Multiplying the cycles per tile operation by the number of tile operations, we obtain the total execution runtime (latency) in terms of the number of cycles as follows:

$$\texttt{RUNTIME} = \left\lceil \frac{k_1 k_2 c}{s_1} \right\rceil \left\lceil \frac{f}{s_2} \right\rceil hwb \tag{5}$$

The utilization of processing elements can be simply calculated as the ratio of the average throughput to the peak throughput. The average throughput (i.e., operations per unit time) is the total number of operations performed during the execution time. Using Eq. 5 and the number of multiply-and-accumulate

---

[1] The ceil function is defined as $\lceil x \rceil = \min\{n \in \mathbb{Z} : n \geq x\}$.

operations required to calculate $\hat{Y}$, which is equal to $hwbk_1k_2cf$, we finally obtain the utilization as follows:

$$\text{UTIL} = \frac{k_1k_2cf}{s_1s_2 \left\lceil \frac{k_1k_2c}{s_1} \right\rceil \left\lceil \frac{f}{s_2} \right\rceil} \tag{6}$$

Consider the case where the convolutional layer's dimensions exactly match the array dimensions: $k_1k_2c = s_1$ and $f = s_2$. Then, Eq. 6 simplifies to a utilization of 1, and the inference platform runs at full capacity. However, if the layer dimensions are slightly increased, for instance $k_1k_2c = s_1 + 1$, the ceil function reveals a significant drop in utilization since $\left\lceil \frac{k_1k_2c}{s_1} \right\rceil = \left\lceil \frac{s_1+1}{s_1} \right\rceil = 2$, resulting in a utilization of about 0.5. In other words, a slight modification in layer dimensions may lead to a significant change in hardware utilization.

To validate the proposed utilization model and to demonstrate the impact of channel dimensions on hardware utilization, we performed dense and convolutional DNN inference with varying numbers of output channels on a Cloud TPU v2 and measured the runtime and utilization values using Google Cloud's XLA op_profiler tool. Figure 3 shows the result of our experiment as well as estimated values with the proposed and roofline [36] models. Because Cloud TPUv2 have an array size of 128×128, we observe significant drops in utilization when the channel dimensions exceed multiples of 128. The roofline model, which accounts only for memory bottleneck, does not capture these drops in utilization, leading to a discrepancy up to 40% between measured and estimated values. The proposed utilization model, however, accounts for the dimension mismatches and is able to estimate the actual utilization value with an error of only up to 2%.

Moreover, hardware utilization also varies significantly across different layer types. For instance, depthwise convolutional layers [31], which are widely used in mobile applications, have only a single filter ($f = 1$) and perform convolution operations channel-by-channel. As a result, depthwise convolutional layers require matrix multiplications with dimensions equal to the $hwb \times k_1k_2$ and $k_1k_2 \times 1$, which is much smaller than the standard convolutional layers. The small matrix dimensions inherent to depthwise convolution often lead to a hardware utilization as low as 1% [7,13], which reduces their inference performance in array-based accelerators. In short, hardware utilization is highly sensitive to both layer type and layer dimensions, and their impact must be accounted for when searching for the optimal DNN architecture.

4   Proposed NAS Framework

Using the proposed utilization model, we introduce a utilization-aware differentiable NAS framework. In this Section, we first explain how we approximate the proposed utilization model, then we formulate our multi-objective loss function, and finally, we describe the NAS algorithm used to search optimal DNN architectures.

4.1   Approximation of the Utilization Function

The ceil function in Eq. 5 is not differentiable and can only be used as a collection of point estimates. This limits the effectiveness of the neural architecture search and allows only for evolutionary or reinforcement learning methods, which require orders of magnitude more computational resources compared to differentiable methods. For this reason, we use the generalised logistic function to obtain a smooth approximation of ceil function:

$$\text{CEIL}_{smooth}(x) = \sum_i \left[1 + \frac{\exp\left(-B(x - w_i)\right)}{C} \right]^{-1/v} \tag{7}$$

where w_i are intervals between zero and a fixed value; C, B, and v are constants that adjust the smoothness of the approximation. We empirically selected $C = 0.2$, $B = 20$, and $v = 0.5$, which leads to a smooth and accurate approximation of the original ceil function. Figure 4 show a comparison between the true utilization, denoted as *hard*, and its smooth counterpart. We verify that both hard and smooth utilization models yield peak utilization values at the same channel dimensions. Therefore, we replace the original utilization model with its smooth approximation in the proposed NAS framework.

4.2   Multi-objective Loss Function

Let $\mathcal{F}$ be the hypothesis class of neural networks that characterizes the search space. The candidate neural network $\alpha \in \mathcal{F}$ implements the function $f_\alpha : \mathcal{X} \to \mathcal{Y}$ where $\mathcal{X}$ and $\mathcal{Y}$ are the domains of the input and the output for our dataset $\mathcal{D}$, respectively. Let $(\boldsymbol{x}, y) \in \mathcal{X} \times \mathcal{Y}$ be a sample. Then the loss function consists of three terms:

$$\mathcal{L}(\boldsymbol{x}, y, \alpha) = \mathcal{L}_{classification}(f_\alpha(\boldsymbol{x}), y) + \lambda \cdot \mathcal{L}_{latency}(\alpha) - \beta \cdot \mathcal{L}_{utilization}(\alpha) \tag{8}$$

where $\lambda > 0$ and $\beta > 0$ determine the tradeoff between the accuracy, latency and utilization. The classification loss corresponds to cross-entropy, while the latency and utilization terms have been discussed in the previous section.

4.3   NAS Algorithm

The search algorithm employs a hierarchical search similar to prior work [25,35]. Concretely, it consists of three stages: microarchitecture search, macro-architecture search and training of the selected architecture $\alpha \in \mathcal{F}$. The first stage searches for layer types and connections using a model of a single cell and fixed channel dimensions. After obtaining the optimal candidate cell, the macroarchitecture stage constructs a model with k sequential cells sequentially and searches for the optimal channel dimensions cell-wise using the Dmasking method [35]. In both stages, each architectural decision (i.e., type of operator in the former and number of channels in the latter) is modelled by a probability simplex of dimension m equal to the number of choices and is parameterized by Gumbel-Softmax [15].

5   Experiments

To evaluate the effectivenes of the proposed method, we perform image classification experiments on the CIFAR10 and ImageNet100 datasets and compare our results with prior work. In this section, we first explain our experimental setup, then analyse the characteristics of the DNN architectures obtained with the proposed method, and finally, report and discuss the performance results of our experiments.

Experimental Setup. We perform experiments on widely used computer vision datasets, namely CIFAR10 [21] and ImageNet100, which is a subset of the Imagenet (ILSVRC 2012) classification dataset [10] with randomly-selected 100 classes. As in prior work [25,37], the optimal-architecture search stage for both datasets is performed on a proxy dataset, namely CIFAR10. We compare the results of our proposed method against three hardware-aware NAS methods that use *FLOPS* [12], *Roofline* [23], and *Blackbox* [37] models to estimate the latency. In FLOPS baseline, we simply calculate the latency as the number of operations required to perform inference divided by the theoretical peak throughput of inference platform assuming full-utilization. In Roofline baseline, we consider two modes, namely memory-bound and compute-bound. While the compute-bound mode is the same as the FLOPS baseline, in memory-bound mode, we calculate the latency as the memory footprint size divided by the off-chip bandwidth. In Blackbox baseline, we fill a lookup table with latency values for all layer types and dimensions with a quantization of 16 obtained with the hardware simulator, and retrieve these values during architecture search using nearest-neighbor interpolation.

Search Space. The cell architecture and search space are inspired by the DARTS architecture [25] with a few minor modifications. In all search and training stages, the candidate architecture consists of a preparatory block, k stack of cells, and a fully connected classifier. Each cell is a multigraph whose edges represent different operators, including depthwise separable, dilated, and standard convolutional layers as well as identity and zero operations corresponding to residual and no connections, respectively. Candidate kernel sizes for all convolutional layers are 3×3 and 5×5. Each cell has two input nodes connected to the output nodes of two previous cells. Each convolution operation has a stride of 1 and is followed by batch normalization and ReLU activation functions. The channel search space corresponds to a dimension range of 64 to 280 with increments of 8. For CIFAR10, we use a stack of three cells ($k = 3$), each of which is followed by a 2×2 maxpooling layer. To accomodate the increased complexity of ImageNet100, we use a stack of nine cells ($k = 9$), where only one of every three cells is followed by maxpooling. More details about the search space are given in appendix.

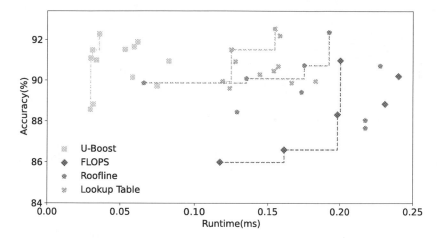

Fig. 5. Experiments on CIFAR10 dataset. Upper left corner is optimal. The dashed lines connect the points in the Pareto Front of each method.

NAS Settings. During the microarchitecture and channel search stages, the first 80% of the batches of each epoch is used to train model weights, while the last 20% is used to train the architectural parameters using a batch size of 64. The weights are optimized with Stochastic Gradient Descent (SGD) with learning rate 0.05, momentum 0.9 and weight decay $3e - 4$, while the architectural parameters use Adam [19] with learning rate 0.1. The microarchitecture and channel search stages last 10 and 30 epochs, respectively. To improve convergence, the temperature parameter τ of the Gumbel-Softmax is annealed exponentially by 0.95 per epoch from the initial value of 1. For fairness, we use the same NAS algorithm and hyperparameters for all baselines and the proposed method. After the search stages are completed, the selected DNN architecture is trained from scratch. In `CIFAR10` experiments, we train the models for 200 epochs with a batch size of 64 using the original image resolution of 32×32. In `ImageNet100` experiments, we train the models for 70 epochs with a batch size of 256 using an input resolution of 128×128. For both datasets, we use a preprocessing stage consisting of normalization, random crop and vertical flip.

Metrics. For all experiments, we report top-1 classification accuracy from the test datasets. Runtime and utilization values are measured by running the DNN models on our custom-made cycle-accurate hardware simulator. Correctness of our hardware simulator is validated against an RTL design of a systolic array architecture. During the hardware simulations, we assumed an array size of 128×128 as in Cloud TPUv4 [17] with a 15 MB on-chip memory and an 80 GB/s off-chip memory bandwidth and 1 GHz clock frequency. To quantify the trade-off between accuracy and latency, we calculate the hypervolume score [43], which is calculated as the volume of the union of axis-aligned rectangles from each point in a Pareto front [11]. We select the reference point to calculate the hypervolume

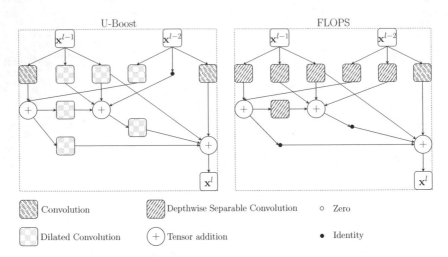

Fig. 6. Visualization of the `CIFAR10` cells obtained from U-Boost and FLOPS models during the microarchitecture search stage.

score as the perfect oracle: 100% accuracy with zero runtime. Consequently, lower scores indicate design points that are close to the ideal.

5.1 `CIFAR10` Experiments

To evaluate the proposed method on `CIFAR10` dataset, we set the utilization coefficient $\beta = 1$ in Eq. 8 and vary the latency coefficient $\lambda \in \{0.1, 0.5, 1, 5\}$ for all baselines to control accuracy-latency trade-off. Figure 5 shows the accuracy and latency of the DNN architectures found by the proposed method and baselines. We observe that U-Boost significantly improves the accuracy-latency Pareto front with a $2.8 - 4\times$ speedup in runtime compared to baseline methods while achieving comparable accuracy. The improvement in the Pareto front is also reflected in the hypervolume metric: U-Boost has a hypervolume of 0.39 whereas FLOPS, Roofline, and Blackbox baselines have hypervolumes of 2.68, 1.86, and 1.47, respectively, corresponding to an improvement in the range of $3.7 - 6.8\times$.

The reason why U-Boost achieves better accuracy-latency Pareto front is mainly because the selected cell microarchitecture and channel dimensions are well-suited for the target inference platform. To validate this insight, we analyze and compare the cell microarchitecture and channel dimensions selected by U-Boost and other baselines. Figure 6 depicts examples of cell microarchitectures selected by U-Boost and FLOPS baseline. We observe that the cell microarchitecture selected by FLOPS baseline mostly consists of depthwise separable convolutional layers because they require a smaller number of operations. However, these layers run at low utilization at the inference platforms, which increases their latency. By contrast, the cell microarchitecture selected by U-Boost consists of standard or dilated convolutional layers because U-Boost is

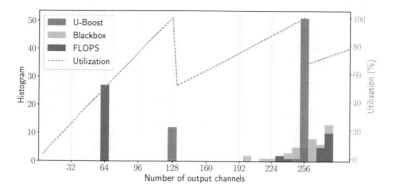

Fig. 7. Histogram of channel dimensions found by U-Boost as well as FLOPS and Blackbox baselines on `CIFAR10` dataset.

utilization-aware and it chooses layers that run at higher utilization in target platforms, reducing the latency.

Besides the cell microarchitecture, we also analyze the channel dimensions selected by the U-Boost and other baselines. Figure 7 shows the histogram of channel dimensions selected by U-Boost, FLOPS, and Blackbox baselines. We observe that the channel dimensions selected by FLOPS and Blackbox baselines are mostly concentrated on each end of the search space, which is bounded by channel dimensions of 64 and 280, rather than dimensions that correspond to high utilization. As a consequence, DNN architectures with such layers run at low utilization in target inference platforms. Unlike FLOPS and Blackbox baselines, we observe that the channel dimensions selected by U-Boost are concentrated on either 128 or 256, which are multiples of the array size and correspond to high utilization. As such, the DNN architectures selected by U-Boost run at high utilization, accelerating the inference at target platforms.

5.2   `ImageNet100` Experiments

To show the effectiveness of the proposed method on a more complex dataset, we also perform a set of experiments on `ImageNet100`. For this set of experiments, we set the latency coefficient $\lambda \in \{0.1, 1.0\}$ to control the accuracy-latency tradeoff. Table 1 reports the results of these experiments. We observe that FLOPS and Roofline baselines result in poor inference hardware utilization ($< 10\%$) as they estimate hardware performance inaccurately during the architecture search. The second best method in terms of utilization, namely Blackbox, improves the hardware utilization to 69% as it can estimate the hardware performance accurately during the search. Still, around 30% of hardware resources remain unutilized during inference as the Blackbox method can not find the optimal channel dimension since it operates on a discrete search space and is unable to exploit gradient information to successfully navigate the search.

Table 1. Experimental results for ImageNet100 experiments. Underlined measurements show best per column (λ), bold show best per metric. Number of parameters reported in millions.

λ	Acc. (%, ↑)		Runt. (ms, ↓)		Util. (%, ↑)		HV (↓)	# Params	
	0.1	1.0	0.1	1.0	0.1	1.0	(across λ)	0.1	1.0
Blackbox	87.5	87.8	4.8	4.05	69.3	68.5	49.4	70.5	55.5
Roofline	86.5	84.0	4.7	3.5	6.8	4.8	72.2	13.7	5.7
FLOPS	87.2	78.4	6.1	3.45	5.5	3.1	108	14.4	3.5
U-Boost	<u>87.8</u>	**87.9**	<u>2.2</u>	**1.05**	**<u>91.1</u>**	<u>78.6</u>	**12.7**	47.3	30.1

By contrast, the proposed U-Boost method, which both estimates the hardware performance accurately and uses the information from gradients to find the optimal cell microarchitecture and channel dimensions, achieves inference hardware utilization up to 91%, which is 1.3× higher than the second best baseline. Consequently, DNN architectures obtained with U-Boost achieve the best top-1 accuracy (87.9%), which is 0.1%, 0.7%, and 1.4% higher than the best of Blackbox, FLOPS, and Roofline baselines, respectively, while achieving speedups of 2.1× and 3.3× compared to the second best baselines across λ values. These results reiterate the importance of incorporating and correctly modeling utilization in hardware-aware NAS for computer vision tasks.

6   Conclusion

In this paper, we have illustrated the importance of resource utilization in runtime characteristics on target inference platforms. We demonstrated that by optimizing DNN architectures in terms of resource utilization as well as task accuracy and latency, we achieve significant improvement in accuracy-latency Pareto front. We proposed a utilization-aware differentiable NAS method, namely U-Boost. We provided an analytical model for resource utilization in widely used array-based hardware accelerators, which allows estimating the utilization efficiently and accurately during the architecture search. Through extensive experiments on popular computer vision datasets and detailed hardware simulations, we showed that the proposed U-Boost NAS method achieves 2.8 − 4× inference latency speedup with similar or improved accuracy, compared to utilization-agnostic NAS methods. This work highlights the importance of a holistic approach for hardware-aware NAS and the proposed method enables the design of DNNs with improved performance in inference accelerators.

Acknowledgements. The work of Ahmet Caner Yüzügüler was supported by the Hasler Foundation (Switzerland) and Nikolaos Dimitriadis was supported by Swisscom (Switzerland) AG.

References

1. Bannon, P., Venkataramanan, G., Sarma, D.D., Talpes, E.: Computer and redundancy solution for the full self-driving computer. In: 2019 IEEE Hot Chips 31 Symposium (HCS), Cupertino, CA, USA, 18–20 August 2019, pp. 1–22. IEEE (2019)
2. Bender, G., Kindermans, P., Zoph, B., Vasudevan, V., Le, Q.V.: Understanding and simplifying one-shot architecture search. In: Proceedings of the 35th International Conference on Machine Learning, ICML 2018, Stockholmsmässan, Stockholm, Sweden, 10–15 July 2018, vol. 80, pp. 549–558. PMLR (2018)
3. Bergstra, J., Bardenet, R., Bengio, Y., Kégl, B.: Algorithms for hyper-parameter optimization. In: Advances in Neural Information Processing Systems 24: 25th Annual Conference on Neural Information Processing Systems 2011. Proceedings of a meeting held 12–14 December 2011, Granada, Spain, pp. 2546–2554 (2011)
4. Cai, H., Zhu, L., Han, S.: Proxylessnas: direct neural architecture search on target task and hardware. In: 7th International Conference on Learning Representations, ICLR 2019, New Orleans, LA, USA, 6–9 May 2019. OpenReview.net (2019)
5. Chang, J., Zhang, X., Guo, Y., Meng, G., Xiang, S., Pan, C.: DATA: differentiable architecture approximation. In: Advances in Neural Information Processing Systems 32: Annual Conference on Neural Information Processing Systems 2019, NeurIPS 2019, 8–14 December 2019, Vancouver, BC, Canada, pp. 874–884 (2019)
6. Chen, Y., Emer, J.S., Sze, V.: Eyeriss: a spatial architecture for energy-efficient dataflow for convolutional neural networks. In: 43rd ACM/IEEE Annual International Symposium on Computer Architecture, ISCA 2016, Seoul, South Korea, 18–22 June 2016, pp. 367–379. IEEE Computer Society (2016)
7. Cho, H.: Risa: a reinforced systolic array for depthwise convolutions and embedded tensor reshaping. ACM Trans. Embed. Comput. Syst. **20**(5s), 53:1-53:20 (2021)
8. Choi, K., Hong, D., Yoon, H., Yu, J., Kim, Y., Lee, J.: DANCE: differentiable accelerator/network co-exploration. In: 58th ACM/IEEE Design Automation Conference, DAC 2021, San Francisco, CA, USA, 5–9 December 2021, pp. 337–342. IEEE (2021)
9. Dai, X., et al.: Chamnet: towards efficient network design through platform-aware model adaptation. In: IEEE Conference on Computer Vision and Pattern Recognition, CVPR 2019, Long Beach, CA, USA, 16–20 June 2019, pp. 11398–11407. Computer Vision Foundation/IEEE (2019)
10. Deng, J., Dong, W., Socher, R., Li, L., Li, K., Fei-Fei, L.: Imagenet: a large-scale hierarchical image database. In: 2009 IEEE Computer Society Conference on Computer Vision and Pattern Recognition (CVPR 2009), 20–25 June 2009, Miami, Florida, USA, pp. 248–255. IEEE Computer Society (2009)
11. Désidéri, J.A.: Multiple-gradient descent algorithm (MGDA) for multiobjective optimization. C.R. Math. **350**, 313–318 (2012)
12. Gordon, A., et al.: Morphnet: fast & simple resource-constrained structure learning of deep networks. In: 2018 IEEE Conference on Computer Vision and Pattern Recognition, CVPR 2018, Salt Lake City, UT, USA, 18–22 June 2018, pp. 1586–1595. Computer Vision Foundation/IEEE Computer Society (2018)
13. Gupta, S., Akin, B.: Accelerator-aware neural network design using automl. CoRR abs/2003.02838 (2020)
14. He, K., Gkioxari, G., Dollár, P., Girshick, R.B.: Mask R-CNN. In: IEEE International Conference on Computer Vision, ICCV 2017, Venice, Italy, 22–29 October 2017, pp. 2980–2988. IEEE Computer Society (2017)

15. Jang, E., Gu, S., Poole, B.: Categorical reparameterization with gumbel-softmax. In: 5th International Conference on Learning Representations, ICLR 2017, Toulon, France, 24–26 April 2017, Conference Track Proceedings. OpenReview.net (2017)
16. Jordà, M., Valero-Lara, P., Peña, A.J.: Performance evaluation of cudnn convolution algorithms on NVIDIA volta gpus. IEEE Access **7**, 70461–70473 (2019)
17. Jouppi, N.P., et al.: Ten lessons from three generations shaped Google's TPUv4i: Industrial product. In: 48th ACM/IEEE Annual International Symposium on Computer Architecture, ISCA 2021, Valencia, Spain, 14–18 June 2021, pp. 1–14. IEEE (2021)
18. Jouppi, N.P., et al.: In-datacenter performance analysis of a tensor processing unit. In: Proceedings of the 44th Annual International Symposium on Computer Architecture, ISCA 2017, Toronto, ON, Canada, 24–28 June 2017, pp. 1–12. ACM (2017)
19. Kingma, D.P., Ba, J.: Adam: a method for stochastic optimization. In: 3rd International Conference on Learning Representations, ICLR 2015, San Diego, CA, USA, 7–9 May 2015, Conference Track Proceedings (2015)
20. Kokkinos, I.: Ubernet: training a universal convolutional neural network for low-, mid-, and high-level vision using diverse datasets and limited memory. In: 2017 IEEE Conference on Computer Vision and Pattern Recognition, CVPR 2017, Honolulu, HI, USA, 21–26 July 2017, pp. 5454–5463. IEEE Computer Society (2017)
21. Krizhevsky, A.: Learning multiple layers of features from tiny images. Tech. rep. (2009)
22. Kung, H.T.: Why systolic architectures? Computer **15**(1), 37–46 (1982)
23. Li, S., et al.: Searching for fast model families on datacenter accelerators. In: IEEE Conference on Computer Vision and Pattern Recognition, CVPR 2021, Virtual, 19–25 June 2021, pp. 8085–8095. Computer Vision Foundation/IEEE (2021)
24. Liu, C., et al.: Progressive neural architecture search. In: Ferrari, V., Hebert, M., Sminchisescu, C., Weiss, Y. (eds.) ECCV 2018. LNCS, vol. 11205, pp. 19–35. Springer, Cham (2018). https://doi.org/10.1007/978-3-030-01246-5_2
25. Liu, H., Simonyan, K., Yang, Y.: DARTS: differentiable architecture search. In: 7th International Conference on Learning Representations, ICLR 2019, New Orleans, LA, USA, 6–9 May 2019. OpenReview.net (2019)
26. Marchisio, A., Massa, A., Mrazek, V., Bussolino, B., Martina, M., Shafique, M.: Nascaps: a framework for neural architecture search to optimize the accuracy and hardware efficiency of convolutional capsule networks. In: IEEE/ACM International Conference On Computer Aided Design, ICCAD 2020, San Diego, CA, USA, 2–5 November 2020, pp. 114:1–114:9. IEEE (2020)
27. Nayman, N., Noy, A., Ridnik, T., Friedman, I., Jin, R., Zelnik-Manor, L.: XNAS: neural architecture search with expert advice. In: Advances in Neural Information Processing Systems 32: Annual Conference on Neural Information Processing Systems 2019, NeurIPS 2019, December, pp. 1975–1985 (2019)
28. Pham, H., Guan, M.Y., Zoph, B., Le, Q.V., Dean, J.: Efficient neural architecture search via parameter sharing. In: Proceedings of the 35th International Conference on Machine Learning, ICML 2018, Stockholmsmässan, Stockholm, Sweden, 10–15 July 2018. Proceedings of Machine Learning Research, vol. 80, pp. 4092–4101. PMLR (2018)

29. Real, E., Aggarwal, A., Huang, Y., Le, Q.V.: Regularized evolution for image classifier architecture search. In: The 33rd AAAI Conference on Artificial Intelligence, AAAI 2019, The 31st Innovative Applications of Artificial Intelligence Conference, IAAI 2019, The 9th AAAI Symposium on Educational Advances in Artificial Intelligence, EAAI 2019, Honolulu, Hawaii, USA, 27 January - 1 February 2019, pp. 4780–4789. AAAI Press (2019)

30. Samajdar, A., Joseph, J.M., Zhu, Y., Whatmough, P.N., Mattina, M., Krishna, T.: A systematic methodology for characterizing scalability of DNN accelerators using scale-sim. In: IEEE International Symposium on Performance Analysis of Systems and Software, ISPASS 2020, Boston, MA, USA, 23–25 August 2020, pp. 58–68. IEEE (2020)

31. Sandler, M., Howard, A.G., Zhu, M., Zhmoginov, A., Chen, L.: Mobilenetv 2: inverted residuals and linear bottlenecks. In: 2018 IEEE Conference on Computer Vision and Pattern Recognition, CVPR 2018, Salt Lake City, UT, USA, 18–22 June 2018, pp. 4510–4520. Computer Vision Foundation/IEEE Computer Society (2018)

32. Smithson, S.C., Yang, G., Gross, W.J., Meyer, B.H.: Neural networks designing neural networks: multi-objective hyper-parameter optimization. In: Proceedings of the 35th International Conference on Computer-Aided Design, ICCAD 2016, Austin, TX, USA, 7–10 November 2016, p. 104. ACM (2016)

33. Stamoulis, D., et al.: Single-path nas: designing hardware-efficient convnets in less than 4 hours. In: Brefeld, U., Fromont, E., Hotho, A., Knobbe, A., Maathuis, M., Robardet, C. (eds.) ECML PKDD 2019. LNCS (LNAI), vol. 11907, pp. 481–497. Springer, Cham (2020). https://doi.org/10.1007/978-3-030-46147-8_29

34. Tan, M., et al.: Mnasnet: platform-aware neural architecture search for mobile. In: IEEE Conference on Computer Vision and Pattern Recognition, CVPR 2019, Long Beach, CA, USA, 16–20 June 2019, pp. 2820–2828. Computer Vision Foundation/IEEE (2019)

35. Wan, A., et al.: Fbnetv2: differentiable neural architecture search for spatial and channel dimensions. In: 2020 IEEE/CVF Conference on Computer Vision and Pattern Recognition, CVPR 2020, Seattle, WA, USA, 13–19 June 2020, pp. 12962–12971. Computer Vision Foundation/IEEE (2020)

36. Williams, S., Waterman, A., Patterson, D.A.: Roofline: an insightful visual performance model for multicore architectures. Commun. ACM **52**(4), 65–76 (2009)

37. Wu, B., et al.: Fbnet: hardware-aware efficient convnet design via differentiable neural architecture search. In: IEEE Conference on Computer Vision and Pattern Recognition, CVPR 2019, Long Beach, CA, USA, 16–20 June 2019, pp. 10734–10742. Computer Vision Foundation/IEEE (2019)

38. Xie, S., Zheng, H., Liu, C., Lin, L.: SNAS: stochastic neural architecture search. In: 7th International Conference on Learning Representations, ICLR 2019, New Orleans, LA, USA, 6–9 May 2019. OpenReview.net (2019)

39. Xiong, Y., et al.: Mobiledets: searching for object detection architectures for mobile accelerators. In: IEEE Conference on Computer Vision and Pattern Recognition, CVPR 2021, virtual, 19–25 June 2021, pp. 3825–3834. Computer Vision Foundation/IEEE (2021)

40. Xu, Y., et al.: PC-DARTS: partial channel connections for memory-efficient architecture search. In: 8th International Conference on Learning Representations, ICLR 2020, Addis Ababa, Ethiopia, 26–30 April 2020. OpenReview.net (2020)

41. Yang, T.-J., et al.: NetAdapt: platform-aware neural network adaptation for mobile applications. In: Ferrari, V., Hebert, M., Sminchisescu, C., Weiss, Y. (eds.) ECCV 2018. LNCS, vol. 11214, pp. 289–304. Springer, Cham (2018). https://doi.org/10. 1007/978-3-030-01249-6_18

42. Zhang, L.L., Yang, Y., Jiang, Y., Zhu, W., Liu, Y.: Fast hardware-aware neural architecture search. In: 2020 IEEE/CVF Conference on Computer Vision and Pattern Recognition, CVPR Workshops 2020, Seattle, WA, USA, 14–19 June 2020, pp. 2959–2967. Computer Vision Foundation/IEEE (2020)

43. Zitzler, E., Thiele, L.: Multiobjective evolutionary algorithms: a comparative case study and the strength pareto approach. IEEE Trans. Evol. Comput. **3**(4), 257–271 (1999)

44. Zoph, B., Le, Q.V.: Neural architecture search with reinforcement learning. In: 5th International Conference on Learning Representations, ICLR 2017, Toulon, France, 24–26 April 2017, Conference Track Proceedings. OpenReview.net (2017)

45. Zoph, B., Vasudevan, V., Shlens, J., Le, Q.V.: Learning transferable architectures for scalable image recognition. In: 2018 IEEE Conference on Computer Vision and Pattern Recognition, CVPR 2018, Salt Lake City, UT, USA, 18–22 June 2018, pp. 8697–8710. Computer Vision Foundation/IEEE Computer Society (2018)

PTQ4ViT: Post-training Quantization for Vision Transformers with Twin Uniform Quantization

Zhihang Yuan[1,3], Chenhao Xue[1], Yiqi Chen[1], Qiang Wu[3], and Guangyu Sun[2(✉)]

[1] School of Computer Science, Peking University, Beijing, China
{yuanzhihang,xch927027}@pku.edu.cn
[2] School of Integrated Circuits, Peking University, Beijing, China
gsun@pku.edu.cn
[3] Houmo AI, Nanjing, China

Abstract. Quantization is one of the most effective methods to compress neural networks, which has achieved great success on convolutional neural networks (CNNs). Recently, vision transformers have demonstrated great potential in computer vision. However, previous post-training quantization methods performed not well on vision transformer, resulting in more than 1% accuracy drop even in 8-bit quantization. Therefore, we analyze the problems of quantization on vision transformers. We observe the distributions of activation values after softmax and GELU functions are quite different from the Gaussian distribution. We also observe that common quantization metrics, such as MSE and cosine distance, are inaccurate to determine the optimal scaling factor. In this paper, we propose the twin uniform quantization method to reduce the quantization error on these activation values. And we propose to use a Hessian guided metric to evaluate different scaling factors, which improves the accuracy of calibration at a small cost. To enable the fast quantization of vision transformers, we develop an efficient framework, PTQ4ViT. Experiments show the quantized vision transformers achieve near-lossless prediction accuracy (less than 0.5% drop at 8-bit quantization) on the ImageNet classification task.

1 Introduction

The self-attention module is the basic building block of the transformer to capture global information [23]. Inspired by the success of transformers [2,6] on natural language processing (NLP) tasks, researchers have brought the self-attention module into computer vision [7,15]. They replaced the convolution layers in convolutional neural networks (CNNs) with self-attention modules and they called these networks vision transformers. Vision transformers are comparable to CNNs on many computer vision tasks and have great potential to be deployed on various applications [11].

Z. Yuan and C. Xue—contribute equally to this paper.

© The Author(s), under exclusive license to Springer Nature Switzerland AG 2022
S. Avidan et al. (Eds.): ECCV 2022, LNCS 13672, pp. 191–207, 2022.
https://doi.org/10.1007/978-3-031-19775-8_12

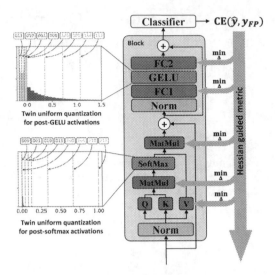

Fig. 1. Overview of the PTQ4ViT.

However, both the CNN and the vision transformer are computationally intensive and consume much energy. The larger and larger scales of neural networks block their deployment on various hardware devices, such as mobile phones and IoT devices, and increase carbon emissions. It is required to compress these neural networks. Quantization is one of the most effective ways to compress neural networks [9]. The floating-point values are quantized to integers with a low bit-width, reducing the memory consumption and the computation cost.

There are two types of quantization methods, quantization-aware training (QAT) [4,28] and post-training quantization (PTQ) [1,5]. Although QAT can generate the quantized network with a lower accuracy drop, the training of the network requires a training dataset, a long optimization time, and the tuning of hyper-parameters. Therefore, QAT is impractical when the training dataset is not available or rapid deployment is required. While PTQ quantizes the network with unlabeled calibration images after training, which enables fast quantization and deployment.

Although PTQ has achieved great success on CNNs, directly bringing it to vision transformer results in more than 1% accuracy drop even with 8-bit quantization [16]. Therefore, we analyze the problems of quantization on vision transformers. We collect the distribution of activation values in the vision transformer and observe there are some special distributions. 1) The values after softmax have a very unbalanced distribution in $[0, 1]$, where most of them are very close to zero. Although the number of large values is very small, they mean high attention between two patches, which is of vital importance in the attention mechanism. This requires a large scaling factor to make the quantization range cover the large value. However, a big scaling factor quantizes the small values to zero, resulting in a large quantization error. 2) The values after the GELU function have an asymmetrical distribution, where the positive values have a large distribution range while the negative values have a very small distribution range. It's difficult to well quantify both the positive values and negative values with uniform quantization. Therefore, we propose the twin uniform quantization, which separately quantifies the values in two ranges. To enable its efficient processing on hardware devices, we design a data format and constrain the scaling factors of the two ranges.

The second problem is that the metric to determine the optimal scaling factor is not accurate on vision transformers. There are various metrics in previous PTQ methods, including MSE, cosine distance, and Pearson correlation coefficient between the layer outputs before and after quantization. However, we observe they are inaccurate to evaluate different scaling factor candidates because only the local information is used. Therefore, we propose to use the Hessian guided metric to determine the quantization parameters, which is more accurate. The proposed methods are demonstrated in Fig. 1.

We develop a post-training quantization framework for vision transformers using twin uniform quantization, PTQ4ViT.[1] Experiments show the quantized vision transformers (ViT, DeiT, and Swin) achieve near-lossless prediction accuracy (less than 0.5% drop at 8-bit quantization) on the ImageNet classification task.

Our contributions are listed as follows:

- We find the problems in PTQ on vision transformers are special distributions of post-softmax and post-GELU activations and the inaccurate metric.
- We propose the twin uniform quantization to handle the special distributions, which can be efficiently processed on existing hardware devices including CPU and GPU.
- We propose to use the Hessian guided metric to determine the optimal scaling factors, which replaces the inaccurate metrics.
- The quantized networks achieve near-lossless prediction accuracy, making PTQ acceptable on vision transformers.

2   Background and Related Work

2.1   Vision Transformer

In the last few years, convolution neural networks (CNNs) have achieved great success in computer vision. The convolution layer is a fundamental component of CNNs to extract features using local information. Recently, the position of CNNs in computer vision is challenged by vision transformers, which take the self-attention modules [23] to make use of the global information. DETR [3] is the first work to replace the object detection head with a transformer, which directly regresses the bounding boxes and achieves comparable results with the CNN-based head. ViT [7] is the first architecture that replaces all convolution layers,

---

[1] Code is in https://github.com/hahnyuan/PTQ4ViT.

which achieves better results on image classification tasks. Following ViT, various vision transformer architectures have been proposed to boost performance [15, 25]. Vision transformers have been successfully applied to downstream tasks [12, 15,24]. They have great potential for computer vision tasks [11].

The input of a transformer is a sequence of vectors. An image is divided into several patches and a linear projection layer is used to project each patch to a vector. These vectors form the input sequence of the vision transformer. We denote these vectors as $X \in R^{N \times D}$, where N is the number of patches and D is the hidden size, which is the size of the vector after linear projection.

A vision transformer contains some blocks. As shown in Fig. 1, each block is composed of a multi-head self-attention module (MSA) and a multi-layer perceptron (MLP). MSA generates the attention between different patches to extract features with global information. Typical MLP contains two fully-connected layers (FC) and the GELU activation function is used after the first layer. The input sequence is first fed into each self-attention head of MSA. In each head, the sequence is linearly projected to three matrices, query $Q = XW^Q$, key $K = XW^K$, and value $V = XW^V$. Then, matrix multiplication QK^T calculates the attention scores between patches. The softmax function is used to normalize these scores to attention probability P. The output of the head is matrix multiplication PV. The process is formulated as Eq. (1):

$$\text{Attention}(Q, K, V) = \text{softmax}(\frac{QK^T}{\sqrt{d}})V, \tag{1}$$

where d is the hidden size of head. The outputs of multiple heads are concatenated together as the output of MSA.

Vision transformers have a large amount of memory, computation, and energy consumption, which hinders their deployment in real-world applications. Researchers have proposed a lot of methods to compress vision transformers, such as patch pruning [21], knowledge distillation [13], and quantization [16].

2.2   Quantization

Network quantization is one of the most effective methods to compress neural networks. The weight values and activation values are transformed from floating-point to integer with lower bit-width, which significantly decreases the memory consumption, data movement, and energy consumption. The uniform symmetric quantization is the most widely used method, which projects a floating-point value x to a k-bit integer value x_q with a scaling factor Δ:

$$x_q = \Psi_k(x, \Delta) = \text{clamp}(\text{round}(\frac{x}{\Delta}), -2^{k-1}, 2^{k-1} - 1), \tag{2}$$

where round projects a value to an integer and clamp constrains the output in the range that k-bit integer can represent. We propose the twin uniform quantization, which separately quantifies the values in two ranges. [8] also uses multiple quantization ranges. However, their method targets CNN and is not suitable for ViT. They use an extra bit to represent which range is used, taking 12.5% more storage than our method. Moreover, they use FP32 computation to align the two ranges, which is not efficient. Our method uses the shift operation, avoiding the format transformation and extra FP32 multiplication and FP32 addition.

There are two types of quantization methods, quantization-aware training (QAT) [4,28] and post-training quantization (PTQ) [1,5]. QAT methods combine quantization with network training. It optimizes the quantization parameters to minimize the task loss on a labeled training dataset. QAT can be used to quantize transformers [18]. Q-BERT [20] uses the Hessian spectrum to evaluate the sensitivity of the different tensors for mixed-precision, achieving 3-bit weight and 8-bit activation quantization. Although QAT achieves lower bit-width, it requires a training dataset, a long quantization time, and hyper-parameter tuning. PTQ methods quantize networks with a small number of unlabeled images, which is significantly faster than QAT and doesn't require any labeled dataset. PTQ methods should determine the scaling factors Δ of activations and weights for each layer. Choukroun et al. [5] proposed to minimize the mean square error (MSE) between the tensors before and after quantization. EasyQuant [27] uses the cosine distance to improve the quantization performance on CNN. Recently, Liu et al. [16] first proposed a PTQ method to quantize the vision transformer. Pearson correlation coefficient and ranking loss are used as the metrics to determine the scaling factors. However, these metrics are inaccurate to evaluate different scaling factor candidates because only the local information is used.

3 Method

In this section, we will first introduce a base PTQ method for vision transformers. Then, we will analyze the problems of quantization using the base PTQ and propose methods to address the problems. Finally, we will introduce our post-training quantization framework, PTQ4ViT.

3.1 Base PTQ for Vision Transformer

Matrix multiplication is used in the fully-connected layer and the computation of QK^T and PV, which is the main operation in vision transformers. In this paper, we formulate it as $O = AB$ and we will focus on its quantization. A and B are quantized to k-bit using the symmetric uniform quantization with scaling factors Δ_A and Δ_B. According to Eq. (2), we have $A_q = \Psi_k(A, \Delta_A)$ and $B_q = \Psi_k(B, \Delta_B)$. In base PTQ, the distance of the output before and after quantization is used as metric to determine the scaling factors, which is formulated as:

$$\min_{\Delta_A, \Delta_B} \text{distance}(O, \hat{O}), \tag{3}$$

where $\hat{O}$ is the output of the matrix multiplication after quantization $\hat{O} = \Delta_A \Delta_B A_q B_q$.

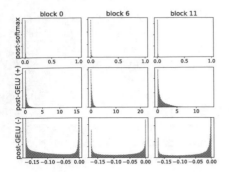
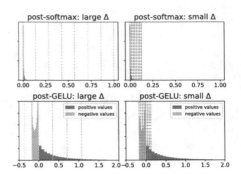
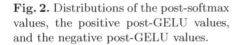

Fig. 2. Distributions of the post-softmax values, the positive post-GELU values, and the negative post-GELU values.

Fig. 3. Demonstration of different scaling factors to quantize the post-softmax and post-GELU activation values.

The same as [27], we use cosine distance as the metric to calculate the distance. We make the search spaces of Δ_A and Δ_B by linearly dividing $[\alpha\frac{A_{max}}{2^{k-1}}, \beta\frac{A_{max}}{2^{k-1}}]$ and $[\alpha\frac{B_{max}}{2^{k-1}}, \beta\frac{B_{max}}{2^{k-1}}]$ to n candidates, respectively. A_{max} and B_{max} are the maximum absolute value of A and B. α and β are two parameters to control the search range. We alternatively search for the optimal scaling factors Δ_A^* and Δ_B^* in the search space. Firstly, Δ_B is fixed, and we search for the optimal Δ_A to minimize distance$(O, \hat{O})$. Secondly, Δ_A is fixed, and we search for the optimal Δ_B to minimize distance$(O, \hat{O})$. Δ_A and Δ_B are alternately optimized for several rounds.

The values of A and B are collected using unlabeled calibration images. We search for the optimal scaling factors of activation or weight layer-by-layer. However, the base PTQ results in more than 1% accuracy drop on quantized vision transformer in our experiments.

3.2 Twin Uniform Quantization

The activation values in CNNs are usually considered Gaussian distributed. Therefore, most PTQ quantization methods are based on this assumption to determine the scaling factor. However, we observe the distributions of post-softmax values and post-GELU values are quite special as shown in Fig. 2. Specifically, (1) The distribution of activations after softmax is very unbalanced, in which most values are very close to zero and only a few values are close to one. (2) The values after the GELU function have a highly asymmetric distribution, in which the unbounded positive values are large while the negative values have a very small distribution range. As shown in Fig. 3, we demonstrate the quantization points of the uniform quantization using different scaling factors.

For the values after softmax, a large value means that there is a high correlation between the two patches, which is important in the self-attention mechanism. A larger scaling factor can reduce the quantization error of these large

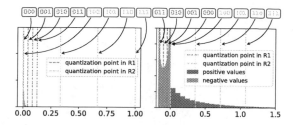

Fig. 4. Demonstration of the 3-bit twin uniform quantization on post-softmax values (left) and post-GELU values (right). We annotate the binary values for different quantization points.

values, which causes smaller values to be quantized to zero. While a small scaling factor makes the large values quantized to small values, which significantly decreases the intensity of attention between two patches. For the values after GELU, it is difficult to quantify both positive and negative values well with symmetric uniform quantization. Non-uniform quantization [10] can be used to solve the problem. It can set the quantization points according to the distribution, ensuring the overall quantization error is small. However, most hardware devices cannot efficiently process the non-uniform quantized values. Acceleration can be achieved only on specially designed hardware.

We propose the twin uniform quantization, which can be efficiently processed on existing hardware devices including CPUs and GPUs. As shown in Fig. 4, twin uniform quantization has two quantization ranges, R1 and R2, which are controlled by two scaling factors Δ_{R1} and Δ_{R2}, respectively. The k-bit twin uniform quantization is be formulated as:

$$T_k(x, \Delta_{R1}, \Delta_{R2}) = \begin{cases} \Psi_{k-1}(x, \Delta_{R1}), x \in R1 \\ \Psi_{k-1}(x, \Delta_{R2}), otherwise \end{cases}. \tag{4}$$

For values after softmax, the values in $R1 = [0, 2^{k-1}\Delta_{R1}^s)$ can be well quantified by using a small Δ_{R1}^s. To avoid the effect of calibration dataset, we keeps Δ_{R2}^s fixed to $1/2^{k-1}$. Therefore, $R2 = [0, 1]$ can cover the whole range, and large values can be well quantified in R2. For activation values after GELU, negative values are located in $R1 = [-2^{k-1}\Delta_{R1}^g, 0]$ and positive values are located in $R2=[0, 2^{k-1}\Delta_{R2}^g]$. We also keep Δ_{R1}^g fixed to make R1 just cover the entire range of negative numbers. Since different quantization parameters are used for positive and negative values respectively, the quantization error can be effectively reduced. When calibrating the network, we search for the optimal Δ_{R1}^s and Δ_{R2}^g.

The uniform symmetric quantization uses the k bit signed integer data format. It consists of one sign bit and $k-1$ bits representing the quantity. In order to efficiently store the twin-uniform-quantized values, we design a new data format. The most significant bit is the range flag to represent which range is used (0 for R1, 1 for R2). The other $k-1$ bits compose an unsigned number to represent

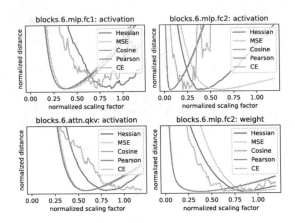

Fig. 5. The distance between the layer outputs before and after quantization and the change of task loss (CE) under different scaling factors on ViT-S/224. The x-axis is the normalized scaling factor by dividing $\frac{A_{max}}{2^k-1}$ or $\frac{B_{max}}{2^k-1}$.

the quantity. Because the sign of values in the same range is the same, the sign bit is removed.

Data in different ranges need to be multiplied and accumulated in matrix multiplication. In order to efficiently process with the twin-uniform-quantized values on CPUs or GPUs, we constrain the two ranges with $\Delta_{R2} = 2^m \Delta_{R1}$, where m is an unsigned integer. Assuming a_q is quantized in R1 and b_q is quantized in R2, the two values can be aligned:

$$a_q \times \Delta_{R1} + b_q \times \Delta_{R2} = (a_q + b_q \times 2^m)\Delta_{R1}. \tag{5}$$

We left shift b_q by m bits, which is the same as multiplying the value by 2^m. The shift operation is very efficient on CPUs or GPUs. Without this constraint, multiplication is required to align the scaling factor, which is much more expensive than shift operations.

3.3   Hessian Guided Metric

Next, we will analyze the metrics to determine the scaling factors of each layer. Previous works [5,16,27] greedily determine the scaling factors of inputs and weights layer by layer. They use various kinds of metrics, such as MSE and cosine distance, to measure the distance between the original and the quantized outputs. The change in the internal output is considered positively correlated with the task loss, so it is used to calculate the distance.

We plot the performance of different metrics in Fig. 5. We observe that MSE, cosine distance, and Pearson correlation coefficient are inaccurate compared with task loss (cross-entropy) on vision transformers. The optimal scaling factors based on them are not consistent with that based on task loss. For instance, on blocks.6.mlp.fc1:activation, they indicate that a scaling factor around $0.4\frac{A_{max}}{2^k-1}$ is the optimal one, while the scaling factor around $0.75\frac{A_{max}}{2^k-1}$ is the optimal according to the task loss. Using these metrics, we get sub-optimal scaling factors, causing the accuracy degradation. The distance between the last layer's output before and after quantization can be more accurate in PTQ. However, using it to determine the scaling factors of internal layers is impractical because it requires executing the network many times to calculate the last layer's output, which consumes too much time.

To achieve high accuracy and quick quantization at the same time, we propose to use the Hessian guided metric to determine the scaling factors. In the classification task, the task loss is $L = \mathrm{CE}(\hat{y}, y)$, where CE is cross-entropy, $\hat{y}$ is the output of the network, and y is the ground truth[2]. When we treat the weights as variables, the expectation of loss is a function of weight $\mathbb{E}[L(W)]$. The quantization brings a small perturbation ϵ on weight $\hat{W} = W + \epsilon$. We can analyze the influence of quantization on task loss by Taylor series expansion.

$$\mathbb{E}[L(\hat{W})] - \mathbb{E}[L(W)] \approx \epsilon^T \bar{g}^{(W)} + \frac{1}{2}\epsilon^T \bar{H}^{(W)}\epsilon, \tag{6}$$

where $\bar{g}^{(W)}$ is the gradients and $\bar{H}^{(W)}$ is the Hessian matrix. The target is to find the scaling factors to minimize the influence: $\min_\Delta (\mathbb{E}[L(\hat{W})] - \mathbb{E}[L(W)])$. Based on the layer-wise reconstruction method in [14], the optimization can be approximated[3] by:

$$\min_\Delta \mathbb{E}[(\hat{O}^l - O^l)^T diag\left(\left(\frac{\partial L}{\partial O_1^l}\right)^2, \dots, \left(\frac{\partial L}{\partial O_{|O^l|}^l}\right)^2\right)(\hat{O}^l - O^l)], \tag{7}$$

where O^l and $\hat{O}^l$ are the outputs of the l-th layer before and after quantization, respectively. As shown in Fig. 5, the optimal scaling factor indicated by Hessian guided metric is closer to that indicated by task loss (CE). Although it still has a gap with the task loss, Hessian guided metric significantly improves the performance. For instance, on blocks.6.mlp.fc1:activation, the optimal scaling factor indicated by Hessian guided metric has less influence on task loss than other metrics.

3.4 PTQ4ViT Framework

To achieve fast quantization and deployment, we develop an efficient post-training quantization framework for vision transformers, PTQ4ViT. Its flow is described in Algorithm 1. It supports the twin uniform quantization and Hessian guided metric. There are two quantization phases. 1) The first phase is to collect the output and the gradient of the output in each layer before quantization. The outputs of the l-th layer O^l are calculated through forward propagation on the calibration dataset. The gradients $\frac{\partial L}{\partial O_1^l}, \dots, \frac{\partial L}{\partial O_a^l}$ are calculated through

---

[2] The ground truth y is not available in PTQ, so we use the prediction of floating-point network y_{FP} to approximate it.

[3] The derivation of it is in Appendix.

Algorithm 1: Searches for the optimal scaling factors of each layer.

1 **for** l *in* 1 *to* L **do**
2 | forward-propagation $O^l \leftarrow A^l B^l$;
3 **end**
4 **for** l *in* L *to* 1 **do**
5 | backward-propagation to get $\frac{\partial L}{\partial O^l}$;
6 **end**
7 **for** l *in* 1 *to* L **do**
8 | initialize $\Delta_{B^l}^* \leftarrow \frac{B_{max}^l}{2^{k}-1}$;
9 | generate search spaces of Δ_{A^l} and Δ_{B^l};
10 | **for** $r = 1$ *to* $\#Round$ **do**
11 | | search for $\Delta_{A^l}^*$ using Eq. (7);
12 | | search for $\Delta_{B^l}^*$ using Eq. (7);
13 | **end**
14 **end**

---

backward propagation. 2) The second phase is to search for the optimal scaling factors layer by layer. Different scaling factors in the search space are used to quantize the activation values and weight values in the l-th layer. Then the output of the layer $\hat{O}^l$ is calculated. We search for the optimal scaling factor Δ^* that minimizes Eq. (7).

In the first phase, we need to store O^l and $\frac{\partial L}{\partial O^l}$, which consumes a lot of GPU memory. Therefore, we transfer these data to the main memory when they are generated. In the second phase, we transfer O^l and $\frac{\partial L}{\partial O^l}$ back to GPU memory and destroy them when the quantization of l-th layer is finished. To make full use of the GPU parallelism, we calculate $\hat{O}^l$ and the influence on loss for different scaling factors in batches.

4 Experiments

In this section, we first introduce the experimental settings. Then we will evaluate the proposed methods on different vision transformer architectures. At last, we will take an ablation study on the proposed methods.

4.1 Experiment Settings

For post-softmax quantization, the search space of Δ_{R1}^s is $[\frac{1}{2^k}, \frac{1}{2^{k+1}}, ..., \frac{1}{2^{k+10}}]$. The search spaces of scaling factors for weight and other activations are the same as that of base PTQ (Sec. 3.1). We set $alpha = 0$, $beta = 1.2$, and $n = 100$. The search round $\#Round$ is set to 3. We experiment on the ImageNet classification task [19]. We randomly select 32 images from the training dataset as calibration images. The ViT models are provided by timm [26].

We quantize all the weights and inputs for the fully-connect layers including the first projection layer and the last prediction layer. We also quantize the two

Table 1. Top-1 accuracy of quantized vision transformers. The result in the bracket is the accuracy drop from floating-point networks. W6A6 means weights are quantized to 6-bit and activations are quantized to 6-bit. The default patch size is 16×16. ViT-S/224/32 means the input resolution is 224×224 and the patch size is 32×32.

Model	FP32	Base PTQ		PTQ4ViT	
		W8A8	W6A6	W8A8	W6A6
ViT-S/224/32	75.99	73.61(2.38)	60.14(15.8)	75.58(0.41)	71.90(4.08)
ViT-S/224	81.39	80.46(0.91)	70.24(11.1)	81.00(0.38)	78.63(2.75)
ViT-B/224	84.54	83.89(0.64)	75.66(8.87)	84.25(0.29)	81.65(2.89)
ViT-B/384	86.00	85.35(0.64)	46.88(39.1)	85.82(0.17)	83.34(2.65)
DeiT-S/224	79.80	77.65(2.14)	72.26(7.53)	79.47(0.32)	76.28(3.51)
DeiT-B/224	81.80	80.94(0.85)	78.78(3.01)	81.48(0.31)	80.25(1.55)
DeiT-B/384	83.11	82.33(0.77)	68.44(14.6)	82.97(0.13)	81.55(1.55)
Swin-T/224	81.39	80.96(0.42)	78.45(2.92)	81.24(0.14)	80.47(0.91)
Swin-S/224	83.23	82.75(0.46)	81.74(1.48)	83.10(0.12)	82.38(0.84)
Swin-B/224	85.27	84.79(0.47)	83.35(1.91)	85.14(0.12)	84.01(1.25)
Swin-B/384	86.44	86.16(0.26)	85.22(1.21)	86.39(0.04)	85.38(1.04)

input matrices for the matrix multiplications in self-attention modules. We use different quantization parameters for different self-attention heads. The scaling factors for W^Q, W^K, and W^V are different. The same as [16], we don't quantize softmax and normalization layers in vision transformers.

4.2 Results on ImageNet Classification Task

We choose different vision transformer architectures, including ViT [7], DeiT [22], and Swin [15]. The results are demonstrated in Table 1. From this table, we observe that base PTQ results in more than 1% accuracy drop on some vision transformers even at the 8-bit quantization. PTQ4ViT achieves less than 0.5% accuracy drop with 8-bit quantization. For 6-bit quantization, base PTQ results in high accuracy drop (9.8% on average) while PTQ4ViT achieves a much smaller accuracy drop (2.1% on average).

We observe that the accuracy drop on Swin is not as significant as ViT and DeiT. The prediction accuracy drops are less than 0.15% on the four Swin transformers at 8-bit quantization. The reason may be that Swin computes the self-attention locally within non-overlapping windows. It uses a smaller number of patches to calculate the self-attention, reducing the unbalance after post-softmax values. We also observe that larger vision transformers are less sensitive to quantization. For instance, the accuracy drops of ViT-S/224/32, ViT-S/224, ViT-B/224, and ViT-B/384 are 0.41, 0.38, 0.29, and 0.17 at 8-bit quantization and 4.08, 2.75, 2.89, and 2.65 at 6-bit quantization, respectively. The reason may be that the larger networks have more weights and generate more activations, making them more robust to the perturbation caused by quantization.

Table 2. Results of different PTQ methods. #ims means the number of calibration images. MP means mixed precision. BC means bias correction.

Model	Method	Bit-width	#ims	Size	Top-1
DeiT-S/224	EasyQuant [27]	W8A8	1024	22.0	76.59
79.80	Liu [16]	W8A8	1024	22.0	77.47
	Liu [16]	W8A8 (MP)	1024	22.2	78.09
	PTQ4ViT	W8A8	32	22.0	**79.47**
	EasyQuant [27]	W6A6	1024	16.5	73.26
	Liu [16]	W6A6	1024	16.5	74.58
	Liu [16]	W6A6 (MP)	1024	16.6	75.10
	PTQ4ViT	W6A6	32	16.5	**76.28**
DeiT-B	EasyQuant [27]	W8A8	1024	86.0	79.36
81.80	Liu [16]	W8A8	1024	86.0	80.48
	Liu [16]	W8A8 (MP)	1024	86.8	81.29
	PTQ4ViT	W8A8	32	86.0	**81.48**
	EasyQuant [27]	W6A6	1024	64.5	75.86
	Liu [16]	W6A6	1024	64.5	77.02
	Liu [16]	W6A6 (MP)	1024	64.3	77.47
	PTQ4ViT	W6A6	32	64.5	**80.25**
	Liu [16]	W4A4 (MP)	1024	43.6	**75.94**
	PTQ4ViT	W4A4	32	43.0	60.91
	PTQ4ViT+BC	W4A4	32	43.0	64.39

Table 2 demonstrates the results of different PTQ methods. EasyQuant [27] is a popular post-training method that alternatively searches for the optimal scaling factors of weight and activation. However, the accuracy drop is more than 3% at 8-bit quantization. Liu et al. [16] proposed using the Pearson correlation coefficient and ranking loss are used as the metrics to determine the scaling factors, which increases the Top-1 accuracy. Since the sensitivity of different layers to quantization is not the same, they also use the mixed-precision technique, achieving good results at 4-bit quantization. At 8-bit quantization and 6-bit quantization, PTQ4ViT outperforms other methods, achieving more than 1% improvement in prediction accuracy on average. At 4-bit quantization, the performance of PTQ4ViT is not good. Although bias correction [17] can improve the performance of PTQ4ViT, the result at 4-bit quantization is lower than the mixed-precision of Liu et al. This indicates that mixed-precision is important for quantization with lower bit-width.

4.3   Ablation Study

Next, we take ablation study on the effect of the proposed twin uniform quantization and Hessian guided metric. The experimental results are shown in Table 3.

Table 3. Ablation study of the effect of the proposed twin uniform quantization and Hessian guided metric. We mark a ✓ if the proposed method is used.

Model	Hessian Guided	Softmax Twin	GELU Twin	Top-1 Accuracy	
				W8A8	W6A6
ViT-S/224				80.47	70.24
81.39	✓			80.93	77.20
	✓	✓		81.11	78.57
	✓		✓	80.84	76.93
		✓	✓	79.25	74.07
	✓	✓	✓	81.00	78.63
ViT-B/224				83.90	75.67
84.54	✓			83.97	79.90
	✓	✓		84.07	80.76
	✓		✓	84.10	80.82
		✓	✓	83.40	78.86
	✓	✓	✓	84.25	81.65
ViT-B/384				85.35	46.89
86.00	✓			85.42	79.99
	✓	✓		85.67	82.01
	✓		✓	85.60	82.21
		✓	✓	84.35	80.86
	✓	✓	✓	85.89	83.19

As we can see, the proposed methods improve the top-1 accuracy of quantized vision transformers. Specifically, using the Hessian guided metric alone can slightly improve the accuracy at 8-bit quantization, and it significantly improves the accuracy at 6-bit quantization. For instance, on ViT-S/224, the accuracy improvement is 0.46% at 8-bit while it is 6.96% at 6-bit. And using them together can further improve the accuracy.

Based on the Hessian guided metric, using the twin uniform quantization on post-softmax activation or post-GELU activation can improve the performance. We observe that using the twin uniform quantization without the Hessian guided metric significantly decreases the top-1 accuracy. For instance, the top-1 accuracy on ViT-S/224 achieves 81.00% with both Hessian guided metric and twin uniform quantization at 8-bit quantization, while it decreases to 79.25% without Hessian guided metric, which is even lower than basic PTQ with 80.47% top-1 accuracy. This is also evidence that the metric considering only the local information is inaccurate.

5   Conclusion

In this paper, we analyzed the problems of post-training quantization for vision transformers. We observed both the post-softmax activations and the post-GELU activations have special distributions. We also found that the common quantization metrics are inaccurate to determine the optimal scaling factor. To solve these problems, we proposed the twin uniform quantization and a Hessian-guided metric. They can decrease the quantization error and improve the prediction accuracy at a small cost. To enable the fast quantization of vision transformers, we developed an efficient framework, PTQ4ViT. The experiments demonstrated that we achieved near-lossless prediction accuracy on the ImageNet classification task, making PTQ acceptable for vision transformers.

Acknowledgements. This work is supported by National Key R&D Program of China (2020AAA0105200), NSF of China (61832020, 62032001, 92064006), Beijing Academy of Artificial Intelligence (BAAI), and 111 Project (B18001).

A   Appendix

A.1   Derivation of Hessian guided metric

Hessian guided metric introduces as small an increment on task loss $L = CE(\hat{y}, y)$ as possible, in which $\hat{y}$ is the prediction of the quantized model and y is the ground truth. Here y is approximated by the prediction of the floating-point model y_{FP}, since no labels of input data are available in PTQ.

Quantization introduces a small perturbation ϵ on weight W, whose effect on task loss $\mathbb{E}[L(W)]$ could be analyzed with Taylor series expansion,

$$\mathbb{E}[L(\hat{W})] - \mathbb{E}[L(W)] \approx \epsilon^T \bar{g}^{(W)} + \frac{1}{2}\epsilon^T \bar{H}^{(W)}\epsilon. \tag{8}$$

Since the pretrained model has converged to a local optimum, The gradients $\bar{g}^{(W)}$ is close to zero and could be ignored. The Hessian matrix $\bar{H}^{(W)}$ on weight could be computed by

$$\frac{\partial^2 L}{\partial w_i \partial w_j} = \frac{\partial}{\partial w_j}(\sum_{k=1}^{m} \frac{\partial L}{\partial O_k} \frac{\partial O_k}{\partial w_i}) = \sum_{k=1}^{m} \frac{\partial L}{\partial O_k} \frac{\partial^2 O_k}{\partial w_i \partial w_j} + \sum_{k,l=1}^{m} \frac{\partial O_k}{\partial w_i} \frac{\partial^2 L}{\partial O_k \partial O_l} \frac{\partial O_l}{\partial w_j}. \tag{9}$$

$O = W^T X \in R^m$ is the output of the layer, and $\dfrac{\partial^2 O_k}{\partial w_i \partial w_j} = 0$. So the first term of Eq. (9) is zero, and $\bar{H}^{(W)} = J_O(W)^T \bar{H}^{(O)} J_O(W)$. Therefore, Eq. 8 could be further written as,

$$\mathbb{E}[L(\hat{W})] - \mathbb{E}[L(W)] \approx \frac{1}{2}(J_O(W)\epsilon)^T \bar{H}^{(O)} J_O(W)\epsilon \approx \frac{1}{2}(\hat{O} - O)^T \bar{H}^{(O)}(\hat{O} - O) \tag{10}$$

Following Liu et al. [14], we use the Diagonal Fisher Information Matrix to substitute $\bar{H}^{(O)}$. The optimization is formulated as:

$$\min_{\Delta_W} \mathbb{E}[(\hat{O} - O)^T \mathrm{diag}((\frac{\partial L}{\partial O_1})^2, \cdots, (\frac{\partial L}{\partial O_m})^2)(\hat{O} - O)]. \tag{11}$$

References

1. Banner, R., Nahshan, Y., Soudry, D.: Post training 4-bit quantization of convolutional networks for rapid-deployment. In: Wallach, H.M., Larochelle, H., Beygelzimer, A., d'Alché-Buc, F., Fox, E.B., Garnett, R. (eds.) Advances in Neural Information Processing Systems 32: Annual Conference on Neural Information Processing Systems 2019, NeurIPS 2019, 8–14 December 2019. Vancouver, BC, Canada, pp. 7948–7956 (2019). https://proceedings.neurips.cc/paper/2019/hash/c0a62e133894cdce435bcb4a5df1db2d-Abstract.html

2. Brown, T.B., et al.: Language models are few-shot learners. In: Larochelle, H., Ranzato, M., Hadsell, R., Balcan, M., Lin, H. (eds.) Advances in Neural Information Processing Systems 33: Annual Conference on Neural Information Processing Systems 2020, NeurIPS 2020, 6–12 December 2020. virtual (2020). https://proceedings.neurips.cc/paper/2020/hash/1457c0d6bfcb4967418bfb8ac142f64a-Abstract.html

3. Carion, N., Massa, F., Synnaeve, G., Usunier, N., Kirillov, A., Zagoruyko, S.: End-to-End object detection with transformers. In: Vedaldi, A., Bischof, H., Brox, T., Frahm, J.-M. (eds.) ECCV 2020. LNCS, vol. 12346, pp. 213–229. Springer, Cham (2020). https://doi.org/10.1007/978-3-030-58452-8_13

4. Choi, J., Wang, Z., Venkataramani, S., Chuang, P.I., Srinivasan, V., Gopalakrishnan, K.: PACT: parameterized clipping activation for quantized neural networks. CoRR abs/1805.06085 (2018). http://arxiv.org/abs/1805.06085

5. Choukroun, Y., Kravchik, E., Yang, F., Kisilev, P.: Low-bit quantization of neural networks for efficient inference. In: 2019 IEEE/CVF International Conference on Computer Vision Workshops, ICCV Workshops 2019, Seoul, Korea (South), 27–28 October 2019, pp. 3009–3018. IEEE (2019). https://doi.org/10.1109/ICCVW.2019.00363

6. Devlin, J., Chang, M., Lee, K., Toutanova, K.: BERT: pre-training of deep bidirectional transformers for language understanding. In: Burstein, J., Doran, C., Solorio, T. (eds.) Proceedings of the 2019 Conference of the North American Chapter of the Association for Computational Linguistics: Human Language Technologies, NAACL-HLT 2019, Minneapolis, MN, USA, June 2–7, 2019, Volume 1 (Long and Short Papers), pp. 4171–4186. Association for Computational Linguistics (2019). https://doi.org/10.18653/v1/n19-1423

7. Dosovitskiy, A., et al.: An image is worth 16×16 words: transformers for image recognition at scale. In: 9th International Conference on Learning Representations, ICLR 2021, Virtual Event, Austria, 3–7 May 2021. OpenReview.net (2021). https://openreview.net/forum?id=YicbFdNTTy

8. Fang, J., Shafiee, A., Abdel-Aziz, H., Thorsley, D., Georgiadis, G., Hassoun, J.H.: Post-training piecewise linear quantization for deep neural networks. In: Vedaldi, A., Bischof, H., Brox, T., Frahm, J.-M. (eds.) ECCV 2020. LNCS, vol. 12347, pp. 69–86. Springer, Cham (2020). https://doi.org/10.1007/978-3-030-58536-5_5

9. Gholami, A., Kim, S., Dong, Z., Yao, Z., Mahoney, M.W., Keutzer, K.: A survey of quantization methods for efficient neural network inference. CoRR abs/2103.13630 (2021). https://arxiv.org/abs/2103.13630

10. Guo, Y., Yao, A., Zhao, H., Chen, Y.: Network sketching: exploiting binary structure in deep cnns. In: 2017 IEEE Conference on Computer Vision and Pattern Recognition, CVPR 2017, Honolulu, HI, USA, 21–26 July 2017, pp. 4040–4048. IEEE Computer Society (2017). https://doi.org/10.1109/CVPR.2017.430, http://doi.ieeecomputersociety.org/10.1109/CVPR.2017.430

11. Han, K., et al.: A survey on visual transformer. CoRR abs/2012.12556 (2020). https://arxiv.org/abs/2012.12556

12. Huang, L., Tan, J., Liu, J., Yuan, J.: Hand-transformer: non-autoregressive structured modeling for 3D hand pose estimation. In: Vedaldi, A., Bischof, H., Brox, T., Frahm, J.-M. (eds.) ECCV 2020. LNCS, vol. 12370, pp. 17–33. Springer, Cham (2020). https://doi.org/10.1007/978-3-030-58595-2_2

13. Jia, D., et al.: Efficient vision transformers via fine-grained manifold distillation. CoRR abs/2107.01378 (2021). https://arxiv.org/abs/2107.01378

14. Li, Y., et al.: BRECQ: pushing the limit of post-training quantization by block reconstruction. In: 9th International Conference on Learning Representations, ICLR 2021, Virtual Event, Austria, 3–7 May 2021. OpenReview.net (2021). https://openreview.net/forum?id=POWv6hDd9XH

15. Liu, Z., et al.: Swin transformer: hierarchical vision transformer using shifted windows. CoRR abs/2103.14030 (2021). https://arxiv.org/abs/2103.14030

16. Liu, Z., Wang, Y., Han, K., Zhang, W., Ma, S., Gao, W.: Post-training quantization for vision transformer, pp. 28092–28103 (2021). https://proceedings.neurips.cc/paper/2021/hash/ec8956637a99787bd197eacd77acce5e-Abstract.html

17. Nagel, M., van Baalen, M., Blankevoort, T., Welling, M.: Data-free quantization through weight equalization and bias correction. In: 2019 IEEE/CVF International Conference on Computer Vision, ICCV 2019, Seoul, Korea (South), 27 October - 2 November 2019, pp. 1325–1334. IEEE (2019). https://doi.org/10.1109/ICCV.2019.00141

18. Prato, G., Charlaix, E., Rezagholizadeh, M.: Fully quantized transformer for machine translation. In: Cohn, T., He, Y., Liu, Y. (eds.) Findings of the Association for Computational Linguistics: EMNLP 2020, Online Event, 16–20 November 2020. Findings of ACL, vol. EMNLP 2020, pp. 1–14. Association for Computational Linguistics (2020). https://doi.org/10.18653/v1/2020.findings-emnlp.1

19. Russakovsky, O., et al.: ImageNet large scale visual recognition challenge. Int. J. Comput. Vision 115(3), 211–252 (2015). https://doi.org/10.1007/s11263-015-0816-y

20. Shen, S., et al.: Q-BERT: hessian based ultra low precision quantization of BERT. In: The 34th AAAI Conference on Artificial Intelligence, AAAI 2020, The 32nd Innovative Applications of Artificial Intelligence Conference, IAAI 2020, The 10th AAAI Symposium on Educational Advances in Artificial Intelligence, EAAI 2020, New York, NY, USA, 7–12 February 2020, pp. 8815–8821. AAAI Press (2020). https://aaai.org/ojs/index.php/AAAI/article/view/6409

21. Tang, Y., et al.: Patch slimming for efficient vision transformers. CoRR abs/2106.02852 (2021). https://arxiv.org/abs/2106.02852

22. Touvron, H., Cord, M., Douze, M., Massa, F., Sablayrolles, A., Jégou, H.: Training data-efficient image transformers & distillation through attention. In: Meila, M., Zhang, T. (eds.) Proceedings of the 38th International Conference on Machine Learning, ICML 2021, 18–24 July 2021, Virtual Event. Proceedings of Machine

Learning Research, vol. 139, pp. 10347–10357. PMLR (2021). http://proceedings.mlr.press/v139/touvron21a.html

23. Vaswani, A., et al.: Attention is all you need. In: Guyon, I., et al. (eds.) Advances in Neural Information Processing Systems 30: Annual Conference on Neural Information Processing Systems 2017, 4–9 December 2017. Long Beach, CA, USA, pp. 5998–6008 (2017). https://proceedings.neurips.cc/paper/2017/hash/3f5ee243547dee91fbd053c1c4a845aa-Abstract.html

24. Wang, H., Zhu, Y., Adam, H., Yuille, A.L., Chen, L.: Max-deeplab: end-to-end panoptic segmentation with mask transformers. In: IEEE Conference on Computer Vision and Pattern Recognition, CVPR 2021, virtual, 19–25 June 2021, pp. 5463–5474. Computer Vision Foundation/IEEE (2021). https://openaccess.thecvf.com/content/CVPR2021/html/Wang_MaX-DeepLab_End-to-End_Panoptic_Segmentation_With_Mask_Transformers_CVPR_2021_paper.html

25. Wang, W., et al.: Pyramid vision transformer: a versatile backbone for dense prediction without convolutions. CoRR abs/2102.12122 (2021). https://arxiv.org/abs/2102.12122

26. Wightman, R.: Pytorch image models (2019). https://doi.org/10.5281/zenodo.4414861, https://github.com/rwightman/pytorch-image-models

27. Wu, D., Tang, Q., Zhao, Y., Zhang, M., Fu, Y., Zhang, D.: Easyquant: post-training quantization via scale optimization. CoRR abs/2006.16669 (2020). https://arxiv.org/abs/2006.16669

28. Zhang, D., Yang, J., Ye, D., Hua, G.: LQ-Nets: learned quantization for highly accurate and compact deep neural networks. In: Ferrari, V., Hebert, M., Sminchisescu, C., Weiss, Y. (eds.) ECCV 2018. LNCS, vol. 11212, pp. 373–390. Springer, Cham (2018). https://doi.org/10.1007/978-3-030-01237-3_23

Bitwidth-Adaptive Quantization-Aware Neural Network Training: A Meta-Learning Approach

Jiseok Youn[1][ID], Jaehun Song[2][ID], Hyung-Sin Kim[2(✉)][ID],
and Saewoong Bahk[1(✉)][ID]

[1] Department of Electrical and Computer Engineering and INMC,
Seoul National University, Seoul, South Korea
jsyoun@netlab.snu.ac.kr, sbahk@snu.ac.kr
[2] Graduate School of Data Science, Seoul National University, Seoul, South Korea
{steve2972,hyungkim}@snu.ac.kr

Abstract. Deep neural network quantization with *adaptive bitwidths* has gained increasing attention due to the ease of model deployment on various platforms with different resource budgets. In this paper, we propose a meta-learning approach to achieve this goal. Specifically, we propose MEBQAT, a simple yet effective way of bitwidth-adaptive quantization-aware training (QAT) where meta-learning is effectively combined with QAT by redefining meta-learning tasks to incorporate bitwidths. After being deployed on a platform, MEBQAT allows the (meta-)trained model to be quantized to any candidate bitwidth with minimal inference accuracy drop. Moreover, in a few-shot learning scenario, MEBQAT can also adapt a model to any bitwidth as well as any *unseen* target classes by adding conventional optimization or metric-based meta-learning. We design variants of MEBQAT to support both (1) a bitwidth-adaptive quantization scenario and (2) a new few-shot learning scenario where both quantization bitwidths and target classes are jointly adapted. Our experiments show that merging bitwidths into meta-learning tasks results in remarkable performance improvement: 98.7% less storage cost compared to bitwidth-dedicated QAT and 94.7% less back propagation compared to bitwidth-adaptive QAT in bitwidth-only adaptation scenarios, while improving classification accuracy by up to 63.6% compared to vanilla meta-learning in bitwidth-class joint adaptation scenarios.

1 Introduction

Recent development in deep learning has provided key techniques for equipping resource-constrained devices with larger networks by reducing neural network computational costs. To this end, several research directions have emerged such as network optimization [23,25], parameter factorization [28,34], network

Supplementary Information The online version contains supplementary material available at https://doi.org/10.1007/978-3-031-19775-8_13.

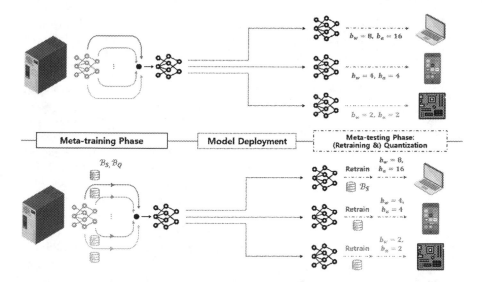

Fig. 1. Overview of MEBQAT on bitwidth-only adaptation (above) and bitwidth-class joint adaptation (below) scenarios.

pruning [33,42], and quantization [5,16,43,45]. In particular, quantization can significantly reduce model size, computational requirements and power consumption by expressing model weights and activations in lower precision. For example, quantizing a model from FP32 to Int8 with devices equipped with fast arithmetic hardware units for low-precision operands can reduce inference delay by up to $5\times$ [20].

However, one challenge associated with quantization is the difficulty of tailoring models to various bitwidths to compensate for platforms with different resource constraints. This is especially important in situations where a quantized model is deployed to platforms with different battery conditions, hardware limitations, or software versions. In order to solve this problem, a recent trend in quantization gave rise to adaptive bitwidths, which allows models to *adapt* to bitwidths of varying precision [2,17,31].

In this paper, we provide a different perspective on this research direction by considering a modified formulation of bitwidth-adaptive quantization-aware training (QAT) with *meta-learning* [9], as shown in Fig. 1. In typical meta-learning scenarios, a meta task is defined as a subset of training data, divided on the basis of class [9] or data configuration [4,12]. With this task definition, the meta-training phase requires a tailored, large-scale dataset for a model to experience many meta tasks while the meta-testing phase needs the model to be retrained with few-shot data for a target task. To apply meta-learning in bitwidth-adaptive QAT, we propose MEBQAT by newly defining a meta task to incorporate a *bitwidth setting*, a model hyperparameter independent of the dataset. Thus, our meta task definition enables *dataset-agnostic* meta-learning: meta-learning without the need for few-shot-learning-specific datasets. In the meta-testing phase, the model is not retrained but quantized immediately with any target bidwidth, resulting in fast adaptation. Experiments show that MEBQAT performs comparably to state-of-the-art bitwidth-adaptive QAT schemes. The results suggest that bitwidth-adaptive QAT can be categorized as a meta learning problem.

In addition, to show that MEBQAT is synergistically combined with typical meta-learning scenarios, we also investigate a new meta-learning scenario where quantization bitwidths and target classes are *jointly adapted*. In this scenario, we define a meta task as a combination of bitwidth setting and target classes. With these modified tasks, we show that MEBQAT can be merged with both optimization-based meta-learning (Model-Agnostic Meta-Learning (MAML) [9]) and metric-based meta-learning (Prototypical Networks [30]) frameworks. Experiments show that MEBQAT produces a model that is not only adaptable to arbitrary bitwidth settings, but also robust to unseen classes when retrained with few-shot data in the meta-testing phase. MEBQAT significantly outperforms both vanilla meta learning and the combination of dedicated QAT and meta learning, demonstrating that MEBQAT successfully merges bitwidth-adaptive QAT and meta learning without losing their own advantages. With this new scenario given by MEBQAT, a model can be deployed on more various platforms regardless of their resource constraints and classification tasks.

To summarize, our contributions are three-fold:

- We propose **ME**ta-learning based **B**itwidth-adaptive **QAT** (MEBQAT), by newly defining meta-learning tasks to include bitwidth settings and averaging gradients approximated for different bitwidths over those tasks to incorporate the essence of quantization-awareness.
- We show that our method can obtain a model robust to various bitwidth settings by conducting extensive experiments on various supervised-learning contexts, datasets, model architectures, and QAT schemes. In the traditional classification problem, MEBQAT shows comparable performance to existing bitwidth-adaptive QAT methods and dedicated training of the model to a given bitwidth, but with higher training efficiency (94.7% less back-propagations required).
- We define a new few-shot classification context for MEBQAT where both bitwidths and classes are jointly adapted using few-shot data. MEBQAT well fits both optimization- and metric-based meta-learning frameworks. In terms of classification accuracy, MEBQAT outperforms vanilla meta-learning by up to 63.6% and a naïve combination of dedicated QAT and meta learning by up to 27.48%, while also adding bitwidth adaptability comparable to state-of-the-art bitwidth-adaptive QAT methods.

2   Related Work

Quantization-Aware Training. Existing approaches to quantization can be broadly split into Post-Training Quantization (PTQ) and QAT. PTQ quantizes a model trained without considering quantization, and requires sophisticated methods such as solving optimization problems [1,3,5,8] and model reconstruction [15,21]. However, given that most platforms that utilize a quantized model are resource constrained, these methods can incur significant computational burden. To reduce post-training computation, we instead focus on QAT, which trains a model to alleviate the drop of accuracy when quantized. QAT methods usually define a formula for approximating a gradient of a quantization function output w.r.t. the input. To this end, recent works suggest integer-arithmetic-only quantization methods [7,16,19,43] and introduce differentiable asymptotic functions for non-differentiable quantization functions [11].

However, conventional QAT is limited in that a model is trained for a single dedicated bitwidth, showing significant performance degradation when quantized to other bitwidths. In other words, supporting multiple bitwidths would require training multiple copies of the model to each bitwidth.

Bitwidth-Adaptive QAT. In order to overcome such shortcomings, some QAT-based approaches aim to train a model only once and use it on various bitwidth settings. A number of studies [1,6,13,29,36–38] use Neural Architecture Search (NAS) to train a *super-network* involving multiple bitwidths in a predefined search space and sample a sub-network quantized with the target bitwidth setting given or searched taking the hardware into account. However, NAS-based approaches usually suffer from difficult training, heavy computation, and collapse on 8-bit precision without special treatment.

AdaBits [17] was the first to propose another research direction, namely the concept of training a single model adaptive to any bitwidth. Specifically, the model is trained via joint quantization and switchable clipping level. As a similar approach, Any-precision DNN [41] enables adaptable bitwidths via knowledge distillation [14] and switchable Batch Normalization (BN) layers. The authors in [31] utilize wavelet decomposition and reconstruction [24] for easy bitwidth adjustment by adjusting hyperparameters. Furthermore, Bit-mixer [2] aims to train a mixed-precision model where its individual layers can be quantized to an arbitrary bitwidth. Although these methods allow a single model to train for multiple bitwidths, some parts of the model (e.g., BN layers) still need to be trained dedicated to each precision candidate which increases the number of parameters w.r.t. the number of bitwidth candidates [6]. Moreover, prior work solely focuses on model quantization and ignores the possibility that users require slightly different tasks that the pretrained model does not support.

Meta Learning. Meta learning has recently attracted much attention in the research community due to its potential to train a model that can flexibly adapt to different tasks, even with a few gradient steps and limited amounts of labeled data, making it ideal for resource-constrained platforms [9,32].

One of the most common approaches to meta learning is optimization-based meta learning that trains a base model from which a model starts to be adapted to a given task by using experience from many different tasks. Model-Agnostic Meta Learning (MAML) [9] suggests to learn from multiple tasks individually, evaluate the overall adaptation performance, and learn to increase it. Many variants of MAML have emerged to improve upon this method [10,22,26,27,35,40,44]. Another approach to meta learning is metric-based meta learning, which attempts to learn an embedding function such that an unseen

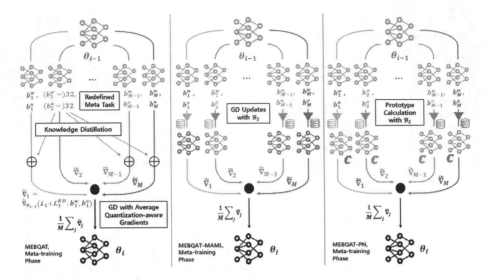

Fig. 2. Illustration of meta-training phase in MEBQAT, MEBQAT-MAML, and MEBQAT-PN. GD stands for Gradient Descent. **c** denotes prototypes. $\mathcal{B}_S$ represents a support set. Illustration of adaptation and inference phase is provided in appendix.

class can be predicted by seeking the label with minimum distance. Prototypical Networks [30] calculates prototypes as a milestone for each label by averaging the corresponding embeddings. While meta learning provides a personalizable model robust to unseen classes, there is still a lack of research concerning the applicability of meta learning in quantization.

To the best of our knowledge, this work is the first to show that bitwidth-adaptive QAT and meta learning can be merged synergistically without sacrificing their own advantages. Specifically, our proposal MEBQAT provides bitwidth-adaptive QAT with zero-copies of any part of the model. In addition, by defining a *meta task* as a combined set of bitwidth setting and target classes, MEBQAT produces a model that quickly adapts to arbitrary bitwidths as well as target classes.

3 Meta-Learning Based Bitwidth-Adaptive QAT

In this section, we introduce **ME**ta-learning based **B**itwidth-adaptive **QAT** (MEBQAT), a once-for-all method that aims to provide a model adaptable to any bitwidth setting by synergistically combining QAT with meta-learning methodologies. Similar to conventional meta-learning schemes, MEBQAT operates in two phases: a meta-training and a meta-testing phase. In the meta-training phase, MEBQAT trains a base model by experiencing various tasks to improve its adaptability. Importantly, the meta-training phase performs QAT with the task definition including bitwidth settings to support bitwidth-adaptive QAT. In the meta-testing phase, the meta-trained base model is deployed at a platform and tailored for a platform-specific target task.

Table 1. Summary of notations.

Notation in Fig. 1	Meaning
$\mathcal{B}_S/\mathcal{B}_Q$	Support data for adaptation / Query data for inference
b_w/b_a	Test bitwidth of weights / activations
Notation in Fig. 2	**Meaning**
$\mathcal{B}_S/\mathcal{B}_Q$	Support data for adaptation / Query data for inference
θ_i	Model parameters after i-th optimization or update
M	Number of inner-loop (task)s per outer-loop (i.e., meta batch size)
b_j^w/b_j^a	Training bitwidth of weights / activations in j-th meta-task $(j = 1, 2, \cdots, M)$
c	Prototype in PN framework, differentiated by colors
$\mathcal{L}_j$	Loss in j-th meta-task
$\mathcal{L}_{j'}^{KD}$	Distillation loss in j'-th meta-task $(j' = 2, \cdots, M)$
$\tilde{\nabla}_j$	Gradients in j-th meta-task, approximated according to the bitwidth (b_j^w, b_j^a)

We consider two practical scenarios for MEBQAT, (1) bitwidth adaptation scenario and (2) bitwidth-class joint adaptation scenario. The former scenario is the main problem that bitwidth-adaptive QAT methods target, and as such, we aim to provide similar performance to state-of-the-art schemes but with less complexity, using our meta-learning-based approach. The bitwidth-class joint adaptation scenario is a new scenario in which a model can adapt to not only an arbitrary target bitwidth but also unseen target classes. To support these scenarios, we provide three variants of MEBQAT, called MEBQAT, -MAML, and -PN, as shown in Fig. 2. Following convention in [9], MEBQAT, -MAML, -PN aim to optimize Eqs. 1–3, respectively.

$$\min_\theta \sum_j \mathcal{L}_j = \sum_j \mathcal{L}(f_{\text{Quantize}(\theta; b_j^w, b_j^a)}) \tag{1}$$

$$\min_\theta \sum_j \mathcal{L}_j^Q = \sum_j \mathcal{L}^Q(f_{\text{Quantize}(\{\theta - \alpha \tilde{\nabla}_\theta \mathcal{L}^S(f_\theta; b_j^w, b_j^a)\}; b_j^w, b_j^a)}) \tag{2}$$

$$\min_\theta \sum_j \mathcal{L}_j^Q = \sum_j \mathcal{L}^Q(f_{\text{Quantize}(\theta; b_j^w, b_j^a)}, c^S(f_{\text{Quantize}(\theta; b_j^w, b_j^a)})) \tag{3}$$

3.1 Bitwidth Adaptation Scenario

In this scenario, we assume that users' target classes are the same as those used for model training; in other words, both meta-training and meta-testing phases have the same classification task. However, each user may have different target bitwidths considering its own resource budget. Therefore in the meta-testing phase, a user immediately quantizes the base model using its own bitwidth setting without the need for fine-tuning.

To support this scenario, we define a bitwidth task set $\mathcal{T}_b$ that consists of various tuples (b^w, b^a) where b^w and b^a are bitwidths for weight quantization and

---

Algorithm 1. MEBQAT, meta-training phase

Initialize base model parameters θ_0, bitwidth task set $\mathcal{T}_b$, training set $\mathcal{D}$ comprising $(\mathbf{x}, y)$, and step size β

for epoch $i = 1$ **to** E **do**

    $\mathcal{B} \leftarrow$ random sample from $\mathcal{D}$

    $\hat{y}_\mathbf{x} \leftarrow f_{\theta_{i-1}}(\mathbf{x})$ for all $\mathbf{x} \in \mathcal{B}$ ▷ Get soft labels using full precision

    for $j = 1$ **to** M **do**

        $(b_j^w, b_j^a) \leftarrow$ random sample from $\mathcal{T}_b$ ▷ Sample a bitwidth task

        $\phi \leftarrow \text{Quantize}(\theta_{i-1}; b_j^w, b_j^a)$

        $\mathcal{L}_j \leftarrow \frac{1}{|\mathcal{B}|} \sum_{(\mathbf{x}, y) \in \mathcal{B}} \mathcal{L}(f_\phi(\mathbf{x}), y)$ ▷ Get task-specific supervised loss

        $\mathcal{L}_j^{KD} \leftarrow \frac{1}{|\mathcal{B}|} \sum_{(\mathbf{x}, y) \in \mathcal{B}} \mathcal{L}(f_\phi(\mathbf{x}), \hat{y}_\mathbf{x})$ ▷ Get task-specific KD loss

        $\tilde{\nabla}_j \leftarrow \tilde{\nabla}_{\theta_{i-1}}(\mathcal{L}_j + \mathcal{L}_j^{KD}; b_j^w, b_j^a)$ ▷ Get task-specific, quant-aware gradient

    end for

    $\theta_i \leftarrow \theta_{i-1} - \frac{\beta}{M} \sum_{j=1}^M \tilde{\nabla}_j$ ▷ Update base model

end for

---

activation quantization, respectively. Assuming that f_θ is a base model parameterized by θ, MEBQAT aims to meta-train θ by experiencing many bitwidth tasks. In contrast to typical meta-learning scenarios, task-specific data samples are not needed because data is decoupled from task definition. Therefore, in each epoch i, MEBQAT samples a (common) batch of data $\mathcal{B}$ from training set $\mathcal{D}$ that all of M sampled tasks share. In addition, before entering into task-specific operation, MEBQAT gets soft labels $\hat{y}_\mathbf{x}$ for the batch $\mathcal{B}$ using the current full precision model parameters θ_{i-1} to utilize knowledge distillation [14] as in Any-precision DNN [41]. The idea is that the full precision model has more information and can teach a quantized model.

MEBQAT samples M bitwidth tasks in each epoch i. For each selected bitwidth tuple (b_j^w, b_j^a), the full precision model θ_{i-1} is quantized to ϕ using the tuple (b_j^w, b_j^a) and two types of task-specific losses are calculated based on the quantized model ϕ: supervised loss $\mathcal{L}_j$ and knowledge distillation loss $\mathcal{L}_j^{KD}$. Note that when (b_j^w, b_j^a) happens to be (FP,FP) (full precision), $\mathcal{L}_j^{KD}$ becomes zero. Using the loss $\mathcal{L}_j + \mathcal{L}_j^{KD}$, task-specific gradients $\tilde{\nabla}_j$ are calculated in a quantization-aware manner. Quantization-aware gradient calculation considers the sample bitwidth (b_j^w, b_j^a) and detailed method depends on which QAT scheme is combined with MEBQAT. Lastly, model parameters are updated to θ_i using gradient descent with step size β Algorithm 1 illustrates this process.

3.2 Bitwidth-Class Joint Adaptation Scenario

In this section, we propose a new meta-learning scenario for bitwidth-class joint adaptation where users may have their own target bitwidths and classification tasks. Each user is assumed to have a small local dataset for their classification tasks, which is used in the meta-testing phase to retrain the base model. Specifically, we consider N-way K-shot tasks where N is the number of target classes and K is the number of data samples per each of N classes. Assuming that $\mathcal{Y}$ is

---

Algorithm 2. MEBQAT-MAML, meta-training phase

Initialize base model parameters θ_0, bitwidth task set $\mathcal{T}_b$, training set $\mathcal{D}$ comprising $(\mathbf{x}, y)$, and step sizes α, β.

for epoch $i = 1$ **to** E **do**

    for $j = 1$ **to** M **do**

        $(b_j^w, b_j^a) \leftarrow$ random sample from $\mathcal{T}_b$ ▷ Sample a bitwidth task

        $\mathcal{Y}_j \leftarrow$ a set of randomly selected N classes   ▷ Sample a data task

        $\mathcal{D}_{\mathcal{Y}_j} \leftarrow$ Subset of $\mathcal{D}$ where $y \in \mathcal{Y}_j$

        $\mathcal{B}_S \leftarrow$ random sample from $\mathcal{D}_{\mathcal{Y}_j}$ ▷ N-way K-shot support set

        $\phi_0 \leftarrow \theta_{i-1}$

        for $u = 1$ **to** U **do**   ▷ Gradient decent with quantization

            $\phi_{u-1}^q \leftarrow$ Quantize$(\phi_{u-1}; b_j^w, b_j^a)$

            $\mathcal{L}_j^S \leftarrow \frac{1}{|\mathcal{B}_S|} \sum_{(\mathbf{x},y) \sim \mathcal{B}_S} \mathcal{L}(f_{\phi_{u-1}^q}(\mathbf{x}), y)$

            $\phi_u \leftarrow \phi_{u-1} - \alpha \tilde{\nabla}_{\phi_{u-1}}(\mathcal{L}_j^S; b_j^w, b_j^a)$

        end for

        $\phi_U^q \leftarrow$ Quantize$(\phi_U; b_j^w, b_j^a)$

        $\mathcal{B}_Q \leftarrow$ random sample from $\mathcal{D}_{\mathcal{Y}_j} \setminus \mathcal{B}_S$ ▷ N-way K-shot query set

        $\mathcal{L}_j^Q \leftarrow \frac{1}{|\mathcal{B}_Q|} \sum_{(\mathbf{x},y) \sim \mathcal{B}_Q} \mathcal{L}(f_{\phi_U^q}(\mathbf{x}), y)$

        $\tilde{\nabla}_j \leftarrow \tilde{\nabla}_{\theta_{i-1}}(\mathcal{L}_i^Q; b_j^w, b_j^a)$ ▷ Get task-specific, quant-aware gradient

    end for

    $\theta_i \leftarrow \theta_{i-1} - \frac{\beta}{M} \sum_{j=1}^M \tilde{\nabla}_j$ ▷ Update base model

end for

---

a set of randomly selected N classes, a single joint task including both bitwidths and classes is defined as $(b_w, b_a, \mathcal{Y})$.

To support this new scenario, we design two types of MEBQAT: (1) MEBQAT-MAML, which adopts a representative optimization-based meta-learning framework MAML [9] and (2) MEBQAT-PN, which adopts a representative metric-based meta-learning framework called Prototypical Networks (PN) [30].

MEBQAT-MAML. The main difference between MEBQAT-MAML and MEBQAT lies in the inner-loop operation for each task. In each iteration j of the inner loop, MEBQAT samples both a bitwidth task (b_j^w, b_j^a) and a data task $\mathcal{Y}_j$. Note that the bitwidth task is newly added to the original MAML operation. Assuming that the current model in epoch i is θ_{i-1}, the model is updated to a task-specific quantized model ϕ_U^q by using U-step gradient decent and a QAT method with the bitwidth setting (b_j^w, b_j^a) and a task-specific support set $\mathcal{B}_S$. Given the task-specific model ϕ_U^q and a query set $\mathcal{B}_Q$, task-specific loss $\mathcal{L}_j^Q$ and gradient $\tilde{\nabla}_j$ are calculated in a quantization-aware manner. Given that $\tilde{\nabla}_j$ requires second-order gradient calculation which is computationally expensive, we instead adopt a first-order approximation of MAML, called FOMAML. Algorithm 2 illustrates the process.

In the meta-testing phase, a user retrains the base model using a local support set of its own classification task and quantizes the model using its own bitwidth

---

Algorithm 3. MEBQAT-PN, meta-training phase

Initialize base model parameters θ_0, bitwidth task set $\mathcal{T}_b$, training set $\mathcal{D}$ comprising $(\mathbf{x}, y)$, and step sizes β.

for epoch $i = 1$ **to** E **do**

    $\mathcal{Y}_i \leftarrow$ a set of randomly selected N classes   ▷ Sample a data task

    $\mathcal{B}_S \leftarrow$ random sample from $\mathcal{D}_{\mathcal{Y}_i}$ ▷ N-way K-shot support set

    $\mathcal{B}_Q \leftarrow$ random sample from $\mathcal{D}_{\mathcal{Y}_i} \setminus \mathcal{B}_S$ ▷ N-way K-shot query set

    for $j = 1$ **to** M **do**

        $(b_j^w, b_j^a) \leftarrow$ random sample from $\mathcal{T}_b$ ▷ Sample a bitwidth task

        $\phi \leftarrow$ Quantize$(\theta_{i-1}; b_j^w, b_j^a)$

        for $n \in \mathcal{Y}_i$ **do**   ▷ Get prototypes using support set

            $\mathbf{c}_n \leftarrow \frac{1}{K}\sum_{(\mathbf{x},y) \in \mathcal{B}_S, y=n} f_\phi(\mathbf{x})$

        end for

        $\mathcal{L}_j^Q = 0$

        for $n \in \mathcal{Y}_i$ **do**   ▷ Get task-specific loss using query set

            for $(\mathbf{x}, y) \in \mathcal{B}_Q$ where $y = n$ **do**

                $\mathcal{L}_j^Q = \mathcal{L}_j^Q + \frac{1}{NK}[d(f_\phi(\mathbf{x}), \mathbf{c}_n) + \log\sum_{n'}\exp(-d(f_\phi(\mathbf{x}), \mathbf{c}_{n'}))]$

            end for

        end for

        $\tilde{\nabla}_j \leftarrow \tilde{\nabla}_{\theta_{i-1}}(\mathcal{L}_j^Q; b_j^w, b_j^a)$.   ▷ Get task-specific, quant-aware gradient

    end for

    $\theta_i \leftarrow \theta_{i-1} - \frac{\beta}{M}\sum_{j=1}^M \tilde{\nabla}_j$ ▷ Update base model

end for

---

setting. Then the model performance is evaluated by inferencing data points in a local query set. Given that a user platform is likely to be resource constrained, the number of gradient decent updates in the meta-testing phase can be smaller than U, as in the original MAML.

MEBQAT-PN. A limitation of MEBQAT-MAML arises from the necessity of gradient decent-based fine-tuning in the meta-testing phase, which can become a computational burden to resource-constrained platforms. In contrast to MAML, Prototypical Network (PN) trains an embedding function such that once data points are converted into embeddings, class prototypes are calculated using a support dataset and query data is classified by using distance from each class prototype. Therefore in the meta-testing phase, MEBQAT-PN does not require gradient descent but simply calculates class prototypes using a local support set, which significantly reduces computation overhead.

Algorithm 3 illustrates MEBQAT-PN's meta-training phase. Unlike original PN, to include a bitwidth setting in a task in each epoch i, MEBQAT samples bitwidths (b_i^w, b_i^a) as well as target classes $\mathcal{Y}_i$, quantizes the current model θ_{i-1} to ϕ using the selected bitwidths, and calculates class prototypes $\mathbf{c}_n$ for $n \in \mathcal{Y}_i$ using the quantized model ϕ and a support set $\mathcal{B}_S$. Then task-specific loss $\mathcal{L}_j^Q$ is calculated using distance between embeddings for query data points in $\mathcal{B}_Q$ and class prototypes. Lastly, task-specific gradient $\tilde{\nabla}_i$ is computed in a quantization-aware manner.

3.3   Implementation

We also include specific implementation details to improve the training process of MEBQAT. First, in each epoch of MEBQAT (Algorithm 1), we fix the bitwidth task in the first inner-loop branch to full-precision (FP,FP) instead of a random sample. This implementation is required for the base model to experience full precision in every epoch, thus improving accuracy. Second, while sampling random bitwidth settings, we exclude unrealistic settings such as (FP,1) and (1,FP) because these settings not only are impractical and improbable, but also hinder convergence. Third, when there are some minor bitwidth settings that a QAT scheme treats differently from other bitwidths (e.g., 1-bit of DoReFa-Net [45]), we sample the minor settings more frequently (e.g. at least once in each epoch).

4   Evaluation

To demonstrate the validity of MEBQAT, we conduct extensive experiments on multiple supervised-learning contexts, datasets, model architectures, and configurations of quantization.

4.1   Experiments on the Bitwidth Adaptation Scenario

In the bitwidth adaptive scenario with shared labels, we compare MEBQAT with (1) (bitwidth-dedicated) QAT and (2) existing bitwidth-adaptive QAT methods (AdaBits [17] and Any-precision DNN (ApDNN) [41]).

MEBQAT adopts multiple quantization configurations depending on the compared scheme. When compared with AdaBits, MEBQAT quantizes a tensor and approximates its gradient using the same Scale-Adjusted Training (SAT) [18] that AdaBits adopts, with $\mathcal{T}_b = \{2, 3, 4, 5, 6, 7, 8, 16, \text{FP}\}$ where FP denotes full-precision Float32. Furthermore, just as in AdaBits, we quantize the first and last layer weights into 8-bits with BN layers remaining full-precision. When compared with Any-precision DNN, MEBQAT quantizes a tensor and approximates its gradient in a DoReFa-Net based manner, with $\mathcal{T}_b = \{1, 2, 3, 4, 5, 6, 7, 8, 16, \text{FP}\}$. Note that we differentiate the formula for 1-bit and other bitwidths as DoReFa-Net [45] does. In this case, we do not quantize the first, last, and BN layers. The number of inner-loop tasks per task is set to 4.

Optimizer and learning rate scheduler settings depend on the model architecture and dataset used. For MobileNet-v2 on CIFAR-10, we use an Adam optimizer for 600 epochs with an initial learning rate 5×10^{-2} and a cosine annealing scheduler without restart. For pre-activation ResNet-20 on CIFAR-10, we use an AdamW optimizer for 400 epochs with an initial learning rate 10^{-3} divided by 10 at epochs $\{150, 250, 350\}$. Finally, for the 8-layer CNN in [41] on SVHN, we use a standard Adam optimizer for 100 epochs with an initial learning rate 10^{-3} divided by 10 at epochs $\{50, 75, 90\}$.

Finally, as in MAML, all BN layers are used in a transductive setting and always use the current batch statistics.

Table 2. Comparison of accuracy (%) with 95% confidence intervals (10 iterations) with bitwidth-dedicated and bitwidth-adaptive QAT methods. † denotes results from [6]. ‡ denotes results from a non-differentiated binarization function. FP stands for 32-bit Full-Precision. '-' denotes results not provided.

(b_w, b_a)	CIFAR-10, MobileNet-v2			CIFAR-10, Pre-activation ResNet-20			SVHN, 8-layer CNN		
	QAT	AdaBits	MEBQAT	QAT	ApDNN	MEBQAT	QAT	ApDNN	MEBQAT
(1, 1)	–	–	–	92.28(±0.116)	92.15‡	91.32(±0.202)	97.27(±0.025)	88.21‡	96.60(±0.060)
(2, 2)	84.40(±0.691)	58.98†	78.50(±0.544)	92.72(±0.146)	93.97	92.52(±0.151)	97.51(±0.043)	94.94	97.25(±0.052)
(3, 3)	90.08(±0.233)	79.30†	88.04(±0.255)	92.61(±0.066)	–	92.65(±0.225)	97.57(±0.024)	–	97.58(±0.041)
(4, 4)	90.44(±0.152)	91.84†	89.30(±0.336)	92.69(±0.195)	93.95	92.77(±0.157)	97.44(±0.068)	96.19	97.62(±0.043)
(5, 5)	90.83(±0.193)	–	89.58(±0.243)	92.64(±0.117)	–	92.80(±0.179)	97.53(±0.028)	–	97.64(±0.050)
(6, 6)	91.10(±0.146)	–	89.46(±0.275)	92.66(±0.120)	–	92.83(±0.188)	97.50(±0.032)	–	97.63(±0.056)
(7, 7)	91.06(±0.138)	–	89.48(±0.303)	92.65(±0.110)	–	92.79(±0.171)	97.56(±0.034)	–	97.64(±0.043)
(8, 8)	91.20(±0.171)	–	89.36(±0.243)	92.57(±0.124)	93.80	92.89(±0.147)	97.52(±0.055)	96.22	97.63(±0.047)
(16, 16)	91.19(±0.145)	–	89.53(±0.209)	92.67(±0.192)	–	92.75(±0.190)	97.51(±0.042)	–	97.65(±0.056)
(FP, FP)	93.00(±0.221)	–	89.24(±0.253)	93.92(±0.107)	93.98	92.90(±0.133)	97.67(±0.079)	96.29	97.40(±0.043)

Table 3. Comparison of training computation and storage costs.

Methods	Training computation cost	Storage cost
Dedicated QAT	1 backprop per update	$T\Theta$
AdaBits/ApDNN	T backprops per update	$(1 - \zeta)\Theta + T\zeta\Theta$
MEBQAT	M backprops per update	Θ

Performance of MEBQAT. Table 2 shows the (meta-)test accuracy after (meta-)trained by QAT/bitwidth-adaptive QAT/MEBQAT in multiple model architectures and datasets. Here, b_w, b_a are bitwidths used during testing. For each bitwidth setting, accuracy is averaged over one test epoch. Results of vanilla QAT come from individually trained models dedicated to a single bitwidth. All other results come from a single adaptable model, albeit with some prior work containing bitwidth-dedicated parts. Results show that MEBQAT achieves performance comparable to or better than the existing methods.

We also tackle the limitations of prior bitwidth-adaptive QAT methods in scalability to the number of target bitwidths. Table 3 shows an overview of training and storage costs of various methods when compared with MEBQAT. Here, $T(\simeq|\mathcal{T}_b|^2)$ represents the number of (test) bitwidths, Θ denotes the total model size, and ζ indicates the ratio of batch normalization layers respective to the entire model. Because MEBQAT is a meta-learning alternative to bitwidth adaptive learning, our method exhibits fast adaptation, requiring only a few train steps M. In evaluation scenarios, $1 < M(= 4) \ll T(= 73$ or $75)$, showing that MEBQAT is up to 18 times more cost-efficient than other methods since it trains a single model with a single batch normalization layer for all different tasks. Note that computation costs are the same for all non-few-shot methods during testing since inference directly follows quantization. In other words, MEBQAT requires zero additional training during inference. Thus, MEBQAT exhibits much more training efficiency than other adaptive methods in non-few-shot scenarios.

Table 4. Comparison of accuracy (%) to vanilla FOMAML and FOMAML + QAT, using 5-layer CNN in [9].

(b_w, b_a)	Omniglot 20-way 1-shot, 5-layer CNN			Omniglot 20-way 5-shot, 5-layer CNN		
	FOMAML	FOMAML+QAT	MEBQAT-MAML	FOMAML	FOMAML+QAT	MEBQAT-MAML
(2, 2)	25.97	62.09	89.57	35.24	84.03	96.94
(3, 3)	75.29	65.24	91.46	83.29	83.29	97.58
(4, 4)	84.43	63.84	91.62	88.19	84.73	97.61
(5, 5)	89.51	67.35	91.65	93.28	97.78	97.61
(6, 6)	91.47	92.95	91.66	96.43	97.53	97.61
(7, 7)	90.94	92.40	91.65	96.61	97.41	97.61
(8, 8)	91.92	93.00	91.66	97.20	97.86	97.61
(16, 16)	93.13	92.82	91.65	97.47	97.41	97.69
(FP, FP)	93.12	93.12	92.39	97.48	97.48	97.88
	MiniImageNet 5-way 1-shot, 5-layer CNN			MiniImageNet 5-way 5-shot, 5-layer CNN		
(2, 2)	34.96	42.35	46.00	47.49	62.25	61.65
(3, 3)	43.89	42.11	47.45	59.02	63.16	63.82
(4, 4)	47.14	48.75	47.56	63.40	64.76	63.54
(5, 5)	48.19	47.07	47.45	64.09	65.54	63.69
(6, 6)	48.56	48.66	47.46	64.31	64.09	63.67
(7, 7)	48.62	48.10	47.43	64.41	64.57	63.72
(8, 8)	48.60	48.26	47.43	64.53	64.65	63.70
(16, 16)	48.65	48.17	47.36	64.48	65.17	63.81
(FP, FP)	48.66	48.66	47.68	64.51	64.51	64.28

4.2   Experiments on the Bitwidth-Class Joint Adaptation Scenario

To the best of our knowledge, there is no prior work on multi-bit quantization in a few-shot context. Therefore, we compare MEBQAT-MAML and MEBQAT-PN to two types of compared schemes: (1) vanilla meta-learning without quantization-awareness and (2) meta-learning combined with bitwidth-dedicated QAT. In (2), by using fake-quantized b-bit models in conventional meta-learning operations, the model shows solid adaptable performance in b-bits. Just as in Sect. 4.1, we conduct experiments with much more various bitwidth candidates than existing QAT-based methods.

When using the MAML framework, there are 16/4 inner-loop tasks using Omniglot/MiniImageNet, respectively. In an inner-loop, the 5-layer CNN in [9] is updated by a SGD optimizer with learning rate $10^{-1}/10^{-2}$ at 5 times with a support set. In an outer-loop, the base model is trained by Adam optimizer with learning rate 10^{-4}. In the meta-testing phase, fine-tuning occurs in 5/10 times, with an optimizer same as inner-loop optimizer in the previous phase. When using the PN framework, a model is optimized by Adam with learning rate 10^{-3}. We use Euclidean distance as a metric for classification. MEBQAT-PN has 4 inner-loop tasks per outer-loop.

Performance of MEBQAT-MAML and MEBQAT-PN. Table 4 shows the meta-testing accuracy after meta-trained by FOMAML, FOMAML + QAT and MEBQAT-MAML. For each bitwidth setting, accuracy is averaged over 600

Table 5. Comparison of accuracy (%) to vanilla PN and PN + QAT, using 4-layer CNN in [30].

(b_w, b_a)	Omniglot 20-way 1-shot, 4-layer CNN			Omniglot 20-way 5-shot, 4-layer CNN		
	PN	PN+QAT	MEBQAT-PN	PN	PN+QAT	MEBQAT-PN
(2, 2)	31.58	95.46	94.87	50.86	98.71	98.32
(3, 3)	81.21	95.90	95.55	93.87	98.77	98.56
(4, 4)	93.73	95.97	95.60	98.37	98.76	98.58
(5, 5)	95.40	95.95	95.60	98.77	98.61	98.58
(6, 6)	95.83	95.82	95.60	98.83	98.65	98.58
(7, 7)	95.84	95.95	95.60	98.87	98.62	98.58
(8, 8)	95.88	95.97	95.60	98.88	98.85	98.58
(16, 16)	96.89	95.55	95.60	98.89	98.93	98.58
(FP, FP)	95.88	95.88	96.06	98.89	98.89	98.70
	MiniImageNet 5-way 1-shot, 4-layer CNN			MiniImageNet 5-way 5-shot, 4-layer CNN		
(2, 2)	26.29	50.06	47.66	30.64	67.45	65.34
(3, 3)	37.51	50.16	48.57	46.74	67.71	66.16
(4, 4)	45.59	50.38	48.38	60.52	67.35	66.22
(5, 5)	48.33	50.18	48.54	64.75	65.95	66.16
(6, 6)	49.77	50.01	48.55	65.81	65.63	66.19
(7, 7)	49.52	49.90	48.55	65.68	66.06	66.19
(8, 8)	49.29	48.35	48.55	65.90	65.86	66.19
(16, 16)	49.75	47.86	48.55	65.94	66.39	66.19
(FP, FP)	49.61	49.61	48.33	65.82	65.82	66.03

different sets of N target classes unseen in the previous phase. It is noteworthy that in some cases, MEBQAT-MAML exceeds the postulated upper bound of accuracy. In other words, although we hypothesized applying bitwidth-dedicated QAT directly to train individual models would have the highest accuracy, we found that in some cases, MEBQAT-MAML achieves performance exceeding the baseline.

Table 5 shows the meta-testing accuracy after meta-trained by PN, PN + QAT and MEBQAT-PN. For each bitwidth setting, accuracy is averaged over 600 different sets of N target classes unseen in the previous phase. The results prove that MEBQAT is also compliant to metric-based meta-learning such that the base model can fit into any target bitwidth as well as target classes without fine-tuning in the test side.

5   Discussion

Although this paper focuses on quantizing the entire model into a single bitwidth, and increasingly growing area of research focuses on quantizing each layer or block of the model into different optimal bitwidths. When MEBQAT is directly applied to this mixed-precision setting, this might require many diverse tasks, which poses heavier computational burdens both during training and when finding an optimal bitwidth for each platform during inference. Development of an efficient meta-learning method for both adaptive- and mixed-precision quantization would be an interesting future work.

A limitation of our current experiments comes from the fact that in our method, QAT does not consist solely of integer-arithmetic-only operations. Moreover, MEBQAT-MAML stipulates fine-tuning at meta-testing phase for adaptation, where the gradient descent during this process is mostly done in full precision. In this case, future work can include applying integer-only methods such as in HAWQ-v3 [39] as a quantization(-aware training) method to further test the feasibility of our method. We can also proceed to use COTS edge devices such as a Coral development board to evaluate the applicability of our method.

Increasing the performance of adaptability of our work is another future work. This is especially true since FOMAML and Prototypical Networks are methods that have been tried and tested for several years. Using other sophisticated meta-learning methods can improve the adaptability performance of our model or reduce the computational complexity of fine-tuning our model at a resource-constrained device.

6 Conclusion

To the best of our knowledge, this paper is the first to attempt training a model with meta-learning which can be independently quantized to any arbitrary bitwidth at runtime. To this end, we investigate the possibility of incorporating bitwidths as an adaptable meta-task, and propose a method by which the model can be trained to adapt into any bitwidth, as well as any target classes in a supervised-learning context. Through experimentation, we found that our proposed method achieves performance greater than or equal to existing work on adaptable bitwidths, showing that incorporating meta-learning could become a viable alternative. We also found that our method is robust to a few-shot learning context, showing better performance than models trained with dedicated meta-learning techniques and quantized using PTQ or QAT. Thus, we demonstrate that MEBQAT can potentially open up an interesting new avenue of research in the field of bitwidth-adaptive QAT.

Acknowledgements. This research was supported in part by the MSIT (Ministry of Science and ICT), Korea, under the ITRC (Information Technology Research Center) support program (IITP-2021-0-02048) supervised by the IITP (Institute of Information & Communications Technology Planning & Evaluation), the National Research Foundation of Korea (NRF) grant (No. 2020R1A2C2101815), and Samsung Research Funding & Incubation Center of Samsung Electronics under Project No. SRFC-TD2003-01.

References

1. Bai, H., Cao, M., Huang, P., Shan, J.: Batchquant: quantized-for-all architecture search with robust quantizer. CoRR abs/2105.08952 (2021)

2. Bulat, A., Tzimiropoulos, G.: Bit-mixer: mixed-precision networks with runtime bit-width selection. In: Proceedings of the IEEE/CVF International Conference on Computer Vision (ICCV), pp. 5188–5197 (2021)

3. Cai, Y., Yao, Z., Dong, Z., Gholami, A., Mahoney, M.W., Keutzer, K.: Zeroq: a novel zero shot quantization framework. In: Proceedings of the IEEE conference on Computer Vision and Pattern Recognition (CVPR), pp. 13166–13175 (2020)

4. Cho, H., Cho, Y., Yu, J., Kim, J.: Camera distortion-aware 3d human pose estimation in video with optimization-based meta-learning. In: Proceedings of the IEEE/CVF International Conference on Computer Vision, pp. 11169–11178 (2021)

5. Choukroun, Y., Kravchik, E., Yang, F., Kisilev, P.: Low-bit quantization of neural networks for efficient inference. In: Proceedings of the IEEE/CVF International Conference on Computer Vision (ICCV) Workshops, pp. 3009–3018 (2019)

6. Cui, Y., et al.: Fully nested neural network for adaptive compression and quantization. In: Proceedings of the International Joint conference on Artificial Intelligence (IJCAI), pp. 2080–2087 (2020)

7. Esser, S.K., McKinstry, J.L., Bablani, D., Appuswamy, R., Modha, D.S.: Learned step size quantization. In: Proceedings of the International conference on Learning Representations (ICLR) (2020)

8. Fang, J., Shafiee, A., Abdel-Aziz, H., Thorsley, D., Georgiadis, G., Hassoun, J.H.: Post-training piecewise linear quantization for deep neural networks. In: Vedaldi, A., Bischof, H., Brox, T., Frahm, J.-M. (eds.) ECCV 2020. LNCS, vol. 12347, pp. 69–86. Springer, Cham (2020). https://doi.org/10.1007/978-3-030-58536-5_5

9. Finn, C., Abbeel, P., Levine, S.: Model-agnostic meta-learning for fast adaptation of deep networks. In: Proceedings of the International Conference on Machine Learning (ICML), vol. 70, pp. 1126–1135 (2017)

10. Finn, C., Xu, K., Levine, S.: Probabilistic model-agnostic meta-learning. In: Proceedings of the advances in Neural Information Processing Systems (NeurIPS), pp. 9537–9548 (2018)

11. Gong, R., et al.: Differentiable soft quantization: bridging full-precision and low-bit neural networks. In: Proceedings of the IEEE/CVF International conference on Computer Vision (ICCV), pp. 4851–4860 (2019)

12. Gong, T., Kim, Y., Shin, J., Lee, S.J.: Metasense: few-shot adaptation to untrained conditions in deep mobile sensing. In: Proceedings of the 17th Conference on Embedded Networked Sensor Systems, pp. 110–123 (2019)

13. Guo, Z., et al.: Single path one-shot neural architecture search with uniform sampling. In: Vedaldi, A., Bischof, H., Brox, T., Frahm, J.-M. (eds.) ECCV 2020. LNCS, vol. 12361, pp. 544–560. Springer, Cham (2020). https://doi.org/10.1007/978-3-030-58517-4_32

14. Hinton, G.E., Vinyals, O., Dean, J.: Distilling the knowledge in a neural network (2015)

15. Hubara, I., Nahshan, Y., Hanani, Y., Banner, R., Soudry, D.: Accurate post training quantization with small calibration sets. In: Proceedings of the International conference on Machine Learning (ICML), vol. 139, pp. 4466–4475 (2021)

16. Jacob, B., et al.: Quantization and training of neural networks for efficient integer-arithmetic-only inference. In: Proceedings of the IEEE Conference on Computer Vision and Pattern Recognition (CVPR), pp. 2704–2713 (2018)

17. Jin, Q., Yang, L., Liao, Z.: Adabits: neural network quantization with adaptive bit-widths. In: Proceedings of the IEEE/CVF conference on Computer Vision and Pattern Recognition (CVPR), pp. 2143–2153 (2020)

18. Jin, Q., Yang, L., Liao, Z., Qian, X.: Neural network quantization with scale-adjusted training. In: Proceedings of the British Machine Vision Conference (BMVC) (2020)

19. Jung, S., et al.: Learning to quantize deep networks by optimizing quantization intervals with task loss. In: Proceedings of the IEEE conference on Computer Vision and Pattern Recognition (CVPR), pp. 4350–4359 (2019)

20. Lee, J.: Fast int8 inference for autonomous vehicles with tensorrt 3 (2017). https://developer.nvidia.com/blog/int8-inference-autonomous-vehicles-tensorrt/

21. Li, Y., et al.: Brecq: pushing the limit of post-training quantization by block reconstruction. In: Proceedings of the International Conference on Learning Representations (ICLR) (2021)

22. Li, Z., Zhou, F., Chen, F., Li, H.: Meta-sgd: learning to learn quickly for few shot learning. CoRR abs/1707.09835 (2017)

23. Liang, T., Glossner, J., Wang, L., Shi, S., Zhang, X.: Pruning and quantization for deep neural network acceleration: a survey. Neurocomputing **461**, 370–403 (2021)

24. Mallat, S.: A theory for multiresolution signal decomposition: the wavelet representation. IEEE Trans. Pattern Anal. Mach. Intell. (TPAMI) **11**(7), 674–693 (1989)

25. Mathieu, M., Henaff, M., LeCun, Y.: Fast training of convolutional networks through ffts. In: Proceedings of the International conference on Learning Representations (ICLR) (2014)

26. Rajeswaran, A., Finn, C., Kakade, S.M., Levine, S.: Meta-learning with implicit gradients. In: Proceedings of the Advances in Neural Information Processing Systems (NeurIPS), pp. 113–124 (2019)

27. Ravi, S., Beatson, A.: Amortized bayesian meta-learning. In: Proceedings of the International conference on Learning Representations (ICLR) (2019)

28. Sainath, T.N., Kingsbury, B., Sindhwani, V., Arisoy, E., Ramabhadran, B.: Low-rank matrix factorization for deep neural network training with high-dimensional output targets. In: Proceedings of the International conference on Acoustics, Speech, and Signal Processing (ICASSP), pp. 6655–6659 (2013)

29. Shen, M., et al.: Once quantization-aware training: high performance extremely low-bit architecture search. CoRR abs/2010.04354 (2020)

30. Snell, J., Swersky, K., Zemel, R.S.: Prototypical networks for few-shot learning. In: Proceedings of the Advances in Neural Information Processing Systems (NeurIPS), pp. 4077–4087 (2017)

31. Sun, Q., et al.: One model for all quantization: a quantized network supporting hot-swap bit-width adjustment (2021)

32. Sung, F., Zhang, L., Xiang, T., Hospedales, T.M., Yang, Y.: Learning to learn: meta-critic networks for sample efficient learning. CoRR abs/1706.09529 (2017)

33. Suzuki, T., et al.: Spectral pruning: compressing deep neural networks via spectral analysis and its generalization error. In: Proceedings of the International Joint conference on Artificial Intelligence (IJCAI), pp. 2839–2846 (2020)

34. Swaminathan, S., Garg, D., Kannan, R., Andres, F.: Sparse low rank factorization for deep neural network compression. Neurocomputing **398**, 185–196 (2020)

35. Triantafillou, E., et al.: Meta-dataset: a dataset of datasets for learning to learn from few examples. In: Proceedings of the International conference on Learning Representations (ICLR) (2020)

36. Wang, K., Liu, Z., Lin, Y., Lin, J., Han, S.: Haq: hardware-aware automated quantization with mixed precision. In: Proceedings of the IEEE Conference on Computer Vision and Pattern Recognition (CVPR), pp. 8612–8620 (2019)

37. Wang, T., Wang, K., Cai, H., Lin, J., Liu, Z., Wang, H., Lin, Y., Han, S.: Apq: joint search for network architecture, pruning and quantization policy. In: Proceedings of the IEEE/CVF Conference on Computer Vision and Pattern Recognition (CVPR), pp. 2075–2084 (2020)
38. Wu, B., Wang, Y., Zhang, P., Tian, Y., Vajda, P., Keutzer, K.: Mixed precision quantization of convnets via differentiable neural architecture search. CoRR abs/1812.00090 (2018)
39. Yao, Z., et al.: HAWQ-V3: dyadic neural network quantization. In: Proceedings of the International Conference on Machine Learning (ICML), vol. 139, pp. 11875–11886 (2021)
40. Yoon, J., Kim, T., Dia, O., Kim, S., Bengio, Y., Ahn, S.: Bayesian model-agnostic meta-learning. In: Proceedings of the Advances in Neural Information Processing Systems (NeurIPS), pp. 7343–7353 (2018)
41. Yu, H., Li, H., Shi, H., Huang, T.S., Hua, G.: Any-precision deep neural networks. In: Proceedings of the AAAI conference on Artificial Intelligence (AAAI), pp. 10763–10771 (2021)
42. Yu, R., et al.: Nisp: pruning networks using neuron importance score propagation. In: Proceedings of the IEEE Conference on Computer Vision and Pattern Recognition (CVPR), pp. 9194–9203 (2018)
43. Zhang, D., Yang, J., Ye, D., Hua, G.: LQ-Nets: learned quantization for highly accurate and compact deep neural networks. In: Ferrari, V., Hebert, M., Sminchisescu, C., Weiss, Y. (eds.) ECCV 2018. LNCS, vol. 11212, pp. 373–390. Springer, Cham (2018). https://doi.org/10.1007/978-3-030-01237-3_23
44. Zhou, P., Zou, Y., Yuan, X., Feng, J., Xiong, C., Hoi, S.C.H.: Task similarity aware meta learning: theory-inspired improvement on maml. In: Proceedings of the Conference on Uncertainty in Artificial Intelligence (UAI), vol. 161, pp. 23–33 (2021)
45. Zhou, S., Ni, Z., Zhou, X., Wen, H., Wu, Y., Zou, Y.: Dorefa-net: training low bitwidth convolutional neural networks with low bitwidth gradients. CoRR abs/1606.06160 (2016)

Understanding the Dynamics of DNNs Using Graph Modularity

Yao Lu[1], Wen Yang[1], Yunzhe Zhang[1], Zuohui Chen[1], Jinyin Chen[1],
Qi Xuan[1(✉)], Zhen Wang[2(✉)], and Xiaoniu Yang[1,3]

[1] Institute of Cyberspace Security, Zhejiang University of Technology,
Hangzhou 310023, China
xsgxlz@live.cn, {chenjinyin,xuanqi}@zjut.edu.cn
[2] School of Artificial Intelligence, Optics and Electronics (iOPEN),
Northwestern Polytechnical University, Xi'an 710072, China
zhenwang0@gmail.com
[3] Science and Technology on Communication Information Security Control
Laboratory, Jiaxing 314033, China

Abstract. There are good arguments to support the claim that deep neural networks (DNNs) capture better feature representations than the previous hand-crafted feature engineering, which leads to a significant performance improvement. In this paper, we move a tiny step towards understanding the dynamics of feature representations over layers. Specifically, we model the process of class separation of intermediate representations in pre-trained DNNs as the evolution of communities in dynamic graphs. Then, we introduce modularity, a generic metric in graph theory, to quantify the evolution of communities. In the preliminary experiment, we find that modularity roughly tends to increase as the layer goes deeper and the degradation and plateau arise when the model complexity is great relative to the dataset. Through an asymptotic analysis, we prove that modularity can be broadly used for different applications. For example, modularity provides new insights to quantify the difference between feature representations. More crucially, we demonstrate that the degradation and plateau in modularity curves represent redundant layers in DNNs and can be pruned with minimal impact on performance, which provides theoretical guidance for layer pruning. Our code is available at https://github.com/yaolu-zjut/Dynamic-Graphs-Construction.

Keywords: Interpretability · Modularity · Layer pruning

1 Introduction

DNNs have gained remarkable achievements in many tasks, from computer vision [21,38,47] to natural language processing [54], which can arguably be attributed

Supplementary Information The online version contains supplementary material available at https://doi.org/10.1007/978-3-031-19775-8_14.

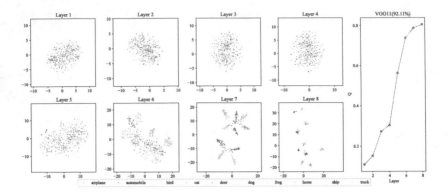

Fig. 1. (Left): t-SNE outputs for CIFAR-10 testing data after each layer in VGG11. (Right): modularity curve of VGG11 on CIFAR-10. Best viewed in color (Color figure online)

to powerful feature representations learned from data [17,48]. Giving insight into DNNs' feature representations is helpful to better understand neural network behavior, which attracts much attention in recent years. Some works seek to characterize feature representations by measuring similarities between the representations of various layers and various trained models [14,25,39,43,46,56,59]. Others visualize the feature representations in intermediate layers for an intuitive understanding, revealing that the feature representations in the shallow layers are relatively general, while those in the deep layers are more specific [7,36,50,58,72]. These studies are insightful, but fundamentally limited, because they ignore the dynamics of DNNs or can only understand the dynamics of DNNs through qualitative visualization instead of quantitative study.

Hence, in this paper, we build upon previous studies and investigate the dynamics of intermediate layers. The left part of Fig. 1 shows t-SNE [34] outputs for 500 CIFAR-10 testing samples after each convolutional layer in VGG11, from which we are able to see how class separation in the feature representations progresses as the layer goes deeper. Inspired by this, we seek to quantify the process of class separation of intermediate representations for better understanding the dynamics of DNNs. Specifically, we treat each sample as a node, and there is an edge between two nodes if their feature representations are similar in the corresponding layer. Then we construct a series of graphs that share the same nodes, which can be modeled as a dynamic graph due to the feature continuity. In this way, we convert quantifying the process of class separation of intermediate representations to investigate the evolution of communities in dynamic graphs. Then, we introduce modularity to quantify the evolution of communities. As shown in the right part of Fig. 1, the value of modularity indeed grows with the depth, which is consistent with the process of class separation shown in the left part of Fig. 1. This indicates that modularity provides a quantifiable interpretation perspective for understanding the dynamics of DNNs.

Then we conduct systematic experiments on exploring how the modularity changes in different scenarios, including the training process, standard and adversarial scenarios. Through further analysis of the modularity, we provide two application scenarios for it: (i) representing the difference of various layers. (ii) providing theoretical guidance for layer pruning. To summarize, we make the following contributions:

- We model the class separation of feature representations from layer to layer as the evolution of communities in dynamic graphs, which provides a novel perspective for researchers to better understand the dynamics of DNNs.
- We leverage modularity to quantify the evolution of communities in dynamic graphs, which tends to increase as the layer goes deeper, but descends or reaches a plateau at particular layers. To preserve the generality of modularity, systematic experiments are conducted in various scenarios, e.g., standard scenarios, adversarial scenarios and training processes.
- Additional experiments show that modularity can also be utilized to represent the difference of various layers, which can provide insights on further theoretical analysis and empirical studies.
- Through further analysis on the degradation and plateau in the modularity curves, we demonstrate that the degradation and plateau reveal the redundancy of DNNs in depth, which provides a theoretical guideline for layer pruning. Extensive experiments show that layer pruning guided by modularity can achieve a considerable acceleration ratio with minimal impact on performance.

2   Related Work

Many researchers have proposed various techniques to analyze certain aspects of DNNs. Hence, in this section, we would like to provide a brief survey of the literature related to our current work.

Understanding Feature Representations. Understanding feature representations of a DNN can obtain more information about the interaction between machine learning algorithms and data than the loss function value alone. Previous works on understanding feature representations can be mainly divided into two categories. One category quantitatively calculates the similarities between the feature representations of different layers and models [14,25,39,43,46,56,59]. For example, Kornblith et al. [25] introduce centered kernel alignment (CKA) to measure the relationship between intermediate representations. Feng et al. [14] propose a new metric, termed as transferred discrepancy, to quantify the difference between two representations based on their downstream-task performance. Compared to previous studies which only utilize feature vectors, Tang et al. [56] leverage both feature vectors and gradients into designing the representations of DNNs. On the basis of [25], Nguyen et al. [43] utilize CKA to explore how varying depth and width affects model feature representations, and find that overparameterized models exhibit the block structure. Through further analysis, they show that some layers exhibit the block structure and can be pruned with minimal impact on performance. Another category attempts to obtain an insightful understanding of feature representations through interpreting feature semantics [7,36,50,58,72]. Wang et al. [58] and Zeiler et al. [72] discover the hierarchical nature of the features in the neural networks. Specifically, shallow layers extract basic and general features while deep layers learn more specifically and globally. Yosinski et al. [68] quantify the degree to which a particular layer is general or specific. Besides, Donahue et al. [11] investigate the transfer of feature representations from the last few layers of a DNN to a novel generic task.

Modularity and Community in DNNs. Previous empirical studies have explored modularity and community in neural networks. Some seek to investigate learned modularity and community structure at the neuron level [8,22,62–64,69] or at the subnetwork level [6,27]. Others train an explicitly modular architecture [2,19] or promote modularity via parameter isolation [24] or regularization [9] during training to develop more modular neural networks. Different from these existing works, in this paper, we explore the evolution of communities at the feature representation level, which provides a new perspective to characterize the dynamics of DNNs.

Layer Pruning. State-of-the-art DNNs often involve very deep networks, which are bound to bring massive parameters and floating-point operations. Therefore, many efforts have been made to design compact models. Pruning is one stream among them, which can be roughly devided into three categories, namely, weight pruning [3,16,31], filter pruning [29,30,57] and layer pruning [5,13,60,61,66,73]. Weight pruning compresses over-parameterized models by dropping redundant individual weights, which has limited applications on general-purpose hardware. Filter pruning seeks to remove entire redundant filters or channels instead of individual weights. Compared to weight pruning and filter pruning, layer pruning removes the entire redundant layers, which is more suitable for general-purpose hardware. Existing layer pruning methods mainly differ in how to determine which layers need to be pruned. For example, Xu et al. [66] first introduce a trainable layer scaling factor to identify the redundant layers during the sparse training. And then they prune the layers with very small factors and retrain the model to recover the accuracy. Elkerdawy et al. [13] leverage imprinting to calculate a per-layer importance score in one-shot and then prune the least important layers and fine-tune the shallower model. Zhou et al. [73] leverage the ensemble view of block-wise DNNs and employ the multi-objective optimization paradigm to prune redundant blocks while avoiding performance degradation. Based on the observations of [1], Chen and Zhao [5], Wang et al. [60], and Wang et al. [61], respectively, utilize linear classifier probes to guide the layer pruning. Specifically, they prune the layers which provide minor contributions on boosting the performance of features.

Adversarial Samples. Although DNNs have gained remarkable achievements in many tasks [21,38,47,54], they have been found vulnerable to adversarial examples, which are born of intentionally perturbing benign samples in a human-imperceptible fashion [18,55]. The vulnerability to adversarial examples hinders

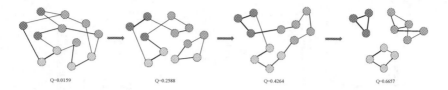

Fig. 2. A dynamic graph with 4 snapshots. Nodes of the same color represent that they are in the same community and the thickness of the line represents the weight of the edge. we use Eq. 1 to calculate the modularity

DNNs from being applied in safety-critical environments. Therefore, attacks and defenses on adversarial examples have attracted significant attention in machine learning. There has been a multitude of work studying methods to obtain adversarial examples [4,18,28,35,37,40,44,53] and to make DNNs robust against adversarial examples [51,65].

3 Methodology

3.1 Preliminary

We start by introducing some concepts in graph theory.

Dynamic graphs are the graphs that change with time [20]. In this paper, we focus on edge-dynamic graphs, i.e., edges may be added or deleted from the graph. Given a dynamic graph: $\mathcal{DG} = \{G_1, G_2, \cdots, G_T\}$, where T is the number of snapshots. Repeatedly leveraging the static methods on each snapshot can collectively look insight into the graph's dynamics. **Communities**, which are quite common in many real networks [23,33,70], are defined as sets of nodes that are more densely connected internally than externally [41,42]. **Ground-truth communities** can be defined as the sets of nodes with common properties, e.g., common attribute, affiliation, role, or function [67]. **Modularity**, as a common measurement to quantify the quality of communities [15,42], carries advantages including intelligibility and adaptivity. In this paper, we adopt modularity Q as following:

$$Q = \frac{1}{2W}\sum\nolimits_{ij} \left(a_{ij} - \frac{s_i s_j}{2W}\right) \delta(c_i, c_j), \tag{1}$$

where a_{ij} is the weight of the edge between node i and node j, $W = \sum_i\sum_j a_{ij}$ is the sum of the weights of all edges, which is used as a normalization factor. $s_i = \sum_j a_{ij}$ and $s_j = \sum_i a_{ij}$ are the strength of nodes i and j, c_i and c_j denote the community that nodes i and j belong to, respectively. $\delta(c_i, c_j)$ is 1 if node i and node j are in the same community and 0 otherwise. In this paper, each node represents an image with the corresponding label. Hence, these nodes can be divided into the corresponding ground-truth communities, which we utilize to calculate the modularity. Figure 2 gives an example to intuitively understand the evolution of communities, from which we find that the communities with a high value of modularity tend to strengthen the intra-community connections and weaken the inter-community connections.

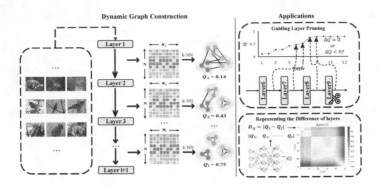

Fig. 3. Pipeline for the dynamic graph construction and the application scenarios of the modularity metric. Best viewed in color, zoomed in for details

3.2   Dynamic Graph Construction

Our proposed dynamic graph construction framework to understand the dynamics of a given DNN is visually summarized in Fig. 3.

Considering a typical image classification problem, the feature representation of the sample is transformed over the course of many layers, to be finally utilized by a classifier at the last layer. Therefore, we can model this process as follows:

$$\tilde{y} = f_{l+1}(f_l(\ldots f_1(x)\ldots)), \tag{2}$$

where $f_1(\cdot), f_2(\cdot), \cdots, f_l(\cdot)$ are n functions to extract feature representations of layers from bottom to top, $f_{l+1}(\cdot)$ and l denote the classifier and the number of layers, respectively. In this paper, we treat a bottleneck building block or a building block as a layer for ResNets [21] and take a sequence of consecutive layers, e.g., Conv-BN-ReLu, as a layer for VGGs [52]. We randomly sample N samples $\mathcal{X} = \{x_1, x_2, \ldots, x_N\}$ with corresponding labels $\mathcal{Y} = \{y_1, y_2, \ldots, y_N\}$ from the test set and feed them into a well-optimized DNN with fixed parameters to obtain an intermediate representation set $\mathcal{R} = \{r_1, r_2, \cdots, r_l\}$, where $r_i \in \mathbb{R}^{N \times C_i \times W_i \times H_i}$ denotes the feature representations in i-th layer, C_i, W_i and H_i are the number of channels, width and height of feature maps in i-th layer, respectively. Then we apply a flatten mapping $f : \mathbb{R}^{N \times C \times W \times H} \rightarrow \mathbb{R}^{N \times M}$ to each element in $\mathcal{R}$, where $M = C \times W \times H$. In order to capture the underlying relationship of samples in feature space, we construct a series of k-nearest neighbor (k-NN) graphs $G_i = (A_i, r_i)$, where A_i is the adjacency matrix of the k-NN graph in i-th layer. Specifically, we calculate the similarity matrix $S = \mathbb{R}^{N \times N}$ among N nodes using Eq. 3.

$$S^i_{jk} = \begin{cases} \dfrac{r_{i,j}^T r_{i,k}}{||r_{i,j}|| \, ||r_{i,k}||}, & \text{if } j \neq k \\ 0, & \text{if } j = k \end{cases}, \tag{3}$$

where $r_{i,j}$ and $r_{i,k}$ are the feature representation vectors of samples j and k in i-th layer. According to the obtained similarity matrix S^i_{jk}, we choose top

k similar node pairs for each node to set edges and corresponding weights, so as to obtain the adjacency matrix A_i. Now, we obtain a series of k-NN graphs $\{G_i = (A_i, r_i)|i = 1, 2, \cdots, l\}$, which reveal the internal relationship between feature representations of different samples in various layers. Due to the continuity of feature representations, i.e., feature representation of the current layer is obtained on the basis of the previous one. Hence, these k-NN graphs are relevant to each other and can be treated as multiple snapshots of the dynamic graph $\mathcal{DG} = \{G_1, G_2, \cdots, G_l\}$ at different time intervals.

Then we revisit the dynamic graph for a better understanding of community evolution. Figure 2 exhibits a demo, from which we can intuitively understand the process of community evolution in the dynamic graph. $\mathcal{DG}$ consists of l snapshots, each snapshot shares same nodes (samples) and reveals the inherent correlations between samples in the corresponding layer. Therefore, our dynamic graph is actually an edge-dynamic graph. Due to the existence of ground-truth label of each sample, we can easily divide samples into K ground-truth communities. Hence, we can calculate the modularity of each snapshot in $\mathcal{DG}$ using Eq. 1 together with the ground-truth communities. According to the obtained modularity of each snapshot, we finally obtain a modularity set $\mathcal{Q} = \{Q_1, Q_2, \cdots, Q_l\}$, which reveals the evolution of communities in the dynamic graph.

4   Experiments

Our goal is to intuitively understand the dynamics in well-optimized DNNs. Reflecting this, our experimental setup consists of a family of VGGs [52] and ResNets [21] trained on standard image classification datasets CIFAR-10 [26], CIFAR-100 [26] and ImageNet [49] (For the sake of simplicity, we choose 50 classes in original ImageNet, termed as ImageNet50). Specifically, we leverage stochastic gradient descent algorithm with an initial learning rate of 0.01 to optimize the model. The batch size, weight decay, epoch and momentum are set to 256, 0.005, 150 and 0.9, respectively. The statistics of pre-trained models and ImageNet50 are shown in Appendix B. All experiments are conducted on two NVIDIA Tesla A100 GPUs. If not special specified, we set $k = 3$, $N = 500$ for CIFAR-10 and ImageNet50, $k = 3$, $N = 1000$ for CIFAR-100 to construct dynamic graphs.

4.1   Understanding the Evolution of Communities

We systematically investigate the modularity curves of various models and repeat each experiment 5 times to obtain the mean and variance of modularity curves. From the results reported in Fig. 4, we have the following observations:

- The modularity roughly tends to increase as the layer goes deeper.
- On the same dataset, the frequency of degradation and plateau existing in the modularity curve gradually increases as models get deeper. Specifically, the modularity curve gradually reaches a plateau on CIFAR-10 and CIFAR-100

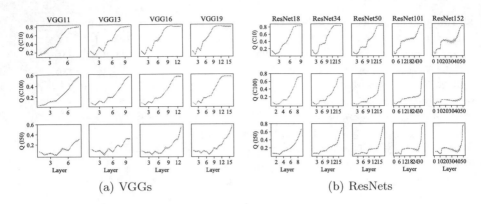

Fig. 4. Modularity curves of different models. C10, C100 and I50 denote CIFAR-10, CIFAR-100 and ImageNet50, respectively. Shaded regions indicate standard deviation

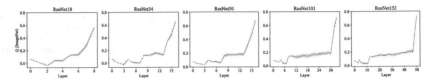

Fig. 5. Modularity curves of ResNets on ImageNet

at the deep layer as VGG gets deeper. Compared to VGGs, the modularity curve of ResNets descends, reaches a plateau, or rises very slowly mostly happening in the repeatable layers.
- According to the modularity curves of VGG16 and VGG19, we can see that as the complexity of the dataset increases, the plateau gradually disappears.

Since previous works [58,72] have shown that shallow layers extract general features while deep layers learn more specifically. Hence, feature representations of the same category are more similar in deep layers than in shallow layers, which can be seen in Fig. 1. Therefore, the samples in the same community (category) tend to connect with each other, i.e., the modularity increases as the layer goes deeper. In this sense, the growth of modularity quantitatively reflects the process of class separation in feature representations inside the DNNs. Besides, the tendency of modularity is consistent with the observation of [1], which utilizes linear classifier probes to measure how suitable the feature representations at every layer are for classification. Compared to [1] that requires training a linear classifier for each layer, our modularity provides a more convenient and effective tool to understand the dynamics of DNNs.

According to the second and third findings, we can draw a conclusion that the degradation and plateau arise when model complexity is great relative to the dataset. With the relative complexity of the model getting greater, a little bit of performance improvement costs nearly doubling the number of layers, e.g., ResNet18 and ResNet34 on CIFAR-10. Many successive layers are essentially

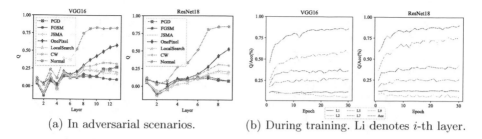

(a) In adversarial scenarios.   (b) During training. Li denotes i-th layer.

Fig. 6. Modularity curves in different scenarios

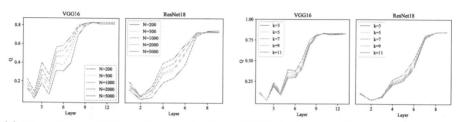

(a) The influence of N on the modularity.   (b) The influence of k on the modularity.

Fig. 7. Hyperparameter sensitivity analysis

performing the same operation over many times, refining feature representations just a little more each time instead of making a more fundamental change of feature representations. Hence, these features representations exhibit a similar degree of class separation, which explains the existence of the plateau and degradation in the modularity curves. To further explore the relationship between the degradation as well as plateau and model relative complexity, we use pre-trained ResNets in torchvision[1] to conduct experiments on ImageNet. From Fig. 5 we find that the tendency of modularity curves is almost consistent with the observation in Fig. 4(b). The main difference lies in the modularity curves of ResNet101 and ResNet152 trained on ImageNet50 exhibit the degradation and plateau in the middle layer, while those trained on ImageNet do not, which further confirms the conclusion we make. Additional experiment on exploring the evolution of communities is presented in Appendix A.1.

4.2 Modularity Curves of Adversarial Samples

In Sect. 4.1, we discuss the modularity curves of normal samples. Here, we would like to explore the modularity curves of adversarial samples. Specifically, we first utilize FGSM [18], PGD [35], JSMA [44], CW [4], OnePixel [53] and Local Search [40] to attack the pre-trained VGG16 and ResNet18 on CIFAR-10 for generating adversarial samples. To make it easier for others to reproduce our results, we

---

[1] https://pytorch.org/vision/stable/index.html.

utilize default parameter settings in AdverTorch [10] to obtain untargeted adversarial samples. The attack success rate is 100% for each attack mode. Then we randomly choose 500 adversarial samples in each attack mode, set $k = 3$ to construct dynamic graphs and plot modularity curves. As shown in Fig. 6(a), we can find that the modularity curve of adversarial samples can reach a smaller peak than normal samples, which can be interpreted as adversarial attacks blur the distinctions among various categories.

4.3 Modularity During Training Time

During the training process, the model is considered to learn valid feature representations. In order to explore the dynamics of DNNs in this process, we conduct experiments on ResNet18 and VGG16. Specifically, in the first 30 training epochs, we save the model file and test accuracy of every epoch. Then we construct a dynamic graph for each model and calculate the modularity of each snapshot in this dynamic graph. Figure 6(b) shows the modularity and accuracy curves on ResNet18 and VGG16, from which we find that the values of modularity in shallow layers nearly keep constant or small fluctuations. We believe that is because feature representations in shallow layers are general [58,72], samples in the same category do not cluster together. Hence, despite learning effective feature representations, the value of modularity does not increase significantly. Compared to shallow layers, feature representations in deep layers are more specific. With the improvement of model performance, the valid feature representations gradually learned by the model make intra-community connections tighter, which explains the modularity curves of deep layers are almost proportional to the accuracy curve. These phenomena may give some information about how the training evolves inside the DNNs and guide the intuition of researchers. Besides, we provide the experiment on randomly initialized models in Appendix A.2.

4.4 Ablation Study

To provide further insight into the modularity, we conduct ablation studies to evaluate the hyperparameter sensitivity of the modularity. The batch size N and the number of edges k that each node connects with are two hyperparameters.

Whether batch size N has a non-negligible impact on the modularity curve? we set $N = 200, 500, 1000, 2000, 5000$, $k = 3$ to repeat experiments on CIFAR-10 with pre-trained VGG16 and ResNet18. In Fig. 7(a), we show that smaller N has relatively lower modularity in early layers but N has less influence on modularity in final layers. Generally speaking, different values of N have the same tendency on modularity curves.

Whether k has a non-negligible impact on the modularity curve? we set $N = 500$, $k = 3, 5, 7, 9, 11$ to repeat the above experiments. Figure 7(b) shows that the different selections of k have a negligible impact on modularity

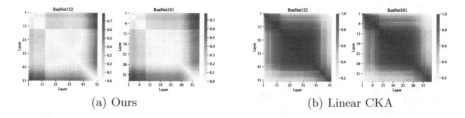

(a) Ours   (b) Linear CKA

Fig. 8. Emergence of the block structure with different methods

because the modularity curves almost overlap together. We observe the similar tendency even when k is large (see Appendix A.3 for details)).

Hence, we can conclude that the modularity is reliable for hyperparameters.

5 Application Scenarios of Modularity

In Sect. 4, we investigate the dynamics of DNNs, from which we gain some insights. On this basis, we provide two application scenarios for the modularity.

5.1 Representing the Difference of Various Layers

In addition to quantifying the degree of class separation of intermediate representations, modularity can also be used to represent the difference of various layers like previous works [14,25,39,43,46,56,59]. Instead of directly calculating the similarity between two feature representations, we compute the difference of modularity of two layers because our modularity reflects the global attribute of the corresponding layer, i.e., the degree of class separation. Specifically, we calculate the difference matrix D, with its element defined as $D_{ij} = |Q_i - Q_j|$. Q_i and Q_j denote the value of modularity in i-th and j-th layer, respectively. We visualize the result as a heatmap, with the x and y axes representing the layers of the model. As shown in Fig. 8(a), the heatmap shows a block structure in representational difference, which arises because representations after residual connections are more different from representations inside ResNet blocks than other post-residual representations. Moreover, we also reproduce the result of linear CKA [25] in Fig. 8(b), from which we can see the same block structure. Note that we measure the difference of various layers while [25] focuses on the similarity, so the darker the color in Fig. 8(b), the less similar.

5.2 Guiding Layer Pruning with Modularity

In Sect. 4.1, we find that the degradation and plateau arise when model complexity is great relative to the dataset. Previous works have demonstrated that DNNs are redundant in depth [71] and overparameterized models exist many consecutive hidden layers that have highly similar feature representations [43]. Informed by these conclusions, we wonder if the degradation and plateau offer

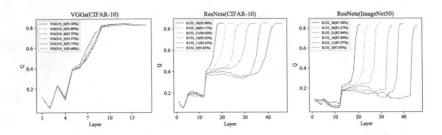

Fig. 9. The modularity curves of VGGs and ResNets of different depths

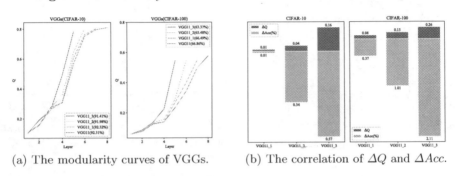

(a) The modularity curves of VGGs.   (b) The correlation of ΔQ and ΔAcc.

Fig. 10. Experiments on pruning the irredundant layers

an intuitive instruction in identifying the redundant layers. Hence, we assume that the plateau and degradation make no contribution or negative contribution to the model. In other words, these layers we consider are redundant and can be pruned with acceptable loss. To verify our assumption, we conduct systematic experiments on CIFAR-10, CIFAR-100 and ImageNet50.

Pruning the Redundant Layers. Since the plateau mostly appears in the last few layers of VGG19, we remove the Conv-BN-ReLu in VGG19 one by one from back to front to obtain a series of variants. As for ResNet152, we remove the bottleneck building blocks from back to front in stage 3 (the degradation and plateau mostly emerge in stage 3) to get variant models. Detailed structures are shown in Appendix B. For example, VGG19_1 denotes VGG19 prunes 1 layer. Then we finetune these variant models with the same hyperparameter setting as Sect. 4.1. The left part of Fig. 9 shows the results of VGGs on CIFAR-10, from which we find that the modularity curves of different VGG almost overlap together. The only difference between these modularity curves is whether the plateau emerges in the last few layers. Specifically, the modularity curve of VGG19_6 does not have the plateau, while VGG19_1 has the obvious plateau. Moreover, the plateau gradually disappears as the layer is removed one by one from back to front. Note that these variant models have similar performance, which proves that the plateau is indeed redundant. The middle part and right part of Fig. 9 show the modularity curves of different ResNet on CIFAR-10 and ImageNet50. These modularity curves almost coincide in the shallow layers, while the plateau gradually narrows with the continuous removal of the middle layers. Consequently, this strongly proves that the plateau can be pruned with minimal impact on performance.

Pruning the Irredundant Layers. In the previous paragraph, we verify that removing redundant layers will not affect the accuracy. Here, we wonder whether removing irredundant layers will result in a significant performance drop. Hence, we conduct further experiments on CIFAR-10 and CIFAR-100 with VGG11 (According to the modularity curves in Fig. 4, we think VGG11 is relatively irredundant on CIFAR-10 and CIFAR-100). We remove the Conv-BN-ReLu in VGG11 one by one from back to front to obtain a series of variants and fine-tune them. Figure 10(a) exhibits the modularity curves of those variants, from which we can see that they can finally reach almost the same peak. Next we calculate the corresponding variation of accuracy brought about by pruning the layer. Specifically, we calculate the variation of modularity ΔQ of the final layer with $\Delta Q = Q_i - Q_{i-1}$, where Q_i denotes the modularity of i-th layer that we want to prune. Finally, we calculate the variation of accuracy ΔAcc using $\Delta Acc = Acc_i - Acc_j$, where Acc_i and Acc_j represent the accuracy of the original model and pruned model, respectively. Figure 10(b) shows the results, from which we can find that on CIFAR-10, removing the layer that modularity increases 0.01 results in nearly no influence (0.01%) on model performance, while pruning the layer that has a 0.16 increment on modularity results in 0.57% degradation. With the complexity of the dataset increasing, this gap becomes more obvious. On CIFAR-100, pruning the layer that modularity goes up 0.08 leads to a 0.37% drop in accuracy, while a variation of 0.26 in modularity causes a variation of 2.11% in accuracy. Hence, we draw a conclusion that the variation of accuracy is proportional to the variation of modularity, which means pruning irredundant layers will result in a more significant drop in performance than removing redundant layers. Besides, This phenomenon becomes more obvious as the complexity of the dataset increases.

Practicality of Layer Pruning by Modularity. According to the above experimental results, we are able to conclude that modularity can be used to provide effective theoretical guidance for layer pruning. Here, we would like to evaluate its practicality. Specifically, we first plot the modularity curve of the original model, then we prune the layer where the curve drops, reaches a plateau or grows slowly, finally we finetune the new model. We adapt number of parameters and required Float Points Operations (denoted as FLOPs), to evaluate model size and computational requirement. We leverage a package in pytorch [45], which terms thop[2] to calculate FLOPs and parameters. Table 1 shows the performance of different layer pruning methods [5,61] on ResNet56 for CIFAR. Compared with Chen et al., our method provides considerable better parameters and FLOPs reductions (43.00% vs. 42.30%, 60.30% vs. 34.80%), while yielding a higher accuracy (93.38% vs. 93.29%). Compared to DBP-0.5, our method shows

---

[2] https://github.com/Lyken17/pytorch-OpCounter.

Table 1. Pruning results of ResNet56 on CIFAR-10. PR is the pruning rate

Method	Top-1%	Params(PR)	FLOPs(PR)
ResNet56	93.27	0%	0%
Chen et al. [5]	93.29	42.30%	34.80%
DBP-0.5 [61]	93.39	/	53.41%
Ours	93.38	43.00%	60.30%

more advantages in FLOPs reduction (60.30% vs. 53.41%), while maintaining a competitive accuracy (93.38% vs. 93.39%).

According to the above experiments, we demonstrate the effectiveness and efficiency of layer pruning guided by modularity.

6   Conclusion and Future Work

In this study, through modeling the process of class separation from layer to layer as the evolution of communities in dynamic graphs, we provide a graph perspective for researchers to better understand the dynamics of DNNs. Then we develop modularity as a conceptual tool and apply it to various scenarios, e.g., the training process, standard and adversarial scenarios, to gain insights into the dynamics of DNNs. Extensive experiments show that modularity tends to rise as the layer goes deeper, which quantitatively reveals the process of class separation in intermediate layers. Moreover, the degradation and plateau arise when model complexity is great relative to the dataset. Through further analysis on the degradation and plateau at particular layers, we demonstrate that modularity can provide theoretical guidance for layer pruning. In addition to guiding layer pruning, modularity can also be used to represent the difference of various layers.

We hope the simplicity of our dynamic graph construction approach could facilitate more research ideas in interpreting DNNs from a graph perspective. Besides, we wish that the modularity presented in this paper can make a tiny step forward in the direction of neural network structure design, layer pruning and other potential applications. Recent work has shown that Vision Transformers can achieve superior performance on image classification tasks [12,32]. In the future, we will further explore the dynamics of Visual Transformers.

Acknowledgments. This work was supported in part by the Key R&D Program of Zhejiang under Grant 2022C01018, by the National Natural Science Foundation of China under Grants U21B2001, 61973273, 62072406, 11931015, U1803263, by the Zhejiang Provincial Natural Science Foundation of China under Grant LR19F030001, by the National Science Fund for Distinguished Young Scholars under Grant 62025602, by the Fok Ying-Tong Education Foundation, China under Grant 171105, and by the Tencent Foundation and XPLORER PRIZE. We also sincerely thank Jinhuan Wang, Zhuangzhi Chen and Shengbo Gong for their excellent suggestions.

References

1. Alain, G., Bengio, Y.: Understanding intermediate layers using linear classifier probes. arXiv preprint arXiv:1610.01644 (2016)
2. Alet, F., Lozano-Pérez, T., Kaelbling, L.P.: Modular meta-learning. In: Conference on Robot Learning, pp. 856–868 (2018)
3. Azarian, K., Bhalgat, Y., Lee, J., Blankevoort, T.: Learned threshold pruning. arXiv preprint arXiv:2003.00075 (2020)
4. Carlini, N., Wagner, D.: Towards evaluating the robustness of neural networks. In: 2017 IEEE Symposium on Security and Privacy (SP), pp. 39–57 (2017)
5. Chen, S., Zhao, Q.: Shallowing deep networks: layer-wise pruning based on feature representations. IEEE Trans. Pattern Anal. Mach. Intell. **41**(12), 3048–3056 (2018)
6. Csordás, R., van Steenkiste, S., Schmidhuber, J.: Are neural nets modular? Inspecting their functionality through differentiable weight masks. In: International Conference on Learning Representations (2021)
7. Das, A., Rad, P.: Opportunities and challenges in explainable artificial intelligence (XAI): a survey. arXiv preprint arXiv:2006.11371 (2020)
8. Davis, B., Bhatt, U., Bhardwaj, K., Marculescu, R., Moura, J.M.: On network science and mutual information for explaining deep neural networks. In: IEEE International Conference on Acoustics, Speech and Signal Processing, pp. 8399–8403 (2020)
9. Delange, M., et al.: A continual learning survey: defying forgetting in classification tasks. IEEE Trans. Pattern Anal. Mach. Intell. **44**, 3366–3385 (2021)
10. Ding, G.W., Wang, L., Jin, X.: AdverTorch v0.1: an adversarial robustness toolbox based on Pytorch. arXiv preprint arXiv:1902.07623 (2019)
11. Donahue, J., et al.: DeCAF: a deep convolutional activation feature for generic visual recognition. In: International Conference on Machine Learning, pp. 647–655. PMLR (2014)
12. Dosovitskiy, A., et al. An image is worth 16x16 words: Transformers for image recognition at scale. In: International Conference on Learning Representations (2021)
13. Elkerdawy, S., Elhoushi, M., Singh, A., Zhang, H., Ray, N.: To filter prune, or to layer prune, that is the question. In: Proceedings of the Asian Conference on Computer Vision (2020)
14. Feng, Y., Zhai, R., He, D., Wang, L., Dong, B.: Transferred discrepancy: quantifying the difference between representations. arXiv preprint arXiv:2007.12446 (2020)
15. Fortunato, S.: Community detection in graphs. CoRR abs/0906.0612 (2009)
16. Frankle, J., Carbin, M.: The lottery ticket hypothesis: finding sparse, trainable neural networks. In: International Conference on Learning Representations (2019)
17. Goh, G., et al.: Multimodal neurons in artificial neural networks. Distill **6**(3), e30 (2021)
18. Goodfellow, I.J., Shlens, J., Szegedy, C.: Explaining and harnessing adversarial examples. In: International Conference on Learning Representations (2015)
19. Goyal, A., et al.: Recurrent independent mechanisms. In: International Conference on Learning Representations (2021)
20. Harary, F., Gupta, G.: Dynamic graph models. Math. Comput. Model. **25**(7), 79–87 (1997)
21. He, K., Zhang, X., Ren, S., Sun, J.: Deep residual learning for image recognition. In: Proceedings of the IEEE/CVF Conference on Computer Vision and Pattern Recognition, pp. 770–778 (2016)

22. Hod, S., Casper, S., Filan, D., Wild, C., Critch, A., Russell, S.: Detecting modularity in deep neural networks. arXiv preprint arXiv:2110.08058 (2021)
23. Jonsson, P.F., Cavanna, T., Zicha, D., Bates, P.A.: Cluster analysis of networks generated through homology: automatic identification of important protein communities involved in cancer metastasis. BMC Bioinform. **7**(1), 1–13 (2006)
24. Kirsch, L., Kunze, J., Barber, D.: Modular networks: Learning to decompose neural computation. In: Advances in Neural Information Processing Systems, pp. 2414–2423 (2018)
25. Kornblith, S., Norouzi, M., Lee, H., Hinton, G.: Similarity of neural network representations revisited. In: International Conference on Machine Learning, pp. 3519–3529 (2019)
26. Krizhevsky, A., Hinton, G., et al.: Learning multiple layers of features from tiny images (2009)
27. Lake, B.M., Ullman, T.D., Tenenbaum, J.B., Gershman, S.J.: Building machines that learn and think like people. Behav. Brain Sci. **40** (2017)
28. Li, J., et al.: Aha! adaptive history-driven attack for decision-based black-box models. In: Proceedings of the IEEE/CVF International Conference on Computer Vision, pp. 16168–16177 (2021)
29. Lin, M., et al.: HRank: filter pruning using high-rank feature map. In: Proceedings of the IEEE/CVF Conference on Computer Vision and Pattern Recognition, pp. 1529–1538 (2020)
30. Lin, S., et al.: Towards optimal structured CNN pruning via generative adversarial learning. In: Proceedings of the IEEE/CVF Conference on Computer Vision and Pattern Recognition, pp. 2790–2799 (2019)
31. Lin, T., Stich, S.U., Barba, L., Dmitriev, D., Jaggi, M.: Dynamic model pruning with feedback. In: International Conference on Learning Representations (2020)
32. Liu, Z., et al.: Swin transformer: hierarchical vision transformer using shifted windows. In: Proceedings of the IEEE/CVF International Conference on Computer Vision, pp. 10012–10022 (2021)
33. Lusseau, D.: The emergent properties of a dolphin social network. Proc. R. Soc. London Ser. B Biol. Sci. **270**(suppl_2), S186–S188 (2003)
34. Van der Maaten, L., Hinton, G.: Visualizing data using t-SNE. J. Mach. Learn. Res. **9**(11) (2008)
35. Madry, A., Makelov, A., Schmidt, L., Tsipras, D., Vladu, A.: Towards deep learning models resistant to adversarial attacks. In: International Conference on Learning Representations (2018)
36. Mahendran, A., Vedaldi, A.: Understanding deep image representations by inverting them. In: Proceedings of the IEEE/CVF Conference on Computer Vision and Pattern Recognition, pp. 5188–5196 (2015)
37. Maho, T., Furon, T., Le Merrer, E.: SurFree: a fast surrogate-free black-box attack. In: Proceedings of the IEEE/CVF Conference on Computer Vision and Pattern Recognition, pp. 10430–10439 (2021)
38. Maqueda, A.I., Loquercio, A., Gallego, G., García, N., Scaramuzza, D.: Event-based vision meets deep learning on steering prediction for self-driving cars. In: Proceedings of the IEEE/CVF Conference on Computer Vision and Pattern Recognition, pp. 5419–5427 (2018)
39. Morcos, A.S., Raghu, M., Bengio, S.: Insights on representational similarity in neural networks with canonical correlation. In: Advances in Neural Information Processing Systems, pp. 5732–5741 (2018)
40. Narodytska, N., Kasiviswanathan, S.P.: Simple black-box adversarial perturbations for deep networks. arXiv preprint arXiv:1612.06299 (2016)

41. Newman, M.E.: Modularity and community structure in networks. Proc. Natl. Acad. Sci. **103**(23), 8577–8582 (2006)
42. Newman, M.E., Girvan, M.: Finding and evaluating community structure in networks. Phys. Rev. E **69**(2), 026113 (2004)
43. Nguyen, T., Raghu, M., Kornblith, S.: Do wide and deep networks learn the same things? uncovering how neural network representations vary with width and depth. In: International Conference on Learning Representations (2021)
44. Papernot, N., McDaniel, P., Jha, S., Fredrikson, M., Celik, Z.B., Swami, A.: The limitations of deep learning in adversarial settings. In: 2016 IEEE European Symposium on Security and Privacy (EuroS&P), pp. 372–387 (2016)
45. Paszke, A., et al.: Automatic differentiation in PyTorch (2017)
46. Raghu, M., Gilmer, J., Yosinski, J., Sohl-Dickstein, J.: SVCCA: singular vector canonical correlation analysis for deep learning dynamics and interpretability. In: Advances in Neural Information Processing Systems, pp. 6076–6085 (2017)
47. Redmon, J., Divvala, S., Girshick, R., Farhadi, A.: You only look once: unified, real-time object detection. In: Proceedings of the IEEE/CVF Conference on Computer Vision and Pattern Recognition, pp. 779–788 (2016)
48. Rumelhart, D.E., Hinton, G.E., Williams, R.J.: Learning internal representations by error propagation. California Univ. San Diego La Jolla Inst. for Cognitive Science, Technical report (1985)
49. Russakovsky, O., et al.: ImageNet large scale visual recognition challenge. Int. J. Comput. Vis. **115**(3), 211–252 (2015)
50. Selvaraju, R.R., Cogswell, M., Das, A., Vedantam, R., Parikh, D., Batra, D.: Gradcam: visual explanations from deep networks via gradient-based localization. In: International Conference on Computer Vision, pp. 618–626 (2017)
51. Shafahi, A., et al.: Adversarial training for free! In: Advances in Neural Information Processing Systems 32 (2019)
52. Simonyan, K., Zisserman, A.: Very deep convolutional networks for large-scale image recognition. In: International Conference on Learning Representations (2015)
53. Su, J., Vargas, D.V., Sakurai, K.: One pixel attack for fooling deep neural networks. IEEE Trans. Evolut. Comput. **23**(5), 828–841 (2019)
54. Sutskever, I., Vinyals, O., Le, Q.V.: Sequence to sequence learning with neural networks. In: Advances in Neural Information Processing Systems, pp. 3104–3112 (2014)
55. Szegedy, C., et al.: Intriguing properties of neural networks. In: International Conference on Learning Representations (2014)
56. Tang, S., Maddox, W.J., Dickens, C., Diethe, T., Damianou, A.: Similarity of neural networks with gradients. arXiv preprint arXiv:2003.11498 (2020)
57. Tang, Y., et al.: Manifold regularized dynamic network pruning. In: Proceedings of the IEEE/CVF Conference on Computer Vision and Pattern Recognition. pp. 5018–5028 (2021)
58. Wang, F., Liu, H., Cheng, J.: Visualizing deep neural network by alternately image blurring and deblurring. Neural Netw. **97**, 162–172 (2018)
59. Wang, L., et al.: Towards understanding learning representations: To what extent do different neural networks learn the same representation. In: Advances in Neural Information Processing Systems, pp. 9607–9616 (2018)
60. Wang, W., et al.: Accelerate CNNs from three dimensions: a comprehensive pruning framework. In: International Conference on Machine Learning, pp. 10717–10726 (2021)

61. Wang, W., Zhao, S., Chen, M., Hu, J., Cai, D., Liu, H.: DBP: discrimination based block-level pruning for deep model acceleration. arXiv preprint arXiv:1912.10178 (2019)

62. Watanabe, C.: Interpreting layered neural networks via hierarchical modular representation. In: International Conference on Neural Information Processing, pp. 376–388 (2019)

63. Watanabe, C., Hiramatsu, K., Kashino, K.: Modular representation of layered neural networks. Neural Netw. **97**, 62–73 (2018)

64. Watanabe, C., Hiramatsu, K., Kashino, K.: Understanding community structure in layered neural networks. Neurocomputing **367**, 84–102 (2019)

65. Wong, E., Rice, L., Kolter, J.Z.: Fast is better than free: revisiting adversarial training. In: International Conference on Learning Representations (2020)

66. Xu, P., Cao, J., Shang, F., Sun, W., Li, P.: Layer pruning via fusible residual convolutional block for deep neural networks. arXiv preprint arXiv:2011.14356 (2020)

67. Yang, J., Leskovec, J.: Defining and evaluating network communities based on ground-truth. Knowl. Inf. Syst. **42**(1), 181–213 (2013). https://doi.org/10.1007/s10115-013-0693-z

68. Yosinski, J., Clune, J., Bengio, Y., Lipson, H.: How transferable are features in deep neural networks? In: Ghahramani, Z., Welling, M., Cortes, C., Lawrence, N.D., Weinberger, K.Q. (eds.) Advances in Neural Information Processing Systems, pp. 3320–3328 (2014)

69. You, J., Leskovec, J., He, K., Xie, S.: Graph structure of neural networks. In: International Conference on Machine Learning, pp. 10881–10891 (2020)

70. Zachary, W.W.: An information flow model for conflict and fission in small groups. J. Anthropol. Res. **33**(4), 452–473 (1977)

71. Zagoruyko, S., Komodakis, N.: Wide residual networks. In: Proceedings of the British Machine Vision Conference (2016)

72. Zeiler, M.D., Fergus, R.: Visualizing and understanding convolutional networks. In: Fleet, D., Pajdla, T., Schiele, B., Tuytelaars, T. (eds.) ECCV 2014. LNCS, vol. 8689, pp. 818–833. Springer, Cham (2014). https://doi.org/10.1007/978-3-319-10590-1_53

73. Zhou, Y., Yen, G.G., Yi, Z.: Evolutionary shallowing deep neural networks at block levels. IEEE Trans. Neural Netw. Learn. Syst. (2021)

Latent Discriminant Deterministic Uncertainty

Gianni Franchi[1]([✉])[iD], Xuanlong Yu[1,2][iD], Andrei Bursuc[3][iD],
Emanuel Aldea[2][iD], Severine Dubuisson[4], and David Filliat[1]

[1] U2IS, ENSTA Paris, Institut polytechnique de Paris, Palaiseau, France
`gianni.franchi@ensta-paris.fr`
[2] SATIE, Paris-Saclay University, Gif-sur-Yvette, France
[3] valeo.ai, Paris, France
[4] CNRS, LIS, Aix Marseille University, Marseille, France

Abstract. Predictive uncertainty estimation is essential for deploying Deep Neural Networks in real-world autonomous systems. However, most successful approaches are computationally intensive. In this work, we attempt to address these challenges in the context of autonomous driving perception tasks. Recently proposed Deterministic Uncertainty Methods (DUM) can only partially meet such requirements as their scalability to complex computer vision tasks is not obvious. In this work we advance a scalable and effective DUM for high-resolution semantic segmentation, that relaxes the Lipschitz constraint typically hindering practicality of such architectures. We learn a discriminant latent space by leveraging a distinction maximization layer over an arbitrarily-sized set of trainable prototypes. Our approach achieves competitive results over Deep Ensembles, the state of the art for uncertainty prediction, on image classification, segmentation and monocular depth estimation tasks. Our code is available at https://github.com/ENSTA-U2IS/LDU.

Keywords: Deep neural networks · Uncertainty estimation · Out-of-distribution detection

1 Introduction

Uncertainty estimation and robustness are essential for deploying Deep Neural Networks (DNN) in real-world systems with different levels of autonomy, ranging from simple driving assistance functions to fully autonomous vehicles. In addition to excellent predictive performance, DNNs are also expected to address different types of uncertainty (noisy, ambiguous or out-of-distribution samples, distribution shift, etc.), while ensuring real-time computational performance. These key

---

G. Franchi and Franchi X. Yu—Equal contribution.

Supplementary Information The online version contains supplementary material available at https://doi.org/10.1007/978-3-031-19775-8_15.

and challenging requirements have stimulated numerous solutions and research directions leading to significant progress in this area [10,27,43,45,57,77]. Yet, the best performing approaches are computationally expensive [45], while faster variants struggle to disentangle different types of uncertainty [23,52,56].

We study a promising new line of methods, termed deterministic uncertainty methods (DUMs) [65], that has recently emerged for estimating uncertainty in a computational efficient manner from a single forward pass [3,48,56,64,74]. In order to quantify uncertainty, these methods rely on some statistical or geometrical properties of the hidden features of the DNNs. While appealing for their good Out-of-Distribution (OOD) uncertainty estimations at low computational cost, they have been used mainly for classification tasks and their specific regularization is often unstable when training deeper DNNs [62]. We then propose a new DUM technique, based on a discriminative latent space that improves both scalability and flexibility. We achieve this by still following the principles of DUMs of learning a sensitive and smooth representation that mirrors well the input distribution, although not by enforcing directly the Lipschitz constraint.

Our DUM, dubbed Latent Discriminant deterministic Uncertainty (LDU), is based on a DNN imbued with a set of prototypes over its latent representations. These prototypes act like a memory that allows to better analyze features from new images in light of the "knowledge" acquired by the DNN from the training data. Various forms of prototypes have been studied for anomaly detection in the past [29] and they often take the shape of a dictionary of representative features. Instead, LDU is trained to learn the optimal prototypes, such that this distance improves the accuracy and the uncertainty prediction. Indeed to train LDU, we introduce a confidence-based loss that learns to predict the error of the DNN given the data. ConfidNet [15] and SLURP [81] have shown that we can train an auxiliary network to predict the uncertainty, at the cost of a more complex training pipeline and more inference steps. Here LDU is lighter, faster and needs only a single forward pass. LDU can be used as a pluggable learning layer on top of DNNs. We demonstrate that LDU avoids feature collapse and can be applied to multiple computer vision tasks. In addition, LDU improves the prediction accuracy of the baseline DNN without LDU.

Contributions. To summarize, our contributions are as follows: **(1)** LDU (Latent Discriminant deterministic Uncertainty): an efficient and scalable DUM approach for uncertainty quantification. **(2)** A study of LDU's properties against feature collapse. **(3)** Evaluations of LDU on a range of computer vision tasks and settings (image classification, semantic segmentation, depth estimation) and the implementation of a set of powerful baselines to further encourage research in this area.

2   Related Work

In this section, we focus on the related works from two perspectives: uncertainty quantification algorithms applied to computer vision tasks and prototype learning on DNNs. In Table 1, we list various uncertainty quantification algorithms according to different computer vision tasks.

Table 1. Summary of the uncertainty estimation methods applied to the specific computer vision tasks.

Uncertainty estimation methods	Computer vision tasks		
	Image classification	Semantic segmentation	1D/2D Regression
Bayesian/Ensemble based methods	Rank-1 BNN [20], PBP [37], Deep Ensembles [45], Bayes by Backprop [10], MultiSWAG [78]	Deep Ensembles [45], Bayes by Backprop [10], MultiSWAG [78]	FlowNetH [40], PBP [37], Deep Ensembles [45], Bayes by Backprop [10], MultiSWAG [78]
Dropout/Sampling based methods	MC-Dropout [27]	MC-Dropout [27], Bayesian SegNet [42]	Infer-perturbations [54], MC-Dropout [27]
Learning distribution/Auxiliary network	ConfidNet [15], Kendall et al. [43]	ConfidNet [15], Kendall et al. [43]	Hu et al. [38], SLURP [81], Mono-Uncertainty [63], Asai et al. [5], Kendall et al. [43], Nix et al. [59]
Deterministic uncertainty methods	SNGP [48], VIB [2], DUM [79],DUE [3], DDU [56], MIR [64], DUQ [74]	MIR [64]	DUE [3], MIR [64]

2.1   Uncertainty Estimation for Computer Vision Tasks

Uncertainty for Image Classification and Semantic Segmentation.
Quantifying uncertainty for classification and semantic segmentation can be done with Bayesian Neural Networks (BNNs) [10,20,37,78], which estimate the posterior distribution of the DNN weights to marginalize the likelihood distribution at inference time. These approaches achieve good performances on image classification, but they do not scale well to semantic segmentation. Deep Ensembles [45] achieve state-of-the-art performance on various tasks. Yet, this approach is computationally costly in both training and inference. Some techniques learn a confidence score as uncertainty [15], but struggle without sufficient negative samples to learn from. MC-Dropout [27] is a generic and easy to deploy approach, however its uncertainty is not always reliable [60] while requiring multiple forward passes. Deterministic Uncertainty Methods (DUMs) [2,3,48,56,64,74,79] are new strategies that allow to quantify epistemic uncertainty in the DNNs with a single forward pass. Yet, except for MIR [64], to the best of our knowledge none of these techniques work on semantic segmentation.

Uncertainty for 1D/2D Regression. Regression in computer vision comprises monocular depth estimation [9,46], optical flow estimation [71,72], or pose estimation [12,69]. One solution for quantifying the uncertainty consists in formalizing the output of a DNN as a parametric distribution and training the DNN to estimate its parameters [43,59]. Multi-hypothesis DNNs [40] consider that the output is a Gaussian distribution and focus on optical flow. Some techniques estimate a confidence score for regression thanks to an auxiliary DNN [63,81]. Deep Ensembles [45] for regression, consider that each DNN outputs the parameters of a Gaussian distribution, to form a mixture of Gaussian distributions. Sampling-based methods [27,54] simply apply dropout or perturbations to some layers during test time to quantify the uncertainty. Yet, their computational cost remains important compared to a single forward pass in the network. Some DUMs [3,64] also work on regression tasks. DUE [3] is applied in a 1D regression task and MIR [64] in monocular depth estimation.

2.2   Prototype Learning in DNNs

Prototype-based learning approaches have been introduced on traditional hand-crafted features [47], and have been recently applied to DNNs as well, for more robust predictions [13,29,76,80]. The center loss [76] can help DNNs to build more discriminative features by compacting intra-class features and dispersing the inter-class ones. Based on this principle, Convolutional Prototype Learning (CPL) [80] with prototype loss also improves the intra-class compactness of the latent features. Chen et al. [13] try to bound the unknown classes by learning reciprocal points for better open set recognition. Similar to [67,75], MemAE [29] learns a memory slot of the prototypes to strengthen the reconstruction error of anomalies in the process of the reconstruction. These prototype-based methods are well suited for classification tasks but are rarely used in semantic segmentation and regression tasks.

3   Latent Discriminant Deterministic Uncertainty (LDU)

3.1   DUM Preliminaries

DUMs arise as a promising line of research for estimating epistemic uncertainty in conventional DNNs in a computationally efficient manner and from a single forward pass. DUM approaches generally focus on learning useful and informative hidden representations of a model [2,3,48,56,64,79] by considering that the distribution of the hidden representation should be representative for the input distribution. Most of the conventional models suffer from the *feature collapse* problem [74] when OOD samples are mapped to similar feature representations as in-distribution ones, thus hindering OOD detection from these representations. DUMs address this issue through various regularization strategies for constraining the hidden representations to mimic distances from the input space. In practice this amounts to striking a balance between *sensitivity* (when the input changes, the feature representation should also change) and *smoothness* (a small change in the input cannot generate major shifts in the feature representation) of the model. To this end, most methods enforce constraints over the Lipschitz constant of the DNN [48,53,74].

Formally, we define $f_\omega(\cdot)$ a DNN with trainable parameters ω, and an input sample $\mathbf{x}$ from a set of images $\mathcal{X}$. Our DNN f_ω is composed of two main blocks: a feature extractor h_ω and a head g_ω, such that $f_\omega(\mathbf{x}) = (g_\omega \circ h_\omega)(\mathbf{x})$. $h_\omega(\mathbf{x})$ computes a latent representation from $\mathbf{x}$, while g_ω is the final layer, that takes $h_\omega(\mathbf{x})$ as input, and outputs the logits of $\mathbf{x}$. The bi-Lipschitz condition implies that for any pair of inputs $\mathbf{x}_1$ and $\mathbf{x}_2$ from $\mathcal{X}$:

$$L_1\|\mathbf{x}_1 - \mathbf{x}_2\| \leq \|h_\omega(\mathbf{x}_1) - h_\omega(\mathbf{x}_2)\| \leq L_2\|\mathbf{x}_1 - \mathbf{x}_2\| \tag{1}$$

where L_1 and L_2 are positive and bounded Lipschitz constants $0 < L_1 < 1 < L_2$. The upper Lipschitz bound enforces the smoothness and is an important condition for the robustness of a DNN by preventing over-sensitivity to perturbations in the input space of $\mathbf{x}$, i.e., the pixel space. The lower Lipschitz bound deals with the sensitivity and strives to preserve distances in the latent space as mappings of distances from the input space, i.e., preventing representations from being too smooth, thus avoiding feature collapse. Liu et al. [48] argue that for residual DNNs [33], we can ensure f_ω to be bi-Lipschitz by forcing its residuals to be Lipschitz and choosing sub-unitary Lipschitz constants.

There are different approaches for imposing the bi-Lipschitz constraint over a DNN, out of which we describe the most commonly used ones in recent works [4, 7, 30, 55]. Wasserstein GAN [4] enforces the Lipschitz constraint by clipping the weights. However, this turns out to be prone to either vanishing or exploding gradients if the clipping threshold is not carefully tuned [30]. An alternative solution from GAN optimization is gradient penalty [30] which is practically an additional loss term that regularizes the L_2 norm of the Jacobian of weight matrices of the DNN. However this can also lead to high instabilities [48, 56] and slower training [56]. Spectral Normalization [7, 55] brings better stability and training speed, however, on the downside, it supports only a fixed pre-defined size for the input, in the same manner as fully connected layers. For computer vision tasks, such as semantic segmentation which is typically performed on high resolution images, constraining the input size is a strong limitation. Moreover, Postels et al. [64] argue that in addition to the architectural constraints, these strategies for avoiding feature collapse risk overfitting epistemic uncertainty to the task of OOD detection. This motivates us to seek a new DUM strategy that does not need the network to comply with the Lipschitz constraint. The recent MIR approach [64] advances an alternative regularization strategy that adds a decoder branch to the network, thus forcing the intermediate activations to better cover and represent the input space. However in the case of high resolution images, reconstruction can be a challenging task and the networks can over-focus on potentially useless and uninformative details at the cost of loss of global information. We detail our strategy below.

3.2   Discriminant Latent Space

An informative latent representation should project similar data samples close and dissimilar ones far away. Yet, it has long been known that in high-dimensional spaces the Euclidean distance and other related p-norms are a very poor indicator of sample similarity as most samples are nearly equally far/close to each other [1, 6]. At the same time, the samples of interest are often not uniformly distributed, and may be projected by means of a learned transform on a lower-dimensional manifold, namely the latent representation space.

Instead of focusing on preserving the potentially uninformative distance in the input space, we can rather attempt to better deal with distances in the lower-dimensional latent space. To this end, we propose to use a distinction maximization (DM) layer [49] that has been recently considered as a replacement for the last layer to produce better uncertainty estimates, in particular for OOD detection [49, 61]. In a DM layer, the units of the classification layer are seen as representative class prototypes and the classification prediction is computed by analyzing the localization of the input sample w.r.t. all class prototypes as indicated by the negative Euclidean distance. Note that a similar idea has been considered in the few-shot learning literature, where DM layers are known as cosine classifiers [28,66,70]. In contrast to all these approaches that use DM as a last layer for classification predictions, we employ it as hidden layer over latent representations. More specifically, we insert DM in the pre-logit layer. We argue that this allows us to better guide learning and preserve the discriminative properties of the latent representations compared to placing DM as last layers where the weights are more specialized for classification decision than for feature representation. We can easily integrate this layer in the architecture without impacting the training pipeline.

Formally, we denote $\mathbf{z} \in \mathbb{R}^n$ the latent representation of dimension n of $\mathbf{x}$, i.e., $\mathbf{z} = h_\omega(\mathbf{x})$, that is given as input to the DM layer. Given a set $\mathbf{p}_\omega = \{\mathbf{p}_i\}_{i=1}^m$, of m vectors ($\mathbf{p}_i \in \mathbb{R}^n$) that are trainable, we define the DM layer as follows:

$$DM_p(\mathbf{z}) = \left[-\|\mathbf{z} - \mathbf{p}_1\|, \dots, -\|\mathbf{z} - \mathbf{p}_m\|\right]^\top \qquad (2)$$

The L_2 distance considered in the DM layer is not bounded, thus when DM is used as intermediate layer, relying on the L_2 distance could cause instability during training. In our proposed approach, we use instead the cosine similarity, $S_c(\cdot, \cdot)$. Our DM layer reads now:

$$DM_p(\mathbf{z}) = \left[S_c(\mathbf{z}, \mathbf{p}_1), \dots, S_c(\mathbf{z}, \mathbf{p}_m)\right]^\top \qquad (3)$$

The vectors $\mathbf{p}_i$ can be seen as a set of prototypes in the latent space that can help in better placing an input sample in the learned representation space using these prototypes as references. This is in contrast to prior works with DM being considered as last layer, where the prototypes represent canonical representations for samples belonging to a class [49,70]. Since hidden layers are used here, we can afford to consider an arbitrary number of prototypes that can define richer latent mapping through a finer coverage of the representation space. DM layers learn the set of weights $\{\mathbf{p}_i\}_{i=1}^m$ such that the cosine similarity (evaluated between $\mathbf{z}$ and the prototypes) is optimal for a given task.

We apply the distinction maximization on this hidden representation, and subsequently use the exponential function as activation function. We consider the exponential function as it can sharpen similarity values and thus facilitates the alignment of the data embedding to the corresponding prototypes in the latent space. Finally, we apply a last fully connected layer for classification on this embedding. Our DNN (see Fig. 1) can be written as:

$$f_\omega(\mathbf{x}) = [g_\omega \circ (\exp(-DM_p(h_\omega)))](\mathbf{x}) \qquad (4)$$

We can see from Eq. (4) that the vector weights $\mathbf{p}_i$ are optimized jointly with the other DNN parameters. We argue that $\mathbf{p}_i$ can work as indicators for analyzing and emphasizing patterns in the latent representation prior to making a classification prediction in the final layers.

3.3   LDU Optimization

Given a DNN f_ω we usually optimize its parameters to minimize a loss $\mathcal{L}^{\text{Task}}$. This can lead to prototypes specialized for solving that task that do not encapsulate uncertainty relevant properties. Hence we propose to enforce the prototypes

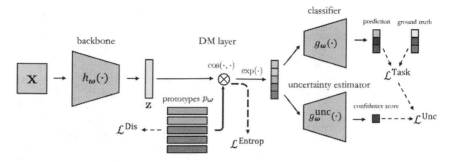

Fig. 1. Overview of LDU: the DNN learns a discriminative latent space thanks to learnable prototypes $\mathbf{p}_\omega$. The DNN backbone computes a feature vector $\mathbf{z}$ for an input $\mathbf{x}$ and then the DM layer matches it with the prototypes. The computed similarities reflecting the position of $\mathbf{z}$ in the learned feature space, are subsequently processed by the classification layer and the uncertainty estimation layer. The dashed arrows point to the loss functions that need to be optimized for training LDU.

to be linked to uncertainty first by avoiding the collapse of all prototypes to a single prototype. Second, we constrain the latent representation $\mathrm{DM}_p(h_\omega)$ of the DNN to not rely only on a single prototype. Finally, we optimize an MLP g_ω^{unc} on the top of the latent representation $\mathrm{DM}_p(h_\omega)$ such that the output of this MLP provides more meaningful information for uncertainty estimation.

First, we add a loss to force the prototypes to be dissimilar:

$$\mathcal{L}^{\mathrm{Dis}} = -\sum_{i<j} \|\mathbf{p}_i - \mathbf{p}_j\|.$$

Then, we also add one loss to constrain the latent representation to stay close to different prototypes. We achieve this with an entropy-like loss:

$$\mathcal{L}^{\mathrm{Entrop}} = \sum_{i=1}^{n} \sigma(\mathrm{DM}_p(h_\omega))_i \cdot \log(\sigma(\mathrm{DM}_p(h_\omega))_i),$$

where σ is the softmax layer, and the subscript index i corresponds to the i-th coefficient of a tensor. Different from per-class prototypes [13,76,80], we obtain more discriminative features by increasing the distance between prototypes and enlarging the dispersion of features corresponding to different prototypes.

We propose to train g_ω^{unc} to predict the error of the DNN, which helps us relate the prototypes to the uncertainty. Formally, given an input data $\mathbf{x}$, its groundtruth y (y can be a scalar or a vector if we deal with regression) and, its loss $\mathcal{L}^{\mathrm{Task}}(g_\omega(\mathbf{x}), y)$, we train g_ω^{unc} by minimizing:

$$\mathcal{L}^{\mathrm{Unc}} = \mathrm{BCE}\big(\big[g_\omega^{\mathrm{unc}} \circ (\exp(-\mathrm{DM}_p(h_\omega)))\big](\mathbf{x}), \mathcal{L}^{\mathrm{Task}}(g_\omega(\mathbf{x}), y)\big),$$

after normalizing $\mathcal{L}^{\mathrm{Task}}(g_\omega(\mathbf{x}), y)$ over the mini-batch such that its maximum value is equal to one and its minimum is equal to zero. BCE stands for the binary

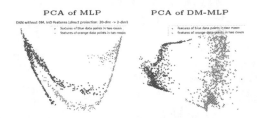

Fig. 2. PCA 2D projection on the left of a standard MLP and on the right of a DM-MLP trained on the two moons dataset. Blue and red points indicate the features of data points of the two classes respectively. As we can see, the representations on the MLP are overlapping between the two classes, leading to a network that will be prone to feature collapse, unlike the DM-MLP. (Color figure online)

cross entropy, which was empirically validated to perform better than common alternatives such as the mean square error and the absolute error.

All these losses combined allow us to have a DNN which can predict uncertainty, avoid feature collapse and have the potential to improve the accuracy of the prediction. To summarize, the following loss function $\mathcal{L}^{\text{total}}$ will be optimized to train a DNN containing a DM layer:

$$\mathcal{L}^{\text{total}} = \mathcal{L}^{\text{Task}} + \lambda(\mathcal{L}^{\text{Entrop}} + \mathcal{L}^{\text{Dis}} + \mathcal{L}^{\text{Unc}}) \tag{5}$$

where λ is a hyper-parameter for the auxiliary losses.

3.4 Addressing Feature Collapse

In order to illustrate the feature collapse problem, we consider a toy example on the two moon dataset. We train on it two MLPs with two hidden layers, each containing 17 neurons. One of the MLP additionally integrates our proposed DM layer and is denoted as DM-MLP, while the standard architecture is called MLP. The two networks reach the same classification performance, about 99% of accuracy. We perform PCA on the pre-logit latent space of both networks after training and visualize PCA projections in Fig. 2. We can observe the feature collapse as the MLP assigns strongly correlated feature representation to both classes which can lead to unreliable uncertainty prediction. However, our DM layer allows a better disentangling of the latent space. Note that, as the networks have the same performance, it is impossible to detect the feature collapse based on the test accuracy alone.

We note that our LDU layer is a Lipschitz function, hence: $\|\exp(-\text{DM}_p(z_1)) - \exp(-\text{DM}_p(z_2))\| \leq k\|z_1 - z_2\|$ with $k \in \mathbb{R}^+$. However, h_ω is not necessarily a Lipschitz function, and we cannot thus guarantee that its features do not entangle ID and OOD data. Yet, using a distance function in the DNN [50,53] can allow it to learn to separate the two data distributions better as illustrated in Figure 2.

Most DUM methods aim for bi-Lipschitz DNNs with small Lipschitz constants. Yet, this is sub-optimal according to the concentration theory. Indeed,

Confidence score | Confidence score
after the first training | after the second training

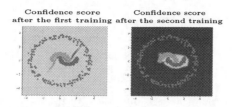

Fig. 3. Illustration of confidence score results on the two moons dataset after the first training (on original data) on the left and with second training (on synthesized outliers) on the right. Orange and blue data points are sampled from two classes in two moons, and the green points are OOD data points. Yellow area indicates high confidence, blue area indicates uncertainty. The left image shows that the uncertain area is between the two classes leading to a confidence score related to aleatoric uncertainty. In the right one, the uncertain area is around the dataset leading to a confidence score related to epistemic uncertainty.

let $\mathbf{X}$ be a set of random vectors of size d i.i.d. from a normal distribution $\mathcal{N}(0, \sigma^2 I_d)$. I_d is the identity matrix of size d. Let $f : \mathbb{R}^d \to \mathbb{R}$ be a Lipschitz function with Lipschitz constant K. The concentration theory ([11], p. 125) stipulates that : $\mathcal{P}(|f(\mathbf{X}) - \mathbb{E}(f(\mathbf{X}))| > t) \leq 2\exp(-\frac{t^2}{2K^2\sigma^2})$ for all $t > 0$. This means that the smaller K is, the more the concentration of the data around their mean increases, leading to increased feature collapse. Hence, it is desirable to have a Lipschitz function that will bring similar data close, but it is at the same time essential to put dissimilar data apart.

3.5 LDU And Epistemic/Aleatoric Uncertainty

We are interested in capturing two types of uncertainty with our DNN: aleatoric and epistemic uncertainty [17,43]. Aleatoric uncertainty is related to the inherent noise and ambiguity in data or in annotations and is often called irreducible uncertainty as it does not fade away with more data in the training set. Epistemic uncertainty is related to the model, the learning process and the amount of training data. Since it can be reduced with more training data, it is also called reducible uncertainty. Disentangling these two sources of uncertainty is generally non-trivial [56] and ensemble methods are usually superior [23,52].

Optional Training with Synthesized Outliers. Due to limited training data and to the penalty enforced by $\mathcal{L}^{\text{Task}}$ being too small, the loss term $\mathcal{L}^{\text{Unc}}$ may potentially force the DNN in some circumstances to overfit the aleatoric uncertainty. Although we did not encounter this behavior on the computer vision tasks given the dataset size, it might occur on more specific data, and among other potential solutions, we propose one relying on synthesized outliers that we illustrate on the two moons dataset as follows. More specifically, we propose to add noise to the data similarly to [19,51], and to introduce an optional step for training g_ω^{unc} on these new samples. We consider a two-stage training scheme. In the first stage we train over data without noise and in the second we optimize only the parameters of g_ω^{unc} over the synthesized outliers. Note that this optional stage would require for vision tasks an adequate OOD synthesizer [8,19] which is beyond the scope of this paper, and that we applied it on the toy dataset. In Fig. 3 we assess the uncertainty estimation performance of this model on the two moons dataset. We can see that the confidence score relates to the aleatoric uncertainty after the first training stage. After the second one, it is linked to the epistemic uncertainty of the model.

Distinguishing between the two sources of uncertainty is essential for various practical applications, such as active learning, monitoring, OOD detection. In the following, we propose two strategies for computing each type of uncertainty.

Aleatoric Uncertainty. For estimating aleatoric uncertainty in classification, maximum class probability (MCP) [36] is a common strategy. The intuition is that a lower MCP can mean a higher entropy, i.e., a potential confusion on the classifier regarding the most likely class of the image. We use this criterion for the aleatoric uncertainty for classification and for semantic segmentation, while for the regression task we use g_ω^{unc} as confidence score.

Epistemic Uncertainty. To estimate epistemic uncertainty, we analyzed the latent representations of the DM layers followed by an exponential activation and found that the maximum value can model well uncertainty. The position of the feature w.r.t. the learned prototypes caries information about the proximity of the current sample with the in-distribution features. Yet, we propose to use the output of g_ω^{unc} as confidence score since we train this criterion for this purpose.

4   Experiments

One major interest of our technique is that it may be seamlessly applied to any computer vision task, be it classification or regression. Thus, we propose to evaluate the quality of uncertainty quantification of different techniques on three major tasks, namely image classification, semantic segmentation and monocular depth estimation. For all the three tasks, we compare our technique against MC Dropout [27] and Deep Ensembles [45]. For image classification, we also compare our technique to relevant DUM techniques for image classification, namely DDU [56], DUQ [74], DUE [3], MIR [64] and SNGP [48].

We evaluate the predictive performance in terms of accuracy for image classification, mIoU [22] for semantic segmentation, and the metrics first introduced in [21] and used in many subsequent works for monocular depth estimation. For image classification and semantic segmentation, we also evaluate the quality of the confidence scores provided by the DNNs via the following metrics: Expected Calibration Error (ECE) [31], AUROC [34,36] and AUPR [34,36]. Note that ECE we use is the confidence ECE defined in [31]. We use the Area Under the Sparsification Error: AUSE-RMSE and AUSE-Absrel similarly to [32,63,81] to better evaluate the uncertainty quantification on monocular depth estimation.

We run all methods ourselves in similar settings using publicly available codes and hyper-parameters for related methods. In the following tables, Top-2 results are highlighted in color.

Table 2. Comparative results for **image classification tasks**. We evaluate on CIFAR-10 for the tasks: in-domain classification, out-of-distribution detection with SVHN. Results are averaged over three seeds.

Method	CIFAR-10			
	Acc ↑	AUC ↑	AUPR ↑	ECE ↓
Baseline (MCP) [36]	88.02	0.8032	0.8713	0.5126
MCP lipz. [7]	88.50	0.8403	0.9058	**0.3820**
Deep Ensembles [45]	**89.96**	0.8513	0.9087	0.4249
SNGP [48]	88.45	0.8447	0.9139	0.4254
DUQ [74]	89.9	0.8446	**0.9144**	0.5695
DUE [3]	87.54	0.8434	0.9082	0.4313
DDU [56]	87.87	0.8199	0.8754	**0.3820**
MIR [64]	87.95	0.7574	0.8556	**0.4004**
LDU #p = 128	87.95	**0.8721**	**0.9147**	0.4933
LDU #p = 64	88.06	0.8625	0.9070	0.5010
LDU #p = 32	87.83	0.8129	0.8900	0.5264
LDU #p = 16	88.33	0.8479	0.9094	0.4975

Image	groundtruth	prediction	MCP	g_ω^{unc}'s prediction

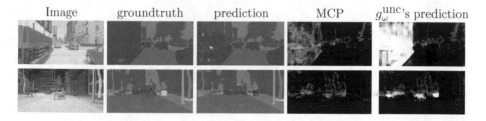

Fig. 4. Illustration of the different confidence scores on one image of MUAD. Note that the class `train`, `bicycle`, `Stand food` and the `animals` are OOD.

4.1 Classification Experiments

To evaluate uncertainty quantification for image classification, we adopt a standard approach based on training on CIFAR-10 [44], and using SVHN [58] as OOD data [25,48,65]. We use ResNet18 [33] as architecture for all methods. Note that for all DNNs, even for DE, we average results over three random seeds for statistical relevance. We follow the corresponding protocol for all DUM techniques (except LDU). For Deep Ensembles, MCP, and LDU, we use the same protocol. Please refer to the appendix for implementation details of LDU. The performances of the different algorithms are shown in Table 2. We can see that LDU has state-of-the-art performances on CIFAR-10. We note that LDU's OOD detection performance improves with the number of prototypes. This can be linked with the fact that the more prototypes we have, the better we can model complex distributions. The ablation studies on sensitivity of the choice of λ and the impact of different losses are provided in the appendix.

4.2 Semantic Segmentation Experiments

Our semantic segmentation study consists of three experiments. The first one is on a new synthetic dataset: MUAD [26]. It comprises a training set and a test set

Table 3. Comparative results for **semantic segmentation on MUAD.**

Evaluation data	normal set		OOD set					low adv. set					high adv. set				
	mIoU ↑	ECE ↓	mIoU ↑	ECE ↓	AUROC ↑	AUPR ↑	FPR ↓	mIoU ↑	ECE ↓	AUROC ↑	AUPR ↑	FPR ↓	mIoU ↑	ECE ↓	AUROC ↑	AUPR ↑	FPR ↓
Baseline (MCP) [36]	68.90%	0.0138	57.32%	0.0607	0.8624	0.2604	0.3943	31.84%	0.3078	0.6349	0.1185	0.6746	18.94%	0.4356	0.6023	0.1073	0.7547
Baseline (MCP) lipz. [7]	53.96%	0.01398	45.97%	0.0601	0.8419	0.2035	0.3940	16.79%	0.3336	0.6303	0.1051	0.7262	7.8%	0.4244	0.5542	0.0901	0.8243
MIR [64]	53.96%	0.01398	45.97%	0.0601	0.6223	0.1469	0.8406	16.79%	0.3336	0.5143	0.1035	0.8708	7.8%	0.4244	0.4470	0.0885	0.9093
MC-Dropout [27]	65.33%	0.0173	55.62%	0.0645	0.8439	0.2225	0.4575	33.38%	0.1329	0.7506	0.1545	0.5807	20.77%	0.3869	0.6864	0.1185	0.6751
Deep Ensembles [45]	69.90%	0.01206	58.29%	0.0588	0.871	0.2802	0.3760	34.91%	0.2447	0.6543	0.1212	0.6425	20.19%	0.4227	0.6101	0.1162	0.7212
LDU (ours)	69.32%	0.01356	58.29%	0.0594	0.8816	0.4418	0.3548	36.12%	0.2674	0.7779	0.2898	0.5381	21.15%	0.4231	0.7107	0.2186	0.6412

Table 4. Comparative results for **semantic segmentation on Cityscapes and Cityscapes-C.**

Evaluation data	Cityscapes		Cityscapes-C lvl 1		Cityscapes-C lvl 2		Cityscapes-C lvl 3		Cityscapes-C lvl 4		Cityscapes-C lvl 5	
	mIoU ↑	ECE ↓	mIoU ↑	ECE ↓	mIoU ↑	ECE ↓	mIoU ↑	ECE ↓	mIoU ↑	ECE ↓	mIoU ↑	ECE ↓
Baseline (MCP) [36]	76.84%	0.1180	51.59%	0.1793	41.45%	0.2291	35.67%	0.2136	30.12%	0.1970	24.84%	0.2131
Baseline (MCP) lipz. [7]	58.38%	0.1037	44.70%	0.1211	38.04%	0.1475	32.70%	0.1802	25.35%	0.2047	18.36%	0.2948
MC-Dropout [27]	71.88%	0.1157	53.61%	0.1501	42.02%	0.2531	35.91%	0.1718	29.52%	0.1947	25.61%	0.2184
Deep Ensembles [45]	77.23%	0.1139	54.98%	0.1422	44.63%	0.1902	38.00%	0.1851	32.14%	0.1602	28.74%	0.1729
LDU (ours)	76.62%	0.0893	52.00%	0.1371	43.02%	0.1314	37.17%	0.1702	32.27%	0.1314	27.30%	0.1712

Table 5. Comparative results obtained on the **OOD detection task on BDD Anomaly** [34] with PSPNet (ResNet50).

OOD technique	mIoU ↑	AUC ↑	AUPR ↑	FPR-95%-TPR ↓
Baseline (MCP) [36]	52.8	86.0	5.4	27.7
MC-Dropout [27]	49.5	85.2	5.0	29.3
Deep Ensembles [45]	57.6	87.0	6.0	25.0
TRADI [25]	52.1	86.1	5.6	26.9
ConfidNET [15]	52.8	85.4	5.1	29.1
LDU (ours)	55.1	87.1	5.8	26.2

without OOD classes and adverse weather conditions. We denote this set **normal set**. MUAD contains three more test sets that we denote **OOD set, low adv. set** and **high adv. set** which contain respectively images with OOD pixels but without adverse weather conditions, images with OOD pixels and weak adverse weather conditions, and for the last set, images with OOD pixels and strong adverse weather conditions. The second experiment evaluates the segmentation precision and uncertainty quality on the Cityscapes [16] and the Cityscapes-C [24,41,68] datasets to assess performance under distribution shift. Finally we analyze OOD detection performance on BDD Anomaly dataset [34] whose test set contains objects unseen during training. We detail the experimental protocol of all datasets in the appendix.

We train a DeepLabV3+ [14] network with ResNet50 encoder [33] on MUAD. Table 3 lists the results from different uncertainty techniques. For this task, we found that enforcing the Lipschitz constraint (see Baseline (MCP) lipz.) has a significant impact. Figure 4 shows a qualitative example of typical uncertainty maps computed on MUAD images.

Similarly to [24,41] we assess predictive uncertainty and robustness under distribution shift using Cityscapes-C, a corrupted version of Cityscapes images with perturbations of varying intensity. We generate Cityscapes-C ourselves from the original Cityscapes images using the code of Hendrycks et al. [35] to apply

Table 6. Comparative results for **monocular depth estimation on KITTI eigen-split validation set.**

Method	Depth performance								Uncertainty performance	
	d1↑	d2↑	d3↑	Abs Rel↓	Sq Rel↓	RMSE↓	RMSE log↓	log10↓	AUSE RMSE↓	AUSE Absrel↓
Baseline	0.955	0.993	0.998	0.060	0.249	2.798	0.096	0.027	-	-
Deep Ensembles [45]	0.956	0.993	0.999	0.060	0.236	2.799	0.094	0.026	0.08	0.21
MC-Dropout [27]	0.945	0.992	0.998	0.072	0.287	2.902	0.107	0.031	0.46	0.50
Single-PU [43]	0.949	0.991	0.998	0.064	0.263	2.796	0.101	0.029	0.08	0.21
Infer-noise [54]	0.955	0.993	0.998	0.060	0.249	2.798	0.096	0.027	0.33	0.48
LDU #p = 5, λ = 1.0	0.954	0.993	0.998	0.063	0.253	2.768	0.098	0.027	0.08	0.21
LDU #p = 15, λ = 0.1	0.954	0.993	0.998	0.062	0.249	2.769	0.098	0.027	0.10	0.28
LDU #p = 30, λ = 0.1	0.955	0.992	0.998	0.061	0.248	2.757	0.097	0.027	0.09	0.26

the different corruptions on the images. Following [35], we apply the following perturbations: Gaussian noise, shot noise, impulse noise, defocus blur, frosted, glass blur, motion blur, zoom blur, snow, frost, fog, brightness, contrast, elastic, pixelate, JPEG. Each perturbation is scaled with five levels of strength. We train a DeepLabV3+ [14] with ResNet50 encoder [33] on Cityscapes. Results in Table 4 show that LDU is closely trailing in accuracy (mIoU score) the much more costly Deep Ensembles [45], while making better calibrated predictions (ECE score).

In order to assess the epistemic uncertainty quantification on real data we used PSPNet [82] with ResNet50 backbone using the experimental protocol in [34]. BDD Anomaly is a subset of BDD dataset, composed of 6688 street scenes for the training set and 361 for the testing set. The training set contains 17 classes, and the test set is composed of the 17 training classes and 2 OOD classes. Results in Table 5 show again that the performances of LDU are close to the ones of Deep Ensembles.

4.3 Monocular Depth Experiments

We set up our experiments on KITTI dataset [73] with Eigen split training and validation set [21] to evaluate and compare the predicted depth accuracy and uncertainty quality. We train BTS [46] with DenseNet161 [39], and we use the default training setting of BTS (number of epochs, weight decay, batch size) to train DNNs for all uncertainty estimation techniques applied on this backbone.

By default, the BTS baseline does not output uncertainty. Similarly to [40,43], we can consider that a DNN may be constructed to find and output the parameters of a parametric distribution (e.g., the mean and variance for a Gaussian distribution). Such networks can be optimized by maximizing their log-likelihood. We denote the result as single predictive uncertainty (Single-PU). We also train a Deep Ensembles [45] by ensembling 3 DNNs, as well as a MC-Dropout [27] with eight forward passes. Without the extra DNNs or training procedures, we also applied Infer-noise [54], which injects Gaussian noise layers to the trained BTS baseline model and propagate eight times to predict the uncertainty.

We have also implemented LDU with the BTS model, but we note however that, in the monocular depth estimation setting and in agreement with previous works [18], the definition of OOD is fundamentally different with respect to the

Table 7. Comparative results for training (forward+backward) and inference wall-clock timings and number of parameters for evaluated methods. Timings are computed per image and averaged over 100 images.

Method	Semantic segmentation			Monocular depth		
	Runtime (ms)	Training time (ms)	#param.	Runtime (ms)	Training time (ms)	#param.
Baseline	14	166.4	39.76	45	92.8	47.00
Deep Ensembles [45]	56	499.2	119.28	133	287.8	141.03
MC-Dropout [27]	199	166.4	39.76	370	92.3	47.00
Single-PU [43]	-	-	-	45	95.6	47.01
LDU (ours)	14	177.8	39.76	45	104.0	47.00

tasks introduced in the prior experiments. Thus, our objective is to investigate whether LDU is robust, can improve the prediction accuracy and still perform well for aleatoric uncertainty estimation. Table 6 lists the depth and uncertainty estimation results on KITTI dataset. Using different settings of #p and λ, the proposed LDU is virtually aligned with the current state-of-the-art, while being significantly lighter computationally (see also Table 7). More ablation results on the influence of #p and λ can be found in the supplementary materials.

5   Discussions and Conclusions

Discussions. In Table 7 we compare the computational cost of LDU and related methods. For each approach we measure the training (forward+backward) and inference time per image on a NVIDIA RTX 3090Ti and report the corresponding number of parameters. We report training and inference wall-clock timings averaged over 100 training and validation. We use the same backbones as mentioned in Sect. 4.2 and Sect. 4.3 for semantic segmentation and monocular depth estimation respectively. We note that the runtime of LDU is almost the same as that of the baseline model (standard single forward model). This underpins the efficiency of our approach during inference, a particularly important requirement for practical applications.

Conclusions. In this work, we propose a simple way to modify a DNN to better estimate its predictive uncertainty. These minimal changes consist in optimizing a set of latent prototypes to learn to quantify the uncertainty by analyzing the position of an input sample in this prototype space. We perform extensive experiments and show that LDU can outperform state-of-the-art DUMs in most tasks and reach results comparable to Deep Ensembles with a significant advantage in terms of computational efficiency and memory requirements.

Along with the current state of the art methods, a limitation of our proposed LDU is that despite the empirical improvements in uncertainty quantification, it does not provide theoretical guarantees on the correctness of the predicted uncertainty. Our perspectives concern further exploration and improvements of the regularization strategies introduced in LDU on the latent feature representation that would allow us to bound the model error while still preserving its main task high performance.

Acknowledgments. This work was performed using HPC resources from GENCI-IDRIS (Grant 2020-AD011011970) and (Grant 2021-AD011011970R1) and Saclay-IA computing platform.

References

1. Aggarwal, C.C., Hinneburg, A., Keim, D.A.: On the surprising behavior of distance metrics in high dimensional space. In: Van den Bussche, J., Vianu, V. (eds.) ICDT 2001. LNCS, vol. 1973, pp. 420–434. Springer, Heidelberg (2001). https://doi.org/10.1007/3-540-44503-X_27
2. Alemi, A.A., Fischer, I., Dillon, J.V.: Uncertainty in the variational information bottleneck. In: UAI (2018)
3. van Amersfoort, J., Smith, L., Jesson, A., Key, O., Gal, Y.: Improving deterministic uncertainty estimation in deep learning for classification and regression. arXiv preprint arXiv:2102.11409 (2021)
4. Arjovsky, M., Chintala, S., Bottou, L.: Wasserstein generative adversarial networks. In: ICLR (2017)
5. Asai, A., Ikami, D., Aizawa, K.: Multi-task learning based on separable formulation of depth estimation and its uncertainty. In: CVPR Workshops (2019)
6. Assent, I.: Clustering high dimensional data. KDD (2012)
7. Behrmann, J., Grathwohl, W., Chen, R.T., Duvenaud, D., Jacobsen, J.H.: Invertible residual networks. In: ICML (2019)
8. Besnier, V., Bursuc, A., Picard, D., Briot, A.: Triggering failures: out-of-distribution detection by learning from local adversarial attacks in semantic segmentation. In: ICCV (2021)
9. Bhat, S.F., Alhashim, I., Wonka, P.: AdaBins: depth estimation using adaptive bins. In: CVPR (2021)
10. Blundell, C., Cornebise, J., Kavukcuoglu, K., Wierstra, D.: Weight uncertainty in neural network. In: ICML (2015)
11. Boucheron, S., Lugosi, G., Massart, P.: Concentration Inequalities: A Nonasymptotic Theory of Independence. Oxford University Press, Oxford (2013)
12. Cao, Z., Hidalgo Martinez, G., Simon, T., Wei, S., Sheikh, Y.A.: OpenPose: real-time multi-person 2D pose estimation using part affinity fields. TPAMI (2019)
13. Chen, G., et al.: Learning open set network with discriminative reciprocal points. In: Vedaldi, A., Bischof, H., Brox, T., Frahm, J.-M. (eds.) ECCV 2020. LNCS, vol. 12348, pp. 507–522. Springer, Cham (2020). https://doi.org/10.1007/978-3-030-58580-8_30
14. Chen, L.-C., Zhu, Y., Papandreou, G., Schroff, F., Adam, H.: Encoder-decoder with atrous separable convolution for semantic image segmentation. In: Ferrari, V., Hebert, M., Sminchisescu, C., Weiss, Y. (eds.) ECCV 2018. LNCS, vol. 11211, pp. 833–851. Springer, Cham (2018). https://doi.org/10.1007/978-3-030-01234-2_49
15. Corbière, C., Thome, N., Bar-Hen, A., Cord, M., Pérez, P.: Addressing failure prediction by learning model confidence. In: NeurIPS (2019)
16. Cordts, M., et al.: The cityscapes dataset for semantic urban scene understanding. In: CVPR (2016)
17. Der Kiureghian, A., Ditlevsen, O.: Aleatory or epistemic? Does it matter? Structural safety (2009)
18. Dijk, T.v., Croon, G.d.: How do neural networks see depth in single images? In: ICCV (2019)

19. Du, X., Wang, Z., Cai, M., Li, Y.: VOS: learning what you don't know by virtual outlier synthesis. In: ICLR (2022)
20. Dusenberry, M., et al.: Efficient and scalable Bayesian neural nets with rank-1 factors. In: ICML (2020)
21. Eigen, D., Puhrsch, C., Fergus, R.: Depth map prediction from a single image using a multi-scale deep network. In: NeurIPS (2014)
22. Everingham, M., et al.: The PASCAL visual object classes challenge: a retrospective. Int. J. Comput. Vis. **111**(1), 98–136 (2014). https://doi.org/10.1007/s11263-014-0733-5
23. Fort, S., Hu, H., Lakshminarayanan, B.: Deep ensembles: a loss landscape perspective. arXiv preprint arXiv:1912.02757 (2019)
24. Franchi, G., Belkhir, N., Ha, M.L., Hu, Y., Bursuc, A., Blanz, V., Yao, A.: Robust semantic segmentation with superpixel-mix. In: BMVC (2021)
25. Franchi, G., Bursuc, A., Aldea, E., Dubuisson, S., Bloch, I.: TRADI: tracking deep neural network weight distributions. In: Vedaldi, A., Bischof, H., Brox, T., Frahm, J.-M. (eds.) ECCV 2020. LNCS, vol. 12362, pp. 105–121. Springer, Cham (2020). https://doi.org/10.1007/978-3-030-58520-4_7
26. Franchi, G., et al.: MUAD: multiple uncertainties for autonomous driving benchmark for multiple uncertainty types and tasks. arXiv preprint arXiv:2203.01437 (2022)
27. Gal, Y., Ghahramani, Z.: Dropout as a Bayesian approximation: representing model uncertainty in deep learning. In: ICML (2016)
28. Gidaris, S., Komodakis, N.: Dynamic few-shot visual learning without forgetting. In: CVPR (2018)
29. Gong, D., et al.: Memorizing normality to detect anomaly: memory-augmented deep autoencoder for unsupervised anomaly detection. In: ICCV (2019)
30. Gulrajani, I., Ahmed, F., Arjovsky, M., Dumoulin, V., Courville, A.: Improved training of Wasserstein GANs. In: NeurIPS (2017)
31. Guo, C., Pleiss, G., Sun, Y., Weinberger, K.Q.: On calibration of modern neural networks. In: ICML (2017)
32. Gustafsson, F.K., Danelljan, M., Schon, T.B.: Evaluating scalable Bayesian deep learning methods for robust computer vision. In: CVPR Workshops (2020)
33. He, K., Zhang, X., Ren, S., Sun, J.: Deep residual learning for image recognition. In: CVPR (2016)
34. Hendrycks, D., Basart, S., Mazeika, M., Mostajabi, M., Steinhardt, J., Song, D.: A benchmark for anomaly segmentation. arXiv preprint arXiv:1911.11132 (2019)
35. Hendrycks, D., Dietterich, T.: Benchmarking neural network robustness to common corruptions and perturbations. In: ICLR (2019)
36. Hendrycks, D., Gimpel, K.: A baseline for detecting misclassified and out-of-distribution examples in neural networks. In: ICLR (2017)
37. Hernández-Lobato, J.M., Adams, R.: Probabilistic backpropagation for scalable learning of Bayesian neural networks. In: ICML (2015)
38. Hu, S., Pezzotti, N., Welling, M.: Learning to predict error for MRI reconstruction. In: de Bruijne, M., et al. (eds.) MICCAI 2021. LNCS, vol. 12903, pp. 604–613. Springer, Cham (2021). https://doi.org/10.1007/978-3-030-87199-4_57
39. Huang, G., Liu, Z., Van Der Maaten, L., Weinberger, K.Q.: Densely connected convolutional networks. In: CVPR (2017)
40. Ilg, E., et al.: Uncertainty estimates and multi-hypotheses networks for optical flow. In: Ferrari, V., Hebert, M., Sminchisescu, C., Weiss, Y. (eds.) ECCV 2018. LNCS, vol. 11211, pp. 677–693. Springer, Cham (2018). https://doi.org/10.1007/978-3-030-01234-2_40

41. Kamann, C., Rother, C.: Benchmarking the robustness of semantic segmentation models with respect to common corruptions. IJCV (2021)

42. Kendall, A., Badrinarayanan, V., Cipolla, R.: Bayesian SegNet: model uncertainty in deep convolutional encoder-decoder architectures for scene understanding. arXiv preprint arXiv:1511.02680 (2015)

43. Kendall, A., Gal, Y.: What uncertainties do we need in Bayesian deep learning for computer vision? In: NeurIPS (2017)

44. Krizhevsky, A., Hinton, G., et al.: Learning multiple layers of features from tiny images. Technical report (2009)

45. Lakshminarayanan, B., Pritzel, A., Blundell, C.: Simple and scalable predictive uncertainty estimation using deep ensembles. In: NeurIPS (2017)

46. Lee, J.H., Han, M.K., Ko, D.W., Suh, I.H.: From big to small: multi-scale local planar guidance for monocular depth estimation. arXiv preprint arXiv:1907.10326 (2019)

47. Liu, C.L., Nakagawa, M.: Evaluation of prototype learning algorithms for nearest-neighbor classifier in application to handwritten character recognition. Pattern Recognit. **34**, 601–615

48. Liu, J.Z., Lin, Z., Padhy, S., Tran, D., Bedrax-Weiss, T., Lakshminarayanan, B.: Simple and principled uncertainty estimation with deterministic deep learning via distance awareness. In: NeurIPS (2020)

49. Macêdo, D., Ren, T.I., Zanchettin, C., Oliveira, A.L., Ludermir, T.: Entropic out-of-distribution detection. In: IJCNN (2021)

50. Macêdo, D., Zanchettin, C., Ludermir, T.: Distinction maximization loss: efficiently improving classification accuracy, uncertainty estimation, and out-of-distribution detection simply replacing the loss and calibrating. arXiv preprint arXiv:2205.05874 (2022)

51. Malinin, A., Gales, M.: Predictive uncertainty estimation via prior networks (2018)

52. Malinin, A., Mlodozeniec, B., Gales, M.: Ensemble distribution distillation. In: ICLR (2020)

53. Mandelbaum, A., Weinshall, D.: Distance-based confidence score for neural network classifiers. arXiv preprint arXiv:1709.09844 (2017)

54. Mi, L., Wang, H., Tian, Y., Shavit, N.: Training-free uncertainty estimation for dense regression: Sensitivity as a surrogate. In: AAAI (2022)

55. Miyato, T., Kataoka, T., Koyama, M., Yoshida, Y.: Spectral normalization for generative adversarial networks. In: ICLR (2018)

56. Mukhoti, J., Kirsch, A., van Amersfoort, J., Torr, P.H., Gal, Y.: Deterministic neural networks with appropriate inductive biases capture epistemic and aleatoric uncertainty. In: ICML Workshops (2021)

57. Mukhoti, J., Kulharia, V., Sanyal, A., Golodetz, S., Torr, P.H., Dokania, P.K.: Calibrating deep neural networks using focal loss. In: NeurIPS (2020)

58. Netzer, Y., Wang, T., Coates, A., Bissacco, A., Wu, B., Ng, A.Y.: Reading digits in natural images with unsupervised feature learning. In: NeurIPS (2011)

59. Nix, D., Weigend, A.: Estimating the mean and variance of the target probability distribution. In: ICNN (1994)

60. Ovadia, Y., et al.: Can you trust your model's uncertainty? Evaluating predictive uncertainty under dataset shift. In: NeurIPS (2019)

61. Padhy, S., Nado, Z., Ren, J., Liu, J., Snoek, J., Lakshminarayanan, B.: Revisiting one-vs-all classifiers for predictive uncertainty and out-of-distribution detection in neural networks. In: ICML Workshops (2020)

62. Pinto, F., Yang, H., Lim, S.N., Torr, P., Dokania, P.K.: Mix-MaxEnt: improving accuracy and uncertainty estimates of deterministic neural networks. In: NeurIPS Workshops (2021)
63. Poggi, M., Aleotti, F., Tosi, F., Mattoccia, S.: On the uncertainty of self-supervised monocular depth estimation. In: CVPR (2020)
64. Postels, J., et al.: The hidden uncertainty in a neural networks activations. In: ICML (2021)
65. Postels, J., Segu, M., Sun, T., Van Gool, L., Yu, F., Tombari, F.: On the practicality of deterministic epistemic uncertainty. arXiv preprint arXiv:2107.00649 (2021)
66. Qi, H., Brown, M., Lowe, D.G.: Low-shot learning with imprinted weights. In: CVPR (2018)
67. Razavi, A., Van den Oord, A., Vinyals, O.: Generating diverse high-fidelity images with VQ-VAE-2. In: NeurIPS (2019)
68. Rebut, J., Bursuc, A., Pérez, P.: StyleLess layer: improving robustness for real-world driving. In: IROS (2021)
69. Rogez, G., Weinzaepfel, P., Schmid, C.: LCR-Net++: multi-person 2D and 3D pose detection in natural images. TPAMI (2019)
70. Snell, J., Swersky, K., Zemel, R.S.: Prototypical networks for few-shot learning. In: NeurIPS (2017)
71. Sun, D., Yang, X., Liu, M.Y., Kautz, J.: PWC-Net: CNNs for optical flow using pyramid, warping, and cost volume. In: CVPR (2018)
72. Teed, Z., Deng, J.: RAFT: recurrent all-pairs field transforms for optical flow. In: Vedaldi, A., Bischof, H., Brox, T., Frahm, J.-M. (eds.) ECCV 2020. LNCS, vol. 12347, pp. 402–419. Springer, Cham (2020). https://doi.org/10.1007/978-3-030-58536-5_24
73. Uhrig, J., Schneider, N., Schneider, L., Franke, U., Brox, T., Geiger, A.: Sparsity invariant CNNs. In: 3DV (2017)
74. Van Amersfoort, J., Smith, L., Teh, Y.W., Gal, Y.: Uncertainty estimation using a single deep deterministic neural network. In: ICML (2020)
75. Van Den Oord, A., Vinyals, O., et al.: Neural discrete representation learning. In: NeurIPS (2017)
76. Wen, Y., Zhang, K., Li, Z., Qiao, Yu.: A discriminative feature learning approach for deep face recognition. In: Leibe, B., Matas, J., Sebe, N., Welling, M. (eds.) ECCV 2016. LNCS, vol. 9911, pp. 499–515. Springer, Cham (2016). https://doi.org/10.1007/978-3-319-46478-7_31
77. Wen, Y., Tran, D., Ba, J.: BatchEnsemble: an alternative approach to efficient ensemble and lifelong learning. In: ICLR (2020)
78. Wilson, A.G., Izmailov, P.: Bayesian deep learning and a probabilistic perspective of generalization. arXiv preprint arXiv:2002.08791 (2020)
79. Wu, M., Goodman, N.: A simple framework for uncertainty in contrastive learning. arXiv preprint arXiv:2010.02038 (2020)
80. Yang, H.M., Zhang, X.Y., Yin, F., Liu, C.L.: Robust classification with convolutional prototype learning. In: CVPR (2018)
81. Yu, X., Franchi, G., Aldea, E.: SLURP: Side learning uncertainty for regression problems. In: BMVC (2021)
82. Zhao, H., Shi, J., Qi, X., Wang, X., Jia, J.: Pyramid scene parsing network. In: CVPR (2017)

Making Heads or Tails: Towards Semantically Consistent Visual Counterfactuals

Simon Vandenhende[(✉)], Dhruv Mahajan, Filip Radenovic,
and Deepti Ghadiyaram

Meta AI, Menlo Park, USA
svandenh@fb.com

Abstract. A visual counterfactual explanation replaces image regions in a query image with regions from a distractor image such that the system's decision on the transformed image changes to the distractor class. In this work, we present a novel framework for computing visual counterfactual explanations based on two key ideas. First, we enforce that the *replaced* and *replacer* regions contain the same semantic part, resulting in more semantically consistent explanations. Second, we use multiple distractor images in a computationally efficient way and obtain more discriminative explanations with fewer region replacements. Our approach is **27%** more semantically consistent and an order of magnitude faster than a competing method on three fine-grained image recognition datasets. We highlight the utility of our counterfactuals over existing works through machine teaching experiments where we teach humans to classify different bird species. We also complement our explanations with the vocabulary of parts and attributes that contributed the most to the system's decision. In this task as well, we obtain state-of-the-art results when using our counterfactual explanations relative to existing works, reinforcing the importance of semantically consistent explanations. Source code is available at github.com/facebookresearch/visual-counterfactuals.

1 Introduction

Explainable AI (XAI) research aims to develop tools that allow lay-users to comprehend the reasoning behind an AI system's decisions [34,61]. XAI tools are critical given the pervasiveness of computer vision technologies in various human-centric applications such as self-driving vehicles, healthcare systems, and facial recognition tools. These tools serve several purposes [2,57]: (i) they help users

---

F. Radenovic and D. Ghadiyaram—Equal contribution.

Supplementary Information The online version contains supplementary material available at https://doi.org/10.1007/978-3-031-19775-8_16.

S. Avidan et al. (Eds.): ECCV 2022, LNCS 13672, pp. 261–279, 2022.
https://doi.org/10.1007/978-3-031-19775-8_16

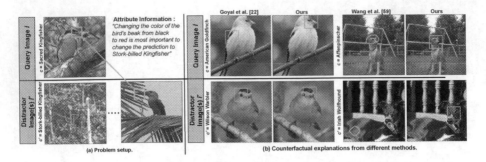

(a) Problem setup.   (b) Counterfactual explanations from different methods.

Fig. 1. Paper overview. (a) Given a query image I (top row) from class c, we provide counterfactual explanations relative to a distractor image I' (bottom row) from class c'. The explanations highlight what regions in I should be replaced from I' for the transformed image to be classified as c'. We also use attribute information to identify the region attributes that contributed the most for a counterfactual. **(b)** Unlike [22,59], our explanations identify regions that are both discriminative and semantically similar.

understand why a decision was reached thereby making systems more transparent, (ii) they allow system developers to improve their system, and (iii) they offer agency to users affected by the system's decision to change the outcome.

One intuitive way to explain a system's decision is through counterfactual explanations [56,57] which describe *in what way* a data instance would need to be different in order for the system to reach an *alternate* conclusion. In this work, we study counterfactual explanations for fine-grained image recognition tasks, where the most confusing classes are often hard to distinguish. The difficulty of this problem makes it a particularly well suited setting to study intuitive and human-understandable explanations. Figure 1-a presents a *query image I* and a *distractor image I'* belonging to the categories *Sacred Kingfisher* (c) and *Stork-billed Kingfisher* (c'), respectively. Given a black-box classification model, a counterfactual explanation aims to answer: "how should the query image I change for the model to predict c' instead of c?" To do this, we utilize the distractor image I' (or a set of distractor images) and identify which regions in I should be replaced with regions from I' for the model's prediction to be c'.

Counterfactual visual explanations are under-explored [22,59], and most popular XAI methods use saliency maps [17,21,38,44,63] or feature importance scores [18,29,33,41,42,51,65] to highlight what image regions or features most contribute to a model's decision. Unlike counterfactual explanations, these methods do not consider alternate scenarios which yield a different result. Additionally, some of these methods [33,41,42] extract explanations via a local model approximation, leading to explanations that are *unfaithful* [3,50], i.e., they misrepresent the model's behavior. By contrast, current counterfactual explanations are faithful by design as they operate on the original model's output to generate explanations. Further, counterfactuals share similarities with how children learn about a concept – by contrasting with other related concepts [9,11]. As studied in [35,56,57], an ideal counterfactual should have the following properties: (i) the highlighted regions in the images I, I' should be <u>discriminative</u> of their respective classes; (ii) the counterfactual should be sensible in that the replaced regions should be <u>semantically consistent</u>, i.e., they correspond to the same object parts; and, (iii) the counterfactual should make as few changes as possible to the query image I as humans find sparse explanations <u>easier to understand</u>.

Prior works [22,59] proposed ways to identify the most discriminative image regions to generate counterfactual explanations. However, naively applying this principle can yield degenerate solutions that are semantically inconsistent. Figure 1-b visualizes such scenarios, where prior works [22,59] replace image regions corresponding to different object parts (e.g., [22] replaces bird's wing in I with a head in I'). Further, these methods rely on a single distractor image I', which often limits the variety of discriminative regions to choose from, leading to explanations that are sometimes less discriminative hence uninformative.

This paper addresses these shortcomings. Specifically, we propose a novel and computationally efficient framework that produces both discriminative and semantically consistent counterfactuals. Our method builds on two key ideas. First, we constrain the identified class-specific image regions that alter a model's decision to allude to the same semantic parts, yielding more semantically consistent explanations. Since we only have access to object category labels, we impose this as a soft constraint in a separate auxiliary feature space learned in a self-supervised way. Second, contrary to prior works, we expand the search space by using multiple distractor images from a given class leading to more discriminative explanations with fewer regions to replace. However, naively extending to multiple distractor images poses a computational bottleneck. We address this by constraining the processing to only the most similar regions by once again leveraging the soft constraint, resulting in an order of magnitude speedup.

Our approach significantly outperforms the s-o-t-a [22,59] across several metrics on three datasets – CUB [58], Stanford-Dogs [28], and iNaturalist-2021 [55] and yields more semantically consistent counterfactuals (Fig. 1-b). While prior work [22] suffers computationally when increasing the number of distractor images, the optimization improvements introduced in our method make it notably efficient. We also study the properties of the auxiliary feature space and justify our design choices. Further, we show the importance of generating semantically consistent counterfactuals via a machine teaching task where we teach lay-humans to recognize bird species. We find that humans perform better when provided with our semantically consistent explanations relative to others [22,59].

We further reinforce the importance of semantically consistent counterfactuals by proposing a method to complement our explanations with the vocabulary of parts and attributes. Consider Fig. 1-a, where the counterfactual changes both the color of the beak and forehead. Under this setup, we provide nameable parts and attributes corresponding to the selected image regions and inform what attributes contributed the most to the model's decision. For example, in Fig. 1-a, our explanation highlights that the beak's color mattered the most. We find that our explanations identify class discriminative attributes – those that belong to class c but not to c', or vice versa – and are more interpretable.

In summary, our contributions are: **(i)** we present a framework to compute semantically consistent and faithful counterfactual explanations by enforcing the model to only replace semantically matching image regions (Sect. 3.2), **(ii)** we leverage multiple distractor images in a computationally efficient way, achieve an order of magnitude speedup, and generate more discriminative and sparse explanations (Sect. 3.3), **(iii)** we highlight the utility of our framework through extensive experiments (Sect. 4.2 and 4.3) and a human-in-the-loop evaluation through machine teaching (Sect. 4.4), **(iv)** we augment visual counterfactuals with nameable part and attribute information (Sect. 5).

2 Related Work

Feature attribution methods [6] rely on the back propagation algorithm [8,40,44–46,62,63] or input perturbations [15,17,20,21,38,65] to identify the image regions that are most important to a model's decision. However, none of these methods can tell how the image should change to get a different outcome. **Counterfactual explanations** [36,39,56,57] transform a query image I of class c such that the model predicts class c' on the transformed image. In computer vision, several works [5,25,26,31,32,43,48,49] used a generative model to synthesize counterfactual examples. However, the difficulties of realistic image synthesis can limit these methods [25,32,43,48] to small-scale problems. A few works [5,26,49] guided the image generation process via pixel-level supervision to tackle more complex scenes. StyleEx [31] uses the latent space of a Style-GAN [27] to identify the visual attributes that underlie the classifier's decision. Despite these efforts, it remains challenging to synthesize realistic counterfactual examples. Our method does not use a generative model but is more related to the works discussed next.

A second group of works [4,22,59] finds the regions or concepts in I that should be changed to get a different outcome. CoCoX [4] identifies visual concepts to add or remove to change the prediction. Still, the most popular methods [22,59] use a distractor image I' from class c' to find and replace the regions in I that change the model's prediction to c'. SCOUT [59] finds these regions via attribute maps. Goyal *et al.* [22] use spatial features of the images to construct counterfactuals. These methods have two key advantages. First, the distractor images are often readily available and thus inexpensive to obtain compared to pixel-level annotations [5,26,49]. Second, these methods fit well with fine-grained recognition tasks, as they can easily identify the distinguishing elements between classes. Our framework follows a similar strategy but differs in two crucial components. First, we enforce that the replaced regions are semantically consistent. Second, our method leverages multiple distractor images in an efficient way.

3   Method

Our key goal is to: (i) generate a counterfactual that selects discriminative and semantically consistent regions in I and I' without using additional annotations, (ii) leverage multiple distractor images efficiently. We first review the foundational method [22] for counterfactual generation that our framework builds on and then introduce our approach, illustrated in Fig. 2.

3.1   Counterfactual Problem Formulation: Preliminaries

Consider a deep neural network with two components: a spatial feature extractor f and a decision network g. Note that any neural network can be divided into such components by selecting an arbitrary layer to split at. In our setup, we split a network after the final down-sampling layer. The spatial feature extractor $f : \mathcal{I} \to \mathbb{R}^{hw \times d}$ maps the image to a $h \times w \times d$ dimensional spatial feature, reshaped to a $hw \times d$ spatial cell matrix, where h and w denote the spatial dimensions and d the number of channels. The decision network $g : \mathbb{R}^{hw \times d} \to \mathbb{R}^{|\mathcal{C}|}$ takes the spatial cells and predicts probabilities over the output space $\mathcal{C}$. Further, let query and distractor image $I, I' \in \mathcal{I}$ with class predictions $c, c' \in \mathcal{C}$.

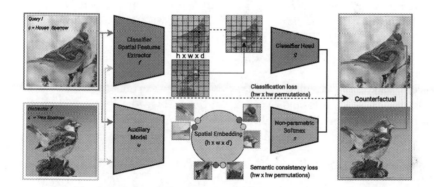

Fig. 2. Our counterfactual explanation identifies regions in a query image I from class c and a distractor image I' from class c' such that replacing the regions in I with the regions in I' changes the model's outcome to c'. Instead of considering actual image regions, we operate on $h \times w$ cells in the spatial feature maps. The cells are selected based upon: (i) a classification loss that increases the predicted probability $g_{c'}$ of class c' and (ii) a semantic consistency loss that selects cells containing the same semantic parts. We use a self-supervised auxiliary model to compute the semantic loss.

Following [22], we construct a counterfactual I^* in the feature space $f(.)$ by replacing spatial cells in $f(I)$ with cells from $f(I')$ such that the classifier predicts c' for I^*. This is done by first rearranging the cells in $f(I')$ to align with $f(I)$ using a permutation matrix $P \in \mathbb{R}^{hw \times hw}$, then selectively replacing entries in $f(I)$ according to a sparse binary gating vector $\mathbf{a} \in \mathbb{R}^{hw}$. Let $\circ$ denote the Hadamard product. The transformed image I^* can be written as:

$$f(I^*) = (\mathbb{1} - \mathbf{a}) \circ f(I) + \mathbf{a} \circ Pf(I') \qquad (1)$$

Classification Loss: Recall that our first goal is to identify class-specific image regions in I and I' such that replacing the regions in I with those in I' increases the predicted probability $g_{c'}(.)$ of class c' for I^*. To avoid a trivial solution where all cells of I are replaced, a sparsity constraint is applied on $\mathbf{a}$ to minimize the number of cell edits (m). Following the greedy approach from [22], we iteratively replace spatial cells in I by repeatedly solving Eq. 2 that maximizes the predicted probability $g_{c'}(\cdot)$ until the model's decision changes.

$$\max_{P,a} g_{c'}((\mathbb{1} - \mathbf{a}) \circ f(I) + \mathbf{a} \circ Pf(I')) \text{ with } ||\mathbf{a}||_1 = 1 \text{ and } a_i \in \{0, 1\} \qquad (2)$$

We evaluate $g_{c'}$ for each of the h^2w^2 permutations constructed by replacing a single cell in $f(I)$ with an arbitrary cell in $f(I')$. The computational complexity is $2 \cdot C_f + mh^2w^2 \cdot C_g$, where C_f and C_g denote the cost of f and g respectively.

Equation 2 does not guarantee that the replaced cells are semantically similar. For example, in the task of bird classification, the counterfactual could replace the wing in I with head in I' (e.g., Fig. 1-b) leading to nonsensical explanations. We address this problem via a semantic consistency constraint, described next.

3.2 Counterfactuals with a Semantic Consistency Constraint

Consider an embedding model $u : \mathcal{I} \to \mathbb{R}^{hw \times d'}$ that brings together spatial cells belonging to the same semantic parts and separates dissimilar cells. Let $u(I)_i$ denote the feature of the i-th cell in I. We estimate the likelihood that cell i of I semantically matches with cell j of I' by:

$$\mathcal{L}_s(u(I)_i, u(I')_j) = \frac{\exp(u(I)_i \cdot u(I')_j / \tau)}{\sum_{j' \in u(I')} \exp(u(I)_i \cdot u(I')_{j'} / \tau)}, \qquad (3)$$

where τ is a temperature hyper-parameter that relaxes the dot product. Equation 3 estimates a probability distribution of a given query cell i over all distractor cells j' using a non-parametric softmax function and indicates what distractor cells are most likely to contain semantically similar regions as the query cell i. Like the classification loss (Eq. 2), we compute the semantic loss for all h^2w^2 cell permutations. Thus, the complexity is $2 \cdot C_u + h^2w^2 \cdot C_{\text{dot}}$, where C_u, C_{dot} denote the cost of the auxiliary model u and the dot-product operation respectively. Empirically, we observe that dot-products are very fast to compute and the semantic loss adds a tiny overhead to the overall computation time. Note that unlike the classification loss which is computed for each edit, $\mathcal{L}_s$ is computed only once in practice, i.e., the cost gets amortized for multiple edits.

Total Loss: We combine both losses to find the single best cell edit:

$$\max_{P,a} \quad \log \quad \underbrace{g_{c'}((1 - a) \circ f(I) + a \circ Pf(I'))}_{\text{Classification loss } \mathcal{L}_c} \quad + \quad \lambda \cdot \log \underbrace{\mathcal{L}_s(a^T u(I), a^T Pu(I'))}_{\text{Semantic consistency loss } \mathcal{L}_s}$$

$$\text{with } P \in \mathbb{R}^{hw \times hw}, \|a\|_1 = 1 \text{ and } a_i \in \{0, 1\}, \text{ and } \lambda \text{ balances } \mathcal{L}_c \text{ and } \mathcal{L}_s.$$

$$(4)$$

We reiterate that $\mathcal{L}_c$ optimizes to find class-specific regions while $\mathcal{L}_s$ ensures that these regions semantically match. We also stress that our explanations are faithful with respect to the underlying deep neural network, since, the proposed auxiliary model, irrespective of the value of λ, only acts as a regularizer and does not affect the class predictions of the transformed images.

Choice of Auxiliary Model: An obvious choice is to use the spatial feature extractor f as the auxiliary model u. We empirically found that since f is optimized for an object classification task, it results in an embedding space that often separates instances of similar semantic parts and is thus unfit to model region similarity. We found that self-supervised models are more appropriate as auxiliary models for two reasons: a) they eliminate the need for part location information, b) several recent studies [14,52,54] showed that self-supervised models based on contrastive learning [16,23,60] or clustering [7,12,13,53] learn richer representations that capture the semantic similarity between local image regions as opposed to task-related similarity in a supervised setup. Such representations have been valuable for tasks such as semantic segment retrieval [52]. Thus, the resulting embedding space inherently brings together spatial cells belonging to the same semantic parts and separates dissimilar cells (see Table 4).

3.3 Using Multiple Distractor Images Through a Semantic Constraint

Recall, the method uses spatial cells from $f(I')$ to iteratively construct $f(I^*)$. Thus, the quality of the counterfactual is sensitive to the chosen distractor image I'. Having to select regions from a single distractor image can limit the variety of discriminative parts to choose from due to factors like pose, scale, and occlusions. We address this limitation by leveraging multiple distractor images from class c'. In this way, we expand our search space in Eq. 4, allowing us to find highly discriminative regions that semantically match, while requiring fewer edits.

However, leveraging (n) multiple distractor images efficiently is not straightforward as it poses a significant computational overhead. This is because, in this new setup, for each edit we can pick any of $n \times hw$ cells from the n distractor images. This makes the spatial cell matrix of the distractor images of shape $nhw \times d$, the matrix P $hw \times nhw$, and $a \in \mathbb{R}^{hw}$. $\mathcal{L}_c$ (Eq. 2) with a single distractor image is already expensive to evaluate due to: (i) its quadratic dependence on hw making the cell edits memory intensive and, (ii) the relatively high cost of evaluating g, involving at least one fully-connected plus zero or more conv layers. This computation gets amplified by a factor n with multiple distractor images.

On the other hand, $\mathcal{L}_s$ (Eq. 3) is computationally efficient as: (i) it does not involve replacing cells and (ii) the dot-product is inexpensive to evaluate. Thus, we first compute $\mathcal{L}_s$ (Eq. 3) to select the top-$k\%$ cell permutations with the lowest loss, excluding the ones that are likely to replace semantically dissimilar cells. Next, we compute $\mathcal{L}_c$ (Eq. 2) only on these selected top-$k\%$ permutations. With this simple trick, we get a significant overall speedup by a factor k (detailed analysis in suppl.). Thus, our overall framework leverages richer information, produces semantically consistent counterfactuals, and is about an order of magnitude faster than [22]. Note that the multi-distractor setup can be extended to [22] but not to SCOUT [59], as the latter was designed for image pairs.

4   Experiments

4.1   Implementation Details and Evaluation Setup

Implementation Details: We evaluate our approach on top of two backbones – VGG-16 [46] for fair comparison with [22] and ResNet-50 [24] for generalizability. As mentioned in Sect. 3.1, we split both networks into components f and g after the final down-sampling layer max_pooling2d_5 in VGG-16 and at conv5_1 in ResNet-50. The input images are of size 224×224 pixels and the output features of f have spatial dimensions 7×7. We examine counterfactual examples for query-distractor class pairs obtained via the confusion matrix – for a query class c, we select the distractor class c' as the class with which images from c are most often confused. This procedure differs from the approach in [22] which uses attribute annotations to select the classes c, c'. Our setup is more generic as it does not use extra annotations. Distractor images are picked randomly from c'.

Auxiliary Model: We adopt the pre-trained ResNet-50 [24] model from Deep-Cluster [13] to measure the semantic similarity of the spatial cells. We remove the final pooling layer and apply up- or down-sampling to match the 7×7 spatial dimensions of features from f. As in [13], we use $\tau = 0.1$ in the non-parametric softmax (Eq. 3). The weight $\lambda = 0.4$ (Eq. 4) is found through grid search. We set $k = 10$ and select top-10% most similar cell pairs to pre-filter.

Evaluation Metrics: We follow the evaluation procedure from [22] and report the following metrics using keypoint part annotations.

- **Near-KP:** measures if the image regions contain object keypoints (KP). This is a proxy for how often we select discriminative cells, i.e., spatial cells that can explain the class differences.
- **Same-KP:** measures how often we select the same keypoints in the query and distractor image, thus measures semantic consistency of counterfactuals.
- **#Edits:** the average number of edits until the classification model predicts the distractor class c' on the transformed image I^*.

Datasets: We evaluate the counterfactuals on three datasets for fine-grained image classification (see Table 1). The CUB dataset [58] consists of images of 200 bird classes. All images are annotated with keypoint locations of 15 bird parts. The iNaturalist-2021 birds dataset [55] contains 1,486 bird classes and more challenging scenes compared to CUB, but lacks keypoint annotations. So we hired raters to annotate bird keypoint locations for 2,060 random val images from iNaturalist-2021 birds and evaluate on this subset. Stanford Dogs [28] contains images of dogs annotated with keypoint locations [10] of 24 parts. The explanations are computed on the validation splits of these datasets.

Table 1. Datasets overview.

Dataset	Statistics			Top-1	
	#Class	#Train	#Val	VGG-16	Res-50
CUB	200	5,994	5,794	81.5	82.0
iNat. (Birds)	1,486	414 k	14,860	78.6	78.8
Stanf. Dogs	120	12 k	8,580	86.7	88.4

4.2   State-of-the-Art Comparison

Table 2 compares our method to other competing methods. We report the results for both (i) the **single edit** found by solving Eq. 4 once and (ii) **all edits** found by repeatedly solving Eq. 4 until the model's decision changes. Our results are directly comparable with [22]. By contrast, SCOUT [59] returns heatmaps that require post-processing. We follow the post-processing from [59] where from the heatmaps, select those regions in I and I' that match the area of a single cell edit to compute the metrics. From Table 2, we observe that our method consistently outperforms prior works across all metrics and datasets. As an example, consider the **all edits** rows for the CUB dataset in Table 2a. The Near-KP metric improved by **13.9%** over [22], indicating that our explanations select more discriminative image regions. More importantly, the Same-KP metric improved by **27%** compared to [22], demonstrating that our explanations are significantly more semantically consistent. The average number of edits have also reduced from 5.5 in [22] to **3.9**, meaning that our explanations require fewer changes to I and are thus sparser, which is a desirable property of counterfactuals [35,57]. Similar performance trends hold on the other two datasets and architectures (Table 2b) indicating the generalizability of the proposed approach. Figure 3 shows a few qualitative examples where we note that our method consistently identifies semantically matched and class-specific image regions, while explanations from [22] and [59] often select regions belonging to different parts.

Table 2. State-of-the-art comparison against our full proposed pipeline.

	Method	CUB-200-2011			INaturalist-2021 Birds			Stanford Dogs Extra		
		Near-KP	Same-KP	# Edits	Near-KP	Same-KP	# Edits	Near-KP	Same-KP	# Edits
(a) Comparison of visual counterfactuals using a VGG-16 model										
Single edit	SCOUT [59]	68.1	18.1	–	74.3	23.1	–	41.7	5.5	–
	Goyal et al. [22]	67.8	17.2	–	78.3	29.4	–	42.6	6.8	–
	Ours	**73.5**	**39.6**	–	**83.6**	**51.0**	–	**49.8**	**23.5**	–
All edits	Goyal et al. [22]	54.6	8.3	5.5	55.2	11.5	5.5	35.7	3.7	**6.3**
	Ours	**68.5**	**35.3**	**3.9**	**70.4**	**36.9**	**4.3**	**37.5**	**16.4**	6.6
(b) Comparison of visual counterfactuals using a ResNet-50 model										
Single edit	SCOUT [59]	43.0	4.4	–	53.9	8.8	–	35.3	3.1	–
	Goyal et al. [22]	61.4	11.5	–	70.5	17.1	–	42.7	6.4	–
	Ours	**71.7**	**36.1**	–	**79.2**	**33.3**	–	**51.2**	**22.6**	–
All edits	Goyal et al. [22]	50.9	6.8	3.6	56.3	10.4	3.3	34.9	3.6	**4.3**
	Ours	**60.3**	**30.2**	**3.2**	**70.9**	**32.1**	**2.6**	**37.2**	**16.7**	4.8

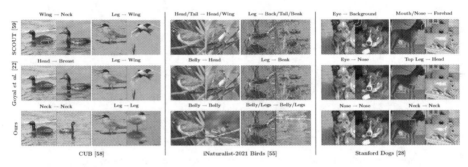

Fig. 3. State-of-the-art comparison of counterfactual explanations (Single Edit - VGG-16). Part labels are included only for better visualization. Image credit: [1].

4.3 Ablation Studies

We now study the different design choices of our framework with [22] as our baseline and use a VGG-16 model for consistent evaluation on CUB.

Analysis of Different Components. Table 3 reports different variants as we add or remove the following components: semantic loss (Sect. 3.2), multiple distractor images (Sect. 3.3), and pre-filtering cells (Sect. 3.3). Our baseline [22] (row 1) establishes a performance limit for the Near-KP and number of edits under the single-distractor setup as the image regions are selected solely based on the predicted class probabilities $g_{c'}(\cdot)$. First, we observe that the semantic loss improves the semantic meaningfulness of the replacements (row 2), i.e., the Same-KP metric increases by **13.7%**. However, the Near-KP slightly decreases by 2.5% and the number of edits increases by 1.3. This may be due to the fact that row 2 considers both the class probabilities $g_{c'}$ and semantic consistency, thereby potentially favoring semantically similar cells over dissimilar cells that yield a larger increase in $g_{c'}$. Second, from rows 1 and 3, we find that allowing multiple distractor images enlarges the search space when solving Eq. 4,

Table 3. Effect of different components of our method: Row 1 is our baseline from [22]. Our method (row 5) uses multiple distractor images combined pre-filtering irrelevant cells and semantic consistency loss. Time measured on a single V-100 GPU.

Row #	Semantic loss	Multi distractor	Filters cells	Near-KP	Same-KP	Time (s)	#Edits
1	✗	✗	✗	54.6	8.3	0.81	5.5
2	✓	✗	✗	52.1 (−2.5)	22.0 (+13.7)	1.02	6.8
3	✗	✓	✗	65.6 (+11.0)	13.8 (+5.5)	9.98	3.5
4	✓	✓	✗	69.2 (+14.6)	36.0 (+23.7)	10.82	3.8
5	✓	✓	✓	68.5 (+13.9)	35.3 (+23.0)	1.15	3.9

resulting in better solutions that are more discriminative (Near-KP ↑), more semantically consistent (Same-KP ↑) and sparser (fewer edits). Combining the semantic loss with multiple distractor images (row 4) further boosts the metrics. However, using multiple distractor images comes at a significant increase in runtime (almost by 10X). We address this by filtering out semantically dissimilar cell pairs. Indeed, comparing rows 4 and 5, we note that the runtime improves significantly while maintaining the performance. Putting everything together, our method outperforms [22] across all metrics (row 1 vs. row 5) and generates explanations that are sparser, more discriminative, and more semantically consistent.

Auxiliary Model: Recall from Sect. 3.2 that representations from self-supervised models efficiently capture richer semantic similarity between local image regions compared to those from supervised models. We empirically verify this by using different pre-training tasks to instantiate the auxiliary model: (i) supervised pre-training with class labels, (ii) self-supervised (SSL) pre-training [12,13,23] with no labels, and (iii) supervised parts detection with keypoint annotations. We train the parts detector to predict keypoint presence in the $h \times w$ spatial cell matrix using keypoint annotations. We stress that the parts detector is used only as an *upperbound* as it uses part ground-truth to model the semantic constraint.

We evaluate each auxiliary model by: (i) measuring the Same-KP metric to study if this model improves the semantic matching, and (ii) measuring clustering accuracy to capture the extent of semantic part disentanglement. To measure the clustering accuracy, we first cluster the d-dimensional cells in a 7×7 spatial matrix from $u(\cdot)$ of all images via K-Means and assign each spatial cell to a cluster. Then, we apply majority voting and associate each cluster with a semantic part using the keypoint annotations. The clustering accuracy measures how often the cells contain the assigned part. From Table 4, we observe that better part disentanglement (high clustering accuracy) correlates with improved semantic matching in the counterfactuals (high Same-KP). Thus, embeddings that disentangle parts are better suited for the semantic consistency constraint via the non-parametric softmax in Eq. 3. The CUB classifier fails to model our constraint because it distinguishes between different types of beaks, wings, etc., to optimize for the classification task (Same-KP drops by 12.1% vs. the

Table 4. Comparison of auxiliary models on CUB: We study the Same-KP metric of the counterfactuals (single distractor) and whether the aux. features can be clustered into parts. [†]Parts detector establishes an upperbound as it uses parts ground-truth.

Auxiliary model	Annotations	Counterfactuals (Same-KP)	Clustering (K-Means Acc.)		
			K = 15	K = 50	K = 250
CUB classifier	Class labels	10.1	18.0	19.3	21.6
IN-1k classifier	Class labels	19.3	42.0	49.5	57.1
IN-1k MoCo [23]	None (SSL)	18.1	33.8	44.1	52.2
IN-1k SWAV [13]	None (SSL)	22.1	45.3	54.2	62.6
IN-1k DeepCluster [12]	None (SSL)	22.0	45.3	54.9	63.5
CUB parts detector[†]	Keypoints	22.2	46.0	59.2	75.4

upperbound). Differently, the SSL features are more generic, making them suitable for our method (Same-KP using DeepCluster drops only 0.2% vs. the upperbound).

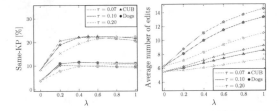

Fig. 4. Influence of temperature τ and weight λ.

Influence of τ and λ: We study how the temperature τ in Eq. 3 and the weight λ parameter in Eq. 4 influence different metrics. Recall that high values of λ favor the semantic loss over the classification loss. Selecting semantically similar cells over dissimilar ones directly improves the Same-KP metric (Fig. 4 (left)), but that comes at a cost of an increased number of edits until the model's decision changes (Fig. 4 (right)). We observe that $\lambda = 0.4$ is a saturation point, after which the Same-KP metric does not notably change. Further, lower values of τ sharpen the softmax distribution making it closer to one-hot, while higher τ yield a distribution closer to a uniform. This has an effect on the number of edits, as a sharper distribution is more selective. We found that for a fixed $\lambda = 0.4$, $\tau = 0.1$ as in [13] is a sweet spot between good Same-KP performance and a small increase in the number of edits. We verified values via 5-fold cross-validation across multiple datasets.

4.4   Online Evaluation Through Machine Teaching

To further demonstrate the utility of high-quality visual counterfactuals, we setup a machine teaching experiment, where humans learn to discern between bird species with the help of counterfactual explanations. Through the experiment detailed below, we verify our hypothesis that humans perform better at this task with more informative and accurate counterfactual explanations.

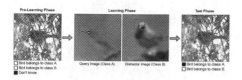

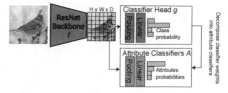

Fig. 5. Machine teaching task phases.   **Fig. 6.** Attribute-based decomposition.

Study Setup: We follow the setup from [59], but differ in two crucial ways: (i) ours is a larger study on 155 query-distractor class pairs, while [59] was done only on one class pair; (ii) we obfuscate the bird class names and replace them with "class A" and "class B". We do this because some class names contain identifiable descriptions (e.g., *Red Headed Woodpecker*) without needing visual cues. The study comprises three phases (simplified visualization in Fig. 5). The **pre-learning phase** gives AMT raters 10 <u>test</u> image examples of 2 bird classes. The raters need to choose one of three options: 'Bird belongs to class A', 'Bird belongs to class B,' or 'Don't know'. The purpose of this stage is for the raters to get familiarized with the user interface, and as in [59] all raters chose 'Don't know' for each example in this stage. Next, during the **learning phase**, we show counterfactual explanations of 10 <u>train</u> image pairs where the query image belongs to class A and the distractor image to class B. We highlight the image content from the counterfactual region, with all other content being blurred (Fig. 5). This ensures that the humans do not perform the classification task based on any other visual cues except the ones identified by a given counterfactual method. Finally, the **test phase** presents to raters 10 <u>test</u> image pairs (same as in the pre-learning stage), and asks to classify them into either class A or B. This time, the option 'Don't know' is not provided. Once the task is done, a different set of bird class pair is selected, and the three stages are repeated.

Task Details: We hired 25 AMT raters, use images from CUB, and compare counterfactuals produced from our method with two baselines: [22] and [59]. For

Table 5. Machine teaching task. The learning phase selects random image pairs (†), or pairs that show the largest improvement in terms of being semantically consistent (∗).

Method	Test acc. (%)	
	(Random)†	(Semantically-acc.)∗
SCOUT [59]	77.4	62.8
Goyal *et al.* [22]	76.7	64.3
Ours	**80.5**	**82.1**

Table 6. Attribute-based counterfactuals. We evaluate whether the top-1 attributes are discriminative of the classes.

Method	Test Acc. (%)
SCOUT [59]	46.7
Goyal *et al.* [22]	67.0
Ours	**74.5**

all three methods, we mine query-distractor classes via the approach mentioned in Sect. 4.1, resulting in 155 unique binary classification tasks. The learning phase visualizes the counterfactual generated from the first edit. To ensure a fair comparison across all methods, we do not use multiple distractor images for generating counterfactuals, use the exact same set of images across all the compared methods, and use the same backbone (VGG-16 [47]) throughout. This controlled setup ensures that any difference in the human study performance can be only due to the underlying counterfactual method. We report results under two setups, which differ in how we select the image pairs (I, I'): **1. random:** we generate explanations from random images using different methods. This is a fair comparison between all methods. **2. semantically-consistent:** we study whether semantically consistent explanations lead to better human teaching. Hence, we exaggerate the differences in Same-KP between our method and [22, 59] by selecting images where our approach has a higher Same-KP metric. If semantic consistency is important in machine teaching, our approach should do much better than 'random', and the baselines should do worse than 'random'.

Results: Table 5 shows that the raters perform better when shown explanations from our method under the 'random' setup. Further, the differences in test accuracy are more pronounced (82.1% vs. 64.3%) when the raters were presented with semantically consistent explanations. This result highlights the importance of semantically consistent visual counterfactuals for teaching humans.

5   Towards Language-Based Counterfactual Explanations

In this section, we propose a novel method to augment visual counterfactual explanations with natural language via the vocabulary of parts and attributes. Parts and attributes bring notable benefits as they enrich the explanations and make them more interpretable [29]. Through this experiment, we further emphasize the importance of semantically consistent counterfactuals and prove them to be a key ingredient towards generating natural-language-based explanations.

Our proof-of-concept experiment uses a ResNet-50 model, where $f(\cdot)$ computes the $h \times w \times d$ spatial feature output of the last conv layer, and $g(\cdot)$ performs a global average pooling operation followed by a linear classifier. We use the CUB [58] dataset with 15 bird parts, where each part (e.g., beak, wing, belly, etc.) is associated with a keypoint location. Additionally, this dataset contains part-attribute annotations (e.g., hooked beak, striped wing, yellow belly, etc.). We perform our analysis on a subset of 77 subsequently denoised part-attributes. Following [30], denoising is performed by majority voting, e.g., if more than 50% of crows have black wings in the data, then all crows are set to have black wings.

In the first step, given a query I from class c and a distractor I' from c', we construct a counterfactual I^*, following our approach from Sect. 3. For fair comparison with [22,59], we limit to single best cell edits. Next, we identify the part corresponding to this best-edit cell in I. We train a parts detector that predicts the top-3 parts for each cell location in the $h \times w$ spatial grid. Note that if the corresponding cell in I' is not semantically consistent with I, the detected

Query Image   Distractor Image   Query Image   Distractor Image   Query Image   Distractor Image

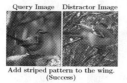

Add striped pattern to the wing.   Remove brown color from the crown.   Add color blue to forehead.
(Success)   (Success)   (Failure)

Fig. 7. Augmenting counterfactuals with part-attributes. We identify the attribute that is most important for changing the model's decision. Best viewed in color.

parts will not match, and the attribute explanations will be nonsensical. Finally, we find *the most important* attribute for the best-edit via the procedure below.

Finding the Best Attribute: We train a part-attribute model A that performs global average pooling followed on the output of $f(.)$ by a linear classifier, thus operating on the same feature space as g. We then use an interpretable basis decomposition [63] to decompose the object classifier weights from $g(\cdot)$ into a weighted sum of part-attribute classifier weights from $A(\cdot)$. A simplified visualization is presented in Fig. 6, see [63] for details. The interpretable basis decomposition yields an importance score s_t for each part-attribute t, and we additionally constrain the part-attributes to only the detected parts in the best-edit cells. E.g., if the detected part is a beak, we only consider the {hooked, long, orange, ...}-beak attribute classifiers. Similarly, we compute an importance score s_t' for the best-edit cell in I'. Finally, we compute the differences of importance scores $s_t' - s_t$, where a positive difference indicates that part-attribute t contributed more to the model's decision being c' compared to c. We select the top-k such part-attributes. Again, note that the difference $s_t' - s_t$ makes sense only if the selected parts are semantically same in I and I' (details in suppl.).

Evaluation: For each class pair (c, c'), we use the available annotations to define part-attributes that belong to class c but not to class c', and vice-versa, as proxy counterfactual ground-truth. Our final explanations are evaluated by measuring how often the top-1 part-attribute, identified via the difference between the estimated importance scores, belongs to the set of ground-truth part-attributes.

Results: Table 6 shows the results using visual counterfactuals from our method and from [22,59]. We observe that our method is significantly better compared to prior work in correctly identifying discriminative part-attributes. Given that all other factors were controlled across the three methods, we argue that this improvement is due to our counterfactuals being semantically consistent. Figure 7 shows the qualitative results. Notice that both the wing's color and pattern are visually distinct in Fig. 7 (left), but the part-attribute explanation points out that the wing's pattern mattered the most to the model while generating the counterfactual. Similarly in Fig. 7 (middle), the part-attribute explanation tells us that the crown color is most important. In both cases, the part-attribute information helps disambiguate the visual explanation. Figure 7 (right) shows a failure case caused by a wrongful prediction from the part-attribute classifiers.

6   Conclusion and Future Work

We presented a novel framework to generate semantically consistent visual counterfactuals. Our evaluation shows that (i) our counterfactuals consistently match semantically similar and class-specific regions, (ii) our proposed method is computationally efficient, and (iii) our explanations significantly outperform the s-o-t-a. Further, we demonstrated the importance of semantically consistent visual counterfactuals via: (i) a machine teaching task on fine-grained bird recognition, and (ii) an approach to augment our counterfactuals with a human interpretable part and attribute vocabulary. Currently, our method greedily searches for one cell replacement at a time. Relaxing this constraint to explore multiple regions in parallel is a fruitful future research problem. Finally, we only scratched the surface in augmenting visual counterfactuals with attribute information. We hope that our work will spark more interest in this worthy topic by the community.

References

1. Authors: Copyright for Figure 3 images from inaturalist-2021, employed for illustration of research work. iNaturalist people: longhairedlizzy: CC BY-NC 4.0, Volker Heinrich: CC BY-NC 4.0, Lee: CC BY-NC 4.0, Jonny Chung: CC BY-NC 4.0, romanvrbicek: CC BY-NC 4.0, poloyellow23: CC BY-NC 4.0. Accessed 02 Mar 2022
2. Adadi, A., Berrada, M.: Peeking inside the black-box: a survey on explainable artificial intelligence (XAI). IEEE Access **6**, 52138–52160 (2018)
3. Adebayo, J., Gilmer, J., Muelly, M., Goodfellow, I., Hardt, M., Kim, B.: Sanity checks for saliency maps. In: NeurIPS (2018)
4. Akula, A., Wang, S., Zhu, S.C.: CoCoX: generating conceptual and counterfactual explanations via fault-lines. In: AAAI (2020)
5. Alipour, K., et al.: Improving users' mental model with attention-directed counterfactual edits. Appl. AI Lett. **2**, e47 (2021)
6. Ancona, M., Ceolini, E., Öztireli, C., Gross, M.: Towards better understanding of gradient-based attribution methods for deep neural networks. In: ICLR (2018)
7. Asano, Y., Rupprecht, C., Vedaldi, A.: Self-labelling via simultaneous clustering and representation learning. In: ICLR (2019)
8. Bach, S., Binder, A., Montavon, G., Klauschen, F., Müller, K.R., Samek, W.: On pixel-wise explanations for non-linear classifier decisions by layer-wise relevance propagation. PloS One **10**, e0130140 (2015)
9. Beck, S.R., Riggs, K.J., Gorniak, S.L.: Relating developments in children's counterfactual thinking and executive functions. Thinking Reason. **15**, 337–354 (2009)
10. Biggs, B., Boyne, O., Charles, J., Fitzgibbon, A., Cipolla, R.: Who left the dogs out? 3D animal reconstruction with expectation maximization in the loop. In: Vedaldi, A., Bischof, H., Brox, T., Frahm, J.-M. (eds.) ECCV 2020. LNCS, vol. 12356, pp. 195–211. Springer, Cham (2020). https://doi.org/10.1007/978-3-030-58621-8_12
11. Buchsbaum, D., Bridgers, S., Skolnick Weisberg, D., Gopnik, A.: The power of possibility: causal learning, counterfactual reasoning, and pretend play. Philos. Trans. Roy. Soc. B: Biol. Sci. **367**, 2202–2212 (2012)

12. Caron, M., Bojanowski, P., Joulin, A., Douze, M.: Deep clustering for unsupervised learning of visual features. In: Ferrari, V., Hebert, M., Sminchisescu, C., Weiss, Y. (eds.) Computer Vision – ECCV 2018. LNCS, vol. 11218, pp. 139–156. Springer, Cham (2018). https://doi.org/10.1007/978-3-030-01264-9_9

13. Caron, M., Misra, I., Mairal, J., Goyal, P., Bojanowski, P., Joulin, A.: Unsupervised learning of visual features by contrasting cluster assignments. In: NeurIPS (2020)

14. Caron, M., et al.: Emerging properties in self-supervised vision transformers. In: ICCV (2021)

15. Chang, C.H., Creager, E., Goldenberg, A., Duvenaud, D.: Explaining image classifiers by counterfactual generation. In: ICLR (2018)

16. Chen, T., Kornblith, S., Norouzi, M., Hinton, G.: A simple framework for contrastive learning of visual representations. In: ICML (2020)

17. Dabkowski, P., Gal, Y.: Real time image saliency for black box classifiers. In: NeurIPS (2017)

18. Datta, A., Sen, S., Zick, Y.: Algorithmic transparency via quantitative input influence: theory and experiments with learning systems. In: IEEE SSP (2016)

19. Deng, J., Dong, W., Socher, R., Li, L.J., Li, K., Fei-Fei, L.: ImageNet: a large-scale hierarchical image database. In: CVPR (2009)

20. Dhurandhar, A., et al.: Explanations based on the missing: towards contrastive explanations with pertinent negatives. In: NeurIPS (2018)

21. Fong, R.C., Vedaldi, A.: Interpretable explanations of black boxes by meaningful perturbation. In: ICCV (2017)

22. Goyal, Y., Wu, Z., Ernst, J., Batra, D., Parikh, D., Lee, S.: Counterfactual visual explanations. In: ICML (2019)

23. He, K., Fan, H., Wu, Y., Xie, S., Girshick, R.: Momentum contrast for unsupervised visual representation learning. In: CVPR (2020)

24. He, K., Zhang, X., Ren, S., Sun, J.: Deep residual learning for image recognition. In: CVPR (2016)

25. Hvilshøj, F., Iosifidis, A., Assent, I.: ECINN: efficient counterfactuals from invertible neural networks. In: BMVC (2021)

26. Jacob, P., Zablocki, É., Ben-Younes, H., Chen, M., Pérez, P., Cord, M.: STEEX: steering counterfactual explanations with semantics. arXiv:2111.09094 (2021)

27. Karras, T., Laine, S., Aittala, M., Hellsten, J., Lehtinen, J., Aila, T.: Analyzing and improving the image quality of stylegan. In: CVPR (2020)

28. Khosla, A., Jayadevaprakash, N., Yao, B., Fei-Fei, L.: Novel dataset for fine-grained image categorization. In: CVPR Workshop (2011)

29. Kim, B., et al.: Interpretability beyond feature attribution: quantitative testing with concept activation vectors (tcav). In: ICML (2018)

30. Koh, P.W., et al.: Concept bottleneck models. In: ICML (2020)

31. Lang, O., et al.: Explaining in style: training a GAN to explain a classifier in stylespace. In: ICCV (2021)

32. Liu, S., Kailkhura, B., Loveland, D., Han, Y.: Generative counterfactual introspection for explainable deep learning. In: GlobalSIP (2019)

33. Lundberg, S.M., Lee, S.I.: A unified approach to interpreting model predictions. In: NeurIPS (2017)

34. Markus, A.F., Kors, J.A., Rijnbeek, P.R.: The role of explainability in creating trustworthy artificial intelligence for health care: a comprehensive survey of the terminology, design choices, and evaluation strategies. JBI **113**, 103655 (2021)

35. Miller, T.: Explanation in artificial intelligence: insights from the social sciences. Artif. Intell. **267**, 1–38 (2019)

36. Mothilal, R.K., Sharma, A., Tan, C.: Explaining machine learning classifiers through diverse counterfactual explanations. In: ACM FAccT (2020)
37. Paszke, A., et al.: PyTorch: an imperative style, high-performance deep learning library. In: NeurIPS (2019)
38. Petsiuk, V., Das, A., Saenko, K.: RISE: randomized input sampling for explanation of black-box models. In: BMVC (2018)
39. Poyiadzi, R., Sokol, K., Santos-Rodriguez, R., De Bie, T., Flach, P.: FACE: feasible and actionable counterfactual explanations. In: AAAI/ACM AIES (2020)
40. Rebuffi, S.A., Fong, R., Ji, X., Vedaldi, A.: There and back again: revisiting back-propagation saliency methods. In: CVPR (2020)
41. Ribeiro, M.T., Singh, S., Guestrin, C.: "Why should i trust you?" explaining the predictions of any classifier. In: SIGKDD (2016)
42. Ribeiro, M.T., Singh, S., Guestrin, C.: Anchors: High-precision model-agnostic explanations. In: AAAI (2018)
43. Rodriguez, P., et al.: Beyond trivial counterfactual explanations with diverse valuable explanations. In: ICCV (2021)
44. Selvaraju, R.R., Cogswell, M., Das, A., Vedantam, R., Parikh, D., Batra, D.: Grad-CAM: visual explanations from deep networks via gradient-based localization. In: ICCV (2017)
45. Shrikumar, A., Greenside, P., Shcherbina, A., Kundaje, A.: Not just a black box: learning important features through propagating activation differences. arXiv:1605.01713 (2016)
46. Simonyan, K., Vedaldi, A., Zisserman, A.: Deep inside convolutional networks: visualising image classification models and saliency maps. arXiv:1312.6034 (2013)
47. Simonyan, K., Zisserman, A.: Very deep convolutional networks for large-scale image recognition. arXiv:1409.1556 (2014)
48. Singla, S., Pollack, B., Chen, J., Batmanghelich, K.: Explanation by progressive exaggeration. In: ICLR (2019)
49. Singla, S., Pollack, B., Wallace, S., Batmanghelich, K.: Explaining the black-box smoothly-a counterfactual approach. arXiv:2101.04230 (2021)
50. Slack, D., Hilgard, S., Jia, E., Singh, S., Lakkaraju, H.: Fooling lime and shap: adversarial attacks on post hoc explanation methods. In: AAAI (2020)
51. Sundararajan, M., Taly, A., Yan, Q.: Axiomatic attribution for deep networks. In: ICML (2017)
52. Van Gansbeke, W., Vandenhende, S., Georgoulis, S., Gool, L.V.: Revisiting contrastive methods for unsupervised learning of visual representations. In: NeurIPS (2021)
53. Van Gansbeke, W., Vandenhende, S., Georgoulis, S., Proesmans, M., Van Gool, L.: SCAN: learning to classify images without labels. In: Vedaldi, A., Bischof, H., Brox, T., Frahm, J.-M. (eds.) ECCV 2020. LNCS, vol. 12355, pp. 268–285. Springer, Cham (2020). https://doi.org/10.1007/978-3-030-58607-2_16
54. Van Gansbeke, W., Vandenhende, S., Georgoulis, S., Van Gool, L.: Unsupervised semantic segmentation by contrasting object mask proposals. In: ICCV (2021)
55. Van Horn, G., Cole, E., Beery, S., Wilber, K., Belongie, S., Mac Aodha, O.: Benchmarking representation learning for natural world image collections. In: CVPR (2021)
56. Verma, S., Dickerson, J., Hines, K.: Counterfactual explanations for machine learning: a review. arXiv:2010.10596 (2020)
57. Wachter, S., Mittelstadt, B., Russell, C.: Counterfactual explanations without opening the black box: automated decisions and the GDPR. Harv. J. Law Technol. **31**, 841 (2018)

58. Wah, C., Branson, S., Welinder, P., Perona, P., Belongie, S.: The caltech-UCSD birds-200-2011 dataset. Technical report, California Institute of Technology (2011)
59. Wang, P., Vasconcelos, N.: SCOUT: self-aware discriminant counterfactual explanations. In: CVPR (2020)
60. Wu, Z., Xiong, Y., Yu, S.X., Lin, D.: Unsupervised feature learning via non-parametric instance discrimination. In: CVPR (2018)
61. Zablocki, É., Ben-Younes, H., Pérez, P., Cord, M.: Explainability of vision-based autonomous driving systems: review and challenges. arXiv:2101.05307 (2021)
62. Zeiler, M.D., Fergus, R.: Visualizing and understanding convolutional networks. In: Fleet, D., Pajdla, T., Schiele, B., Tuytelaars, T. (eds.) ECCV 2014. LNCS, vol. 8689, pp. 818–833. Springer, Cham (2014). https://doi.org/10.1007/978-3-319-10590-1_53
63. Zhou, B., Khosla, A., Lapedriza, A., Oliva, A., Torralba, A.: Learning deep features for discriminative localization. In: CVPR (2016)
64. Zhou, B., Sun, Y., Bau, D., Torralba, A.: Interpretable basis decomposition for visual explanation. In: Ferrari, V., Hebert, M., Sminchisescu, C., Weiss, Y. (eds.) ECCV 2018. LNCS, vol. 11212, pp. 122–138. Springer, Cham (2018). https://doi.org/10.1007/978-3-030-01237-3_8
65. Zintgraf, L.M., Cohen, T.S., Adel, T., Welling, M.: Visualizing deep neural network decisions: prediction difference analysis. In: ICLR (2017)

HIVE: Evaluating the Human Interpretability of Visual Explanations

Sunnie S. Y. Kim$^{(\boxtimes)}$, Nicole Meister , Vikram V. Ramaswamy ,
Ruth Fong , and Olga Russakovsky

Princeton University, Princeton, NJ 08544, USA
{sunniesuhyoung,nmeister,vr23,ruthfong,olgarus}@princeton.edu

Abstract. As AI technology is increasingly applied to high-impact, high-risk domains, there have been a number of new methods aimed at making AI models more human interpretable. Despite the recent growth of interpretability work, there is a lack of systematic evaluation of proposed techniques. In this work, we introduce HIVE (Human Interpretability of Visual Explanations), a novel human evaluation framework that assesses the utility of explanations to human users in AI-assisted decision making scenarios, and enables falsifiable hypothesis testing, cross-method comparison, and human-centered evaluation of visual interpretability methods. To the best of our knowledge, this is the first work of its kind. Using HIVE, we conduct IRB-approved human studies with nearly 1000 participants and evaluate four methods that represent the diversity of computer vision interpretability works: GradCAM, BagNet, ProtoPNet, and ProtoTree. Our results suggest that explanations engender human trust, even for incorrect predictions, yet are not distinct enough for users to distinguish between correct and incorrect predictions. We open-source HIVE to enable future studies and encourage more human-centered approaches to interpretability research. HIVE can be found at https://princetonvisualai.github.io/HIVE.

Keywords: Interpretability · Explainable AI (XAI) · Human studies · Evaluation framework · Human-centered AI

1 Introduction

With the growing adoption of AI in high-impact, high-risk domains, there have been a surge of efforts aimed at making AI models more interpretable. Motivations for interpretability include allowing human users to trace through a model's reasoning process (accountability, transparency), verify that the model is basing its predictions on the right reasons (fairness, ethics), and assess their level

Supplementary Information The online version contains supplementary material available at https://doi.org/10.1007/978-3-031-19775-8_17.

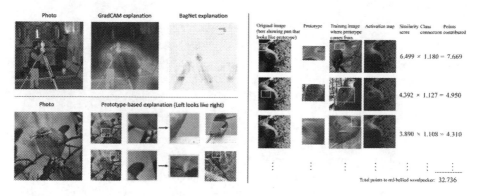

Fig. 1. Different forms of explanation. (Top left) Heatmap explanations (Grad-CAM [61], BagNet [10]) highlight decision-relevant image regions. (Bottom left) Prototype-based explanations (ProtoPNet [15], ProtoTree [48]) match image regions to prototypical parts learned during training. This schematic is much simpler than actual explanations. (Right) Actual ProtoPNet explanation example from the original paper. **While existing evaluation methods typically apply to only one explanation form, HIVE evaluates and compares diverse interpretability methods.**

of confidence in the model (trustworthiness). The *interpretability* research field tackles these questions and is comprised of diverse works, including those that provide explanations of the behavior and inner workings of complex AI models [6,7,25,27,50,61,64,73,77], those that design inherently interpretable models [10,13–15,17,18,38,48,53], and those that seek to understand what is easy and difficult for these models [3,68,75] to make their behavior more interpretable.

Despite much methods development, there is a relative lack of standardized evaluation methods for proposed techniques. Existing evaluation methods for computer vision interpretability methods are focused on feature attribution heatmaps that highlight "important" image regions for a model's prediction. Since we lack ground-truth knowledge about which regions are *actually* responsible for the prediction, different evaluation metrics use different proxy tasks for verifying these important regions (e.g., measuring the impact of deleting regions or the overlap between ground-truth objects and highlighted regions) [26,34,50,51,70,74]. However, these automatic evaluation metrics are disconnected from downstream use cases of explanations; they don't capture how useful end-users find heatmaps in their decision making. Further, these metrics don't apply to other forms of explanations, such as prototype-based explanations produced by some of the recent interpretable-by-design models [15,17,48].

In part due to these challenges, the interpretability of a proposed method is often argued through a few exemplar explanations that highlight how a method is more interpretable than a baseline model. However, recent works suggest that some methods are not as interpretable as originally imagined and may engender over-trust in automated systems [1,19,32,33,46,47,49,62]. They caution against an over-reliance on intuition-based justifications and raise awareness for the need of falsifiable hypotheses [44] and proper evaluation in interpretability research.

Our Contributions. As more diverse interpretability methods are being proposed, it is more important than ever to have a standardized and rigorous evaluation framework that allows for falsifiable hypothesis testing, cross-method comparison, and human-centered evaluation. To this end, we develop HIVE (Human Interpretability of Visual Explanations). HIVE evaluates diverse visual interpretability methods by evaluating all methods on a common task. We carefully design the tasks to reduce the effect of confirmation bias and human prior knowledge in interpretability evaluation, and assess the utility of explanations in AI-assisted decision making scenarios. HIVE also examines how well interpretable-by-design models' reasoning process aligns with that of humans, and how human users tradeoff interpretability and accuracy.

To demonstrate the extensibility and applicability of HIVE, we conduct IRB-approved human studies with nearly 1000 participants and evaluate four existing methods that represent different streams of interpretability work (e.g., post-hoc explanations, interpretable-by-design models, heatmaps, and prototype-based explanations): GradCAM [61], BagNet [10], ProtoPNet [15], ProtoTree [48]. To the best of our knowledge, we are the first to compare interpretability methods with different explanation forms (see Fig. 1) and the first to conduct human studies of the evaluated interpretable-by-design models [10,15,48].

We obtain a number of insights through our studies:

- When provided explanations, participants tend to believe that the model predictions are correct, revealing an issue of *confirmation bias*. For example, our participants found 60% of the explanations for *incorrect* model predictions convincing. Prior work has made similar observations for non-visual interpretability methods [52]; we substantiate them for visual explanations and demonstrate a need for rigorous evaluation of proposed methods.
- When given multiple model predictions and explanations, participants struggle to distinguish between correct and incorrect predictions based on the explanations (e.g., achieving only 40% accuracy on a multiple-choice task with four options). This result suggests that interpretability methods need to be improved to be reliably useful for AI-assisted decision making.
- There exists a gap between the similarity judgments of humans and prototype-based models [15,48] which can hurt the quality of their interpretability.
- Participants prefer to use a model with explanations over a baseline model without explanations. To switch their preference, they require the baseline model to have +6.2% to +10.9% higher accuracy.

As interpretability is fundamentally a human-centric concept, it needs to be evaluated in a human-centric way. We hope our work helps pave the way towards human evaluation becoming commonplace, by presenting and analyzing a human study design, demonstrating its effectiveness and informativeness for interpretability evaluation, and open-sourcing the code to enable future work.

2   Related Work

Interpretability Landscape in Computer Vision. Interpretability research can be described along several axes: first, whether a method is post-hoc or interpretable-by-design; second, whether it is global or local; and third, the form of an explanation (see [4,11,16,24,28,30,57,59] for surveys). *Post-hoc explanations* focus on explaining predictions made by already-trained models, whereas *interpretable-by-design (IBD)* models are intentionally designed to possess a more explicitly interpretable decision-making process [10,13–15,17,18,38,48,53]. Furthermore, explanations can either be *local explanations* of a single input-output example or *global explanations* of a network (or its component parts). Local, post-hoc methods include heatmap [25,50,61,63,64,73,77], counterfactual explanation [29,65,69], approximation [56], and sample importance [37,71] methods. In contrast, global, post-hoc methods aim to understand global properties of CNNs, often by treating them as an object of scientific study [6,7,27,36] or by generating class-level explanations [55,78]. Because we focus on evaluating the utility of explanations in AI-assisted decision making, we do not evaluate global, post-hoc methods. *IBD* models can provide local and/or global explanations, depending on the model type. Lastly, explanations can take a variety of forms: two more popular ones we study are *heatmaps* highlighting important image regions and *prototypes* (i.e., image patches) from the training set that form interpretable decisions. In our work, we investigate four popular methods that span these types of interpretability work: GradCAM [61] (post-hoc, heatmap), BagNet [10] (IBD, heatmap), ProtoPNet [15] (IBD, prototypes), and ProtoTree [48] (IBD, prototypes). See Fig. 1 for examples of their explanations.

Evaluating Heatmaps. Heatmap methods are arguably the most-studied class of interpretability work. Several automatic evaluation metrics have been proposed [5,26,34,50,51,70,74], however, there is a lack of consensus on how to evaluate these methods. Further, the authors of [1,2] and BAM [70] highlight how several methods fail basic "sanity checks" and call for more comprehensive metrics. Complementing these works, we use HIVE to study how useful heatmaps are to human users in AI-assisted decision making scenarios and demonstrate insights that cannot be gained from automatic evaluation metrics.

Evaluating Interpretable-by-Design Models. In contrast, there has been relatively little work on assessing interpretable-by-design models. Quantitative evaluations of these methods typically focus on demonstrating their competitive performance with a baseline CNN, while the quality of their interpretability is often demonstrated through qualitative examples. Recently, a few works revisited several methods' interpretability claims. Hoffmann et al. [33] highlight that prototype similarity of ProtoPNet [15] does not correspond to semantic similarity and that this disconnect can be exploited. Margeloiu et al. [47] analyze concept bottleneck models [38] and demonstrate that learned concepts fail to correspond to real-world, semantic concepts. In this work, we conduct the first human study of three popular interpretable-by-design models [10,15,48] and quantify prior work's [33,48] anecdotal observation on the misalignment between prototype-based models [15,48] and humans' similarity judgment.

Evaluating Interpretability with Human Studies. Outside the computer vision field, human studies are commonly conducted for models trained on tabular datasets [40,41,43,52,76]; however, these do not scale to the complexity of modern vision models. Early human studies for visual explanations have been limited in scope: They typically ask participants which explanation they find more reasonable or which model they find more trustworthy based on explanations [35,61]. Recently, more diverse human studies have been conducted [8,9,23,49,62,63,80].

Closest to our work are [23,49,62]. Shen and Huang [62] ask users to select incorrectly predicted labels with or without showing explanations; Nguyen et al. [49] ask users to decide whether model predictions are correct based on explanations; Fel et al. [23] ask users to predict model outputs in a concurrent work. Regarding [49,62], our *distinction* task also investigates how useful explanations are in distinguishing correct and incorrect predictions. However, different from these works, we ask users to select the correct prediction out of multiple predictions to reduce the effect of confirmation bias and don't show class labels to prevent users from relying their prior knowledge. Regarding [23], we also ask users to predict model outputs, but mainly as a supplement to our *distinction* task. Further, we ask users to identify the model output out of multiple predictions based on the explanations, whereas [23] first trains users to be a meta-predictor of the model by showing example model predictions and explanations, and then at test time asks users to predict the model output for a given image without showing any explanation. Most importantly, different from [23,49,62], we evaluate interpretability methods beyond heatmaps and conduct cross-method comparison. Our work is similar in spirit to work by Zhou et al. [79] on evaluating generative models with human perception. For general guidance on running human studies in computer vision, refer to work by Bylinskii et al. [12].

3   HIVE Design Principles

In this work, we focus on AI-assisted decision making scenarios, in particular those that involve an image classification model. For a given input image, a user is shown a model's prediction along with an associated explanation, and is asked to make a decision about whether the model's prediction is correct or more generally about whether to use the model. In such a scenario, explanations are provided with several goals in mind: help the user identify if the model is making an error, arrive at a more accurate prediction, understand the model's reasoning process, decide how much to trust the model, etc.

To study whether and to what extent different visual interpretability methods are useful for AI-assisted decision making, we develop a novel human evaluation framework named HIVE (Human Interpretability of Visual Explanations). In particular, we design HIVE to allow for *falsifiable hypothesis testing* regarding the usefulness of explanations for identifying model errors, *cross-method comparison* between different explanation approaches, and *human-centered evaluation* for understanding the practical effectiveness of interpretability.

3.1 Falsifiable Hypothesis Testing

We join a growing body of work that cautions against intuition-based justification and subjective self-reported ratings in interpretability evaluation [1,39,44, 60] and calls for objective assessment with behavior indicators [42,52,72,76]. To this end, we design two evaluation tasks, the *agreement* and *distinction* tasks, that enable *falsifiable hypothesis testing* about the evaluated interpretability method.

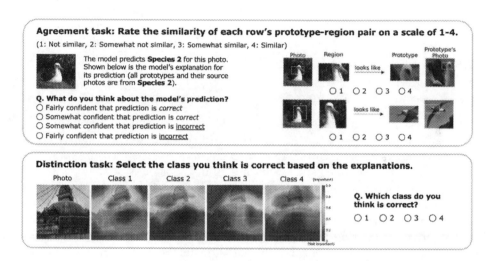

Fig. 2. Study user interfaces (UIs). We show simplified UIs for evaluating ProtoPNet [15] on the *agreement* task (top) and GradCAM [61] on the *distinction* task (bottom). Full UI snapshots are in supp. mat. See Sect. 3 for description of the tasks

In the *agreement* task, we present participants with one prediction-explanation pair at a time and ask how confident they are in the model's prediction based on the explanation. We evaluate methods on this task in part because it is closer to existing interpretability evaluation schemes that consider a model's top-1 prediction and its explanation [61], and also because it allows us to quantify the degree to which participants believe in model predictions based on explanations.

The *agreement* task measures the amount of *confirmation bias* that arises for a given interpretability method. However, it doesn't measure the utility of explanations in distinguishing correct and incorrect predictions, a crucial functionality of explanations in AI-assisted decision making. Hence, we design and use the *distinction* task as our main evaluation task. Here we simultaneously show four predictions and their associated explanations for a given input image and ask users to identify the correct prediction based on the provided explanations. The *distinction* task also mitigates the effect of confirmation bias in interpretability evaluation, as participants now have to reason about multiple explanations at once. See Fig. 2 for the evaluation task UIs.

One concern with this setup is ensuring that participants use the provided explanations rather than their knowledge to complete the task. We take two measures to remove the effect of *human prior knowledge* in our evaluations. First, we evaluate all interpretability methods in the context of fine-grained bird species classification [66], which is a challenging task for non-bird experts. Second, as a more general measure, we omit the semantic class labels of the predictions. This measure is particularly important when evaluating interpretability methods in easier contexts, e.g., coarse-grained object classification with ImageNet [58], because the task becomes too easy otherwise (i.e., participants can select the correct prediction based on the class labels instead of using the explanations). Note that ground-truth class labels are also omitted to simulate a realistic decision making scenario where users do not have access to the ground truth.

3.2   Cross-method Comparison

Existing evaluation methods typically apply to only one explanation form (e.g., heatmaps are compared against each other). In contrast, HIVE enables *cross-method comparison* between different explanation forms by focusing on downstream uses of explanations and evaluating all methods on a common task.

However, there remains a number of practical roadblocks. First, different methods may have been developed for different scenarios (e.g., fine-grained vs. coarse-grained classification), requiring us to carefully analyze the effect of the particular setting during evaluation. Second, different methods may be more or less digestible to the users. While this is an inherent part of what we are trying to evaluate, we also want to ensure that the evaluation task is doable by study participants with limited machine learning background, given most human studies in the field are run through Amazon Mechanical Turk. Hence, we actualize a specific evaluation setup for each interpretability method by creating an individual evaluation UI that respects the method's characteristics (e.g., its explanation form, dataset used for model training). We briefly describe the four methods we evaluate in this work (see Fig. 1 for example explanations) and their evaluation setups. When making any adaptations, we tried to present each method in as favorable of a way as possible. More details are in supp. mat.

GradCAM [61]. GradCAM is a post-hoc method that produces a heatmap that highlights important regions in an input image that contribute to a model's prediction. We evaluate GradCAM on ImageNet [58], which it was originally developed for, as well as on CUB [66], for which we train a standard CNN model to use as the underlying model for generating GradCAM heatmaps.

BagNet [10]. In contrast, BagNet is an interpretable-by-design model that collects evidence for a class from small regions of an image. For each class, BagNet creates a heatmap where higher values (i.e., darker red in our visualizations) imply stronger evidence for the class. BagNet then sums the values in each heatmap and predicts the class with the highest sum. We evaluate BagNet on ImageNet, for which it was originally designed, as well as on CUB, for which we train a new BagNet model using the authors' code.

ProtoPNet [15]. The next two methods reason with *prototypes*, which are small image patches from the training set that these models deem as representative for a certain class. At test time, ProtoPNet compares a given image to the set of prototypes it learned during training and finds regions in the image that are the most similar to each prototype. It computes a similarity score between each prototype-region pair, then predicts the class with the highest weighted sum of the similarity scores. The ProtoPNet model for CUB learns 10 prototypes for each of the 200 bird species (2,000 total) and produces one of the most complex explanations. Its explanation for a single prediction consists of 10 prototypes and their source images, heatmaps that convey the similarity between matched image regions and prototypes, continuous and unnormalized similarity scores, and weights multiplied to the scores (see Fig. 1 right). In our evaluation, we abstract away most technical details based on our pilot studies, and focus on showing the most crucial component of ProtoPNet's reasoning process: the prototype-image region matches. We also ask participants to rate the similarity of each match (see Fig. 2 top) to assess how well the model's similarity judgment aligns with that of humans. See supp. mat. for the task and explanation modification details.

ProtoTree [48]. Finally, the ProtoTree model learns a tree structure along with the prototypes. Each node in the tree contains a prototype from a training image. At each node, the model compares a given test image to the node's prototype and produces a similarity score. If the score is above some threshold, the model judges that the prototype is present in the image and absent if not. The model then proceeds to the next node and repeats this process until it reaches a leaf node, which corresponds to a class. The ProtoTree model for CUB trained by the authors has 511 decision nodes and up to 10 decision steps, and our pilot studies revealed that is too overwhelming for participants. Thus in our evaluation, we significantly simplify the decision process. Participants are shown the model's decisions until the penultimate decision node, and then are asked to make decisions for only the final two nodes of the tree by judging whether the prototype in each node is present or absent in the image. This leads the participants to select one of the four (2^2) classes as the final prediction. One additional challenge is that participants may not be familiar with decision trees and thus may have trouble following the explanation. To help understanding, we introduce a simple decision tree model with two levels, walk through an example, and present two warm up exercises so that participants can get familiar with decision trees before encountering ProtoTree. See supp. mat. for more information.

3.3   Human-Centered Evaluation

HIVE complements existing algorithmic evaluation methods by bringing humans back into the picture and taking a *human-centered* approach to interpretability evaluation. The design of HIVE, particularly the inclusion/exclusion of class labels in Sect. 3.1 and careful actualization of the evaluation setup in Sect. 3.2, is focused on making this evaluation tractable for the participants and as fair as possible with respect to different interpretability methods. We also went through multiple iterations of UI design to present visual explanations in digestible bits so as to not overwhelm participants with their complexity. Despite the challenges, there is a very important payoff from human studies. We are able to evaluate different interpretability methods through participants' 1) ability to *distinguish* between correct and incorrect predictions based on the provided explanations, simulating a more realistic AI-assisted decision-making setting, and 2) level of *alignment* with the model's intermediate reasoning process in the case of prototype-based, interpretable-by-design models. We also gain a number of valuable insights that can only be obtained through human studies.

3.4   Generalizability and Scalability

In closing we discuss two common concerns about human studies: generalizability and scalability. We have shown HIVE's *generalizability* by using it to evaluate a variety of methods (post-hoc explanations, interpretable-by-design models, heatmaps, prototype-based explanations) in two different settings (coarse-grained object recognition with ImageNet, fine-grained bird recognition with CUB). Further, a recent work by Ramaswamy et al. [54] uses HIVE to set up new human studies, for evaluating example-based explanations and finding the ideal complexity of concept-based explanations, demonstrating that HIVE can be easily generalized to new methods and tasks. Regarding *scalability*, human study costs are not exorbitant contrary to popular belief and can be budgeted for like we budget for compute. For example, our GradCAM distinction study cost $70 with 50 participants compensated at $12/hr. The real obstacles are typically the time, effort, and expertise required for study design and UI development; with HIVE open-sourced, these costs are substantially mitigated.

4   HIVE Study Design

In this section, we describe our IRB-approved study design. See supp. mat. and https://princetonvisualai.github.io/HIVE for UI snapshots and code.

Introduction. For each participant, we first introduce the study and receive their informed consent. We also request optional demographic data regarding gender identity, race and ethnicity, and ask about the participant's experience with machine learning; however, no personally identifiable information was collected. Next we explain the evaluated interpretability method in simple terms by avoiding technical jargon (i.e., replacing terms like "image" and "training set" to "photo" and "previously-seen photos"). We then show a preview of the evaluation task and provide example explanations for one correct and one incorrect prediction made by the model to give the participant appropriate references. The participant can access the method description at any time during the task.

Objective Evaluation Tasks. Next we evaluate the interpretability method on a behavioral task (*distinction* or *agreement*) introduced in Sect. 3.1 and Fig. 2. Detailed task descriptions are available in supp. mat.

Subjective Evaluation Questions. While the core of HIVE is in the objective evaluation tasks, we also ask subjective evaluation questions to make the most out of the human studies. Specifically, we ask the participant to self-rate their level of understanding of the evaluated method before and after completing the task, to investigate if the participant's self-rated level of understanding undergoes any changes during the task. After the task completion, we disclose the participant's performance on the task and ask the question one last time.

Interpretability-Accuracy Tradeoff Questions. While interpretability methods offer useful insights into a model's decision, some explanations come at the cost of lower model accuracy. Hence in the final part of the study, we investigate the *interpretability-accuracy tradeoff* participants are willing to make when comparing an interpretable method against a baseline model that doesn't come with any explanation. In high-risk scenarios a user may prefer to maximize model performance over interpretability. However, another user may prefer to prioritize interpretability in such settings so that there would be mechanisms for examining the model's predictions. To gain insight into the tradeoff users are willing to make, we present three scenarios: low-risk (e.g., bird species recognition for scientific or educational purposes), medium-risk (e.g., object recognition for automatic grocery checkout), and high-risk (e.g., scene understanding for autonomous driving). For each scenario, we then ask the participant to input the minimum accuracy of the baseline model that would convince them to use it over the model with explanations and also describe the reason for their choices.

5   Experiments

5.1   Experimental Details

Datasets and Models. We evaluate all interpretability methods on classification tasks and use images from the CUB [66] test set and the ImageNet [58] validation set to generate model predictions and explanations. On CUB, we evaluate all four methods: GradCAM [61], BagNet [10], ProtoPNet [15], ProtoTree [48]. On ImageNet, we evaluate GradCAM and BagNet. See supp. mat. for details.

Human Studies. For each study, i.e., an evaluation of one interpretability method on one task (*distinction* or *agreement*), we recruited 50 participants through Amazon Mechanical Turk (AMT). In total, we conducted 19 studies with 950 participants; see supp. mat. for the full list. The self-reported machine learning experience of the participants was 2.5 ± 1.0, between "2: have heard about..." and "3: know the basics..." The mean study duration was 6.9 minutes for GradCAM, 6.6 for BagNet, 13.6 for ProtoPNet, and 10.4 for ProtoTree. Participants were compensated based on the state-level minimum wage of $12/hr.

Statistical Analysis. For each study, we report the mean task accuracy and standard deviation of the participants' performance which captures the variability between individual participants' performance. We also compare the study result to random chance and compute the p-value from a 1-sample t-test.[1] When comparing results between two groups, we compute the p-value from a 2-sample t-test. Results are deemed statistically significant under $p < 0.05$ conditions.

5.2   The Issue of Confirmation Bias

Let us first examine how the four methods perform on the *agreement* task, where we present participants with one prediction-explanation pair at a time and ask how confident they are in the model's prediction. Results are summarized in Table 1. On CUB, participants found 72.4% of correct predictions convincing for GradCAM, 75.6% for BagNet, 73.2% for ProtoPNet, and 66.0% ProtoTree. However, they also thought 67.2% of incorrect predictions were correct for Grad-CAM, 57.6% for BagNet, 53.6% for ProtoPNet, and 62.8% for ProtoTree. Similarly on ImageNet, participants found 70.8% of correct predictions convincing for GradCAM and 66.0% for BagNet, yet also believed in 55.2% and 64.4% of incorrect predictions, respectively. These results reveal an issue of *confirmation bias*: When given explanations, participants tend to believe model predictions are correct, even if they are wrong. Still, the confidence ratings are overall higher for correct predictions than incorrect predictions, suggesting there is some difference between their explanations. More results and discussion are in supp. mat.

Table 1. Agreement task results. For each study, we show mean accuracy, standard deviation of the participants' performance, and mean confidence rating in parentheses. *Italics* denotes methods with accuracy not statistically significantly different from 50% random chance ($p > 0.05$); **bold** denotes the highest performing method in each group. **In all studies, participants leaned towards believing that model predictions are correct when provided explanations, regardless of if they are actually correct.** For example, for GradCAM on CUB, participants thought 72.4% of correct predictions were correct and $100 - 32.8 = 67.2\%$ of incorrect predictions were correct. These results reveal an issue of *confirmation bias*. See Sect. 5.2 for a discussion.

CUB	GradCAM [61]	BagNet [10]	ProtoPNet [15]	ProtoTree [48]
Correct	72.4% ± 21.5 (2.9)	**75.6% ± 23.4 (3.0)**	73.2% ± 24.9 (3.0)	66.0% ± 33.8 (2.8)
Incorrect	32.8% ± 24.3 (2.8)	*42.4% ± 28.7 (2.7)*	*46.4% ± 35.9 (2.4)*	37.2% ± 34.4 (2.7)
ImageNet	GradCAM [61]	BagNet [10]	–	–
Correct	**70.8% ± 26.6 (2.9)**	66.0% ± 27.2 (2.8)	–	–
Incorrect	*44.8% ± 31.6 (2.7)*	35.6% ± 26.9 (2.7)	–	–

---

[1] We compare our results to chance performance instead of a baseline without explanations because we omit semantic class labels to remove the effect of human prior knowledge (see Sect. 3.1); so such a baseline would contain no relevant information.

5.3   Objective Assessment of Interpretability

Next we discuss findings from our main evaluation task, the *distinction* task, where we ask participants to select the correct prediction out of four options based on the provided explanations. Results are summarized in Table 2.

Participants Perform Better on Correctly Predicted Samples. On correctly predicted samples from CUB, the mean task accuracies are 71.2% on GradCAM, 45.6% on BagNet, 54.5% on ProtoPNet and 33.8% on ProtoTree, all above the 25% chance baseline. That is, participants can identify which of the four explanations correspond to the ground-truth class correctly predicted by the model. On incorrect predictions, however, the accuracies drop from 71.2% to 26.4% for GradCAM and from 45.6% to 32.0% for BagNet, and we observe a similar trend in the ImageNet studies. These results suggest that explanations for correct predictions may be more coherent and convincing than those for incorrect predictions. Even so, all accuracies are far from 100%, indicating that the evaluated methods are not yet reliably useful for AI-assisted decision making.

Participants Struggle to Identify The Model's Prediction. For Grad-CAM and BagNet, we ask participants to select the class they think the model predicts (*output prediction*) in addition to the class they think is correct (*distinction*). For BagNet, this is a straightforward task where participants just need to identify the most activated (most red, least blue) heatmap among the four options, as BagNet by design predicts the class with the most activated heatmap. However, accuracy is not very high, only marginally above the *distinction* task accuracy. This result suggests that BagNet heatmaps for the top-4 (or top-3 plus

Table 2. Distinction and output prediction task results. For each study, we report the mean accuracy and standard deviation of the participants' performance. *Italics* denotes methods that do not statistically significantly outperform 25% random chance ($p > 0.05$); **bold** denotes the highest performing method in each group. In the top half, we show the results of all four methods on CUB. In the bottom half, we show GradCAM and BagNet results on ImageNet, without vs. with ground-truth class labels. **Overall, participants struggle to identify the correct prediction or the model output based on explanations.** See Sect. 5.3 for a discussion.

CUB		GradCAM [61]	BagNet [10]	ProtoPNet [15]	ProtoTree [48]
Distinction	Correct	**71.2% ± 33.3**	45.6% ± 28.0	54.5% ± 30.3	33.8% ± 15.9
	Incorrect	*26.4% ± 19.8*	**32.0% ± 20.8**	–	–
Output prediction	Correct	**69.2% ± 32.3**	50.4% ± 32.8	–	–
	Incorrect	**53.6% ± 27.0**	*30.0% ± 24.1*	–	–
ImageNet		GradCAM [61]	With labels	BagNet [10]	With labels
Distinction	Correct	**51.2% ± 24.7**	49.2% ± 30.8	38.4% ± 28.0	34.8% ± 27.7
	Incorrect	*30.0% ± 22.4*	*27.2% ± 20.3*	*26.0% ± 18.4*	*27.2% ± 18.7*
Output prediction	Correct	**48.0% ± 28.3**	**48.0% ± 35.6**	46.8% ± 29.0	42.8% ± 27.4
	Incorrect	**35.6% ± 24.1**	33.2% ± 25.2	34.0% ± 24.1	32.8% ± 25.5

ground-truth) classes look similar to the human eye, and may not be suitable for assisting humans with tasks that involve distinguishing one class from another. For GradCAM, participants also struggle on this task but to a lesser degree.

Showing Ground-Truth Labels Hurts Performance. For GradCAM and BagNet, we also investigate the effect of showing ground-truth class labels for the presented images. We have not been showing them to simulate a realistic decision making scenario where users don't have access to the ground truth. However, since the task may be ambiguous for datasets like ImageNet whose images may contain several objects, we run a second version of the ImageNet studies showing ground-truth class labels on the same set of images and compare results. Somewhat surprisingly, we find that accuracy decreases, albeit by a small amount, with class labels. One possible explanation is that class labels implicitly bias participants to value heatmaps with better localization properties, which could be a suboptimal signal for the *distinction* and *output prediction* tasks.

Automatic Evaluation Metrics Correlate Poorly with Human Study Results. We also analyze GradCAM results using three automatic metrics that evaluate the localization quality of post-hoc attribution maps: pointing game [74], energy-based pointing game [67], and intersection-over-union [77]. In the *agreement* studies, we find near-zero correlation between participants' confidence in the model prediction and localization quality of heatmaps. In the *distinction* studies, we also do not see meaningful relationships between the participants' choices and these automatic metrics. These observations are consistent with the findings of [23, 49], i.e., automatic metrics poorly correlate with human performance in post-hoc attribution heatmap evaluation. See supp. mat. for details.

5.4   A Closer Examination of Prototype-Based Models

We are the first to conduct human studies of ProtoPNet and ProtoTree which produce some of the most complex visual explanations. As such, we take a closer look at their results to better understand how human users perceive them.

A Gap Exists Between Similarity Ratings of ProtoPNet and ProtoTree and Those of Humans. We quantify prior work's [33, 48] anecdotal observation that there exists a gap between model and human similarity judgment. For ProtoTree, the Pearson correlation coefficient between the participants' similarity ratings and the model similarity scores is 0.06, suggesting little to no relationship. For ProtoPNet, whose similarity scores are not normalized across images, we compute the Spearman's rank correlation coefficient ($\rho = -0.25, p = 0.49$ for *distinction* and $\rho = -0.52$, $p = 0.12$ for *agreement*). There is no significant negative correlation between the two, indicating a gap in similarity judgment that may hurt the models' interpretability. See supp. mat. for more discussion.

Participants Perform Relatively Poorly on ProtoTree, But They Understand How a Decision Tree Works. Since the previously described ProtoTree *agreement* study does not take into account the model's inherent tree structure, we run another version of the study where, instead of asking participants to rate each prototype's similarity, we ask them to select the first step they disagree with in the model's explanation. The result of this study ($52.8\%\pm19.9\%$) is similar to that of the original study ($53.6\%\pm15.2\%$); in both cases, we cannot conclude that participants outperform 50% random chance ($p = 0.33$, $p = 0.10$). To ensure participants understand how decision trees work, we provided a simple decision tree example and subsequent questions asking participants if the decision tree example makes a correct or incorrect prediction. Participants achieved 86.5% performance on this task, implying that the low task accuracy for ProtoTree is not due to a lack of comprehension of decision trees. See supp. mat. for details.

5.5   Subjective Evaluation Of Interpretability

To complement the objective evaluation tasks, we asked participants to self-rate their level of method understanding three times. The average ratings are 3.7 ± 0.9 after the method explanation, 3.8 ± 0.9 after the task, and 3.5 ± 1.0 after seeing their task performance, which all lie between the fair (3) and good (4) ratings. Interestingly, the rating tends to *decrease* after participants see their task performance ($p < 0.05$). Several participants indicated that their performance was lower than what they expected, whereas no one suggested the opposite, suggesting that participants might have been disappointed in their task performance, which in turn led them to lower their self-rated level of method understanding.

5.6   Interpretability-Accuracy Tradeoff

In the final part of our studies, we asked participants for the minimum accuracy of a baseline model they would require to use it over the evaluated interpretable model with explanations for its predictions. Across all studies, participants require the baseline model to have a higher accuracy than the model that comes with explanations, and by a greater margin for higher-risk settings. On average, participants require the baseline model to have +6.2% higher accuracy for low-risk, +8.2% for medium-risk, and +10.9% for high-risk settings. See supp. mat. for the full results and the participants' reasons for their choices.

6   Conclusion

In short, we introduce and open-source HIVE, a novel human evaluation framework for evaluating diverse visual interpretability methods, and use it to evaluate four existing methods: GradCAM, BagNet, ProtoPNet, and ProtoTree.

There are a few limitations of our work: First, we use a relatively small sample size of 50 participants for each study due to our desire to evaluate four methods, some under multiple conditions. Second, while HIVE takes a step towards use case driven evaluation, our evaluation setup is still far from real-world uses of interpretability methods. An ideal evaluation would be contextually situated and conducted with domain experts and/or end-users of a real-world application (e.g., how would bird experts choose to use one method over another when given multiple interpretability methods for a bird species recognition model).

Nonetheless, we believe our work will facilitate more user studies and encourage human-centered interpretability research [20–22,45], as our human evaluation reveals several key insights about the field. In particular, we find that participants generally believe model predictions are correct when given explanations for them. Humans are naturally susceptible to confirmation bias; thus, interpretable explanations will likely engender trust from humans, even if they are incorrect. Our findings underscore the need for evaluation methods that fairly and rigorously assess the usefulness and effect of explanations. We hope our work helps shift the field's objective from focusing on method development to also prioritizing the development of high-quality evaluation methods.

Acknowledgments. This material is based upon work partially supported by the National Science Foundation (NSF) under Grant No. 1763642. Any opinions, findings, and conclusions or recommendations expressed in this material are those of the author(s) and do not necessarily reflect the views of the NSF. We also acknowledge support from the Princeton SEAS Howard B. Wentz, Jr. Junior Faculty Award (OR), Princeton SEAS Project X Fund (RF, OR), Open Philanthropy (RF, OR), and Princeton SEAS and ECE Senior Thesis Funding (NM). We thank the authors of [10,15,31,33,48,61] for open-sourcing their code and/or trained models. We also thank the AMT workers who participated in our studies, anonymous reviewers who provided thoughtful feedback, and Princeton Visual AI Lab members (especially Dora Zhao, Kaiyu Yang, and Angelina Wang) who tested our user interface and provided helpful suggestions.

References

1. Adebayo, J., Gilmer, J., Muelly, M., Goodfellow, I., Hardt, M., Kim, B.: Sanity checks for saliency maps. In: NeurIPS (2018)
2. Adebayo, J., Muelly, M., Liccardi, I., Kim, B.: Debugging tests for model explanations. In: NeurIPS (2020)
3. Agarwal, C., D'souza, D., Hooker, S.: Estimating example difficulty using variance of gradients. In: CVPR (2022)
4. Arrieta, A.B., et al.: Explainable artificial intelligence (XAI): concepts, taxonomies, opportunities and challenges toward responsible AI. Inf. Fus. **58**, 82–115 (2020)
5. Bach, S., Binder, A., Montavon, G., Klauschen, F., Müller, K.R., Samek, W.: On pixel-wise explanations for non-linear classifier decisions by layer-wise relevance propagation. PLoS ONE **10**, e0130140 (2015)
6. Bau, D., Zhou, B., Khosla, A., Oliva, A., Torralba, A.: Network dissection: quantifying interpretability of deep visual representations. In: CVPR (2017)
7. Bau, D., et al.: Seeing what a GAN cannot generate. In: ICCV (2019)
8. Biessmann, F., Refiano, D.I.: A psychophysics approach for quantitative comparison of interpretable computer vision models (2019)
9. Borowski, J., et al.: Exemplary natural images explain CNN activations better than state-of-the-art feature visualization. In: ICLR (2021)

10. Brendel, W., Bethge, M.: Approximating CNNs with bag-of-local-features models works surprisingly well on ImageNet. In: ICLR (2019)
11. Brundage, M., et al.: Toward trustworthy AI development: mechanisms for supporting verifiable claims (2020)
12. Bylinskii, Z., Herman, L., Hertzmann, A., Hutka, S., Zhang, Y.: Towards better user studies in computer graphics and vision. arXiv (2022)
13. Böhle, M., Fritz, M., Schiele, B.: Convolutional dynamic alignment networks for interpretable classifications. In: CVPR (2021)
14. Böhle, M., Fritz, M., Schiele, B.: B-Cos networks: alignment is all we need for interpretability. In: CVPR (2022)
15. Chen, C., Li, O., Tao, D., Barnett, A., Rudin, C., Su, J.K.: This looks like that: deep learning for interpretable image recognition. In: NeurIPS (2019)
16. Chen, V., Li, J., Kim, J.S., Plumb, G., Talwalkar, A.: Towards connecting use cases and methods in interpretable machine learning. In: ICML Workshop on Human Interpretability in Machine Learning (2021)
17. Donnelly, J., Barnett, A.J., Chen, C.: Deformable ProtoPNet: an interpretable image classifier using deformable prototypes. In: CVPR (2022)
18. Dubey, A., Radenovic, F., Mahajan, D.: Scalable interpretability via polynomials. arXiv (2022)
19. Dzindolet, M.T., Peterson, S.A., Pomranky, R.A., Pierce, L.G., Beck, H.P.: The role of trust in automation reliance. In: IJHCS (2003)
20. Ehsan, U., Riedl, M.O.: Human-centered explainable AI: towards a reflective sociotechnical approach. In: Stephanidis, C., Kurosu, M., Degen, H., Reinerman-Jones, L. (eds.) HCII 2020. LNCS, vol. 12424, pp. 449–466. Springer, Cham (2020). https://doi.org/10.1007/978-3-030-60117-1_33
21. Ehsan, U., et al.: Operationalizing human-centered perspectives in explainable AI. In: CHI Extended Abstracts (2021)
22. Ehsan, U., et al.: Human-centered explainable AI (HCXAI): beyond opening the black-box of AI. In: CHI Extended Abstracts (2022)
23. Fel, T., Colin, J., Cadène, R., Serre, T.: What I cannot predict, I do not understand: a human-centered evaluation framework for explainability methods (2021)
24. Fong, R.: Understanding convolutional neural networks. Ph.D. thesis, University of Oxford (2020)
25. Fong, R., Patrick, M., Vedaldi, A.: Understanding deep networks via extremal perturbations and smooth masks. In: ICCV (2019)
26. Fong, R., Vedaldi, A.: Interpretable explanations of black boxes by meaningful perturbation. In: ICCV (2017)
27. Fong, R., Vedaldi, A.: Net2Vec: quantifying and explaining how concepts are encoded by filters in deep neural networks. In: CVPR (2018)
28. Gilpin, L.H., Bau, D., Yuan, B.Z., Bajwa, A., Specter, M., Kagal, L.: Explaining explanations: an overview of interpretability of machine learning. In: DSAA (2018)
29. Goyal, Y., Wu, Z., Ernst, J., Batra, D., Parikh, D., Lee, S.: Counterfactual visual explanations. In: ICML (2019)
30. Gunning, D., Aha, D.: DARPA's explainable artificial intelligence (XAI) program. AI Mag. **40**, 44–58 (2019)
31. He, K., Zhang, X., Ren, S., Sun, J.: Deep residual learning for image recognition. In: CVPR (2016)
32. Herlocker, J.L., Konstan, J.A., Riedl, J.: Explaining collaborative filtering recommendations. In: CSCW (2000)

33. Hoffmann, A., Fanconi, C., Rade, R., Kohler, J.: This looks like that... does it? Shortcomings of latent space prototype interpretability in deep networks. In: ICML Workshop on Theoretic Foundation, Criticism, and Application Trend of Explainable AI (2021)

34. Hooker, S., Erhan, D., Kindermans, P.J., Kim, B.: A benchmark for interpretability methods in deep neural networks. In: NeurIPS (2019)

35. Jeyakumar, J.V., Noor, J., Cheng, Y.H., Garcia, L., Srivastava, M.: How can I explain this to you? An empirical study of deep neural network explanation methods. In: NeurIPS (2020)

36. Kim, B., Reif, E., Wattenberg, M., Bengio, S., Mozer, M.C.: Neural networks trained on natural scenes exhibit gestalt closure. Comput. Brain Behav. **4**, 251–263 (2021). https://doi.org/10.1007/s42113-021-00100-7

37. Koh, P.W., Liang, P.: Understanding black-box predictions via influence functions. In: ICML (2017)

38. Koh, P.W., Nguyen, T., Tang, Y.S., Mussmann, S., Pierson, E., Kim, B., Liang, P.: Concept bottleneck models. In: ICML (2020)

39. Kunkel, J., Donkers, T., Michael, L., Barbu, C.M., Ziegler, J.: Let me explain: Impact of personal and impersonal explanations on trust in recommender systems. In: CHI (2019)

40. Lage, I., Chen, E., He, J., Narayanan, M., Kim, B., Gershman, S.J., Doshi-Velez, F.: Human evaluation of models built for interpretability. In: HCOMP (2019)

41. Lage, I., Ross, A.S., Kim, B., Gershman, S.J., Doshi-Velez, F.: Human-in-the-loop interpretability prior. In: NeurIPS (2018)

42. Lai, V., Tan, C.: On human predictions with explanations and predictions of machine learning models: a case study on deception detection. In: FAccT (2019)

43. Lakkaraju, H., Bach, S.H., Leskovec, J.: Interpretable decision sets: a joint framework for description and prediction. In: KDD (2016)

44. Leavitt, M.L., Morcos, A.S.: Towards falsifiable interpretability research. In: NeurIPS Workshop on ML Retrospectives, Surveys & Meta-Analyses (2020)

45. Liao, Q.V., Varshney, K.R.: Human-centered explainable AI (XAI): from algorithms to user experiences. arXiv (2021)

46. Lipton, Z.C.: The mythos of model interpretability: in machine learning, the concept of interpretability is both important and slippery. Queue **16**, 31–57 (2018)

47. Margeloiu, A., Ashman, M., Bhatt, U., Chen, Y., Jamnik, M., Weller, A.: Do concept bottleneck models learn as intended? In: ICLR Workshop on Responsible AI (2021)

48. Nauta, M., van Bree, R., Seifert, C.: Neural prototype trees for interpretable fine-grained image recognition. In: CVPR (2021)

49. Nguyen, G., Kim, D., Nguyen, A.: The effectiveness of feature attribution methods and its correlation with automatic evaluation scores. In: NeurIPS (2021)

50. Petsiuk, V., Das, A., Saenko, K.: RISE: Randomized input sampling for explanation of black-box models. In: BMVC (2018)

51. Poppi, S., Cornia, M., Baraldi, L., Cucchiara, R.: Revisiting the evaluation of class activation mapping for explainability: a novel metric and experimental analysis. In: CVPR Workshop on Responsible Computer Vision (2021)

52. Poursabzi-Sangdeh, F., Goldstein, D.G., Hofman, J.M., Wortman Vaughan, J.W., Wallach, H.: Manipulating and measuring model interpretability. In: CHI (2021)

53. Radenovic, F., Dubey, A., Mahajan, D.: Neural basis models for interpretability. arXiv (2022)

54. Ramaswamy, V.V., Kim, S.S.Y., Fong, R., Russakovsky, O.: Overlooked factors in concept-based explanations: dataset choice, concept salience, and human capability. arXiv (2022)
55. Ramaswamy, V.V., Kim, S.S.Y., Meister, N., Fong, R., Russakovsky, O.: ELUDE: generating interpretable explanations via a decomposition into labelled and unlabelled features. arXiv (2022)
56. Ribeiro, M.T., Singh, S., Guestrin, C.: "Why should I trust you?": explaining the predictions of any classifier. In: KDD (2016)
57. Rudin, C., Chen, C., Chen, Z., Huang, H., Semenova, L., Zhong, C.: Interpretable machine learning: fundamental principles and 10 grand challenges. In: Statistics Surveys (2021)
58. Russakovsky, O., et al.: ImageNet large scale visual recognition challenge. Int. J. Compu. Vis. **115**(3), 211–252 (2015). https://doi.org/10.1007/s11263-015-0816-y
59. Alber, M.: Software and application patterns for explanation methods. In: Samek, W., Montavon, G., Vedaldi, A., Hansen, L.K., Müller, K.-R. (eds.) Explainable AI: Interpreting, Explaining and Visualizing Deep Learning. LNCS (LNAI), vol. 11700, pp. 399–433. Springer, Cham (2019). https://doi.org/10.1007/978-3-030-28954-6_22
60. Schaffer, J., O'Donovan, J., Michaelis, J., Raglin, A., Höllerer, T.: I can do better than your AI: expertise and explanations. In: IUI (2019)
61. Selvaraju, R.R., Cogswell, M., Das, A., Vedantam, R., Parikh, D., Batra, D.: Grad-CAM: visual explanations from deep networks via gradient-based localization. In: ICCV (2017)
62. Shen, H., Huang, T.H.K.: How useful are the machine-generated interpretations to general users? A human evaluation on guessing the incorrectly predicted labels. In: HCOMP (2020)
63. Shitole, V., Li, F., Kahng, M., Tadepalli, P., Fern, A.: One explanation is not enough: structured attention graphs for image classification. In: NeurIPS (2021)
64. Simonyan, K., Vedaldi, A., Zisserman, A.: Deep inside convolutional networks: Visualising image classification models and saliency maps. In: ICLR Workshops (2014)
65. Vandenhende, S., Mahajan, D., Radenovic, F., Ghadiyaram, D.: Making heads or tails: Towards semantically consistent visual counterfactuals. In: Farinella T. (ed.) ECCV 2022. LNCS, vol. 13672, pp. 261–279 (2022)
66. Wah, C., Branson, S., Welinder, P., Perona, P., Belongie, S.: The caltech-UCSD birds-200-2011 dataset. Technical report CNS-TR-2011-001, California Institute of Technology (2011)
67. Wang, H., et al.: Score-CAM: score-weighted visual explanations for convolutional neural networks. In: CVPR Workshops (2020)
68. Wang, P., Vasconcelos, N.: Towards realistic predictors. In: Ferrari, V., Hebert, M., Sminchisescu, C., Weiss, Y. (eds.) ECCV 2018. LNCS, vol. 11217, pp. 37–53. Springer, Cham (2018). https://doi.org/10.1007/978-3-030-01261-8_3
69. Wang, P., Vasconcelos, N.: SCOUT: self-aware discriminant counterfactual explanations. In: CVPR (2020)
70. Yang, M., Kim, B.: Benchmarking attribution methods with relative feature importance (2019)
71. Yeh, C.K., Kim, J., Yen, I.E.H., Ravikumar, P.K.: Representer point selection for explaining deep neural networks. In: NeurIPS (2018)
72. Yin, M., Wortman Vaughan, J., Wallach, H.: Understanding the effect of accuracy on trust in machine learning models. In: CHI (2019)

73. Zeiler, M.D., Fergus, R.: Visualizing and understanding convolutional networks. In: Fleet, D., Pajdla, T., Schiele, B., Tuytelaars, T. (eds.) ECCV 2014. LNCS, vol. 8689, pp. 818–833. Springer, Cham (2014). https://doi.org/10.1007/978-3-319-10590-1_53

74. Zhang, J., Lin, Z., Brandt, J., Shen, X., Sclaroff, S.: Top-down neural attention by excitation backprop. In: Leibe, B., Matas, J., Sebe, N., Welling, M. (eds.) ECCV 2016. LNCS, vol. 9908, pp. 543–559. Springer, Cham (2016). https://doi.org/10.1007/978-3-319-46493-0_33

75. Zhang, P., Wang, J., Farhadi, A., Hebert, M., Parikh, D.: Predicting failures of vision systems. In: CVPR (2014)

76. Zhang, Y., Liao, Q.V., Bellamy, R.K.E.: Effect on confidence and explanation on accuracy and trust calibration in AI-assisted decision making. In: FAccT (2020)

77. Zhou, B., Khosla, A., Lapedriza, A., Oliva, A., Torralba, A.: Learning deep features for discriminative localization. In: CVPR (2016)

78. Zhou, B., Sun, Y., Bau, D., Torralba, A.: interpretable basis decomposition for visual explanation. In: Ferrari, V., Hebert, M., Sminchisescu, C., Weiss, Y. (eds.) ECCV 2018. LNCS, vol. 11212, pp. 122–138. Springer, Cham (2018). https://doi.org/10.1007/978-3-030-01237-3_8

79. Zhou, S., Gordon, M.L., Krishna, R., Narcomey, A., Fei-Fei, L., Bernstein, M.S.: HYPE: A benchmark for human eye perceptual evaluation of generative models. In: NeurIPS (2019)

80. Zimmermann, R.S., Borowski, J., Geirhos, R., Bethge, M., Wallis, T.S.A., Brendel, W.: How well do feature visualizations support causal understanding of CNN activations? In: NeurIPS (2021)

BayesCap: Bayesian Identity Cap for Calibrated Uncertainty in Frozen Neural Networks

Uddeshya Upadhyay[1][(✉)], Shyamgopal Karthik[1], Yanbei Chen[1],
Massimiliano Mancini[1], and Zeynep Akata[1,2]

[1] University of Tübingen, Tübingen, Germany
uddeshya.upa@gmail.com
[2] Max Planck Institute for Intelligent Systems, Stuttgart, Germany

Abstract. High-quality calibrated uncertainty estimates are crucial for numerous real-world applications, especially for deep learning-based deployed ML systems. While Bayesian deep learning techniques allow uncertainty estimation, training them with large-scale datasets is an expensive process that does not always yield models competitive with non-Bayesian counterparts. Moreover, many of the high-performing deep learning models that are already trained and deployed are non-Bayesian in nature and do not provide uncertainty estimates. To address these issues, we propose `BayesCap` that learns a Bayesian identity mapping for the frozen model, allowing uncertainty estimation. `BayesCap` is a memory-efficient method that can be trained on a small fraction of the original dataset, enhancing pretrained non-Bayesian computer vision models by providing calibrated uncertainty estimates for the predictions without (i) hampering the performance of the model and (ii) the need for expensive retraining the model from scratch. The proposed method is agnostic to various architectures and tasks. We show the efficacy of our method on a wide variety of tasks with a diverse set of architectures, including image super-resolution, deblurring, inpainting, and crucial application such as medical image translation. Moreover, we apply the derived uncertainty estimates to detect out-of-distribution samples in critical scenarios like depth estimation in autonomous driving. Code is available at https://github.com/ExplainableML/BayesCap.

Keywords: Uncertainty estimation · Calibration · Image translation

1 Introduction

Image enhancement and translation tasks like super-resolution [36], deblurring [30,31], inpainting [74], colorization [25,78], denoising [48,58], medical image syn-

U. Upadhyay and S. Karthik—Equal contribution.

Supplementary Information The online version contains supplementary material available at https://doi.org/10.1007/978-3-031-19775-8_18.

S. Avidan et al. (Eds.): ECCV 2022, LNCS 13672, pp. 299–317, 2022.
https://doi.org/10.1007/978-3-031-19775-8_18

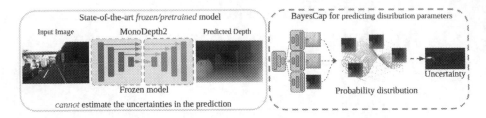

Fig. 1. Computer vision models for image enhancements and translations deterministically map input to output without producing the uncertainty in the latter (example on the right shows depth estimation using MonoDepth2 [21]). BayesCap approximates the underlying distribution and adds uncertainty to the predictions of pretrained models efficiently, details in Sect. 3.3

thesis [1,11,57,59,60,63,81], monocular depth estimation in autonomous driving [16,21], etc., have been effectively tackled using deep learning methods generating high-fidelity outputs. But, the respective state-of-the-art models usually learn a *deterministic* one-to-one mapping between the input and the output, without modeling the uncertainty in the prediction. For instance, a depth estimation model predicts a depth map from the input RGB image (Fig. 1-(Left)), without providing uncertainty. In contrast, learning a probabilistic mapping between the input and the output yields the underlying distribution and provides uncertainty estimates for the predictions. This is a vital feature in safety-critical applications such as autonomous driving and medical imaging. For instance, well-calibrated uncertainty estimates can be used to trigger human/expert intervention in highly uncertain predictions, consequently preventing fatal automated decision making [10,53,54]. The conventional approach for obtaining uncertainty estimates is to train Bayesian models *from scratch*. However, Bayesian deep learning techniques are difficult to train and are not scalable to large volumes of high-dimensional data [12]. Moreover, they cannot be easily integrated with sophisticated deterministic architectures and training schemes tailored for specific tasks, vital to achieving state-of-the-art in vision applications [14,20].

To address the above challenges, we enhance the predictions of pretrained state-of-the-art non-Bayesian deterministic deep models with uncertainty estimation while preserving their strong model performances. There is limited literature tackling the similar problem [3,12,66] but these methods do not yield well-calibrated uncertainty estimates or do not scale to high-dimensional cases such as image synthesis, translation, and enhancement.

In this work, we propose BayesCap, shown in Fig. 1-(Right), an architecture agnostic, plug-and-play method to generate uncertainty estimates for pretrained models. The key idea is to train a Bayesian autoencoder over the output images of the pretrained network, approximating the underlying distribution for the output images. Due to its Bayesian design, in addition to reconstructing the input, BayesCap also estimates the parameters of the underlying distribution, allowing us to compute the uncertainties. BayesCap is highly data-efficient and can be trained on a small fraction of the original dataset. For instance, BayesCap is 3–5× faster to train as compared to a Bayesian model from scratch, while still achieving uncertainty estimates that are better calibrated than the baselines.

To summarize, we make the following contributions. (1) We propose BayesCap, a simple method for generating post-hoc uncertainty estimates, by learning a Bayesian identity mapping, over the outputs of image synthesis/ translation tasks with deterministic pretrained models. (2) BayesCap leads to calibrated uncertainties while retaining the performance of the underlying state-of-the-art pretrained network on a variety of tasks including super-resolution, deblurring, inpainting, and medical imaging. (3) We also show that quantifying uncertainty using BayesCap can help in downstream tasks such as Out-of-Distribution (OOD) detection in critical applications like autonomous driving.

2 Related Works

Image Enhancement and Translations. Advances in computer vision led to tackle challenging problems such as super-resolution [13,36], denoising [48,58], deblurring [30,31,44], inpainting [46,74], depth estimation [16,21] among others. Such problems are tackled using a diverse set of architectures and learning schemes. For instance, the popular method for super-resolution involves training a conditional *generative adversarial networks* (GANs), where the generator is conditioned with a low-resolution image and employs a pretrained VGG network [55] to enforce the content loss in the feature space along with the adversarial term from the discriminator [36]. Differently, for the inpainting task, [74] uses a conditional GAN with contextual attention and trains the network using spatially discounted reconstruction loss. In the case of monocular depth estimation, recent works exploit the left-right consistency as a cue to train the model in an unsupervised fashion [20]. While these methods are highly diverse in their architectures, training schemes, supervisory signals, etc., they typically focus on providing a deterministic one-to-one mapping which may not be ideal in many critical scenarios such as autonomous driving [42] and medical imaging [61–63]. Our BayesCap preserves the high-fidelity outputs provided by such deterministic pretrained models while approximating the underlying distribution of the output of such models, allowing uncertainty estimation.

Uncertainty Estimation. Bayesian deep learning models are capable of estimating the uncertainties in their prediction [27,32]. Uncertainties can be broadly divided into two categories; (1) Epistemic uncertainty which is the uncertainty due to the model parameters [6,9,12,17,22,32,70]. (2) Aleatoric uncertainty which is the underlying uncertainty in the measurement itself, often estimated by approximating the per-pixel residuals between the predictions and the ground-truth using a *heteroscedastic* distribution whose parameters are predicted as the output of the network which is trained *from scratch* to maximize the likelihood of the system [4,27,33,38,66,67]. While epistemic uncertainty is important in low-data regimes as parameter estimation becomes noisy, however, this is often not the case in computer vision settings with large scale datasets where aleatoric uncertainty is the critical quantity [27]. However, it is expensive to train these models and they often perform worse than their deterministic counterparts [12,45,50]. Unlike these works, BayesCap is a fast and efficient method to estimate uncertainty over the predictions of a pretrained deterministic model.

Post-hoc Uncertainty Estimation. While this has not been widely explored, some recent works [12,15] have tried to use the Laplace approximation for this purpose. However, these methods computes the Hessian which is not feasible for high-dimensional modern problems in computer vision [16,30,46,48,78]. Another line of work to tackle this problem is test-time data augmentation [3,66] that perturbs the inputs to obtain multiple outputs leading to uncertainties. However, these estimates are often poorly calibrated [18]. It is of paramount importance that the uncertainty estimates are well calibrated [23,29,33,35,47,77]. In many high-dimensional computer vision problems the per-pixel output is often a continuous value [30,36,46,78], i.e., the problem is regression in nature. Recent works focused on *Uncertainty Calibration Error* that generalizes to high dimensional regression [29,33,35,37]. Unlike prior works [3,12,15,66], BayesCap scales to high-dimensional tasks, providing well-calibrated uncertainties.

3   Methodology: BayesCap - Bayesian Identity Cap

We first describe the problem formulation in Sect. 3.1, and preliminaries on uncertainty estimation in Sect. 3.2. In Sect. 3.3, we describe construction of BayesCap that models a probabilistic identity function capable of estimating the high-dimensional complex distribution from the frozen deterministic model, estimating calibrated uncertainty for the predictions.

3.1   Problem Formulation

Let $\mathcal{D} = \{(\mathbf{x}_i, \mathbf{y}_i)\}_{i=1}^N$ be the training set with pairs from domain $\mathbf{X}$ and $\mathbf{Y}$ (i.e., $\mathbf{x}_i \in \mathbf{X}, \mathbf{y}_i \in \mathbf{Y}, \forall i$), where $\mathbf{X}, \mathbf{Y}$ lies in $\mathbb{R}^m$ and $\mathbb{R}^n$, respectively. While our proposed solution is valid for data of arbitrary dimension, we present the formulation for images with applications for image enhancement and translation tasks, such as super-resolution, inpainting, etc. Therefore, $(\mathbf{x}_i, \mathbf{y}_i)$ represents a pair of images, where $\mathbf{x}_i$ refers to the input and $\mathbf{y}_i$ denotes the transformed/enhanced output. For instance, in super-resolution $\mathbf{x}_i$ is a low-resolution image and $\mathbf{y}_i$ its high-resolution version. Let $\mathbf{\Psi}(\cdot; \theta) : \mathbb{R}^m \to \mathbb{R}^n$ represent a Deep Neural Network parametrized by θ that maps images from the set $\mathbf{X}$ to the set $\mathbf{Y}$, e.g. from corrupted to the non-corrupted/enhanced output images.

We consider a real-world scenario, where $\mathbf{\Psi}(\cdot; \theta)$ has already been trained using the dataset $\mathcal{D}$ and it is in a *frozen state* with parameters set to the learned optimal parameters θ^*. In this state, given an input $\mathbf{x}$, the model returns a point estimate of the output, i.e., $\hat{\mathbf{y}} = \mathbf{\Psi}(\mathbf{x}; \theta^*)$. However, point estimates do not capture the distributions of the output ($\mathcal{P}_{\mathbf{Y}|\mathbf{X}}$) and thus the uncertainty in the prediction that is crucial in many real-world applications [27]. Therefore, we propose to estimate $\mathcal{P}_{\mathbf{Y}|\mathbf{X}}$ for the pretrained model in a fast and cheap manner, quantifying the uncertainties of the output without re-training the model itself.

3.2   Preliminaries: Uncertainty Estimation

To understand the functioning of our BayesCap that produces uncertainty estimates for the *frozen or pretrained* neural networks, we first consider a model trained from scratch to address the target task and estimate uncertainty. Let us denote this model by $\boldsymbol{\Psi}_s(\cdot; \zeta) : \mathbb{R}^m \to \mathbb{R}^n$, with a set of trainable parameters given by ζ. To capture the *irreducible* (i.e., aleatoric) uncertainty in the output distribution $\mathcal{P}_{Y|X}$, the model must estimate the parameters of the distribution. These are then used to maximize the likelihood function. That is, for an input $\mathbf{x}_i$, the model produces a set of parameters representing the output given by, $\{\hat{\mathbf{y}}_i, \hat{\nu}_i \dots \hat{\rho}_i\} := \boldsymbol{\Psi}_s(\mathbf{x}_i; \zeta)$, that characterizes the distribution $\mathcal{P}_{Y|X}(\mathbf{y}; \{\hat{\mathbf{y}}_i, \hat{\nu}_i \dots \hat{\rho}_i\})$, such that $\mathbf{y}_i \sim \mathcal{P}_{Y|X}(\mathbf{y}; \{\hat{\mathbf{y}}_i, \hat{\nu}_i \dots \hat{\rho}_i\})$. The likelihood $\mathcal{L}(\zeta; \mathcal{D}) := \prod_{i=1}^N \mathcal{P}_{Y|X}(\mathbf{y}_i; \{\hat{\mathbf{y}}_i, \hat{\nu}_i \dots \hat{\rho}_i\})$ is then maximized in order to estimate the optimal parameters of the network. Moreover, the distribution $\mathcal{P}_{Y|X}$ is often chosen such that uncertainty can be estimated using a closed form solution $\mathscr{F}$ depending on the estimated parameters of the neural network, i.e.,

$$\{\hat{\mathbf{y}}_i, \hat{\nu}_i \dots \hat{\rho}_i\} := \boldsymbol{\Psi}_s(\mathbf{x}_i; \zeta) \tag{1}$$

$$\zeta^* := \underset{\zeta}{\operatorname{argmax}}\ \mathscr{L}(\zeta; \mathcal{D}) = \underset{\zeta}{\operatorname{argmax}} \prod_{i=1}^N \mathcal{P}_{Y|X}(\mathbf{y}_i; \{\hat{\mathbf{y}}_i, \hat{\nu}_i \dots \hat{\rho}_i\}) \tag{2}$$

$$\mathrm{Uncertainty}(\hat{\mathbf{y}}_i) = \mathscr{F}(\hat{\nu}_i \dots \hat{\rho}_i) \tag{3}$$

It is common to use a *heteroscedastic* Gaussian distribution for $\mathcal{P}_{Y|X}$ [27,66], in which case $\boldsymbol{\Psi}_s(\cdot; \zeta)$ is designed to predict the *mean* and *variance* of the Gaussian distribution, i.e., $\{\hat{\mathbf{y}}_i, \hat{\sigma}_i^2\} := \boldsymbol{\Psi}_s(\mathbf{x}_i; \zeta)$, and the predicted *variance* itself can be treated as uncertainty in the prediction. The optimization problem becomes,

$$\zeta^* = \underset{\zeta}{\operatorname{argmax}} \prod_{i=1}^N \frac{1}{\sqrt{2\pi\hat{\sigma}_i^2}} e^{-\frac{|\hat{\mathbf{y}}_i - \mathbf{y}_i|^2}{2\hat{\sigma}_i^2}} = \underset{\zeta}{\operatorname{argmin}} \sum_{i=1}^N \frac{|\hat{\mathbf{y}}_i - \mathbf{y}_i|^2}{2\hat{\sigma}_i^2} + \frac{\log(\hat{\sigma}_i^2)}{2} \tag{4}$$

$$\mathrm{Uncertainty}(\hat{\mathbf{y}}_i) = \hat{\sigma}_i^2. \tag{5}$$

The above equation models the per-pixel residual (between the prediction and the ground-truth) as a Gaussian distribution. However, this may not always be fit, especially in the presence of outliers and artefacts, where the residuals often follow heavy-tailed distributions. Recent works such as [61,63] have shown that heavy-tailed distributions can be modeled as a heteroscedastic generalized Gaussian distribution, in which case $\boldsymbol{\Psi}_s(\cdot; \zeta)$ is designed to predict the *mean* $(\hat{\mathbf{y}}_i)$, *scale* $(\hat{\alpha}_i)$, and *shape* $(\hat{\beta}_i)$ as trainable parameters, i.e., $\{\hat{\mathbf{y}}_i, \hat{\alpha}_i, \hat{\beta}_i\} := \boldsymbol{\Psi}_s(\mathbf{x}_i; \zeta)$,

$$\zeta^* := \underset{\zeta}{\operatorname{argmax}}\ \mathscr{L}(\zeta) = \underset{\zeta}{\operatorname{argmax}} \prod_{i=1}^N \frac{\hat{\beta}_i}{2\hat{\alpha}_i \Gamma(\frac{1}{\hat{\beta}_i})} e^{-(|\hat{\mathbf{y}}_i - \mathbf{y}_i|/\hat{\alpha}_i)^{\hat{\beta}_i}} = \underset{\zeta}{\operatorname{argmin}} -\log \mathscr{L}(\zeta)$$

$$= \underset{\zeta}{\operatorname{argmin}} \sum_{i=1}^N \left(\frac{|\hat{\mathbf{y}}_i - \mathbf{y}_i|}{\hat{\alpha}_i}\right)^{\hat{\beta}_i} - \log \frac{\hat{\beta}_i}{\hat{\alpha}_i} + \log \Gamma(\frac{1}{\hat{\beta}_i}) \tag{6}$$

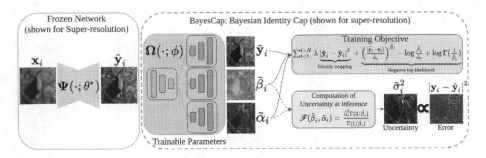

Fig. 2. BayesCap ($\Omega(\cdot; \phi)$) in tandem with the pretrained network with frozen parameters ($\Psi(\cdot; \theta^*)$) (details in Sect. 3.3). While the pretrained network cannot estimate the uncertainty, the proposed BayesCap feeds on the output of the pretrained network and maps it to the underlying probability distribution that allows computation of well calibrated uncertainty estimates

$$\text{Uncertainty}(\hat{\mathbf{y}}_i) = \frac{\hat{\alpha}_i^2 \Gamma(\frac{3}{\tilde{\beta}_i})}{\Gamma(\frac{1}{\tilde{\beta}_i})}. \tag{7}$$

Here $\Gamma(z) = \int_0^\infty x^{z-1} e^{-x} dx, \forall z > 0$, represents the Gamma function [2]. While the above formulation (Eq. (4)–(7)) shows the dependence of various predicted distribution parameters on one another when maximizing the likelihood, it requires training the model from scratch, that we want to avoid. In the following, we describe how we address this problem through our BayesCap.

3.3   Constructing BayesCap

In the above, $\Psi_s(\cdot; \zeta)$ was trained from scratch to predict all the parameters of distribution and does *not* leverage the *frozen* model $\Psi(\cdot; \theta^*)$ estimating $\mathbf{y}_i$ using $\hat{\mathbf{y}}_i$ in a deterministic fashion. To circumvent the training from scratch, we notice that one only needs to estimate the remaining parameters of the underlying distribution. Therefore, to augment the frozen point estimation model , we learn a Bayesian identity mapping represented by $\Omega(\cdot; \phi): \mathbb{R}^n \to \mathbb{R}^n$, that reconstructs the output of the frozen model $\Psi(\cdot; \theta^*)$ and also produces the parameters of the distribution modeling the reconstructed output. We refer to this network as BayesCap (schematic in Fig. 2). As in Eq. (7), we use heteroscedastic generalized Gaussian to model output distribution, i.e.,

$$\Omega(\hat{\mathbf{y}}_i = \Psi(\mathbf{x}_i; \theta^*); \phi) = \{\tilde{\mathbf{y}}_i, \tilde{\alpha}_i, \tilde{\beta}_i\}, \text{ with } \mathbf{y}_i \sim \frac{\tilde{\beta}_i}{2\tilde{\alpha}_i \Gamma(\frac{1}{\tilde{\beta}_i})} e^{-(|\tilde{\mathbf{y}}_i - \mathbf{y}_i|/\tilde{\alpha}_i)^{\tilde{\beta}_i}} \tag{8}$$

To enforce the identity mapping, for every input $\mathbf{x}_i$, we regress the reconstructed output of the BayesCap ($\tilde{\mathbf{y}}_i$) with the output of the pretrained base network ($\hat{\mathbf{y}}_i$). This ensures that, the distribution predicted by BayesCap for an input $\mathbf{x}_i$, i.e., $\Omega(\Psi(\mathbf{x}_i; \theta^*); \phi)$, is such that the point estimates $\tilde{\mathbf{y}}_i$ match the point estimates of the pretrained network $\hat{\mathbf{y}}_i$. Therefore, as the quality of the reconstructed output improves, the uncertainty estimated by $\mathbf{\Omega}(\cdot; \phi)$ also approximates the uncertainty for the prediction made by the pretrained $\mathbf{\Psi}(\cdot; \theta^*)$, i.e.,

$$\tilde{\mathbf{y}}_i \rightarrow \hat{\mathbf{y}}_i \implies \tilde{\sigma}_i^2 = \frac{\tilde{\alpha}_i^2 \Gamma(3/\tilde{\beta}_i)}{\Gamma(1/\tilde{\beta}_i)} \rightarrow \hat{\sigma}_i^2 \tag{9}$$

To train $\mathbf{\Omega}(\cdot; \phi)$ and obtain optimal parameters (ϕ^*), we minimize the fidelity term between $\tilde{\mathbf{y}}_i$ and $\hat{\mathbf{y}}_i$, along with the negative log-likelihood for $\mathbf{\Omega}(\cdot; \phi)$, i.e.,

$$\phi^* = \underset{\phi}{\arg\min} \sum_{i=1}^{N} \lambda \underbrace{|\tilde{\mathbf{y}}_i - \hat{\mathbf{y}}_i|^2}_{\text{Identity mapping}} + \underbrace{\left(\frac{|\tilde{\mathbf{y}}_i - \mathbf{y}_i|}{\tilde{\alpha}_i}\right)^{\tilde{\beta}_i} - \log\frac{\tilde{\beta}_i}{\tilde{\alpha}_i} + \log\Gamma(\frac{1}{\tilde{\beta}_i})}_{\text{Negative log-likelihood}} \tag{10}$$

Here λ represents the hyperparameter controlling the contribution of the fidelity term in the overall loss function. Extremely high λ will lead to improper estimation of the $(\tilde{\alpha})$ and $(\tilde{\beta})$ parameters as other terms are ignored. Equation (10) allows BayesCap to estimate the underlying distribution and uncertainty.

Equation (8) and (10) show that the construction of $\mathbf{\Omega}$ is independent of $\mathbf{\Psi}$ and the $\mathbf{\Omega}$ always performs the Bayesian identity mapping regardless of the task performed by $\mathbf{\Psi}$. This suggests that task specific tuning will have minimal impact on the performance of $\mathbf{\Omega}$. In our experiments, we employed the same network architecture for BayesCap with the same set of hyperparameters, across different tasks and it achieves state-of-the-art uncertainty estimation results without task specific tuning (as shown in Sect. 4), highlighting that BayesCap is not sensitive towards various design choices including architecture, learning-rate, etc.

4 Experiments

We first describe our experimental setup (i.e., datasets, tasks, and evaluation metrics) in Sect. 4.1. We compare our model to a wide variety of state-of-the-art methods quantitatively and qualitatively in Sect. 4.2. Finally in Sect. 4.3, we provide an ablation analysis along with a real world application of BayesCap for detecting out-of-distribution samples.

4.1 Tasks and Datasets

We show the efficacy of our BayesCap on various image enhancement and translation tasks including super-resolution, deblurring, inpainting, and MRI translation, as detailed below. In general, image enhancement and translation tasks are highly ill-posed problems as an *injective function* between input and output may not exist [39,61], thereby necessitating the need to learn a probabilistic mapping to quantify the uncertainty and indicate poor reconstruction of the output images. For each task we choose a well established deterministic pretrained network, for which we estimate uncertainties.

Super-Resolution. The goal is to map low-resolution images to their high-resolution counterpart. We choose pretrained SRGAN [36] as our base model $\Psi(\cdot; \theta^*)$. The BayesCap model $\Omega(\cdot; \phi)$ is trained on ImageNet patches sized 84×84 to perform $4\times$ super-resolution. The resulting combination of SRGAN and BayesCap is evaluated on the Set5 [5], Set14 [76], and BSD100 [41] datasets.

Deblurring. The goal is to remove noise from images corrupted with blind motion. We use the pretrained DeblurGANv2 [31] which shows improvements over the original DeblurGAN [30]. The BayesCap model is evaluated on the GoPro dataset [44], using standard train/test splits.

Inpainting. The goal is to fill masked regions of an input image. We use pretrained DeepFillv2 [75], that improves over DeepFill [74], as the base model for inpainting. Both the original base model and the BayesCap are trained and tested on the standard train/test split of Places365 dataset [79].

MRI Translation. We predict the T2 MRI from T1 MRI, an important problem in medical imaging as discussed in [7,8,24,72,73]. We use the pretrained deterministic UNet as base model [52,63]. Both the base model and BayesCap are trained and tested on IXI datatset [51] following [63].

Baselines. For all tasks, we compare BayesCap against 7 methods in total, out of which 6 baselines can estimate uncertainty of a pretrained model without re-training and one baseline modifies the base network and train it from *scratch* to estimate the uncertainty. The first set of baselines belong to *test-time data augmentation* (**TTDA**) technique [3,65,66], where we generate multiple perturbed copies of the input and use the set of corresponding outputs to compute the uncertainty. We consider three different ways of perturbing the input, (i) per-pixel noise perturbations **TTDAp** [3,65], (ii) affine transformations **TTDAa** [3,65] and (iii) random corruptions from Gaussian blurring, contrast enhancement, and color jittering (**TTDAc**) [3,65]. As additional baseline, we also consider **TTDApac** that generates the multiple copies by combining pixel-level perturbations, affine transformations, and corruptions as described above.

Another set of baselines uses dropout [17,26,34,40,56] before the final predictions. This is possible even for the models *that are not originally trained with dropout*. We refer to this model as DO. In addition, we consider a baseline that combines dropout with test-time data augmentation (**DOpac**). Finally, we also compare against a model trained from scratch to produce the uncertainty as described in [27]. We refer to this as **Scratch**. **Metrics.** We evaluate the performance of various models on two kinds of metrics (i) image reconstruction quality and (ii) predicted uncertainty calibration quality. To measure reconstruction quality, we use SSIM [69] and PSNR. For inpainting we also show mean ℓ_1 and ℓ_2 error, following the convention in original works [74,75]. We emphasize

Fig. 3. Input (LR,**x**) and output of pretrained SRGAN (SR,$\hat{\mathbf{y}}$) along with output of BayesCap ($\{\tilde{\mathbf{y}}, \tilde{\alpha}, \tilde{\beta}\}$). Spatially varying parameters ($\tilde{\alpha}, \tilde{\beta}$) lead to well-calibrated uncertainty $\tilde{\sigma}^2$, highly correlated with the SRGAN error, $|\mathbf{y} - \hat{\mathbf{y}}|^2$

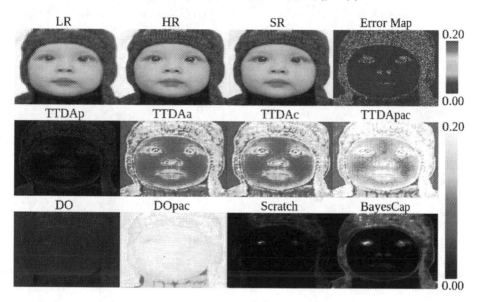

Fig. 4. Qualitative example showing the results of the pre-trained SRGAN model along with the uncertainty maps produced by BayesCap and the other methods. Uncertainty derived from BayesCap has better correlation with the error

that all the methods, *except* Scratch, can use the output of the pretrained base model and only derive the uncertainty maps using different estimation techniques described above. Therefore image reconstruction quality metrics like SSIM, PSNR, mean ℓ_1 and ℓ_2 error remain the same as that of base network. However, Scratch method *does not* have access to the pretrained model, therefore it has to use its own predicted output and uncertainty estimates.

To quantify the quality of the uncertainty, we use the *uncertainty calibration error* (UCE) as described in [23,35] for regression tasks. It measures the discrepancy between the predictive error and predictive uncertainty, given by, UCE $:= \sum_{m=1}^{M} |B_m|/N |\mathrm{err}(B_m) - \mathrm{uncer}(B_m)|$, where B_m is one of the uniformly separated bins, $\mathrm{err}(B_m) := 1/|B_m| \sum_{i \in B_m} ||\hat{\mathbf{y}}_i - \mathbf{y}_i||^2$, and $\mathrm{uncer}(B_m) := 1/|B_m| \sum_{i \in B_m} \hat{\sigma}_i^2$. We also use *correlation coefficient* (C.Coeff.) between the error and the uncertainty, as high correlation is desirable.

Implementation Details. We optimize Eq. (10) using the Adam optimizer [28] and a batch size of 2 with images that are resized to 256×256. During training we exponentially anneal the hyperparameter λ that is initially set to 10. This guides the BayesCap to learn the identity mapping in the beginning, and gradually learn the optimal parameters of the underlying distribution via maximum likelihood estimation.

4.2   Results

Super-Resolution. Table 1 shows the image reconstruction performance along with the uncertainty calibration performance on the Set5, Set14, and the BSD100 datasets for all the methods. We see that BayesCap significantly outperforms all other methods in terms of UCE and C.Coeff while retaining the image reconstruction performance of the base model. For instance, across all the 3 sets, the correlation coefficient is always greater than 0.4 showing that the error and the uncertainty are correlated to a high degree. The model trained from scratch has a correlation coefficient between 0.22–0.31 which is lower than BayesCap. The baselines based on dropout and test-time data augmentation show nearly no correlation between the uncertainty estimates and the error (C.Coeff. of 0.03–0.17). A similar trend can be seen with the UCE where BayesCap has the best UCE scores followed by the model trained from scratch, while the test-time data augmentation and the dropout baselines have a very high UCE (between 0.33 and 0.83) suggesting poorly calibrated uncertainty estimates.

Qualitatively, Fig. 3 shows the prediction of the pretrained SRGAN along with the predictions of the BayesCap showing per-pixel estimated distribution parameters along with the uncertainty map on a sample from Set5 dataset. High correlation between the per-pixel predictive error of SRGAN and uncertainties from BayesCap suggests that BayesCap produces well-calibrated uncertainty estimates. Moreover, Fig. 4 shows that uncertainty produced by other baselines are

Table 1. Quantitative results showing the performance of pretrained SRGAN in terms of PSNR and SSIM, along with the quality of of uncertaintiy maps obtained by BayesCap and other baselines, in terms of UCE and Correlation Coefficient (C.Coeff). All results on 3 datasets including Set5, Set14, and BSD100

D	Metrics	SRGAN	TTDAp	TTDAa	TTDAc	TTDApac	DO	DOpac	Scratch	BayesCap
Set5	PSNR↑	29.40	29.40	29.40	29.40	29.40	29.40	29.40	27.83	29.40
	SSIM↑	0.8472	0.8472	0.8472	0.8472	0.8472	0.8472	0.8472	0.8166	0.8472
	UCE↓	**NA**	0.39	0.40	0.42	0.47	0.33	0.36	0.035	**0.014**
	C.Coeff↑	**NA**	0.17	0.13	0.08	0.03	0.05	0.07	0.28	**0.47**
Set14	PSNR↑	26.02	26.02	26.02	26.02	26.02	26.02	26.02	25.31	26.02
	SSIM↑	0.7397	0.7397	0.7397	0.7397	0.7397	0.7397	0.7397	0.7162	0.7397
	UCE↓	**NA**	0.57	0.63	0.61	0.69	0.48	0.52	0.048	**0.017**
	C.Coeff↑	**NA**	0.07	0.04	0.04	0.06	0.08	0.04	0.22	**0.42**
BSD100	PSNR↑	25.16	25.16	25.16	25.16	25.16	25.16	25.16	24.39	25.16
	SSIM↑	0.6688	0.6688	0.6688	0.6688	0.6688	0.6688	0.6688	0.6297	0.6688
	UCE↓	**NA**	0.72	0.77	0.81	0.83	0.61	0.64	0.057	**0.028**
	C.Coeff↑	**NA**	0.13	0.09	0.11	0.09	0.10	0.08	0.31	**0.45**

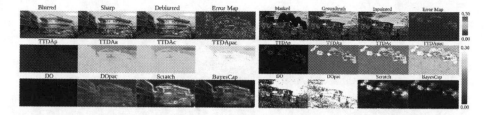

Fig. 5. Qualitative example showing the results of the pretrained DeblurGANv2 and DeepFillv2 on image deblurring (left) and inpainting (right) tasks along with the uncertainty maps produced by different methods

Table 2. Results showing the performance of pretrained DeblurGANv2 in terms of PSNR and SSIM, along with the quality of of uncertaintiy obtained by `BayesCap` and other methods, in terms of UCE and C.Coeff on GoPro dataset

D	Metrics	DeblurGANv2	TTDAp	TTDAa	TTDAc	TTDApac	DO	DOpac	Scratch	BayesCap
GoPro	PSNR↑	29.55	29.55	29.55	29.55	29.55	29.55	29.55	26.16	29.55
	SSIM↑	0.9340	0.9340	0.9340	0.9340	0.9340	0.9340	0.9340	0.8136	0.9340
	UCE↓	**NA**	0.44	0.45	0.49	0.53	0.52	0.59	0.076	**0.038**
	C.Coeff↑	**NA**	0.17	0.13	0.08	0.03	0.05	0.07	0.21	**0.32**

Table 3. Performance of pretrained DeepFillv2 in terms of mean ℓ_1 error, mean ℓ_2 error, PSNR and SSIM, along with the quality of uncertainty obtained by `BayesCap` and other methods, in terms of UCE and C.Coeff on Places365 dataset

D	Metrics	DeepFillv2	TTDAp	TTDAa	TTDAc	TTDApac	DO	DOpac	Scratch	BayesCap
Places365	m. ℓ_1 err.↓	9.1%	9.1%	9.1%	9.1%	9.1%	9.1%	9.1%	15.7%	9.1%
	m. ℓ_2 err.↓	1.6%	1.6%	1.6%	1.6%	1.6%	1.6%	1.6%	5.8%	1.6%
	PSNR↑	18.34	18.34	18.34	18.34	18.34	18.34	18.34	17.24	18.34
	SSIM↑	0.6285	0.6285	0.6285	0.6285	0.6285	0.6285	0.6285	0.6032	0.6285
	UCE↓	**NA**	0.63	0.88	0.87	0.93	1.62	1.49	0.059	**0.011**
	C.Coeff↑	**NA**	0.26	0.11	0.12	0.08	0.09	0.12	0.44	**0.68**

not in agreement with the error (e.g., `TTDAp` does not show high uncertainty within the eye, where error is high) indicating that they are poorly calibrated.

Deblurring. We report the results on the GoPro dataset in Table 2. Deblur-GANv2 achieves significantly better results than `Scratch` (29.55 vs 26.16 PSNR). In terms of UCE, `BayesCap` outperforms all the methods by achieving a low score of 0.038. While the `Scratch` is close and achieves a UCE of 0.076, all the other methods have a UCE that is nearly 10 times higher suggesting that `BayesCap` estimates the most calibrated uncertainty. This is also visible in Fig. 5-(left) where the uncertainties provided by `BayesCap` is correlated with the error (C.Coeff. of 0.32) unlike methods from `TTDA` and `DO` class that have very low correlation between the uncertainty and the error (C.Coeff of 0.03–0.17). While `Scratch` achieves a reasonable score, second only to `BayesCap`, in terms of UCE (0.076 vs. 0,038) and C.Coeff (0.21 vs. 0.32), it has much poorer image reconstruction output with a PSNR of 26.16 and SSIM of 0.8136. The poor

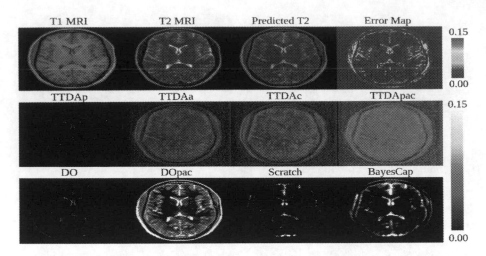

Fig. 6. Qualitative example showing the results of the pretrained UNet for T1 to T2 MRI translation along with the uncertainty produced by different methods.

Table 4. Performance of pretrained UNet for MRI translation in terms of PSNR and SSIM, along with the quality of of uncertainty obtained by `BayesCap` and other methods, in terms of UCE and C.Coeff on IXI Dataset.

D	Metrics	UNet	TTDAp	TTDAa	TTDAc	TTDApac	DO	DOpac	Scratch	BayesCap
IXI	PSNR↑	25.70	25.70	25.70	25.70	25.70	25.70	25.70	25.50	25.70
	SSIM↑	0.9272	0.9272	0.9272	0.9272	0.9272	0.9272	0.9272	0.9169	0.9272
	UCE↓	**NA**	0.53	0.46	0.41	0.44	0.38	0.40	0.036	**0.029**
	C.Coeff↑	**NA**	0.05	0.14	0.16	0.08	0.13	0.47	0.52	**0.58**

reconstruction also justifies the relatively higher uncertainty values for `Scratch` (as seen in Fig. 5-(left)) when compared to `BayesCap`.

Inpainting. Table 3 shows the results on Places365 dataset. The pretrained base model DeepFillv2 [75] achieves a mean L1 error of 9.1%, however `Scratch` is much worse, achieving a mean L1 error of 15.7%. This again demonstrates that training a Bayesian model from scratch often does not replicate the performance of deterministic counterparts. Also, `BayesCap` retains the reconstruction performance of DeepFillv2 [75] and provides well-calibrated uncertainties, as demonstrated by highest C.Coeff. (0.68), and the lowest UCE (0.011). Methods belonging to `TTDA` and `DO` classes are unable to provide good uncertainties (C.Coeff. of 0.08–0.26). The example in Fig. 5-(right) also illustrates an interesting phenomenon. Although, the uncertainties are predicted for the entire image, we see that `BayesCap` automatically learns to have extremely low uncertainty values outside the masked region which is perfectly reconstructed. Within the masked region, uncertainty estimates are highly correlated with the error.

MRI Translation. We perform T1 to T2 MRI translation as described in [61, 63] which has an impact in clinical settings by reducing MRI acquisition times [7,

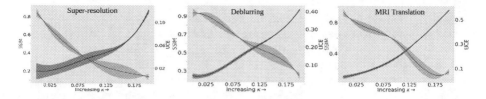

Fig. 7. *Impact of the identity mapping.* Degrading the quality of the identity mapping (SSIM) at inference, leads to poorly calibrated uncertainty (UCE). κ represents the magnitude of noise used for degrading the identity mapping.

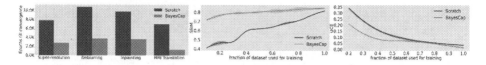

Fig. 8. BayesCap can be trained to achieve optimal performance in fewer epochs (left), while being more data-efficient (achieves better results with fewer samples) as compared to Scratch (middle and right), shown for super-resolution.

8, 24, 72, 73]. The quantitative results on the IXI dataset are show in Table 4. The pretrained base model employing U-Net architecture, achieves a SSIM score of 0.9272 which is nearly matched by Scratch (0.9169). However, we see that BayesCap performs better than Scratch in terms of UCE (0.036 vs. 0.029) and C.Coeff (0.52 vs. 0.58). Other methods are poorly calibrated, as indicated by high UCE and low C.Coeff. This is also evident from Fig. 6, indicating high correlation between the BayesCap uncertainty and error, while low correlation of the same for other methods.

4.3 Ablation Studies

As discussed in Sect. 3.3, BayesCap can help in estimating uncertainty for the frozen model only if BayesCap provides perfect reconstruction. To study this, we design an experiment that deteriorates the identity mapping learned by BayesCap, leading to poor uncertainty estimates. Moreover, we also demonstrate that BayesCap is much more data and compute efficient when compared to Scratch (i.e., model capable of estimating the uncertainty, trained from scratch).

Figure 7 shows that preserving the identity mapping is essential to providing well-calibrated post-hoc uncertainty estimates using BayesCap. Here, we gradually feed increasingly noisy samples (corrupted by zero-mean Gaussian with variance given by κ^2) to the model that leads to poor reconstruction by BayesCap and as a result degrades the identity mapping. As the quality of the identity mapping degrades (decreasing SSIM), we see that quality of uncertainty also degrades (increasing UCE). For instance, for super-resolution, with zero noise in the input the reconstruction quality of the BayesCap is at its peak (SSIM, 0.8472) leading to almost identical mapping. This results in well-calibrated uncertainty maps derived from BayesCap (UCE, 0.014). However, with $\kappa = 0.15$, the reconstruction quality decreases sharply (SSIM of 0.204) leading to poorly calibrated uncertainty (UCE of 0.0587), justifying the need for the identity mapping.

We also show that BayesCap is more efficient than training a Bayesian model from scratch both in terms of time and data required to train the model in Fig. 8-(Left). On all the 4 tasks, BayesCap can be trained 3–5× faster than Scratch. For super-resolution, we show that BayesCap can achieve competitive results even when trained on a fraction of the dataset, whereas the performance of Scratch reduces sharply in low data regime, as shown in Fig. 8-(Middle and Right). For instance, with just 50% of training data, BayesCap performs 33% better than Scratch in terms of SSIM. This is because BayesCap learns the autoencoding task, whereas Scratch learns the original translation task.

4.4 Application: Out-of-Distribution Analysis

The proposed BayesCap focuses on estimating well-calibrated uncertainties of pretrained image regression models in a post-hoc fashion. This can be crucial in many critical real-world scenarios such as autonomous driving and navigation [27,43,68,71]. We consider monocular depth estimation (essential for autonomous driving) and show that the estimated uncertainty can help in detecting *out-of-distribution* (OOD) samples. We take MonoDepth2 [21] that is trained on the KITTI dataset [19]. The resulting model with BayesCap is evaluated on the validation sets of the KITTI dataset, the India Driving Dataset(IDD) [64] as well as the Places365 dataset [79]. The KITTI dataset captures images from German cities, while the India Driving Dataset has images from Indian roads. Places365 consists of images for scene recognition and is vastly different from driving datasets. Intuitively, both IDD and Places365 represent OOD as there is a shift in distribution when compared to training data (KITTI), this is captured in Fig. 9-(a,b,c,d), representing degraded depth and uncertainty on OOD images. The uncertainties also reflect this with increasingly higher values for OOD samples as also shown in the bar plot (Fig. 9-(d)). This suggests that mean uncertainty value for an image can be used to detect if it belongs to OOD.

To quantify this, we plot the ROC curve for OOD detection on a combination of the KITTI (in-distribution) samples and IDD and Places365 (out-of-distribution) samples in Fig. 9-(e). Additionally, we compare using uncertainty for OOD detection against (i) using intermediate features from the pretrained model [49] and (ii) using features from the bottleneck of a trained autoencoder [80]. Samples are marked OOD if the distance between the features of the samples and mean feature of the KITTI validation set is greater than a threshold. We clearly see that using the mean uncertainty achieves far superior results (AUROC of 0.96), against the pretrained features (AUROC of 0.82) and autoencoder based approach (AUROC of 0.72). Despite not being specifically tailored for OOD detection, BayesCap achieves strong results. This indicates the benefits of its well-calibrated uncertainty estimates on downstream tasks.

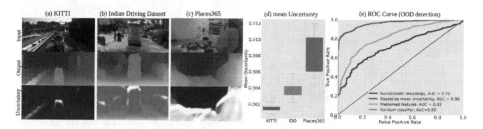

Fig. 9. BayesCap with MonoDepth2 [21] for depth estimation in autonomous driving. Trained on KITTI and evaluated on (a) KITTI, (b) Indian Driving Dataset, and (c) Places365. (d) and (e) Plots show mean uncertainty values and ROC curve for OOD detection respectively, as described in Sect. 4.4.

5 Conclusions

We proposed BayesCap, a fast and cheap post-hoc uncertainty estimation method for pretrained deterministic models. We show that our method consistently produces well-calibrated uncertainty estimates across a wide variety of image enhancement and translation tasks without hampering the performance of the pretrained model. This is in sharp contrast to training a Bayesian model from scratch that is more expensive and often not competitive with deterministic counterparts. We demonstrate that derived calibrated uncertainties can be used in critical scenarios for detecting OOD and helping decision-making. One limitation of our model is the assumption that the input is sufficiently contained in the output of the target task. Future works may address this limitation, as well as extending BayesCap to discrete predictions.

Acknowledgment. This work has been partially funded by the ERC (853489 - DEXIM), by the DFG (2064/1 - Project number 390727645). The authors thank the International Max Planck Research School for Intelligent Systems (IMPRS-IS) for supporting Uddeshya Upadhyay and Shyamgopal Karthik.

References

1. Armanious, K., et al.: MedGAN: medical image translation using GANs. Comput. Med. Imaging Graph. **79**, 101684 (2020)
2. Artin, E.: The Gamma Function. Courier Dover Publications (2015)
3. Ayhan, M.S., Berens, P.: Test-time data augmentation for estimation of heteroscedastic aleatoric uncertainty in deep neural networks. In: MIDL (2018)
4. Bae, G., Budvytis, I., Cipolla, R.: Estimating and exploiting the aleatoric uncertainty in surface normal estimation. In: IEEE ICCV (2021)
5. Bevilacqua, M., Roumy, A., Guillemot, C., Alberi-Morel, M.L.: Low-complexity single-image super-resolution based on nonnegative neighbor embedding. In: BMVC (2012)
6. Blundell, C., Cornebise, J., Kavukcuoglu, K., Wierstra, D.: Weight uncertainty in neural network. In: ICML (2015)

7. Bowles, C., Qin, C., Ledig, C., Guerrero, R., Gunn, R., Hammers, A., Sakka, E., Dickie, D.A., Hernández, M.V., Royle, N., Wardlaw, J., Rhodius-Meester, H., Tijms, B., Lemstra, A.W., van der Flier, W., Barkhof, F., Scheltens, P., Rueckert, D.: Pseudo-healthy image synthesis for white matter lesion segmentation. In: Tsaftaris, S.A., Gooya, A., Frangi, A.F., Prince, J.L. (eds.) SASHIMI 2016. LNCS, vol. 9968, pp. 87–96. Springer, Cham (2016). https://doi.org/10.1007/978-3-319-46630-9_9

8. Chartsias, A., Joyce, T., Giuffrida, M.V., Tsaftaris, S.A.: Multimodal MR synthesis via modality-invariant latent representation. IEEE TMI **37**, 803–814(2017)

9. Chen, T., Fox, E., Guestrin, C.: Stochastic gradient Hamiltonian Monte Carlo. In: ICML. PMLR (2014)

10. Coglianese, C., Lehr, D.: Regulating by robot: administrative decision making in the machine-learning era. Geo, LJ **105**, 1147 (2016)

11. Cohen, J.P., Luck, M., Honari, S.: Distribution matching losses can hallucinate features in medical image translation. In: Frangi, A.F., Schnabel, J.A., Davatzikos, C., Alberola-López, C., Fichtinger, G. (eds.) MICCAI 2018. LNCS, vol. 11070, pp. 529–536. Springer, Cham (2018). https://doi.org/10.1007/978-3-030-00928-1_60

12. Daxberger, E., Kristiadi, A., Immer, A., Eschenhagen, R., Bauer, M., Hennig, P.: Laplace redux-effortless Bayesian deep learning. In: NeurIPS (2021)

13. Dong, C., Loy, C.C., He, K., Tang, X.: Image super-resolution using deep convolutional networks. IEEE TPAMI **38**, 295–307(2015)

14. Dosovitskiy, A., et al.: FlowNet: learning optical flow with convolutional networks. In: IEEE ICCV (2015)

15. Eschenhagen, R., Daxberger, E., Hennig, P., Kristiadi, A.: Mixtures of laplace approximations for improved Post-Hoc uncertainty in deep learning. In: NeurIPS Workshop on Bayesian Deep Learning (2021)

16. Fu, H., Gong, M., Wang, C., Batmanghelich, K., Tao, D.: Deep ordinal regression network for monocular depth estimation. In: IEEE CVPR (2018)

17. Gal, Y., Ghahramani, Z.: Dropout as a Bayesian approximation: representing model uncertainty in deep learning. In: ICML (2016)

18. Gawlikowski, J., et al.: A survey of uncertainty in deep neural networks. arXiv preprint arXiv:2107.03342 (2021)

19. Geiger, A., Lenz, P., Stiller, C., Urtasun, R.: Vision meets robotics: the KITTI dataset. Int. J. Robot. Res. **32**, 1231–1237 (2013)

20. Godard, C., Mac Aodha, O., Brostow, G.J.: Unsupervised monocular depth estimation with left-right consistency. In: IEEE CVPR (2017)

21. Godard, C., Mac Aodha, O., Firman, M., Brostow, G.J.: Digging into self-supervised monocular depth estimation. In: IEEE ICCV (2019)

22. Graves, A.: Practical variational inference for neural networks. In: NIPS (2011)

23. Guo, C., Pleiss, G., Sun, Y., Weinberger, K.Q.: On calibration of modern neural networks. In: ICML. PMLR (2017)

24. Iglesias, J.E., Konukoglu, E., Zikic, D., Glocker, B., Van Leemput, K., Fischl, B.: Is synthesizing MRI Contrast useful for inter-modality analysis? In: Mori, K., Sakuma, I., Sato, Y., Barillot, C., Navab, N. (eds.) MICCAI 2013. LNCS, vol. 8149, pp. 631–638. Springer, Heidelberg (2013). https://doi.org/10.1007/978-3-642-40811-3_79

25. Iizuka, S., Simo-Serra, E., Ishikawa, H.: Let there be color! Joint end-to-end learning of global and local image priors for automatic image colorization with simultaneous classification. ACM Trans. Graph. (ToG) **35**, 1–11 (2016)

26. Kendall, A., Badrinarayanan, V., Cipolla, R.: Bayesian SegNet: model uncertainty in deep convolutional encoder-decoder architectures for scene understanding. arXiv preprint arXiv:1511.02680 (2015)
27. Kendall, A., Gal, Y.: What uncertainties do we need in Bayesian deep learning for computer vision? In: NIPS (2017)
28. Kingma, D.P., Ba, J.: Adam: a method for stochastic optimization. In: ICLR (2015)
29. Kuleshov, V., Fenner, N., Ermon, S.: Accurate uncertainties for deep learning using calibrated regression. In: ICML (2018)
30. Kupyn, O., Budzan, V., Mykhailych, M., Mishkin, D., Matas, J.: DeblurGAN: blind motion deblurring using conditional adversarial networks. In: IEEE CVPR (2018)
31. Kupyn, O., Martyniuk, T., Wu, J., Wang, Z.: DeblurGAN-V2: deblurring (orders-of-magnitude) faster and better. In: IEEE ICCV (2019)
32. Lakshminarayanan, B., Pritzel, A., Blundell, C.: Simple and scalable predictive uncertainty estimation using deep ensembles. arXiv preprint arXiv:1612.01474 (2016)
33. Laves, M.H., Ihler, S., Fast, J.F., Kahrs, L.A., Ortmaier, T.: Well-calibrated regression uncertainty in medical imaging with deep learning. In: MIDL (2020)
34. Laves, M.H., Ihler, S., Kortmann, K.P., Ortmaier, T.: Well-calibrated model uncertainty with temperature scaling for dropout variational inference. arXiv preprint arXiv:1909.13550 (2019)
35. Laves, M.H., Ihler, S., Kortmann, K.P., Ortmaier, T.: Calibration of model uncertainty for dropout variational inference. arXiv preprint arXiv:2006.11584 (2020)
36. Ledig, C., et al.: Photo-realistic single image super-resolution using a generative adversarial network. In: IEEE CVPR (2017)
37. Levi, D., Gispan, L., Giladi, N., Fetaya, E.: Evaluating and calibrating uncertainty prediction in regression tasks. arXiv preprint arXiv:1905.11659 (2019)
38. Litjens, G., et al.: A survey on deep learning in medical image analysis. Med. Image Anal. **42**, 60–88 (2017)
39. Liu, M.Y., Breuel, T., Kautz, J.: Unsupervised image-to-image translation networks. In: NIPS (2017)
40. Maddox, W.J., Izmailov, P., Garipov, T., Vetrov, D.P., Wilson, A.G.: A simple baseline for Bayesian uncertainty in deep learning. In: NeurIPS (2019)
41. Martin, D., Fowlkes, C., Tal, D., Malik, J.: A database of human segmented natural images and its application to evaluating segmentation algorithms and measuring ecological statistics. In: IEEE ICCV (2001)
42. McAllister, R., et al.: Concrete problems for autonomous vehicle safety: advantages of Bayesian deep learning. In: IJCAI (2017)
43. Michelmore, R., Kwiatkowska, M., Gal, Y.: Evaluating uncertainty quantification in end-to-end autonomous driving control. arXiv preprint arXiv:1811.06817 (2018)
44. Nah, S., Hyun Kim, T., Mu Lee, K.: Deep multi-scale convolutional neural network for dynamic scene deblurring. In: IEEE CVPR (2017)
45. Osawa, K., et al.: Practical deep learning with Bayesian principles. In: NeurIPS (2019)
46. Pathak, D., Krahenbuhl, P., Donahue, J., Darrell, T., Efros, A.A.: Context encoders: feature learning by inpainting. In: IEEE CVPR (2016)
47. Phan, B., Salay, R., Czarnecki, K., Abdelzad, V., Denouden, T., Vernekar, S.: Calibrating uncertainties in object localization task. arXiv preprint arXiv:1811.11210 (2018)
48. Plotz, T., Roth, S.: Benchmarking denoising algorithms with real photographs. In: IEEE CVPR (2017)

49. Reiss, T., Cohen, N., Bergman, L., Hoshen, Y.: Panda: adapting pretrained features for anomaly detection and segmentation. In: IEEE CVPR (2021)
50. Riquelme, C., Tucker, G., Snoek, J.: Deep Bayesian bandits showdown: an empirical comparison of Bayesian deep networks for Thompson sampling. arXiv preprint arXiv:1802.09127 (2018)
51. Robinson, E.C., Hammers, A., Ericsson, A., Edwards, A.D., Rueckert, D.: Identifying population differences in whole-brain structural networks: a machine learning approach. In: NeuroImage (2010)
52. Ronneberger, O., Fischer, P., Brox, T.: U-Net: convolutional networks for biomedical image segmentation. In: Navab, N., Hornegger, J., Wells, W.M., Frangi, A.F. (eds.) MICCAI 2015. LNCS, vol. 9351, pp. 234–241. Springer, Cham (2015). https://doi.org/10.1007/978-3-319-24574-4_28
53. van der Schaar, M., Alaa, A.M., Floto, A., Gimson, A., Scholtes, S., Wood, A., McKinney, E., Jarrett, D., Lio, P., Ercole, A.: How artificial intelligence and machine learning can help healthcare systems respond to COVID-19. Mach. Learn. **110**, 1–14 (2021). https://doi.org/10.1007/s10994-020-05928-x
54. Schwarting, W., Alonso-Mora, J., Rus, D.: Planning and decision-making for autonomous vehicles. Ann. Rev. Control Robot. Auton. Syst. **1**, 187–210 (2018)
55. Simonyan, K., Zisserman, A.: Very deep convolutional networks for large-scale image recognition. In: ICLR (2015)
56. Srivastava, N., Hinton, G., Krizhevsky, A., Sutskever, I., Salakhutdinov, R.: Dropout: a simple way to prevent neural networks from overfitting. JMLR **15**, 1929–1958 (2014)
57. Sudarshan, V.P., Upadhyay, U., Egan, G.F., Chen, Z., Awate, S.P.: Towards lower-dose pet using physics-based uncertainty-aware multimodal learning with robustness to out-of-distribution data. Med. Image Anal. **73**, 102187 (2021)
58. Tian, C., Fei, L., Zheng, W., Xu, Y., Zuo, W., Lin, C.W.: Deep learning on image denoising: an overview. Neural Netw. **131**, 251–275 (2020)
59. Upadhyay, U., Awate, S.P.: A mixed-supervision multilevel GAN framework for image quality enhancement. In: Shen, D., Liu, T., Peters, T.M., Staib, L.H., Essert, C., Zhou, S., Yap, P.-T., Khan, A. (eds.) MICCAI 2019. LNCS, vol. 11768, pp. 556–564. Springer, Cham (2019). https://doi.org/10.1007/978-3-030-32254-0_62
60. Upadhyay, U., Awate, S.P.: Robust super-resolution GAN, with manifold-based and perception loss. In: IEEE International Symposium on Biomedical Imaging (2019)
61. Upadhyay, U., Chen, Y., Akata, Z.: Robustness via uncertainty-aware cycle consistency. In: NeurIPS (2021)
62. Upadhyay, U., Chen, Y., Hepp, T., Gatidis, S., Akata, Z.: uncertainty-guided progressive GANs for medical image translation. In: de Bruijne, M., Cattin, P.C., Cotin, S., Padoy, N., Speidel, S., Zheng, Y., Essert, C. (eds.) MICCAI 2021. LNCS, vol. 12903, pp. 614–624. Springer, Cham (2021). https://doi.org/10.1007/978-3-030-87199-4_58
63. Upadhyay, U., Sudarshan, V.P., Awate, S.P.: Uncertainty-aware GAN with adaptive loss for robust MRI image enhancement. In: IEEE ICCV Workshop (2021)
64. Varma, G., Subramanian, A., Namboodiri, A., Chandraker, M., Jawahar, C.: IDD: a dataset for exploring problems of autonomous navigation in unconstrained environments. In: IEEE WACV (2019)
65. Wang, G., Li, W., Aertsen, M., Deprest, J., Ourselin, S., Vercauteren, T.: Test-time augmentation with uncertainty estimation for deep learning-based medical image segmentation. In: MIDL (2018)

66. Wang, G., Li, W., Aertsen, M., Deprest, J., Ourselin, S., Vercauteren, T.: Aleatoric uncertainty estimation with test-time augmentation for medical image segmentation with convolutional neural networks. Neurocomputing **338**, pp. 34–45 (2019)
67. Wang, X., Aitchison, L.: Bayesian OOD detection with aleatoric uncertainty and outlier exposure. In: Fourth Symposium on Advances in Approximate Bayesian Inference (2021)
68. Wang, Y., et al.: Pseudo-lidar from visual depth estimation: bridging the gap in 3D object detection for autonomous driving. In: IEEE CVPR (2019)
69. Wang, Z., Bovik, A.C., Sheikh, H.R., Simoncelli, E.P.: Image quality assessment: from error visibility to structural similarity. In: IEEE TIP (2004)
70. Welling, M., Teh, Y.W.: Bayesian learning via stochastic gradient langevin dynamics. In: ICML (2011)
71. Xu, W., Pan, J., Wei, J., Dolan, J.M.: Motion planning under uncertainty for on-road autonomous driving. In: IEEE ICRA (2014)
72. Ye, D.H., Zikic, D., Glocker, B., Criminisi, A., Konukoglu, E.: Modality propagation: coherent synthesis of subject-specific scans with data-driven regularization. In: Mori, K., Sakuma, I., Sato, Y., Barillot, C., Navab, N. (eds.) MICCAI 2013. LNCS, vol. 8149, pp. 606–613. Springer, Heidelberg (2013). https://doi.org/10.1007/978-3-642-40811-3_76
73. Yu, B., Zhou, L., Wang, L., Shi, Y., Fripp, J., Bourgeat, P.: EA-GANs: edge-aware generative adversarial networks for cross-modality MR image synthesis. IEEE TMI **38**, 1750–1762 (2019)
74. Yu, J., Lin, Z., Yang, J., Shen, X., Lu, X., Huang, T.S.: Generative image inpainting with contextual attention. In: IEEE CVPR (2018)
75. Yu, J., Lin, Z., Yang, J., Shen, X., Lu, X., Huang, T.S.: Free-form image inpainting with gated convolution. In: IEEE CVPR (2019)
76. Zeyde, R., Elad, M., Protter, M.: On single image scale-up using sparse-representations. In: Boissonnat, J.-D., Chenin, P., Cohen, A., Gout, C., Lyche, T., Mazure, M.-L., Schumaker, L. (eds.) Curves and Surfaces 2010. LNCS, vol. 6920, pp. 711–730. Springer, Heidelberg (2012). https://doi.org/10.1007/978-3-642-27413-8_47
77. Zhang, J., Kailkhura, B., Han, T.Y.J.: Mix-n-match: Ensemble and compositional methods for uncertainty calibration in deep learning. In: ICML (2020)
78. Zhang, R., Isola, P., Efros, A.A.: Colorful image colorization. In: Leibe, B., Matas, J., Sebe, N., Welling, M. (eds.) ECCV 2016. LNCS, vol. 9907, pp. 649–666. Springer, Cham (2016). https://doi.org/10.1007/978-3-319-46487-9_40
79. Zhou, B., Lapedriza, A., Khosla, A., Oliva, A., Torralba, A.: Places: a 10 million image database for scene recognition. IEEE TPAMI **40**, 1452–1464 (2017)
80. Zhou, C., Paffenroth, R.C.: Anomaly detection with robust deep autoencoders. In: ACM KDD (2017)
81. Zhu, Y., Tang, Y., Tang, Y., Elton, D.C., Lee, S., Pickhardt, P.J., Summers, R.M.: Cross-domain medical image translation by shared latent gaussian mixture model. In: Martel, A.L., Abolmaesumi, P., Stoyanov, D., Mateus, D., Zuluaga, M.A., Zhou, S.K., Racoceanu, D., Joskowicz, L. (eds.) MICCAI 2020. LNCS, vol. 12262, pp. 379–389. Springer, Cham (2020). https://doi.org/10.1007/978-3-030-59713-9_37

SESS: Saliency Enhancing with Scaling and Sliding

Osman Tursun$^{(\boxtimes)}$ ⓘ, Simon Denman ⓘ, Sridha Sridharan ⓘ,
and Clinton Fookes ⓘ

SAIVT Lab, Queensland University of Technology, Brisbane, Australia
{osman.tursun,s.denman,s.sridharan,c.fookes}@qut.edu.au

Abstract. High-quality saliency maps are essential in several machine learning application areas including explainable AI and weakly supervised object detection and segmentation. Many techniques have been developed to generate better saliency using neural networks. However, they are often limited to specific saliency visualisation methods or saliency issues. We propose a novel saliency enhancing approach called **SESS** (**S**aliency **E**nhancing with **S**caling and **S**liding). It is a method and model agnostic extension to existing saliency map generation methods. With SESS, existing saliency approaches become robust to scale variance, multiple occurrences of target objects, presence of distractors and generate less noisy and more discriminative saliency maps. SESS improves saliency by fusing saliency maps extracted from multiple patches at different scales from different areas, and combines these individual maps using a novel fusion scheme that incorporates channel-wise weights and spatial weighted average. To improve efficiency, we introduce a pre-filtering step that can exclude uninformative saliency maps to improve efficiency while still enhancing overall results. We evaluate SESS on object recognition and detection benchmarks where it achieves significant improvement. The code is released publicly to enable researchers to verify performance and further development. Code is available at https://github.com/neouyghur/SESS.

1 Introduction

Approaches that generate saliency or importance maps based on the decision of deep neural networks (DNNs) are critical in several machine learning application areas including explainable AI and weakly supervised object detection and semantic segmentation. High-quality saliency maps increase the understanding and interpretability of a DNN's decision-making process, and can increase the accuracy of segmentation and detection results.

Since the development of DNNs, numerous approaches have been proposed to efficiently produce high-quality saliency maps. However, most methods have

Supplementary Information The online version contains supplementary material available at https://doi.org/10.1007/978-3-031-19775-8_19.

limited transferability and versatility. Existing methods are designed for DNN models with specific structures (i.e. a global average pooling layer), for certain types of visualisation (for details refer to Sect. 2), or to address a specific limitation. For instance, CAM [24] requires a network with global average pooling. Guided backpropagation (Guided-BP) [18] is restricted to gradient-based approaches. Score-CAM [20] seeks to reduce the method's running-time, while SmoothGrad [17] aims to generate saliency maps with lower noise.

In this work, we propose Saliency Enhancing with Scaling and Sliding (SESS), a model and method agnostic black-box extension to existing saliency visualisation approaches. SESS is only applied to the input and output spaces, and thus does not need to access the internal structure and features of DNNs, and is not sensitive to the design of the base saliency method. It also addresses multiple limitations that plague existing saliency methods. For example, in Fig. 1, SESS shows improvements when applied to three different saliency methods. The saliency map extracted with the gradient-based approach (Guided-BP) is discriminative but noisy. Saliency maps generated by the activation-based method Grad-CAM [14] and perturbation-based method RISE [12] generate smooth saliency maps, but lack detail around the target object and fail to precisely separate the target object from the scene. With SESS, the results of all three methods become less noisy and a more discriminative boundary around the target is obtained.

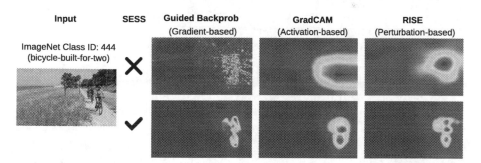

Fig. 1. Example results of three well-known deep neural network visualisation methods with and without SESS. Each of these methods represents one type of saliency map extraction technique. With SESS, all methods generate less noisy and more discriminative saliency maps. The results are extracted with ResNet50, and layer4 is used for Grad-CAM. Target ImageNet class ID is 444 (bicycle-built-for-two).

SESS addresses the following limitations of existing approaches:

- **Weak scale invariance:** Several studies claim that generated saliency maps are inconsistent when there are scale differences [7,21], and we also observe that generated saliency maps are less discriminative when the target objects are comparatively small (see Figs. 1 and 8).
- **Inability to detect multiple occurrences:** Some deep visualisation methods (i.e., Grad-CAM) fail to capture multiple occurrences of the same object in a scene [1,11] (see Fig. 8).

- **Impacted by distractors:** Extracted saliency maps frequently incorrectly highlight regions when distractors exist. This is especially true when the class of the distractor returns a high confidence score, or is correlated with the target class.
- **Noisy results:** Saliency maps extracted with gradient based visualisation approaches [15,19] appear visually noisy as shown in Fig. 1.
- **Less discriminative results:** Activation based approaches (e.g., Grad-CAM) tend to be less discriminative, often highlighting large regions around the target such that background regions are often incorrectly captured as being salient.
- **Fixed input size requirements:** Neural networks with fully-connected layers like VGG-16 [16] require a fixed input size. Moreover, models perform better when the input size at inference is the same as the input size during training. As such, most visualisation methods resize the input to a fixed size. This impacts the resolution and aspect ratio, and may cause poor visualisation results [21].

SESS is a remedy for all of the limitations mentioned above. SESS extracts multiple equally sized (i.e., 224×224) patches from different regions of multiple scaled versions of an input image through resizing and sliding window operations. This step ensures that it is robust to scale variance and multiple occurrences. Moreover, since each extracted patch is equal in size to the default input size of the model, SESS takes advantage of high-resolution inputs and respects the aspect ratio of the input image. Each extracted patch will contribute to the final saliency map, and the final saliency map is the fusion of the saliency maps extracted from patches. In the fusion step, SESS considers the confidence score of each patch, which serves to reduce noise and the impact of distractors while increasing SESS's discriminative power.

The increased performance of SESS is achieved countered a reduction in efficiency due to the use of multiple patches. Quantitative ablation studies show using more scales and denser sliding windows are beneficial, but increase computational costs. To reduce this cost, SESS uses a pre-filtering step that filters out background regions with low target class activation scores. Compared to saliency extraction, the inference step is efficient as it only requires a single forward pass and can exploit parallel computation and batch processing. As such, SESS obtains improved saliency masks with a small increase in run-time requirements. Ablation studies show that the proposed method outperforms its base saliency methods when using pre-filtering with a high pre-filter ratio. In a Pointing Game experiment [22] all methods with SESS achieved significant improvements, despite of a pre-filter ratio of 99% that excludes the majority of extracted patches from saliency generation.

We quantitatively and qualitatively evaluate SESS and conduct ablation studies regarding multiple scales, pre-filtering and fusion. All experimental results show that SESS is a useful and versatile extension to existing saliency methods.

To summarize, the main contributions of this work are as follows:

- We propose, SESS, a model and method agnostic black-box extension to existing saliency methods which is simple and efficient.

– We demonstrate that SESS increases the visual quality of saliency maps, and improves their performance on object recognition and localisation tasks.

2   Related Work

Deep Saliency Methods: Numerous deep neural network-based visualisation methods have been developed in recent years. Based on how the saliency map is extracted, they can be broadly categorised into three groups: gradient-based [15,17,19], class activation-based [14,20,23,24], and perturbation-based [2,5,12] methods.

Gradient-based methods interpret the gradient with respect to the input image as a saliency map. They are efficient as they only require a single forward and backward propagation operation. However, saliency maps generated from raw gradients are visually noisy. Activation-based methods aggregate target class activations of a selected network layer to generate saliency maps. Compared with gradient-based methods, activation-based methods are less noisy, but are also less discriminative and will often incorrectly show strong activations in nearby background regions. Perturbation-based methods generate saliency maps by measuring the changes in the output when the input is perturbed. Perturbation-based methods are slow when compared to most gradient- and activation-based approaches, as they require multiple queries.

Methods can also be split into black-box and white-box according to whether they access the model architecture and parameters. Except for some perturbation-based methods [5,12], saliency methods are all white-box in nature [14,15,17,19,24]. White-box methods are usually more computationally efficient than black-box methods, and require a single forward and backward pass through the network. However, black-box methods are model agnostic, while white-box methods may only work on models with specific architectural features.

Approaches can also be one-shot or multi-shot in nature. One-shot approaches require a single forward and backward pass. Most gradient- and activation-based methods are single-shot. However, multi-shot variants are developed to obtain further improvements. For example, SmoothGrad [17] generates a sharper visualisations through multiple passes across noisy samples of the input image. Integrated Gradients (IG) [19] addresses the "gradient saturation" problem by taking the average of the gradients across multiple interpolated images. Augmented Grad-CAM [10] generates high-resolution saliency maps through multiple low-resolution saliency maps extracted from augmented variants of the input. Smooth Grad-CAM++ [11] utilises the same idea proposed in Smooth-Grad to generate sharper saliency maps. To the best of our knowledge, all perturbation methods are multi-shot in nature, as they require multiple queries to the model, each of which has a different perturbation.

Attempts have been made to make multi-shot approaches more efficient. Most such approaches seek to create perturbation masks in an efficient way. Dabkowski *et al.* [2] generates a perturbation mask with a second neural network. Score-CAM [20] uses class activation maps (CAM) as masks; and Group-CAM [23] follows a similar idea, but further reduces the number of masks through the merging of adjacent maps.

The proposed SESS is a method and model agnostic saliency extension. It can be a "plug-and-play" extension for any saliency methods. However, like perturbation methods, it requires multiple queries. As such, for the sake of efficiency, single-shot and efficient multi-shot approaches are most appropriate for use with SESS.

Enhancing Deep Saliency Visualisations: Many attempts have been made to generate discriminative and low-noise saliency maps. Early Gradient-based methods are visually noisy, and several methods have been proposed to address this. Guided-BP [18] ignores zero gradients during backpropagation by using a RELU as the activation unit. SmoothGrad [18] takes the average gradient of noisy samples [17] to generate cleaner results.

The first of the activation-based methods, CAM, is model sensitive. It requires the model apply a global average pooling over convolutional feature map channels immediately prior to the classification layer [12]. Later variants such as Grad-CAM relax this restriction by using average channel gradients as weights. However, Grad-CAM [14] is also less discriminative, and is unable to locate multiple occurrences of target objects. Grad-CAM++ [1] uses positive partial derivatives of features maps as weights. Smooth Grad-CAM++ [11] combines techniques from both Grad-CAM++ and SmoothGrad to generate sharper visualisations.

Perturbation methods are inefficient, as they send multiple queries to the model. For example, RISE [12] sends 8000 queries to the model to evaluate the importance of regions covered by 8000 randomly selected masks. Recent works reduce the number of masks by using channels in CAMs as masks. For instance, Score-CAM uses all channels in CAMs, while Group-CAM further minimises the number of masks by grouping the channels of CAMs.

All the aforementioned methods have successfully improved certain issues relating to saliency methods, but have limited transferability and versatility. In comparison, SESS is a model and method agnostic extension, which can be applied to any existing saliency approach (though we note that single-pass or efficient multi-pass methods are most suitable). Moreover, SESS is robust to scale-variance, noise, multiple occurrences and distractors. SESS can generate clean and focused saliency maps, and significantly improves the performance of saliency methods for image recognition and detection tasks.

3   Saliency Enhancing with Scaling and Sliding (SESS)

In this section, we introduce SESS. A system diagram is shown in Fig. 2, and the main steps are described in Algorithm 1. The implementation of SESS is simple and includes six steps: *multi-scaling, sliding window, pre-filtering, saliency extraction, saliency fusion, and smoothing*. The first four steps are applied to the input space, and the last two steps are applied at the output space. SESS

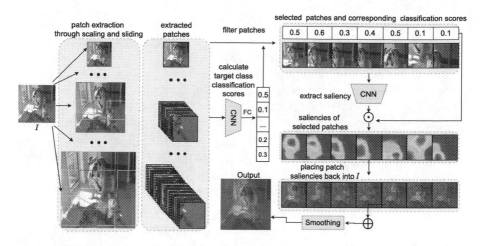

Fig. 2. The SESS process: SESS includes six major steps: multi-scaling, sliding window, pre-filtering, saliency extraction, saliency fusion and smoothing.

is therefore a black-box extension and a model and method agnostic approach. Each of these steps will be discussed in detail in this section.

Algorithm 1. SESS

Input: Image I, Model f, Target class c, Scale n, Window size (w, h), Pre-filtering ratio r

Output: Saliency map L_{sess}^c

1: $M, P \leftarrow []$
2: **for** $i \in [1, \ldots, n]$ **do**   ▷ Scaling
3:     M.append(resize(I, $224 + 64 \times (i - 1)$))
4: **end for**
5: **for** $m \in M$ **do**   ▷ Extracting patches
6:     P.append(sliding-window(m, w, h))
7: **end for**
8: $B \leftarrow$ batchify(P)
9: $S^c \leftarrow f(B, c)$ ▷ S^c as activation scores of class c
10: $S_{fil.}^c, P_{fil.} \leftarrow$ pre-filtering(S^c, P, r)   ▷ filter out patches whose class c activation score is lower than top $(100 - r)\%$
11: $A \leftarrow$ saliency_extraction($P_{fil.}$, f, c)   ▷ get saliency maps of patches after pre-filtering
12: $L \leftarrow$ calibration($P_{fil.}$, A)   ▷ L is a tensor with shape $n \times w \times h$
13: $L' \leftarrow L \otimes S_{fil.}^c$ ▷ Apply channel-wise weight
14: $L_{sess}^c =$ weighted_average(L') ▷ Apply binary weights to obtain the average of the non-zero values
15: **return** L_{sess}^c

Multi-scaling: Generating multiple scaled versions of the input image I is the first step of SESS. In this study, the number of scales, n, ranges from 1 to 12. The set of sizes of all scales is equal to $\{224 + 64 * (i-1)|i \in \{1, 2, \ldots, n\}\}$. The smallest size is equal to the default size of pre-trained models, and the largest size is approximately four times the smallest size. The smaller side of I is resized to the given scale, while respecting the original aspect ratio. M represents the set of all Is at different scales. Benefits of multi-scaling include:

- Most saliency extraction methods are scale-variant. Thus saliency maps generated at different scales are inconsistent. By using multiple scales and combining the saliency results from these, scale-invariance is achievable.
- Small objects will be distinct and visible in salience maps after scaling.

Sliding Window: For efficiency the sliding window step occurs after multi-scaling, which calls n resizing operations. A sliding window is applied to each image in M to extract patches. The width w and height h of the sliding window is set to 224. Thus patch sizes are equal to the default input size of pre-trained models in PyTorch[1]. The sliding operation starts from the top-left corner of the given image, and slides through from top to bottom and left to right. By default, for efficiency, the step-size of the sliding window is set to 224, in other words there is no overlap between neighbouring windows. However, patches at image boundaries are allowed to overlap with their neighbours to ensure that the entire image is sampled. The minimum number of generated patches is $\sum_{i=1}^{n} \lceil 0.25i + 0.75 \rceil^2$. When I has equal width and height and $n = 1$, only one patch of size 224×224 will be extracted, and SESS will return the same results as it's base saliency visualisation method. Thus, SESS can be viewed as a generalisation of existing saliency extraction methods.

Pre-filtering: To increase the efficiency of SESS, a pre-filtering step is introduced. Generating saliency maps for each extracted patch is computationally expensive. Generally, only a few patches are extracted that contain objects which belong to the target class, and they have comparatively large target class activation scores. Calculating target class activation scores requires only a forward pass, and can be sped-up by exploiting batch operations. After sorting the patches based on activation scores, only patches that have a score in the top $(100 - r)\%$ of patches are selected to generate saliency maps. Here, we denote r the pre-filter ratio. When $r = 0$, no pre-filter is applied. As shown in Fig. 3, when r increases only the region which covers the target object remains, and the number of patches is greatly reduced. For instance, only four patches from an initial set of 303 patches are retained after applying a pre-filter with $r = 99$, and these patches are exclusively focussed on the target object. Of course, a large pre-filter ratio i.e., $r > 50$ will decrease the quality of the generated saliency maps as shown in Fig. 3. Note we use notation S_{fil}^c to represent the class "c" activation scores of the remaining patches after filtering.

Saliency Extraction: The saliency maps for the patches retained after pre-filtering are extracted with a base saliency extraction method. Any saliency

---

[1] https://pytorch.org.

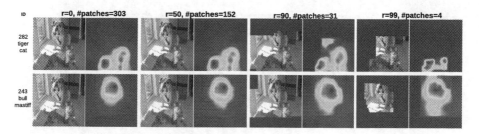

Fig. 3. Visualisation of regions and saliency maps after pre-filtering, when computing saliency maps for the target classes "tiger cat" (top row) and "bull mastiff" (bottom row). All patches that overlap with the red region are removed after pre-filtering. (Color figure online)

extraction method is suitable; however real-time saliency extraction methods including Grad-CAM, Guided-BP and Group-CAM are recommended for efficiency. Each extracted saliency map is normalised with Min-Max normalisation.

Saliency Fusion: Since each patch is extracted from a different position or a scaled version of I, a calibration step is applied before fusion. Each saliency map is overlayed on a zero mask image which has the same size as the scaled I from which it was extracted. Then all masks are resized to the same size as I. Here, notation L represents the channel-wise concatenation of all masks. L has n channels of size $w \times h$. Before fusion, a channel-wise weight is applied. S_{fil}^c, the activation scores of patches after filtering, is used as the weight. The weighted L' is then obtained using,

$$L' = L \otimes S_{fil}^c. \tag{1}$$

Finally, a weighted average that excludes non-zero values is applied at each spatial position for fusion. The modified weighted average is used over uniform average to ignore the zero saliency values introduced during the calibration step. Thus, the saliency value at (i, j) of the final saliency map becomes,

$$L_{sess}(i, j) = \frac{\sum_{i=1}^{n} L'(n, i, j) * \sigma(L'(n, i, j))}{\sum_{i=1}^{n} \sigma(L'(n, i, j))}, \tag{2}$$

where $\sigma(x) = 1$ if $x > \theta$, else $\sigma(x) = 0$, and $\theta = 0$. A Min-Max normalisation is applied after fusion.

Smoothing: Visual artefacts typically remain between patches after fusion, as shown in Fig. 4. Gaussian filtering is applied to eliminate these artefacts. This paper sets the kernel size to 11 and $\sigma = 5$.

4 Experiments

In this section, we first conduct a series of ablation studies to find the optimal hyper-parameters and show the significance of the steps in our approach. Then,

(a) input   (b) before smoothing   (c) after smoothing

Fig. 4. An example of the effect of the smoothing step. After the smoothing step, edge artefacts are removed and the generated saliency is more visually pleasing.

we qualitatively and quantitatively evaluate the efficiency and effectiveness of SESS compared to other widely used saliency methods.

4.1   Experimental Setup

All experiments are conducted on the validation split of the three publicly available datasets: ImageNet-1k [13], PASCAL VOC07 [3] and MSCOCO2014 [9]. Pre-trained VGG-16 (layer: Feature.29) [16] and ResNet-50 (layer: layer4) [6] networks are used as backbones in our experiments. We used Grad-CAM [14], Guided-BP [18] and Group-CAM [23] as base saliency extraction methods. Grad-CAM and Guided-BP are selected as widely used representations of activation-based and gradient-based approaches. We selected Group-CAM as a representative perturbation-based method given it's efficiency.

In qualitative experiments, the number of scales and the pre-filter ratio are set to 12 and 0, and smoothing is applied. In quantitative experiments, to reduce computation time and fair comparison, we employ fewer scales and higher pre-filtering ratios, and omit the smoothing step.

4.2   Ablation Studies

We conduct ablation studies on 2000 random images selected from the validation split of ImageNet-1k [13]. ImageNet pre-trained VGG-16 and ResNet-50 networks are used during the ablation study, and Grad-CAM is used as the base saliency method. Insertion and deletion scores [12] are used as evaluation metrics. The intuition behind this metric is that deletion/insertion of pixels with high saliency will cause a sharp drop/increase in the classification score of the target class. The area under the classification score curve (AUC) is used as the quantitative indicator of the insertion/deletion score. A lower deletion score and a higher insertion score indicates a high-quality saliency map. We also reported the overall score as in [23], where the overall score is defined as $AUC(insertion) - AUC(deletion)$. The implementation is the same as [23]. 3.6% of pixels are gradually deleted from the original image in the deletion test, while

3.6% of pixels are recovered from a highly blurred version of the original image in the insertion test.

Scale: To study the role of multi-scale inputs, we tested different numbers of scales with insertion and deletion tests. As Fig. 5 shows, for both VGG-16 and ResNet-50, when the number of scales increases, the insertion scores increase and the deletion scores decrease. Improvements begin to plateau once five scales are used, and converge once ten scales are used. Overall, the improvement is clear even when using images from the ImageNet dataset, where the main object typically covers the majority of the image, and the role of scaling is less apparent.

Pre-filtering Ratio: To find a high pre-filtering ratio which increases efficiency whilst retaining high performance, we test 10 different global filters from 0 to 0.9. The insertion score decreases as the pre-filter ratio increases, while the deletion scores fluctuate just slightly until the pre-filter ratio reaches 0.6, after which they increase sharply. This shows pre-filter ratio can be set to 0.5 for both high quality and efficiency. However, we used a pre-filter ratio larger than 0.9 in quantitative experiments (Fig. 6).

Channel-Wise Weights: In the fusion step of SESS, channel-wise weights are applied. Figure 7 qualitatively shows the role of the channel-wise weights. With the channel-wise weights, the extracted saliency maps are more discriminative, better highlighting relevant image regions. Without the channel-wise weights, background regions are more likely to be detected as salient.

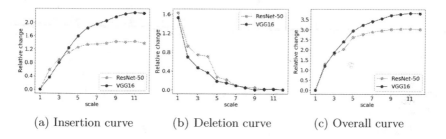

(a) Insertion curve   (b) Deletion curve   (c) Overall curve

Fig. 5. Ablation study considering scale factor in terms of deletion (lower AUC is better), insertion (higher AUC is better), and overall score (higher AUC is better) on the ImageNet-1k validation split (on a randomly selected set of 2k images).

4.3   Image Recognition Evaluation

Here, we also use the insertion and deletion metrics to evaluate the performance of the proposed SESS. We evaluate SESS with three base saliency extraction approaches (Grad-CAM, Guided-BP and Group-CAM) and two backbones (VGG-16 and ResNet-50) on 5000 randomly selected images from the ImageNet-1k validation split. Considering the efficiency, the number of scales and pre-filter

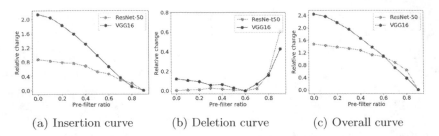

(a) Insertion curve   (b) Deletion curve   (c) Overall curve

Fig. 6. Ablation study considering the pre-filtering operation in terms of deletion (lower is better), insertion (higher is better), and overall (higher is better) scores on the ImageNet-1k validation split (on a randomly selected set of 2k images).

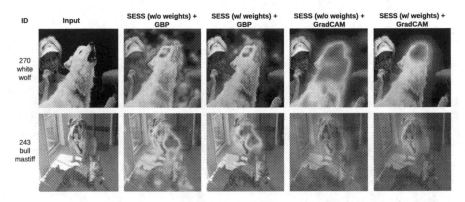

Fig. 7. Impact of the channel-wise weights: The use of the channel-wise weights suppresses activations in background regions, and results in a more focussed saliency map.

ratio are set to 10 and 0.9. With SESS, all three methods with two different backbones achieve improvements, especially Guided-BP whose overall score increases by nearly 5% (Table 1).

4.4 Qualitative Results

This section qualitatively illustrates how much visual improvement SESS brings over the base visualisation approaches such as Grad-CAM [14] and Guided-BP [18]. As a baseline, we selected five visualisation approaches: Guided-BP, SmoothGrad [17], RISE [12], Score-CAM [20] and Grad-CAM. ResNet-50 is selected as the backbone for all methods. We selected challenging cases for queries, including instances with multiple occurrences of the target classes, the presence of distractors, small targets, and curved shapes (Fig. 8).

Table 1. Comparison in terms of deletion (lower is better), insertion (higher is better), and the overall (higher is better) scores on a randomly selected set of 5000 images from the ImageNet-1k validation split.

Method	Model/layer	SESS	Insertion (↑)	Deletion (↓)	Over-all (↑)
Grad-CAM [14]	ResNet-50		68.1	12.1	56.0
		✓	68.6	11.3	57.3
	VGG-16		60.6	9.1	51.5
		✓	60.3	8.1	52.2
Guided-BP [18]	ResNet-50		47.8	11.0	36.8
		✓	53.0	12.0	41.0
	VGG-16		38.8	6.8	32.0
		✓	44.3	6.9	37.4
Group-CAM [23]	ResNet-50		68.2	12.1	56.2
		✓	68.8	11.3	57.4
	VGG-16		61.1	8.8	52.3
		✓	61.1	8.1	53.1

As shown in Fig. 8, visualisation results with SESS are more discriminative and contain less noise. SESS reduces noise and suppresses distractors from the saliency maps, while making the Grad-CAM maps more discriminative and robust to small scales and multiple occurrences.

4.5   Running Time

We calculated the average running time of Grad-CAM, Guided-BP and Group-CAM with/without SESS on a randomly selected set of 5000 images from the ImageNet-1k validation split. Since SESS's running time is decided by the pre-filter ratio and the number of scales, we calculated SESS's running time with two scales (6 and 12) and three pre-filtering ratios (0, 50%, 99%). For comparison, we also calculated the average running time of RISE [12], Score-CAM [20] and XRAI [8]. These experiments are conducted with an NVIDIA T4 Tensor Core GPU and four Intel Xeon 6140 CPUs. Results are given in Table 2. With SESS, the average computation time increased, though it is substantially reduced by using a higher pre-filter ratio and a lower number of scales. Compared to perturbation-based methods, activation/gradient-based methods with SESS are efficient. For instance, in the worst case, SESS requires 16.66 s which is still over twice as fast as RISE and XRAI, which require more than 38 s for a single enquiry.

4.6   Localisation Evaluation

In this section, we evaluate SESS using the Pointing Game introduced in [22]. This allows us to evaluate the performance of the generated saliency maps on

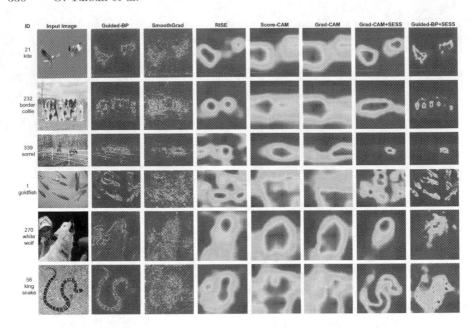

Fig. 8. Qualitative comparison with SOTA saliency methods. Target class IDs and inputs images are shown in the first two columns. Later columns show saliency maps for (from left to right) Guided-BP, SmoothGrad, RISE, Score-CAM, Grad-CAM, Grad-CAM with SESS, and Guided-BP with SESS.

Table 2. Comparison of average run-time (in seconds) of a single query from a set of 5000 randomly selected images from the ImageNet-1K validation split.

Method	Without SESS	With SESS		
		Pre-filter = 0% Scale = 6/12	Pre-filter = 50% Scale = 6/12	Pre-filter = 99% Scale = 6/12
Grad-CAM [14]	0.03	1.23/3.96	0.84/2.36	0.42/0.70
Guided-BP [18]	0.04	1.29/4.09	0.86/2.37	0.41/0.73
Group-CAM [23]	0.13	4.42/16.66	2.54/8.54	0.51/0.92
RISE [12]	38.25	–	–	–
Score-CAM [20]	2.47	–	–	–
XRAI [8]	42.17	–	–	–

weakly-supervised object localisation tasks. The localisation accuracy of each class is calculated using $Acc = \#Hits/(\#Hits + \#Misses)$. $\#Hit$ is increased by one if the highest saliency point is within the ground-truth bounding box of the target class, otherwise $\#Misses$ is incremented. The overall performance is measured by computing the mean accuracy across different categories. Higher accuracy indicates better localisation performance.

We conduct the Pointing Game on the test split of PASCAL VOC07 and the validation split of MSCOCO2014. VGG16 and ResNet50 networks are used as backbones, and are initialised with the pre-trained weights provided by [22]. For implementation, we adopt the TorchRay[2] library. Grad-CAM, Guided-BP and Group-CAM are chosen as base saliency methods. For efficiency and fair comparison, we set the pre-filter ratio to 99% and the number of scales to 10. As per [22], the results of both the "all" and "difficult" sets are reported. The "difficult" set includes images with small objects (covering less than 1/4 of the image) and distractors.

Results are shown in Table 3. With SESS, all three methods achieved significant improvements, especially on the "difficult" set. The average improvement for all cases is 11.2%, and the average improvement for difficult cases is 19.8%. Grad-CAM with SESS achieved SOTA results. The results further demonstrate that the multi-scaling and sliding window steps of SESS are beneficial when scale variance and distractors exist.

Table 3. Comparative evaluation on the Pointing Game [22].

Method	SESS	VOC07 Test (All/Diff)		COCO Val. (All/Diff)	
		VGG16	ResNet50	VGG16	ResNet50
Grad-CAM [14]		86.6/74.0	90.4/82.3	54.2/49.0	57.3/52.3
	✓	**90.4/80.8**	**93.0/86.1**	**62.0/57.8**	**67.0/63.2**
Guided-BP [18]		75.9/53.0	77.2/59.4	39.1/31.4	42.1/35.3
	✓	79.4/64.2	86.0/75.7	39.5/34.5	44.0/39.4
Group-CAM [23]		80.2/64.9	84.2/71.0	47.4/41.1	48.6/42.4
	✓	89.5/79.8	92.4/85.3	61.2/56.9	66.2/62.3
RISE [12]		86.9/75.1	86.4/78.8	50.8/45.3	54.7/50.0
EBP [22]		77.1/56.6	84.5/70.8	39.8/32.8	49.6/43.9
EP [4]		88.0/76.1	88.9/78.7	51.5/45.9	56.5/51.5

5   Conclusion

In this work, we proposed SESS, a novel model and method agnostic extension for saliency visualisation methods. As qualitative results show, with SESS, the generated saliency maps are more visually pleasing and discriminative. Improved quantitative experimental results on object recognition and detection tasks demonstrate that SESS is beneficial for weakly supervised object detection and recognition tasks.

---

[2] https://facebookresearch.github.io/TorchRay.

References

1. Chattopadhay, A., Sarkar, A., Howlader, P., Balasubramanian, V.N.: Grad-CAM++: generalized gradient-based visual explanations for deep convolutional networks. In: 2018 IEEE Winter Conference on Applications of Computer Vision (WACV), pp. 839–847. IEEE (2018)
2. Dabkowski, P., Gal, Y.: Real time image saliency for black box classifiers. Adv. Neural Inf. Process. Syst. **30**, 1–9 (2017)
3. Everingham, M., Van Gool, L., Williams, C.K., Winn, J., Zisserman, A.: The pascal visual object classes (VOC) challenge. Int. J. Comput. Vision **88**(2), 303–338 (2010)
4. Fong, R., Patrick, M., Vedaldi, A.: Understanding deep networks via extremal perturbations and smooth masks. In: Proceedings of the IEEE/CVF International Conference on Computer Vision, pp. 2950–2958 (2019)
5. Fong, R.C., Vedaldi, A.: Interpretable explanations of black boxes by meaningful perturbation. In: Proceedings of the IEEE International Conference on Computer Vision, pp. 3429–3437 (2017)
6. He, K., Zhang, X., Ren, S., Sun, J.: Deep residual learning for image recognition. In: Proceedings of the IEEE Conference on Computer Vision and Pattern Recognition, pp. 770–778 (2016)
7. Jo, S., Yu, I.J.: Puzzle-CAM: improved localization via matching partial and full features. In: 2021 IEEE International Conference on Image Processing (ICIP), pp. 639–643. IEEE (2021)
8. Kapishnikov, A., Bolukbasi, T., Viégas, F., Terry, M.: XRAI: better attributions through regions. In: Proceedings of the IEEE/CVF International Conference on Computer Vision, pp. 4948–4957 (2019)
9. Lin, T.-Y., et al.: Microsoft COCO: common objects in context. In: Fleet, D., Pajdla, T., Schiele, B., Tuytelaars, T. (eds.) ECCV 2014. LNCS, vol. 8693, pp. 740–755. Springer, Cham (2014). https://doi.org/10.1007/978-3-319-10602-1_48
10. Morbidelli, P., Carrera, D., Rossi, B., Fragneto, P., Boracchi, G.: Augmented Grad-CAM: heat-maps super resolution through augmentation. In: ICASSP 2020–2020 IEEE International Conference on Acoustics, Speech and Signal Processing (ICASSP), pp. 4067–4071. IEEE (2020)
11. Omeiza, D., Speakman, S., Cintas, C., Weldermariam, K.: Smooth Grad-CAM++: an enhanced inference level visualization technique for deep convolutional neural network models. arXiv preprint arXiv:1908.01224 (2019)
12. Petsiuk, V., Das, A., Saenko, K.: Rise: randomized input sampling for explanation of black-box models. arXiv preprint arXiv:1806.07421 (2018)
13. Russakovsky, O., et al.: ImageNet large scale visual recognition challenge. Int. J. Comput. Vision **115**(3), 211–252 (2015)
14. Selvaraju, R.R., Cogswell, M., Das, A., Vedantam, R., Parikh, D., Batra, D.: Grad-CAM: visual explanations from deep networks via gradient-based localization. In: Proceedings of the IEEE International Conference on Computer Vision, pp. 618–626 (2017)
15. Simonyan, K., Vedaldi, A., Zisserman, A.: Deep inside convolutional networks: visualising image classification models and saliency maps. arXiv preprint arXiv:1312.6034 (2013)
16. Simonyan, K., Zisserman, A.: Very deep convolutional networks for large-scale image recognition. arXiv preprint arXiv:1409.1556 (2014)
17. Smilkov, D., Thorat, N., Kim, B., Viégas, F., Wattenberg, M.: SmoothGrad: removing noise by adding noise. arXiv preprint arXiv:1706.03825 (2017)

18. Springenberg, J., Dosovitskiy, A., Brox, T., Riedmiller, M.: Striving for simplicity: the all convolutional net. In: ICLR (Workshop Track) (2015)
19. Sundararajan, M., Taly, A., Yan, Q.: Axiomatic attribution for deep networks. In: International Conference on Machine Learning, pp. 3319–3328. PMLR (2017)
20. Wang, H.,et al.: Score-CAM: score-weighted visual explanations for convolutional neural networks. In: Proceedings of the IEEE/CVF Conference on Computer Vision and Pattern Recognition Workshops, pp. 24–25 (2020)
21. Wang, Y., Zhang, J., Kan, M., Shan, S., Chen, X.: Self-supervised equivariant attention mechanism for weakly supervised semantic segmentation. In: Proceedings of the IEEE/CVF Conference on Computer Vision and Pattern Recognition, pp. 12275–12284 (2020)
22. Zhang, J., Bargal, S.A., Lin, Z., Brandt, J., Shen, X., Sclaroff, S.: Top-down neural attention by excitation backprop. Int. J. Comput. Vision **126**(10), 1084–1102 (2018)
23. Zhang, Q., Rao, L., Yang, Y.: Group-CAM: group score-weighted visual explanations for deep convolutional networks. arXiv preprint arXiv:2103.13859 (2021)
24. Zhou, B., Khosla, A., Lapedriza, A., Oliva, A., Torralba, A.: Learning deep features for discriminative localization. In: Proceedings of the IEEE conference on computer vision and pattern recognition, pp. 2921–2929 (2016)

No Token Left Behind: Explainability-Aided Image Classification and Generation

Roni Paiss$^{(\boxtimes)}$, Hila Chefer, and Lior Wolf

The Blavatnik School of Computer Science, Tel Aviv University, Tel Aviv, Israel
paiss.roni@gmail.com

Abstract. The application of zero-shot learning in computer vision has been revolutionized by the use of image-text matching models. The most notable example, CLIP, has been widely used for both zero-shot classification and guiding generative models with a text prompt. However, the zero-shot use of CLIP is unstable with respect to the phrasing of the input text, making it necessary to carefully engineer the prompts used. We find that this instability stems from a selective similarity score, which is based only on a subset of the semantically meaningful input tokens. To mitigate it, we present a novel explainability-based approach, which adds a loss term to ensure that CLIP focuses on all relevant semantic parts of the input, in addition to employing the CLIP similarity loss used in previous works. When applied to one-shot classification through prompt engineering, our method yields an improvement in the recognition rate, without additional training or fine-tuning. Additionally, we show that CLIP guidance of generative models using our method significantly improves the generated images. Finally, we demonstrate a novel use of CLIP guidance for text-based image generation with spatial conditioning on object location, by requiring the image explainability heatmap for each object to be confined to a pre-determined bounding box. Our code is available at https://github.com/apple/ml-no-token-left-behind.

1 Introduction

State-of-the-art computer vision models are often trained as task-specific models that infer a fixed number of labels. In contrast, [29] have demonstrated that by training an image-text matching model that employs Transformers for encoding each modality, tens of downstream tasks can be performed without further training ("zero-shot"), with comparable accuracy to the state of the art [29]. Due to its zero shot capabilities and its semantic latent space, CLIP [29] has been widely used in recent research to guide pretrained generative networks to create images according to a text prompt [10,15,23] and edit an input image according to a text description [20,26,28].

While CLIP shows great promise in zero-shot image classification tasks and generative network guidance, it suffers from instabilities that often lead to

Supplementary Information The online version contains supplementary material available at https://doi.org/10.1007/978-3-031-19775-8_20.

biased similarity scores between non-matching text and image pairs. To mitigate these instabilities, [29] suggest a prompt engineering technique that averages the embeddings of multiple textual templates. Since CLIP summarizes the relation between a given image and a given text with a single similarity score, it can present a myopic behavior and focus on specific elements within the sentence and/or the image. In order to alleviate this issue, it is necessary to rely on an additional signal. In this work, we propose using the explainability maps to steer the optimization process towards solutions that rely on the relevant parts of the input, and away from solutions that focus on irrelevant parts, or on a small subset of the relevant parts. We explore two domains in which we guide CLIP to account for the important tokens in an input prompt: one-shot classification and zero-shot text-based image generation. For one-shot classification, we incorporate a loss based on the explainability scores of the class name in the prompt engineering method proposed by [42]. Our results demonstrate that guiding the prompt engineering process using explainability improves performance in both the class-agnostic and the class-specific cases. In the domain of image editing guided by text, we employ a similar explainability-based loss. This loss allows the generative network to avoid local minima caused by focusing on irrelevant words in the input text, by requiring the explainability scores of important tokens to be high. We demonstrate that our method significantly improves the generated images. Additionally, by applying similar principles to the image-side relevancy map, we use the obtained heatmaps to facilitate CLIP-guided text-based image generation with spatial conditioning. As far as we can ascertain, we are the first to present a spatial layout to image method using CLIP. As we demonstrate, a straightforward application of the similarity score over a requested bounding box does not guarantee that the entire object will be contained within that bounding box. When relying on the explainability heatmap, our method helps ensure that the object does not deviate from the provided bounding box.

2   Related Work

Zero-Shot Classification. Zero-shot classification in computer vision usually refers to a model's ability to generalize to unseen labels. While several works used weakly labeled Google images as training data [3,9,14,32,35,38,40] the method of [21] was perhaps the first to study zero-shot transfer learning to unseen datasets, which is a broader approach to zero-shot classification. This approach was adopted by CLIP [29], which trains an image-text matching engine using an image encoder and a text encoder, via contrastive learning. The image encoder architectures used are either ViT [12] or ResNet [16], and the text encoder is based on a Transformer [37] with the modifications of [30]. CLIP was trained on a set of $400M$ matching image-text pairs, and showed remarkable zero-shot capabilities on the ImageNet dataset [11]. Following CLIP, [42] proposed few-shot prompt engineering to enhance CLIP's classification accuracy on unseen datasets. Their approach opts to learn the textual templates fed to CLIP rather than manually engineering them, as originally done by [29]. As we show, CLIP-guided optimization methods such as CoOp [42] tend to focus on a sparse set of tokens in the input, and often neglect important parts of it. Our work attempts to mitigate this issue by applying an explainability-based loss.

CLIP-Guided Generation. Following the success of CLIP in zero-shot classification, several works have used the similarity scores produced by CLIP to guide pretrained generative networks. These methods usually construct a similarity-based loss, encouraging a generator to produce the output with the desired semantic attributes. Some of the applications of CLIP for guiding generative networks include image manipulation [28], image essence transfer [6], style transfer [15,43], 3D object style editing [25], and image captioning [36]. While demonstrating great capabilities, we show that methods such as StyleCLIP and VQGAN+CLIP are limited by the tendency of CLIP to sometimes ignore meaningful parts of the text. Our explainability-based loss addresses this issue.

Transformer Explainability. Many methods were suggested for generating a heatmap that indicates local relevancy, given an input image and a CNN [4, 24,33,34]. However, the literature on Transformer explainability is relatively sparse and most methods focus on pure self-attention architectures [1,8]. The recent method of [7], which we employ in this work, is the first to also offer a comprehensive method for bi-modal networks.

3   Method

We begin by describing how to produce relevance values for each word and image patch using CLIP. We then describe how our method is applied to one-shot classification via prompt engineering and to zero-shot image generation.

3.1   Explainability

Given a pair of text t and image i, CLIP produces a score $\text{CLIP}(t, i)$, which determines how semantically similar the textual description t is to the image i. The goal of the explainability method is to produce a relevance score for each input text token and image patch in the computation of the similarity score $\text{CLIP}(t, i)$. The score of each token should reflect its impact on the similarity score, and input tokens with the greatest influence on the similarity score should receive a high relevancy score and vice versa. We employ the method described in [7] (details in the supplementary) to produce a relevance score for each image patch and text token, given the calculated similarity score. As the relevance scores are calculated per token, we define the relevance score of a word to be the maximal relevance score among its tokens. For each word $W = w_1, ..., w_n$, where $w_1, ...w_n$ are the tokens it contains, we define its relevance score $\mathcal{R}_{expl}(W)$ as $\mathcal{R}_{expl}(W) = \max_{k \in w_1, ..., w_n} \mathbf{R}[k]$, where $\mathbf{R}[k]$ is the relevance score of [7].

3.2   Prompt Engineering

Image classification is the task of assigning a label from a set of possible classes $c \in C$ to an input image i. In order to adapt CLIP to different classification tasks, [29] propose employing prompt templates with each possible class $c \in C$ inserted, e.g. "A photo of a {label}." These templates are necessary because in the process of CLIP's pre-training most textual inputs are full sentences rather than single words. Let i be the input image to be classified, and let t denote the textual template. $t(c)$ denotes the template t, where the {label} placeholder was replaced by the class name c. CLIP scores per class are obtained using the similarity score between the input image and the class-template as follows:

$$\Pr(\text{output} = c|i) = \frac{e^{\text{CLIP}(t(c),i)}}{\sum_{c' \in C} e^{\text{CLIP}(t(c'),i)}}. \tag{1}$$

[42] replace the manual selection of the textual templates with a few-shot learning of it. Given the desired template length M, the template $t(\text{label}) = v_1, ..., v_i, \{\text{label}\}, v_{i+1}, ..., v_M$ is optimized with a cross-entropy loss using Eq. 1 to extract the distribution. Note that the learned templates are prone to overfitting, due to the small number of example images for each label, which can result in prompts that describe distinctive parts of the images that are irrelevant to their class, yielding a similarity score $\text{CLIP}(t(c), i)$ that is not based on the class name. This problem is most prominent in the one-shot scenario where the prompts are optimized based on a single image per class. To help avoid this phenomenon, our method employs a novel explainability-based loss. For each class $c \in C$ and image i, the similarity score $\text{CLIP}(t(c), i)$ is produced, and then a normalized explainability score is calculated. This score reflects the relevance of the class $c \in C$ to the similarity of the template and the image:

$$\mathcal{S}_{expl}(c) = \frac{\max_{W \in c} \mathcal{R}_{expl}(W)}{\sum_{U \in t(c)/c} \mathcal{R}_{expl}(U)} \tag{2}$$

where, as above, $W \in c$ are the words comprising the label $c \in C$. The score $\mathcal{S}_{expl}(c)$ encapsulates the impact that the class name c has on the calculated similarity score $\text{CLIP}(t(c), i)$, in comparison to all other words in the sentence. Our explainability-based loss is, therefore:

$$\mathcal{L}_{expl} = \lambda_{expl} \left(-\mathcal{S}_{expl}(c_{gt}) + \sum_{c \neq c_{gt} \in C} \mathcal{S}_{expl}(c) \right)$$

where c_{gt} is the ground truth label, and λ_{expl} is a hyperparameter. The first term, $-\mathcal{S}_{expl}(c_{gt})$ encourages the similarity score for the ground truth class to be based on the class name tokens, in order to avoid focusing on other, irrelevant tokens. The second term, $\sum_{c \neq c_{gt} \in C} \mathcal{S}_{expl}(c)$ encourages the similarity score for all the counterfactual classes to be based on tokens that are not the false class name, since the class name does not correspond to the image.

3.3 Zero-shot Text-Guided Image Manipulation

Recent research [20,28] demonstrates that CLIP can be effective for guiding generative networks that synthesize and edit images, by maximizing the similarity

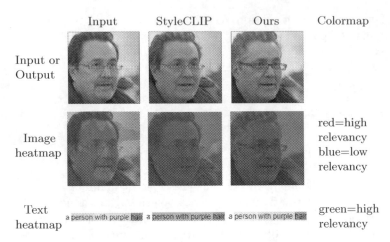

Fig. 1. Manipulations for "A person with purple hair". StyleClip [28] produces a manipulation that is not consistent with the semantic meaning of the prompt, and the color of the person's shirt and the background are altered. Our method generates an output that is faithful to the input text query, and the high values of the explainability heatmaps are much more correlated with the prompt.

score between the desired text and image. As pointed out by [23], methods that integrate a pre-trained generator with the CLIP score to allow text-based editing suffer from instabilities, since similarity-based optimization often reaches local minima that do not reflect the desired semantic meaning of the input query.

As shown in Fig. 1, this shortcoming is often manifested in the explainability scores, and is caused by similarity scores relying only on a partial subset of the semantic tokens in the text query. Thus, our method leverages the explainability scores, to ensure that the output similarity scores are derived from all of the tokens that determine the semantic meaning of the input text. Given a pre-trained generator G (a mapping from a latent vector z to an output image), an input image i, and an input text t, the goal of the optimization algorithm A is to find a vector $A(G, i, t) = z$ such that $G(z)$ combines the visual appearance of image i with the description in the text query t. In order to assess the correlation between the manipulation $G(z)$ and the desired semantic property in t, the algorithm A uses the CLIP similarity score between the manipulated image $G(z)$ and the textual prompt t as a loss term $\mathcal{L}_{similarity} = -\text{CLIP}(G(z), t)$. This loss is applied in addition to other loss terms that lead to a high visual similarity between i and $G(z)$.

As mentioned, $\mathcal{L}_{similarity}$ can produce biased similarity scores, which do not reflect the semantic meaning in t, due to focusing only on a subset of the semantically important words in t. Our method remedies this bias by adding a loss term that encourages CLIP to attend to all semantic words in t.

Let S be the set of semantic words in t. Since the textual prompts for image editing are of the format: "a person/ man/ woman with {*description*}" or of the format "a {*description*} person/ man/ woman", the set S is considered to be the words that comprise the description. Our method adds the following explainability-based loss term to the optimization algorithm A:

$$\mathcal{L}_{expl} = -\lambda_{expl}\frac{1}{|S|}\left(\sum_{s \in S}\mathcal{R}_{expl}(s)\right), \tag{3}$$

where λ_{expl} is a hyperparameter. For example, in Fig. 1, the set of semantic words is defined to be: $S = \{$"purple", "hair"$\}$. This helps the optimization process to favor results where the similarity score is based on the hair color of the subject of the image. As can be seen in the figure, when our loss is not applied, the optimization results in coloring the shirt and the background.

Choosing λ_{expl}. Our modified optimization algorithm has an additional hyperparameter λ_{expl}. Since CLIP-based optimization is sensitive to the choice of hyperparameters, it is better to set them based specifically on the input image i, input text t, and generator G. In order to provide an automatic mechanism for choosing λ_{expl}, we consider a range of possible λ_{expl}, and choose the value of λ_{expl} for which the similarity predicted by CLIP for the generated image and the input text is maximal. Note that after applying our method, the similarity scores become more stable, as they consider all semantic tokens in the input.

3.4   Zero-Shot Text-to-Image with Spatial Conditioning

While the textual descriptions provided to CLIP can include the spatial positioning of objects, images generated by optimizing CLIP similarity score with such texts tend not to follow these spatial restrictions, as shown in Fig 4. We attribute this to the nature of the task CLIP was trained on, which is predicting how similar a given image is to a given text. The existence of matching entities in both inputs is a stronger indication of their similarity than the positions of the entities. This intuition is reflected in the distribution by speech parts (POS) of the explainability scores calculated for CLIP, as shown in the supplementary.

To alleviate this shortcoming of providing spatial positioning with textual description, we add spatial conditioning as an additional input. As far as we can ascertain, CLIP has not been used before for image generation conditioned on spatial masks. The somewhat related task of CLIP-based zero-shot image inpainting was recently successfully performed by [2,26], who point out that a simple masking of the input image presented to CLIP, such that the similarity score is only predicted based on a specific region of the image, does not guarantee that the generated object will not deviate from the provided region.

Preventing the generator from producing objects outside the designated spatial locations requires applying additional losses on the background or restricting the optimization process such that only the parts of the latent vector that affect the desired region of the image are optimized. These methods limit the spatial conditioning to applications that receive an input image to be used as unaltered background. However, since the explainability maps produced for CLIP indicate the location of the objects, we can effectively limit the location of generated objects using explainability.

Algorithm 1. Obtain IoU loss from masks and image.

Input : (i) $m_1, ..., m_k$- bounding boxes of the objects to be generated, (ii) $t_1, ..., t_k$-
textual descriptions of the objects we wish to generate, (iii) C- a pre-trained CLIP
model. (iv) the input image i, (v) a threshold T over relevancy scores (vi) $temp$ - a
temperature for the sigmoid operation (vii) $expl$ - the image explainability algorithm
for CLIP, which outputs relevance scores for the image patches for each pair of image
and text.

Output : $\mathcal{L}_{IoU}$- an explainability-based IoU loss for the input masks $m_1, ..., m_k$, and
input image i.

1: $\mathcal{L}_{IoU} \leftarrow 0$
2: $for\ j \in \{1, ..., k\},:$
3:     $R_j \leftarrow expl(i, t_j)$
4:     $R_j \leftarrow R_j / R_j.max()$
5:     $pred\_mask_j \leftarrow sigmoid((R_j - T) * temp)$
6:     $intersection \leftarrow \sum_{p \in i}(pred\_mask_j[p] \cdot m_j[p])$
7:     $\mathcal{L}_{IoU} \leftarrow \mathcal{L}_{IoU} + \frac{2*intersection}{\sum_{p \in i} pred\_mask_j[p] + \sum_{p \in i} m_j[p]}$

Our method employs an IoU-inspired loss based on the explainability of the
image modality. Algorithm 1 describes how we produce the loss $\mathcal{L}_{IoU}$ given the
input spatial conditioning masks $m_1, ..., m_k$ and the input image i. For each
bounding box m_j and the text t_j describing the object we wish to generate in
that location, we generate the explainability for CLIP with the entire image i and
text t_j (L.3). This explainability map represents the location in which the object
is currently found in the image by CLIP. We then transform the explainability
map into a semi-binary mask (L.5) by substracting a threshold value T and
passing the output through a sigmoid function with high temperature $temp$. This
predicted mask is then used to calculate a Dice Loss with respect to the ground
truth object mask (L.7). After calculating the IoU-based loss, we incorporate
the similarity-based loss $\mathcal{L} = -\lambda_{expl} \cdot \mathcal{L}_{IoU} - \sum_{j=1}^{k} \text{CLIP}(i, t_j)$, where λ_{expl}
is a hyperparameter, and the sum calculates the CLIP similarity between the
image and each object we wish to generate, in order to ensure that all objects in
$\{t_1, ..., t_k\}$ appear in i. λ_{expl}, T and $temp$ are chosen empirically, using examples
from the MSCOCO [22] validation set.

4   Experiments

We evaluate our method in various contexts, including one-shot prompt engineering for image classification based on [42], zero-shot text-guided image
manipulation based on [28], and zero-shot text-guided image generation with
spatial conditioning based on CLIP-guidance for VQGAN [10, 13].

4.1   One-Shot Prompt Engineering

We compare the classification accuracy of CLIP using the prompts optimized
with CoOp [42] and with our method, as described in Sect. 3.2. Following [42],

Original image | Ground truth class label CoOp [42]   Ours | Counter factual class label CoOp [42]   Ours

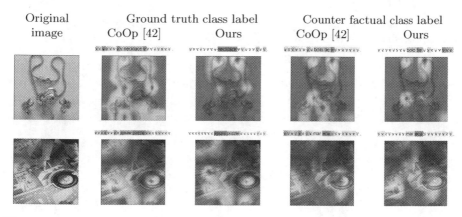

Fig. 2. A qualitative comparison of prompt engineering using CoOp [42] with and without our method on 2 exemplary samples from ImageNetV2 [31]. We present the relevance maps for the ground truth class chosen by our method ("necklace", "jigsaw"), and the counterfactual class chosen by CoOp ("bolo tie", "maraca"). The learned vectors for the prompt are annotated by the letter "v" in the textual explainability maps, since the vectors do not represent actual tokens. As can be seen, for the ground truth classes "necklace" and "jigsaw", our prompts encourage CLIP to focus on the class name in the input text, while CoOp leads CLIP to consider unrelated tokens. This can cause CLIP to produce biased similarity scores based on the engineered prompts.

we evaluate the methods on ImageNet [11] test set, ImageNetV2 [31], ImageNet-Sketch [39], ImageNet-A [18], and Imagenet-R [17].

Following [42], two scenarios are tested: unified prompt engineering and class-specific prompt engineering. In the unified scenario, a single prompt is optimized for all class names. In the class-specific (CSC) case, a different prompt is optimized per class. Note that for all datasets, the prompts are optimized using labeled examples only from the ImageNet training set, in order to test the robustness of the optimized prompts on different ImageNet variations.

For both methods we test different backbones for the visual encoder of CLIP (see Table 1), including variations of ViT [12] and of ResNet [16]. Following [42], we optimize a template with $M = 16$ tokens. We also include results for $M = 4$, as it was noted to sometimes achieve superior results on ImageNet.

Two options for positioning the class name tokens in the prompt were reported in [42], with similar outcomes. The first has the class name located in the middle of the prompt, i.e.: $t = v_1, ..., v_8, \{label\}, v_9, ..., v_{16}$, where $v_1, ..., v_{16}$ are the prompt tokens, and the second has the class name located at the end, i.e.: $t = v_1, ..., v_{16}, \{label\}$. In the main text we report the results of the former; for the latter, see the supplementary. We use $\lambda_{expl} = 1$ for experiments that use ViT-B/16 as backbone and $\lambda_{expl} = 3$ for all other backbones.

Table 1 shows the 1-shot accuracy of CoOp and our method, in addition to 0-shot manual prompt selection and linear probing of the image embedding produced by CLIP, which are the baselines used by CoOp [42]. 2-shot and

Table 1. 1-shot accuracy (in percentage) of linear probing (LP) and CLIP [29] with prompts produced by the method of [42] (CoOp) or with our explainability-guided variant, with various image backbones. All methods are trained on ImageNet and evaluated on several variants. Unified stands for training a single prompt for all classes, and CSC (class-specific) stands for optimizing a prompt for each class name. Results are averaged over 3 random seeds.

Image backbone		ImageNet Unified	CSC	ImageNetV2 Unified	CSC	INet-Sketch Unified	CSC	ImageNet-A Unified	CSC	ImageNet-R Unified	CSC
ResNet-50	0-shot	58.18	-	51.34	-	**33.32**	-	21.65	-	56.00	-
	LP	21.70	-	17.78	-	5.57	-	0.11	-	0.07	-
	CoOp M=16	54.45	28.40	47.11	23.92	28.12	11.80	19.97	10.39	50.38	26.83
	CoOp M=4	57.63	35.88	50.34	30.49	30.18	16.28	21.43	13.45	53.53	32.06
	Ours M=16	58.13	31.90	51.30	26.82	32.49	13.52	22.12	11.77	**57.73**	29.26
	Ours M=4	**59.05**	**38.79**	**52.33**	**33.58**	32.59	**18.25**	**22.74**	**14.33**	57.15	**34.63**
ResNet-101	0-shot	61.62	-	54.81	-	**38.71**	-	28.05	-	64.38	-
	LP	26.41	-	21.75	-	9.61	-	0.08	-	0.07	-
	CoOp M=16	57.84	33.51	51.25	26.98	33.80	15.78	26.82	14.28	59.02	32.40
	CoOp M=4	60.41	38.96	53.68	33.43	36.71	21.19	27.94	16.91	61.08	40.27
	Ours M=16	61.76	36.01	55.02	31.07	37.96	18.70	29.56	15.97	63.92	36.02
	Ours M=4	**62.31**	**40.77**	**55.65**	**35.18**	38.51	**21.88**	**30.07**	**17.80**	**65.33**	**40.44**
ViT-B/16	0-shot	66.73	-	60.83	-	46.15	-	47.77	-	73.76	-
	LP	32.26	-	27.33	-	16.48	-	0.10	-	0.08	-
	CoOp M=16	63.66	38.86	56.53	33.55	40.96	22.59	43.93	23.30	69.33	42.76
	CoOp M=4	66.93	46.20	60.14	40.28	44.97	28.26	47.44	31.87	72.12	51.16
	Ours M=16	67.09	40.78	60.28	35.25	45.71	23.77	48.29	25.03	74.9	44.43
	Ours M=4	**67.62**	**48.74**	**61.07**	**42.58**	**46.33**	**30.34**	**49.46**	**34.08**	**75.66**	**53.75**
ViT-B/32	0-shot	62.05	-	54.79	-	**40.82**	-	29.57	-	65.99	-
	LP	27.03	-	22.38	-	11.32	-	0.12	-	0.08	-
	CoOp M=16	57.64	33.42	50.24	28.39	35.12	17.63	27.53	13.84	59.46	34.30
	CoOp M=4	61.48	40.66	54.01	34.52	38.26	22.76	29.56	18.58	63.11	41.12
	Ours M=16	62.55	38.63	55.14	33.23	40.40	21.08	31.22	16.8	**67.22**	39.64
	Ours M=4	**63.69**	**42.98**	**55.84**	**37.21**	40.23	**24.26**	30.78	**20.48**	66.49	**44.22**

4-shot results are available in the supplementary. As can be seen, both linear probing and CoOp are heavily overfitting and actually achieve significantly lower accuracy than 0-shot results. Using the explainability-based loss, our method is consistently able to improve upon CoOp, leading to higher accuracy across all backbones, all datasets, and both scenarios (unified and CSC).

A Sensitivity analysis for λ_{expl} is presented in Fig. 5, showing that the improvement in accuracy is consistent across a large range of λ_{expl} values. Figure 2 presents a qualitative comparison of using CoOp with and without our method, see caption for a detailed description.

4.2 Zero-Shot Text-Guided Image Manipulation

Next, we compare our explainability-based optimization (Sect. 3.3) with the optimization presented in [28]. There are three methods for text-based image editing using StyleGAN [19] presented by [28] - latent optimization, mapper training, and global directions extraction. We focus on latent optimization, since

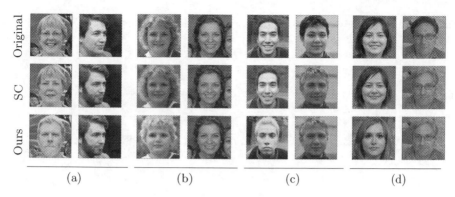

Original
SC
Ours

(a)   (b)   (c)   (d)

Fig. 3. A qualitative comparison between StyleCLIP (SC) and our method on 4 different textual prompts. (a) "A man with a beard", (b) "A person with purple hair", (c) "A blond man", (d) "A person with grey hair". For each prompt we present examples where StyleCLIP is successful (right column), and unsuccessful (left column). For the failure cases, the optimization in StyleCLIP hardly modifies the original image, leading to a high identity preservation score when no semantic change was applied. When StyleCLIP is successful, our method produces similar or identical results.

our focus is on zero-shot methods and the other two methods employ additional training. As described in Sect. 3.3, we add the explainability-based loss from Eq. 3 to the optimization algorithm of [28]. We choose the set of hyperparameters for our explainability-based loss from the set: $\lambda_{expl} = \{0, 0.5, 1, 1.5, 2, 2.5, 3, 3.5\}$, and use the best value for λ_{expl} according to the highest CLIP similarity score.

Since the optimization in [28] requires a different hyperparameter setting for each prompt, we select prompts that appear in the paper or official code, and use the same hyperparameters (in other words, we do not manually select the hyperparameters for our method). Next, we choose 20 random seeds to be used across all our experiments to generate the images to be edited, i. For each image i, and text prompt t we produce the edited image with StyleCLIP's optimization, and with our modified optimization.

For evaluation, we extract all examples where our method produces a different output than StyleCLIP, i.e., all cases where the automatic procedure selected $\lambda_{expl} \neq 0$, and conduct a user study among 46 users. Users were asked to evaluate each manipulation by the similarity between the manipulated image and the input prompt t and by the loss of identity between the manipulated image and the original image i, both on a scale of $1 - 5$ (higher is better).

Figure 3 presents sample results from our user study (See the supplementary for full results). Notice that for challenging manipulations, such as using the prompt "a man with a beard" on a woman, StyleCLIP tends to leave the input image i almost unchanged. In contrast, in many of these cases, our method compels the optimization to reach a solution that fits the required semantic edit. We present the results of the user study for each prompt in Table 2 (see results with standard deviation in the supplementary). As can be seen, our method

Table 2. A user study comparing text-based image editing with StyleCLIP (SC) and our method on 4 different textual prompts: "A man with a beard", "A person with purple hair", "A blond man", "A person with grey hair". Quality refers to the similarity between the prompt and the manipulation; Identity refers to the identity preservation of the manipulation. Scores are averaged across 20 random seeds, on a scale of 1–5 (higher is better).

Method	A man with a beard		A person with purple hair		A blond man		A person with grey hair	
	Quality	Identity	Quality	Identity	Quality	Identity	Quality	Identity
SC	2.92	**3.61**	1.17	**4.13**	3.93	**3.67**	2.59	**3.82**
Ours	**4.28**	2.23	**2.29**	3.51	**4.28**	2.63	**3.27**	3.10

produces results that are consistently rated by users as more similar to the target text. However, StyleCLIP, which, as can be seen in Fig. 3, often leaves the images unchanged, obtains a higher identity preservation score. Evidently, the gap in the identity score is much bigger for the prompts "A **man** with a beard" and "A blond **man**". These prompts modify the gender of the subject of the image i, thereby requiring a more substantial identity change.

4.3   Zero-Shot Text-to-Image with Spatial Conditioning

We use CLIP-guided VQGAN as implemented by [10]. Since, as far as we can ascertain, there is no previous literature on zero-shot CLIP-guided text-to-image generation with spatial conditioning on the location of the generated objects, we use two variations of a similarity-based CLIP loss to create baselines without explainability conditioning. The first baseline employs the loss $\mathcal{L}_{masked} = \sum_{t \in \{t_1, \dots, t_k\}} \sum_{m \in \{m_1, \dots, m_k\}} -CLIP(i_m, t)$, where i_m is the image i masked according to bounding box m, i.e. for each mask m, we black out all pixels outside m, in order to ensure that the objects identified by CLIP reside within the bounding boxes, and t is a prompt of the form "a photo of {label}" where "label" is the target class to be generated in bounding box m. This masking technique has also been employed in previous works [2,26], for CLIP-guided image inpainting. The second similarity-based baseline we consider employs the loss $\mathcal{L}_{masked}$ in addition to the similarity loss in the unmasked image $\mathcal{L}_{similarity} = \sum_{t \in \{t_1, \dots, t_k\}} -CLIP(i, t)$. This baseline uses the loss: $\mathcal{L} = \mathcal{L}_{similarity} + \mathcal{L}_{masked}$, which considers both the information inside the bounding boxes and the information in the entire image.

As mentioned in Sect. 3.4, since a simple similarity-based loss has no spatial restrictions, the baselines produce objects outside the input bounding box (see Fig. 4), while our loss produces objects within the bounding box, thanks to spatial conditioning based on explainability. Moreover, the examples in Fig. 4 demonstrate the ability of our method to generate images in a variety of cases, including multiple bounding boxes with varying heights and widths (see the supplementary for additional examples and visualizations of the explainability maps). As smaller bounding boxes require stronger supervision, we set λ_{expl_i} for object i to be $\lambda_{expl_i} = \frac{0.15}{\sqrt{r(m_i)}}$, where m_i is the bounding box assigned to object

Input conditioning	Textual conditioning	Similarity-based	Similarity-based 2	ours

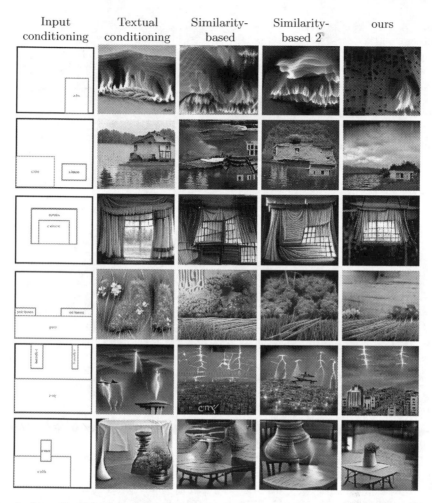

Fig. 4. A qualitative comparison between the two similarity-based baselines and our method for CLIP-guided zero-shot text-based image generation with spatial conditioning. Textual conditioning refers to specifying the spatial positioning of objects within the text prompts, for example "a vase on a table". Additional examples are presented in the supplementary.

i and $r(m_i)$ is the ratio between the area of the mask and the area of the entire image. The threshold T is set to 0.1 and $temp$ is set to 20.

In order to provide quantitative metrics for our spatially conditioned text-to-image generation, we use the validation set from MSCOCO [22] which contains bounding boxes for each object in the image. In order to ensure a varying number and size of objects, while maintaining enough background to allow object-free generation, which is challenging for CLIP-guided VQGAN, we filter the layout as follows: we keep the k largest bounding boxes whose commutative area is less than 50% of the image, where adding the next largest bounding box would result

Table 3. Precision, recall, F1, average precision, and average recall for spatially conditioned image generation with our method, and two similarity-based baselines (results in percentage). Metrics were averaged across 100 random samples from the MSCOCO [22] validation set and four random seeds. Average precision and average recall are calculated using DETR [5].

	Similarity-based	Similarity-based 2	Ours
Precision	46.4	26.9	**71.7**
Recall	48.3	30.5	**63.4**
F1	40.5	24.28	**62.6**
AP	8	5.4	**26.2**
AR	21.6	19	**40**
$AP_{0.5}$	18	15.4	**56.5**

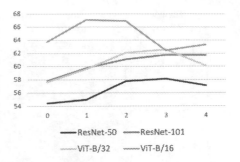

Fig. 5. 1-shot accuracy (in percentage) on the ImageNet test set for different choices of λ_{expl} for all visual backbones of CLIP. The accuracy achieved by the baselines is denoted as $\lambda_{expl} = 0$.

in occupying more than 50% of the image. By focusing on the largest objects we also help ensure that the size of each bounding box suffices for the CLIP encoder. For our first experiment, we sample 100 MSCOCO images at random, and use our method and the similarity-based baselines to generate images corresponding to the annotated spatial layout. We then produce an explainability map for each text description t, as described in Algorithm 1 (L.2-3). We use these maps as soft semantic segmentation, binarize them using thresholds produced with Otsu's method [27], and calculate the precision, recall, and F1 scores of the binarized maps with the ground truth bounding boxes $m_1, .., m_k$. Note that both precision and recall are limited and cannot reach 100% due to the square shape of the bounding boxes, which is not suited to non-square objects. Precision is also limited because images often contain more than one instance of a specific class, leading to a high explainability score for the other occurrences as well. As can be seen in Table 3, our method significantly outperforms the baselines.

Next, we use object detection to evaluate the quality of the generated objects, as well as the overlap between their location and the target spatial condition. DETR [5] is used to produce bounding boxes for each object. These bounding boxes are evaluated against the input spatial conditioning masks using the average precision and average recall scores. As can be seen in Table 3, our method greatly outperforms the baselines in this evaluation as well, implying that the explainability signal is indeed indicative enough to enforce spatial restrictions over an image.

5   Discussion

In our experiments, we presented a generic application of explainability to improve classification, image editing, and image synthesis. There are also specific situations in which a limited view of the input is detrimental and where explainability can help ensure a more even distribution of information pooling. One such case, studied in the supplementary, is that of compound nouns, e.g. "apple juice" or "blueberry muffin". As we show, state-of-the-art zero-shot text-to-image generation engines might overly emphasize or ignore some of the textual input, leading to misinterpretation of the text. The method we present for equalizing the contributions to avoid such neglect not only leads to considerably better image outputs for such cases, but also slightly improves the FID score for other sentences. See the supplementary for full details of the method implementation, visual examples, and the results of a user study conducted against results obtained with a state-of-the-art method. In order to demonstrate the wide applicability of our approach, we have modified multiple zero-shot and one-shot approaches. While the baseline approaches are impressive, we do note that they are not yet ready to replace supervised methods. Prompt engineering is not yet competitive with supervised methods, CLIP-guided VQGAN often generates substandard images, and StyleCLIP optimization method often fails and requires different parameters for each prompt. Therefore, other signals need to be considered to allow zero-shot applications to compete against fully supervised ones. Explainability, as we show, is an example of such beneficial signal.

6 Conclusions

While explainability methods are constantly improving, their use as a feedback mechanism to improve classification or generation methods is still relatively unexplored. As far as we can ascertain, their utilization as such a building block is currently limited to weakly supervised segmentation [7,41]. In this work, we show how explainability can help overcome the neglect problem of bi-modal transformers, which have become a cornerstone in the current rapid evolution of zero-shot methods. We demonstrate how preventing neglect, as reflected through the lens of the explainability score, helps improve one-shot classification, zero-shot image editing, and zero-shot layout-conditioned image generation. In the first two domains, neglect is prevented in the text domain, while in the latter, the constraint on the heatmap is placed in the image domain.

Acknowledgments. This project has received funding from the European Research Council (ERC) under the European Unions Horizon 2020 research and innovation programme (grant ERC CoG 725974). We thank Ariel Landau for his assistance.

References

1. Abnar, S., Zuidema, W.: Quantifying attention flow in transformers. arXiv preprint arXiv:2005.00928 (2020)
2. Bau, D., et al.: Paint by word. arXiv preprint arXiv:2103.10951 (2021)
3. Berg, T., Forsyth, D.: Animals on the web. In: CVPR (2006)
4. Binder, A., Montavon, G., Lapuschkin, S., Müller, K.-R., Samek, W.: Layer-wise relevance propagation for neural networks with local renormalization layers. In: Villa, A.E.P., Masulli, P., Pons Rivero, A.J. (eds.) ICANN 2016. LNCS, vol. 9887, pp. 63–71. Springer, Cham (2016). https://doi.org/10.1007/978-3-319-44781-0_8

5. Carion, N., Massa, F., Synnaeve, G., Usunier, N., Kirillov, A., Zagoruyko, S.: End-to-end object detection with transformers. arXiv preprint arXiv:2005.12872 (2020)

6. Chefer, H., Benaim, S., Paiss, R., Wolf, L.: Image-based clip-guided essence transfer (2021)

7. Chefer, H., Gur, S., Wolf, L.: Generic attention-model explainability for interpreting bi-modal and encoder-decoder transformers. In: Proceedings of the IEEE/CVF International Conference on Computer Vision (ICCV), pp. 397–406, October 2021

8. Chefer, H., Gur, S., Wolf, L.: Transformer interpretability beyond attention visualization. In: Proceedings of the IEEE/CVF Conference on Computer Vision and Pattern Recognition (CVPR), pp. 782–791, June 2021

9. Chen, X., Gupta, A.K.: Webly supervised learning of convolutional networks. In: 2015 IEEE International Conference on Computer Vision (ICCV), pp. 1431–1439 (2015)

10. Crowson, K.: VQGAN+CLIP (2021). https://colab.research.google.com/drive/1L8oL-vLJXVcRzCFbPwOoMkPKJ8-aYdPN

11. Deng, J., Dong, W., Socher, R., Li, L.J., Li, K., Fei-Fei, L.: ImageNet: a large-scale hierarchical image database. In: 2009 IEEE Conference on Computer Vision and Pattern Recognition, pp. 248–255 (2009)

12. Dosovitskiy, A., et al.: An image is worth 16x16 words: transformers for image recognition at scale. arXiv preprint arXiv:2010.11929 (2020)

13. Esser, P., Rombach, R., Ommer, B.: Taming transformers for high-resolution image synthesis. In: Proceedings of the IEEE/CVF Conference on Computer Vision and Pattern Recognition (CVPR), pp. 12873–12883, June 2021

14. Fergus, R., Fei-Fei, L., Perona, P., Zisserman, A.: Learning object categories from internet image searches. Proc. IEEE **98**(8), 1453–1466 (2010). https://doi.org/10.1109/JPROC.2010.2048990

15. Gal, R., Patashnik, O., Maron, H., Chechik, G., Cohen-Or, D.: StyleGAN-NADA: CLIP-guided domain adaptation of image generators. ACM Trans. Graph. **41**, 1–3 (2021)

16. He, K., Zhang, X., Ren, S., Sun, J.: Deep residual learning for image recognition. In: Proceedings of the IEEE Conference on Computer Vision and Pattern Recognition, pp. 770–778 (2016)

17. Hendrycks, D., et al.: The many faces of robustness: a critical analysis of out-of-distribution generalization. In: Proceedings of the IEEE/CVF International Conference on Computer Vision (ICCV), pp. 8340–8349, October 2021

18. Hendrycks, D., Zhao, K., Basart, S., Steinhardt, J., Song, D.: Natural adversarial examples. In: Proceedings of the IEEE/CVF Conference on Computer Vision and Pattern Recognition (CVPR), pp. 15262–15271, June 2021

19. Karras, T., Laine, S., Aittala, M., Hellsten, J., Lehtinen, J., Aila, T.: Analyzing and improving the image quality of StyleGAN. In: CVPR, pp. 8110–8119 (2020)

20. Kim, G., Ye, J.C.: DiffusionCLIP: text-guided image manipulation using diffusion models (2021)

21. Li, A., Jabri, A., Joulin, A., van der Maaten, L.: Learning visual n-grams from web data. In: Proceedings of the IEEE International Conference on Computer Vision (ICCV), October 2017

22. Lin, T.-Y., et al.: Microsoft COCO: common objects in context. In: Fleet, D., Pajdla, T., Schiele, B., Tuytelaars, T. (eds.) ECCV 2014. LNCS, vol. 8693, pp. 740–755. Springer, Cham (2014). https://doi.org/10.1007/978-3-319-10602-1_48

23. Liu, X., Gong, C., Wu, L., Zhang, S., Su, H., Liu, Q.: FuseDream: training-free text-to-image generation with improved CLIP+GAN space optimization. arXiv:abs/2112.01573 (2021)

24. Lundberg, S.M., Lee, S.I.: A unified approach to interpreting model predictions. In: Advances in Neural Information Processing Systems, pp. 4765–4774 (2017)

25. Michel, O.J., Bar-On, R., Liu, R., Benaim, S., Hanocka, R.: Text2mesh: text-driven neural stylization for meshes. arXiv:abs/2112.03221 (2021)

26. Omri Avrahami, D.L., Friedn, O.: Blended diffusion for text-driven editing of natural images. arXiv preprint arxiv:2111.14818 (2021)

27. Otsu, N.: A threshold selection method from gray-level histograms. IEEE Trans. Syst. Man Cybern. **9**(1), 62–66 (1979)

28. Patashnik, O., Wu, Z., Shechtman, E., Cohen-Or, D., Lischinski, D.: StyleCLIP: text-driven manipulation of StyleGAN imagery. In: Proceedings of the IEEE/CVF International Conference on Computer Vision, pp. 2085–2094 (2021)

29. Radford, A., et al.: Learning transferable visual models from natural language supervision. arXiv preprint arXiv:2103.00020 (2021)

30. Radford, A., et al.: Language models are unsupervised multitask learners. OpenAI Blog **1**(8), 9 (2019)

31. Recht, B., Roelofs, R., Schmidt, L., Shankar, V.: Do ImageNet classifiers generalize to ImageNet? In: Chaudhuri, K., Salakhutdinov, R. (eds.) Proceedings of the 36th International Conference on Machine Learning. Proceedings of Machine Learning Research, vol. 97, pp. 5389–5400. PMLR, 09–15 June 2019. https://proceedings. mlr.press/v97/recht19a.html

32. Rubinstein, M., Joulin, A., Kopf, J., Liu, C.: Unsupervised joint object discovery and segmentation in internet images. In: 2013 IEEE Conference on Computer Vision and Pattern Recognition, pp. 1939–1946 (2013). https://doi.org/10.1109/ CVPR.2013.253

33. Selvaraju, R.R., Cogswell, M., Das, A., Vedantam, R., Parikh, D., Batra, D.: Grad-CAM: visual explanations from deep networks via gradient-based localization. In: Proceedings of the IEEE International Conference on Computer Vision, pp. 618–626 (2017)

34. Shrikumar, A., Greenside, P., Shcherbina, A., Kundaje, A.: Not just a black box: learning important features through propagating activation differences. arXiv preprint arXiv:1605.01713 (2016)

35. Tang, K., Joulin, A., Li, L.J., Fei-Fei, L.: Co-localization in real-world images. In: 2014 IEEE Conference on Computer Vision and Pattern Recognition, pp. 1464–1471 (2014). https://doi.org/10.1109/CVPR.2014.190

36. Tewel, Y., Shalev, Y., Schwartz, I., Wolf, L.: Zero-shot image-to-text generation for visual-semantic arithmetic. CoRR abs/2111.14447 (2021). arXiv:abs/2111.14447

37. Vaswani, A., et al.: Attention is all you need. In: Advances in Neural Information Processing Systems, pp. 5998–6008 (2017)

38. Vijayanarasimhan, S., Grauman, K.: Keywords to visual categories: multiple-instance learning for weakly supervised object categorization. In: 2008 IEEE Conference on Computer Vision and Pattern Recognition, pp. 1–8 (2008). https://doi. org/10.1109/CVPR.2008.4587632

39. Wang, H., Ge, S., Lipton, Z., Xing, E.P.: Learning robust global representations by penalizing local predictive power. In: Wallach, H., Larochelle, H., Beygelzimer, A., d'Alché-Buc, F., Fox, E., Garnett, R. (eds.) Advances in Neural Information Processing Systems, vol. 32. Curran Associates, Inc. (2019). https://proceedings. neurips.cc/paper/2019/file/3eefceb8087e964f89c2d59e8a249915-Paper.pdf

40. Wang, X.J., Zhang, L., Li, X., Ma, W.Y.: Annotating images by mining image search results. IEEE Trans. Pattern Anal. Mach. Intell. **30**(11), 1919–1932 (2008). https://doi.org/10.1109/TPAMI.2008.127

41. Zabari, N., Hoshen, Y.: Semantic segmentation in-the-wild without seeing any segmentation examples. arXiv:abs/2112.03185 (2021)
42. Zhou, K., Yang, J., Loy, C.C., Liu, Z.: Learning to prompt for vision-language models. arXiv preprint arXiv:2109.01134 (2021)
43. Zhu, P., Abdal, R., Femiani, J.C., Wonka, P.: Mind the gap: domain gap control for single shot domain adaptation for generative adversarial networks. arXiv:abs/2110.08398 (2021)

Interpretable Image Classification with Differentiable Prototypes Assignment

Dawid Rymarczyk[1,2]([✉])(ID), Łukasz Struski[1](ID), Michał Górszczak[1](ID),
Koryna Lewandowska[3](ID), Jacek Tabor[1](ID), and Bartosz Zieliński[1,2](ID)

[1] Faculty of Mathematics and Computer Science, Jagiellonian University,
Kraków, Poland
[2] Ardigen SA, Kraków, Poland
dawid.rymarczyk@student.uj.edu.pl
[3] Department of Cognitive Neuroscience and Neuroergonomics, Institute of Applied
Psychology, Jagiellonian University, Kraków, Poland

Abstract. Existing prototypical-based models address the black-box
nature of deep learning. However, they are sub-optimal as they often
assume separate prototypes for each class, require multi-step optimization, make decisions based on prototype absence (so-called negative reasoning process), and derive vague prototypes. To address those shortcomings, we introduce ProtoPool, an interpretable prototype-based model
with positive reasoning and three main novelties. Firstly, we reuse prototypes in classes, which significantly decreases their number. Secondly, we
allow automatic, fully differentiable assignment of prototypes to classes,
which substantially simplifies the training process. Finally, we propose a
new focal similarity function that contrasts the prototype from the background and consequently concentrates on more salient visual features. We
show that ProtoPool obtains state-of-the-art accuracy on the CUB-200-
2011 and the Stanford Cars datasets, substantially reducing the number
of prototypes. We provide a theoretical analysis of the method and a
user study to show that our prototypes capture more salient features
than those obtained with competitive methods. We made the code available at https://github.com/gmum/ProtoPool.

Keywords: Deep learning · Interpretability · Case-based reasoning

1 Introduction

The broad application of deep learning in fields like medical diagnosis [3] and
autonomous driving [53], together with current law requirements (such as GDPR
in EU [21]), enforces models to explain the rationale behind their decisions.
That is why explainers [6,23,29,39,44] and self-explainable [4,7,58] models are
developed to justify neural network predictions. Some of them are inspired by

Supplementary Information The online version contains supplementary material
available at https://doi.org/10.1007/978-3-031-19775-8_21.

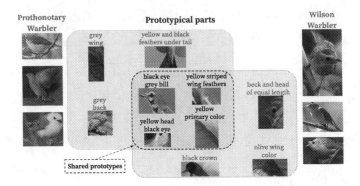

Fig. 1. Automatically discovered prototypes (Names of prototypical parts were generated based on the annotations from CUB-200-2011 dataset (see details in Supplementary Materials)) for two classes, *Prothonotary Warbler* and *Wilson Warbler* (each class represented by three images on left and right side). Three prototypical parts on the blue and green background are specific for a *Prothonotary Warbler* and *Wilson Warbler*, respectively (they correspond to heads and wings feathers). At the same time, four prototypes shared between those classes (related to yellow feathers) are presented in the intersection. Prototypes sharing reduces their amount, leads to a more interpretable model, and discovers classes similarities.

mechanisms used by humans to explain their decisions, like matching image parts with memorized prototypical features that an object can poses [8,27,34,43].

Recently, a self-explainable model called Prototypical Part Network (ProtoPNet) [8] was introduced, employing feature matching learning theory [40,41]. It focuses on crucial image parts and compares them with reference patterns (prototypical parts) assigned to classes. The comparison is based on a similarity metric between the image activation map and representations of prototypical parts (later called prototypes). The maximum value of similarity is pooled to the classification layer. As a result, ProtoPNet explains each prediction with a list of reference patterns and their similarity to the input image. Moreover, a global explanation can be obtained for each class by analyzing prototypical parts assigned to particular classes.

However, ProtoPNet assumes that each class has its own separate set of prototypes, which is problematic because many visual features occur in many classes. For instance, both *Prothonotary Warbler* and *Wilson Warbler* have yellow as a primary color (see Fig. 1). Such limitation of ProtoPNet hinders the scalability because the number of prototypes grows linearly growing number of classes. Moreover, a large number of prototypes makes ProtoPNet hard to interpret by the users and results in many background prototypes [43].

To address these limitations, ProtoPShare [43] and ProtoTree [34] were introduced. They share the prototypes between classes but suffer from other drawbacks. ProtoPShare requires previously trained ProtoPNet to perform the merge-pruning step, which extends the training time. At the same time, ProtoTree builds a decision tree and exploits the negative reasoning process that may result in explanations based only on prototype absence. For example, a model can predict a *sparrow* because an image does not contain red feathers, a long beak, and wide wings. While this characteristic is true in the case of a *sparrow*, it also matches many other species.

To deal with the above shortcomings, we introduce ProtoPool, a self-explainable prototype model for fine-grained images classification. ProtoPool introduces significantly novel mechanisms that substantially reduce the number of prototypes and obtain higher interpretability and easier training. Instead of using hard assignment of prototypes to classes, we implement the soft assignment represented by a distribution over the set of prototypes. This distribution is randomly initialized and binarized during training using the Gumbel-Softmax trick. Such a mechanism simplifies the training process by removing the pruning step required in ProtoPNet, ProtoPShare, and ProtoTree. The second novelty is a focal similarity function that focuses the model on the salient features. For this purpose, instead of maximizing the global activation, we widen the gap between the maximal and average similarity between the image activation map and prototypes (see Fig. 4). As a result, we reduce the number of prototypes and use the positive reasoning process on salient features, as presented in Figs. 2 and 10.

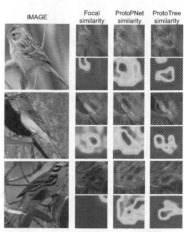

Fig. 2. Focal similarity focuses the prototype on a salient visual feature. While the other similarity metrics are more distributed through the image, making the interpretation harder to comprehend. It is shown with three input images, the prototype activation map, and its overlay

We confirm the effectiveness of ProtoPool with theoretical analysis and exhaustive experiments, showing that it achieves the highest accuracy among models with a reduced number of prototypes. What is more, we discuss interpretability, perform a user study, and discuss the cognitive aspects of the Proto Pool over existing methods.

The main achievements of the paper can be summarized as follows:

- We construct ProtoPool, a case-based self-explainable method that shares prototypes between data classes without any predefined concept dictionary.
- We introduce fully differentiable assignments of prototypes to classes, allowing the end-to-end training.
- We define a novel similarity function, called focal similarity, that focuses the model on the salient features.
- We increase interpretability by reducing prototypes number and providing explanations in a positive reasoning process.

2   Related Works

Attempts to explain deep learning models can be divided into the post hoc and self-explainable [42] methods. The former approaches assume that the reasoning process is hidden in a black box model and a new explainer model has to be created to reveal it. Post hoc methods include a saliency map [31,38,44–46] generating a heatmap of crucial image parts, or Concept Activation Vectors (CAV) explaining the internal network state as user-friendly concepts [9,14,23,25,55]. Other methods provide counterfactual examples [1,15,33,36,52] or analyze the networks' reaction to the image perturbation [6,11,12,39]. Post hoc methods are easy to implement because they do not interfere with the architecture, but they can produce biased and unreliable explanations [2]. That is why more focus is recently put on designing self-explainable models [4,7] that make the decision process directly visible. Many interpretable solutions are based on the attention [28,48,54,57–59] or exploit the activation space [16,37], e.g. with adversarial autoencoder. However, most recent approaches built on an interpretable method introduced in [8] (ProtoPNet) with a hidden layer of prototypes representing the activation patterns.

ProtoPNet inspired the design of many self-explainable models, such as Tes-Net [51] that constructs the latent space on a Grassman manifold without prototypes reduction. Other models like ProtoPShare [43] and ProtoTree [34] reduce the number of prototypes used in the classification. The former introduces data-dependent merge-pruning that discovers prototypes of similar semantics and joins them. The latter uses a soft neural decision tree that may depend on the negative reasoning process. Alternative approaches organize the prototypes hierarchically [17] to classify input at every level of a predefined taxonomy or transform prototypes from the latent space to data space [27]. Moreover, prototype-based solutions are widely adopted in various fields such as medical imaging [3,5,24,47], time-series analysis [13], graphs classification [56], and sequence learning [32].

3   ProtoPool

In this section, we describe the overall architecture of ProtoPool presented in Fig. 3 and the main novelties of ProtoPool compared to the existing models, including the mechanism of assigning prototypes to slots and the focal similarity. Moreover, we provide a theoretical analysis of the approach.

Overall Architecture. The architecture of ProtoPool, shown in Fig. 3, is generally inspired by ProtoNet [8]. It consists of convolutional layers f, a prototype pool layer g, and a fully connected layer h. Layer g contains a pool of M trainable prototypes $P = \{p_i \in \mathbb{R}^D\}_{i=1}^M$ and K slots for each class. Each slot is implemented as a distribution $q_k \in \mathbb{R}^M$ of prototypes available in the pool, where successive values of q_k correspond to the probability of assigning successive prototypes to slot k ($\|q_k\| = 1$). Layer h is linear and initialized to enforce

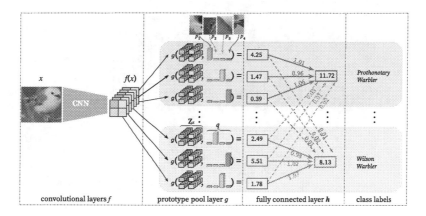

Fig. 3. The architecture of our ProtoPool with a prototype pool layer g. Layer g contains a pool of prototypes $p_1 - p_4$ and three slots per class. Each slot is implemented as a distribution $q \in \mathbb{R}^4$ of prototypes from the pool, where successive values of q correspond to the probability of assigning successive prototypes to the slot. In this example, p_1 and p_2 are assigned to the first slot of *Prothonotary Warbler* and *Wilson Warbler*, respectively. At the same time, the shared prototypes p_3 and p_4 are assigned to the second and third slots of both classes.

the positive reasoning process, i.e. weights between each class c and its slots are initialized to 1 while remaining weights of h are set to 0.

Given an input image $x \in X$, the convolutional layers first extract image representation $f(x)$ of shape $H \times W \times D$, where H and W are the height and width of representation obtained at the last convolutional layer for image x, and D is the number of channels in this layer. Intuitively, $f(x)$ can be considered as a set of $H \cdot W$ vectors of dimension D, each corresponding to a specific location of the image (as presented in Fig. 3). For the clarity of description, we will denote this set as $Z_x = \{z_i \in f(x) : z_i \in \mathbb{R}^D, i = 1, ..., H \cdot W\}$. Then, the prototype pool layer is used on each k-th slot to compute the aggregated similarity $g_k = \sum_{i=1}^{M} q_k^i g_{p_i}$ between Z_x and all prototypes considering the distribution q_k of this slot, where g_p is defined below. Finally, the similarity scores (K values per class) are multiplied by the weight matrix w_h in the fully connected layer h. This results in the output logits, further normalized using softmax to obtain a final prediction.

Focal Similarity. In ProtoPNet [8] and other models using prototypical parts, the similarity of point z to prototype p is defined as[1] $g_p(z) = \log(1 + \frac{1}{\|z-p\|^2})$, and the final activation of the prototype p with respect to image x is given by $g_p = \max_{z \in Z_x} g_p(z)$. One can observe that such an approach has two possible disadvantages. First, high activation can be obtained when all the elements in Z_x are similar to a prototype. It is undesirable because the prototypes can then

---

[1] The following regularization is used to avoid numerical instability in the experiments: $g_p(z) = \log(\frac{\|z-p\|^2+1}{\|z-p\|^2+\varepsilon})$, with a small $\varepsilon > 0$.

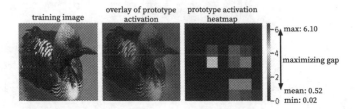

Fig. 4. Our focal similarity limits high prototype activation to a narrow area (corresponding to white and black striped wings). It is obtained by widening the gap between the maximal and average activation (equal 6.10 and 0.52, respectively). As a result, our prototypes correspond to more salient features (according to our user studies described in Sect. 5).

concentrate on the background. The other negative aspect concerns the training process, as the gradient is passed only through the most active part of the image.

To prevent those behaviors, in ProtoPool, we introduce a novel focal similarity function that widens the gap between maximal and average activation

$$g_p = \max_{z \in Z_x} g_p(z) - \text{mean}_{z \in Z_x}\, g_p(z), \tag{1}$$

as presented in Fig. 4. The maximal activation of focal similarity is obtained if a prototype is similar to only a narrow area of the image x (see Fig. 2). Consequently, the constructed prototypes correspond to more salient features (according to our user studies described in Sect. 5), and the gradient passes through all elements of Z_x.

Assigning One Prototype per Slot. Previous prototypical methods use the hard predefined assignment of the prototypes to classes [8,43,51] or nodes of a tree [34]. Therefore, no gradient propagation is needed to model the prototypes assignment. In contrast, our ProtoPool employs a soft assignment based on prototypes distributions to use prototypes from the pool optimally. To generate prototype distribution q, one could apply softmax on the vector of size $\mathbb{R}^M$. However, this could result in assigning many prototypes to one slot and consequently could decrease the interpretability. Therefore, to obtain distributions with exactly one probability close to 1, we require a differentiable arg max function. A perfect match, in this case is the Gumbel-Softmax estimator [20], where for $q = (q^1, \ldots, q^M) \in \mathbb{R}^M$ and $\tau \in (0, \infty)$

$$\text{Gumbel}-\text{softmax}(q, \tau) = (y^1, \ldots, y^M) \in \mathbb{R}^M,$$

where $y^i = \frac{\exp\big((q^i + \eta_i)/\tau\big)}{\sum_{m=1}^{M} \exp((q^m + \eta_m)/\tau)}$ and η_m for $m \in 1, .., M$ are samples drawn from standard Gumbel distribution. The Gumbel-Softmax distribution interpolates between continuous categorical densities and discrete one-hot-encoded categorical distributions, approaching the latter for low temperatures $\tau \in [0.1, 0.5]$ (see Fig. 5).

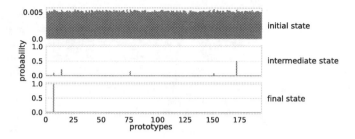

Fig. 5. A sample distribution (slot) at the initial, middle, and final step of training. In the beginning, all prototypes are assigned with a probability of 0.005. Then, the distribution binarizes, and finally, one prototype is assigned to this slot with a probability close to 1.

Slots Orthogonality. Without any additional constraints, the same prototype could be assigned to many slots of one class, wasting the capacity of the prototype pool layer and consequently returning poor results. Therefore, we extend the loss function with

$$\mathcal{L}_{orth} = \sum_{i<j}^{K} \frac{\langle q_i, q_j \rangle}{\|q_i\|_2 \cdot \|q_j\|_2}, \tag{2}$$

where $q_1, .., q_K$ are the distributions of a particular class. As a result, successive slots of a class are assigned to different prototypes.

Prototypes Projection. Prototypes projection is a step in the training process that allows prototypes visualization. It replaces each abstract prototype learned by the model with the representation of the nearest training patch. For prototype p, it can be expressed by the following formula

$$p \leftarrow \arg\min_{z \in Z_C} \|z - p\|_2, \tag{3}$$

where $Z_C = \{z : z \in Z_x \text{ for all } (x, y) : y \in C\}$. In contrast to [8], set C is not a single class but the set of classes assigned to prototype p.

Theoretical Analysis. Here, we theoretically analyze why ProtoPool assigns one prototype per slot and why each prototype does not repeat in a class. For this purpose, we provide two observations.

Observation 1. Let $q \in [0, 1]^M$, $\sum q_i = 1$ be a distribution (slot) of a particular class. Then, the limit of Gumbel-softmax(q, τ), as τ approaches zero, is the canonical vector $e_i \in \mathbb{R}^M$, i.e. for q there exists $i = 1, .., M$ such that $\lim_{\tau \to 0}$ Gumbel-softmax$(q, \tau) = e_i$.

The temperature parameter $\tau > 0$ controls how closely the new samples approximate discrete one-hot vectors (the canonical vector). From paper [20] we know that as $\tau \to 0$, the softmax computation smoothly approaches the arg max, and the sample vectors approach one-hot q distribution (see Fig. 5).

Table 1. Comparison of ProtoPool with other prototypical methods trained on the CUB-200-2011 and Stanford Cars datasets, which considers a various number of prototypes and types of convolutional layers f. In the case of the CUB-200-2011 dataset, ProtoPool achieves the highest accuracy than other models, even those containing ten times more prototypes. Moreover, the ensemble of three ProtoPools surpasses the ensemble of five TesNets with 17 times more prototypes. On the other hand, in the case of Stanford Cars, ProtoPool achieves competitive results with significantly fewer prototypes. Please note that the results are first sorted by backbone network and then by the number of prototypes, R stands for ResNet, iN means pretrained on iNaturalist, and Ex is an ensemble of three or five models.

CUB-200-2011				Stanford Cars			
Model	Arch.	Proto. #	Acc [%]	Model	Arch.	Proto. #	Acc [%]
ProtoPool (ours)		202	80.3±0.2	ProtoPool (ours)		195	89.3±0.1
ProtoPShare [43]	R34	400	74.7	ProtoPShare [43]	R34	480	86.4
ProtoPNet [8]		1655	79.5	ProtoPNet [8]		1960	86.1±0.2
TesNet [51]		2000	82.7±0.2	TesNet [51]		1960	92.6±0.3
ProtoPool (ours)		202	81.5±0.1	ProtoPool (ours)	R50	195	88.9±0.1
ProtoPShare [43]	R152	1000	73.6	ProtoTree [34]		195	86.6±0.2
ProtoPNet [8]		1734	78.6	ProtoPool (ours)	Ex3	195×3	91.1
TesNet [51]		2000	82.8±0.2	ProtoTree [34]		195×3	90.5
ProtoPool (ours)	iNR50	202	85.5±0.1	ProtoPool (ours)		195×5	91.6
ProtoTree [34]		202	82.2±0.7	ProtoTree [34]	Ex5	195×5	91.5
ProtoPool (ours)	Ex3	202×3	87.5	ProtoPNet [8]		1960×5	91.4
ProtoTree [34]		202×3	86.6	TesNet [51]		1960×5	**93.1**
ProtoPool (ours)		202×5	**87.6**				
ProtoTree [34]	Ex5	202×5	87.2				
ProtoPNet [8]		2000×5	84.8				
TesNet [51]		2000×5	86.2				

Observation 2. Let $K \in \mathbb{N}$ and $q_1, .., q_K$ be the distributions (slots) of a particular class. If $\mathcal{L}_{orth}$ defined in Eq. (2) is zero, then each prototype from a pool is assigned to only one slot of the class.

It follows the fact that $\mathcal{L}_{orth} = 0$ only if $\langle q_i, q_j \rangle = 0$ for all $i < j \leq K$, i.e. only if q_i, q_j have non-zero values for different prototypes.

4   Experiments

We train our model on CUB-200-2011 [50] and Stanford Cars [26] datasets to classify 200 bird species and 196 car models, respectively. As the convolutional layers f of the model, we take ResNet-34, ResNet-50, ResNet-121 [18], DenseNet-121, and DenseNet-161 [19] without the last layer, pretrained on ImageNet [10]. The one exception is ResNet-50 used with CUB-200-2011 dataset, which we pretrain on iNaturalist2017 [49] for fair comparison with ProtoTree

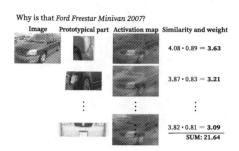

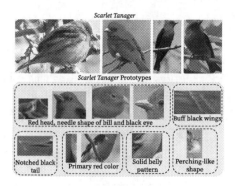

Fig. 6. Sample explanation of *Ford Freestar Minivan 2007* predictions. Except for an image, we present a few prototypical parts of this class, their activation maps, similarity function values, and the last layer weights. Moreover, we provide the sum of the similarities multiplied by the weights. ProtoPool returns the class with the largest sum as a prediction.

Fig. 7. Samples of *Scarlet Tanager* and prototypical parts assigned to this class by our ProtoPool model. Prototypes correspond, among others, to primary red color of feathers, black eye, perching-like shape, black notched tail, and black buff wings (Names of prototypical parts were generated based on the annotations from CUB-200-2011 dataset (see details in Supplementary Materials).

model [34]. In the testing scenario, we make the prototype assignment hard, i.e. we set all values of a distribution q higher than 0.5 to 1, and the remaining values to 0 otherwise. We set the number of prototypes assigned to each class to be at most 10 and use the pool of 202 and 195 prototypical parts for CUB-200-2011 and Stanford Cars, respectively. Details on experimental setup and results for other backbone networks are provided in the Supplementary Materials.

Comparison with Other Prototypical Models. In Table 1 we compare the efficiency of our ProtoPool with other models based on prototypical parts. We report the mean accuracy and standard error of the mean for 5 repetitions. Additionally, we present the number of prototypes used by the models, and we use this parameter to sort the results. We compare ProtoPool with ProtoPNet [8], ProtoPShare [43], ProtoTree [34], and TesNet [51].

One can observe that ProtoPool achieves the highest accuracy for the CUB-200-2011 dataset, surpassing even the models with a much larger number of prototypical parts (TesNet and ProtoPNet). For Stanford Cars, our model still performs better than other models with a similarly low number of prototypes, like ProtoTree and ProtoPShare, and slightly worse than TesNet, which uses ten times more prototypes. The higher accuracy of the latter might be caused by prototype orthogonality enforced in training. Overall, our method achieves competitive results with significantly fewer prototypes. However, ensemble ProtoPool or TesNet should be used if higher accuracy is preferred at the expense of interpretability.

Table 2. Characteristics of prototypical methods for fine-grained image classification that considers the number of prototypes, reasoning type, and prototype sharing between classes. ProtoPool uses 10% of ProtoPNet's prototypes but only with positive reasoning. It shares the prototypes between classes but, in contrast to ProtoPShare, is trained in an end-to-end, fully differentiable manner. Please notice that 100% corresponds to 2000 and 1960 of prototypes for CUB-200-2011 and Stanford Cars datasets, respectively.

Model	ProtoPool	ProtoTree	ProtoPShare	ProtoPNet	TesNet
Portion of prototypes	~10%	~10%	[20%;50%]	100%	100%
Reasoning type	+	+/−	+	+	+
Prototype sharing	Direct	Indirect	Direct	None	None

Fig. 8. Sample prototype of a *convex tailgate* (left top corner) shared by nine classes. Most of the classes correspond to luxury cars, but some exceptions exist, such as *Fiat 500*.

5 Interpretability

In this section, we analyze the interpretability of the ProtoPool model. Firstly, we show that our model can be used for local and global explanations. Then, we discuss the differences between ProtoPool and other prototypical approaches, and investigate its stability. Then, we perform a user study on the similarity functions used by the ProtoPNet, ProtoTree, and ProtoPool to assess the saliency of the obtained prototypes. Lastly, we consider ProtoPool from the cognitive psychology perspective.

Local and Global Interpretations. Except for local explanations that are similar to those provided by the existing methods (see Fig. 6), ProtoPool can provide a global characteristic of a class. It is presented in Fig. 7, where we show the prototypical parts of *Scarlet Tanager* that correspond to the visual features of this species, such as red feathers, a puffy belly, and a short beak. Moreover, similarly to ProtoPShare, ProtoPool shares the prototypical parts between data classes. Therefore, it can describe the relations between classes relying only on the positive reasoning process, as presented in Fig. 1 (in contrast, ProtoTree also uses negative reasoning). In Fig. 8, we further provide visualization of the prototypical part shared by nine classes. More examples are provided in Supplementary Materials.

Differences Between Prototypical Methods. In Table 2, we compare the characteristics of various prototypical-based methods. Firstly, ProtoPool and

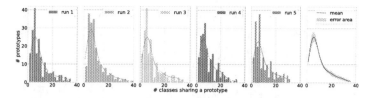

Fig. 9. Distribution presenting how many prototypes are shared by the specific number of classes (an estimation plot is represented with a dashed line). Each color corresponds to a single ProtoPool training on Stanford Cars dataset with ResNet50 as a backbone network. The right plot corresponds to the mean and standard deviation for five training runs. One can observe that the distribution behaves stable between runs.

ProtoTree utilize fewer prototypical parts than ProtoPNet and TesNet (around 10%). ProtoPShare also uses fewer prototypes (up to 20%), but it requires a trained ProtoPNet model before performing merge-pruning. Regarding class similarity, it is directly obtained from ProtoPool slots, in contrast to ProtoTree, which requires traversing through the decision tree. Moreover, ProtoPNet and TesNet have no mechanism to detect inter-class similarities. Finally, ProtoTree depends, among others, on *negative* reasoning process, while in the case of ProtoPool, it relies only on the positive reasoning process, which is a desirable feature according to [8].

Stability of Shared Prototypes. The natural question that appears when analyzing the assignment of the prototypes is: *Does the similarity between two classes hold for many runs of ProtoPool training?* To analyze this behavior, in Fig. 9 we show five distributions for five training runs. They present how many prototypes are shared by the specific number of classes. One can observe that difference between runs is negligible. In all runs, most prototypes are shared by five classes, but there exist prototypes shared by more than thirty classes. Moreover, on average, a prototype is shared by 2.73 ± 0.51 classes. A sample inter-class similarity graph is presented in the Supplementary Materials.

User Study on Focal Similarity. To validate if using focal similarity results in more salient prototypical parts, we performed a user study where we asked the participants to answer the question: *"How salient is the feature pointed out by the AI system?"*. The task was to assign a score from 1 to 5 where 1 meant *"Least salient"* and 5 meant *"Most salient"*. Images were generated using prototypes obtained for ProtoPool with ProtoPNets similarity or with focal similarity and from a trained ProtoTree[2]. To perform the user study, we used Amazon Mechanical Turk (AMT) system[3]. To assure the reliability of the answers, we required the users to be masters according to AMT. 40 workers participated in our study

---

[2] ProtoTree was trained using code from https://github.com/M-Nauta/ProtoTree and obtained accuracy similar to [34]. For ProtoPNet similarity, we used code from https://github.com/cfchen-duke/ProtoPNet.

[3] https://www.mturk.com.

and answered 60 questions (30 per dataset) presented in a random order, which resulted in 2400 answers. Each question contained an original training image and the same image with overlayed activation map, as presented in Fig. 2.

Results presented in Fig. 10 show that ProtoPool obtains mostly scores from 3 to 5, while other methods often obtain lower scores. We obtained a mean value of scores equal to 3.66, 2.87, and 2.85 for ProtoPool, ProtoTree, and ProtoPool without focal similarity, respectively. Hence, we conclude that ProtoPool with focal similarity generated more salient prototypes than the reference models, including ProtoTree. See Supplementary Materials for more information about a user study, detailed results, and a sample questionnaire.

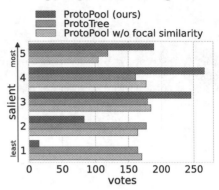

Fig. 10. Distribution of scores from user study on prototypes obtained for ProtoPool without and with focal similarity and for ProtoTree. One can observe that ProtoPool with focal similarity generates more salient prototypes than the other models.

ProtoPool in the Context of Cognitive Psychology. ProtoPool can be described in terms of parallel or simultaneous information processing, while ProtoTree may be characterized by serial or successive processing, which takes more time [22,30,35]. More specifically, human cognition is marked with the speed-accuracy trade-off. Depending on the perceptual situation and the goal of a task, the human mind can apply a categorization process (simultaneous or successive) that is the most appropriate in a given context, i.e. the fastest or the most accurate. Both models have their advantages. However, ProtoTree has a specific shortcoming because it allows for a categorization process to rely on an absence of features. In other words, an object characterized by none of the enlisted features is labeled as a member of a specific category. This type of reasoning is useful when the amount of information to be processed (i.e. number of features and categories) is fixed and relatively small. However, the time of object categorization profoundly elongates if the number of categories (and therefore the number of features to be crossed out) is high. Also, the chance of miscategorizing completely new information is increased.

6   Ablation Study

In this section, we analyze how the novel architectural choices, the prototype projection, and the number of prototypes influence the model performance.

Table 3. The influence of prototype projection on ProtoPool performance for CUB-200-2011 and Stanford Cars datasets is negligible. Note that for CUB-200-2011, we used ResNet50 pretrained on iNaturalist.

	CUB-200-2011		Stanford cars	
Architecture	Acc [%] before	Acc [%] after	Acc [%] before	Acc [%] after
ResNet34	80.8 ± 0.2	80.3 ± 0.2	89.1 ± 0.2	89.3 ± 0.1
ResNet50	85.9 ± 0.1	85.5 ± 0.1	88.4 ± 0.1	88.9 ± 0.1
ResNet152	81.2 ± 0.2	81.5 ± 0.1	—	—

Table 4. The influence of novel architectural choices on ProtoPool performance for CUB-200-2011 and Stanford Cars datasets is significant. We consider training without orthogonalization loss, with softmax instead of Gumbel-Softmax, and with similarity from ProtoPNet instead of focal similarity. One can observe that the mix of the proposed mechanisms (i.e. ProtoPool) obtains the best accuracy.

	CUB-200-2011	Stanford cars
Model	Acc [%]	Acc [%]
ProtoPool	**85.5**	**88.9**
w/o $\mathcal{L}_{orth}$	82.4	86.8
w/o Gumbel-Softmax trick	80.3	64.5
w/o Gumbel-Softmax trick and $\mathcal{L}_{orth}$	65.1	30.8
w/o focal similarity	85.3	88.8

Influence of the Novel Architectural Choices. Additionally, we analyze the influence of the novel components we introduce on the final results. For this purpose, we train ProtoPool without orthogonalization loss, with softmax instead of Gumbel-Softmax trick, and with similarity from ProtoPNet instead of focal similarity. Results are presented in Table 4 and in Supplementary Materials. We observe that the Gumbel-Softmax trick has a significant influence on the model performance, especially for the Stanford Cars dataset, probably due to lower inter-class similarity than in CUB-200-2011 dataset [34]. On the other hand, the focal similarity does not influence model accuracy, although as presented in Sect. 5, it has a positive impact on the interpretability. When it comes to orthogonality, it slightly increases the model accuracy by forcing diversity in slots of each class. Finally, the mix of the proposed mechanisms gets the best results.

Before and After Prototype Projection. Since ProtoPool has much fewer prototypical parts than other models based on a positive reasoning process, applying projection could result in insignificant prototypes and reduced model performance. Therefore, we decided to test model accuracy before and after the projection (see Table 3), and we concluded that differences are negligible.

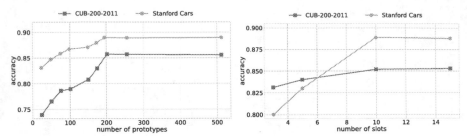

(a) Accuracy depending on the number of prototypes. One can observe that the model reaches a plateau for around 200 prototypical parts, and there is no gain in further increase of prototype number.

(b) Accuracy depending on the number of slots. One can observe that the model reaches a plateau for around 10 slots per class.

Fig. 11. ProtoPool accuracy with ResNet50 backbone depending on the number of prototypes and slots for CUB-200-2011 (blue square) and Stanford Cars (orange circle) datasets. (Color figure online)

Number of Prototypes and Slots vs Accuracy. Finally, in Fig. 11 we investigate how the number of prototypical parts or slots influences accuracy for the CUB-200-2011 and Stanford Cars datasets. We observe that up to around 200 prototypical parts, the accuracy increases and reaches the plateau. Therefore, we conclude that the amount of prototypes optimal for ProtoTree is also optimal for ProtoPool. Similarly, in the case of slots, ProtoPool accuracy increases till the 10 slots and then reaches the plateau.

7   Conclusions

We presented ProtoPool, a self-explainable method that incorporates the paradigm of prototypical parts to explain its predictions. This model shares the prototypes between classes without pruning operations, reducing their number up to ten times. Moreover, it is fully differentiable. To efficiently assign the prototypes to classes, we apply the Gumbel-Softmax trick together with orthogonalization loss. Additionally, we introduced focal similarity that focuses on salient features. As a result, we increased the interpretability while maintaining high accuracy, as we showed through theoretical analysis, multiple experiments, and user study.

Acknowledgments. The research of J. Tabor, K. Lewandowska and D. Rymarczyk were carried out within the research project "Bio-inspired artificial neural network" (grant no. POIR.04.04.00-00-14DE/18-00) within the Team-Net program of the Foundation for Polish Science co-financed by the European Union under the European Regional Development Fund. The work of Ł. Struski and B. Zieliński were supported by the National Centre of Science (Poland) Grant No. 2020/39/D/ST6/01332, and 2021/41/B/ST6/01370, respectively.

References

1. Abbasnejad, E., Teney, D., Parvaneh, A., Shi, J., Hengel, A.: Counterfactual vision and language learning. In: Proceedings of the IEEE/CVF Conference on Computer Vision and Pattern Recognition, pp. 10044–10054 (2020)
2. Adebayo, J., Gilmer, J., Muelly, M., Goodfellow, I., Hardt, M., Kim, B.: Sanity checks for saliency maps. In: Bengio, S., Wallach, H., Larochelle, H., Grauman, K., Cesa-Bianchi, N., Garnett, R. (eds.) Advances in Neural Information Processing Systems. vol. 31. Curran Associates, Inc. (2018). www.proceedings.neurips.cc/paper/2018/file/294a8ed24b1ad22ec2e7efea049b8737-Paper.pdf
3. Afnan, M.A.M., et al.: Interpretable, not black-box, artificial intelligence should be used for embryo selection. Human Reprod. Open. **2021**, 1–8 (2021)
4. Alvarez Melis, D., Jaakkola, T.: Towards robust interpretability with self-explaining neural networks. In: Bengio, S., Wallach, H., Larochelle, H., Grauman, K., Cesa-Bianchi, N., Garnett, R. (eds.) Advances in Neural Information Processing Systems. vol. 31. Curran Associates, Inc. (2018). www.proceedings.neurips.cc/paper/2018/file/3e9f0fc9b2f89e043bc6233994dfcf76-Paper.pdf
5. Barnett, A.J., et al.: IAIA-BL: a case-based interpretable deep learning model for classification of mass lesions in digital mammography. arXiv preprint arXiv:2103.12308 (2021)
6. Basaj, D., et al.: Explaining self-supervised image representations with visual probing. In: International Joint Conference on Artificial Intelligence (2021)
7. Brendel, W., Bethge, M.: Approximating CNNs with bag-of-local-features models works surprisingly well on ImageNet. In: International Conference on Learning Representations (2019). www.openreview.net/forum?id=SkfMWhAqYQ
8. Chen, C., Li, O., Tao, D., Barnett, A., Rudin, C., Su, J.K.: This looks like that: deep learning for interpretable image recognition. In: NeurIPS, pp. 8930–8941 (2019)
9. Chen, Z., Bei, Y., Rudin, C.: Concept whitening for interpretable image recognition. Nat. Mach. Intell. **2**(12), 772–782 (2020)
10. Deng, J., Dong, W., Socher, R., Li, L.J., Li, K., Fei-Fei, L.: ImageNet: a large-scale hierarchical image database. In: 2009 IEEE Conference on Computer Vision and Pattern Recognition, pp. 248–255. IEEE (2009)
11. Fong, R., Patrick, M., Vedaldi, A.: Understanding deep networks via extremal perturbations and smooth masks. In: Proceedings of the IEEE/CVF International Conference on Computer Vision, pp. 2950–2958 (2019)
12. Fong, R.C., Vedaldi, A.: Interpretable explanations of black boxes by meaningful perturbation. In: Proceedings of the IEEE International Conference on Computer Vision, pp. 3429–3437 (2017)
13. Gee, A.H., Garcia-Olano, D., Ghosh, J., Paydarfar, D.: Explaining deep classification of time-series data with learned prototypes. In: CEUR Workshop Proceedings, vol. 2429, p. 15. NIH Public Access (2019)
14. Ghorbani, A., Wexler, J., Zou, J.Y., Kim, B.: Towards automatic concept-based explanations. In: Wallach, H., Larochelle, H., Beygelzimer, A., d' Alché-Buc, F., Fox, E., Garnett, R. (eds.) Advances in Neural Information Processing Systems, vol. 32. Curran Associates, Inc. (2019). www.proceedings.neurips.cc/paper/2019/file/77d2afcb31f6493e350fca61764efb9a-Paper.pdf
15. Goyal, Y., Wu, Z., Ernst, J., Batra, D., Parikh, D., Lee, S.: Counterfactual visual explanations. In: International Conference on Machine Learning, pp. 2376–2384. PMLR (2019)

16. Guidotti, R., Monreale, A., Matwin, S., Pedreschi, D.: Explaining image classifiers generating exemplars and counter-exemplars from latent representations. Proc. AAAI Conf. Artif. Intell. **34**(09), 13665–13668 (2020). https://doi.org/10.1609/aaai.v34i09.7116
17. Hase, P., Chen, C., Li, O., Rudin, C.: Interpretable image recognition with hierarchical prototypes. In: Proceedings of the AAAI Conference on Human Computation and Crowdsourcing, vol. 7, pp. 32–40 (2019)
18. He, K., Zhang, X., Ren, S., Sun, J.: Deep residual learning for image recognition. In: Proceedings of the IEEE Conference on Computer Vision and Pattern Recognition, pp. 770–778 (2016)
19. Huang, G., Liu, Z., Van Der Maaten, L., Weinberger, K.Q.: Densely connected convolutional networks. In: Proceedings of the IEEE Conference on Computer Vision and Pattern Recognition, pp. 4700–4708 (2017)
20. Jang, E., Gu, S., Poole, B.: Categorical reparameterization with gumbel-softmax. arXiv:1611.01144 (2016)
21. Kaminski, M.E.: The right to explanation, explained. In: Research Handbook on Information Law and Governance. Edward Elgar Publishing (2021)
22. Kesner, R.: A neural system analysis of memory storage and retrieval. Psychol. Bull. **80**(3), 177 (1973)
23. Kim, B., et al.: Interpretability beyond feature attribution: quantitative testing with concept activation vectors (TCAV). In: International Conference on Machine Learning, pp. 2668–2677. PMLR (2018)
24. Kim, E., Kim, S., Seo, M., Yoon, S.: XProtoNet: diagnosis in chest radiography with global and local explanations. In: Proceedings of the IEEE/CVF Conference on Computer Vision and Pattern Recognition, pp. 15719–15728 (2021)
25. Koh, P.W., et al.: Concept bottleneck models. In: III, H.D., Singh, A. (eds.) Proceedings of the 37th International Conference on Machine Learning. Proceedings of Machine Learning Research, vol. 119, pp. 5338–5348. PMLR, 13–18 July 2020. www.proceedings.mlr.press/v119/koh20a.html
26. Krause, J., Stark, M., Deng, J., Fei-Fei, L.: 3d object representations for fine-grained categorization. In: Proceedings of the IEEE International Conference on Computer Vision Workshops, pp. 554–561 (2013)
27. Li, O., Liu, H., Chen, C., Rudin, C.: Deep learning for case-based reasoning through prototypes: a neural network that explains its predictions. In: Proceedings of the AAAI Conference on Artificial Intelligence, vol. 32 (2018)
28. Liu, N., Zhang, N., Wan, K., Shao, L., Han, J.: Visual saliency transformer. In: Proceedings of the IEEE/CVF International Conference on Computer Vision, pp. 4722–4732 (2021)
29. Lundberg, S.M., Lee, S.I.: A unified approach to interpreting model predictions. In: Proceedings of the 31st International Conference on Neural Information Processing Systems, pp. 4768–4777 (2017)
30. Luria, A.: The origin and cerebral organization of man's conscious action. In: Children with Learning Problems: Readings in a Developmental-interaction, pp. 109–130. New York, Brunner/Mazel (1973)
31. Marcos, D., Lobry, S., Tuia, D.: Semantically interpretable activation maps: what-where-how explanations within CNNs. In: 2019 IEEE/CVF International Conference on Computer Vision Workshop (ICCVW), pp. 4207–4215. IEEE (2019)
32. Ming, Y., Xu, P., Qu, H., Ren, L.: Interpretable and steerable sequence learning via prototypes. In: Proceedings of the 25th ACM SIGKDD International Conference on Knowledge Discovery & Data Mining, pp. 903–913 (2019)

33. Mothilal, R.K., Sharma, A., Tan, C.: Explaining machine learning classifiers through diverse counterfactual explanations. In: Proceedings of the 2020 Conference on Fairness, Accountability, and Transparency, pp. 607–617 (2020)
34. Nauta, M., et al.: Neural prototype trees for interpretable fine-grained image recognition. In: CVPR, pp. 14933–14943 (2021)
35. Neisser, U.: Cognitive Psychology (New York: Appleton). Century, Crofts (1967)
36. Niu, Y., Tang, K., Zhang, H., Lu, Z., Hua, X.S., Wen, J.R.: Counterfactual VQA: a cause-effect look at language bias. In: Proceedings of the IEEE/CVF Conference on Computer Vision and Pattern Recognition, pp. 12700–12710 (2021)
37. Puyol-Antón, E., et al.: Interpretable deep models for cardiac resynchronisation therapy response prediction. In: Martel, A.L., et al. (eds.) MICCAI 2020. LNCS, vol. 12261, pp. 284–293. Springer, Cham (2020). https://doi.org/10.1007/978-3-030-59710-8_28
38. Rebuffi, S.A., Fong, R., Ji, X., Vedaldi, A.: There and back again: revisiting back-propagation saliency methods. In: Proceedings of the IEEE/CVF Conference on Computer Vision and Pattern Recognition, pp. 8839–8848 (2020)
39. Ribeiro, M.T., Singh, S., Guestrin, C.: "why should i trust you?" Explaining the predictions of any classifier. In: Proceedings of the 22nd ACM SIGKDD International Conference on Knowledge Discovery and Data Mining, pp. 1135–1144 (2016)
40. Rosch, E.: Cognitive representations of semantic categories. J. Exp. Psychol. Gener. **104**(3), 192 (1975)
41. Rosch, E.H.: Natural categories. Cogn. Psychol. **4**(3), 328–350 (1973)
42. Rudin, C.: Stop explaining black box machine learning models for high stakes decisions and use interpretable models instead. Nat. Mach. Intell. **1**(5), 206–215 (2019)
43. Rymarczyk, D., et al.: Protopshare: prototypical parts sharing for similarity discovery in interpretable image classification. In: SIGKDD, pp. 1420–1430 (2021)
44. Selvaraju, R.R., Cogswell, M., Das, A., Vedantam, R., Parikh, D., Batra, D.: Grad-CAM: visual explanations from deep networks via gradient-based localization. In: Proceedings of the IEEE International Conference On Computer Vision, pp. 618–626 (2017)
45. Selvaraju, R.R., et al.: Taking a hint: leveraging explanations to make vision and language models more grounded. In: Proceedings of the IEEE/CVF International Conference on Computer Vision, pp. 2591–2600 (2019)
46. Simonyan, K., Vedaldi, A., Zisserman, A.: Deep inside convolutional networks: Visualising image classification models and saliency maps. In: In Workshop at International Conference on Learning Representations. Citeseer (2014)
47. Singh, G., Yow, K.C.: These do not look like those: an interpretable deep learning model for image recognition. IEEE Access **9**, 41482–41493 (2021)
48. Sundararajan, M., Taly, A., Yan, Q.: Axiomatic attribution for deep networks. In: International Conference on Machine Learning, pp. 3319–3328. PMLR (2017)
49. Van Horn, G., et al.: The inaturalist species classification and detection dataset. In: Proceedings of the IEEE Conference on Computer Vision and Pattern Recognition, pp. 8769–8778 (2018)
50. Wah, C., Branson, S., Welinder, P., Perona, P., Belongie, S.: The Caltech-UCSD birds-200-2011 dataset (2011)
51. Wang, J., et al.: Interpretable image recognition by constructing transparent embedding space. In: ICCV, pp. 895–904 (2021)
52. Wang, P., Vasconcelos, N.: Scout: Self-aware discriminant counterfactual explanations. In: Proceedings of the IEEE/CVF Conference on Computer Vision and Pattern Recognition, pp. 8981–8990 (2020)

53. Wiegand, G., Schmidmaier, M., Weber, T., Liu, Y., Hussmann, H.: I drive-you trust: explaining driving behavior of autonomous cars. In: Extended abstracts of the 2019 CHI conference on human factors in computing systems, pp. 1–6 (2019)
54. Xiao, T., Xu, Y., Yang, K., Zhang, J., Peng, Y., Zhang, Z.: The application of two-level attention models in deep convolutional neural network for fine-grained image classification. In: Proceedings of the IEEE Conference on Computer Vision and Pattern Recognition, pp. 842–850 (2015)
55. Yeh, C.K., Kim, B., Arik, S., Li, C.L., Pfister, T., Ravikumar, P.: On completeness-aware concept-based explanations in deep neural networks. In: Larochelle, H., Ranzato, M., Hadsell, R., Balcan, M.F., Lin, H. (eds.) Advances in Neural Information Processing Systems, vol. 33, pp. 20554–20565. Curran Associates, Inc. (2020). www.proceedings.neurips.cc/paper/2020/file/ecb287ff763c169694f682af52c1f309-Paper.pdf
56. Zhang, Z., Liu, Q., Wang, H., Lu, C., Lee, C.: ProtGNN: towards self-explaining graph neural networks (2022)
57. Zheng, H., Fu, J., Mei, T., Luo, J.: Learning multi-attention convolutional neural network for fine-grained image recognition. In: Proceedings of the IEEE International Conference on Computer Vision, pp. 5209–5217 (2017)
58. Zheng, H., Fu, J., Zha, Z.J., Luo, J.: Looking for the devil in the details: learning trilinear attention sampling network for fine-grained image recognition. In: Proceedings of the IEEE/CVF Conference on Computer Vision and Pattern Recognition, pp. 5012–5021 (2019)
59. Zhou, B., Sun, Y., Bau, D., Torralba, A.: Interpretable basis decomposition for visual explanation. In: Ferrari, V., Hebert, M., Sminchisescu, C., Weiss, Y. (eds.) ECCV 2018. LNCS, vol. 11212, pp. 122–138. Springer, Cham (2018). https://doi.org/10.1007/978-3-030-01237-3_8

Contributions of Shape, Texture, and Color in Visual Recognition

Yunhao Ge$^{(\boxtimes)}$, Yao Xiao, Zhi Xu, Xingrui Wang, and Laurent Itti

University of Southern California, Los Angeles, USA
yunhaoge@usc.edu
https://github.com/gyhandy/Humanoid-Vision-Engine

Abstract. We investigate the contributions of three important features of the human visual system (HVS)—shape, texture, and color—to object classification. We build a humanoid vision engine (HVE) that explicitly and separately computes shape, texture, and color features from images. The resulting feature vectors are then concatenated to support the final classification. We show that HVE can summarize and rank-order the contributions of the three features to object recognition. We use human experiments to confirm that both HVE and humans predominantly use some specific features to support the classification of specific classes (e.g., texture is the dominant feature to distinguish a zebra from other quadrupeds, both for humans and HVE). With the help of HVE, given any environment (dataset), we can summarize the most important features for the whole task (task-specific; e.g., color is the most important feature overall for classification with the CUB dataset), and for each class (class-specific; e.g., shape is the most important feature to recognize boats in the iLab-20M dataset). To demonstrate more usefulness of HVE, we use it to simulate the open-world zero-shot learning ability of humans with no attribute labeling. Finally, we show that HVE can also simulate human imagination ability with the combination of different features.

1 Introduction

The human vision system (HVS) is the gold standard for many current computer vision algorithms, on various challenging tasks: zero/few-shot learning [31,35,40,48,50], meta-learning [2,29], continual learning [43,52,57], novel view imagination [16,59], etc. Understanding the mechanism, function, and decision pipeline of HVS becomes more and more important. The vision systems of humans and other primates are highly differentiated. Although HVS provides us a unified image of the world around us, this picture has multiple facets or features, like shape, depth, motion, color, texture, etc. [15,22]. To understand the

---

Y. Ge and Y. Xiao—Contributed equally.

Supplementary Information The online version contains supplementary material available at https://doi.org/10.1007/978-3-031-19775-8_22.

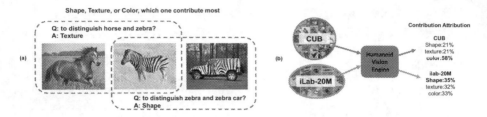

Fig. 1. (a): Contributions of Shape, Texture, and Color may be different among different scenarios/tasks. Here, texture is most important to distinguish zebra from horse, but shape is most important for zebra vs. zebra car. (b): Humanoid Vision Engine takes dataset as input and summarizes how shape, texture, and color contribute to the given recognition task in a pure learning manner (E.g., In ImageNet classification, shape is the most discriminative feature and contributes most to visual recognition).

contributions of the most important three features—shape, texture, and color—in visual recognition, some research compares the HVS with an artificial convolutional Neural Network (CNN). A widely accepted intuition about the success of CNNs on perceptual tasks is that CNNs are the most predictive models for the human ventral stream object recognition [7,58]. To understand which feature is more important for CNN-based recognition, recent paper shows promising results: ImageNet-trained CNNs are biased towards texture while increasing shape bias improves accuracy and robustness [32].

Due to the superb success of HVS on various complex tasks [2,18,35,43,59], human bias may also represent the most efficient way to solve vision tasks. And it is likely task-dependent (Fig. 1). Here, inspired by HVS, we wish to find a general way to understand how shape, texture, and color contribute to a recognition task by pure data-driven learning. The summarized feature contribution is important both for the deep learning community (guide the design of accuracy-driven models [6,14,21,32]) and for the neuroscience community (understanding the contributions or biases in human visual recognition) [33,56].

It has been shown by neuroscientists that there are separate neural pathways to process these different visual features in primates [1,11]. Among the many kinds of features crucial to visual recognition in humans, the shape property is the one that we primarily rely on in static object recognition [15]. Meanwhile, some previous studies show that surface-based cues also play a key role in our vision system. For example, [20] shows that scene recognition is faster for color images compared with grayscale ones and [36,38] found a special region in our brain to analyze textures. In summary, [8,9] propose that shape, color and texture are three separate components to identify an object.

To better understand the task-dependent contributions of these features, we build a Humanoid Vision Engine (HVE) to simulate HVS by explicitly and separately computing shape, texture, and color features to support image classification in an objective learning pipeline. HVE has the following key contributions: (1) Inspired by the specialist separation of the human brain on different features

[1,11], for each feature among shape, texture, and color, we design a specific feature extraction pipeline and representation learning model. (2) To summarize the contribution of features by end-to-end learning, we design an interpretable humanoid Neural Network (HNN) that aggregates the learned representation of three features and achieves object recognition, while also showing the contribution of each feature during decision. (3) We use HVE to analyze the contribution of shape, texture, and color on three different tasks subsampled from ImageNet. We conduct human experiments on the same tasks and show that both HVE and humans predominantly use some specific features to support object recognition of specific classes. (4) We use HVE to explore the contribution, relationship, and interaction of shape, texture, and color in visual recognition. Given any environment (dataset), HVE can summarize the most important features (among shape, texture, and color) for the whole task (task-specific) and for each class (class-specific). To the best of our knowledge, we provide the first fully objective, data-driven, and indeed first-order, quantitative measure of the respective contributions. (5) HVE can help guide accuracy-driven model design and performs as an evaluation metric for model bias. For more applications, we use HVE to simulate the open-world zero-shot learning ability of humans which needs no attribute labels. HVE can also simulate human imagination ability across features.

2   Related Works

In recent years, more and more researchers focus on the interpretability and generalization of computer vision models like CNN [23,46] and vision transformer [12]. For CNN, many researchers try to explore what kind of information is most important for models to recognize objects. Some paper show that CNNs trained on the ImageNet are more sensitive to texture information [6,14,21]. But these works fail to quantitatively explain the contribution of shape, texture, color as different features, comprehensively in various datasets and situations. While most recent studies focus on the bias of Neural Networks, exploring the bias of humans or a humanoid learning manner is still under-explored and inspiring.

Besides, many researchers contribute to the generalization of computer vision models and focus on zero/few-shot learning [10,17,31,35,48,54], novel view imagination [16,19,59], open-world recognition [3,26,27], etc. Some of them tackled these problems by feature learning—representing an object by different features, and made significant progress in this area [37,53,59]. But, there still lacks a clear definition of what these properties look like or a uniform design of a system that can do humanoid tasks like generalized recognition and imagination.

3   Humanoid Vision Engine

The goal of the humanoid vision engine (HVE) is to summarize the contribution of shape, texture, and color in a given task (dataset) by separately computing the three features to support image classification, similar to humans' recognizing

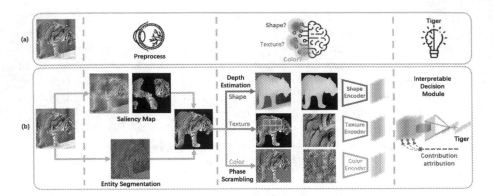

Fig. 2. Pipeline for humanoid vision engine (HVE). (a) shows how will humans' vision system deal with an image. After humans' eyes perceive the object, the different parts of the brain will be activated. The human brain will organize and summarize that information to get a conclusion. (b) shows how we design HVE to correspond to each part of the human's vision system.

objects. During the pipeline and model design, we borrow the findings of neuroscience on the structure, mechanism and function of HVS [1,11,15,20,36,38]. We use end-to-end learning with backpropagation to simulate the learning process of humans and to summarize the contribution of shape, texture, and color. The advantage of end-to-end training is that we can avoid human bias, which may influence the objective of contribution attribution (e.g., we avoid handcrafted elementary shapes as done in Recognition by Components [4]). We only use data-driven learning, a straightforward way to understand the contribution of each feature from an effectiveness perspective, and we can easily generalize HVE to different tasks (datasets). As shown in Fig. 2, HVE consists of (1) **a humanoid image preprocessing pipeline**, (2) **feature representation** for shape, texture, and color, and (3) **a humanoid neural network** that aggregates the representation of each feature and achieves interpretable object recognition.

3.1   Humanoid Image Preprocessing and Feature Extraction

As shown in Fig. 2 (a), humans (or primates) can localize an object intuitively in a complex scene before we recognize what it is [28]. Also, there are different types of cells or receptors in our primary visual cortex extracting specific information (like color, shape, texture, shading, motion, etc.) information from the image [15]. In our HVE (Fig. 2 (b)), for an input raw image $I \in \mathbb{R}^{H \times W \times C}$, we first parse the object from the scene as preprocessing and then extract our defined shape, texture, and color features I_s, I_t, I_c, for the following humanoid neural network.

Image Parsing and Foreground Identification. As shown in the preprocessing part of Fig. 2 (b), we use the entity segmentation method [39] to simulate the process of parsing objects from a scene in our brain. Entity segmentation is an open-world model and can segment the object from the image without labels. This method aligns with human behavior, which can (at least in some cases; e.g., autostereograms [28]) segment an object without deciding what it is. After we get the segmentation of the image, we use a pre-trained CNN and GradCam [45] to find the foreground object among all masks. (More details in appendix.)

We design three different feature extractors after identifying the foreground object segment: shape, texture, and color extractor, similar to the separate neural pathways in the human brain which focus on specific property [1, 11]. The three extractors focus only on the corresponding features, and the extracted features, shape I_s, texture I_t, and color I_c, are disentangled from each other.

Shape Feature Extractor. For the shape extractor, we want to keep both 2D and 3D shape information while eliminating the information of texture and color. We first use a 3D depth prediction model [41, 42] to obtain the 3D depth information of the whole image. After element-wise multiplying the 3D depth estimation and 2D mask of the object, we obtain our shape feature I_s. We can notice that this feature only contains 2D shape and 3D structural information (the 3D depth) and without color or texture information (Fig. 2(b)).

Texture Feature Extractor. In texture extractor, we want to keep both local and global texture information while eliminating shape and color information. Figure 3 visualizes the extraction process. First, to remove the color information, we convert the RGB object segmentation to a grayscale image. Next, we cut this image into several square patches with an adaptive strategy (the patch size and location are adaptive with object sizes to cover more texture information). If the overlap ratio between the patch and the original 2D object segment is larger than a threshold τ, we add that patch to a patch pool (we set τ to be 0.99 in our experiments, which means the over 99% of the area of the patch belongs to the object). Since we want to extract both local (one patch) and global (whole image) texture information, we randomly select 4 patches from the patch pool and concatenate them into a new texture image (I_t). (More details in appendix.)

Color Feature Extractor. To represent the color feature for I. We use phase scrambling, which is popular in psychophysics and in signal processing [34, 51]. Phase scrambling transforms the image into the frequency domain using the fast Fourier transform (FFT). In the frequency domain, the phase of the signal is

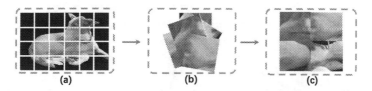

Fig. 3. Pipeline for extracting texture feature: (a) Crop images and compute the overlap ratio between 2D mask and patches. Patches with overlap > 0.99 are shown in a green shade. (b) add the valid patches to a patch pool. (c) randomly choose 4 patches from pool and concatenate them to obtain a texture image I_t. (Color figure online)

then randomly scrambled, which destroys shape information while preserving color statistics. Then we use IFFT to transfer back to image space and get $I_c \in \mathbb{R}^{H \times W \times C}$. I_c and I have the same distribution of pixel color values (Fig. 2(b)).

3.2   Humanoid Neural Network

After preprocessing, we have three features, i.e. shape I_s, texture I_t, color I_c of an input image I. To simulate the separate neural pathways in humans' brains for different feature information [1,11], we design three feature representation encoders for shape, texture, and color, respectively. Shape feature encoder E_s takes a 3D shape feature I_s as input and outputs the shape representation ($V_s = E_s(I_s)$). Similarly, texture encoder E_t and color encoder E_c take the texture patch image I_t or color phase scrambled image I_c as input, after embedded by E_t (or E_c), we get the texture feature V_t and color feature V_c. We use ResNet-18 [23] as the backbone for all feature encoders to project the three types of features to the corresponding well-separated embedding spaces. It is hard to define the ground-truth label of the distance between features. Given that the objects from the same class are relatively consistent in shape, texture, and color, the encoders can be trained in the classification problem independently instead, with the supervision of class labels. After training our encoders as classifiers, the feature map of the last convolutional layer will serve as the final feature representation. To aggregate separated feature representations and conduct object recognition, we freeze the three encoders and train a contribution interpretable aggregation module Aggr_θ, which is composed of two fully-connected layers (Fig. 2 (b) right). We concatenate V_s, V_t, V_c and send it to Aggr_θ. The output is denoted as $p \in \mathbb{R}^n$, where n is the number of classes. So we have $p = \text{Aggr}_\theta (\text{concat}(V_s, V_t, V_c))$. (More details and exploration of our HNN are in appendix.)

  We also propose a gradient-based *contribution attribution* method to interpret the contributions of shape, texture, and color to the classification decision, respectively. Take the shape feature as an example, given a prediction p and the probability of class k, namely p^k, we compute the gradient of p^k with respect to the shape feature V^s. We define the gradient as shape importance weights α_s^k, i.e. $\alpha_s^k = \frac{\partial p^k}{\partial V_s}, \alpha_t^k = \frac{\partial p^k}{\partial V_t}, \alpha_c^k = \frac{\partial p^k}{\partial V_c}$. Then we calculate element-wise product between V_s and α_s^k to get the final shape contribution S_s^k, i.e. $S_s^k = \text{ReLU}\left(\sum \alpha_s^k V_s\right)$. In other words, S_s^k represents the "contribution" of shape feature to classifying this image as class k. We can do the same thing to get texture contribution S_t^k and color contribution S_c^k. After getting the feature contributions for each image, we can calculate the average value of all images in this class to assign feature contributions to each class (class-specific bias) and the average value of all classes to assign feature contributions to the whole dataset (task-specific bias).

4   Experiments

In this section, we first show the effectiveness of feature encoders on representation learning (Sect. 4.1); then we show the contribution interpretation

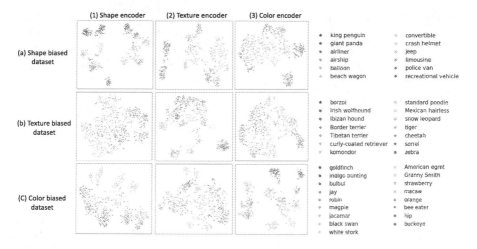

Fig. 4. T-SNE results of feature encoders on their corresponding biased datasets

performance of Humanoid NN on different feature-biased datasets in ImageNet (Sect. 4.2); We use human experiments to confirm that both HVE and humans predominantly use some specific features to support the classification of specific classes (Sect. 4.3); Then we use HVE to summarize the contribution of shape, texture, and color on different datasets (CUB [55] and iLab-20M [5]) (Sect. 4.4).

4.1 Effectiveness of Feature Encoders

To show that our three feature encoders focus on embedding their corresponding sensitive features, we handcrafted three subsets of ImageNet [30]: shape-biased dataset (D_{shape}), texture-biased dataset (D_{texture}), and color-biased dataset (D_{color}). **Shape-biased dataset** containing 12 classes, where the classes were chosen which intuitively are strongly determined by shape (e.g., vehicles are defined by shape more than color). **Texture-biased dataset** uses 14 classes which we believed are more strongly determined by texture. **Color-biased dataset** includes 17 classes. The intuition of class selection of all three datasets will be verified by our results in Table 1 with further illustration in Sect. 4.2. All these datasets are randomly selected as around 800 training images and 200 testing images. The class details of biased datasets are shown in Fig. 4.

If our feature extractors actually learned their *feature-constructive* latent spaces, their T-SNE results will show clear clusters in the feature-biased datasets. "Bias" here means we can classify the objects based on the biased feature easily, but it is more difficult to make decisions based on the other two features.

After pre-processing the original images and getting their feature images, we input the feature images into feature encoders and get the T-SNE results shown in Fig. 4. Each row represents one feature-biased dataset and each column is bounded with one feature encoder, each image shows the results of one combination. T-SNE results are separated perfectly on corresponding datasets

Table 1. "Original" column means the accuracy of Resnet18 on the original images as our upper bound. Shape, texture and color columns represent the accuracy of feature nets. "all" means results of our HNN that combines the 3 feature nets. It approaches the upper bound, suggesting that the split into 3 feature nets preserved most information needed for image classification.

Accuracy	Original	Shape	Texture	Color	All
Shape biased dataset	**97%**	**90%**	84%	71%	**95%**
Texture biased dataset	**96%**	64%	**81%**	65%	**91%**
Color biased dataset	**95%**	70%	73%	**82%**	**92%**

(diagonal) but not as well on others' datasets (off-diagonal), which shows that our feature encoders are predominantly sensitive to the corresponding features.

4.2   Effectiveness of Humanoid Neural Network

We can use feature encoders to serve as classifiers after adding fully-connected layers. As these classifiers classify images based on corresponding feature representation, we call them *feature nets*. We tested the accuracy of feature nets on these three biased datasets. As shown in Table 1, a ResNet-18 trained on the original segmented images (without explicit separated features, e.g. Fig. 2 (b) tiger without background) provided an upper bound for the task. We find that feature net consistently obtains the best performance on their own biased dataset (e.g., on the shape-biased dataset, shape net classification performance is better than that of the color net or texture net). If we combine these three feature nets with the interpretable aggregation module, the classification accuracy is very close to the upper bound, which means our vision system can classify images based on these three features almost as well as based on the full original color images. This demonstrates that we can obtain most information of original images by our feature nets, and our aggregation and interpretable decision module actually learned how to combine those three features by end-to-end learning.

Table 2a shows the quantitative contribution summary results of Humanoid NN (Sect. 3.2). For task-specific bias, shape plays a dominant role in shape-biased tasks, and texture, color also contribute most to their related biased tasks.

4.3   Human Experiments

Intuitively, we expect that humans may rely on different features to classify different objects (Fig. 1). To show this, we designed human experiments that asked participants to classify reduced images with only shape, texture, or color features. If an object is mainly recognizable based on shape for humans, we could then check whether it is also the same for HVE, and also for color and texture.

Experiments Design. Three datasets in Table 1 have a clear bias towards corresponding features (Fig. 4). We asked the participants to classify objects in

Table 2. Contributions of features from HVE and humans' recognition accuracy.

(a) Contributions of features for different biased datasets summarized by HVE.

Contribution ratio	Shape	Texture	Color
Shape biased dataset	**47%**	34%	19%
Texture biased dataset	5%	**65%**	30%
Color biased dataset	11%	19%	**70%**

(b) Humans' accuracy of different feature images on different biased datasets.

Accuracy	Shape	Texture	Color
Shape biased dataset	**90.0%**	49.0%	16.8%
Texture biased dataset	33.1%	**40.0%**	11.1%
Color biased dataset	32.3%	19.7%	**46.5%**

each dataset based on one single feature image computed by one of our feature extractors (Fig. 5). Participants were asked to choose the correct class label for the reduced image (from 12/14/17 classes in shape/texture/color datasets).

Human Performance Results. The results here are based on 3270 trials, 109 participants. The accuracy for different feature questions on different biased datasets can be seen in Table 2b. Human performance is similar to our feature nets' performance (compare Table 1 with Table 2b). On shape-biased dataset, both human and feature nets attain the highest accuracy with shape. The same for the color and texture biased datasets. Both HVE and humans predominantly use some specific features to support recognition of specific classes. Interestingly, humans can perform not badly on all three biased datasets with shape features.

4.4 Contributions Attribution in Different Tasks

With our vision system, we can summarize the task-specific bias and class-specific bias for any dataset. This enables several applications: (1) Guide accuracy-driven model design [6,14,21,32]; Our method provides objective summarization of dataset bias. (2) Evaluation metric for model bias. Our method can help correct an initially wrong model bias on some datasets (e.g., that most CNN trained on ImageNet are texture biased [21,32]). (3) Substitute human intuition to obtain more objective summarization with end-to-end learning. We implemented the biased summarization experiments on two datasets, CUB [55] and iLab-20M [5]. Figure 1(b) shows the task-specific biased results. Since CUB is a dataset of birds, which means all the classes in CUB have a similar shape

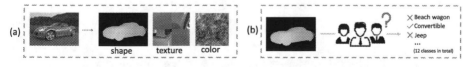

Fig. 5. Sample question for the human experiment. (a) A test image (left) is first converted into shape, color, and texture images using our feature extractors. (b) On a given trial, human participants are presented with one shape, color, or texture image, along with 2 reference images for each class in the corresponding dataset (not shown here, see appendix. For a screenshot of an experiment trial). Participants are asked to guess the correct object class from the feature image.

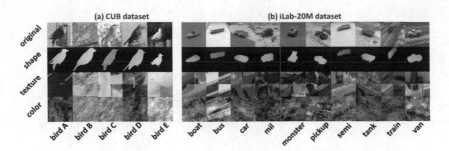

Fig. 6. Processed CUB and iLab-20M dataset examples

with feather textures, hence color may indeed be the most discriminative feature (Fig. 6 (a)).

As for iLab (Fig. 6 (b)), we also conduct the class-specific biased experiments on iLab and summarize the class biases in Table 3. It is interesting to find that the dominant feature is different for different classes. For instance, boat is shape-biased while military vehicle (mil) is color-biased. (More examples in appendix.)

5   More Humanoid Applications with HVE

To further explore more applications with HVE, we use HVE to simulate the visual reasoning process of humans and propose a new solution for conducting open-world zero-shot learning without predefined attribute labels (Sect. 5.1). We also use HVE to simulate human imagination ability through cross-feature retrieval and imagination (Sect. 5.2).

5.1   Open-World Zero-Shot Learning with HVE

Zero-shot learning needs to classify samples from classes never seen during training. Most current methods [13,31,35] need humans to provide detailed attribute labels for each image, which is costly in time and energy. However, given an image from an unseen class, humans can still *describe* it with their learned knowledge. For example, we may use horse-like shape, panda-like color, and tiger-like texture to describe an unseen class zebra. In this section, we show how our HVE can simulate this feature-wise open-world image description by feature retrieval

Table 3. Class-specific bias for each class in iLab-20M

Ratio	Boat	Bus	Car	Mil	Monster	Pickup	Semi	Tank	Train	Van
Shape	**40%**	35%	**44%**	18%	36%	28%	**40%**	36%	31%	**40%**
Texture	32%	31%	**40%**	30%	34%	20%	31%	32%	34%	27%
Color	28%	34%	16%	**52%**	30%	**53%**	29%	32%	35%	33%

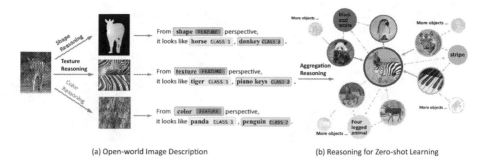

(a) Open-world Image Description   (b) Reasoning for Zero-shot Learning

Fig. 7. The zero-shot learning method with HVE. We first describe the novel image in the perspective of shape, texture, and color. Then we use ConceptNet as common knowledge to reason and predict the label.

and ranking. And based on these image descriptions, we propose a feature-wise open-world zero-shot learning pipeline with the help of ConceptNet [49], like the *reasoning or consulting* process of humans. The whole process shows in Fig. 7.

Step 1: Description. We use HVE to provide feature-wise descriptions for any unseen class images without predefined attribute labels. First, to represent learnt knowledge, we use trained three feature extractors (described in Sect. 3.2) to get the shape, texture, and color representation image of seen class k. Then, given an unseen class image I_{un}, we use the same feature extractors to get its feature-wise representation. To retrieve learnt classes as descriptions, we calculate the average distance between I_{un} and images of other class k in the latent space on shape, texture, and color features. In this way, we can find the top K closest classes of I_{un} from the perspective of each feature, and we call these K classes "roots" of each feature. Now, we can describe I_{un} using our three sets of roots. For example, as shown in Fig. 7(a), for the unseen class zebra, we can describe its shape by {horse, donkey}, texture by {tiger, piano keys}, and color by {panda}.

Step 2: Open-world classification. To further predict the actual class of I_{un} based on the feature-wise description, we use ConceptNet as common knowledge to conduct reasoning. As shown in Fig. 7(b), for every feature roots, we retrieve their common attribute in ConceptNet, (e.g., stripe the is common attribute root of {tiger, piano keys}). We form a reasoning root pool R^* consisting of classes from feature roots obtained during image description and shared attribute roots. The reasoning roots will be our *evidence* for reasoning. For every root in R^*, we can search its neighbors in ConceptNet, which are treated as possible candidate classes for I_{un}. All candidates form a possible candidate pool P, which contains all hypothesis classes. Now we have two pools, root pool R^* and candidate pool P. For every candidate $p_i \in P$ and $r_i \in R^*$, we calculate the ranking score of p_i as: $\bar{S}(p_i) = \sum_{r_j \in R^*} \cos(\mathcal{E}(p_i), \mathcal{E}(r_j))$. where $\mathcal{E}(\cdot)$ is the word embedding in ConceptNet and $\cos(A, B)$ means cosine similarity between A and B.

Table 4. Open-world zero-shot accuracy and FID of cross-features imagination.

(a) Accuracy of unseen class for zero-shot learning. One-shot on Prototype and zero-shot on ours

Method	Fowl	Zebra	Wolf	Sheep	Apple
Prototype	19%	16%	17%	21%	74%
Ours	78%	87%	63%	72%	98%

(b) Cross-features imagination quality comparison. We compare HVE methods with three pix2pix GANs as baselines

FID (↓)	Shape input	Texture input	Color input
Baselines	123.915	188.854	203.527
Ours	**96.871**	**105.921**	**52.846**

We choose the candidate with the highest score as our predicted label. In our prototype zero-shot learning dataset, we select 34 seen classes as the training set and 5 unseen classes as the test set, with 200 images per class. We calculate the accuracy of the test set (Table 4a). As a comparison, we conduct prototypical networks [47] using its one-shot setting. More details are in the appendix.

5.2   Cross Feature Imagination with HVE

We show HVE has the potential to simulate human imagination ability. Humans can intuitively imagine an object when seeing one aspect of a feature, especially when this feature is prototypical (contribute most to classification). For instance, we can imagine a zebra when seeing its stripe (texture). This process is similar but harder than the classical image generation task since the input features modality here is *dynamic* which can be any feature among shape, texture, or color. To solve this problem, using HVE, we separate this procedure into two steps: (1) **cross feature retrieval** and (2) **cross feature imagination**. Given any feature (shape, texture, or color) as input, cross-feature retrieval finds the most possible two other features. Cross-feature imagination then generate a whole object based on a group of shapes, textures, and color features.

Cross Feature Retrieval. We learn a feature agnostic encoder that projects the three features into one same feature space and makes sure that the features belonging to the same class are in the nearby regions.

As shown in Fig. 8(a), during training, the shape I_s, texture I_t and color I_c are first sent into the corresponding frozen encoders E_s, E_t, E_c, which are the same encoders in Sect. 3.2. Then all of the outputs are projected into a cross-feature embedding space by a feature agnostic net $\mathcal{M}$, which contains three convolution layers. We also add a fully connected layer to predict the class labels of the features. We use cross-entropy loss $\mathcal{L}_{cls}$ to regularize the prediction label and a triplet loss $\mathcal{L}_{triplet}$ [44] to regularize the projection of $\mathcal{M}$. For any input feature x (e.g., a bird A shape), positive sample x_{pos} are either same class same modality (another bird A shape) or same class different feature modality (a bird A texture or color); negative sample x_{neg} are any features from different class. $\mathcal{L}_{triplet}$ pulls the embedding of x closer to that of the positive sample x_{pos}, and pushes it apart from the embedding of the negative sample x_{neg}. The triplet

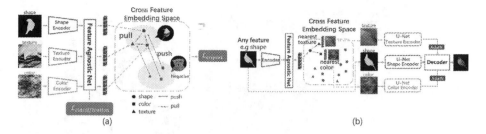

Fig. 8. (a) The structure and training process of the cross-feature retrieval model. E_s, E_t, E_c are the same encoders in Sect. 3.2. The feature agnostic net then projects them to shared feature space for retrieval. (b) The process of cross-feature imagination. After retrieval, we design a cross-feature pixel2pixel GAN model to generate the final image.

loss is defined as $\mathcal{L}_{\text{triplet}} = \max(\|\mathcal{F}(x) - \mathcal{F}(x_{\text{pos}})\|_2 - \|\mathcal{F}(x) - \mathcal{F}(x_{\text{neg}})\|_2 + \alpha, 0)$, where $\mathcal{F}(\cdot) := \mathcal{M}(E(\cdot))$, E is one of the feature encoders. α is the margin size in the feature space between classes, $\|\cdot\|_2$ represents ℓ_2 norm.

We test the retrieval model in all three biased datasets (Fig. 4) separately. During retrieval, given any feature of any object, we can map it into the cross feature embedding space by the corresponding encoder net and the feature agnostic net. Then we apply the ℓ_2 norm to find the other two features closest to the input one as output. The output is correct if they belong to the same class as the input. For each dataset, we retrieve the three features pair by pair (accuracy in appendix). The retrieval performs better when the input feature is the dominant of the dataset, which again verifies the feature bias in each dataset.

Cross Feature Imagination. To stimulate imagination, we propose a cross-feature imagination model to generate plausible final images with the input and retrieved features. The procedure of imagination is shown in Fig. 8(b). Inspired by the pixel2pixel GAN [25] and AdaIN [24], we design a cross-feature pixel2pixel GAN model to generate the final image. The GAN model is trained and tested on the three biased datasets. In Fig. 9, we show more results of the generation, which show that our model satisfyingly generates the object from a single feature. From the comparison between (c) and (e), we can clearly find that they are alike from the view of the corresponding input feature, but the imagination results preserve the retrieval features. The imagination variance also shows the feature contributions from a generative view: if the given feature is the dominant feature of a class (contribute most in classification. e.g., the stripe of zebra), then the retrieved features and imagined images have smaller variance (most are zebras); While non-dominant given feature (shape of zebra) lead to large imagination variance (can be any horse-like animals). We create a baseline generator by using three pix2pix GANs where each pix2pix GAN is responsible for one specific feature (take one modality of feature as input and imagine the raw image). The FID comparison is in Table 4b. More details are in the appendix.

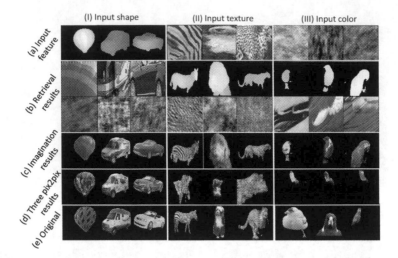

Fig. 9. Imagination with shape, texture, and color feature input (columns I, II, III). Line (a): input feature. Line (b): retrieved features given (a). Line (c): imagination results with HVE and our GAN model. Line (d): results of baseline 3 pix2pix GANs. Line (e): original images to which the input features belong. Our model can reasonably "imagine" the object given a single feature.

6   Conclusion

To explore the task-specific contribution of shape, texture, and color features in human visual recognition, we propose a humanoid vision engine (HVE) that explicitly and separately computes these features from images and then aggregates them to support image classification. With the proposed contribution attribution method, given any task (dataset), HVE can summarize and rank-order the task-specific contributions of the three features to object recognition. We use human experiments to show that HVE has a similar feature contribution to humans on specific tasks. We show that HVE can help simulate more complex and humanoid abilities (e.g., open-world zero-shot learning and cross-feature imagination) with promising performance. These results are the first step towards better understanding the contributions of object features to classification, zero-shot learning, imagination, and beyond.

Acknowledgments. This work was supported by C-BRIC (one of six centers in JUMP, a Semiconductor Research Corporation (SRC) program sponsored by DARPA), DARPA (HR00112190134) and the Army Research Office (W911NF2020053). The authors affirm that the views expressed herein are solely their own, and do not represent the views of the United States government or any agency thereof.

References

1. Amir, Y., Harel, M., Malach, R.: Cortical hierarchy reflected in the organization of intrinsic connections in macaque monkey visual cortex. J. Comp. Neurol. **334**(1), 19–46 (1993)
2. Andrychowicz, M., et al.: Learning to learn by gradient descent by gradient descent. In: Advances in Neural Information Processing Systems, pp. 3981–3989 (2016)
3. Bendale, A., Boult, T.: Towards open world recognition. In: Proceedings of the IEEE Conference on Computer Vision and Pattern Recognition, pp. 1893–1902 (2015)
4. Biederman, I.: Recognition-by-components: a theory of human image understanding. Psychol. Rev. **94**(2), 115 (1987)
5. Borji, A., Izadi, S., Itti, L.: ilab-20m: A large-scale controlled object dataset to investigate deep learning. In: Proceedings of the IEEE Conference on Computer Vision and Pattern Recognition, pp. 2221–2230 (2016)
6. Brendel, W., Bethge, M.: Approximating CNNs with bag-of-local-features models works surprisingly well on imagenet. In: 7th International Conference on Learning Representations, ICLR 2019, New Orleans, LA, USA, 6–9 May 2019. OpenReview.net (2019). www.openreview.net/forum?id=SkfMWhAqYQ
7. Cadieu, C.F., et al.: Deep neural networks rival the representation of primate it cortex for core visual object recognition. PLoS Comput. Biolo. **10**(12), e1003963 (2014)
8. Cant, J.S., Goodale, M.A.: Attention to form or surface properties modulates different regions of human occipitotemporal cortex. Cereb. Cortex **17**(3), 713–731 (2007)
9. Cant, J.S., Large, M.E., McCall, L., Goodale, M.A.: Independent processing of form, colour, and texture in object perception. Perception **37**(1), 57–78 (2008)
10. Cheng, H., Wang, Y., Li, H., Kot, A.C., Wen, B.: Disentangled feature representation for few-shot image classification. CoRR abs/2109.12548 (2021). arxiv.org/abs/2109.12548
11. DeYoe, E.A., et al.: Mapping striate and extrastriate visual areas in human cerebral cortex. Proc. Natl. Acad. Sci. **93**(6), 2382–2386 (1996)
12. Dosovitskiy, A., et al.: An image is worth 16x16 words: transformers for image recognition at scale. In: 9th International Conference on Learning Representations, ICLR 2021, Virtual Event, Austria, 3–7 May 2021. OpenReview.net (2021). www.openreview.net/forum?id=YicbFdNTTy
13. Fu, Y., Xiang, T., Jiang, Y.G., Xue, X., Sigal, L., Gong, S.: Recent advances in zero-shot recognition: toward data-efficient understanding of visual content. IEEE Sig. Process. Mag. **35**(1), 112–125 (2018)
14. Gatys, L.A., Ecker, A.S., Bethge, M.: Texture and art with deep neural networks. Curr. Opin. Neurobiol. **46**, 178–186 (2017)
15. Gazzaniga, M.S., Ivry, R.B., Mangun, G.: Cognitive Neuroscience. The Biology of the Mind (2014) (2006)
16. Ge, Y., Abu-El-Haija, S., Xin, G., Itti, L.: Zero-shot synthesis with group-supervised learning. In: 9th International Conference on Learning Representations, ICLR 2021, Virtual Event, Austria, 3–7 May 2021. OpenReview.net (2021). www.openreview.net/forum?id=8wqCDnBmnrT
17. Ge, Y., Xiao, Y., Xu, Z., Li, L., Wu, Z., Itti, L.: Towards generic interface for human-neural network knowledge exchange (2021)

18. Ge, Y., et al.: A peek into the reasoning of neural networks: interpreting with structural visual concepts. In: Proceedings of the IEEE/CVF Conference on Computer Vision and Pattern Recognition, pp. 2195–2204 (2021)
19. Ge, Y., Xu, J., Zhao, B.N., Itti, L., Vineet, V.: Dall-e for detection: language-driven context image synthesis for object detection. arXiv preprint arXiv:2206.09592 (2022)
20. Gegenfurtner, K.R., Rieger, J.: Sensory and cognitive contributions of color to the recognition of natural scenes. Curr. Biol. **10**(13), 805–808 (2000)
21. Geirhos, R., Rubisch, P., Michaelis, C., Bethge, M., Wichmann, F.A., Brendel, W.: Imagenet-trained CNNs are biased towards texture; increasing shape bias improves accuracy and robustness. In: 7th International Conference on Learning Representations, ICLR 2019, New Orleans, LA, USA, 6–9 May 2019. OpenReview.net (2019). www.openreview.net/forum?id=Bygh9j09KX
22. Grill-Spector, K., Malach, R.: The human visual cortex. Annu. Rev. Neurosci. **27**, 649–677 (2004)
23. He, K., Zhang, X., Ren, S., Sun, J.: Deep residual learning for image recognition. In: Proceedings of the IEEE Conference on Computer Vision and Pattern Recognition, pp. 770–778 (2016)
24. Huang, X., Belongie, S.: Arbitrary style transfer in real-time with adaptive instance normalization. In: Proceedings of the IEEE International Conference on Computer Vision, pp. 1501–1510 (2017)
25. Isola, P., Zhu, J.Y., Zhou, T., Efros, A.A.: Image-to-image translation with conditional adversarial networks. In: Proceedings of the IEEE Conference on Computer Vision and Pattern Recognition, pp. 1125–1134 (2017)
26. Jain, L.P., Scheirer, W.J., Boult, T.E.: Multi-class open set recognition using probability of inclusion. In: Fleet, D., Pajdla, T., Schiele, B., Tuytelaars, T. (eds.) ECCV 2014. LNCS, vol. 8691, pp. 393–409. Springer, Cham (2014). https://doi.org/10.1007/978-3-319-10578-9_26
27. Joseph, K., Khan, S., Khan, F.S., Balasubramanian, V.N.: Towards open world object detection. In: Proceedings of the IEEE/CVF Conference on Computer Vision and Pattern Recognition, pp. 5830–5840 (2021)
28. Julesz, B.: Binocular depth perception without familiarity cues: random-dot stereo images with controlled spatial and temporal properties clarify problems in stereopsis. Science **145**(3630), 356–362 (1964)
29. Khodadadeh, S., Bölöni, L., Shah, M.: Unsupervised meta-learning for few-shot image classification. In: Wallach, H.M., Larochelle, H., Beygelzimer, A., d'Alché-Buc, F., Fox, E.B., Garnett, R. (eds.) Advances in Neural Information Processing Systems 32: Annual Conference on Neural Information Processing Systems 2019, NeurIPS 2019, 8–14 December 2019, Vancouver, BC, Canada, pp. 10132–10142 (2019). www.proceedings.neurips.cc/paper/2019/hash/fd0a5a5e367a0955d81278062ef37429-Abstract.html
30. Krizhevsky, A., Sutskever, I., Hinton, G.E.: Imagenet classification with deep convolutional neural networks. Adv. Neural. Inf. Process. Syst. **25**, 1097–1105 (2012)
31. Lampert, C.H., Nickisch, H., Harmeling, S.: Learning to detect unseen object classes by between-class attribute transfer. In: 2009 IEEE Conference on Computer Vision and Pattern Recognition, pp. 951–958. IEEE (2009)
32. Li, Y., et al.: Shape-texture debiased neural network training. arXiv preprint arXiv:2010.05981 (2020)
33. Oliva, A., Schyns, P.G.: Diagnostic colors mediate scene recognition. Cogn. Psychol. **41**(2), 176–210 (2000). https://doi.org/10.1006/cogp.1999.0728

34. Oppenheim, A.V., Lim, J.S.: The importance of phase in signals. Proc. IEEE **69**(5), 529–541 (1981)
35. Palatucci, M., Pomerleau, D., Hinton, G.E., Mitchell, T.M.: Zero-shot learning with semantic output codes. In: Bengio, Y., Schuurmans, D., Lafferty, J.D., Williams, C.K.I., Culotta, A. (eds.) Advances in Neural Information Processing Systems 22: 23rd Annual Conference on Neural Information Processing Systems 2009. Proceedings of a meeting held 7–10 December 2009, Vancouver, British Columbia, Canada, pp. 1410–1418. Curran Associates, Inc. (2009). www.proceedings.neurips.cc/paper/2009/hash/1543843a4723ed2ab08e18053ae6dc5b-Abstract.html
36. Peuskens, H., Claeys, K.G., Todd, J.T., Norman, J.F., Van Hecke, P., Orban, G.A.: Attention to 3-D shape, 3-D motion, and texture in 3-D structure from motion displays. J. Cogn. Neurosci. **16**(4), 665–682 (2004)
37. Prabhudesai, M., Lal, S., Patil, D., Tung, H., Harley, A.W., Fragkiadaki, K.: Disentangling 3D prototypical networks for few-shot concept learning. In: 9th International Conference on Learning Representations, ICLR 2021, Virtual Event, Austria, 3–7 May 2021. OpenReview.net (2021). www.openreview.net/forum?id=-Lr-u0b42he
38. Puce, A., Allison, T., Asgari, M., Gore, J.C., McCarthy, G.: Differential sensitivity of human visual cortex to faces, letterstrings, and textures: a functional magnetic resonance imaging study. J. Neurosci. **16**(16), 5205–5215 (1996)
39. Qi, L., et al.: Open-world entity segmentation. arXiv preprint arXiv:2107.14228 (2021)
40. Rahman, S., Khan, S., Porikli, F.: A unified approach for conventional zero-shot, generalized zero-shot, and few-shot learning. IEEE Trans. Image Process. **27**(11), 5652–5667 (2018)
41. Ranftl, R., Bochkovskiy, A., Koltun, V.: Vision transformers for dense prediction. arXiv preprint (2021)
42. Ranftl, R., Lasinger, K., Hafner, D., Schindler, K., Koltun, V.: Towards robust monocular depth estimation: mixing datasets for zero-shot cross-dataset transfer. IEEE Trans. Pattern Anal. Mach. Intell. (TPAMI) (2020)
43. Schlimmer, J.C., Fisher, D.: A case study of incremental concept induction. In: AAAI, vol. 86, pp. 496–501 (1986)
44. Schroff, F., Kalenichenko, D., Philbin, J.: FaceNet: a unified embedding for face recognition and clustering. In: Proceedings of the IEEE Conference on Computer Vision and Pattern Recognition, pp. 815–823 (2015)
45. Selvaraju, R.R., Cogswell, M., Das, A., Vedantam, R., Parikh, D., Batra, D.: Gradcam: visual explanations from deep networks via gradient-based localization. In: Proceedings of the IEEE International Conference on Computer Vision, pp. 618–626 (2017)
46. Simonyan, K., Zisserman, A.: Very deep convolutional networks for large-scale image recognition. In: Bengio, Y., LeCun, Y. (eds.) 3rd International Conference on Learning Representations, ICLR 2015, San Diego, CA, USA, 7–9 May 2015, Conference Track Proceedings (2015). www.arxiv.org/abs/1409.1556
47. Snell, J., Swersky, K., Zemel, R.: Prototypical networks for few-shot learning. In: Advances in Neural Information Processing Systems, vol. 30 (2017)
48. Snell, J., Swersky, K., Zemel, R.S.: Prototypical networks for few-shot learning. In: Guyon, I., et al. (eds.) Advances in Neural Information Processing Systems 30: Annual Conference on Neural Information Processing Systems 2017, 4–9 December 2017, Long Beach, CA, USA, pp. 4077–4087 (2017). www.proceedings.neurips.cc/paper/2017/hash/cb8da6767461f2812ae4290eac7cbc42-Abstract.html

49. Speer, R., Chin, J., Havasi, C.: Conceptnet 5.5: an open multilingual graph of general knowledge. In: Thirty-First AAAI Conference on Artificial Intelligence (2017)
50. Sung, F., Yang, Y., Zhang, L., Xiang, T., Torr, P.H., Hospedales, T.M.: Learning to compare: relation network for few-shot learning. In: Proceedings of the IEEE Conference on Computer Vision and Pattern Recognition, pp. 1199–1208 (2018)
51. Thomson, M.G.: Visual coding and the phase structure of natural scenes. Netw. Comput. Neural Syst. **10**(2), 123 (1999)
52. Thrun, S., Mitchell, T.M.: Lifelong robot learning. Robot. Auton. Syst. **15**(1–2), 25–46 (1995)
53. Tokmakov, P., Wang, Y.X., Hebert, M.: Learning compositional representations for few-shot recognition. In: Proceedings of the IEEE/CVF International Conference on Computer Vision, pp. 6372–6381 (2019)
54. Vinyals, O., Blundell, C., Lillicrap, T., Wierstra, D., et al.: Matching networks for one shot learning. Adv. Neural. Inf. Process. Syst. **29**, 3630–3638 (2016)
55. Wah, C., Branson, S., Welinder, P., Perona, P., Belongie, S.: The Caltech-UCSD birds-200-2011 dataset (2011)
56. Walther, D.B., Chai, B., Caddigan, E., Beck, D.M., Fei-Fei, L.: Simple line drawings suffice for functional MRI decoding of natural scene categories. Proc. Natl. Acad. Sci. **108**(23), 9661–9666 (2011). https://doi.org/10.1073/pnas.1015666108, https://www.pnas.org/doi/abs/10.1073/pnas.1015666108
57. Wen, S., Rios, A., Ge, Y., Itti, L.: Beneficial perturbation network for designing general adaptive artificial intelligence systems. IEEE Trans. Neural Netw. Learn. Syst. (2021)
58. Yamins, D.L., Hong, H., Cadieu, C.F., Solomon, E.A., Seibert, D., DiCarlo, J.J.: Performance-optimized hierarchical models predict neural responses in higher visual cortex. Proc. Natl. Acad. Sci. **111**(23), 8619–8624 (2014)
59. Zhu, J.Y., et al.: Visual object networks: image generation with disentangled 3D representations. In: Bengio, S., Wallach, H., Larochelle, H., Grauman, K., Cesa-Bianchi, N., Garnett, R. (eds.) Advances in Neural Information Processing Systems, vol. 31. Curran Associates, Inc. (2018). www.proceedings.neurips.cc/paper/2018/file/92cc227532d17e56e07902b254dfad10-Paper.pdf

STEEX: Steering Counterfactual Explanations with Semantics

Paul Jacob[1], Éloi Zablocki[1(✉)], Hédi Ben-Younes[1], Mickaël Chen[1], Patrick Pérez[1], and Matthieu Cord[1,2]

[1] Valeo.ai, Paris, France
eloi.zablocki@gmail.com
[2] Sorbonne University, Paris, France

Abstract. As deep learning models are increasingly used in safety-critical applications, explainability and trustworthiness become major concerns. For simple images, such as low-resolution face portraits, synthesizing visual counterfactual explanations has recently been proposed as a way to uncover the decision mechanisms of a trained classification model. In this work, we address the problem of producing cotual explanations for high-quality images and complex scenes. Leveraging recent semantic-to-image models, we propose a new generative counterfactual explanation framework that produces plausible and sparse modifications which preserve the overall scene structure. Furthermore, we introduce the concept of "region-targeted counterfactual explanations", and a corresponding framework, where users can guide the generation of counterfactuals by specifying a set of semantic regions of the query image the explanation must be about. Extensive experiments are conducted on challenging datasets including high-quality portraits (CelebAMask-HQ) and driving scenes (BDD100k). Code is available at: https://github.com/valeoai/STEEX.

Keywords: Explainable AI · Counterfactual analysis · Visual explanations · Region-targeted counterfactual explanation

1 Introduction

Deep learning models are now used in a wide variety of application domains, including safety-critical ones. As the underlying mechanisms of these models remain opaque, explainability and trustworthiness have become major concerns. In computer vision, *post-hoc* explainability often amounts to producing saliency maps, which highlight regions on which the model grounded the most its decision [2,13,38,40,43,54,59]. While these explanations show *where* the regions of interest for the model are, they fail to indicate *what* specifically in these regions leads to the obtained output. A desirable explanation should not only be *region*-based

Supplementary Information The online version contains supplementary material available at https://doi.org/10.1007/978-3-031-19775-8_23.

S. Avidan et al. (Eds.): ECCV 2022, LNCS 13672, pp. 387–403, 2022.
https://doi.org/10.1007/978-3-031-19775-8_23

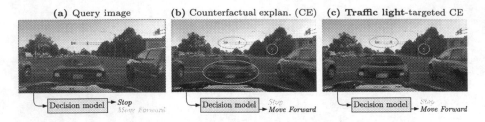

Fig. 1. Overview of counterfactual explanations generated by our framework STEEX. Given a trained model and a query image (a), a counterfactual explanation is an answer to the question *"What other image, slightly different and in a meaningful way, would change the model's outcome?"* In this example, the 'Decision model' is a binary classifier that predicts whether or not it is possible to move forward. On top of explaining decisions for large and complex images (b), we propose 'region-targeted counterfactual explanations' (c), where produced counterfactual explanations only target specified semantic regions. Green ellipses are manually provided to highlight details (Color figure online)

but also *content*-based by expressing in some way how the content of a region influences the outcome of the model. For example, in autonomous driving, while it is useful to know that a stopped self-driving car attended the traffic light, it is paramount to know that the red color of the light was decisive in the process.

In the context of simple tabular data, *counterfactual explanations* have recently been introduced to provide fine content-based insights on a model's decision [8,47,48]. Given an input query, a counterfactual explanation is a version of the input with *minimal* but *meaningful* modifications that change the output decision of the model. *Minimal* means that the new input must be as similar as possible to the query input, with only sparse changes or in the sense of some distance to be defined. *Meaningful* implies that changes must be semantic, i.e., human-interpretable. This way, a counterfactual explanation points out in an understandable way *what* is important for the decision of the model by presenting a close hypothetical reality that contradicts the observed decision. As they are contrastive and as they usually focus on a small number of feature changes, counterfactuals can increase user's trust in the model [39,53,56]. Moreover, these explanations can also be leveraged by machine learning engineers, as they can help to identify spurious correlations captured by a model [36,45,55]. Despite growing interest, producing visual counterfactual explanations for an image classification model is especially challenging as naively searching for small input changes results in adversarial perturbations [6,14,18,33,44]. To this date, there only exists a very limited number of counterfactual explanation methods able to deal with image classifiers [19,36,41,50]. Yet, these models present significant limitations, as they either require a target image of the counterfactual class [19,50] or can only deal with classification settings manipulating simple images such as low-resolution face portraits [36,41].

In this work, we tackle the generation of counterfactual explanations for deep classifiers operating on large images and/or visual scenes with complex structures. Dealing with such images comes with unique challenges, beyond technical issues. Indeed, because of scene complexity, it is likely that the model's decision can be changed by many admissible modifications in the input. For a driving action classifier, it could be for instance modifying the color of traffic lights, the road markings or the visibility conditions, but also adding new elements to the scene such as pedestrians and traffic lights, or even replacing a car on the road with an obstacle. Even if it was feasible to provide an exhaustive list of counterfactual explanations, the task of selecting which ones in this large collection are relevant would fall on the end-user, hindering the usability of the method. To limit the space of possible explanations while preserving sufficient expressivity, we propose that the overall structure of the query image remains untouched when creating the counterfactual example. Accordingly, through semantic guidance, we impose that a generated counterfactual explanation respects the original layout of the query image.

Our model, called STEEX for STEering counterfactual EXplanations with semantics, leverages recent breakthroughs in semantic-to-real image synthesis [32,37,60]. A pre-trained encoder network decomposes the query image into a spatial layout structure and latent representations encoding the content of each semantic region. By carefully modifying the latent codes towards a different decision, STEEX is able to generate meaningful counterfactuals with relevant semantic changes and a preserved scene layout, as illustrated in Fig. 1b. Additionally, we introduce a new setting where users can guide the generation of counterfactuals by specifying which semantic region of the query image the explanation must be about. We coin "region-targeted counterfactual explanations" such generated explanations where only a subset of latent codes is allowed to be modified. In other words, such explanations are answers to questions such as *"How should the traffic lights change to switch the model's decision?"*, as illustrated in Fig. 1d. To validate our claims, extensive experiments of STEEX are conducted on a variety of image classification models trained for different tasks, including self-driving action decision on the BDD100k dataset, and high-quality face recognition networks trained on CelebAMask-HQ. Besides, we investigate how explanations for different decision models can hint at their distinct and specific behaviors.

To sum up, our contributions are as follows:

- We tackle the generation of visual counterfactual explanations for classifiers dealing with large and/or complex images.
- By leveraging recent semantic-to-image generative models, we propose a new framework capable of generating counterfactual explanations that preserve the semantic layout of the image.
- We introduce the concept of "region-targeted counterfactual explanations" to target specified semantic regions in the counterfactual generation process.
- We validate the quality, plausibility and proximity to their query, of obtained explanations with extensive experiments, including classification models for high-quality face portraits and complex urban scenes.

2 Related Work

The black-box nature of deep neural networks has led to the recent development of many explanation methods [1,3,12,16]. In particular, our work is grounded within the *post-hoc* explainability literature aiming at explaining a trained model, which contrasts with approaches building interpretable models *by design* [9,57]. Post-hoc methods can be either be *global* if they seek to explain the model in its entirety, or *local* when they explain the prediction of the model for a specific instance. Global approaches include model translation techniques, that distill the black-box model into a more interpretable one [15,20], or the more recent disentanglement methods that search for latent dimensions of the input space that are, at the level of the dataset, correlated with the output variations of the target classifier [25,28]. Instead, in this paper, we focus on *local* methods that provide explanations, tailored to a given image.

Usually, *post-hoc local* explanations of vision models are given in the form of saliency maps, which attribute the output decision to image regions. Gradient-based approaches compute this attribution using the gradient of the output with respect to input pixels or intermediate layers [4,34,38,43]. Differently, perturbation-based approaches [13,49,54,58] evaluate how sensitive to input variations is the prediction. Other explainability methods include locally fitting a more interpretable model such as a linear function [35] or measuring the effect of including a feature with game theory tools [30]. However, these methods only provide information on *where* are the regions of interest for the model but do not tell *what* in these regions is responsible for the decision.

Counterfactual explanations [48], on the other hand, aim to inform a user on why a model M classifies a specific input x into class y instead of a *counter class* $y' \neq y$. To do so, a *counterfactual example* x' is constructed to be similar to x but classified as y' by M. Seminal methods have been developed in the context of low-dimensional input spaces, like the ones involved in credit scoring tasks [48]. Naive attempts to scale the concept to higher-dimensional input spaces, such as natural images, face the problem of producing adversarial examples [6,18,31,44], that is, *imperceptible* changes to the query image that switch the decision. While the two problems have similar formulations, their goals are in opposition [14,33] since counterfactual explanations must be understandable, achievable, and informative for a human. Initial attempts to counterfactual explanations of vision models would explain a decision by comparing the image x to one or several real instances classified as y' [19,21,50]. However, these discriminative counterfactuals do not produce natural images as explanations, and their interpretability is limited when many elements vary from one image to another.

To tackle these issues, generative methods leverage deep generative models to produce counterfactual explanations. For instance, DiVE [36] is built on β-TCVAE [11] and takes advantage of its disentangled latent space to discover such meaningful sparse modifications. With this method, it is also possible to generate multiple orthogonal changes that correspond to different valid counterfactual examples. Progressive Exaggeration (PE) [41], instead, relies on a Generative Adversarial Network (GAN) [17] conditioned on a perturbation value that is introduced as input in the generator via conditional batch normalization. PE modifies the query image so that the prediction of the decision model is shifted by this perturbation value towards the counter class. By applying this modification multiple times, and by showing the progression, PE highlights adjustments that would change the decision model's output. Unfortunately, none of these previous works is designed to handle complex scenes. The β-TCVAE used in DiVE hardly scales beyond small centered images, requiring specifically-designed enhancement methods [27,42], and PE performs style-based manipulations that are unsuited images with multiple small independent objects of interest. Instead, our method relies on segmentation-to-image GANs [32,37,60], that have demonstrated good generative capabilities on high-quality images containing multiple objects.

3   Model STEEX

We now describe our method to obtain counterfactual explanations with semantic guidance. First, we formalize the generative approach for visual counterfactual explanations in Sect. 3.1. Within this framework, we then incorporate a semantic guidance constraint in Sect. 3.2. Next we propose in Sect. 3.3 a new setting where the generation targets specified semantic regions. Finally, Sect. 3.4 details the instantiation of each component. An overview of STEEX is presented in Fig. 2.

3.1   Visual Counterfactual Explanations

Consider a trained differentiable machine learning model M, which takes an image $x^I \in \mathcal{X}$ from an input space $\mathcal{X}$ and outputs a prediction $y^I = M(x^I) \in \mathcal{Y}$. A counterfactual explanation for the obtained decision y^I is an image x which is as close to the image x^I as possible, but such that $M(x) = y$ where $y \neq y^I$ is another class. This problem can be formalized and relaxed as follows:

$$\operatorname{argmin}_{x \in \mathcal{X}} L_{\text{decision}}(M(x), y) + \lambda L_{\text{dist}}(x^I, x), \qquad (1)$$

where L_{decision} is a classification loss, L_{dist} measures the distance between images, and the hyperparameter λ balances the contribution of the two terms.

In computer vision applications where input spaces are high-dimensional, additional precautions need to be taken to avoid ending up with adversarial examples [6,14,33,44]. To prevent those uninterpretable perturbations, which leave the data manifold by adding imperceptible high-frequency patterns, counterfactual methods impose that visual explanations lie in the original input domain $\mathcal{X}$. Incorporating this in-domain constraint can be achieved by using a deep generator network as an implicit prior [5,46]. Consider a generator $G : z \mapsto x$ that maps vectors z in latent space $\mathcal{Z}$ to in-distribution images x. Searching images only in the output space of such a generator would be sufficient to satisfy the in-domain constraint, and the problem now reads:

$$\operatorname{argmin}_{z \in \mathcal{Z}} L_{\text{decision}}(M(G(z)), y) + \lambda L_{\text{dist}}(x^I, G(z)). \qquad (2)$$

Eq. 2 formalizes practices introduced in prior works [36,41] that also aim to synthesize counterfactual explanations for images.

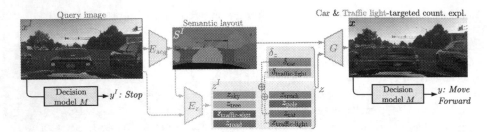

Fig. 2. Overview of STEEX. The query image x^I is first decomposed into a semantic map S^I and $z^I = (z_c)_{c=1}^N$, a collection of N semantic embeddings which encode each the aspect of their corresponding semantic category c. The perturbation δ_z is optimized such that the generated image $x = G(S^I, z^I + \delta_z)$ is classified as y by the decision model M, while staying small. As the generator uses the semantic layout S^I of the query image x^I, the generated counterfactual explanation x retains the original image structure. The figure specifically illustrates the region-targeted setting, where only the subset {'car', 'traffic light'} of the semantic style codes is targeted

Furthermore, assuming that a latent code z^I exists and can be recovered for the image x^I, we can express the distance loss directly in the latent space $\mathcal{Z}$:

$$\text{argmin}_{z \in \mathcal{Z}} \, L_{\text{decision}}(M(G(z)), y) + \lambda L_{\text{dist}}(z^I, z). \tag{3}$$

By searching for an optimum in a low-dimensional latent space rather than in the raw pixel space, we operate over inputs that have a higher-level meaning, which is reflected in the resulting counterfactual examples.

3.2   Semantic-Guided Counterfactual Generation

The main objective of our model is to scale counterfactual image synthesis to large and complex scenes involving multiple objects within varied layouts. In such a setting, identifying and interpreting the modifications made to the query image is a hurdle to the usability of counterfactual methods. Therefore we propose to generate counterfactual examples that preserve the overall structure of the query and, accordingly, design a framework that optimizes under a fixed semantic layout. Introducing semantic masks for counterfactual explanations comes with additional advantages. First, we can leverage semantic-synthesis GANs that are particularly well-suited to generate diverse complex scenes [32,37,60]. Second, it provides more control over the counterfactual explanation we wish to synthesize, allowing us to target the changes to a specific set of semantic regions, as we detail in Sect. 3.3. To do so, we adapt the generator G and condition it on a semantic mask S that associates each pixel to a label indicating its semantic category (for instance, in the case of a driving scene, such labels can be cars, road, traffic signs, etc.). The output of the generator $G : (S, z) \mapsto x$ is now restricted to follow the layout indicated by S. We can then find a counterfactual example for image x^I that has an associated semantic mask S^I by optimizing the following objective:

$$\text{argmin}_{z \in \mathcal{Z}} \, L_{\text{decision}}(M(G(S^I, z)), y) + \lambda L_{\text{dist}}(z^I, z). \tag{4}$$

This formulation guarantees that the semantic mask S^I of the original scene is kept as is in the counterfactuals.

3.3 Region-Targeted Counterfactual Explanations

We introduce a new setting enabling finer control in the generation of counterfactuals. In this setup, a user specifies a set of semantic regions that the explanation must be about. For example, in Fig. 2, the user selects 'car' and 'traffic light', and the resulting counterfactual is only allowed to alter these regions. Such a selection allows studying the influence of different semantic concepts in the image for the target model's behavior. In practice, given a semantic mask S with N classes, we propose to decompose z into N vectors, $z = (z_c)_{c=1}^N$, where each z_c is a latent vector associated with one class in S. With such a formulation, it becomes possible to target a subset $C \subset \{1, \ldots, N\}$ for the counterfactual explanation. Region-targeted counterfactuals only optimize on the specified components $(z_c)_{c \in C}$, and all other latent codes remain unmodified.

3.4 Instantiation of STEEX

We now present the modeling choices we make for each part of our framework.

Generator G. The generator G can be any of the recent segmentation-to-image GANs [32,37,60] that transform a latent code z and a segmentation layout S into an image x. As such generators typically allow for a different vector z_c to be used for each class in the semantic mask [37,60], the different semantic regions can be modified independently in the output image. This property enables STEEX to perform region-targeted counterfactual explanations as detailed in Sect. 3.3.

Obtaining the Code z^I. To recover the latent code z^I from the image x^I, we exploit the fact that in aforementioned frameworks [37,60], the generator G can be trained jointly, in an auto-encoding pipeline, with an encoder E_z that maps an image x^I and its associated segmentation layout S^I into a latent code z^I. Such a property ensures that we can efficiently compute this image-to-latent mapping and that there is indeed a semantic code that corresponds to each image, leading to an accurate reconstruction in the first place.

Obtaining the Mask S^I. As query images generally have no associated annotated segmentation masks S^I, these need to be inferred. To do so, we add a segmentation network E_{seg} in the pipeline: we first obtain the map $S^I = E_{seg}(x^I)$ and then use the encoder: $z^I = E_z(x^I, S^I)$, so STEEX is applicable to any image.

Loss Functions. The decision loss L_{dist} ensures that the output image x is classified as y by the decision model M. It is thus set as the negative log-likelihood of the targeted counter class y for $M(G(z))$:

$$L_{decision}(M(G(z)), y) = -\mathcal{L}(M(G(z))|y). \tag{5}$$

The distance loss L_{dist} is the sum of squared L2 distance between each semantic component of z^I and z:

$$L_{\text{dist}}(z^I, z) = \sum_{c=1}^{N} \|z_c^I - z_c\|_2^2. \tag{6}$$

We stress that Eq. 4 is optimized on the code z only. All of the network parameters (G, E_z and E_{seg}) remain frozen.

4   Experiments

We detail in Sect. 4.1 our experimental protocol to evaluate different aspects of generated counterfactuals: the plausibility and perceptual quality (Sect. 4.2) as well as the proximity to query images (Sect. 4.3). We then present in Sect. 4.4 region-targeted counterfactual explanations. In Sect. 4.5, we use STEEX to explain different decision models for the same task, and show that produced explanations hint at the specificities of each model. Finally, we present an ablation study in Sect. 4.6. Our code and pretrained models will be made available.

4.1   Experimental Protocol

We evaluate our method on five decision models across three different datasets. We compare against two recently proposed visual counterfactual generation frameworks, Progressive Exaggeration (PE) [41] and DiVE [36], previously introduced in Sect. 2. We report scores directly from their paper when available (CelebA) and used the public and official implementation to evaluate them otherwise (CelebAMask-HQ and BDD100k). We now present each dataset and the associated experimental setup.

BDD100k [52]. The ability of STEEX to explain models handling complex visual scenes is evaluated on the driving scenes of BDD100k. Most images of this dataset contain diversely-positioned objects that can have fine relationships with each other, and small details in size can be crucial for the global understanding of the scene (e.g., traffic light colors). The decision model to be explained is a *Move Forward* vs. *Stop/Slow down* action classifier trained on BDD-OIA [51], a 20,000-scene extension of BDD100k annotated with binary attributes representing the high-level actions that are allowed in a given situation. The image resolution is 512×256. The segmentation model E_{seg} is a DeepLabV3 [10] trained on a subset of 10,000 images annotated with semantic masks that cover 20 classes (e.g., road, truck, car, tree, etc.). On the same set, the semantic encoder E_z and the generator G are jointly trained within a SEAN framework [60]. Counterfactual scores are computed on the validation set of BDD100k.

CelebAMask-HQ [26]. CelebAMask-HQ contains 30,000 high-quality face portraits with semantic segmentation annotation maps including 19 semantic classes (e.g., skin, mouth nose, etc.). The portraits are also annotated with identity and 40 binary attributes, allowing us to perform a quantitative evaluation for high-quality images. Decision models to be explained are two DenseNet121 [23] binary classifiers trained to respectively recognize *Smile* and *Young* attributes.

To obtain semantic segmentation masks for the query images, we instantiate E_{seg} with a DeepLabV3 [10] pre-trained on the 28,000-image training split. On the same split, the semantic encoder E_z and generator G are jointly learned within a SEAN framework [60]. Counterfactual explanations are computed on the 2000-image validation set, with images rescaled to the resolution 256×256.

CelebA [29]. CelebA contains 200,000 face portraits, annotated with identity and 40 binary attributes, but of smaller resolution (128×128 after processing) and of lower quality compared to CelebAMask-HQ. STEEX is designed to handle more complex and larger images, but we include this dataset for the sake of completeness as previous works [36,41] use it as their main benchmark. We report their score directly from their respective papers and align our experiment protocol with the one described in [36]. As in previous works, we explain two decision models: a *Smile* classifier and a *Young* classifier, both with DenseNet121 architecture [23]. We obtain E_{seg} with a DeepLabV3 [10] trained on CelebAMask-HQ images. Then, we jointly train the semantic encoder E_z and generator G with a SEAN architecture [60] on the training set of CelebA. Explanations are computed on the 19,868-image validation split of CelebA.

Optimization Scheme. As M and G are differentiable, we optimize z using ADAM [24] with a learning rate $1 \cdot 10^{-2}$ for 100 steps with $\lambda = 0.3$. Hyperparameters have been found on the training splits of the datasets.

4.2 Quality of the Counterfactual Explanations

We first ensure that the success rate of STEEX, i.e., the fraction of explanations that are well classified into the counter class, is higher than 99.5% for all of the five tested classifiers. Then, as STEEX's counterfactuals must be realistic and informative, we evaluate their perceptual quality.

Similarly with previous works [36,41], we use the Fréchet Inception Distance (FID) [22] between all explanations and the set of query images, and report this metric in Table 1. For each classifier, STEEX outperforms the baselines by a large margin, meaning that our explanations are more realistic-looking, which verifies that they belong to the input domain of the decision model.

Generating realistic counterfactuals for classifiers that deal with large and complex images is difficult, as reflected by large FID discrepancies between

Table 1. Perceptual quality, measured with FID ↓. Five attribute classifiers are explained, across three datasets. Results of PE and DiVE are reported from original papers on CelebA. For CelebAMask-HQ and BDD100k, their models are retrained using their code. DiVE does not converge on BDD100k

FID ↓	CelebA		CelebAM-HQ		BDD100k
	Smile	Young	Smile	Young	Move For.
PE [41]	35.8	53.4	52.4	60.7	141.6
DiVE [36]	29.4	33.8	107.0	107.5	—
STEEX	**10.2**	**11.8**	**21.9**	**26.8**	**58.8**

Counterfactual explanation for Smile   Counterfactual explanation for Young

Fig. 3. Counterfactual explanations on CelebAMask-HQ, generated by STEEX (ours), PE, and DiVE. Explanations are generated for two binary classifiers, on *Smile* and *Young* attributes, at resolution 256×256. Other examples in the Supplementary

CelebA, CelebAMask-HQ and BDD100k. Scaling the generation of counterfactual explanations from 128×128 (CelebA) to 256×256 (CelebAMask-HQ) face portraits is not trivial as a significant drop in performance can be observed for all models, especially for DiVE. Despite our best efforts to train DiVE on BDD100k, we were unable to obtain satisfying 512×256 explanations, as all reconstructions were nearly uniformly gray. As detailed in Sect. 2, VAE-based models are indeed usually limited to images with a fairly regular structure, and they struggle to deal with the diversity of driving scenes.

We display examples of STEEX's counterfactual explanations on CelebAMask-HQ in Fig. 3, compared with PE [41] and DiVE [36]. For the *Smile* classifier, STEEX explains positive (top-row) and negative (bottom-row) smile predictions through sparse and photo-realistic modifications of the lips and the skin around the mouth and the eyes. Similarly, for the *Young* classifier, STEEX explain decisions by adding or removing facial wrinkles. In comparison, PE introduces high-frequency artifacts that harm the realism of generated examples. DiVE generates blurred images and applies large modifications so that it becomes difficult to identify the most crucial changes for the target model. Figure 4 shows other samples for the action classifier on the BDD100k dataset, where we

Fig. 4. Counterfactual explanations on BDD100k. Explanations are generated for a binary classifier for the action *Move Forward*, with images at resolution 512×256. Our method finds interpretable, sparse and meaningful semantic modifications to the query image. Other examples are available in the Supplementary

overlay green ellipses to point the reader's attention to significant region changes. STEEX finds sparse but highly semantic modifications to regions that strongly influence the output decision, such as the traffic light colors or the brake lights of a leading vehicle. Finally, the semantic guidance leads to a fine preservation of the scene structure in STEEX's counterfactuals, achieving both global coherence and high visual quality.

4.3   Proximity to the Query Image

We now verify the *proximity* of counterfactuals to query images, as well as the *sparsity* of changes.

We first compare STEEX to previous work with respect to the **Face Verification Accuracy (FVA)**. The FVA is the percentage of explanations that preserve the person's identity, as revealed by a cosine similarity above 0.5 between features of the counterfactual and the query. Following previous works [36,41], features are computed by a pre-trained re-identification network on VGGFace2 [7]. As shown in Table 2, even if STEEX is designed for high-quality or complex scenes image classifiers, it reaches high FVA on the low-quality CelebA dataset. Moreover, STEEX significantly outperforms PE and DiVE on CelebAMask-HQ, showing its ability to scale up to higher image sizes. Again, DiVE suffers from the poor capacities of $\beta-$TCVAE to reconstruct high-quality images Sect. 2. To support this claim, we compute the FVA between query images and reconstructions with the $\beta-$TCVAE of DiVE and obtain 45.9%, which indicates a low reconstruction capacity.

We then measure the sparsity of explanations using the **Mean Number of Attributes Changed (MNAC)**. This metric averages the number of facial attributes that differ between the query image and its associated counterfactual explanation. As STEEX successfully switches the model's decision almost every time, explanations that obtain a low MNAC are likely to have altered only the necessary elements to build a counterfactual. Following previous work [36], we use an oracle ResNet pretrained on VGGFace2 [7], and fine-tuned on 40 attributes provided in CelebA/CelebAMask-HQ. As reported in Table 2, STEEX has a lower MNAC than PE and DiVE on both CelebA and CelebAMask-HQ. Conditioning the counterfactual generation on semantic masks helps obtaining small variations that are meaningful enough for the model to switch its decision. This property makes STEEX useful in practice and well-suited to explain image classifiers.

Table 2. Face Verification Accuracy (FVA ↑) (%) and Mean Number of Attributes Changed (MNAC ↓), on CelebA and CelebAMask-HQ. For PE and DiVE, CelebA scores come from the original papers, and we re-train their models using official implementations for CelebAMask-HQ

FVA ↑	CelebA		CelebAM-HQ		MNAC ↓	CelebA		CelebAM-HQ	
	Smile	Young	Smile	Young		Smile	Young	Smile	Young
PE [41]	85.3	72.2	79.8	76.2	PE [41]	—	3.74	7.71	8.51
DiVE [36]	**97.3**	**98.2**	35.7	32.3	DiVE [36]	—	4.58	7.41	6.76
STEEX	96.9	97.5	**97.6**	**96.0**	**STEEX**	4.11	3.44	5.27	5.63

Fig. 5. Semantic region-targeted counterfactual explanations on BDD100k.
Explanations are generated for a binary classifier trained on the attribute *Move Forward*,
at resolution 512×256. Each row shows explanations where we restrict the optimization
process to one specific semantic region, on two examples: one where the model initially
goes forward, and one where it initially stops. Significant modifications are highlighted
within the green ellipses. Note that even when targeting specific regions, others may still
slightly differ from the original image: this is mostly due to small errors in the reconstruction $G(S^I, z^I) \approx x^I$ (more details in the Supplementary) (Color figure online)

4.4 Region-Targeted Counterfactual Explanations

As can be seen in Figs. 1b and 4, when the query image is complex, the counterfactual explanations can encompass multiple semantic concepts at the same
time. In Fig. 1b for instance, in order to switch the decision of the model to *Move
Forward*, the traffic light turns green and the car's brake lights turn off. It raises
ambiguity about how these elements compound to produce the decision. In other
words, "Are both changes necessary, or changing only one region is sufficient to
switch the model's decision?".

To answer this question, we generate *region-targeted* counterfactual explanations, as explained in Sect. 3.3. In Fig. 1d, we observe that targeting the traffic
light region can switch the decision of the model, despite the presence of a
stopped car blocking the way. Thereby, region-targeted counterfactuals can help
to identify potentially safety-critical issues with the decision model.

More generally, region-targeted counterfactual explanations empower the
user to separately assess how different concepts impact the decision. We show in
Fig. 5 qualitative examples of such region-targeted counterfactual explanations
on the *Move Forward* classifier. On the one hand, we can verify that the decision
model relies on cues such as the color of the traffic lights and brake lights of cars,
as changing them often successfully switch the decision. On the other hand, we
discover that changes in the appearance of buildings can flip the model's decision. Indeed, we see that green or red gleams on facades can fool the decision

Fig. 6. Counterfactual explanations on CelebAMask-HQ for three different *Young* classifiers, namely M_{top}, M_{mid}, and M_{bot} that respectively only attend to the top, mid, and bottom parts of the image. Other examples are available in the Supplementary

model into predicting *Move Forward* or *Stop* respectively, suggesting that the model could need further investigation before being safely deployed.

4.5   Analyzing Decision Models

An attractive promise of explainable AI is the possibility to detect and characterize biases or malfunctions of explained decision models. In this section, we investigate how specific are explanations to different decision models and if the explanations can point at the particularity of each model. In practice, we consider three decision models, namely M_{top}, M_{mid}, and M_{bot}, that were trained on images with masked out pixels except the for the top, middle, and bottom parts of the input respectively. Figure 6 reports qualitative results, and we can identify that M_{top} has based its decisions mainly on the color of the hair, while M_{mid} uses the wrinkles on the face, and M_{bot} focuses on facial hair and the neck.

We also measure how much each semantic region has been modified to produce the counterfactual. Accordingly, we assess the impact of a semantic class c in the decision with the average value of $\|\delta_{z_c}\|_2 = \|z_c^I - z_c\|_2$ aggregated over the validation set. Note that while the absolute values of δ_{z_c} can be compared across the studied decision models, they cannot be directly compared across different semantic classes, as the z_c can be at different scales for different values of c in the generative model. To make this comparison in Table 3, we instead compute the

Table 3. Most and least impactful semantic classes for a decision model relatively to others. The impact of a class for a given model has been determined as the average value of $\|\delta_{z_c}\|_2 = \|z_c^I - z_c\|_2$ for each semantic class c, relatively to the same value averaged for other models

Model	Most impactful	Least impactful
M_{top}	Hat, hair, background	Necklace, eyes, lips
M_{mid}	Nose, glasses, eyes	Necklace, neck, hat
M_{bot}	Neck, necklace, cloth	Eyes, brows, glasses

Table 4. Ablation study measuring the role of the distance loss L_{dist} in Eq. 4 and upper bound results that would be achieved with ground-truth segmentation masks

	Smile		Young	
	FID ↓	FVA ↑	FID ↓	FVA ↑
STEEX	21.9	97.6	26.8	96.0
without L_{dist}	29.7	65.2	45.7	37.0
with ground-truth segmentation	21.2	98.9	25.7	98.2

value of δ_{z_c} for the target model *relatively* to the average value for all models. The semantic classes of most impact in Table 3 indicate how each decision model is biased towards a specific part of the face and ignores cues that are important for the other models.

4.6   Ablation Study

We propose an ablation study on CelebAMask-HQ, reported in Table 4, to assess the role of the distance loss L_{dist} and the use of predicted segmentation masks.

First, we evaluate turning off the distance loss by setting $\lambda = 0$, such that the latent codes z_c are no longer constrained to be close to z_c^I. Doing so, for both *Young* and *Smile* classifiers, the FVA and FID of STEEX degrade significantly, which respectively indicate that the explanation proximity to the real images is deteriorated and that the counterfactuals are less plausible. The distance loss is thus an essential component for STEEX.

Second, we investigate if the segmentation network E_{seg} is a bottleneck in STEEX. To do so, we replace the segmenter's outputs with ground-truth masks and generate counterfactual explanations with these. The fairly similar scores of both settings indicate that STEEX works well with inferred layouts.

5   Conclusion

In this work, we present STEEX, a method to generate counterfactual explanations for complex scenes, by steering the generative process using predicted semantics. To our knowledge, we provide the first framework for complex scenes where numerous elements can affect the decision of the target network. Experiments on driving scenes and high-quality portraits show the capacity of our method to finely explain deep classification models. For now, STEEX is designed to generate explanations that preserve the semantic structure. While we show the merits of this property, future work can consider how, within our framework, to handle operations such as shifting, removing, or adding objects, while keeping the explanation simple to interpret. Finally, we hope that the setup we propose in Sect. 4.5, when comparing explanations for multiple decision models with known behaviors, can serve as a basis to measure the interpretability of an explanation method.

References

1. Adadi, A., Berrada, M.: Peeking inside the black-box: a survey on explainable artificial intelligence (XAI). IEEE Access (2018)
2. Bach, S., Binder, A., Montavon, G., Klauschen, F., Müller, K.R., Samek, W.: On pixel-wise explanations for non-linear classifier decisions by layer-wise relevance propagation. PloS ONE (2015)
3. Beaudouin, V., et al.: Flexible and context-specific AI explainability: a multidisciplinary approach. CoRR abs/2003.07703 (2020)
4. Bojarski, M., et al.: Visualbackprop: efficient visualization of CNNs for autonomous driving. In: ICRA (2018)
5. Bora, A., Jalal, A., Price, E., Dimakis, A.G.: Compressed sensing using generative models. In: ICML (2017)
6. Browne, K., Swift, B.: Semantics and explanation: why counterfactual explanations produce adversarial examples in deep neural networks. CoRR abs/2012.10076 (2020)
7. Cao, Q., Shen, L., Xie, W., Parkhi, O.M., Zisserman, A.: Vggface2: a dataset for recognising faces across pose and age. In: FG (2018)
8. Chang, C., Creager, E., Goldenberg, A., Duvenaud, D.: Explaining image classifiers by counterfactual generation. In: ICLR (2019)
9. Chen, C., Li, O., Tao, D., Barnett, A., Rudin, C., Su, J.: This looks like that: deep learning for interpretable image recognition. In: NeurIPS (2019)
10. Chen, L., Papandreou, G., Schroff, F., Adam, H.: Rethinking atrous convolution for semantic image segmentation. CoRR abs/1706.05587 (2017)
11. Chen, R.T.Q., Li, X., Grosse, R., Duvenaud, D.: Isolating sources of disentanglement in variational autoencoders. In: NeurIPS (2018)
12. Das, A., Rad, P.: Opportunities and challenges in explainable artificial intelligence. (XAI), a survey. CoRR (2020)
13. Fong, R.C., Vedaldi, A.: Interpretable explanations of black boxes by meaningful perturbation. In: ICCV (2017)
14. Freiesleben, T.: Counterfactual explanations & adversarial examples - common grounds, essential differences, and potential transfers. CoRR abs/2009.05487 (2020)
15. Frosst, N., Hinton, G.E.: Distilling a neural network into a soft decision tree. In: Workshop on Comprehensibility and Explanation in AI and ML @AI*IA (2017)
16. Gilpin, L.H., Bau, D., Yuan, B.Z., Bajwa, A., Specter, M., Kagal, L.: Explaining explanations: an overview of interpretability of machine learning. In: DSSA (2018)
17. Goodfellow, I.J., et al.: Generative adversarial nets. In: Ghahramani, Z., Welling, M., Cortes, C., Lawrence, N.D., Weinberger, K.Q. (eds.) NeurIPS (2014)
18. Goodfellow, I.J., Shlens, J., Szegedy, C.: Explaining and harnessing adversarial examples. In: ICLR (2015)
19. Goyal, Y., Wu, Z., Ernst, J., Batra, D., Parikh, D., Lee, S.: Counterfactual visual explanations. In: ICML (2019)
20. Harradon, M., Druce, J., Ruttenberg, B.E.: Causal learning and explanation of deep neural networks via autoencoded activations. CoRR (2018)
21. Hendricks, L.A., Hu, R., Darrell, T., Akata, Z.: Grounding visual explanations. In: Ferrari, V., Hebert, M., Sminchisescu, C., Weiss, Y. (eds.) ECCV 2018. LNCS, vol. 11206, pp. 269–286. Springer, Cham (2018). https://doi.org/10.1007/978-3-030-01216-8_17

22. Heusel, M., Ramsauer, H., Unterthiner, T., Nessler, B., Hochreiter, S.: GANs trained by a two time-scale update rule converge to a local NASH equilibrium. In: NeurIPS (2017)

23. Huang, G., Liu, Z., van der Maaten, L., Weinberger, K.Q.: Densely connected convolutional networks. In: CVPR (2017)

24. Kingma, D.P., Ba, J.: Adam: a method for stochastic optimization. In: ICLR (2015)

25. Lang, O., et al.: Explaining in style: training a GAN to explain a classifier in stylespace. In: ICCV (2021)

26. Lee, C.H., Liu, Z., Wu, L., Luo, P.: MaskGAN: towards diverse and interactive facial image manipulation. In: CVPR (2020)

27. Lee, W., Kim, D., Hong, S., Lee, H.: High-fidelity synthesis with disentangled representation. In: Vedaldi, A., Bischof, H., Brox, T., Frahm, J.-M. (eds.) ECCV 2020. LNCS, vol. 12371, pp. 157–174. Springer, Cham (2020). https://doi.org/10. 1007/978-3-030-58574-7_10

28. Li, Z., Xu, C.: Discover the unknown biased attribute of an image classifier. In: In: The IEEE International Conference on Computer Vision (ICCV) (2021)

29. Liu, Z., Luo, P., Wang, X., Tang, X.: Deep learning face attributes in the wild. In: ICCV (2015)

30. Lundberg, S.M., Lee, S.: A unified approach to interpreting model predictions. In: NeurIPS (2017)

31. Moosavi-Dezfooli, S., Fawzi, A., Frossard, P.: DeepFool: a simple and accurate method to fool deep neural networks. In: CVPR (2016)

32. Park, T., Liu, M., Wang, T., Zhu, J.: Semantic image synthesis with spatially-adaptive normalization. In: CVPR (2019)

33. Pawelczyk, M., Joshi, S., Agarwal, C., Upadhyay, S., Lakkaraju, H.: On the connections between counterfactual explanations and adversarial examples. CoRR abs/2106.09992 (2021)

34. Rebuffi, S., Fong, R., Ji, X., Vedaldi, A.: There and back again: revisiting back-propagation saliency methods. In: CVPR (2020)

35. Ribeiro, M.T., Singh, S., Guestrin, C.: "why should I trust you?": explaining the predictions of any classifier. In: SIGKDD (2016)

36. Rodríguez, P., et al.: Beyond trivial counterfactual explanations with diverse valuable explanations. In: ICCV (2021)

37. Schönfeld, E., Sushko, V., Zhang, D., Gall, J., Schiele, B., Khoreva, A.: You only need adversarial supervision for semantic image synthesis. In: ICLR (2021)

38. Selvaraju, R.R., Cogswell, M., Das, A., Vedantam, R., Parikh, D., Batra, D.: Grad-CAM: visual explanations from deep networks via gradient-based localization. In: ICCV (2017)

39. Shen, Y., et al.: To explain or not to explain: a study on the necessity of explanations for autonomous vehicles. CoRR (2020)

40. Shrikumar, A., Greenside, P., Kundaje, A.: Learning important features through propagating activation differences. In: ICML (2017)

41. Singla, S., Pollack, B., Chen, J., Batmanghelich, K.: Explanation by progressive exaggeration. In: ICLR (2020)

42. Srivastava, A., et al.: Improving the reconstruction of disentangled representation learners via multi-stage modelling. CoRR abs/2010.13187 (2020)

43. Sundararajan, M., Taly, A., Yan, Q.: Axiomatic attribution for deep networks. In: ICML (2017)

44. Szegedy, C., et al.: Intriguing properties of neural networks. In: ICLR (2014)

45. Tian, Y., Pei, K., Jana, S., Ray, B.: DeeptEST: automated testing of deep-neural-network-driven autonomous cars. In: ICSE (2018)

46. Ulyanov, D., Vedaldi, A., Lempitsky, V.S.: Deep image prior. IJCV (2020)
47. Verma, S., Dickerson, J.P., Hines, K.: Counterfactual explanations for machine learning: a review. CoRR abs/2010.10596 (2020)
48. Wachter, S., Mittelstadt, B., Russell, C.: Counterfactual explanations without opening the black box: automated decisions and the GDPR. Harvard J. Law Technol. (2017)
49. Wagner, J., Köhler, J.M., Gindele, T., Hetzel, L., Wiedemer, J.T., Behnke, S.: Interpretable and fine-grained visual explanations for convolutional neural networks. In: CVPR (2019)
50. Wang, P., Vasconcelos, N.: SCOUT: self-aware discriminant counterfactual explanations. In: CVPR (2020)
51. Xu, Y., et al.: Explainable object-induced action decision for autonomous vehicles. In: CVPR (2020)
52. Yu, F., et al.: BDD100K: a diverse driving dataset for heterogeneous multitask learning. In: CVPR (2020)
53. Zablocki, É., Ben-Younes, H., Pérez, P., Cord, M.: Explainability of vision-based autonomous driving systems: review and challenges. CoRR abs/2101.05307 (2021)
54. Zeiler, M.D., Fergus, R.: Visualizing and understanding convolutional networks. In: Fleet, D., Pajdla, T., Schiele, B., Tuytelaars, T. (eds.) ECCV 2014. LNCS, vol. 8689, pp. 818–833. Springer, Cham (2014). https://doi.org/10.1007/978-3-319-10590-1_53
55. Zhang, M., Zhang, Y., Zhang, L., Liu, C., Khurshid, S.: DeepRoad: GAN-based metamorphic testing and input validation framework for autonomous driving systems. In: IEEE ASE (2018)
56. Zhang, Q., Yang, X.J., Robert, L.P.: Expectations and trust in automated vehicles. In: CHI (2020)
57. Zhang, Q., Wu, Y.N., Zhu, S.: Interpretable convolutional neural networks. In: CVPR (2018)
58. Zhou, B., Khosla, A., Lapedriza, À., Oliva, A., Torralba, A.: Object detectors emerge in deep scene CNNs. In: ICLR (2015)
59. Zhou, B., Khosla, A., Lapedriza, À., Oliva, A., Torralba, A.: Learning deep features for discriminative localization. In: CVPR (2016)
60. Zhu, P., Abdal, R., Qin, Y., Wonka, P.: SEAN: image synthesis with semantic region-adaptive normalization. In: CVPR (2020)

Are Vision Transformers Robust to Patch Perturbations?

Jindong Gu[1(✉)], Volker Tresp[1], and Yao Qin[2]

[1] University of Munich, Munich, Germany
jindong.gu@outlook.com
[2] Google Research, Mountain View, USA

Abstract. Recent advances in Vision Transformer (ViT) have demonstrated its impressive performance in image classification, which makes it a promising alternative to Convolutional Neural Network (CNN). Unlike CNNs, ViT represents an input image as a sequence of image patches. The patch-based input image representation makes the following question interesting: How does ViT perform when individual input image patches are perturbed with natural corruptions or adversarial perturbations, compared to CNNs? In this work, we study the robustness of ViT to patch-wise perturbations. Surprisingly, we find that ViTs are more robust to naturally corrupted patches than CNNs, whereas they are more vulnerable to adversarial patches. Furthermore, we discover that the attention mechanism greatly affects the robustness of vision transformers. Specifically, the attention module can help improve the robustness of ViT by effectively ignoring natural corrupted patches. However, when ViTs are attacked by an adversary, the attention mechanism can be easily fooled to focus more on the adversarially perturbed patches and cause a mistake. Based on our analysis, we propose a simple temperature-scaling based method to improve the robustness of ViT against adversarial patches. Extensive qualitative and quantitative experiments are performed to support our findings, understanding, and improvement of ViT robustness to patch-wise perturbations across a set of transformer-based architectures.

Keywords: Understanding Vision Transformer · Adversarial Robustness

1 Introduction

Recently, Vision Transformer (ViT) has demonstrated impressive performance [7, 8, 10, 14, 15, 25, 47, 49, 50], which makes it become a potential alternative to convolutional neural networks (CNNs). Meanwhile, the robustness of ViT has also received great attention [5, 20, 38, 39, 41, 42, 45]. On the one hand, it is important to improve its robustness for safe deployment in the real world. On the other

---

Supplementary Information The online version contains supplementary material available at https://doi.org/10.1007/978-3-031-19775-8_24.

hand, diagnosing the vulnerability of ViT can also give us a deeper understanding of its underlying working mechanisms. Existing works have intensively studied the robustness of ViT and CNNs when the whole input image is perturbed with natural corruptions or adversarial perturbations [2,3,5,28,41]. Unlike CNNs, ViT processes the input image as a sequence of image patches. Then, a self-attention mechanism is applied to aggregate information from all patches. Based on the special patch-based architecture of ViT, we mainly focus on studying the robustness of ViT to patch-wise perturbations.

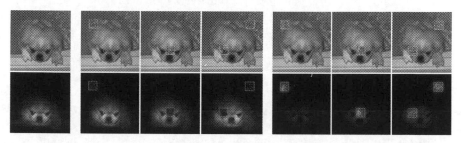

(a) Clean Image (b) with Naturally Corrupted Patch   (c) with Adversarial Patch

Fig. 1. Images with patch-wise perturbations (top) and their corresponding attention maps (bottom). The attention mechanism in ViT can effectively ignore the naturally corrupted patches to maintain a correct prediction in Fig. b, whereas it is forced to focus on the adversarial patches to make a mistake in Fig. c. The images with corrupted patches (Fig. b) are all correctly classified. The images with adversary patches (Fig. c) are misclassified as *dragonfly*, *axolotl*, and *lampshade*, respectively.

In this work, two typical types of perturbations are considered to compare the robustness between ViTs and CNN (e.g., ResNets [16]). One is natural corruptions [17], which is to test models' robustness under distributional shift. The other is adversarial perturbations [13,44], which are created by an adversary to specifically fool a model to make a wrong prediction. Surprisingly, we find ViT does *not always* perform more robustly than ResNet. When individual image patches are naturally corrupted, ViT is more robust compared to ResNet. However, when input image patch(s) are adversarially attacked, ViT shows a higher vulnerability than ResNet.

Digging down further, we revealed that ViT's stronger robustness to natural corrupted patches and higher vulnerability against adversarial patches are both caused by the attention mechanism. Specifically, the self-attention mechanism of ViT can effectively ignore the natural patch corruption, while it's also easy to manipulate the self-attention mechanism to focus on an adversarial patch. This is well supported by rollout attention visualization [1] on ViT. As shown in Fig. 1 (a), ViT successfully attends to the class-relevant features on the clean image, *i.e.*, the head of the dog. When one or more patches are perturbed with natural corruptions, shown in Fig. 1 (b), ViT can effectively ignore the corrupted patches and still focus on the main foreground to make a correct prediction. In Fig. 1 (b), the attention weights on the positions of naturally corrupted patches are much smaller even when the patches appear on the foreground. In contrast, when the patches are perturbed with adversarial perturbations by an adversary, ViT is successfully fooled to make a wrong prediction, as shown in Fig. 1 (c). This is because the attention of ViT is misled to focus on the adversarial patch instead.

Based on this understanding that the attention mechanism leads to the vulnerability of ViT against adversarial patches, we propose a simple Smoothed Attention to discourage the attention mechanism to a single patch. Specifically, we use a temperature to smooth the attention weights computed by a *softmax* operation in the attention. In this way, a single patch can hardly dominate patch embeddings in the next layer, which can effectively improve the robustness of ViT against adversarial patch attacks.

Our main contributions can be summarized as follows:

- **Finding:** Based on a fair comparison, we discover that ViT is more robust to natural patch corruption than ResNet, whereas it is more vulnerable to adversarial patch perturbation.
- **Understanding:** We reveal that the self-attention mechanism can effectively ignore natural corrupted patches to maintain a correct prediction but be easily fooled to focus on adversarial patches to make a mistake.
- **Improvement:** Inspired by our understanding, we propose Smoothed Attention, which can effectively improve the robustness of ViT against adversarial patches by discouraging the attention to a single patch.

2   Related Work

Robustness of Vision Transformer. The robustness of ViT have achieved great attention due to its great success [2–5,18,28–30,33,33,34,38,39,41,51]. On the one hand, [5,36] show that vision transformers are more robust to natural corruptions [17] compared to CNNs. On the other hand, [5,36,41] demonstrate that ViT achieves higher adversarial robustness than CNNs under adversarial attacks. These existing works, however, mainly focus on investigating the robustness of ViT when a whole image is naturally corrupted or adversarially perturbed. Instead, our work focuses on patch perturbation, given the patch-based architecture trait of ViT. The patch-based attack [12,20] and defense [32,42] methods have also been proposed recently. Different from their work, we aim to understand the robustness of patch-based architectures under patch-based natural corruption and adversarial patch perturbation.

Adversarial Patch Attack. The seminal work [35] shows that adversarial examples can be created by perturbing only a small amount of input pixels. Further, [6,24] successfully creates universal, robust, and targeted adversarial patches. These adversarial patches therein are often placed on the main object in the images. The works [11,31] shows that effective adversarial patches can be created without access to the target model. However, both universal patch attacks and black-box attacks are weak to be used for our study. They can only

Table 1. Comparison of popular ResNet and ViT models. The difference in model robustness can not be blindly attributed to the model architectures. It can be caused by different training settings. WS, GN and WD correspond to Weight Standardization, Group Normalization and Weight Decay, respectively.

Model	Pretraining	DataAug	Input Size	WS	GN	WD
ResNet [16]	N	N	224	N	N	Y
BiT [22]	Y	N	480	Y	Y	N
ViT [10]	Y	N	224/384	N	N	N
DeiT [47]	N	Y	224/384	N	N	N

achieve very low fooling rates when a single patch of ViT (only 0.5% of image) is attacked. In contrast, the white-box attack [21,23,26,37,48] can fool models by attacking only a very small patch. In this work, we apply the most popular adversarial patch attack in [21] to both ViT and CNNs for our study.

3 Experimental Settings to Compare ViT and ResNet

Fair Base Models. We list the state-of-the-art ResNet and ViT models and part of their training settings in Table 1. The techniques applied to boost different models are different, *e.g.*, pretraining. A recent work [3] points out the necessity of a fair setting. Our investigation finds weight standardization and group normalization also have a significant impact on model robustness (More in Appendix A). This indicates that the difference in model robustness can not be blindly attributed to the model architectures if models are trained with different settings. Hence, we build fair models to compare ViT and ResNet as follows.

First, we follow [47] to choose two pairs of fair model architectures, DeiT-small vs. ResNet50 and DeiT-tiny vs. ResNet18. The two models of each pair (*i.e.* DeiT and its counter-part ResNet) are of similar model sizes. Further, we train ResNet50 and ResNet18 using the **exactly same setting** as DeiT-small and Deit-tiny in [47]. In this way, we make sure the two compared models, *e.g.*, DeiT-small and ResNet50, have similar model sizes, use the same training techniques, and achieve similar test accuracy (See Appendix A). The two fair base model pairs are used across this paper for a fair comparison.

Adversarial Patch Attack. We now introduce adversarial patch attack [21] used in our study. The first step is to specify a patch position and replace the original pixel values of the patch with random initialized noise δ. The second step is to update the noise to minimize the probability of ground-truth class, *i.e.* maximize the cross-entropy loss via multi-step gradient ascent [27]. The adversary patches are specified to align with input patches of DeiT.

Evaluation Metric. We use the standard metric **Fooling Rate (FR)** to evaluate the model robustness. First, we collect a set of images that are correctly classified by both models that we compare. The number of these collected images is denoted as P. When these images are perturbed with natural patch corruption or adversarial patch attack, we use Q to denoted the number of images that are misclassified by the model. The Fooling Rate is then defined as FR $= \frac{Q}{P}$. The lower the FR is, the more robust the model is.

Table 2. Fooling Rates (in %) are reported. DeiT is more robust to naturally corrupted patches than ResNet, while it is significantly more vulnerable than ResNet against adversarial patches. Bold font is used to mark the lower fooling rate, which indicates the higher robustness.

Model	# Naturally Corrupted Patches				# Adversarial Patches			
	32	96	160	196	1	2	3	4
ResNet50	3.7	18.2	43.4	49.8	**30.6**	**59.3**	**77.1**	**87.2**
DeiT-small	**1.8**	**7.4**	**22.1**	**38.9**	61.5	95.4	99.9	100
ResNet18	6.8	31.6	56.4	61.3	**39.4**	**73.8**	**90.0**	**96.1**
DeiT-tiny	**6.4**	**14.6**	**35.8**	**55.9**	63.3	95.8	99.9	100

4   ViT Robustness to Patch-Wise Perturbations

Following the setting in [47], we train the models DeiT-small, ResNet50, DeiT-tiny, and ResNet18 on ImageNet 1k training data respectively. Note that no distillation is applied. The input size for training is $H = W = 224$, and the patch size is set to 16. Namely, there are 196 image patches totally in each image. We report the clean accuracy in Appendix A where DeiT and its counter-part ResNet show similar accuracy on clean images.

4.1   Patch-Wise Natural Corruption

First, we investigate the robustness of DeiT and ResNet to patch-based natural corruptions. Specifically, we randomly select 10k test images from ImageNet-1k validation dataset [9] that are correctly classified by both DeiT and ResNet. Then for each image, we randomly sample n input image patches x_i from 196 patches and perturb them with natural corruptions. As in [17], 15 types of natural corruptions with the highest level are applied to the selected patches, respectively. The fooling rate of the patch-based natural corruption is computed over all the test images and all corruption types. We test DeiT and ResNet with the same naturally corrupted images for a fair comparison.

We find that both DeiT and ResNet hardly degrade their performance when a small number of patches are corrupted (*e.g.*, 4). When we increase the number of patches, the difference between two architectures emerges: DeiT achieves a lower FR compared to its counter-part ResNet (See Table 2). This indicates that

DeiT is more robust against naturally corrupted patches than ResNet. The same conclusion holds under the extreme case when the number of patches $n = 196$. That is: the whole image is perturbed with natural corruptions. This is aligned with the observation in the existing work [5] that vision transformers are more robust to ResNet under distributional shifts. More details on different corruption types are in Appendix B.

In addition, we also increase the patch size of the perturbed patches, $e.g.$, if the patch size of the corrupted patch is 32×32, it means that it covers 4 continuous and independent input patches as the input patch size is 16×16. As shown in Fig. 2 (Left), even when the patch size of the perturbed patches becomes larger, DeiT (marked with red lines) is still more robust than its counter-part ResNet (marked with blue lines) to natural patch corruption.

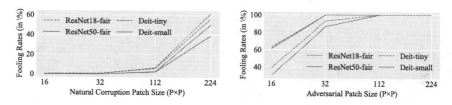

Fig. 2. DeiT with red lines shows a smaller FR to natural patch corruption and a larger FR to adversarial patch of different sizes than counter-part ResNet. (Color figure online)

4.2   Patch-Wise Adversarial Attack

In this section, we follow [21] to generate adversarial patch attack and then compare the robustness of DeiT and ResNet against adversarial patch attack. We first randomly select the images that are correctly classified by both models from imagenet-1k validation dataset. Following [21], the ℓ_∞-norm bound, the step size, and the attack iterations are set to $255/255$, $2/255$, and 10K respectively. Each reported FR score is averaged over 19.6k images.

As shown in Table 2, DeiT achieves much higher fooling rate than ResNet when one of the input image patches is perturbed with adversarial perturbation. This consistently holds even when we increase the number of adversarial patches, sufficiently supports that DeiT is more vulnerable than ResNet against patch-wise adversarial perturbation. When more than 4 patches ($\sim$2% area of the input image) are attacked, both DeiT and ResNet can be successfully fooled with almost 100% FR.

When we attack a large continuous area of the input image by increasing the patch size of adversarial patches, the FR on DeiT is still much larger than counter-part ResNet until both models are fully fooled with 100% fooling rate. As shown in Fig. 2 (Right), DeiT (marked with red lines) consistently has higher FR than ResNet under different adversarial patch sizes.

Taking above results together, we discover that DeiT is more robust to natural patch corruption than ResNet, whereas it is significantly more vulnerable to adversarial patch perturbation.

(a) on ResNet50 under Adversary Patch Attack

(b) on DeiT-small under Adversary Patch Attack

Fig. 3. Gradient Visualization. The clean image, the images with adversarial patches, and their corresponding gradient maps are visualized. We use a blue box on the gradient map to mark the location of the adversarial patch. The adversary patch on DeiT attracts attention, while the one on ResNet hardly do. (Color figure online)

5   Understanding ViT Robustness to Patch Perturbation

In this section, we design and conduct experiments to analyze the robustness of ViT. Especially, we aim to obtain deep understanding of how ViT performs when its input patches are perturbed with natural corruption or adversary patches.

5.1   How ViT Attention Changes Under Patch Perturbation?

We visualize and analyze models' attention to understand the different robustness performance of DeiT and ResNet against patch-wise perturbations. Although there are many existing methods, *e.g.*, [40,43,53], designed for CNNs to generate saliency maps, it is not clear yet how suitable to generalize them to vision transformers. Therefore, we follow [21] to choose the **model-agnostic** vanilla gradient visualization method to compare the gradient (saliency) map [52] of DeiT and ResNet. Specifically, we consider the case where DeiT and ResNet are attacked by adversarial patches. The gradient map is created as follow: we obtain the gradients of input examples towards the predicted classes, sum the absolute values of the gradients over three input channels, and visualize them by mapping the values into gray-scale saliency maps.

Qualitative Evaluation. As shown in Fig. 3 (a), when we use adversarial patch to attack a ResNet model, the gradient maps of the original images and the images with adversarial patch are similar. The observation is consistent with the one made in the previous work [21]. In contrast to the observation on ResNet, the adversarial patch can change the gradient map of DeiT by attracting more attention. As shown in Fig. 3 (b), even though the main attention of DeiT is still on the object, part of the attention is misled to the adversarial patch. More visualizations are in Appendix C.

Quantitative Evaluation. We also measure our observation on the attention changes with the metrics in [21]. In each gradient map, we score each patch according to (1) the maximum absolute value within the patch (MAX); and (2) the sum of the absolute values within the patch (SUM). We first report the percentage of patches where the MAX is also the maximum of the whole gradient map. Then, we divide the SUM of the patch by the SUM of the all gradient values and report the percentage.

Table 3. Quantitative Evaluation. Each cell lists the percent of patches in which the maximum gradient value inside the patches is also the maximum of whole gradient map. SUM corresponds to the sum of element values inside patch divided by the sum of values in the whole gradient map. The average over all patches is reported.

	Towards ground-truth Class				Towards misclassified Class			
	SUM		MAX		SUM		MAX	
Patch Size	16	32	16	32	16	32	16	32
ResNet50	0.42	1.40	0.17	0.26	0.55	2.08	0.25	0.61
DeiT-small	**1.98**	**5.33**	**8.3**	**8.39**	**2.21**	**6.31**	**9.63**	**12.53**
ResNet18	0.24	0.74	0.01	0.02	0.38	1.31	0.05	0.13
DeiT-tiny	**1.04**	**3.97**	**3.67**	**5.90**	**1.33**	**4.97**	**6.49**	**10.16**

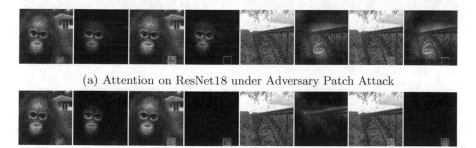

(a) Attention on ResNet18 under Adversary Patch Attack

(b) Attention on DeiT-tiny under Adversary Patch Attack

Fig. 4. Attention Comparison between ResNet and DeiT under Patch Attack. The clean image, the adversarial images, and their corresponding attention are visualized. The adversary patch on DeiT attract attention, while the ones on ResNet hardly do.

As reported in Table 3, the pixel with the maximum gradient value is more likely to fall inside the adversarial patch on DeiT, compared to that on ResNet. Similar behaviors can be observed in the metric of SUM. The quantitative experiment also supports our claims above that adversarial patches mislead DeiT by attracting more attention.

Besides the gradient analysis, another popular tool used to visualize ViT is Attention Rollout [1]. To further confirm our claims above, we also visualize DeiT with Attention Rollout in Fig. 4. The rollout attention also shows that the attention of DeiT is attracted by adversarial patches. The attention rollout is not applicable to ResNet. As an extra check, we visualize and compare the feature maps of classifications on ResNet. The average of feature maps along the channel dimension is visualized as a mask on the original image. The visualization also supports the claims above. More visualizations are in Appendix D. Both qualitative and quantitative analysis verifies our claims that the adversarial patch can mislead the attention of DeiT by attacthing it.

However, the gradient analysis is not available to compare ViT and ResNet on images with natural corrupted patches. When a small number of patch of input images are corrupted, both Deit and ResNet are still able to classify them correctly. The slight changes are not reflected in vanilla gradients since they are noisy. When a large area of the input image is corrupted, the gradient is very noisy and semantically not meaningful. Due to the lack of a fair visualization tool to compare DeiT and ResNet on naturally corrupted images, we apply Attention Rollout to DeiT and Feature Map Attention visualization to ResNet for comparing the their attention.

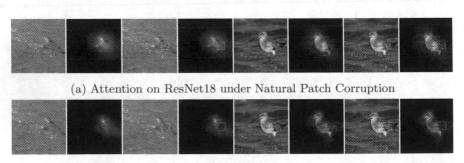

(a) Attention on ResNet18 under Natural Patch Corruption

(b) Attention on DeiT-tiny under Natural Patch Corruption

Fig. 5. Attention Comparison between ResNet and DeiT under Natural Patch Corruption. The clean image, the naturally corrupted images, and their corresponding attention are visualized. The patch corruptions on DeiT are ignored by attending less to the corrupted patches, while the ones on ResNet are treated as normal patches.

The attention visualization of these images is shown in Fig. 5. We can observe that ResNet treats the naturally corrupted patches as normal ones. The attention of ResNet on naturally patch-corrupted images is almost the same as that on the clean ones. Unlike CNNs, DeiT attends less to the corrupted patches when they cover the main object. When the corrupted patches are placed in the background, the main attention of DeiT is still kept on the main object. More figures are in Appendix E.

5.2   How Sensitive is ViT Vulnerability to Attack Patch Positions?

To investigate the sensitivity against the location of adversarial patch, we visualize the FR on each patch position in Fig. 6. We can clearly see that adversarial patch achieves higher FR when attacking DeiT-tiny than ResNet18 in different patch positions. Interestingly, we find that the FRs in different patch positions of DeiT-tiny are similar, while the ones in ResNet18 are center-clustered. A similar pattern is also found on DeiT-small and ResNet50 in Appendix F.

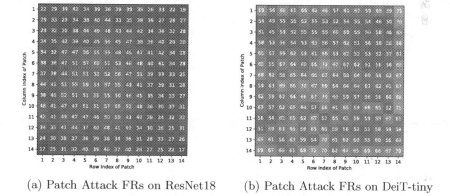

(a) Patch Attack FRs on ResNet18   (b) Patch Attack FRs on DeiT-tiny

Fig. 6. Patch Attack FR (in %) in each patch position is visualized. FRs in different patch positions of DeiT-tiny are similar, while the ones in ResNet18 are center-clustered.

(a) Corner-biased Images

(b) Center-biased Images

Fig. 7. Collection of two sets of biased data. The fist set contains only images with corner-biased object(s), and the other set contains center-biased images.

Considering that ImageNet are center-biased where the main objects are often in the center of the images, we cannot attribute the different patterns to the model architecture difference without further investigation.

Hence, we design the following experiments to disentangle the two factors, *i.e.*, model architecture and data bias. Specifically, we select two sets of correctly classified images from ImageNet 1K validation dataset. As shown in Fig. 7a, the first set contains images with corner bias where the main object(s) is in the image corners. In contrast, the second set is more center-biased where the main object(s) is exactly in the central areas, as shown in Fig. 7b.

We apply patch attack to corner-biased images (*i.e.*, the first set) on ResNet. The FRs of patches in the center area are still significantly higher than the ones in the corner (See Appendix G). Based on this, we can conclude that such a relation of FRs to patch position on ResNet is caused by ResNet architectures instead of data bias. The reason behind this might be that pixels in the center can affect more neurons of ResNet than the ones in corners.

Similarly, we also apply patch attack to center-biased images (the second set) on DeiT. We observe that the FRs of all patch positions are still similar even the input data are highly center-biased (See Appendix H). Hence, we draw the conclusion that DeiT shows similar sensitivity to different input patches regardless of the content of the image. We conjecture it can be explained by the architecture trait of ViT, in which each patch equally interact with other patches regardless of its position.

Table 4. Transferability of adversarial patch across different patch positions of the the image. Translation X/Y stands for the number of pixels shifted in rows or columns. When they are shifted to cover other patches exactly, adversarial patches transfer well, otherwise not.

Trans-(X,Y)	(0, 1)	**(0, 16)**	**(0, 32)**	(1, 0)	**(16, 0)**	**(32, 0)**	(1, 1)	**(16, 16)**
ResNet50	0.06	0.31	0.48	0.06	0.18	0.40	0.08	0.35
DeiT-small	0.27	**8.43**	**4.26**	0.28	**8.13**	**3.88**	0.21	**4.97**
ResNet18	0.22	0.46	0.56	0.19	0.49	0.68	0.15	0.49
DeiT-tiny	2.54	**29.15**	**18.19**	2.30	**28.37**	**17.32**	2.11	**21.23**

5.3   Are Adversarial Patches on ViT Still Effective When Shifted?

The work [21] shows that the adversarial patch created on an image on ResNet is not effective anymore even if a single pixel is shifted away. Similarly, we also find that the adversarial patch perturbation on DeiT does not transfer as well when shifting a single pixel away. However, when an adversarial patch is shifted to exactly match another input patch, it remains highly effective, as shown in Table 4. This mainly because the attention can still be misled to focus on the adversarial patch as long as it is perfectly aligned with the input patch. In contrast, if a single pixel is shifted away, the structure of the adversarial pertur-bation is destroyed due to the misalignment between the input patch of DeiT and the constructed adversarial patch. Additionally, We find that the adversarial patch perturbation can hardly transfer across images or models regardless of the alignment. Details can be found in Appendix I.

6   Improving ViT Robustness to Adversarial Patch

Given an input image $x \in \mathbb{R}^{H \times W \times C}$, ViT [10] first reshapes the input x into a sequence of image patches $\{x_i \in \mathbb{R}^{(\frac{H}{P} \cdot \frac{W}{P}) \times (P^2 \cdot C)}\}_{i=1}^N$ where P is the patch size and N is the number of patches. A class-token patch x_0 is concatenated to the patch sequence. A set of self-attention blocks is applied to obtain patch embeddings of the l-th block $\{x_i^l\}_{i=0}^N$. The class-token patch embedding of the last block is mapped to the output.

The patch embedding of the i-th patch in the l-th layer is the weighted sum of all patch embedding $\{x_j^{l-1}\}_{j=0}^N$ of the previous layer. The weights are the attention weights obtained from the attention module. Formally, the patch embedding x_i^l is computed with following equation

$$x_i^l = \sum_{j=0}^N \alpha_{ij} \cdot x_j^{l-1}, \qquad \alpha_{ij} = \frac{\exp(Z_{ij})}{\sum_{j=0}^N \exp(Z_{ij})} \qquad (1)$$

where α_{ij} is the attention weight that stands for the attention of the i-th patch of the l-th layer to the j-th patch of the $(l\text{-}1)$-th layer. Z_{ij} is the scaled dot-product between the key of the j-th patch and the query of the i-th patch in the $(l\text{-}1)$-th layer, i.e., the logits before *softmax* attention.

Given a classification task, we denote the patch embedding of the clean image as x_i^{*l}. When the k-th patch is attacked, the patch embedding of the i-th patch in the l-th layer deviates from x_i^{*l}. The deviation distance is described as

$$d(x_i^l, x_i^{*l}) = \sum_{j=0}^N \alpha_{ij} \cdot x_j^{l-1} - \sum_{j=0}^N \alpha_{ij}^* \cdot x_j^{l-1}, \qquad (2)$$

where α_{ij}^* is the attention weight corresponding to the clean image. Our analysis shows that the attention is misled to focus on the attacked patch. In other words, α_{ik} is close to 1, and other attention weights are close to zero.

To address this, we replace the original attention with smoothed attention using temperature scaling in the *softmax* operation. Formally, the smoothed attention is defined as

$$\alpha_{ij}^\diamond = \frac{\exp(Z_{ij}/T)}{\sum_{j=0}^N \exp(Z_{ij}/T)}, \qquad (3)$$

where $T(> 1)$ is the hyper-parameter that determines the smoothness of the proposed attention. With the smoothed attention, the deviation of the patch embedding from the clean patch embedding is smaller.

$$d(x_i^{\diamond l}, x_i^{*l}) = \sum_{j=0}^N \alpha_{ij}^{\diamond l} \cdot x_j^{l-1} - \sum_{j=0}^N \alpha_{ij}^* \cdot x_j^{l-1} < d(x_i^l, x_i^{*l}) \qquad (4)$$

We can see that the smoothed attention naturally encourages self-attention
not to focus on a single patch. To validate
if ViT becomes more robust to adversarial
patches, we apply the method to ViT and
report the results in Fig. 8. Under different temperatures, the smoothed attention
can improve the adversarial robustness
of ViT to adversarial patches and rarely
reduce the clean accuracy. In addition, the
effectiveness of smoothed attention also
verifies our understanding of the robustness of ViT in Sect. 5: it is the attention
mechanism that causes the vulnerability
of ViT against adversarial patch attacks.

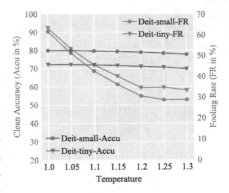

Fig. 8. The robustness of ViT can be
improved with Smoothed Attention.

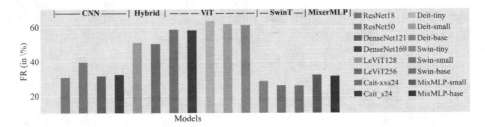

Fig. 9. We report Fooling Rates on different versions of ViT, CNN as well as Hybrid
architectures under adversarial patch attacks.

7   Discussion

In previous sections, we mainly focus on studying the state-of-the-art patch
attack methods on the most primary ViT architecture and ResNet. In this
section, we further investigate different variants of model architectures as well
as adversarial patch attacks.

Different Model Architectures. In addition to DeiT and ResNet, we also
investigate the robustness of different versions of ViT [10,25,47], CNN [16,19]
as well as Hybrid architectures [14] under adversarial patch attacks. Following
the experimental setting in Sect. 3, we train all the models and report fooling
rate on each model in Fig. 9. Four main conclusions can be drawn from the figure.

1. CNN variants are consistently more robust than ViT models.
2. The robustness of LeViT model [14] with hybrid architecture (*i.e.,* Conv
   Layers + Self-Attention Blocks) lives somewhere between ViT and CNNs.

3. Swin Transformers [25] are as robust as CNNs. We conjecture this is because attention cannot be manipulated by a single patch due to hierarchical attention and the shifted windows therein. Specifically, the self-attention in Swin Transformers is conducted on patches within a local region rather than the whole image. In addition, a single patch will interact with patches from different groups in different layers with shifted windows. This makes effective adversarial patches challenging.
4. Mixer-MLP [46] uses the same patch-based architecture as ViTs and has no attention module. Mixer-base with FR (31.36) is comparable to ResNet and more robust than ViTs. The results further confirm that the vulnerability of ViT can be attributed to self-attention mechanism.

Our proposed attention smoothing by temperature scaling can effectively improve the robustness of DeiT and Levit. However, the improvement on Swin Transformers is tiny due to its architecture design.

Different Patch Attacks. Other than adversarial patch attacks studied previously, we also investigate the robustness of ViT and ResNet against the following variants of adversarial patch attacks.

Imperceptible Patch Attack. In previous sections, we use unbounded local patch attacks where the pixel intensity can be set to any value in the image range $[0, 1]$. The adversarial patches are often visible, as shown in Fig. 1. In this section, we compare DeiT and ResNet under a popular setting where the adversarial perturbation is imperceptible to humans, bounded by $8/225$. In the case of a single patch attack, the attacker achieves FR of 2.9% on ResNet18 and 11.2% on DeiT-tiny (see Appendix J for more results). That is: DeiT is still more vulnerable than ResNet when attacked with imperceptible patch perturbation.

Targeted Patch Attack. We also compare DeiT and ResNet under targeted patch attacks, which can be achieved by maximizing the probability of the target class. Specifically, we randomly select a target class other than the ground-truth class for each image. Under a single targeted patch attack, the FR is 15.4% for ResNet18 vs. 32.3% for DeiT-tiny, 7.4% for ResNet50 vs. 24.9% for DeiT-small. The same conclusion holds: DeiT is more vulnerable than ResNet. Visualization of adversarial patches is in Appendix K.

Patch Attack Generated with Different Iterations. Following [21], we generate adversarial patch attacks with 10k iterations. In this section, we further study the minimum iterations required to successfully attack the classifier, which is averaged over all patch positions of the misclassified images. We find that the minimum attack iterations on DeiT-tiny is much smaller than that on ResNet18 (65 vs. 342). Similar results on DeiT-small and ResNet50 (294 vs. 455). This further validates DeiT is more vulnerable than ResNet.

ViT-Agnostic Patch Attack. In this section, we study ViT-agnostic patch attack where the adversarial patch of the same size as an input patch is placed to a random area of the image. The covered area can involve pixels from multiple input patches. We find that DeiT becomes less vulnerable to adversarial patch attack, *e.g.*, the FR on DeiT-small decreases from 61.5% to 47.9%. When the adversarial patch is not aligned with the input patch, *i.e.*, only part of patch pixels can be manipulated, the attention of DeiT is less likely to be misled. Under such ViT-agnostic patch attack, ViT is still more vulnerable than ResNet.

8   Conclusion

This work starts with an interesting observation on the robustness of ViT to patch perturbations. Namely, vision transformer (e.g., DeiT) is more robust to natural patch corruption than ResNet, whereas it is significantly more vulnerable against adversarial patches. Further, we discover the self-attention mechanism of ViT can effectively ignore natural corrupted patches but be easily misled to adversarial patches to make mistakes. Based on our analysis, we propose attention smoothing to improve the robustness of ViT to adversarial patches, which further validates our developed understanding. We believe this study can help the community better understand the robustness of ViT to patch perturbations.

References

1. Abnar, S., Zuidema, W.: Quantifying attention flow in transformers. In: Annual Meeting of the Association for Computational Linguistics (ACL) (2020)
2. Aldahdooh, A., Hamidouche, W., Deforges, O.: Reveal of vision transformers robustness against adversarial attacks. arXiv:2106.03734 (2021)
3. Bai, Y., Mei, J., Yuille, A., Xie, C.: Are transformers more robust than CNNs? arXiv:2111.05464 (2021)
4. Benz, P., Ham, S., Zhang, C., Karjauv, A., Kweon, I.S.: Adversarial robustness comparison of vision transformer and MLP-mixer to CNNs. arXiv preprint arXiv:2110.02797 (2021)
5. Bhojanapalli, S., Chakrabarti, A., Glasner, D., Li, D., Unterthiner, T., Veit, A.: Understanding robustness of transformers for image classification. arXiv:2103.14586 (2021)
6. Brown, T.B., Mané, D., Roy, A., Abadi, M., Gilmer, J.: Adversarial patch. arXiv:1712.09665v1 (2017)
7. Chen, C.F., Fan, Q., Panda, R.: CrossVit: cross-attention multi-scale vision transformer for image classification. arXiv:2103.14899 (2021)
8. Chen, Z., Xie, L., Niu, J., Liu, X., Wei, L., Tian, Q.: VisFormer: the vision-friendly transformer. arXiv:2104.12533 (2021)
9. Deng, J., Dong, W., Socher, R., Li, L.J., Li, K., Fei-Fei, L.: ImageNet: a large-scale hierarchical image database. In: IEEE Conference on Computer Vision and Pattern Recognition (CVPR) (2009)
10. Dosovitskiy, A., et al.: An image is worth 16x16 words: transformers for image recognition at scale. arXiv:2010.11929 (2020)
11. Fawzi, A., Frossard, P.: Measuring the effect of nuisance variables on classifiers. In: Proceedings of the British Machine Vision Conference (BMVC) (2016)

12. Fu, Y., Zhang, S., Wu, S., Wan, C., Lin, Y.: Patch-fool: are vision transformers always robust against adversarial perturbations? In: International Conference on Learning Representations (2021)

13. Goodfellow, I.J., Shlens, J., Szegedy, C.: Explaining and harnessing adversarial examples. arXiv:1412.6572 (2014)

14. Graham, B., et al.: Levit: a vision transformer in convnet's clothing for faster inference. arXiv:2104.01136 (2021)

15. Han, K., Xiao, A., Wu, E., Guo, J., Xu, C., Wang, Y.: Transformer in transformer. arXiv:2103.00112 (2021)

16. He, K., Zhang, X., Ren, S., Sun, J.: Deep residual learning for image recognition. In: IEEE Conference on Computer Vision and Pattern Recognition (CVPR) (2016)

17. Hendrycks, D., Dietterich, T.: Benchmarking neural network robustness to common corruptions and perturbations. In: International Conference on Learning Representations (ICLR) (2019)

18. Hu, H., Lu, X., Zhang, X., Zhang, T., Sun, G.: Inheritance attention matrix-based universal adversarial perturbations on vision transformers. IEEE Sig. Process. Lett. **28**, 1923–1927 (2021)

19. Huang, G., Liu, Z., Van Der Maaten, L., Weinberger, K.Q.: Densely connected convolutional networks. In: Proceedings of the IEEE Conference on Computer Vision and Pattern Recognition, pp. 4700–4708 (2017)

20. Joshi, A., Jagatap, G., Hegde, C.: Adversarial token attacks on vision transformers. arXiv:2110.04337 (2021)

21. Karmon, D., Zoran, D., Goldberg, Y.: Lavan: localized and visible adversarial noise. In: International Conference on Machine Learning (ICML) (2018)

22. Kolesnikov, A., et al.: Big transfer (BiT): general visual representation learning. In: Vedaldi, A., Bischof, H., Brox, T., Frahm, J.-M. (eds.) ECCV 2020. LNCS, vol. 12350, pp. 491–507. Springer, Cham (2020). https://doi.org/10.1007/978-3-030-58558-7_29

23. Liu, A., et al.: Perceptual-sensitive GAN for generating adversarial patches. In: AAAI (2019)

24. Liu, A., Wang, J., Liu, X., Cao, B., Zhang, C., Yu, H.: Bias-based universal adversarial patch attack for automatic check-out. In: Vedaldi, A., Bischof, H., Brox, T., Frahm, J.-M. (eds.) ECCV 2020. LNCS, vol. 12358, pp. 395–410. Springer, Cham (2020). https://doi.org/10.1007/978-3-030-58601-0_24

25. Liu, Z., et al.: Swin transformer: hierarchical vision transformer using shifted windows. arXiv:2103.14030 (2021)

26. Luo, J., Bai, T., Zhao, J.: Generating adversarial yet inconspicuous patches with a single image (student abstract). In: Proceedings of the AAAI Conference on Artificial Intelligence, vol. 35, pp. 15837–15838 (2021)

27. Madry, A., Makelov, A., Schmidt, L., Tsipras, D., Vladu, A.: Towards deep learning models resistant to adversarial attacks. In: arXiv:1706.06083 (2017)

28. Mahmood, K., Mahmood, R., Van Dijk, M.: On the robustness of vision transformers to adversarial examples. In: Proceedings of the IEEE/CVF International Conference on Computer Vision, pp. 7838–7847 (2021)

29. Mao, X., et al.: Towards robust vision transformer. arXiv:2105.07926 (2021)

30. Mao, X., Qi, G., Chen, Y., Li, X., Ye, S., He, Y., Xue, H.: Rethinking the design principles of robust vision transformer. arXiv:2105.07926 (2021)

31. Metzen, J.H., Finnie, N., Hutmacher, R.: Meta adversarial training against universal patches. arXiv preprint arXiv:2101.11453 (2021)

32. Mu, N., Wagner, D.: Defending against adversarial patches with robust self-attention. In: ICML 2021 Workshop on Uncertainty and Robustness in Deep Learning (2021)
33. Naseer, M., Ranasinghe, K., Khan, S., Hayat, M., Khan, F.S., Yang, M.H.: Intriguing properties of vision transformers. arXiv:2105.10497 (2021)
34. Naseer, M., Ranasinghe, K., Khan, S., Khan, F.S., Porikli, F.: On improving adversarial transferability of vision transformers. arXiv:2106.04169 (2021)
35. Papernot, N., McDaniel, P., Jha, S., Fredrikson, M., Celik, Z.B., Swami, A.: The limitations of deep learning in adversarial settings. In: 2016 IEEE European Symposium on Security and Privacy (EuroS&P) (2016)
36. Paul, S., Chen, P.Y.: Vision transformers are robust learners. arXiv:2105.07581 (2021)
37. Qian, Y., Wang, J., Wang, B., Zeng, S., Gu, Z., Ji, S., Swaileh, W.: Visually imperceptible adversarial patch attacks on digital images. arXiv preprint arXiv:2012.00909 (2020)
38. Qin, Y., Zhang, C., Chen, T., Lakshminarayanan, B., Beutel, A., Wang, X.: Understanding and improving robustness of vision transformers through patch-based negative augmentation. arXiv preprint arXiv:2110.07858 (2021)
39. Salman, H., Jain, S., Wong, E., Madry, A.: Certified patch robustness via smoothed vision transformers. arXiv:2110.07719 (2021)
40. Selvaraju, R.R., Cogswell, M., Das, A., Vedantam, R., Parikh, D., Batra, D.: Gradcam: visual explanations from deep networks via gradient-based localization. In: ICCV (2017)
41. Shao, R., Shi, Z., Yi, J., Chen, P.Y., Hsieh, C.J.: On the adversarial robustness of visual transformers. arXiv:2103.15670 (2021)
42. Shi, Y., Han, Y.: Decision-based black-box attack against vision transformers via patch-wise adversarial removal. arXiv preprint arXiv:2112.03492 (2021)
43. Shrikumar, A., Greenside, P., Kundaje, A.: Learning important features through propagating activation differences. In: International Conference on Machine Learning (ICML) (2017)
44. Szegedy, C., Zaremba, W., Sutskever, I., Bruna, J., Erhan, D., Goodfellow, I., Fergus, R.: Intriguing properties of neural networks. In: International Conference on Learning Representations (ICLR) (2014)
45. Tang, S., et al.: Robustart: benchmarking robustness on architecture design and training techniques. arXiv preprint arXiv:2109.05211 (2021)
46. Tolstikhin, I., et al.: MLP-mixer: an all-MLP architecture for vision. In: arXiv:2105.01601 (2021)
47. Touvron, H., Cord, M., Douze, M., Massa, F., Sablayrolles, A., Jégou, H.: Training data-efficient image transformers & distillation through attention. In: International Conference on Machine Learning (ICML) (2021)
48. Wang, J., Liu, A., Bai, X., Liu, X.: Universal adversarial patch attack for automatic checkout using perceptual and attentional bias. IEEE Trans. Image Process. **31**, 598–611 (2021)
49. Wu, B., et al.: Visual transformers: token-based image representation and processing for computer vision. arXiv:2006.03677 (2020)
50. Xiao, T., Singh, M., Mintun, E., Darrell, T., Dollár, P., Girshick, R.: Early convolutions help transformers see better. arXiv:2106.14881 (2021)
51. Yu, Z., Fu, Y., Li, S., Li, C., Lin, Y.: Mia-former: efficient and robust vision transformers via multi-grained input-adaptation. arXiv preprint arXiv:2112.11542 (2021)

52. Zeiler, M.D., Fergus, R.: Visualizing and understanding convolutional networks. In: Fleet, D., Pajdla, T., Schiele, B., Tuytelaars, T. (eds.) ECCV 2014. LNCS, vol. 8689, pp. 818–833. Springer, Cham (2014). https://doi.org/10.1007/978-3-319-10590-1_53
53. Zhou, B., Khosla, A., Lapedriza, A., Oliva, A., Torralba, A.: Learning deep features for discriminative localization. In: IEEE Conference on Computer Vision and Pattern Recognition (CVPR) (2016)

A Dataset Generation Framework for Evaluating Megapixel Image Classifiers and Their Explanations

Gautam Machiraju[✉][iD], Sylvia Plevritis[iD], and Parag Mallick[iD]

Stanford University, Stanford, USA
gmachi@stanford.edu

Abstract. Deep learning-based megapixel image classifiers have exceptional prediction performance in a number of domains, including clinical pathology. However, extracting reliable, human-interpretable model explanations has remained challenging. Because real-world megapixel images often contain latent image features highly correlated with image labels, it is difficult to distinguish correct explanations from incorrect ones. Furthering this issue are the flawed assumptions and designs of today's classifiers. To investigate classification and explanation performance, we introduce a framework to (a) generate synthetic control images that reflect common properties of megapixel images and (b) evaluate average test-set correctness. By benchmarking two commonplace Convolutional Neural Networks (CNNs), we demonstrate how this interpretability evaluation framework can inform architecture selection beyond classification performance—in particular, we show that a simple Attention-based architecture identifies salient objects in all seven scenarios, while a standard CNN fails to do so in six scenarios. This work carries widespread applicability to any megapixel imaging domain.

Keywords: eXplainable AI · Interpretable ML · Salient object detection · Model selection · Synthetic data · Coarse supervision · Megapixel imagery

1 Introduction

Megapixel image datasets are increasingly common in multiple scientific and human-centered application domains (*e.g.*, histopathology [39,64], autonomous systems [117], remote sensing and atmospheric sciences [24], and cosmology [79]), but pose unique analytical challenges that are not present in standard image datasets (*i.e.*, those containing $<10^6$ pixels per image). Firstly, such datasets contain images that are often described as either coarsely labeled, weakly labeled,

Supplementary Information The online version contains supplementary material available at https://doi.org/10.1007/978-3-031-19775-8_25.

or *weakly annotated data* (WAD) [85], meaning that each image is only paired with an image-level label and lacks sub-image annotations. The WAD characterization also implies sufficiently high resolution to contain semantically distinct objects, or visual concepts [44], at multiple scales. However, such images are not typically annotated due to costs and required domain expertise. Secondly, these datasets typically have smaller sample sizes (of labeled images) than standard image datasets. Thirdly, megapixel image datasets now include multiplexed or multispectral images (*i.e.*, with high channel-wise dimensionality) generated in scientific domains including histopathology [10, 11, 22, 36, 37, 42, 72, 86, 106, 111]. Finally, due to the scarcity of these datasets and the diversity of their channel spaces, the creation of pre-trained and foundation models [12] is uncertain. These four unique challenges, coupled with memory constraints of today's GPUs, have necessitated new machine learning approaches for classifying and understanding spatial systems imaged at high-resolution, megapixel scale.

While megapixel image classification has seen significant success with recent deep learning approaches, model explanations are largely still unreliable and uninterpretable. Recent studies have shifted classifiers toward end-to-end deep learning from traditional machine learning of hand-crafted (i.e. hypothesis-driven or keypoint-derived) featurization [39]. In particular, *Patch-based Convolutional Neural Networks* (*PatchCNNs*) [43,54] have reached state-of-the-art performances in domains such as cancer diagnostics and prognostics via histopathology [16, 17, 20, 26, 30, 40, 43, 52, 60, 72, 77, 78, 107] and remote sensing of geoeconomic indicators via multispectral satellite imagery [49, 113, 115, 121]. Despite modeling success on deployment datasets—as commonly defined by classification performance statistics, *e.g.*, area under the Receiver Operating Characteristic (AUROC) and area under the Precision-Recall Curve (AUPRC)—studies may report qualitative and anecdotal assessment of model explanations or report a lack of interpretability [17, 20, 52, 78]. To carry out this cursory assessment, Explanation Maps are commonly used to identify input-specific salient objects and regions-of-interest (ROIs) in test sets. However, Explanation Maps are rarely quantitatively assessed for correctness due to a lack of ground truth pixel-level annotations in WAD settings. Thus, challenging questions remain for megapixel image classifiers: Are models learning and explaining truly salient, human-interpretable objects? Or are those objects latent features or spurious correlations [120]? Are current modeling choices [88] impeding human-interpretable explanations? Which choices lead to enhanced interpretability? How should we assess interpretability? Conversely, can model explanations reveal learning mechanisms and behaviors [68,81], and if so, what mechanisms are desirable for megapixel imagery?

A growing focus on eXplainable Artificial Intelligence (XAI) and Interpretable Machine Learning (IML) [28, 29, 67, 68, 75, 82] in human-centered and scientific application domains [4, 52, 105] has reframed interpretability as a priority for model development and selection. *Post-hoc model interpretability*—or a model's ability to make human-interpretable, input-specific explanations—is desired for decision-making but is often untenable due to discordance between optimized model objectives (*e.g.*, predictive performance) and the end user's real-world objectives (*e.g.*, identifying salient ROIs) [68]. In order to quantify interpretability as a form of explanation correctness, a classification-adjacent task has emerged: *weakly supervised Salient Object Detection* (*wsSOD*). The wsSOD task can be conceptualized as a form of image segmentation targeting salient objects used for classification, but without annotated regions-of-interest (ROIs) as training inputs [13,23,109]. While Salient Object Detection (SOD) is the goal task evaluated via *post-hoc* Explanation Mapping, classification is the workhorse task used to define the learning objective and evaluation scheme. Using neural networks' *in situ* explanations, wsSOD presents an opportunity to both identify and validate salient objects held as ground truth, as well as discover objects *de novo* [30,33,44,109]. This opportunity is especially of interest in megapixel imagery to better understand large-scale spatial systems. This interest is compounded and even necessitated in multiplexed imagery (*e.g.*, spatial proteomics [61]) since Explanation Maps scale in channel-space. Alas, the lack of ground truth salient objects in such settings poses a roadblock to quantify interpretability and motivates wsSOD-based architectural evaluation prior to deployment.

As wsSOD gains popularity in megapixel imagery, the computer vision community needs benchmarking datasets and a quantitative evaluation framework for assessing classifiers and their explanations. To address this need, we present **MISO**, a novel dataset generator that creates **M**egapixel **I**mages with **S**alient **O**bjects for the wsSOD task. MISO creates synthetic control images with differentially expressed, class-specific properties and predefined ground truths to assess classification and wsSOD. We developed MISO and its derivative datasets to help understand the impact of modeling assumptions and design configurations (architectures, hyperparameters, etc.) [88] on these tasks, perform architecture selection, and coax out learning mechanisms. Finally, we demonstrate architecture selection through head-to-head comparisons between commonplace architectures and explanation methods. In summary, our contributions are as follows:

- A unifying and generalized framework for designing PatchCNNs, a popular approach to megapixel image classification
- MISO, a dataset generator and framework for creating synthetic megapixel images with multiple training and testing scenarios that simulate common properties of such datasets
- MISO-1, a benchmark dataset of synthetic controls generated by MISO
- MISO-2, a benchmark dataset derived from real-world histopathology data
- Interpretability Report Cards, a quantitative framework for measuring the average input-specific correctness of predictions and explanations

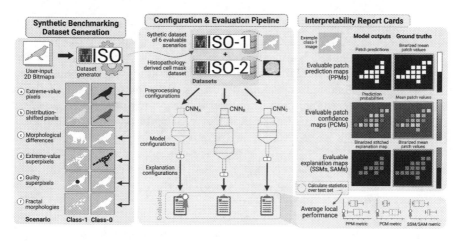

Fig. 1. Overview of the MISO generator and benchmarking datasets, MISO-1 and MISO-2. MISO-1 and MISO-2 alike evaluate models' abilities to detect salient objects and ROIs. Scenarios (a)–(f) of MISO-1 test models' abilities to detect differentially expressed properties and are trained and tested on separately. MISO-2 combines tested properties of MISO-1 to offer a more challenging wsSOD task.

2 Related Work

2.1 Modern Megapixel Image Classifiers

Megapixel image classification has shifted toward deep learning-based approaches, driven by a well-posed data assumption. Implicit in the WAD characterization of megapixel images is the assumption of differentially expressed image features, often formulated as the *Multiple Instance (MI) assumption* [5,18,53,63,122]: that is, there exists at least one class-specific *instance* (*i.e.*, ≥ 1 pixels) within a class-1 image that corresponds with the class-1 image-level label. Inversely, a class-0 image reflects the absence of such instances. These "guilty" instances (*i.e.*, salient objects) are not known *a priori*, but the MI assumption can be encoded into a classifier's design [17,46,72,113]. This intuitively shifts the traditional classification task toward object detection and segmentation, albeit without ground truth ROI annotations (*e.g.*, bounding boxes). This assumption arises frequently in clinical settings: pathologists search for instances of cancer cells in healthy tissue (to determine cancer diagnosis and stage) and for anaplastic cancer cell morphology in tumor tissue (to determine tumor grade) [52,76].

In accordance with the MI assumption and a push to scalably process megapixel images as inputs to neural networks, classifiers have shifted the unit of analysis toward small sub-image croppings of source images. These croppings, often referred to as *patches* (as depicted in Fig. 2), sample source images and offer a granular, frequentist view of their heterogeneity with predictive utility in various settings [6,50,60]. With patches as inputs, popular models of today often take a disjoint two-stage, Transfer Learning approach to image classification: (1) patch-level predictions through the use of a Convolutional Neural

Network (CNN) followed by (2) patch aggregation and image-level prediction. The family of models in Stage-1 is sometimes referred to as a *Patch-based CNN* or *PatchCNN* [43,54] and can include Attention modules [48,58,94,103] as seen in recent architectures [46,51,72]. The implementation of Stage-2 has spanned a variety of models, ranging from simple decision rules to more sophisticated functions trained on any combinations of hidden vector representations, independent patch-level predictions, and the spatial arrangement of patch predictions (*i.e.*, a *Patch Prediction Map*, or *PPM*) and their probabilities [17,40] (*i.e.*, a *Patch Confidence Map*, or *PCM*). In summary, this two-stage strategy (Fig. 2) and its input patches are used for interconnected reasons: to operate under the memory constraints of GPUs, to significantly amplify the labeled dataset size while preserving image richness, and to perform fine-grained analyses.

However, PatchCNNs in their simplest forms make multiple inherently flawed modeling assumptions. Firstly, (I) patches are assumed and treated as statistically independent and identically distributed (IID), making image-level classification a combination of independent patch predictions. This is inherently false due to spatial autocorrelation between neighboring patches. Based on patch size, features captured may only represent local contextual information [13]. Secondly, because of the aforementioned WAD constraint and MI assumption, (II) patch labeling is often conducted via image-level label inheritance (ILI). ILI is a labeling strategy that labels all constituent patches of an image with their associated image-level label. Despite ILI being a noisy extension of the MI assumption through its "guilt by association" clause, this coarse supervision strategy has shown surprising classification success in the histopathology application domain [17,43]. Thirdly, as discussed, (III) PatchCNNs usually appear in decoupled, two-stage modeling pipelines. Thus, image-level predictions are not part of the PatchCNN learning objective when training on patches, meaning image-level features are not learned when constructing patch-level representations. This design implies (and enforces) that any learned salient objects are self-contained within IID patches. Despite operating on at least one of these limiting modeling assumptions, PatchCNN applications in patient prognostics via histopathology have achieved state-of-the-art classification performances with at AUROCs approaching 1.0 [17,40,43,77,107]. However, while qualitative analyses have been conducted for salient objects *in situ* [89], few examples quantitatively evaluate them [98] or objectively assess architectural or model interpretability.

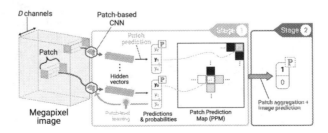

Fig. 2. Generalized schematic of the PatchCNN pipeline. Stage-1 operates on patches independently while Stage-2 operates on all patches per image.

2.2   Explanation Maps and Salient Object Detection

Current explanation methods for neural network classifiers are predominantly *post-hoc* and input-specific (*i.e.*, local) [2,28,82], lending themselves to anecdotal and qualitative interpretations. These explanations often take the form of Explanation Maps (*e.g.*, Saliency Maps [62,95,100], Class Activation Maps [91], and Visual Attention Maps [51]). However, neural networks often fail to provide reliable and human-interpretable explanations through Explanation Mapping [57]. This discordance has been quantitatively studied in standard image domains (*e.g.*, applications in clinical decision support [8]), but is largely missing for megapixel image domains to our knowledge. Studies often omit large-scale assessment of model explanations and instead typically show meaningful ROIs for cherry-picked examples. Ultimately, quantifying explanation *correctness* is required to quantify a model's *post-hoc* interpretability—and is currently unobtainable without human-derived ground truth salient objects. Due to expensive annotation costs beyond image-level labels, wsSOD is an important task due to its wide applicability in assessing model explanations [108,109]. Conventional SOD, much like conventional classification approaches, uses hypothesis-driven, hand-crafted featurizers or keypoint detectors [1,9] to extract domain-specific image features (*e.g.*, intensity, color, texture) and identify salient objects. In contrast, deep learning-based wsSOD is often driven by an end-to-end classification task followed by Explanation Mapping to identify salient objects or ROIs [38,108].

2.3   Benchmarks for Salient Object Detection

While datasets for SOD (and wsSOD) exist for standard image domains, they do not exist for megapixel image domains to our knowledge. Current SOD datasets span a wide range of scenarios—simpler scenarios can contain a few objects overlaid on a solid background, while more complicated scenarios contain multiple objects with varying backgrounds [109]. While these datasets all have ground truths for salient regions (*e.g.*, pixel-wise annotated masks, bounding boxes, human eye fixation locations, etc.), they usually lack class structure and thereby any differentially expressed, class-specific objects for (weakly) supervised learning. Additionally, to our knowledge, no SOD datasets contain megapixel images and thus do not contain any data-relevant properties encountered in real-world datasets (Sect. 3.1). On the other hand, there exist very few real-world megapixel image datasets that have ground truth annotations to compare against any detected salient objects. Typically used for tasks such as ROI segmentation and classification (*e.g.*, in histopathology [69]), such datasets only test models' abilities to identify true positive expert annotations as defined by limited *a priori* domain knowledge (albeit often with low inter-rater reliability [87,97]), but fail to test detected false negatives beyond its scope. Realistic data synthesized from generative models can fall short for similar reasons—defining ground truth salient objects is difficult to condition on, define, and verify, and (ii) even if successful, such salient objects will still only ever reflect domain knowledge. In reality, many real-world scientific settings will likely never have truly exhaustive annotations for ground truth salience. Finally, to our knowledge, there exists no dataset and evaluation framework to systematically evaluate a patch-based model's textural (*i.e.*, pixel-value heterogeneity), morphological (*i.e.*, edge shape between foreground and background), and contextual awareness (*i.e.*, patch co-localization) at patch and image scales.

3   Proposed Benchmarking Datasets

3.1   Data-Relevant Properties for Megapixel Imagery

Megapixel images often contain differentially expressed properties that classifiers should be attuned to. One property is differential (A) pixel values, where pixel intensities can be indicative of class membership. An analogy from histopathology includes channel expression indicative of a phenotype (*e.g.*, benign versus malignant tumor tissue [74]). Additionally, (B) local abnormalities in intensity, texture, and morphology can exist, or relatively small objects found within larger ones (*i.e.*, MI assumption). An illustrative example from histopathology includes cancer cells in healthy tissue [59]. Another property we may see is differential (C) large-scale morphologies, *i.e.*, patterning and different edge shapes. In histopathology, this can be exemplified by invasive cancer cells altering the stromal patterning in tissue [55]. Finally, the global texture or degree of (D) object clustering (*i.e.*, the size, number, and density of objects) can differ between classes. This is demonstrated by the tightly packed organization of cells in healthy tissue versus cellular anaplasticity in high-grade tumors (*e.g.*, Gleason grade [99]).

3.2   MISO Dataset Generator and MISO-1 Benchmark

We present **MISO** to offer a principled approach to systematically evaluate megapixel image classifiers and their explanations. MISO is a dataset generator and generalizable framework for creating synthetic grayscale control images. Generated using standard techniques in image processing (Appendix A.3), images are partitioned into six datasets, i.e. scenarios, that each simulate one or more differentially expressed data properties (A)-(D) between classes. While trivial to classify at the image-level, a scenario's images may pose difficulty in determining class membership at the patch-level. Each scenario is intended to be trained and tested on individually. We also present **MISO-1**, a benchmarking dataset generated by MISO for the wsSOD task (Fig. 1, Fig. 3). Each of MISO-1's six scenarios contain $n \approx 100$ weakly annotated megapixel images for each of the training and test sets (Fig. 3). Dataset specifications are outlined in Appendix A.3. Because we define ground truth salient objects in the creation of MISO-1, we can test models' capabilities to identify *all* differentially expressed salient objects—a necessity for selecting interpretable models when ground truth annotations are unavailable or non-exhaustive. Regardless of scale, Scenarios (a), (b), and (d) test awareness of pixel value, while (c), (e), and (f) test awareness of morphology:

(a) **Extreme-value pixels (EVP):** This scenario tests data property (A). Images in this scenario contain 0- or 1-valued objects of varying large-scale morphologies for their respective 0-class and 1-class images. Since image textures are relatively smooth (with "salt or peppered" fuzziness in some cases), this scenario tests a patch-based model's ability to map patches to classes via their pixel values alone, regardless of what parts of the image's objects are contained within the patches.

(b) **Distribution-shifted pixels (DSP):** This scenario also tests data property (A), albeit with increased difficulty than the EVP scenario. Images in this scenario contain objects with pixel values that are sampled from non-overlapping class distributions. More concretely, there exist varying large-scale morphologies with pixels ranging from [0.6,1] for 1-class images and [0,0.4] for 0-class images. Similar to the EVP scenario, this scenario tests a patch-based model's ability to map patches to classes by pixel values alone.

(c) **Morphological differences (MD):** This scenario tests data property (C). Images in class-0 have a shared large-scale morphology (Fig. 3). Images in class-1 take on all other large-scale morphologies derived from the other 23 input bitmaps. Because data property (A) is tested by both EVP and DSP scenarios, MD has 1-valued objects for both classes. This scenario tests a patch-based model's contextual and morphological awareness.

(d) **Extreme-value superpixels (EVSP):** This scenario tests data property (A) like the EVP scenario, but with added difficulty—this scenario simulates settings with varying large-scale morphologies comprised of many small objects. To achieve this, we took coordinates from in-house cell segmentation data (masks provided with MISO-1) and overlaid 0- and 1-valued circles (of radius 10 pixels) within our large-scale morphologies for 0- and 1-class images, respectively. This scenario also tests a patch-based model's ability to map patch pixel values to classes.

(e) **Guilty superpixels (GSP):** This scenario tests data property (B). Class-0 images contain 1-valued objects with varying large-scale morphologies and class-1 images contain those objects but with randomly placed 0-valued circles of (randomly selected) radii between [100] pixels within them. This tests a patch-based model's textural and contextual awareness.

(f) **Fractal morphologies (FM):** This scenario tests data property (D). Class-0 images contain 1-valued objects of varying large-scale morphologies, while class-1 images contain mosaics of those same objects but shrunken, repeatedly tiled, and mapped within their original morphology boundaries. This approach creates sufficient class balance between class-0 and class-1 patches. This tests a patch-based model's textural, morphological, and contextual awareness.

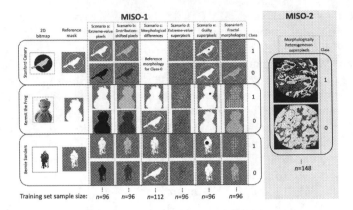

Fig. 3. MISO-1 example images for three reference masks. MISO-2 example images.

3.3 MISO-2 Benchmark

To evaluate PatchCNNs in a more realistic setting, we also present MISO-2, a binary image dataset derived from histopathology data. This dataset includes morphologically heterogeneous superpixels in the form of segmented cell masks from $n = 184$ images. The masks span two histologic types of lung cancer, defining our class structure for image-level labels: 92 adenocarcinoma samples as class-0 and 92 squamous cell carcinoma samples as class-1. Histologic type is used as a label for its ability to describe superpixel patterns represented throughout MISO-2's images. Cell morphology heterogeneity arises from both intra-sample heterogeneity (via multiple cell types, *i.e.*, stromal and cancer cells) and inter-sample heterogeneity (via histologic type). We split the data using an 80–20 split, resulting in 148 training images and 36 test images. This benchmark simultaneously tests data properties (B)–(D).

4   Experiments and Results

4.1   Baseline Methods and Evaluable Outputs

To establish baseline classifiers for MISO-1 and MISO-2, we chose two commonplace PatchCNN architectures to perform the wsSOD task. We used *VGG-19* [96] and *VGG with Attention* (*VGG-Att*) with pre-pooled modules [51] with shared patch sizes (96 × 96 pixels for MISO-1 and 224 × 224 pixels for MISO-2) and background filtration. Both architectures were trained (over 10 epochs for MISO-1 and 20 epochs for MISO-2) and tested on all scenarios separately, resulting in fourteen evaluable models. Both architectures output PPMs and PCMs, which spatially contextualize patch predictions and prediction probabilities per image. For explanations, we also chose standard baseline methods: Saliency and Attention Maps. VGG-19 can compute Saliency Maps per patch, while VGG-Att can compute both Saliency and Attention Maps per patch. To

speed up run-time and smooth out ROIs, we averaged absolute saliency or attention scores [3] per patch. We then concatenated scores to construct an array per
image that we refer to as *Stitched Saliency Maps* (*SSMs*) and *Stitched Attention
Maps* (*SAMs*). Binarization of the SSMs and SAMs was performed with the popular adaptive threshold of double the mean saliency or attention score [13]. All
other design configurations (*e.g.*, hyperparameters) for preprocessing, modeling,
and explanations are shared between architectures. Refer to Appendix A.4 for a
summary of configurations that define benchmarks for architecture selection.

4.2   Evaluation Framework: Interpretability Report Cards

Because ground truths are defined by the control images in MISO-1 and MISO-2,
quantitative evaluation includes the correctness of spatially resolved predictions
and explanations. To evaluate both model predictions and explanations, we use
performance statistics from classification, image segmentation, and saliency analysis [13,38,65,109,116] to score outputs against ground truths (Fig. 1). Three
types of analyses are conducted in this section: (i) independent patch predictions (of their constituent image's class) over the whole patch dataset and the
resulting image-level predictions based on decision rules; (ii) PPMs and PCMs
to assess the correctness of predictions and prediction probabilities in space; and
(iii) SSMs and SAMs to assess wsSOD capabilities, *i.e.*, correctness of explanations. We define correctness using the notion of *explanation plausibility*, *i.e.*,
the quality of alignment between model explanations and human interpretations
[47], and use it as our proxy for interpretability. Thus, we compute similarity scores between an explanation and its corresponding ground truth salient
objects. We do not assess *explanation faithfulness*, *i.e.*, how accurately an explanation reflects a model's true reasoning process [47], due to its relative difficulty
to evaluate quantitatively. To summarize evaluation, we provide *Interpretability
Report Cards* per model and per scenario, which consist of average statistics and
·confidence intervals over test set images and visualizations of example predictions
and explanations (Fig. 4, Appendix A.6). Specifically for Explanation Mapping,
this strategy moves toward a global measure of *post-hoc* model interpretability through the aggregate evaluation of local, or input-specific, explanations.
It should be noted that the following statistics are all inflated by background
patch filtration, a commonly performed preprocessing step in several application
domains.

Classification Performance Statistics: For independent patch predictions,
ground truth annotations are derived from ILI labeling. Patch-level AUROC,
AUPRC, and Average Precision (AP) are computed over all IID-assumed
patches, thus reflecting class membership prediction of individual patches.
Image-level AUROC, AUPRC, and AP are also computed over all image-level
labels using patch aggregation functions (*i.e.*, image-level decision rules described
in Appendix A.4). Image-level prediction probabilities are generated with each
decision rule's pooling strategy. Results are found in Appendix A.5 Table 3.

Patch Prediction and Confidence Maps: Ground truth annotations for PCMs are derived from patch means, while ground truth annotations for PPMs are derived from patch means and an applied manual binarization threshold specific to image label and scenario (Fig. 1). To assess PPMs on an IID patch-level, we calculate set-theoretic statistics including F_β-measure (with $\beta^2 = 0.3$ [13]), Dice (*i.e.*, F_1-measure), Jaccard, and Overlap coefficients, as well as Sensitivity, Specificity, and Mean Absolute Error (MAE). To assess PPMs structurally, we use the E-measure [32] and also introduce a new metric called the *Scagnostics Distance* (*ScagDist*). ScagDist simply featurizes binary masks by their topological properties using techniques from computational geometry and graph theory [112] and takes the cosine distance between them. It should be noted that we tallied default ScagDist values of 1.0 for blank SSM or SAM outputs. We assess PCMs on a patch-level via the MAE and structurally via the Structural Similarity Index Measure (SSIM) [110]. Results are shown in Table 2 (and Appendix A.5 Table 4). It should be noted that PPMs and PCMs were not evaluated for the GSP scenario due to the difficulty in defining a single notion of correctness for ground truths, given the ILI labeling scheme used for training (Appendix A.6 Fig. 10).

Stitched Saliency and Attention Maps: Due to the differentially expressed nature of the image properties in MISO-1 and MISO-2, ground truth annotations for SSMs and SAMs (salient ROIs) are conveniently the same as those constructed for PPMs (patch labels)—they are derived from patch means and an applied manual binarization threshold specific to image labels and scenarios (Fig. 1, Appendix A.3 Fig. 6). For ease, we only assess class-1 test images. To assess binarized SSMs and SAMs on an IID patch-level, we again calculate

Table 1. Average statistics over test-set PPMs. A **boldface** result indicates a superior score between architectures for a given scenario (directionality denoted by $\uparrow, \downarrow$). [†]Generated mean value required image sampling to avoid runtime or memory issues. ♣Denotes binary map evaluation. ♢Denotes structural evaluation. ♡Denotes IID patch-level evaluation.

Scenario	Sensitivity♣♡↑	Specificity♣♡↑	ScagDist♣♢↓	F_β-measure♣♡↑	E-measure♣♢↑	MAE♣♡↓
VGG-19						
EVP	0.999 ± 0.000	0.926 ± 0.016	0.011 ± 0.006	0.931 ± 0.016	0.952 ± 0.005	0.038 ± 0.008
DSP	0.500 ± 0.100	0.929 ± 0.016	0.138 ± 0.035	0.499 ± 0.100	0.666 ± 0.036	0.246 ± 0.056
MD	0.500 ± 0.093	0.938 ± 0.011	$\mathbf{0.004} \pm 0.001$	0.499 ± 0.092	0.837 ± 0.015	0.137 ± 0.025
EVSP	0.498 ± 0.100	0.923 ± 0.019	0.535 ± 0.094	0.321 ± 0.068	0.514 ± 0.028	0.160 ± 0.024
FM	0.500 ± 0.100	$\mathbf{0.923} \pm 0.018$	0.508 ± 0.099	0.429 ± 0.086	0.575 ± 0.033	0.205 ± 0.039
MISO-2	$\mathbf{0.557}^{\dagger} \pm 0.167$	$\mathbf{0.890}^{\dagger} \pm 0.040$	$\mathbf{0.171}^{\dagger} \pm 0.094$	$0.572^{\dagger} \pm 0.171$	$\mathbf{0.628}^{\dagger} \pm 0.072$	$\mathbf{0.323}^{\dagger} \pm 0.121$
VGG-Att						
EVP	0.999 ± 0.000	0.926 ± 0.016	0.011 ± 0.006	0.931 ± 0.016	0.952 ± 0.005	0.038 ± 0.008
DSP	$\mathbf{0.999} \pm 0.000$	$\mathbf{0.928} \pm 0.016$	$\mathbf{0.013}^{\dagger} \pm 0.006$	$\mathbf{0.927} \pm 0.018$	$\mathbf{0.951} \pm 0.005$	$\mathbf{0.038} \pm 0.008$
MD	$\mathbf{0.721} \pm 0.048$	0.938 ± 0.011	$0.008^{\dagger} \pm 0.001$	$\mathbf{0.719} \pm 0.051$	$\mathbf{0.908} \pm 0.008$	$\mathbf{0.099} \pm 0.017$
EVSP	$\mathbf{0.998} \pm 0.001$	0.923 ± 0.019	$\mathbf{0.035} \pm 0.012$	$\mathbf{0.819} \pm 0.041$	$\mathbf{0.888} \pm 0.013$	$\mathbf{0.057} \pm 0.013$
FM	$\mathbf{0.986} \pm 0.009$	0.922 ± 0.017	$\mathbf{0.009} \pm 0.004$	$\mathbf{0.924} \pm 0.016$	$\mathbf{0.944} \pm 0.006$	$\mathbf{0.045} \pm 0.009$
MISO-2	$0.400^{\dagger} \pm 0.086$	$0.865^{\dagger} \pm 0.040$	$0.317^{\dagger} \pm 0.071$	$\mathbf{0.601}^{\dagger} \pm 0.095$	$0.461^{\dagger} \pm 0.040$	$0.437^{\dagger} \pm 0.068$

set similarity coefficients as described in the previous subsection. To assess binarized maps structurally, we again use the E-measure and ScagDist. We use MAE to assess non-binarized maps on a patch-level. Finally, to assess non-binarized maps structurally, we use the S-measure [31] and SSIM. Results are shown in Table 2 (and Appendix A.5 Table 5). Testing every patch explanation against its patch-level ground truth helps measure image-level awareness of differential expression.

4.3   Analysis of Results

Are models learning and explaining truly salient, human-interpretable objects? Yes, our wsSOD experiments suggest that a subset of models are able to do so despite the aforementioned flawed modeling assumptions. However, our experiments highlight divergent patterns of classification and explanation performance over a number of evaluation statistics. *Are current modeling choices impeding human-interpretable explanations? Which choices lead to enhanced interpretability?* For MISO-1, while both architectures perform fairly well with patch- and image-level class predictions (Table 1, Appendix A.5 Table 3 and Table 4), VGG-Att still generally outperforms VGG-19 in both generating accurate class predictions and human-interpretable explanations. VGG-Att's relatively high performance is increasingly clear as the scenarios become intuitively

Table 2. Average statistics over test set Explanation Maps. Stitched Saliency Maps (SSMs) are constructed for both models, while Stitched Attention Maps (SAMs) are also constructed for VGG-Att. A **boldface** result indicates a superior score between architectures for a given scenario (directionality denoted by ↑, ↓). *Indicates that SAMs yielded a top score for VGG-Att. †Generated mean value required image sampling to avoid runtime or memory issues. ♣Denotes binary map evaluation. ♠Denotes non-binary map evaluation. ◇Denotes structural evaluation. ♡Denotes IID patch-level evaluation.

Scenario	MAE♠◇↓	F_β-measure♣◇↑	ScagDist♣◇↓	S-measure♠◇↑	E-measure♣◇↑	SSIM♠◇↑
VGG-19						
EVP	**0.427** ± 0.050	0.475 ± 0.082	0.214 ± 0.047	**0.289** ± 0.025	0.503 ± 0.044	**0.270** ± 0.063
DSP	0.413 ± 0.051	0.000 ± 0.000	1.000 ± 0.000	0.293 ± 0.026	0.250 ± 0.000	0.280 ± 0.064
MD	0.394 ± 0.049	0.000 ± 0.000	1.000 ± 0.000	0.303 ± 0.025	0.250 ± 0.000	0.279 ± 0.055
EVSP	0.206 ± 0.043	0.000 ± 0.000	1.000 ± 0.000	0.397 ± 0.022	0.250 ± 0.000	0.384 ± 0.066
GSP	0.020 ± 0.004	0.000 ± 0.000	1.000† ± 0.000	**0.529** ± 0.021	0.250 ± 0.000	0.913 ± 0.014
FM	0.331 + 0.058	0.000 + 0.000	1.000 ± 0.000	0.335 ± 0.029	0.250 ± 0.000	0.314 ± 0.071
MISO-2	0.660 ± 0.023	0.000 ± 0.000	1.000 ± 0.000	0.170 ± 0.012	0.250 ± 0.000	0.001 ± 0.001
VGG-Att						
EVP	0.430 ± 0.050	**0.609** ± 0.083	**0.179** ± 0.056	0.285 ± 0.025	**0.581** ± 0.043	0.265 ± 0.063
DSP	**0.412** ± 0.051	**0.585*** ± 0.098	**0.170*†** ± 0.068	**0.541*** ± 0.016	**0.610*** ± 0.045	**0.412*** ± 0.054
MD	**0.391** ± 0.049	**0.627** ± 0.030	**0.135*†** ± 0.045	**0.306** ± 0.025	**0.657*** ± 0.042	**0.282** ± 0.055
EVSP	**0.205** ± 0.043	**0.691*** ± 0.047	**0.076*** ± 0.028	**0.446*** ± 0.013	**0.781*** ± 0.021	**0.435*** ± 0.053
GSP	0.020 ± 0.004	**0.350** ± 0.072	0.266† ± 0.044	0.490 ± 0.002	**0.656** ± 0.040	0.913 ± 0.014
FM	**0.313** ± 0.055	**0.664*** ± 0.046	**0.051*** ± 0.014	**0.363** ± 0.026	**0.645*** ± 0.031	**0.339** ± 0.068
MISO-2	0.660 ± 0.023	**0.390*** ± 0.078	**0.398†** ± 0.046	0.170 ± 0.012	**0.258*** ± 0.015	**0.002** ± 0.001

more challenging—regarding PPM correctness in Table 1, VGG-19 has low sensitivity (near 0.5) for all scenarios other than EVP, thus pointing to an average of near-random classification of foreground patches per image. Regarding SSM and SAM correctness (Table 2), VGG-19 often failed to identify any salient objects in scenarios other than EVP. Results from MISO-2 show us that while VGG-19, on average, tends to make more correct class predictions than VGG-Att (Table 1), VGG-Att is far superior at identifying salient objects on average (Table 2). *Are some salient objects simply latent features or spurious correlations?* Regarding latent features, VGG-19 is learning highly predictive patch representations almost all scenarios Table 1, but systemically struggles to create interpretable explanations. This discrepancy points to the architecture's potential reliance on latent features. Unfortunately, we are limited in our ability to characterize spurious correlations from this work alone. For simplicity and to probe high-supervision performance on MISO-1 and MISO-2, we conducted complete patch filtering of image backgrounds. This choice thereby limits any learning of spurious correlations and subsequently constrained learning and evaluation to foreground objects.

Can model explanations reveal learning mechanisms and behaviors? What mechanisms are desirable for megapixel imagery? While model explanations on synthetic datasets do not directly reveal learning mechanisms, they can shed light on resulting emergent behaviors and capabilities that certain architectures afford us. Firstly, VGG-Att seems to have a stronger ability to learn ranges of pixel values between classes and to identify relatively small salient morphologies, respectively supported by DSP and FM scenarios (*e.g.*, near-doubled PPM

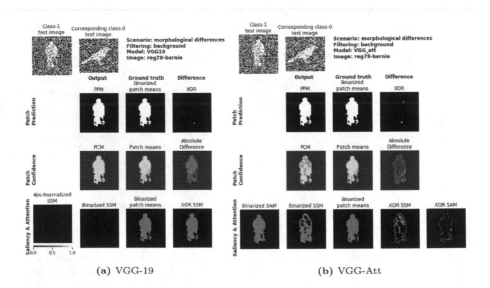

(a) VGG-19   (b) VGG-Att

Fig. 4. Example Report Card for morphological differences (MD). Test-set examples, PPMs, PCMs, and SSMs (and SAMs if applicable) are displayed in rows 1–4, respectively.

sensitivity and F_β-measure in Table 1). Additionally, while VGG-Att has lower patch-level classification performance scores for GSP (Appendix A.5 Table 3), the scores actually reflect the model's ability to work beyond the confines of fuzzy (*i.e.*, ILI) patch labeling and classify "guilty" regions as the sole class-1 patches. VGG-Att's under-reliance on fuzzy patch labeling is also apparent in its explanations. Curiously, it identified salient "guilty" ROIs in larger morphologies (Appendix A.6 Fig. 10, Table 2 ScagDist and E-measure) despite flawed modeling assumptions: despite only having access to IID-assumed patches and ILI labels, solely computing Attention scores within each patch, and without any form of global Attention across patches. This finding hints at superior contextual awareness, such as capabilities for Gestalt Closure [56] and foreground-background edge detection. VGG-Att's heightened edge detection is supported by properly modulated patch prediction probabilities in the MD scenario's PCMs (Fig. 4b, Table 4)—by assigning lower prediction probabilities to interior foreground patches, it appears to use edges as the primary discernment between large-scale morphologies. Interestingly, MISO-2's explanation results (Table 2) could point to VGG-19's heightened global textural awareness (a potential bias for CNNs [34]), despite overall lower SSM performance than VGG-Att (Table 1). The relative complexity of MISO-2 as a dataset may point to VGG-Att's limited bias toward textures, but could also reflect the need for increased training iterations or sample size. Even so, VGG-Att's dominance in interpretability points to its ability to construct representations without solely relying on latent features. VGG-Att's overall modeling capabilities are most likely granted by its model-intrinsic explanations [29], but don't necessarily imply greater utility of SAMs over SSMs—both were generally accurate, but also excelled in different scenarios (*e.g.*, SSMs outperformed SAMs in all statistics in the EVP scenario). These findings reiterate Attention-based models' (*e.g.*, *Vision Transformers*) starkly different learning mechanisms for visual recognition than traditional CNNs, their fine-grained attentiveness, and their subsequent human-interpretability in standard image domains [14,19,21,80].

How should we assess interpretability? To quantify and assess interpretability, we recommend measuring explanation plausibility. Specifically for megapixel imagery, classifiers should be assessed via wsSOD to investigate their abilities to explain predictions with long-range dependencies. Even though our evaluation framework only tests plausibility, it allows us to interrogate model behaviors, hypothesize learning mechanisms, and even sufficiently differentiate between two comparable, but architecturally distinct baselines. Because this form of analysis requires complete salient object annotations, synthetic controls (with differentially expressed properties) provides automated evaluation in deployment settings where exhaustive annotations are infeasible or impossible. Furthermore, because predictive performance and explanation plausibility are not necessarily correlated, we support similar frameworks for architectural (*i.e.*, model family) evaluation, benchmarking, debugging and testing workflows [118], and selection prior to model deployment. In order to push models toward XAI and IML, we must promote interpretability as a quantifiable criterion in the design process to ultimately build inherently interpretable architectures and trained models.

While MISO-1 lacks realism, its scenarios independently and systematically test megapixel imagery's common properties to reveal architectural behaviors. Even with MISO-2's real-world origins and combined elements of MISO-1 scenarios (*i.e.*, MD, GSP, FM), VGG-19's low-plausibility explanations for MISO-1 recur for MISO-2 (Table 2). While this consistency hints at generalized behavior, generalizability should be assessed further in settings with greater data complexity (as discussed in Limitations section Appendix A.7). For these reasons, we recommend that MISO-1, MISO-2, and any other custom MISO-generated datasets be used *in tandem* with domain-specific segmentation datasets (*i.e.*, where annotated ROIs are withheld during training and evaluated against). This multi-pronged strategy can respectively test (A) a model's full set of plausible explanations in settings without exhaustive ground truths and (B) true positive salient objects in real-world environments. We believe the proposed benchmarks can act as a community resource similar to MNIST [15], but customized for megapixel imagery and assessing interpretability. These datasets act as debugging sanity checks that *any* interpretable model should be able to pass—especially before deployment in low-annotation settings.

5 Conclusion

In summary, our primary goals are to both provide synthetic datasets that reflect one or more common, differentially expressed data properties, as well as systematically probe the interpretability of megapixel image classifiers. We show the utility of this approach for evaluating classifiers and their explanation plausibility via their propensities to perform wsSOD. Through experimentation, we also put commonplace PatchCNN architectures into question. While current modeling paradigms lack interpretability, extensions toward context-aware, Attention-based architectures have great potential as salient object detectors aligned with human interpretation. This work has widespread applicability for megapixel image and application domains, and can even provide an groundwork for interpretability evaluation in standard image domains.

References

1. Achanta, R., Estrada, F., Wils, P., Süsstrunk, S.: Salient region detection and segmentation. In: Gasteratos, A., Vincze, M., Tsotsos, J.K. (eds.) ICVS 2008. LNCS, vol. 5008, pp. 66–75. Springer, Heidelberg (2008). https://doi.org/10.1007/978-3-540-79547-6_7
2. Adadi, A., Berrada, M.: Peeking inside the Black-Box: a survey on explainable artificial intelligence (XAI). IEEE Access **6**, 52138–52160 (2018)
3. Adebayo, J., Gilmer, J., Muelly, M., Goodfellow, I., Hardt, M., Kim, B.: Sanity checks for saliency maps (2020)
4. Amann, J., Blasimme, A., Vayena, E., Frey, D., Madai, V.I.: Precise4Q consortium: explainability for artificial intelligence in healthcare: a multidisciplinary perspective. BMC Med. Inform. Decis. Mak. **20**(1), 310 (2020)

5. Amores, J.: Multiple instance classification: review, taxonomy and comparative study. Artif. Intell. **201**, 81–105 (2013)
6. Anonymous: Patches are all you need? In: Submitted to The Tenth International Conference on Learning Representations (2022). https://openreview.net/forum?id=TVHS5Y4dNvM. under review
7. Arazo, E., Ortego, D., Albert, P., O'Connor, N.E., McGuinness, K.: Pseudo-labeling and confirmation bias in deep semi-supervised learning. arXiv:1908.02983 [cs], June 2020. arXiv: 1908.02983
8. Arun, N., et al.: Assessing the (un)trustworthiness of saliency maps for localizing abnormalities in medical imaging. bioRxiv, July 2020
9. Bartol, K., Bojanić, D., Pribanić, T., Petković, T., Donoso, Y.D., Mas, J.S.: On the comparison of classic and deep keypoint detector and descriptor methods. arXiv, July 2020
10. Berry, S., et al.: Analysis of multispectral imaging with the AstroPath platform informs efficacy of PD-1 blockade. Science **372**(6547) (2021)
11. Black, S., et al.: CODEX multiplexed tissue imaging with DNA-conjugated antibodies. Nat. Protoc. **16**, 3802–3835 (2021)
12. Bommasani, R., et al.: On the opportunities and risks of foundation models. arXiv, August 2021
13. Borji, A., Cheng, M.-M., Hou, Q., Jiang, H., Li, J.: Salient object detection: a survey. Comput. Vis. Media **5**(2), 117–150 (2019). https://doi.org/10.1007/s41095-019-0149-9
14. Brunner, G., Liu, Y., Pascual, D., Richter, O., Ciaramita, M., Wattenhofer, R.: On identifiability in transformers. arXiv, August 2019
15. Burges, C.J.C.: MNIST handwritten digit database, Yann LeCun, Corinna Cortes and Chris Burges. https://yann.lecun.com/exdb/mnist/. Accessed 20 July 2022
16. Bándi, P., et al.: From detection of individual metastases to classification of lymph node status at the patient level: the CAMELYON17 challenge. IEEE Trans. Med. Imaging **38**(2), 550–560 (2019). https://doi.org/10.1109/TMI.2018.2867350
17. Campanella, G., et al.: Clinical-grade computational pathology using weakly supervised deep learning on whole slide images. Nat. Med. **25**(8), 1301 (2019)
18. Carbonneau, M.A., Cheplygina, V., Granger, E., Gagnon, G.: Multiple instance learning: a survey of problem characteristics and applications. arXiv, December 2016
19. Caron, M., et al.: Emerging properties in self-supervised vision transformers. arXiv, April 2021
20. Chan, L., Hosseini, M., Rowsell, C., Plataniotis, K., Damaskinos, S.: HistoSeg-Net: semantic segmentation of histological tissue type in whole slide images. In: 2019 IEEE/CVF International Conference on Computer Vision (ICCV). IEEE, October 2019
21. Chefer, H., Gur, S., Wolf, L.: Transformer interpretability beyond attention visualization. arXiv, December 2020
22. Chevrier, S., et al.: An immune atlas of clear cell renal cell carcinoma. Cell **169**(4), 736-749.e18 (2017). https://doi.org/10.1016/j.cell.2017.04.016
23. Choe, J., Oh, S.J., Lee, S., Chun, S., Akata, Z., Shim, H.: Evaluating weakly supervised object localization methods right. arXiv, January 2020
24. Coffey, V.C.: Multispectral imaging moves into the mainstream. Opt. Photonics News **23**(4), 18 (2012)
25. Cohen, T.S., Welling, M.: Group equivariant convolutional networks. arXiv, February 2016

26. Cruz-Roa, A., Arévalo, J., Judkins, A., Madabhushi, A., González, F.: A method for medulloblastoma tumor differentiation based on convolutional neural networks and transfer learning. In: 11th International Symposium on Medical Information Processing and Analysis, vol. 9681, p. 968103. International Society for Optics and Photonics, December 2015. https://doi.org/10.1117/12.2208825. https://www.spiedigitallibrary.org/conference-proceedings-of-spie/9681/968103/A-method-for-medulloblastoma-tumor-differentiation-based-on-convolutional-neural/10.1117/12.2208825.short

27. Dalal, N., Triggs, B.: Histograms of oriented gradients for human detection. In: 2005 IEEE Computer Society Conference on Computer Vision and Pattern Recognition (CVPR 2005), vol. 1, pp. 886–893 (2005). https://doi.org/10.1109/CVPR.2005.177

28. Das, A., Rad, P.: Opportunities and challenges in explainable artificial intelligence (XAI). A survey. arXiv, June 2020

29. Du, M., Liu, N., Hu, X.: Techniques for interpretable machine learning. arXiv, July 2018

30. Echle, A., Rindtorff, N.T., Brinker, T.J., Luedde, T., Pearson, A.T., Kather, J.N.: Deep learning in cancer pathology: a new generation of clinical biomarkers. Br. J. Cancer **124**(4), 686–696 (2020)

31. Fan, D.P., Cheng, M.M., Liu, Y., Li, T., Borji, A.: Structure-measure: a new way to evaluate foreground maps. arXiv, August 2017

32. Fan, D.P., Gong, C., Cao, Y., Ren, B., Cheng, M.M., Borji, A.: Enhanced-alignment measure for binary foreground map evaluation. arXiv, May 2018

33. Frintrop, S., García, G.M., Cremers, A.B.: A cognitive approach for object discovery. In: 2014 22nd International Conference on Pattern Recognition, pp. 2329–2334, August 2014

34. Geirhos, R., Rubisch, P., Michaelis, C., Bethge, M., Wichmann, F.A., Brendel, W.: ImageNet-trained CNNs are biased towards texture; increasing shape bias improves accuracy and robustness. arXiv, November 2018

35. Ghafoorian, M., et al.: Location sensitive deep convolutional neural networks for segmentation of white matter hyperintensities. Sci. Rep. **7**(1), 5110 (2017)

36. Giesen, C., et al.: Highly multiplexed imaging of tumor tissues with subcellular resolution by mass cytometry. Nat. Methods **11**(4), 417–422 (2014). https://doi.org/10.1038/nmeth.2869

37. Goltsev, Y., et al.: Deep profiling of mouse splenic architecture with CODEX multiplexed imaging. Cell **174**(4), 968–981.e15 (2018)

38. Gupta, A.K., Seal, A., Prasad, M., Khanna, P.: Salient object detection techniques in computer vision-a survey. Entropy **22**(10) (2020)

39. Gurcan, M.N., Boucheron, L., Can, A., Madabhushi, A., Rajpoot, N., Yener, B.: Histopathological image analysis: a review. IEEE Rev. Biomed. Eng. **2**, 147–171 (2009). https://doi.org/10.1109/RBME.2009.2034865. https://www.ncbi.nlm.nih.gov/pmc/articles/PMC2910932/

40. Halicek, M., et al.: Head and neck cancer detection in digitized Whole-Slide histology using convolutional neural networks. Sci. Rep. **9**(1), 14043 (2019)

41. Harris, C.R., et al.: Quantifying and correcting slide-to-slide variation in multiplexed immunofluorescence images, July 2021

42. Hickey, J.W., et al.: Spatial mapping of protein composition and tissue organization: a primer for multiplexed antibody-based imaging. arXiv, July 2021

43. Hou, L., Samaras, D., Kurc, T.M., Gao, Y., Davis, J.E., Saltz, J.H.: Patch-based convolutional neural network for whole slide tissue image classification. arXiv:1504.07947 [cs], March 2016. arXiv: 1504.07947

44. Huang, H., Chen, Z., Rudin, C.: SegDiscover: visual concept discovery via unsupervised semantic segmentation. arXiv, April 2022
45. Huttenlocher, D., Klanderman, G., Rucklidge, W.: Comparing images using the Hausdorff distance. IEEE Trans. Pattern Anal. Mach. Intell. **15**(9), 850–863 (1993). https://doi.org/10.1109/34.232073
46. Ilse, M., Tomczak, J.M., Welling, M.: Attention-based deep multiple instance learning. arXiv, February 2018
47. Jacovi, A., Goldberg, Y.: Towards faithfully interpretable NLP systems: how should we define and evaluate faithfulness? arXiv, April 2020
48. Jaderberg, M., Simonyan, K., Zisserman, A., Kavukcuoglu, K.: Spatial transformer networks. arXiv:1506.02025 [cs], February 2016. arXiv: 1506.02025
49. Jean, N., Burke, M., Xie, M., Davis, W.M., Lobell, D.B., Ermon, S.: Combining satellite imagery and machine learning to predict poverty. Science **353**(6301), 790–794 (2016)
50. Jean, N., Wang, S., Samar, A., Azzari, G., Lobell, D., Ermon, S.: Tile2Vec: unsupervised representation learning for spatially distributed data. arXiv, May 2018
51. Jetley, S., Lord, N.A., Lee, N., Torr, P.H.S.: Learn to pay attention. arXiv, April 2018
52. Jiang, Y., Yang, M., Wang, S., Li, X., Sun, Y.: Emerging role of deep learning-based artificial intelligence in tumor pathology. Cancer Commun. **40**(4), 154–166 (2020)
53. Kandemir, M., Hamprecht, F.A.: Computer-aided diagnosis from weak supervision: a benchmarking study. Comput. Med. Imaging Graph. **42**, 44–50 (2015)
54. Kao, P.Y., et al.: Improving patch-based convolutional neural networks for MRI brain tumor segmentation by leveraging location information. Front. Neurosci. **13**, 1449 (2019)
55. Kawamura, Y., et al.: Histological and immunohistochemical evaluation of stroma variations and their correlation with the KI-67 index and expressions of glucose transporter 1 and monocarboxylate transporter 1 in canine thyroid C-cell carcinomas. J. Vet. Med. Sci. **78**(4), 607–612 (2016)
56. Kim, B., Reif, E., Wattenberg, M., Bengio, S., Mozer, M.C.: Neural networks trained on natural scenes exhibit gestalt closure. arXiv, March 2019
57. Kim, B., Seo, J., Jeon, S., Koo, J., Choe, J., Jeon, T.: Why are saliency maps noisy? Cause of and solution to noisy saliency maps. arXiv, February 2019
58. Kitaev, N., Kaiser, L., Levskaya, A.: Reformer: the efficient transformer. arXiv:2001.04451 [cs, stat], February 2020. arXiv: 2001.04451
59. Klevesath, M.B., Bobrow, L.G., Pinder, S.E., Purushotham, A.D.: The value of immunohistochemistry in sentinel lymph node histopathology in breast cancer. Br. J. Cancer **92**(12), 2201–2205 (2005)
60. van der Laak, J., Ciompi, F., Litjens, G.: No pixel-level annotations needed. Nat. Biomed. Eng. **3**(11), 855–856 (2019)
61. Lähnemann, D., et al.: Eleven grand challenges in single-cell data science. Genome Biol. **21**(1), 31 (2020)
62. LeCun, Y., Bottou, L., Bengio, Y., Ha, P.: Gradient-based learning applied to document recognition. Proc. IEEE **86**, 46 (1998)
63. Lerousseau, M., Vakalopoulou, M., Deutsch, E., Paragios, N.: SparseconvMIL: sparse convolutional context-aware multiple instance learning for whole slide image classification. In: COMPAY 2021: The Third MICCAI Workshop on Computational Pathology (2021). https://openreview.net/forum?id=3byhkJb8FUj
64. Lewis, S.M., et al.: Spatial omics and multiplexed imaging to explore cancer biology. Nat. Methods **18**, 997–1012 ((2021)

65. Li, X.H., et al.: Quantitative evaluations on saliency methods: an experimental study. arXiv, December 2020
66. Liebel, L., Körner, M.: Auxiliary tasks in multi-task learning. arXiv, May 2018
67. Linardatos, P., Papastefanopoulos, V., Kotsiantis, S.: Explainable AI: a review of machine learning interpretability methods. Entropy **23**(1) (2020)
68. Lipton, Z.C.: The mythos of model interpretability. arXiv, June 2016
69. Litjens, G., et al.: 1399 H&E-stained sentinel lymph node sections of breast cancer patients: the CAMELYON dataset. Gigascience **7**(6) (2018)
70. Liu, Y., Zhuang, B., Shen, C., Chen, H., Yin, W.: Auxiliary learning for deep multi-task learning. arXiv, September 2019
71. Liu, Z., et al.: Swin transformer: hierarchical vision transformer using shifted windows. arXiv, March 2021
72. Lu, M.Y., Williamson, D.F.K., Chen, T.Y., Chen, R.J., Barbieri, M., Mahmood, F.: Data-efficient and weakly supervised computational pathology on whole-slide images. Nat. Biomed. Eng. **55**, 555–570 (2021)
73. Marcos, D., Volpi, M., Tuia, D.: Learning rotation invariant convolutional filters for texture classification. arXiv, April 2016
74. Matos, L.L.D., Trufelli, D.C., de Matos, M.G.L., da Silva Pinhal, M.A.: Immuno-histochemistry as an important tool in biomarkers detection and clinical practice. Biomark. Insights **5**, 9–20 (2010)
75. Molnar, C., Casalicchio, G., Bischl, B.: Interpretable machine learning - a brief history, State-of-the-Art and challenges. arXiv, October 2020
76. Nathanson, S.D.: Insights into the mechanisms of lymph node metastasis. Cancer **98**(2), 413–423 (2003). https://doi.org/10.1002/cncr.11464
77. Nazeri, K., Aminpour, A., Ebrahimi, M.: Two-stage convolutional neural network for breast cancer histology image classification. arXiv:1803.04054 [cs] 10882, pp. 717–726 (2018). https://doi.org/10.1007/978-3-319-93000-881. arXiv: 1803.04054
78. Nguyen, A., Yosinski, J., Clune, J.: Deep neural networks are easily fooled: high confidence predictions for unrecognizable images. In: 2015 IEEE Conference on Computer Vision and Pattern Recognition (CVPR), pp. 427–436. IEEE, Boston, June 2015. https://doi.org/10.1109/CVPR.2015.7298640. https://ieeexplore.ieee.org/document/7298640/
79. Pesenson, M.Z., Pesenson, I.Z., McCollum, B.: The data big bang and the expanding digital universe: high-dimensional, complex and massive data sets in an inflationary epoch. Adv. Astron. **2010**, 1–16 (2010). https://doi.org/10.1155/2010/350891. arXiv: 1003.0879
80. Raghu, M., Unterthiner, T., Kornblith, S., Zhang, C., Dosovitskiy, A.: Do vision transformers see like convolutional neural networks? arXiv, August 2021
81. Rahwan, I., et al.: Machine behaviour. Nature **568**(7753), 477–486 (2019)
82. Ras, G., Xie, N., van Gerven, M., Doran, D.: Explainable deep learning: a field guide for the uninitiated. arXiv, April 2020
83. Ratner, A., De Sa, C., Wu, S., Selsam, D., Ré, C.: Data programming: creating large training sets, quickly. Adv. Neural. Inf. Process. Syst. **29**, 3567–3575 (2016)
84. Ribeiro, M.T., Singh, S., Guestrin, C.: "Why should I trust you?": explaining the predictions of any classifier. arXiv, February 2016
85. Robinson, J., Jegelka, S., Sra, S.: Strength from weakness: fast learning using weak supervision. arXiv, February 2020
86. Rost, S., Giltnane, J., Bordeaux, J.M., Hitzman, C., Koeppen, H., Liu, S.D.: Multiplexed ion beam imaging analysis for quantitation of protein expression in cancer tissue sections. Lab. Invest. **97**(8), 992–1003 (2017)

87. Sakamoto, T., et al.: A narrative review of digital pathology and artificial intelligence: focusing on lung cancer. Transl. Lung Cancer Res. **9**(5), 2255–2276 (2020)

88. Salvi, M., Acharya, U.R., Molinari, F., Meiburger, K.M.: The impact of pre- and post-image processing techniques on deep learning frameworks: a comprehensive review for digital pathology image analysis. Comput. Biol. Med. **128**, 104129 (2021)

89. Schaumberg, A.J., et al.: Interpretable multimodal deep learning for real-time pan-tissue pan-disease pathology search on social media. Mod. Pathol. **33**(11), 2169–2185 (2020)

90. Selvaraju, R.R., Cogswell, M., Das, A., Vedantam, R., Parikh, D., Batra, D.: Grad-CAM: visual explanations from deep networks via gradient-based localization. arXiv, October 2016

91. Selvaraju, R.R., Cogswell, M., Das, A., Vedantam, R., Parikh, D., Batra, D.: Grad-CAM: visual explanations from deep networks via gradient-based localization. Int. J. Comput. Vis. **128**(2), 336–359 (2020). https://doi.org/10.1007/s11263-019-01228-7. arXiv:1610.02391

92. Shaban, M., et al.: Context-aware convolutional neural network for grading of colorectal cancer histology images. IEEE Trans. Med. Imaging **39**(7), 2395–2405 (2020)

93. Shamir, R.R., Duchin, Y., Kim, J., Sapiro, G., Harel, N.: Continuous dice coefficient: a method for evaluating probabilistic segmentations. arXiv, June 2019

94. Sharma, S., Kiros, R., Salakhutdinov, R.: Action recognition using visual attention. arXiv:1511.04119 [cs], February 2016. arXiv: 1511.04119

95. Simonyan, K., Vedaldi, A., Zisserman, A.: Deep inside convolutional networks: visualising image classification models and saliency maps. arXiv:1312.6034 [cs], April 2014. arXiv: 1312.6034

96. Simonyan, K., Zisserman, A.: Very deep convolutional networks for Large-Scale image recognition. arXiv, September 2014

97. Sooriakumaran, P., Lovell, D.P., Henderson, A., Denham, P., Langley, S.E.M., Laing, R.W.: Gleason scoring varies among pathologists and this affects clinical risk in patients with prostate cancer. Clin. Oncol. **17**(8), 655–658 (2005)

98. Palatnik de Sousa, I., Maria Bernardes Rebuzzi Vellasco, M., Costa da Silva, E.: Local interpretable model-agnostic explanations for classification of lymph node metastases. Sensors (Basel, Switzerland) **19**(13) (2019). https://doi.org/10.3390/s19132969. https://www.ncbi.nlm.nih.gov/pmc/articles/PMC6651753/

99. Stamey, T.A., McNeal, J.E., Yemoto, C.M., Sigal, B.M., Johnstone, I.M.: Biological determinants of cancer progression in men with prostate cancer. JAMA **281**(15), 1395–1400 (1999). https://doi.org/10.1001/jama.281.15.1395. http://jamanetwork.com/journals/jama/fullarticle/189523 Association 7

100. Szegedy, C., Vanhoucke, V., Ioffe, S., Shlens, J., Wojna, Z.: Rethinking the inception architecture for computer vision. arXiv:1512.00567 [cs], December 2015. arXiv: 1512.00567

101. Thompson, G.Z., Maitra, R.: CatSIM: a categorical image similarity metric. arXiv, April 2020

102. Tourniaire, P., Ilie, M., Hofman, P., Ayache, N., Delingette, H.: Attention-based multiple instance learning with mixed supervision on the camelyon16 dataset. In: COMPAY 2021: the third MICCAI Workshop on Computational Pathology (2021). https://openreview.net/forum?id=Z_L9j0HW3QM

103. Vaswani, A., et al.: Attention is all you need. arXiv, June 2017

104. Veeling, B.S., Linmans, J., Winkens, J., Cohen, T., Welling, M.: Rotation equivariant CNNs for digital pathology. arXiv:1806.03962 [cs, stat], June 2018. arXiv: 1806.03962
105. Vinuesa, R., Sirmacek, B.: Interpretable deep-learning models to help achieve the sustainable development goals. Nat. Mach. Intell. **3**(11), 926–926 (2021)
106. Wagner, J., et al.: A single-cell atlas of the tumor and immune ecosystem of human breast cancer. Cell **177**(5), 1330-1345.e18 (2019). https://doi.org/10.1016/j.cell.2019.03.005
107. Wang, D., Khosla, A., Gargeya, R., Irshad, H., Beck, A.H.: Deep learning for identifying metastatic breast cancer. arXiv:1606.05718 [cs, q-bio], June 2016. arXiv: 1606.05718
108. Wang, L., et al.: Learning to detect salient objects with image-level supervision. In: 2017 IEEE Conference on Computer Vision and Pattern Recognition (CVPR). IEEE, July 2017
109. Wang, W., Lai, Q., Fu, H., Shen, J., Ling, H., Yang, R.: Salient object detection in the deep learning era: an in-depth survey. IEEE Trans. Pattern Anal. Mach. Intell. **44**, 3239–3259 (2021)
110. Wang, Z., Bovik, A., Sheikh, H., Simoncelli, E.: Image quality assessment: from error visibility to structural similarity. IEEE Trans. Image Process. **13**(4), 600–612 (2004). https://doi.org/10.1109/TIP.2003.819861
111. Wen, S., et al.: Comparison of different classifiers with active learning to support quality control in nucleus segmentation in pathology images. AMIA Jt Summits Transl. Sci. Proc. **2017**, 227–236 (2018)
112. Wilkinson, L., Anand, A., Grossman, R.: Graph-theoretic scagnostics. In: IEEE Symposium on Information Visualization 2005, INFOVIS 2005, pp. 157–164, October 2005
113. Xie, J., Xu, K., Li, Z., Bi, Q., Qin, K.: Building scene recognition based on deep multiple instance learning convolutional neural network using high resolution remote sensing image. In: Proceedings of the 2019 International Conference on Video, Signal and Image Processing, VSIP 2019, pp. 60–63. Association for Computing Machinery, New York, October 2019
114. Xu, H., Jiang, C., Liang, X., Li, Z.: Spatial-aware graph relation network for large-scale object detection. In: 2019 IEEE/CVF Conference on Computer Vision and Pattern Recognition (CVPR). IEEE, June 2019
115. Yeh, C., et al.: Using publicly available satellite imagery and deep learning to understand economic well-being in Africa. Nat. Commun. **11**(1), 2583 (2020)
116. Yildirim, G., Sen, D., Kankanhalli, M., Süsstrunk, S.: Evaluating salient object detection in natural images with multiple objects having multi-level saliency. arXiv, March 2020
117. Yurtsever, E., Lambert, J., Carballo, A., Takeda, K.: A survey of autonomous driving: common practices and emerging technologies. IEEE Access **8**, 58443–58469 (2020)
118. Zhang, J.M., Harman, M., Ma, L., Liu, Y.: Machine learning testing: survey, landscapes and horizons. arXiv, June 2019
119. Zhang, J., Yu, X., Li, A., Song, P., Liu, B., Dai, Y.: Weakly-Supervised salient object detection via scribble annotations. arXiv, March 2020
120. Zhang, M., Sohoni, N.S., Zhang, H.R., Finn, C., Ré, C.: Correct-N-contrast: a contrastive approach for improving robustness to spurious correlations (2021)
121. Zhao, W., Du, S.: Learning multiscale and deep representations for classifying remotely sensed imagery. ISPRS J. Photogramm. Remote. Sens. **113**, 155–165 (2016)
122. Zhou, Z.H.: Multi-instance learning: a survey (2016)

Cartoon Explanations of Image Classifiers

Stefan Kolek[1]([⊠]), Duc Anh Nguyen[1], Ron Levie[3], Joan Bruna[4],
and Gitta Kutyniok[1,2]

[1] Ludwig Maximilian University of Munich, Munich, Germany
{kolek,danguyen,kutyniok}@math.lmu.de
[2] University of Tromsø, Tromsø, Norway
[3] Technion-Israel Institute of Technology, Haifa, Israel
levieron@technion.ac.il
[4] New York University, New York City, NY, USA
bruna@cims.nyu.edu

Abstract. We present *CartoonX* (Cartoon Explanation), a novel model-agnostic explanation method tailored towards image classifiers and based on the rate-distortion explanation (RDE) framework. Natural images are roughly piece-wise smooth signals—also called cartoon-like images—and tend to be sparse in the wavelet domain. CartoonX is the first explanation method to exploit this by requiring its explanations to be sparse in the wavelet domain, thus extracting the *relevant piece-wise smooth* part of an image instead of relevant pixel-sparse regions. We demonstrate that CartoonX can reveal novel valuable explanatory information, particularly for misclassifications. Moreover, we show that CartoonX achieves a lower distortion with fewer coefficients than state-of-the-art methods.

1 Introduction

Powerful machine learning models such as deep neural networks are inherently opaque, which has motivated numerous explanation methods over the last decade (see for example the survey by [4]). A significant fraction of the research literature has focused on explaining image classifications due to both the practical relevance of computer vision tasks and the ease at which heatmaps can communicate explanatory information. Despite the great variety in methods and explanation philosophies, all current methods share the following characteristic: they operate in pixel space. Roughly speaking, existing explanation methods for image classifiers either allocate additive attribution scores to each (super)pixel or optimize a deletion mask on the pixel coefficients to mark a relevant set of pixels. The result is typically a pixel-sparse and jittery explanation. We challenge the conventional approach to explain in pixel space by successfully applying the rate-distortion explanation (RDE) framework [10,18] in the wavelet domain of images. Our novel explanation method, *CartoonX*, extracts the relevant piece-

S. Avidan et al. (Eds.): ECCV 2022, LNCS 13672, pp. 443–458, 2022.
https://doi.org/10.1007/978-3-031-19775-8_26

wise smooth part of an image. Instead of demanding sparsity in pixel space, as in [2,18], CartoonX demands sparsity in the wavelet domain, which produces piece-wise smooth explanations. Piece-wise smooth images are also known as *cartoon-like images* [14]—a class of 2D signals that has been well studied, and for which wavelets provides an efficient representation system [24]. Our work makes the following contributions.

Reformulation and reinterpretation of the RDE framework: We reformulate the RDE framework in a more general manner with enhanced flexibility in the input representation to accommodate complex interpretation queries such as "What is the piece-wise smooth part of the input signal that leads to its model decision?". Thereby, we reinterpret RDE as a simplification of the input signal, which is interpretable to humans and adheres to a meaningful interpretation query. The simplification is achieved by demanding sparsity in a suitable representation system, which sparsely represents the class of explanations that are desirable for the interpretation query.

CartoonX, a novel explanation method tailored to image classifiers: CartoonX is the first explanation method to extract the relevant piece-wise smooth part of an image instead of relevant pixel sparse regions. This is achieved by demanding sparsity in the wavelet domain of images, where sparsity translates into piece-wise smooth images. We demonstrate that our piece-wise smooth explanations can reveal relevant piece-wise smooth patterns that are not easily visible with existing pixel-based methods. Quantitatively, we also corroborate that CartoonX achieves a lower distortion in the model output using fewer coefficients than other state-of-the-art methods.

2   Related Work

The Rate-Distortion Explanation (RDE) framework was first introduced in [18], and extended in [10], as a mathematically well-founded and intuitive explanation framework. RDEs are model-agnostic explanations and inspired by rate-distortion theory, which studies lossy-data compression. An explanation in RDE consists of a relatively sparse mask over the input features, highlighting the relevant set of features. The mask is optimized to produce low distortion in the model output after applying perturbations to the unselected features in the input while remaining relatively sparse. The authors of [10] also applied RDE to non-canonical input representations to explain model decisions in challenging domains such as audio classification [7] and radio-map estimation [15,16]. The explanation principle of optimizing a mask $s \in [0,1]^n$ was first proposed by [8] who explained image classification decisions by considering one of the two "deletion games": (1) optimizing for the smallest deletion mask that causes the class score to drop significantly or (2) optimizing for the largest deletion mask that has no significant effect on the class score. The original RDE approach [18] is based on the second deletion game. We decided to work within the RDE framework, due to its flexible mathematical formulation. However, we note that other viable mask-based explanation frameworks such as RISE [22], which does not assume access to the model gradient, exist. Other explanation methods developed by the research community are typically either (1) gradient-based such as Smoothgrad [29], Integrated Gradients [31], and Grad-CAM [25], (2) surrogate models such as LIME [23], (3) based on propagation of activations in neurons such as LRP [1,27], and DeepLIFT [27], (4) based on Shapely values from game-theory [17], (6) concept-based such as Concept Activation Vectors [12], or (7) based on generative causal explanations [21]. Also related are methods that were developed to explain individual neurons such as in [6,20]. To our knowledge, all existing explainability methods operate in pixel space and all methods looking for sparse explanations demand sparsity in pixel space [2,8,18].

3   Background: RDE

In this section, we review the rate-distortion explanation (RDE) framework, which was introduced by [18] and later extended by [10] by applying RDE to non-canonical input representations. Suppose $\Phi : \mathbb{R}^n \to \mathbb{R}^m$ is a pre-trained model, *e.g.*, a classifier (with m class labels) or a regression model (with m-dimensional output), where n denotes the dimension of the model input. RDE produces an explanation for a model decision $\Phi(x)$ with $x \in \mathbb{R}^n$ as a relatively sparse mask $s \in \{0, 1\}^n$ marking the relevant input features in x. More precisely, RDE aims to solve the following constrained optimization problem over a mask $s \in \{0, 1\}^n$:

$$\min_{s \in \{0,1\}^n : \|s\|_0 \leq \ell} \; \mathbb{E}_{v \sim \mathcal{V}} \left[d\Big(\Phi(x), \Phi(x \odot s + (1 - s) \odot v) \Big) \right] \tag{1}$$

where $\odot$ denotes the Hadamard product (element-wise multiplication), $d(\Phi(x), \cdot)$ is a measure of distortion (*e.g.*, $d(\Phi(x), \cdot) = \|\Phi(x) - \cdot\|_2$), $\mathcal{V}$ is a distribution over input perturbations $v \in \mathbb{R}^n$, and $\ell \in \{1, ..., n\}$ is a given sparsity level for the explanation mask s. A solution s^* to the optimization problem (1) masks relatively few components in the model input x that suffice to approximately retain the model output $\Phi(x)$. This approach is in the spirit of rate-distortion theory, which deals with lossy compression of data. Therefore, [18] coined such explanations *rate-distortion explanations* (RDEs).

In practice, the RDE optimization problem is relaxed to continuous masks $s \in [0, 1]^n$ solving:

$$\min_{s \in [0,1]^n} \; \mathbb{E}_{v \sim \mathcal{V}} \left[d\Big(\Phi(x), \Phi(x \odot s + (1 - s) \odot v) \Big) \right] + \lambda \|s\|_1 \tag{2}$$

In the relaxed optimization problem, the sparsity level of the mask is determined by $\lambda > 0$ and an approximate solution can be found with stochastic gradient descent in $s \in [0, 1]^n$ if Φ is differentiable. The authors of [18] applied the RDE method as described above to image classifiers in the pixel domain of images, where each mask entry $s_i \in [0, 1]$ corresponds to the i-th pixel values. We refer to this method as *Pixel RDE* throughout this work.

4   RDE Reformulated and Reinterpreted

Instead of applying RDE to the standard input representation $x = [x_1 \dots x_n]^T$, we can apply RDE to a different representation of x to answer a particular interpretation query. For example, consider a 1D-signal $x \in \mathbb{R}^n$: if we ask "What is the smooth part in the signal x that leads to the model decision $\Phi(x)$?", then we can apply RDE in the Fourier basis of x. Since frequency-sparse signals are smooth, applying RDE in the Fourier basis of x extracts the relevant smooth part of the signal. To accommodate such interpretation queries, we reformulate RDE in Sect. 4.1. Finally, based on the reformulation, we reinterpret RDE in Sect. 4.2. Later in Sect. 5, we use our reformulation and reinterpretation of RDE to derive and motivate CartoonX as a special case and novel explanation method tailored towards image classifiers.

4.1   General Formulation

An input signal $x = [x_1, \dots, x_n]^T$ is represented in a basis $\{b_1, \dots, b_n\}$ as a linear combination $\sum_{i=1}^n h_i b_i$ with coefficients $[h_i]_{i=1}^n$. As we argued above and demonstrate later on, some choices for a basis may be more suitable than others to explain a model decision $\Phi(x)$. Therefore, we define the RDE mask not only on the canonical input representation $[x_i]_{i=1}^n$ but also on a different representation $[h_i]_{i=1}^n$ with respect to a choice of basis $\{b_1, \dots, b_n\}$. Examples of non-canonical choices for a basis include the Fourier basis and the wavelet basis. This work is centered around CartoonX, which applies RDE in the wavelet basis, i.e., a linear data representation. Nevertheless, there also exist other domains and interpretation queries where applying RDE to a non-linear data representation can make sense (see the interpretation query "Is phase or magnitude more important for an audio classifier?" in [10]). Therefore, we formulate RDE in terms of a data representation function $f : \prod_{i=1}^k \mathbb{R}^c \to \mathbb{R}^n$, $f(h_1, \dots, h_k) = x$, which does not need to be linear and allows to mask c channels in the input at once. In the important linear case and $c = 1$, we have $f(h_1, \dots, h_k) = \sum_{i=1}^k h_i b_i$, where $\{b_i, \dots, b_k\} \subset \mathbb{R}^n$ are k fixed vectors that constitute a basis. The case $c > 1$ is useful when one wants to mask out several input channels at once, e.g., all color channels of an image, to reduce the number of entries in the mask that will operate on $[h_i]_{i=1}^k$. In the following, we introduce the important definitions of *obfuscations, expected distortion, the RDE mask,* and *RDE's ℓ_1-relaxation,* which generalize the RDE framework of [18] to abstract input representations.

Definitions. The first two key concepts in RDE are *obfuscations* and *expected distortions,* which are defined below.

Definition 1 (Obfuscations and expected distortions). *Let $\Phi : \mathbb{R}^n \to \mathbb{R}^m$ be a model and $x \in \mathbb{R}^n$ a data point with a data representation $x = f(h_1, ..., h_k)$ as discussed above. For every mask $s \in [0, 1]^k$, let $\mathcal{V}$ be a probability distribution over $\prod_{i=1}^k \mathbb{R}^c$. Then the* obfuscation *of x with respect to s and $\mathcal{V}$ is defined as the*

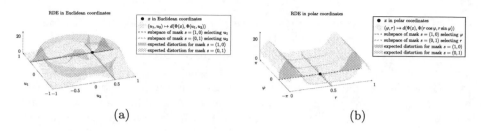

(a)   (b)

Fig. 1. RDE for a hypothetical toy-example in (a) Euclidean coordinates and (b) polar coordinates. Here, the RDE mask can find low expected distortion in polar coordinates but not in Euclidean coordinates. Therefore, in this example, polar coordinates are more appropriate to explain $\Phi(x)$, and RDE would determine that the angle φ, not the magnitude r, is relevant for $\Phi(x)$.

random vector $y := f(s \odot h + (1-s) \odot v)$, where $v \sim \mathcal{V}$, $(s \odot h)_i = s_i h_i \in \mathbb{R}^c$ and $((1-s) \odot v)_i = (1 - s_i) v_i \in \mathbb{R}^c$, for $i \in \{1, \ldots, k\}$. A choice for the distribution $\mathcal{V}$ is called obfuscation strategy. *Furthermore, the expected distortion of x with respect to the mask s and the perturbation distribution $\mathcal{V}$ is defined as*

$$D(x, s, \mathcal{V}, \Phi) := \mathop{\mathbb{E}}_{v \sim \mathcal{V}} \Big[d\Big(\Phi(x), \Phi(y)\Big) \Big],$$

where $d : \mathbb{R}^m \times \mathbb{R}^m \to \mathbb{R}_+$ is a measure of distortion between two model outputs.

In the RDE framework, the explanation is given by a mask that minimizes distortion while remaining relatively sparse. The rate-distortion explanation mask is defined as follows.

Definition 2 (The RDE mask). *In the setting of Definition 1, we define the RDE mask as a solution $s^*(\ell)$ to the minimization problem*

$$\min_{s \in \{0,1\}^k} \quad D(x, s, \mathcal{V}, \Phi) \quad s.t. \quad \|s\|_0 \leq \ell, \tag{3}$$

where $\ell \in \{1, \ldots, k\}$ is the desired level of sparsity.

Geometrically, the RDE mask s is associated with a particular subspace. The complement mask $(1 - s)$ can be seen as selecting a large stable subspace of Φ, where each point represents a possible perturbation in unselected coefficients in h. The RDE mask minimizes the expected distortion along its associated subspace, which requires non-local information of Φ. We illustrate this geometric view of RDE in Fig. 1 with a toy example for a hypothetical classifier $\Phi : \mathbb{R}^2 \to \mathbb{R}^m$ and two distinct input representations: (1) Euclidean coordinates, *i.e.*, f is the identity in $x = f(h)$, and (2) polar coordinates, *i.e.*, $f(h) = (h_2 \cos h_1, h_2 \sin h_1) = x$. In the example, we assume $\mathcal{V}$ to be a uniform distribution on $[-1, 1]^2$ in the Euclidean representation and a uniform distribution on $[-\pi, \pi] \times [0, 1]$ in the polar representation. The expected distortion associated with the masks $s = (1, 0)$ and $s = (0, 1)$ is given by the red and green shaded area, respectively. The RDE mask aims for low expected distortion, and hence, in polar coordinates, the RDE mask would be the green subspace, *i.e.*, $s = (0, 1)$. On the other hand, in Euclidean coordinates, neither $s = (1, 0)$ nor $s = (0, 1)$ produces a particularly low expected distortion, making the Euclidean explanation less meaningful than the polar explanation. The example illustrates why certain input representations can yield more meaningful explanatory insight for a given classifier than others—an insight that underpins our novel CartoonX method. Moreover, the plot in polar coordinates illustrates why the RDE mask cannot be simply chosen with local distortion information, *e.g.*, with the lowest eigenvalue of the Hessian of $h \mapsto d(\Phi(x), \Phi(f(h)))$: the lowest eigenvalue in polar coordinates belongs to the red subspace and does not see the large distortion on the tails.

As was shown by [18], the RDE mask from Definition 2 cannot be computed efficiently for non-trivial input sizes. Nevertheless, one can find an approximate solution by considering continuous masks $s \in [0, 1]^k$ and encouraging sparsity through the ℓ_1-norm.

Definition 3 (RDE's ℓ_1-relaxation). *In the setting of Definition 1, we define RDE's ℓ_1-relaxation as a solution $s^*(\lambda)$ to the minimization problem*

$$\min_{s \in [0,1]^k} \quad D(x, s, \mathcal{V}, \Phi) + \lambda \|s\|_1, \tag{4}$$

where $\lambda > 0$ is a hyperparameter for the sparsity level.

The ℓ_1-relaxation above can be solved with stochastic gradient descent (SGD) over the mask s while approximating $D(x, s, \mathcal{V}, \Phi)$ with i.i.d. samples from $v \sim \mathcal{V}$.

Obfuscation Strategies. An obfuscation strategy is defined by the choice of the perturbation distribution $\mathcal{V}$. Common choices are Gaussian noise [8,18], blurring [8], constants [8], and inpainting GANs [2,10]. Inpainting GANs train a generator $G(s, z, h)$ (z denotes random latent factors) such that for samples $v \sim G(s, z, h)$ the obfuscation $f(s \odot h + (1-s) \odot v)$ remains in the data manifold. In our work, we refrain from using an inpainting GAN due to the following reason: it is hard to tell whether a GAN-based mask did not select coefficients because they are unimportant or because the GAN can easily inpaint them from a biased context (*e.g.*, a GAN that always inpaints a car when the mask shows a traffic light). We want to explain a black-box method transparently, which is why we opt for a simple distribution on the price of not accurately representing the data distribution. We choose a simple and well-understood obfuscation strategy, which we call *Gaussian adaptive noise*. It works as follows: Let $A_1, ..., A_j$ be a pre-defined choice for a partition of $\{1, ..., k\}$. For $i = 1, ..., j$, we compute the empirical mean and empirical standard deviation for each A_i:

$$\mu_i := \frac{\sum_{a \in A_i, t=1,...,d_a} h_{at}}{\sum_{a \in A_i} d_a}, \quad \sigma_i := \sqrt{\frac{1}{\sum_{a \in A_i} d_a} \sum_{a \in A_i, t=1,...,d_a} (\mu_i - h_{at})^2} \tag{5}$$

The adaptive Gaussian noise strategy then samples $v_{at} \sim \mathcal{N}(\mu_i, \sigma_i^2)$ for all members $a \in A_i$ and channels $t = 1, ..., d_a$. We write $v \sim \mathcal{N}(\mu, \sigma^2)$ for the resulting Gaussian random vector $v \in \prod_{i=1}^{k} \mathbb{R}^c$. For Pixel RDE, we only use one set $A_1 = \{1, ..., k\}$ for all k pixels. In CartoonX, which represents input signals in the discrete wavelet domain, we partition $\{1, ..., k\}$ along the scales of the discrete wavelet transform.

Measures of distortion. There are various choices for the measure of distortion $d(\Phi(x), \Phi(y))$. For example, one can take the squared distance in the post-softmax probability of the predicted label for x, i.e., $d(\Phi(x), \Phi(y)) :=$ $\left(\Phi_{j^*}(x) - \Phi_{j^*}(y)\right)^2$, where $j^* := \arg\max_{i=1,...,m} \Phi_i(x)$ and $\Phi(x)$ is assumed to be the post-softmax probabilities of a neural net. Alternatively, one could also choose $d(\Phi(x), \Phi(y))$ as the ℓ_2-distance or the KL-Divergence in the post-softmax layer of Φ. In our experiments for CartoonX, we found that these choices had no significant effect on the explanation (see Fig. 8d).

4.2 Interpretation

The philosophy of the generalized RDE framework is that an explanation for a decision $\Phi(x)$ on a generic input signal $x = f(h)$ should be some simplified version of the signal, which is interpretable to humans. The simplification is achieved by demanding sparsity in a suitable representation system h, *which sparsely represents the class of explanations that are desirable for the interpretation query*. This philosophy is the fundamental premise of CartoonX, which aims to answer the interpretation query *"What is the relevant piece-wise smooth part of the image for a given image classifier?"*. CartoonX first employs RDE on a representation system $x = f(h)$ that sparsely represents piece-wise smooth images and finally visualizes the relevant piece-wise smooth part as an image back in pixel space. In the following section, we explain why wavelets provide a suitable representation system in CartoonX, discuss the CartoonX implementation, and evaluate CartoonX qualitatively and quantitatively on ImageNet.

5 CartoonX

The focus of this paper is *CartoonX*, a novel explanation method—tailored to image classifications—that we obtain as a special case of our generalized RDE framework formulated in Sect. 4. CartoonX first performs RDE in the discrete wavelet position-scale domain of an image x, and finally, visualizes the wavelet mask s as a piece-wise smooth image in pixel space. Wavelets provide optimal representations for piece-wise smooth 1D functions [5], and represent 2D piece-wise smooth images, also called *cartoon-like images* [14], efficiently as well [24]. In particular, sparse vectors in the wavelet coefficient space encode cartoon-like images reasonably well [19]—certainly better than sparse pixel representations. Moreover, wavelets constitute an established tool in image processing [19].

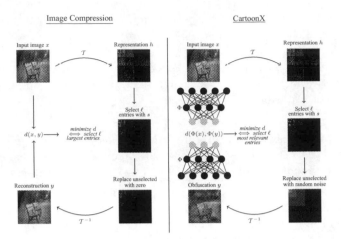

Fig. 2. CartoonX shares many interesting parallels to wavelet-based image compression. Distortion is denoted as d, Φ is an image classifier, h denotes the discrete wavelet coefficients, $\mathcal{T}$ is the discrete wavelet transform, and ℓ is the coefficient budget.

The optimization process underlying CartoonX produces sparse vectors in the discrete wavelet coefficient space, which results in cartoon-like images as explanations. This is the fundamental difference to Pixel RDE, which produces rough, jittery, and pixel-sparse explanations. Cartoon-like images provide a natural model of simplified images. Since the goal of the RDE framework is to generate an easy to interpret simplified version of the input signal, we argue that CartoonX explanations are more appropriate for image classification than Pixel RDEs. Previous work, such as Grad-CAM [25], produces smooth explanations, which also avoid jittery explanations. CartoonX produces *roughly piecewise smooth explanations and not smooth explanations*, which we believe to be more appropriate for images, since smooth explanations cannot preserve edges well. Moreover, we believe that CartoonX enforces piece-wise smoothness in a mathematically more natural manner than explicit smoothness regularization (as in [9]) because wavelets sparsely represent piece-wise smooth signals well. Therefore, CartoonX does not rely on additional smoothness hyperparameters.

CartoonX exhibits interesting parallels to wavelet-based image compression. In image compression, distortion is minimized in the image domain, which is equivalent to selecting the ℓ largest entries in the discrete wavelet transform (DWT) coefficients. CartoonX minimizes distortion in the model output of Φ, which translates to selecting the ℓ most relevant entries in the DWT coefficients. The objective in image compression is efficient data representation, *i.e.*, producing minimal data distortion with a budget of ℓ entries in the DWT coefficients. Conversely, in CartoonX, the objective is extracting the relevant piece-wise smooth part, *i.e.*, producing minimal model distortion with a budget of ℓ entries in the DWT coefficients. We illustrate this connection in Fig. 2—highlighting once more the *rate-distortion* spirit of the RDE framework.

Fig. 3. Visualization of the DWT coefficients for five scales. Three L-shaped sub-images describe coefficients for details in vertical, horizontal, and diagonal orientation at a particular scale. The largest sub-images (the outer L-shape) belong to the lowest scale, *i.e.*, the highest resolution. The smaller L-shaped sub-images gradually build up to higher scales, *i.e.*, lower resolution features.

5.1   Implementation

An image $x \in [0,1]^n$ with $c \in \{1,3\}$ channels, $k \in \mathbb{N}$ pixels can be represented in a wavelet basis by computing its DWT, which is defined by the number of scales $J \in \{1, \ldots, \lfloor \log_2 k \rfloor\}$, the padding mode, and a choice of the mother wavelet (*e.g.*, Haar or Daubechies). For images, the DWT computes four types of coefficients: details in (1) horizontal, (2) vertical, and (3) diagonal orientation at scale $j \in \{1, \ldots, J\}$, and (4) coefficients of the image at the very coarsest resolution. We briefly illustrate the DWT for an example image in Fig. 3.

CartoonX, as described in Algorithm 1, computes the RDE mask in the wavelet domain of images. More precisely, for the data representation $x = f(h)$, we choose h as the concatenation of all the DWT coefficients along the channels, *i.e.*, $h_i \in \mathbb{R}^c$. The representation function f is then the discrete inverse wavelet transform, *i.e.*, the summation of the DWT coefficients times the DWT basis vectors. We optimize the mask $s \in [0,1]^k$ on the DWT coefficients $[h_1, \ldots, h_k]^T$ to minimize RDE's ℓ_1-relaxation from Definition 3. For the obfuscation strategy $\mathcal{V}$, we use adaptive Gaussian noise with a partition by the DWT scale (see Sect. 4.1), *i.e.*, we compute the empirical mean and standard deviation per scale. To visualize the final DWT mask s as a piece-wise smooth image in pixel space, we multiply the mask with the DWT coefficients of the greyscale image $\hat{x}$ of x before inverting the product back to pixel space with the inverse DWT. The pixel values of the inversion are finally clipped into $[0,1]$ as are obfuscations during the RDE optimization to avoid overflow (we assume here the pixel values in x are normalized into $[0,1]$). The clipped inversion in pixel space is the final CartoonX explanation.

5.2   Experiments

We compare CartoonX to the closely related Pixel RDE [18] and several other state-of-the-art explanation methods, *i.e.*, Integrated Gradients [31], Smoothgrad [29], Guided Backprop [30], LRP [1], Guided Grad-CAM [26], Grad-CAM [26], and LIME [23]. Our experiments use the pre-trained ImageNet classifiers MobileNetV3-Small [11] (67.668% top-1 acc.) and VGG16 [28] (71.592% top-1

Algorithm 1: CartoonX

Data: Image $x \in [0,1]^n$ with c channels and k pixels, pre-trained classifier Φ.
Initialization: Initialize mask $s := [1, ..., 1]^T$ on
DWT coefficients $h = [h_1, ..., h_k]^T$ with $x = f(h)$, where f is the inverse DWT.
Choose sparsity level $\lambda > 0$, number of steps N, number of noise samples L, and
measure of distortion d.

for $i \leftarrow 1$ **to** N **do**

Sample L adaptive Gaussian noise samples $v^{(1)}, ..., v^{(L)} \sim \mathcal{N}(\mu, \sigma^2)$;
Compute obfuscations $y^{(1)}, ..., y^{(L)}$ with $y^{(i)} := f(h \odot s + (1-s) \odot v^{(i)})$;
Clip obfuscations into $[0,1]^n$;
Approximate expected distortion $\hat{D}(x, s, \Phi) := \sum_{i=1}^{L} d(\Phi(x), \Phi(y^{(i)}))^2 / L$;
Compute loss for the mask, *i.e.*, $\ell(s) := \hat{D}(x, s, \Phi) + \lambda \|s\|_1$;
Update mask s with gradient descent step using $\nabla_s \ell(s)$ and clip s back to
$[0,1]^k$;

end

Get DWT coefficients $\hat{h}$ for greyscale image $\hat{x}$ of x;
Set $\mathcal{E} := f(\hat{h} \odot s)$ and finally clip $\mathcal{E}$ into $[0,1]^k$;

---

acc.). Images were preprocessed to have 256×256 pixel values in $[0,1]$. Throughout our experiments with CartoonX and Pixel RDE, we used the Adam optimizer [13], a learning rate of $\epsilon = 0.001$, $L = 64$ adaptive Gaussian noise samples, and $N = 2000$ steps. Several different sparsity levels were used. We specify the sparsity level in terms of the number of mask entries k, *i.e.*, by choosing the product λk. Pixel RDE typically requires a smaller sparsity level than CartoonX. We chose $\lambda k \in [20, 80]$ for CartoonX and $\lambda k \in [3, 20]$ for Pixel RDE. The obfuscation strategy for Pixel RDE was chosen as Gaussian adaptive noise with mean and standard deviation computed for all pixel values (see Sect. 4.1). We implemented the DWT for CartoonX with the Pytorch Wavelets package, which is compatible with PyTorch gradient computations, and chose the Daubechies 3 wavelet system with $J = 5$ scales and zero-padding. For the Integrated Gradients method, we used 100 steps, and for the Smoothgrad method, we used 10 samples and a standard deviation of 0.1.

Interpreting CartoonX. In order to correctly interpret CartoonX, we briefly review important properties of the DWT. To cover a large area in an image with a constant value or slowly and smoothly changing gray levels, it suffices to select very few high-scale wavelet coefficients. Hence, for the wavelet mask in CartoonX, it is cheap to cover large image regions with constant or blurry values. Conversely, one needs many high-scale wavelet coefficients to produce fine details such as edges in an image, so fine details are expensive for CartoonX. Hence, the fine details present in the CartoonX are important features for the outcome of the classifier, and fine image features that are replaced by smooth areas in CartoonX are not important for the classifier. It is important to keep in mind that the final CartoonX explanation is a visualization of the wavelet mask

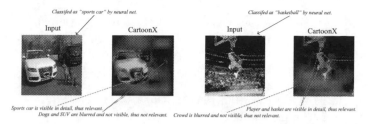

Fig. 4. The CartoonX explanation is an image that suffices to retain the classification decision. For the sports car, CartoonX blurs out the SUV and the dogs. This means the dog and the SUV are irrelevant. For the basketball, the crowd is blurred out. This means the crowd is not relevant since the player and the basket with the crowd blurred out retains the classification as "basketball". The left example also shows that CartoonX is class-discriminative since it blurs out the dogs and the SUV, which belong to other classes.

in pixel space, and *should not be interpreted as a pixel-mask or ordinal pixel-attribution.* CartoonX is not a saliency-map or heatmap but an explanation that is to be interpreted as an image that suffices to retain the classification decision. We illustrate this point in Fig. 4 with two examples.

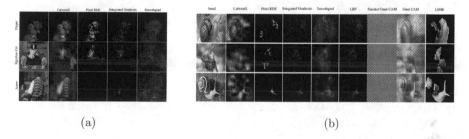

        (a)   (b)

Fig. 5. (a) Each row compares CartoonX explanations of misclassifications by MobileNetV3-Small. The predicted label is depicted next to each misclassified image. (b) Comparing CartoonX explanations for VGG16 for three different images of correctly classified snails.

Qualitative Evaluation. In practice, explaining misclassifications is particularly relevant since good explanations can pinpoint model biases and causes for model failures. In Fig. 5a, we illustrate how CartoonX can help explain misclassified examples by revealing classifier-relevant piece-wise smooth patterns that are not easily visible in other pixel-based methods. In the first row in Fig. 5a, the input image shows a man holding a dog that was classified as a "diaper". CartoonX shows the man not holding a dog but a baby, possibly revealing that the neural net associated diapers with babies and babies with the pose with which the man is holding the dog. In the second row, the input image shows a dog sitting on a chair with leopard patterns. The image was classified as an "Egyptian Cat", which can exhibit leopard-like patterns. CartoonX exposes the Egyptian cat by connecting the dog's head to parts of the armchair forming a cat's torso and legs. In the last row, the input image displays the backside of a man wearing a striped sweater that was classified as a "screw". CartoonX reveals how the stripe patterns look like a screw to the neural net.

Figure 5b further compares CartoonX explanations of correct classifications by VGG16. We also compare CartoonX on random ImageNet samples in Fig. 6a to provide maximal transparency and fair qualitative comparison. In Fig. 6b, we also show failures of CartoonX. These are examples of explanations that are not interpretable and seem to fail at explaining the model prediction. Notably, most failure examples are also not particularly well explained by other state-of-the-art methods. It is challenging to state with certainty the underlying reason for the CartoonX failures since there it is always possible that the neural net bases its decision on non-interpretable grounds.

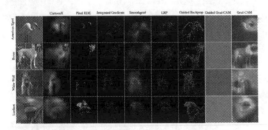
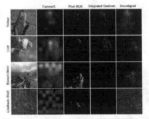
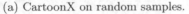

(a) CartoonX on random samples.   (b) Examples of CartoonX failures.

Fig. 6. On random Imagenet samples, CartoonX consistently produces interpretable explanations. Established explanation methods tend to also be difficult to interpret on CartoonX's failure examples.

Quantitative Evaluation. To compare CartoonX quantitatively against other explanation methods, we computed explanations for 100 random ImageNet samples and ordered the image coefficients (for CartoonX the wavelet coefficients) by their respective relevance score. Figure 7a plots the rate-distortion curve, *i.e.*, the distortion achieved in the model output (measured as the ℓ_2-norm in the post-softmax layer) when keeping the most relevant coefficients and randomizing the others. We expect a good explanation to have the most rapid decaying rate-distortion curve for low rates (non-randomized components), which is the case for CartoonX. Note that the random baseline in the wavelet representation is not inherently more efficient than the random baseline in the pixel representation. Moreover, Fig. 7b plots the achieved distortion versus the fraction of randomized relevant components. Here, we expect a good explanation to have the sharpest

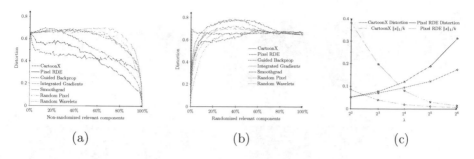

(a)   (b)   (c)

Fig. 7. In (a) the best explanation exhibits steepest early decay. In (b) best explanation exhibits sharpest early increase. In (c) best explanation exhibits lowest distortion and lowest normalized ℓ_1-norm of mask (*i.e.*, highest sparsity).

early increase, which CartoonX again realizes. Lastly, Fig. 7c plots the distortion and non-sparsity (measured as the normalized ℓ_1-norm) of the RDE mask for Pixel RDE and CartoonX at different λ values. The plot underscores the efficiency advantage of CartoonX over Pixel RDE since CartoonX achieves lower distortion and higher sparsity throughout all λ values. For all three plots, random perturbations were drawn from the adaptive Gaussian distribution described in Sect. 4.1.

Sensitivity to Hyperparameters. We compare qualitatively CartoonX's sensitivity to its primary hyperparameters. Figure 8a plots CartoonX explanations and Pixel RDEs for increasing λ. We conistently find that CartoonX is less sensitive than Pixel RDE to λ. In practice, this means one can find a suitable λ faster for CartoonX than for Pixel RDE. Note that for $\lambda = 0$, Pixel RDE is entirely yellow because the mask is initialized as $s = [1 \dots 1]^T$ and $\lambda = 0$ provides no incentive to make s sparser. For the same reason, CartoonX is simply the greyscale image when $\lambda = 0$. Figure 8b plots CartoonX explanations for two choices of $\mathcal{V}$: (1) Gaussian adaptive noise (see Sect. 4.1) and (2) constant zero perturbations. We observe that the Gaussian adaptive noise gives much more meaningful explanations than the simple zero baseline perturbations. Figure 8d plots CartoonX explanations for four choices of $d(\Phi(x), \Phi(y))$, where x is the original input, y is the RDE obfuscation, and Φ outputs post-softmax probabilities: (1) squared ℓ_2 in probability of predicted label j^*, (2) $d(\Phi(x), \cdot) = \|\Phi_{j^*}(x) - 1\|$, *i.e.*, distance that maximizes probability of predicted label, (3) ℓ_2 in post-softmax, (4) KL-Divergence in post-softmax. We do not observe a significant effect by the distortion measure on the explanation. Finally, in Fig. 8c we compare the effect of the mother wavelet in the DWT on the CartoonX explanation. All choices of mother wavelets (labeled as in the Pytorch Wavelets package) provide consistent explanations except for the Haar wavelet, which produces images built of large square pixels.

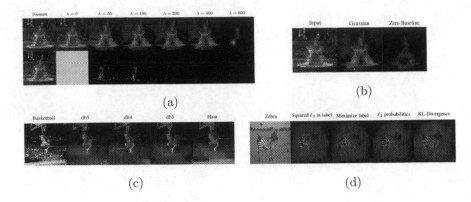

Fig. 8. (a) Top row depicts CartoonX, and the bottom row depicts Pixel RDE, for increasing values of λ. CartoonX for different (b) perturbation distributions, (c) mother wavelets, (d) distortion measures.

Limitations. For MobileNetV3-Small, an image of 256×256 pixels, 16 noise samples, and 2000 optimization steps, we reported a runtime of 45.10 s for CartoonX and 34.09 s for Pixel RDE on the NVIDIA Titan RTX GPU. CartoonX is only slightly slower than Pixel RDE. However, like other perturbation-based methods, CartoonX is significantly slower than gradient or propagation-based methods, which only compute a single or few forward and backward passes and are very fast (Integrated Gradients computes an explanation in 0.48 s for the same image, model, and hardware). We acknowledge that the runtime for CartoonX in its current form constitutes a considerable limitation for many critical applications. However, we are confident that we can significantly reduce the runtime in future work by either learning a strong initial wavelet mask with a neural net or even learning the final wavelet mask with a neural net, similar to the real time image saliency work in [3]. Finally, solving RDE's ℓ_1-relaxation requires access to the model's gradients. Hence, CartoonX is limited to differentiable models.

6   Conclusion

CartoonX is the first explanation method for differentiable image classifiers based on wavelets. We corroborated experimentally that CartoonX can reveal novel explanatory insight and achieves a better rate-distortion than state-of-the-art methods. Nonetheless, CartoonX is still computationally quite expensive, like other perturbation-based explanation methods. In the future, we hope to devise new techniques to speed up the runtime for CartoonX and study the effect of using inpainting GANs for perturbations. We believe CartoonX is a valuable new explanation method for practitioners and potentially a great source of inspiration for future explanation methods tailored to specific data domains.

Acknowledgments. GK was supported in part by the ONE Munich Strategy Forum as well as by Grant DFG-SFB/TR 109, Project C09 and DFG-SPP-2298, KU 1446/31-1 and KU 1446/32-1.

References

1. Bach, S., Binder, A., Montavon, G., Klauschen, F., Müller, K.R., Samek, W.: On pixel-wise explanations for non-linear classifier decisions by layer-wise relevance propagation. PLoS ONE **10**(7), e0130140 (2015)
2. Chang, C., Creager, E., Goldenberg, A., Duvenaud, D.: Explaining image classifiers by counterfactual generation. In: Proceedings of the 7th International Conference on Learning Representations, ICLR (2019)
3. Dabkowski, P., Gal, Y.: Real time image saliency for black box classifiers. In: NIPS (2017)
4. Das, A., Rad, P.: Opportunities and challenges in explainable artificial intelligence (XAI): a survey. ArXiv abs/2006.11371 (2020)
5. DeVore, R.A.: Nonlinear approximation. Acta Numer **7**, 51–150 (1998)
6. Dhamdhere, K., Sundararajan, M., Yan, Q.: How important is a neuron. In: International Conference on Learning Representations (2019)
7. Engel, J., et al.: Neural audio synthesis of musical notes with WaveNet autoencoders. In: Proceedings of the 34th International Conference on Machine Learning, ICML, vol. 70, pp. 1068–1077 (2017)
8. Fong, R.C., Vedaldi, A.: Interpretable explanations of black boxes by meaningful perturbation. In: Proceedings of 2017 IEEE International Conference on Computer Vision (ICCV), pp. 3449–3457 (2017)
9. Fong, R., Patrick, M., Vedaldi, A.: Understanding deep networks via extremal perturbations and smooth masks. In: Proceedings of the IEEE/CVF International Conference on Computer Vision (ICCV), October 2019
10. Heiß, C., Levie, R., Resnick, C., Kutyniok, G., Bruna, J.: In-distribution interpretability for challenging modalities. Preprint arXiv:2007.00758 (2020)
11. Howard, A., et al.: Searching for MobileNetV3. In: Proceedings of the 2019 IEEE/CVF International Conference on Computer Vision (ICCV), pp. 1314–1324 (2019)
12. Kim, B., et al.: Interpretability beyond feature attribution: quantitative testing with concept activation vectors (TCAV). In: ICML (2018)
13. Kingma, D., Ba, J.: Adam: a method for stochastic optimization. In: International Conference on Learning Representations (2014)
14. Kutyniok, G., Lim, W.Q.: Compactly supported shearlets are optimally sparse. J. Approx. Theory **163**(11), 1564–1589 (2011)
15. Levie, R., Yapar, C., Kutyniok, G., Caire, G.: Pathloss prediction using deep learning with applications to cellular optimization and efficient D2D link scheduling. In: ICASSP 2020–2020 IEEE International Conference on Acoustics, Speech and Signal Processing (ICASSP), pp. 8678–8682 (2020). https://doi.org/10.1109/ICASSP40776.2020.9053347
16. Levie, R., Yapar, C., Kutyniok, G., Caire, G.: RadioUNet: fast radio map estimation with convolutional neural networks. IEEE Trans. Wireless Commun. **20**(6), 4001–4015 (2021)
17. Lundberg, S.M., Lee, S.: A unified approach to interpreting model predictions. In: Proceedings of the 31st International Conference on Neural Information Processing Systems, NeurIPS, pp. 4768–4777 (2017)

18. Macdonald, J., Wäldchen, S., Hauch, S., Kutyniok, G.: A rate-distortion framework for explaining neural network decisions. Preprint arXiv:1905.11092 (2019)
19. Mallat, S.: A Wavelet Tour of Signal Processing (Third Edition), chap. 11.3. Academic Press, third edition edn. (2009)
20. Nguyen, A., Dosovitskiy, A., Yosinski, J., Brox, T., Clune, J.: Synthesizing the preferred inputs for neurons in neural networks via deep generator networks. In: Advances in Neural Information Processing Systems (NIPS) (2016)
21. O'Shaughnessy, M., Canal, G., Connor, M., Rozell, C., Davenport, M.: Generative causal explanations of black-box classifiers. In: Larochelle, H., Ranzato, M., Hadsell, R., Balcan, M.F., Lin, H. (eds.) Advances in Neural Information Processing Systems, vol. 33, pp. 5453–5467. Curran Associates, Inc. (2020)
22. Petsiuk, V., Das, A., Saenko, K.: Rise: randomized input sampling for explanation of black-box models. In: BMVC (2018)
23. Ribeiro, M.T., Singh, S., Guestrin, C.: "Why should I trust you?": explaining the predictions of any classifier. In: Proceedings of the 22nd International Conference on Knowledge Discovery and Data Mining, ACM SIGKDD, pp. 1135–1144. Association for Computing Machinery (2016)
24. Romberg, J.K., Wakin, M.B., Baraniuk, R.G.: Wavelet-domain approximation and compression of piecewise smooth images. IEEE Trans. Image Process. **15**, 1071–1087 (2006)
25. Selvaraju, R.R., Cogswell, M., Das, A., Vedantam, R., Parikh, D., Batra, D.: GradCAM: visual explanations from deep networks via gradient-based localization. In: 2017 IEEE International Conference on Computer Vision (ICCV), pp. 618–626 (2017). https://doi.org/10.1109/ICCV.2017.74
26. Selvaraju, R.R., Das, A., Vedantam, R., Cogswell, M., Parikh, D., Batra, D.: GradCAM: visual explanations from deep networks via gradient-based localization. Int. J. Comput. Vision **128**, 336–359 (2019)
27. Shrikumar, A., Greenside, P., Kundaje, A.: Learning important features through propagating activation differences. In: Proceedings of the 34th International Conference on Machine Learning, ICML. vol. 70, pp. 3145–3153 (2017)
28. Simonyan, K., Zisserman, A.: Very deep convolutional networks for large-scale image recognition. In: International Conference on Learning Representations (2015)
29. Smilkov, D., Thorat, N., Kim, B., Viégas, F., Wattenberg, M.: SmoothGrad: removing noise by adding noise. In: Workshop on Visualization for Deep Learning, ICML (2017)
30. Springenberg, J., Dosovitskiy, A., Brox, T., Riedmiller, M.: Striving for simplicity: the all convolutional net. In: ICLR (Workshop Track) (2015)
31. Sundararajan, M., Taly, A., Yan, Q.: Axiomatic attribution for deep networks. In: Proceedings of the 34th International Conference on Machine Learning, ICML, vol. 70, pp. 3319–3328 (2017)

Shap-CAM: Visual Explanations for Convolutional Neural Networks Based on Shapley Value

Quan Zheng[1,2,3], Ziwei Wang[1,2,3], Jie Zhou[1,2,3], and Jiwen Lu[1,2,3]([✉])

[1] Department of Automation, Tsinghua University, Beijing, China
{zhengq20,wang-zw18}@mails.tsinghua.edu.cn,
{jzhou,lujiwen}@tsinghua.edu.cn
[2] State Key Lab of Intelligent Technologies and Systems, Beijing, China
[3] Beijing National Research Center for Information Science and Technology,
Beijing, China

Abstract. Explaining deep convolutional neural networks has been recently drawing increasing attention since it helps to understand the networks' internal operations and why they make certain decisions. Saliency maps, which emphasize salient regions largely connected to the network's decision-making, are one of the most common ways for visualizing and analyzing deep networks in the computer vision community. However, saliency maps generated by existing methods cannot represent authentic information in images due to the unproven proposals about the weights of activation maps which lack solid theoretical foundation and fail to consider the relations between each pixels. In this paper, we develop a novel post-hoc visual explanation method called Shap-CAM based on class activation mapping. Unlike previous gradient-based approaches, Shap-CAM gets rid of the dependence on gradients by obtaining the importance of each pixels through Shapley value. We demonstrate that Shap-CAM achieves better visual performance and fairness for interpreting the decision making process. Our approach outperforms previous methods on both recognition and localization tasks.

Keywords: CNNs · Explainable AI · Interpretable ML · Neural network interpretability

1 Introduction

The dramatic advance of machine learning within the form of deep neural networks has opened up modern Artificial Intelligence (AI) capabilities in real-world applications. Deep learning models achieve impressive results in tasks like object detection, speech recognition, machine translation, which offer tremendous benefits. However, the connectionist approach of deep learning is fundamentally

© The Author(s), under exclusive license to Springer Nature Switzerland AG 2022
S. Avidan et al. (Eds.): ECCV 2022, LNCS 13672, pp. 459–474, 2022.
https://doi.org/10.1007/978-3-031-19775-8_27

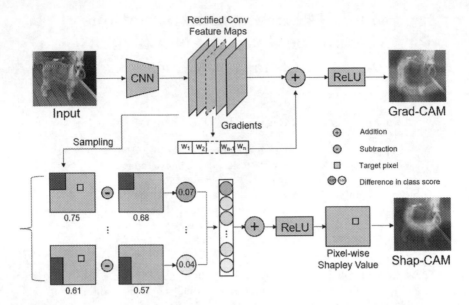

Fig. 1. Comparison between the conventional gradient-based CAM methods and our proposed Shap-CAM. The gradient-based CAM methods, taking Grad-CAM for example, combine the rectified convolutional feature maps and the gradients via backpropagation to compute the saliency map which represents where the model has to look to make the particular decision. Our Shap-CAM introduces Shapley value to estimate the marginal contribution of pixels. For a given pixel in the feature map (viewed in color), we sample various pixel combinations and compute the score difference if the given pixel is added. The score differences are synthesized to obtain pixel-wise Shapley value, which generates the finally Shap-CAM saliency map.

different from earlier AI systems where the predominant reasoning methods are logical and symbolic. These early systems can generate a trace of their inference steps, which at that point serves as the basis for explanation. On the other hand, the usability of today's intelligent systems is limited by the failure to explain their decisions to human users. This issue is particularly critical for risk-sensitive applications such as security, clinical decision support or autonomous navigation.

For this gap, various methods have been proposed by researchers over the last few years to figure out what knowledge is hidden in the layers and connections when utilizing deep learning models. While encouraging development has been carrying this field forward, existing efforts are restricted and the goal of explainable deep learning still has a long way to go, given the difficulty and wide range of issue scopes.

In the context of understanding Convolutional Neural Networks (CNNs), Zhou et al. proposed a technique called CAM (Class Activation Mapping), and demonstrated that various levels of the CNN functioned as unsupervised object detectors [34]. They were able to obtain heat maps that illustrate which regions of an input image were looked at by the CNN for assigning a label by employing a global average pooling layer and showing the weighted combination of the resulting feature maps at the penultimate (pre-softmax) layer. However, this technique was architecture-sensitive and involved retraining a linear classifier for each class. Similar methods were examined with different pooling layers such as global max pooling and log-sum-exp pooling [19,22]. After that, Selvaraju et al. developed Grad-CAM, an efficient version of CAM that combines the class-conditional property of CAM with current pixel-space gradient visualization techniques like Guided Back-propagation and Deconvolution to emphasize fine-grained elements on the image [24]. Grad-CAM improved the transparency of CNN-based models by displaying input regions with high resolution details that are critical for prediction. The variations of Grad-CAM, such as Grad-CAM++ [7], introduce more reliable expressions for pixel-wise weighting of the gradients. However, gradient-based CAM methods cannot represent authentic information in images due to the unproven proposals about the weights of activation maps [1,2,6,8,17,18]. Adversarial model manipulation methods fool the explanations by manipulating the gradients without noticeable modifications to the original images [15], proving that the gradient-based CAM methods are not robust and reliable enough.

To this end, Wang et al. proposed Score-CAM which got rid of the dependence on gradients by obtaining the weight of each activation map through its forward passing score on target class [31]. Though Score-CAM discarded gradients for generating explanations, it still suffered from self designed expression of score which lacked solid theoretical foundation and failed to take the relationship between pixels into consideration. In this work, we present a new post-hoc visual explanation method, named Shap-CAM, where the importance of pixels is derived from their marginal contribution to the model output utilizing Shapley value. Our contributions are:

- We propose a novel gradient-free visual explanation method, Shap-CAM, which introduces Shapley value in the cooperative game theory to estimate the marginal contribution of pixels. Due to the superiority of Shapley value and the consideration of relationship between pixels, more rational and accurate contribution of each pixel is obtained.
- We quantitatively evaluate the generated saliency maps of Shap-CAM on recognition and localization tasks and show that Shap-CAM better discovers important features.
- We show that in a constrained teacher-student setting, it is possible to achieve an improvement in the performance of the student by using a specific loss function inspired from the explanation maps generated by Shap-CAM, which indicates that our explanations discover authentic semantic information mined in images.

The remainder of the paper is organized as follows. In Sect. 2, we introduce the related work about visual explanations and Shapley value. In Sect. 3, we develop our Shap-CAM for the generation of visual explanations based on Shapley value. In Sect. 4, we present some experimental results on recognition and localization tasks and show the effectiveness of our proposed method. We finish the paper with final conclusions and remarks.

2   Related Work

2.1   Visual Explanations

We give a summary of related attempts in recent years to understand CNN predictions in this part. Zeiler et al. provided one of the earliest initiatives in this field, developing a deconvolution approach to better grasp what the higher layers of a given network have learned [33]. Springenberg et al. extended this work to guided backpropagation, which allowed them to better comprehend the impact of each neuron in a deep network on the input image [29]. From a different perspective, Ribeiro et al. introduced LIME (Local Interpretable Model-Agnostic Explanations), an approach that uses smaller interpretable classifiers like sparse linear models or shallow decision trees to make a local approximation to the complex decision surface of any deep model [23]. Shrikumar et al. presented DeepLift, which approximates the instantaneous gradients (of the output with respect to the inputs) with discrete gradients to determine the relevance of each input neuron for a given decision [26]. Al-Shedivat et al. presented Contextual Explanation Networks (CENs), a class of models that learns to anticipate and explain its decision simultaneously [3]. Unlike other posthoc model-explanation tools, CENs combine deep networks with context-specific probabilistic models to create explanations in the form of locally-correct hypotheses.

Class Activation Mapping (CAM) [34] is a technique for discovering discriminative regions and giving understandable explanations of deep models across domains. In CAM, the authors demonstrate that a CNN with a Global Average Pooling (GAP) layer after the last convolutional layer shows localization capabilities despite not being explicitly trained to do so. The CAM explanation regards the importance of each channel as the weight of fully connected layer connecting the global average pooling and the output probability distribution. However, an obvious limitation of CAM is the requirements of a GAP penultimate layer and retraining of an additional fully connected layer. To resolve this problem, Grad-CAM [24] extends the CAM explanation and regards the importance of each channel as the gradient of class confidence w.r.t. the activation map. In Grad-CAM, the authors naturally regard gradients as the importance of each channel towards the class probability, which avoids any retraining or model modification. Variations of Grad-CAM, like Grad-CAM++ [7], use different combinations of gradients and revise the weights for adapting the explanations to different conditions.

However, gradient-based CAM methods do not have solid theoretical foundation and receive poor performances when the gradients are not reliable. Hoe et al. explored whether the neural network interpretation methods can be fooled via adversarial model manipulation, a model fine-tuning step that aims to radically alter the explanations without hurting the accuracy of the original models [15]. They showed that the state-of-the-art gradient-based interpreters can be easily fooled by manipulating the gradients with no noticeable modifications to the original images, proving that the gradient-based CAM methods are not robust and reliable enough. Score-CAM gets rid of the dependence of gradients and introduces channel-wise increase of confidence as the importance of each channel [31]. Score-CAM obtains the weight of each activation map through its forward passing score on target class, the final result is obtained by a linear combination of weights and activation maps. This approach however suffers from self-designed expression of score which lacked solid theoretical foundation. Besides, it fails to consider the relationship between different pixels.

2.2   Shapley Value

One of the most important solution concepts in cooperative games was defined by Shapley [25]. This solution concept is now known as the Shapley value. The Shapley value is useful when there exists a need to allocate the worth that a set of players can achieve if they agree to cooperate. Although the Shapley value has been widely studied from a theoretical point of view, the problem of its calculation still exists. In fact, it can be proved that the problem of computing the Shapley value is an NP-complete problem [9].

Several authors have been trying to find algorithms to calculate the Shapley value precisely for particular classes of games. In Bilbao et al. for example, where a special class of voting game is examined, theoretical antimatriod concepts are used to polynomially compute the Shapley value [11]. In Granot et al. a polynomial algorithm is developed for a special case of an operation research game [13]. In Castro et al., it is proved that the Shapley value for an airport game can be computed in polynomial time by taking into account that this value is obtained using the serial cost sharing rule [4].

Considering the wide application of game theory to real world problems, where exact solutions are often not possible, a need exists to develop algorithms that facilitate this approximation. Although the multilinear extension defined by Owen is an exact method for simple games [20], the calculation of the corresponding integral is not a trivial task. So, when this integral is approximated (using the central limit theorem) this methodology could be considered as an approximation method. In Fatima et al. , a randomized polynomial method for determining the approximate Shapley value is presented for voting games [10]. Castro et al. develop an efficient algorithm that can estimate the Shapley value for a large class of games [5]. They use sampling to estimate the Shapley value and any semivalues. These estimations are efficient if the worth of any coalition can be calculated in polynomial time.

In this work, we propose a new post-hoc visual explanation method, named Shap-CAM, where the importance of pixels is derived from their marginal contribution to the model output utilizing Shapley value. Due to the superiority of Shapley value and the consideration of relationship between pixels, more rational and accurate explanations are obtained.

3   Approach

In this section, we first present the prelimilaries of visual explanations and the background the CAM methods. Then we introduce the theory of Shapley Value [25], a way to quantify the marginal contribution of each player in the cooperative game theory. We apply this theory to our problem and propose the definition of Shap-CAM. Finally, we clarify the estimation of Shapley value in our method. The comparison between the conventional gradient-based CAM methods and our proposed Shap-CAM is illustrated in Fig. 1.

3.1   Prelimilaries

Let function $Y = f(X)$ be a CNN which takes X as an input data point and outputs a probability distribution Y. We denote Y^c as the probability of class c. For the last convolutional layer, A^k denotes the feature map of the k-th channel.

In CAM, the authors demonstrate that a CNN with a Global Average Pooling (GAP) layer after the last convolutional layer shows localization capabilities despite not being explicitly trained to do so. However, an obvious limitation of CAM is the requirements of a GAP penultimate layer and retraining of an additional fully connected layer. To resolve this problem, Grad-CAM [24] extends the CAM explanation and regards the importance of each channel as the gradient of class confidence Y w.r.t. the activation map A, which is defined as:

$$L^c_{ij, \ Grad-CAM} = ReLU\left(\sum_k w^c_k A^k_{ij} \right) \tag{1}$$

where

$$w^c_k = \frac{1}{Z} \sum_i \sum_j \frac{\partial Y^c}{\partial A^k_{ij}} \tag{2}$$

Constant Z stands for the number of pixels in the activation map. In Grad-CAM, the explanation is a weighted summation of the activation maps A^k, where the gradients are regarded as the importance of each channel towards the class probability, which avoids any retraining or model modification. Variations of Grad-CAM, like Grad-CAM++ [7], use different combinations of gradients and revise w^c_k in Eq. (1) for adapting the explanations to different conditions.

However, gradient-based CAM methods do not have solid theoretical foundation and can be easily fooled by adversarial model manipulation methods [15]. Without noticeable modifications to the original images or hurting the accuracy of the original models, the gradient-based explanations can be radically altered by manipulating the gradients, proving that the gradient-based CAM methods are not robust and reliable enough. Score-CAM [31] gets rid of the dependence of gradients and introduces channel-wise increase of confidence as the importance score of each channel. This approach however suffers from self designed expression of the score which fails to consider the relationship between different pixels.

3.2   Definition of Shap-CAM

In order to obtain more accurate and rational estimation of the marginal contribution of each pixel to the model output, we turn to the cooperative game theory. The Shapley value [25] is useful when there exists a need to allocate the worth that a set of players can achieve if they agree to cooperate. Consider a set of n players $\mathbb{P}$ and a function $f(\mathbb{S})$ which represents the worth of the subset of s players $\mathbb{S} \subseteq \mathbb{P}$. The function $f : 2^{\mathbb{P}} \to \mathbb{R}$ maps each subset to a real number, where $2^{\mathbb{P}}$ indicates the power set of $\mathbb{P}$. Shapley Value is one way to quantify the marginal contribution of each player to the result $f(\mathbb{P})$ of the game when all players participate. For a given player i, its Shapley value can be computed as:

$$Sh_i(f) = \sum_{\mathbb{S} \subseteq \mathbb{P}, i \notin \mathbb{S}} \frac{(n-s-1)!s!}{n!} \left[f(\mathbb{S} \cup \{i\}) - f(\mathbb{S}) \right] \tag{3}$$

The Shapley value for player i defined above can be interpreted as the average marginal contribution of player i to all possible coalitions $\mathbb{S}$ that can be formed without it. Notably, it can be proved that Shapley value is the only way of assigning attributions to players that satisfies the following four properties:

Null Player. If the class probability does not depend on any pixels, then its attribution should always be zero. It ensures that a pixel has no contribution if it does not bring any score changes to every possible coalition.

Symmetry. If the class probability depends on two pixels but not on their order (i.e. the values of the two pixels could be swapped, never affecting the probability), then the two pixels receive the same attribution. This property, also called anonymity , is arguably a desirable property for any attribution method: if two players play the exact same role in the game, they should receive the same attribution.

Linearity. If the function f can be seen as a linear combination of the functions of two sub-networks (i.e. $f = af_1 + bf_2$), then any attribution should also be a linear combination, with the same weights, of the attributions computed on the sub-networks, i.e. $Sh_i(\boldsymbol{x}|f) = a \cdot Sh_i(\boldsymbol{x}|f_1) + b \cdot Sh_i(\boldsymbol{x}|f_2)$. Intuitively, this is justified by the need for preserving linearities within the network.

Efficiency. An attribution method satisfies efficiency when attributions sum up to the difference between the value of the function evaluated at the input, and the value of the function evaluated at the baseline, i.e. $\sum_{i=1}^{n} Sh_i = \Delta f = f(\boldsymbol{x}) - f(\boldsymbol{0})$. In our problem, this property indicates that all the attributions of the pixels sum up to the difference between the output probability of the original feature map and the output of the feature map where no original pixels remain. This property, also called completeness or conservation, has been recognized by previous works as desirable to ensure the attribution method is comprehensive in its accounting. If the difference $\Delta f > 0$, there must exist some pixels assigned a non-zero attribution, which is not necessarily true for gradient-based methods.

Back to our problem on class activation mapping, we consider each pixel (i, j) in the feature map of the last convolutional layer $\boldsymbol{A}$ as a player in the cooperative game. Let $\mathbb{P} = \{(i,j)|i = 1, \ldots, h; j = 1, \ldots, w\}$ be the set of pixels in the feature map $\boldsymbol{A}$, where h, w stand for the height and width of the feature map. Let $n = h \cdot w$ be the number of pixels in the activation map. We then define the worth function f in Eq. (3) as the class confidence Y^c, where c is the class of interest. For each subset $\mathbb{S} \subseteq \mathbb{P}$, $Y^c(\mathbb{S})$ represents the output probability of class c when only the pixels in the set $\mathbb{S}$ remain and the others are set to the average value of the whole feature map. By the symbolization above, the original problem turns to an n-player game $(\mathbb{P}, Y^c)$. Naturally, the Shapley Value of the pixel (i,j) represents its marginal contribution to the class confidence. Thus, we define the Shapley Value as the saliency map of our Shap-CAM:

$$
\begin{aligned}
L^c_{ij,\ Shap-CAM} &= Sh_{(i,j)}(Y^c) \\
&= \sum_{\mathbb{S} \subseteq \mathbb{P}, (i,j) \notin \mathbb{S}} \frac{(n-s-1)!s!}{n!} \left[Y^c(\mathbb{S} \cup \{(i,j)\}) - Y^c(\mathbb{S}) \right]
\end{aligned}
\tag{4}
$$

The obtained heatmap is then upsampled to the size of the original image, as it is done in Grad-CAM.

The contribution formula that uniquely satisfies all these properties is that a pixel's contribution is its marginal contribution to the class confidence of every subset of the original feature map. Most importantly, this formula takes into account the interactions between different pixels. As a simple example, suppose there are two pixels that improve the class confidence only if they are both present or absent and harm the confidence if only one is present. The equation considers all these possible settings. This is one of the few methods that take such interactions into account and is inspired by similar approaches in Game Theory. Shapley value is introduced as an equitable way of sharing the group reward among the players where equitable means satisfying the aforementioned properties. It's possible to make a direct mapping between our setting and a cooperative game; therefore, proving the uniqueness of Shap-CAM.

3.3   Estimation of Shapley Value

Exactly computing Eq. (3) would require $\mathcal{O}(2^n)$ evaluations. Intuitively, this is required to evaluate the contribution of each activation with respect to all possible subsets that can be enumerated with the other ones. Clearly, the exact computation of Shapley values is computationally unfeasible for real problems. Sampling is a process or method of drawing a representative group of individuals or cases from a particular population. Sampling and statistical inference are used in circumstances in which it is impractical to obtain information from every member of the population. Taking this into account, we use sampling in this paper to estimate the Shapley value and any semivalues. These estimations are efficient if the worth of any coalition can be calculated in polynomial time. Here we use a sampling algorithm to estimate Shapley value which reduces the complexity to $\mathcal{O}(mn)$, where m is the number of samples taken [5].

Following the definition of Shapley value in Eq. (3), an alternative definition of the Shapley value can be expressed in terms of all possible orders of the players.

Let $O : \{1, \ldots, n\} \rightarrow \{1, \ldots, n\}$ be a permutation that assigns to each position k the player $O(k)$. Let us denote by $\pi(\mathbb{P})$ the set of all possible permutations with player set $\mathbb{P}$. Given a permutation O, we denote by $Pre^i(O)$ the set of predecessors of the player i in the order O, i.e. $Pre^i(O) = \{O(1), \ldots, O(k-1)\}$, if $i = O(k)$.

It can be proved that the Shapley value in Eq. (3) can be expressed equivalently in the following way:

$$Sh_i(f) = \frac{1}{n!} \sum_{O \in \pi(\mathbb{P})} \left[f(Pre^i(O) \cup \{i\}) - f(Pre^i(O)) \right], \quad i = 1, \ldots, n \quad (5)$$

In estimation, we randomly take m samples of player order O from $\pi(\mathbb{P})$, calculate the marginal contribution of the players in the order O, which is defined in the summation of the equation above, and finally average the marginal contributions as the approximation.

Then we will obtain, in polynomial time, an estimation of the Shapley value with some desirable properties. To estimate the Shapley value, we will use a unique sampling process for all players. The sampling process is defined as follows:

- The population of the sampling process P will be the set of all possible orders of n players. The vector parameter under study is $Sh = (Sh_1, \ldots, Sh_n)$.
- The characteristics observed in each sampling unit are the marginal contributions of the players in the order O, i.e.

$$\chi(O) = \{\chi(O_i)\}_{i=1}^n, \quad where \ \chi(O)_i = f(Pre^i(O) \cup \{i\}) - f(Pre^i(O)) \quad (6)$$

- The estimate of the parameter will be the mean of the marginal contributions over the sample M, i.e.

$$\hat{Sh} = (\hat{Sh}_1, \ldots, \hat{Sh}_n), \quad where \quad \hat{Sh}_i = \frac{1}{m} \sum_{O \in M} \chi(O)_i. \quad (7)$$

4   Experiments

In this section, we conduct experiments to evaluate the effectiveness of the proposed explanation method. We first introduce the datasets for evaluation and our implementation details. Then we assess the fairness of the explanation (the significance of the highlighted region for the model's decision) qualitatively via visualization in Sect. 4.2 and quantatively on image recognition in Sect.4.3. In Sect. 4.4 we show the effectiveness for class-conditional localization of objects in a given image. The knowledge distillation experiment is followed in Sect. 4.5.

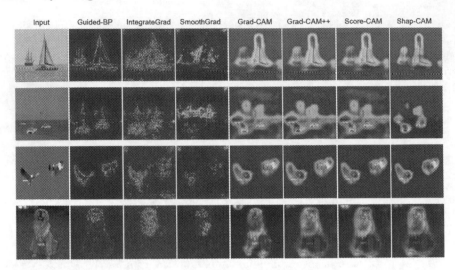

Fig. 2. Visualization results of guided backpropagation [29], SmoothGrad [28], Integrate-Grad [30], Grad-CAM [24], Grad-CAM++ [7], Score-CAM [31] and our proposed Shap-CAM.

4.1 Datasets and Implementation Details

We first detail the datasets that we carried out experiments on: The ImageNet (ILSVRC2012) dataset consists of about 1.2 million and 50k images from 1,000 classes for training and validation respectively. We conducted the following experiments on the validation split of ImageNet. The PASCAL VOC dataset consists of 9,963 natural images from 20 different classes. We used the PASCAL VOC 2007 trainval sets which contained 5,011 images for the recognition evaluation.

For both the ImageNet and the PASCAL VOC datasets, all images are resized to $224 \times 224 \times 3$, transformed to the range $[0, 1]$, and then normalized using mean vector $[0.485, 0.456, 0.406]$ and standard deviation vector $[0.229, 0.224, 0.225]$. In the following experiments, we use pre-trained VGG16 network [27] from the Pytorch model zoo as a base model. As for the calculation of Shapley value in Eq. (4), only the pixels in the set $\mathbb{S}$ are preserved and the others are set to the average value of the whole feature map. Unless stated otherwise, the sampling number for estimating Shapley value is set to 10^4. For a fair comparison, all saliency maps are upsampled with bilinear interpolate to 224×224.

4.2 Qualitative Evaluation via Visualization

We qualitatively compare the saliency maps produced by recently SOTA methods, including gradient-based methods (Guided Backpropagation [29], IntegrateGrad [30], SmoothGrad [28]), and activation-based methods (Grad-CAM [24], Grad-CAM [24], Grad-CAM++ [7]) to validate the effectiveness of Shap-CAM.

Table 1. Recognition evaluation results on the ImageNet (ILSVRC2012) validation set (lower is better in average drop, higher is better in average increase).

Method	Mask	RISE	GradCAM	GradCAM++	ScoreCAM	ShapCAM
Avr. Drop(%)	63.5	47.0	47.8	45.5	31.5	**28.0**
Avr. Increase(%)	5.29	14.0	19.6	18.9	30.6	**31.8**

Table 2. Recognition evaluation results on the PASCAL VOC 2007 validation set (lower is better in Average Drop, higher is better in Average Increase).

Method	Mask	RISE	GradCAM	GradCAM++	ScoreCAM	ShapCAM
Avr. Drop (%)	45.3	31.3	28.5	19.5	15.6	**13.2**
Avr. Increase (%)	10.7	18.2	21.4	19.0	28.9	**32.7**

As shown in Fig. 2, results in Shap-CAM, random noises are much less than that other methods. In addition, Shap-CAM generates smoother saliency maps comparing with gradient-based methods. Due to the introduction of Shapley value, our Shap-CAM is able to estimate the marginal contribution of the pixels and take the relationship between pixels into consideration, and thus can obtain more rational and accurate saliency maps.

4.3   Faithfulness Evaluation via Image Recognition

The faithfulness evaluations are carried out as depicted in Grad-CAM++ [7] for the purpose of object recognition. The original input is masked by point-wise multiplication with the saliency maps to observe the score change on the target class. In this experiment, rather than do point-wise multiplication with the original generated saliency map, we slightly modify by limiting the number of positive pixels in the saliency map. Two metrics called Average Drop, Average Increase In Confidence are introduced:

Average Drop: The Average Drop refers to the maximum positive difference in the predictions made by the prediction using the input image and the prediction using the saliency map. It is given as: $\sum_{i=1}^{N} \frac{\max(0, Y_i^c - O_i^c)}{Y_i^c} \times 100\%$. Here, Y_i^c refers to the prediction score on class c using the input image i and O_i^c refers to the prediction score on class c using the saliency map produced over the input image i. A good explanation map for a class should highlight the regions that are most relevant for decision-making. It is expected that removing parts of an image will reduce the confidence of the model in its decision, as compared to its confidence when the full image is provided as input.

Increase in Confidence: Complementary to the previous metric, it would be expected that there must be scenarios where providing only the explanation map region as input (instead of the full image) rather increases the confidence in the prediction (especially when the context is distracting). In this metric, we measure the number of times in the entire dataset, the model's confidence increased when providing only the explanation map regions as input. The Average Increase in Confidence is denoted as:$\sum_{i=1}^{N} \dfrac{sign(Y_i^c < O_i^c)}{N} \times 100\%$, where $sign$ presents an indicator function that returns 1 if input is True.

Our comparison extends with state-of-the-art methods, namely gradient-based, perturbation-based and CAM-based methods, including Mask [12], RISE [21], Grad-CAM [24] and Grad-CAM++ [7]. Experiment conducts on the ImageNet (ILSVRC2012) validation set, 2000 images are randomly selected. Results are reported in Table 1. Results on the PASCAL VOC 2007 validation set are reported in Table 2.

As shown in Table 1 and Table 2, Shap-CAM outperforms other perturbation-based and CAM-based methods. Shap-CAM can successfully locate the most distinguishable part of the target item, rather than only determining what humans think is important, based on its performance on the recognition challenge. Results on the recognition task show that Shap-CAM can more accurately reveal the decision-making process of the original CNN model than earlier techniques. Previous methods lack solid theoretical foundation and suffer from self-designed explanations, which are prone to the manipulation of gradients or fail to consider the relationship between different pixels. Our Shap-CAM instead is able to obtain more rational and accurate contribution of each pixel due to the superiority of Shapley value and the consideration of relationship between pixels.

Furthermore, for a more comprehensive comparison, we also evaluate our method on deletion and insertion metrics which are proposed in [21]. In our experiment, we simply remove or introduce pixels from an image by setting the pixel values to zero or one with step 0.01 (remove or introduce 1% pixels of the whole image each step). Example are shown in Fig. 3, where our approach achieves better performance on both metrics compared with SOTA methods.

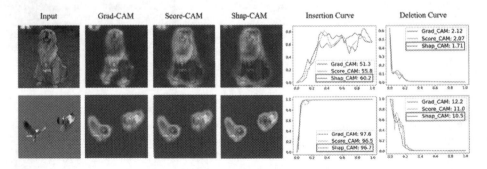

Fig. 3. Grad-CAM, Score-CAM and Shap-CAM generated saliency maps for representative images in terms of deletion and insertion curves. In the insertion curve, a better explanation is expected that the prediction score to increase quickly, while in the deletion curve, it is expected the classification confidence to drop faster.

4.4 Localization Evaluation

In this section, we measure the quality of the generated saliency map through localization ability. Bounding box evaluations are accomplished. We employ the similar metric, as specified in Score-CAM, called the Energy-based pointing game. Here, the amount of energy of the saliency map is calculated by finding out how much of the saliency map falls inside the bounding box. Specifically, the input is binarized with the interior of the bounding box marked as 1 and the region outside the bounding box as 0. Then, this input is multiplied with the generated saliency map and summed over to calculate the proportion ratio, which given as

$$Proportion = \frac{\sum L^c_{(i,j)\in bbox}}{\sum L^c_{(i,j)\in bbox} + \sum L^c_{(i,j)\notin bbox}}. \tag{8}$$

Two pre-trained models, namely VGG-16 [27], ResNet18 [14], are used to conduct the energy-based pointing game on the 2000 randomly chosen images from the ILSVRC 2012 Validation set.

We randomly select images from the validation set by removing images where object occupies more than 50% of the whole image. For convenience, we only consider these images with only one bounding box for target class. We experiment on 500 random selected images from the ILSVRC 2012 validation set. Evaluation results are reported in Table 3, which show that our method outperforms previous works. This also confirms that the Shap-CAM-generated saliency map has fewer noises. As is shown in the previous research [15], the state-of-the-art gradient-based interpreters can be easily fooled by manipulating the gradients with no noticeable modifications to the original images, proving that the SOTA methods are not robust enough to the noise. On the contrary, our Shap-CAM can alleviate the effect of this problem by estimating the importance of each pixel more accurately.

Table 3. Localization evaluations of proportion (%) using energy-based pointing game (higher the better).

Method	Grad-CAM	Grad-CAM++	Score-CAM	Shap-CAM
VGG-16	39.95	40.16	40.10	**40.45**
ResNet18	40.90	40.85	40.76	**41.28**

4.5 Learning from Explanations: Knowledge Distillation

Following the knowledge distillation experiment settings in Grad-CAM++ [7], we show that in a constrained teacher-student learning setting, knowledge transfer to a shallow student (commonly called knowledge distillation) is possible from the explanation of CNN decisions generated by CAM methods. We use Wide

Table 4. Test error rate (%) for knowledge distillation to train a student from a deeper teacher network. L_{CE} is the normal cross entropy loss function. The Column 2–6 refer to the modified loss function L_{stu} where the explanations for images are generated using the corresponding interpreter.

Loss function	L_{CE}	GradCAM	GradCAM++	ScoreCAM	ShapCAM
w/o L_{KD}	6.78	6.86	6.74	6.75	**6.69**
w/ L_{KD}	5.68	5.80	5.56	5.42	**5.37**

ResNets [32] for both the student and teacher networks. We train a WRN-40-2 teacher network (2.2 M parameters) on the CIFAR-10 dataset. In order to train a student WRN-16-2 network (0.7 M parameters), we introduce a modified loss L_{stu}, which is a weighted combination of the standard cross entropy loss L_{CE} and an interpretability loss L_{interp}.

$$L_{stu}(c, W_s, W_t, I) = L_{CE}(c, W_s(I)) + \alpha \|L_s^c(W_s(I)) - L_t^c(W_t(I))\|_2^2 \quad (9)$$

where the first term represents the cross entropy loss and the second term represents the interpretability loss L_{interp}. In the above equations, I indicates the input image and c stands for the corresponding output class label. L^c is the explanations given by a certain interpreter. α is a hyper parameter that controls the importance given to the interpretability loss. W_s denotes the weights of the student network, and W_t the weights of the teacher network. This equation above forces the student network not only minimize standard cross-entropy loss for classification, but also learn from the most relevant parts of a given image used for making a decision from the teacher network, which is the influence of the interpretability loss L_{interp} term.

Table 4 shows the results for this experiment. L_{CE} is the normal cross entropy loss function, i.e. the student network is trained independently on the dataset without any intervention from the expert teacher. The following four columns refer to loss functions defined in Eq. (9) where the explanations for image I are generated using the corresponding interpreter. We further also included L_{KD}, the knowledge distillation loss introduced by Hinton et al. with temperature parameter set to 4 [16]. It is indicated from these results that knowledge distillation can be improved by considering the explanations of the teacher. The results also show that Shap-CAM provides better explanation-based knowledge distillation than existing CAM-based methods.

5   Conclusion

We propose Shap-CAM, a novel CAM variant, for visual explanations of deep convolutional networks. We introduce Shapley value to represent the marginal contribution of each pixel to the model output. Due to the superiority of Shapley value and the consideration of relationship between pixels, more rational and accurate explanations are obtained. We present evaluations of the generated saliency maps on recognition and localization tasks and show that Shap-CAM better discovers important features. In a constrained teacher-student setting, our Shap-CAM provides better explanation-based knowledge distillation than the state-of-the-art explanation approaches.

Acknowledgement. This work was supported in part by the National Key Research and Development Program of China under Grant 2017YFA0700802, in part by the National Natural Science Foundation of China under Grant 62125603 and Grant U1813218, in part by a grant from the Beijing Academy of Artificial Intelligence (BAAI).

References

1. Adebayo, J., Gilmer, J., Goodfellow, I., Kim, B.: Local explanation methods for deep neural networks lack sensitivity to parameter values. arXiv preprint arXiv:1810.03307 (2018)
2. Adebayo, J., Gilmer, J., Muelly, M., Goodfellow, I., Hardt, M., Kim, B.: Sanity checks for saliency maps. In: Advances in Neural Information Processing Systems (2018)
3. Al-Shedivat, M., Dubey, A., Xing, E.P.: Contextual explanation networks. arXiv preprint arXiv:1705.10301 (2017)
4. Castro, J., Gomez, D., Tejada, J.: A polynomial rule for the problem of sharing delay costs in pert networks. In: Computers and Operation Research (2008)
5. Castro, J., Gomez, D., Tejada, J.: Polynomial calculation of the shapley value based on sampling. Comput. Oper. Res. **36**, 1726–1730 (2009)
6. Chang, C.H., Creager, E., Goldenberg, A., Duvenaud, D.: Explaining image classifiers by counterfactual generation. arXiv preprint arXiv:1807.08024 (2018)
7. Chattopadhay, A., Sarkar, A., Howlader, P., Balasubramanian, V.N.: Grad-cam++: generalized gradient-based visual explanations for deep convolutional networks. In: IEEE Winter Conference on Applications of Computer Vision, pp. 839–847 (2018)
8. Dabkowski, P., Gal, Y.: Real time image saliency for black box classifiers. In: Advances in Neural Information Processing Systems (2017)
9. Deng, X., Papadimitriou, C.: On the complexity of cooperative solution concepts. Math. Oper. Res. **19** (1994)
10. Fatima, S.S., Wooldridge, M., Jennings, N.R.: An analysis of the shapley value and its uncertainty for the voting game. In: La Poutré, H., Sadeh, N.M., Janson, S. (eds.) AMEC/TADA -2005. LNCS (LNAI), vol. 3937, pp. 85–98. Springer, Heidelberg (2006). https://doi.org/10.1007/11888727_7
11. Fernández, J., et al.: Generating functions for computing the myerson value. Ann. Oper. Res. **109**(1), 143–158 (2002)
12. Fong, R.C., Vedaldi, A.: Interpretable explanations of black boxes by meaningful perturbation. In: ICCV (2017)
13. Granot, D., Kuipers, J., Chopra, S.: Cost allocation for a tree network with heterogeneous customers. Math. Oper. Res. **27**, 647–661 (2002)
14. He, K., Zhang, X., Ren, S., Sun, J.: Deep residual learning for image recognition. In: CVPR (2016)
15. Heo, J., Joo, S., Moon, T.: Fooling neural network interpretations via adversarial model manipulation. In: NeurIPS (2019)

16. Hinton, G., Vinyals, O., Dean, J.: Distilling the knowledge in a neural network. arXiv preprint arXiv:1503.02531 (2015)
17. Lin, M., Chen, Q., Yan, S.: Network in network. arXiv preprint arXiv:1312.4400 (2013)
18. Omeiza, D., Speakman, S., Cintas, C., Weldermariam, K.: Smooth Grad-CAM++: an enhanced inference level visualization technique for deep convolutional neural network models. arXiv preprint arXiv:1908.01224 (2019)
19. Oquab, M., Bottou, L., Laptev, I., Sivic, J.: Is object localization for free? weakly-supervised learning with convolutional neural networks. In: CVPR, pp. 685–694 (2015)
20. Owen, G.: Multilinear extensions of games. In: Management Science Series B-Application, vol. 18, No. 5, Theory Series, Part 2, Game Theory and Gaming, January 1972, pp. P64–P79. INFORMS (1972)
21. Petsiuk, V., Das, A., Saenko, K.: Rise: Randomized input sampling for explanation of black-box models. arXiv preprint arXiv:1806.07421 (2018)
22. Pinheiro, P.O., Collobert, R.: From image-level to pixel-level labeling with convolutional networks. In: CVPR, pp. 1713–1721 (2015)
23. Ribeiro, M.T., Singh, S., Guestrin, C.: Why should I trust you?: Explaining the predictions of any classifier. In: Proceedings of the 22nd ACM SIGKDD International Conference on Knowledge Discovery and Data Mining, pp. 1135–1144 (2016)
24. Selvaraju, R.R., Das, A., Vedantam, R., Cogswell, M., Parikh, D., Batrat, D.: Grad-CAM: why did you say that? Visual explanations from deep networks via gradient-based localization. arXiv preprint arXiv:1610.02391 (2016)
25. Shapley, L.S.: A value for n-person games. Contribut. Theory Games 2(28), 307–317 (1953)
26. Shrikumar, A., Greenside, P., Kundaje, A.: Learning important features through propagating activation differences. arXiv preprint arXiv:1704.02685 (2017)
27. Simonyan, K., Zisserman, A.: Very deep convolutional networks for large-scale image recognition. arXiv preprint arXiv:1409.1556 (2014)
28. Smilkov, D., Thorat, N., Kim, B., Viegas, F., Wattenberg, M.: SmoothGrad: removing noise by adding noise. arXiv preprint arXiv:1706.03825 (2017)
29. Springenberg, J.T., Dosovitskiy, A., Brox, T., Riedmiller, M.: Striving for simplicity: the all convolutional net. arXiv preprint arXiv:1412.6806 (2014)
30. Sundararajan, M., Taly, A., Yan, Q.: Axiomatic attribution for deep networks. In: Proceedings of the 34th International Conference on Machine Learning (2017)
31. Wang, H., et al.: Score-cam: Score-weighted visual explanations for convolutional neural networks. In: CVPR (2020)
32. Zagoruyko, S., Komodakis, N.: Wide residual networks. arXiv preprint arXiv:1605.07146 (2016)
33. Zeiler, M.D., Fergus, R.: Visualizing and understanding convolutional networks. In: ECCV,. pp. 818–833 (2014)
34. Zhou, B., Khosla, A., Lapedriza, A., Oliva, A., Torralba, A.: Learning deep features for discriminative localization. In: CVPR, pp. 2921–2929 (2016)

Privacy-Preserving Face Recognition with Learnable Privacy Budgets in Frequency Domain

Jiazhen Ji[1]([⊠])[iD], Huan Wang[2][iD], Yuge Huang[1][iD], Jiaxiang Wu[1][iD], Xingkun Xu[1][iD], Shouhong Ding[1][iD], ShengChuan Zhang[2][iD], Liujuan Cao[2], and Rongrong Ji[2]

[1] Youtu Lab, Tencent, Shanghai, China
{royji,yugehuang,willjxwu,xingkunxu,ericding}@tencent.com
[2] Xiamen University, Xiamen, China
hanawh@stu.xmu.edu.cn, {zsc_2016,caoliujuan,rrji}@xmu.edu.cn

Abstract. Face recognition technology has been used in many fields due to its high recognition accuracy, including the face unlocking of mobile devices, community access control systems, and city surveillance. As the current high accuracy is guaranteed by very deep network structures, facial images often need to be transmitted to third-party servers with high computational power for inference. However, facial images visually reveal the user's identity information. In this process, both untrusted service providers and malicious users can significantly increase the risk of a personal privacy breach. Current privacy-preserving approaches to face recognition are often accompanied by many side effects, such as a significant increase in inference time or a noticeable decrease in recognition accuracy. This paper proposes a privacy-preserving face recognition method using differential privacy in the frequency domain. Due to the utilization of differential privacy, it offers a guarantee of privacy in theory. Meanwhile, the loss of accuracy is very slight. This method first converts the original image to the frequency domain and removes the direct component termed DC. Then a privacy budget allocation method can be learned based on the loss of the back-end face recognition network within the differential privacy framework. Finally, it adds the corresponding noise to the frequency domain features. Our method performs very well with several classical face recognition test sets according to the extensive experiments. Code will be available at https://github.com/Tencent/TFace/tree/master/recognition/tasks/dctdp.

Keywords: Privacy-preserving · Face recognition · Differential privacy

Supplementary Information The online version contains supplementary material available at https://doi.org/10.1007/978-3-031-19775-8_28.

1   Introduction

With the rapid development of deep learning, face recognition models based on convolutional neural networks have gained a remarkable breakthrough in recent years. The extremely high accuracy rate has led to its application in many daily life scenarios. However, due to the privacy sensitivity of facial images and the unauthorized collection and use of data by some service providers, people have become increasingly concerned about the leakage of their face privacy. In addition to these unregulated service providers, malicious users and hijackers pose a significant danger to privacy leakage. It is necessary to apply some privacy-preserving mechanisms to face recognition.

Homomorphic encryption [11] is an encryption method that encrypts the original data and performs inference on the encrypted data. It protects data privacy and maintains a high level of recognition accuracy. However, the introduction of the encryption process requires a great additional computation. Therefore, it is not suitable for large-scale and interactive scenarios.

PEEP [6] is a very typical approach to privacy protection that makes use of differential privacy. It first converts the original image to a projection on the eigenfaces. Then, it adds noise to it by utilizing the concept of differential privacy to provide better privacy guarantees than other privacy models. PEEP has a low computational complexity, not to slow down the inference speed. However, it significantly reduces the accuracy of face recognition. As shown in Fig. 1, the current face recognition privacy-preserving methods all perform poorly with large data sets and have relatively mediocre privacy-preserving capabilities.

This paper aims to limit the face recognition service provider to learn only the classification result (e.g., identity) with a certain level of confidence but does not have access to the original image (even by some automatic recovery techniques). We propose a privacy-preserving framework to tackle the privacy issues in deep face recognition. Inspired by frequency analysis performing well in selecting the

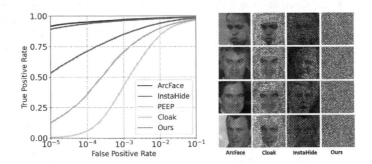

Fig. 1. Comparison with different privacy-preserving methods. Left: ROC curves under IJB-C dataset, the closer to the upper left area, the better the performance. **Right**: Visualization of processed images using LFW dataset as examples. Compared with other methods, our method has good privacy-preserving performance.

signal of interest, we deeply explored the utility of frequency domain privacy preservation. We use block discrete cosine transform (DCT) to transfer the raw facial image to the frequency domain. It is a prerequisite for separating information critical to visualization from information critical to identification. Next, we remove the direct component (DC) channel because it aggregates most of the energy and visualization information in the image but is not essential for identification. Meanwhile, we consider that elements at different frequencies of the input image do not have the same importance for the identification task. Therefore it is not reasonable to set precisely the same privacy budget for all elements. We propose a method taking into account the importance of elements for identification. It only needs to set the average privacy budget to obtain the trade-off between privacy and accuracy. Then the distribution of privacy budgets over all the elements will be learned according to the loss of the face recognition model. Compared with PEEP, our approach to switching to the frequency domain is simpler, faster, and easier to deploy. Moreover, because we learn different privacy budgets for different features from the loss of face recognition, our recognition accuracy is also far better than SOTAs. The contributions of this paper are summarized as follows:

- We propose a framework for face privacy protection based on the differential privacy method. The method is fast and efficient and adjusts the privacy-preserving capability according to the choice of privacy budget.
- We design a learnable privacy budget allocation structure for image representation in the differential privacy framework, which can protect privacy while reducing accuracy loss.
- We design various privacy experiments that demonstrate the high privacy-preserving capability of our approach with marginal loss of accuracy.
- Our method can transform the original face recognition dataset into a privacy-preserving dataset while maintaining high availability.

2   Related Work

2.1   Face Recognition

The state-of-the-art (SOTA) research on face recognition mainly improve from the perspective of softmax-based loss function which aims to maximize the inter-class discrepancy and minimize the intra-class variance for better recognition accuracy [8, 18, 22, 29, 33]. However, existing margin-based methods do not consider potential privacy leakage. In actual applications, raw face images of users need to be delivered to remote servers with GPU devices, which raises the user's concern about the abuse of their face images and potential privacy leakage during the transmission process. Our method takes masked face images rather than raw face images as the face recognition model's inputs, which reduces the risk of misuse of user images.

2.2 Frequency Domain Learning

Frequency analysis has always been widely used in signal processing, which is a powerful tool for filtering signals of interest. In recent years, Some works that introduced frequency-domain analysis have been proposed to tackle the various aspects of the problem.

For instance, [13] trained CNNs directly on the blockwise discrete cosine transform (DCT) coefficients. [36] made an in-depth analysis of the selection of DCT coefficients on three different high-level vision tasks such as image classification, detection, and segmentation. [16] presented a frequency space domain randomization technique that achieved superior segmentation performance. In the field of deepfake, frequency is an essential tool to distinguish real from synthetic images (or videos) [10]. As for face recognition, [20] presents a new approach through a feature-level fusion of face and iris traits extracted by polar fast fourier transform. [34] splits the frequency domain channels of the image and selects only some of them for subsequent tasks. On the basis of these works, we firstly and deeply explored the utility of privacy-preserving in the frequency domain.

2.3 Privacy Preserving

Privacy-preserving research can be broadly categorized based on how to process input data, i.e., encryption or perturbation. The majority of data encryption methods fall under the Homomorphic Encryption (HE) and Secure Multiparty Computation (SMC) [2,12,19,21]. However, these data encryption methods are unsuitable for current SOTA face recognition systems due to their prohibitive computation cost. Data perturbation, in contrast, avoids the high cost of encryption by applying a perturbation to raw inputs. Differential privacy (DP) is a common-used perturbation approach. [6] applied perturbation to eigenfaces utilizing differential privacy that equally split the privacy budget to every eigenface resulting in a great loss of accuracy. In addition to DP, there are other methods for data disturbance. For instance, [17] used the Mixup [37] method to perturb data while it has been hacked successfully [5]. [25] presented a Gaussian noise disturbance method to suppress unimportant pixels before sending them to the cloud. Recently, some privacy-preserving methods have also appeared in the field of face recognition. [1] proposed federated face recognition to train face recognition models using multi-party data via federated learning to avoid privacy risks. K-same [27] is a de-identification approach by using K-anonymity for face images. Our method maintains accuracy to the greatest extent based on a learnable privacy budget.

3 Method

In this section, we describe the framework of our proposed privacy-preserving face recognition method. Our method consists of three main modules, a frequency

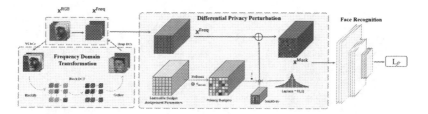

Fig. 2. Overview of our proposed method. It consists of three modules: frequency-domain transformation, differential privacy perturbation, and face recognition. The frequency-domain transformation module transforms the facial image to frequency domain features. The differential privacy perturbation module adds perturbations to frequency domain features. The face recognition module takes the perturbed features as input and performs face recognition.

domain transformation module, a perturbation module that utilizes differential privacy, and a face recognition module, as shown in Fig. 2. Each input image is first converted to frequency domain features by the frequency domain transformation module. Furthermore, the differential privacy perturbation module will generate the corresponding noise and add it to the frequency domain features. Finally, the perturbed frequency domain features will be transferred to the face recognition model. Because the size of the perturbed frequency domain features is $[H, W, C]$, we only need to change the input channels of the face recognition model from 3 to C to suit our input. In Sect. 3.1, we describe in detail the specific process of frequency-domain conversion. In Sect. 3.2, we first introduce the background knowledge of differential privacy and then describe the specific improvements in our differential privacy perturbation module.

3.1 Frequency-Domain Transformation Module

[32] discovered that humans rely only on low-frequency information for image recognition while neural networks use low-and high-frequency information. Therefore, using low-frequency information as little as possible is very effective for image privacy protection. DCT transformation can be beneficial in separating the low-frequency information that is important for visualization from the high-frequency information that is important for identification. In the frequency domain transformation module, inspired by the compression operation in JPEG, we utilized block discrete cosine transform (BDCT) as our basis of frequency-domain transformation. For each input image, we first convert it from RGB color spaces to YCbCr color spaces. We then adjust its value range to $[-128, 127]$ to meet the requirement of BDCT input and then split it into $\frac{H}{8} \times \frac{W}{8}$ blocks with a size of 8×8. For a fairer comparison and as little adjustment as possible to the structure of the recognition network, we perform an 8-fold up-sampling on the facial images before BDCT. Then a normalized, two-dimensional type-II DCT is used to convert each block into 8×8 frequency-domain coefficients.

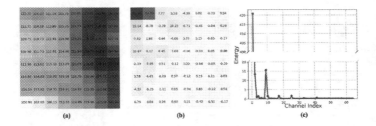

Fig. 3. (a) One block of the Original image with size 8×8. (b) Frequency-domain features of the image block after DCT transformation. (c) Energies among different channels. DC channels take up most of the energy

At this time, we can see that the element in the upper left corner in Fig. 3(b) has an extreme value. It defines the basic tone of the entire block, which we call the DC component. For elements in the same position in each block, we collect them together in the same channel and arrange them according to the relative position of the block in the original image. Here we have converted the original image with size $[H, W, 3]$ into a frequency domain representation with size $[H, W, 8, 8, 3]$. As we can see in Fig. 3(c), the most energy (91.6%) of facial images is concentrated in the DC channel. According to our experiment results in Table 3, DC is crucial for visualization, while it has little impact on recognition. Thus it is removed before it enters the next module.

3.2 Differential Privacy (DP)

Definition. DP [24] is known to be a privacy model that provides maximum privacy by minimizing the possibility of personal records being identified. By adding noise to the query information of the database, DP helps to ensure the security of personal information while obtaining comprehensive statistical information. For any two databases that differ by only one individual, we say they are adjacent.

A randomized algorithm $\mathcal{A}$ satisfies ϵ-DP if for every adjacent pairs of database D_1 and D_2 and all $\mathcal{Q} \subseteq Range(\mathcal{A})$, Eq. 1 holds.

$$Pr(\mathcal{A}(D_1) \in \mathcal{Q}) \leq e^\epsilon Pr(\mathcal{A}(D_2) \in \mathcal{Q}) \tag{1}$$

DP for Generative Models. Since our goal is to protect facial privacy by scrambling the images after BDCT, we need to move from the database domain to the generative model representations domain. Under the generative model representation domain, query sensitivity and adjacency in the database domain are no longer applicable. Each input image should correspond to a separate database. Some methods [7] apply DP generalization to arbitrary secrets, where a secret is any numerical representation of the data.

In our approach, we consider the BDCT representation of the facial image as a secret. The distance between secrets replaces the notion of adjacency between databases. We can control the noise by the distance metric to make similar (in visualization) secrets indistinguishable while keeping very different secrets that remain distinguishable. Thus, the recoverability is minimized while ensuring as much identifiability as possible. Therefore the choice of distance metric for secrets is critical. In our approach, each image is transformed to a BDCT representation with size $[H, W, C]$. Let $R_{i,j,k} = [r^{i,j,k}_{min}, r^{i,j,k}_{max}]$ be the sensitivity of the element in the $[i, j, k]$ position of representation. Then we define an element-wise distance as follows:

$$d_{i,j,k}(x_1, x_2) = \frac{|x_1 - x_2|}{r^{i,j,k}_{max} - r^{i,j,k}_{min}} \qquad \forall x_1, x_2 \in R_{i,j,k} \tag{2}$$

Moreover, the distance between the whole representations are defined as follow:

$$d(X_1, X_2) = \max_{i,j,k}(d_{i,j,k}(x_1, x_2))$$
$$\forall X_1, X_2 \in \mathbb{R}^{H,W,C} \tag{3}$$

Thus, the ϵ-DP protection can be guaranteed for mechanism $\mathcal{K}$ if for any representation X_1, X_2 and $E \in \mathbb{R}^{H,W,C}$. Equation 4 satisfied:

$$Pr(\mathcal{K}(X_1) = E) \leq e^{\epsilon d(X_1, X_2)} Pr(\mathcal{K}(X_2) = E)$$
$$\forall X_1, X_2, E \in \mathbb{R}^{H,W,C} \tag{4}$$

With the definition of distance between representations and DP for representations, we can claim

Lemma 1. *Any BDCT representation of image* $X \in \mathbb{R}^{H,W,C}$ *can be protected by* ϵ-*DP though the addition of a vector* $Y \in \mathbb{R}^{H,W,C}$ *where each* $Y_{i,j,k}$ *is an independent random variable following a Laplace distribution with a scaling parameter*

$$\sigma_{i,j,k} = \frac{r^{i,j,k}_{max} - r^{i,j,k}_{min}}{\epsilon_{i,j,k}}, \qquad where \sum_{i,j,k} \epsilon_{i,j,k} = \epsilon \tag{5}$$

With the setting in Lemma 1, Eq. 4 can be guaranteed. The proof has been attached in Appendix.

Differential Privacy Perturbation Module. The size of the output of the frequency domain transformation module is $[H, W, C]$. We need a noise matrix of the same size to mask the frequency features. Due to the properties of DCT, most energies are gathered in small areas. Thus, the frequency features have very different importance for face recognition. In order to protect face privacy with the best possible face recognition accuracy, we use a learnable privacy budgets allocation method in the module. It allows us to assign more privacy budgets to locations that are important for face recognition.

To achieve the idea of learnable privacy budgets, we utilize the setting in Lemma 1. We first initialize learnable privacy budget assignment parameters with the same size as the frequency features and put them into the softmax layer. Then, we multiply each element of the output of the softmax layer by ϵ. Because the sum of the output after the softmax layer is 1, no matter how the learnable budget assignment parameters are changed, the total privacy budget always equals ϵ. According to Lemma 1, if we add a vector $Y \in \mathbb{R}^{H,W,C}$ to the frequency features where each $Y_{i,j,k}$ is an independent random variable following a Laplace distribution with a scaling parameter $\sigma_{i,j,k} = \frac{r_{max}^{i,j,k} - r_{min}^{i,j,k}}{\epsilon_{i,j,k}}$, then we can ensure that this feature is protected by $\epsilon - DP$. To get the sensitivities of the frequency features of facial images, we transferred all the images in VGGFace2 [4] and refined MS1MV2 into the frequency domain. Then we obtained the maximum values and minimum values at each position. Sensitivities will equal the value of MAX - MIN. By now, we have prepared all the scaling parameters of Laplace noise. We sample the Laplace distribution according to the parameters and add them to the frequency features. The masked frequency features will be the output of the differential privacy perturbation module and transmitted to the face recognition model. The learnable budget assignment parameters are learned based on the loss function of the face recognition model and get the best allocation scheme that guarantees recognition accuracy. For face recognition module, we use ArcFace [8] as loss function and ResNet50 [14] as backbone.

4   Experiment

4.1   Datasets

We use VGGFace2 [4] that contains about 3.31M images of 9131 subjects for training. We extensively test our method on several popular face recognition benchmarks, including five small testing datasets and two general large-scale benchmarks. LFW [15] contains 13233 web-collected images from 5749 different identities. CFP-FP [30], CPLFW [38], CALFW [38] and AgeDB [26] utilize the similar evaluation metric of LFW to test face recognition with various challenges, such as cross pose, cross age.IJB-B [35] and IJB-C [23] are two general large-scale benchmarks. The IJB-B dataset contains $12,115$ templates with $10,270$ genuine matches and 8M impostor matches. The IJB-C dataset is a further extension of IJB-B, having $23,124$ templates with $19,557$ genuine matches and $15,639$K impostor matches.

4.2   Implementation Details

Each input face is resized to 112×112. After the frequency domain transformation, the input channel C is set to 189. We set the same random seed in all experiments. Unless otherwise stated, we use the following setting. We train the baseline model on ResNet50 [14] backbone. For our proposed model, we first convert the raw input RGB image to BDCT coefficients using some functions in

TorchJPEG [9]. Secondly, we process the BDCT coefficients using the proposed method. We calculate the sensitivity among the whole training dataset. The initial values of the learnable budget allocation parameters are set to be 0 so that the privacy budget of each pixel is equal in the initial stage. The whole model is trained from scratch using the SGD algorithm for 24 epochs, and the batch size is set to be 512. The learning rate of the learnable budget allocation parameters and the backbone parameters is 0.1. The momentum and the weight-decay are set to be 0.9 and 5e−4, respectively. We conducted all the experiments on 8 NVIDIA Tesla V100 GPU with the PyTorch framework. We divide the learning rate by 10 at 10, 18, 22 epochs. For ArcFace [8], we set $s = 64$ and $m = 0.4$. For CosFace [31], we set $s = 64$ and $m = 0.35$.

4.3   Comparisons with SOTA Methods

Settings for Other Methods. To evaluate the effectiveness of the model, we compare it with five baselines: **(1) ArcFace** [8]: The model is ResNet50 equipped with ArcFace, which is the simplest baseline with the original RGB image as input and introduces an additive angular margin inside the target cosine similarity. **(2) CosFace** [31]: The model is ResNet50 equipped with CosFace, which is another baseline with the original RGB image as input and subtracts a positive value from the target cosine similarity. **(3) PEEP** [6]: This method is the first to use DP in privacy-preserving face recognition. We reproduce it and run it on our benchmarks. Due to a large amount of training data, half of the data is selected for each ID to calculate the eigenface. The privacy budget ϵ is set to 5. **(4) Cloak** [25]: We run its official code. Note that when training large datasets, the privacy-accuracy parameter is adjusted according to our experiment setting, which is set to 100. **(5) InstaHide** [17]: This method incorporates mix-up to solve the privacy-preserving problem. We adapt it to face privacy-preserving problems. We set k to 2 and adopt an inside-dataset scheme that mixes each training image with random images within the same private training dataset.

Results on LFW, CFP-FP, CPLFW, CALFW, AgeDB. The results of the comparison with other SOTA methods can be seen in Table 1. For our method, the accuracy on all five datasets is close. Our method has a similar performance to the baselines on LFW and CALFW, with only an average drop of 0.14% and 0.65%. For CFP-FP, AgeDB, and CPLFW, our method has an average drop of 2.5% compared with the baseline. We believe this is because the images in these datasets have more complex poses and are therefore inherently less robust and more susceptible to interference from noise. However, our method still has a considerable lead in performance on these datasets compared to other SOTA privacy-preserving methods. In particular, on the CFP-FP dataset, other privacy-preserving methods have accuracy losses of more than 10%, but we still perform well.

Results on IJB-B and IJB-C. We also compare our method with baseline and other SOTA privacy-preserving methods over IJB-B and IJB-C. As shown

Table 1. Comparison of the face recognition accuracy among different privacy-preserving face recognition methods.

Method (%)	Privacy-Preserving	LFW	CFP-FP	AgeDB-30	CALFW	CPLFW	IJB-B(1e−4)	IJB-C(1e−4)
ArcFace (Baseline)	No	99.60	98.32	95.88	94.16	92.68	91.02	93.25
CosFace (Baseline)	No	99.63	98.52	95.83	93.96	93.30	90.79	93.14
PEEP	Yes	98.41	74.47	87.47	90.06	79.58	5.82	6.02
Cloak	Yes	98.91	87.97	92.60	92.18	83.43	33.58	33.82
InstaHide	Yes	96.53	83.20	79.58	86.24	81.03	61.88	69.02
Ours, Arcface ($\epsilon_{mean} = 0.5$)	Yes	**99.48**	**97.20**	**94.37**	**93.47**	90.6	89.33	**91.22**
Ours, Cosface ($\epsilon_{mean} = 0.5$)	Yes	99.47	97.16	94.13	93.36	**90.88**	**89.37**	91.21

in Table 1 and Fig. 1, our method has a very similar performance compared to the baseline. Under different false positive rates, the true positive rate of our method is still at a high level. However, the other SOTA methods do not perform well in this area. Their true-positive rate is much lower than the baseline with the lower false-positive rate.

4.4 Privacy Attack

White-Box Attacking Experiments. We assume that the attacker already knows all our operations in the white-box attack section. Therefore he will perform an inverse DCT (IDCT) operation on the transmitted data. Since our operation to remove DCs is practically irreversible, we assume he fills all DCs to 0. We set the privacy budget to 0.5 and 100, respectively, to better demonstrate the effect. However, in practice, we do not recommend setting the privacy budget as large as 100. Further, we denoise the images after IDCT. Here we use non-local means denoising [3] as the denoising method. As shown in Fig. 4, after losing the information of DC, it is difficult to show the original facial information in the recovered figure of the white box attack. Even if the privacy budget is as large as 100, we cannot obtain valid information about the user's facial features from the recovered image. Moreover, at this time, the denoising method also cannot work effectively because the IDCT image is full of noise.

To further demonstrate our approach's inability to provide valid information about the user's facial features, we assume that the attacker knows this data originates from a specific dataset. He makes certain guesses about the user to whom the data belongs and adds the DC of the guessed user to the data. The corresponding results are shown in Fig. 5. We assumed that the attacker guessed exactly the corresponding user in the above row. However, with the privacy budget set to 0.5, the recovery image still does not reveal any facial information. In the following row, we assume that the data belong to another user. However, the attacker still guesses that the data belongs to the same user as in the above

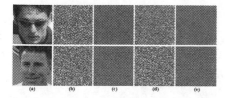

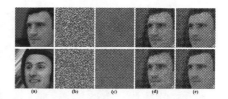

Fig. 4. White-box attack for our method. (a) Raw image. (b) IDCT with $\epsilon=0.5$. (c)Denoising image of (b). (d) IDCT with $\epsilon=100$. (e) Denoising image of (d).

Fig. 5. White-box attack for our method with guessed DC. For both line, we add the DC of the upper user. (a) Raw image. (b) IDCT with $\epsilon=0.5$. (c) Denoising image of (b). (d) IDCT with $\epsilon=100$. (e) Denoising image of (d).

row. In other words, the attacker guesses the wrong data source and adds the wrong DC to it. The recovery image does not give any information to the attacker about whether he guesses the right user to whom the data belongs. In this way, we can say that our method can protect the user's privacy well.

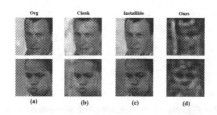

Fig. 6. Visualization of different privacy-preserving methods under black-box attack.

Metric	Method	LFW	CFP-FP	CPLFW	AgeDB	CALFW
	Cloak	20.292	20.069	19.705	19.946	20.327
PSNR(db)	InstaHide	23.519	22.807	22.546	23.271	23.010
	Ours	**14.281**	**13.145**	**13.515**	**13.330**	**13.330**
	Cloak	0.564	0.464	0.578	0.574	0.526
Similarity	InstaHide	0.729	0.649	0.732	0.737	0.693
	Ours	**0.214**	**0.175**	**0.264**	**0.250**	**0.202**

Fig. 7. PSNRs and similarities between the original images and the ones recovered from output of different privacy-preserving methods. The lower the value is, the better the privacy-preserving method is.

Black-Box Attacking Experiments In this section, we analyze the privacy-preserving reliability of the proposed method from the perspective of a black-box attack. A black box attack means that the attacker does not know the internal structure and parameters of the proposed model. However, attackers can collect large-scale face images from websites or other public face datasets. They can obtain the processed inputs by feeding those data to the model. Subsequently, they can train a decoder to map the processed inputs to the original face images. Finally, attackers can employ the trained decoder to recover the user's face image. Under these circumstances, we use UNet [28] as our decoder to reconstruct original images from processed images. In the training phase, we use an SGD optimizer with a learning rate of 0.1 with 10 epochs, and the batch size is set to 512. As shown in Fig. 6, we compare our method with other methods on reconstructed images. For Cloak, the face is still evident since the added

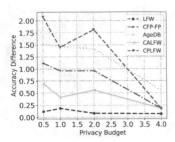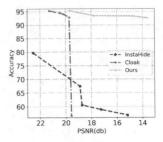

Fig. 8. Left: The recognition accuracy among different settings of privacy budget under different data set. The lower the value is, the less impact the method has on recognition accuracy. Right: Trade-off abilities of different approaches. The position in the upper right-hand corner indicates greater resilience to recovery while maintaining accuracy.

noise only affects the face's background. For InstaHide, most encrypted images can be recovered under our setting. For our method, the reconstructed images have been blurred. The facial structure of recovered images has been disrupted with an average privacy budget of 0.5.

In addition, Fig. 7 shows some quantitative results to illustrate the effectiveness of our method further. The reported results correspond to the results in Table 1. We first compare the PSNR between the original image and the one reconstructed. PSNR is often used to measure the reconstruction quality of lossy compression. It is also a good measure of image similarity. As we can find in the first row of Fig. 7, compared with other methods, our method has lower PSNR and higher recognition accuracy. Furthermore, we also evaluate the average Feature Similarity across five small data sets in order to show the privacy-preserving ability of the model at the feature level, as shown in the second row of Fig. 7. Precisely, we feed recovered RGB images to a pre-trained Arcface backbone. Then we can get new feature embedding to perform feature similarity calculation with origin embedding. The smaller the similarity, the more recovery-resistant the privacy-preserving method is at the feature level. As we can see, when the average privacy budget is chosen to be lower than 1, the similarities are much smaller than the other methods.

We further experimented with the trade-off ability of different approaches for privacy and accuracy, as shown in Fig. 8. Cloak does not protect the privacy of the face to a high degree, so the PSNR of the recovered image under blackbox attack and the original image can hardly be lower than 19. In contrast, InstaHide can protect face privacy quite well as the number of mixed images increases. However, the resulting loss of accuracy is very heavy. For our approach, depending on the choice of the average privacy budget, the degree of face privacy protection can be freely chosen, and the loss of accuracy can be controlled to a small extent. In summary, our method is more resistant to recovery than other methods, which means a more robust privacy-preserving ability.

4.5   Ability to Protect the Privacy of Training Data

Unlike other privacy-preserving methods, we can protect privacy during the inference and training stages. After the first training, we can get fixed privacy budget allocation parameters. We perform a DCT operation on the existing raw face recognition dataset and add noise according to the privacy budget allocation parameters. Thus, the original privacy leaked dataset is transformed into a privacy-protected dataset. Moreover, we experimentally demonstrate that the transformed dataset can still be used to train face recognition tasks with high accuracy. As shown in Table 2, the model trained using privacy-preserving datasets still has high accuracy. Notably, even though we trained fixed privacy assignment parameters using VGGFace2 and transformed MS1M using it, the transformed privacy-preserving MS1M retained high usability. It demonstrated the versatility and separability of the privacy budget allocation parameters. The corresponding results are shown in the fourth row.

Table 2. Comparison of recognition accuracy training with different datasets. **VGGFace2/MS1M**: Training with the original VGGFace2/MS1M. **Privacy-Preserved VGGFace2/MS1M**: Training with the dataset that transformed from the original VGGFace2/MS1M. They used the same fixed privacy budget allocation parameters learned from VGGFace2 and $\epsilon = 2$.

Method (%)	LFW	CFP-FP	AgeDB-30	CALFW	CPLFW
VGGFace2	99.60	98.32	95.88	94.16	92.68
MS1M	99.76	97.94	98.00	96.10	92.30
Privacy-Preserved VGGFace2	99.68	97.88	95.85	93.97	92.11
Privacy-Preserved MS1M	99.73	96.94	97.96	95.96	91.73

4.6   Ablation Study

Effects of Transformation to the Frequency Domain. To demonstrate the effects of transformation to the frequency domain, we directly input the raw RGB facial image to the differential privacy perturbation module. Here we choose $\epsilon_{mean} = 0.5$. As shown in Table 3, the model without transformation to frequency domain has a much lower accuracy even with the same setting for other parts.

Fig. 9. From left to right, the original image, the IDCT image without DC removal and the IDCT image with dc removal.

Effects of Removing DCs. To show the effects of removing DCs, we train models with and without removing DCs for comparison. We chose a larger privacy budget to make the difference between removing DC and not removing DC more visible. Here we set the budget equal to 20. By comparing the IDCT visualization of masked images without removing DCs shown in Fig. 9, we can see that the removal of DCs has a good effect on facial information protection.

Table 3. Comparison of the face recognition accuracy among methods with or without privacy protection. **BDCT-DC/BDCT-NoDC**: The baseline model trained on BDCT coefficients with/without DC. **RGB**: The model trained on the original image with learnable privacy budgets.

Method (%)	LFW	CFP-FP	AgeDB-30	CALFW	CPLFW
ArcFace (Baseline)	99.60	98.32	95.88	94.16	92.68
BDCT-DC	99.68	98.3	95.93	94.25	93.28
BDCT-NoDC	99.48	98.39	95.73	94.35	92.92
RGB ($\epsilon_{mean} = 0.5$)	84.78	63.87	73.66	76.85	63.08
Ours ($\epsilon_{mean} = 0.5$)	**99.48**	**97.20**	**94.37**	**93.47**	**90.6**

Effects of Using Learnable DP Budgets. To demonstrate the effects of using learnable DP budgets, we compare the accuracy of using learnable DP budgets and using the same DP budget for all elements. We set the privacy budget to be 0.5 at every element and test its accuracy using a pre-trained model with noiseless frequency domain features as input. The accuracy over LFW is only 65.966%, which is much lower than the one with learnable DP budgets, shown in Table 3.

Effects of Choosing Different DP Budgets. To show the effect of choosing different privacy budgets, we chose different privacy budgets and tested their loss of accuracy relative to the baseline. The results of the tests are presented in Fig. 8. As we have seen, the smaller the privacy budget is set, the higher the accuracy loss will be. This result is in line with the theory of differential privacy that a smaller privacy budget means more privacy-preserving and has a correspondingly higher accuracy loss. In our approach, the smaller privacy budget also provides stronger privacy protection, which is proven in Sect. 4.4.

5   Conclusions

In this paper, we propose a privacy-preserving approach for face recognition. It provides a privacy-accuracy trade-off capability. Our approach preserves image privacy by transforming the image to the frequency domain and adding random perturbations. It utilizes the concept of differential privacy to provide strong protection of face privacy. It regulates the ability of privacy protection by changing the average privacy budget. It is lightweight and easily compatible to be easily added to existing face recognition models. Moreover, it is shown experimentally that our method can fully protect face privacy under white-box attacks and maintain similar accuracy as the baseline. The masked images can defend against the black-box recovery attack of UNet. These show that our method performs far better than other SOTA face recognition protection methods.

References

1. Bai, F., Wu, J., Shen, P., Li, S., Zhou, S.: Federated face recognition. arXiv preprint arXiv:2105.02501 (2021)
2. Boemer, F., Cammarota, R., Demmler, D., Schneider, T., Yalame, H.: Mp2ml: a mixed-protocol machine learning framework for private inference. In: Proceedings of the 15th International Conference on Availability, Reliability and Security, pp. 1–10 (2020)
3. Buades, A., Coll, B., Morel, J.M.: A non-local algorithm for image denoising. In: 2005 IEEE Computer Society Conference on Computer Vision and Pattern Recognition (CVPR'05), vol. 2, pp. 60–65. IEEE (2005)
4. Cao, Q., Shen, L., Xie, W., Parkhi, O.M., Zisserman, A.: VGGFace2: a dataset for recognising faces across pose and age. In: 2018 13th IEEE International Conference on Automatic Face & Gesture Recognition (FG 2018), pp. 67–74. IEEE (2018)
5. Carlini, N., et al.: An attack on instahide: is private learning possible with instance encoding? arXiv preprint arXiv:2011.05315 (2020)
6. Chamikara, M.A.P., Bertók, P., Khalil, I., Liu, D., Camtepe, S.: Privacy preserving face recognition utilizing differential privacy. Comput. Secur. **97**, 101951 (2020)
7. Croft, W.L., Sack, J.R., Shi, W.: Obfuscation of images via differential privacy: from facial images to general images. Peer-to-Peer Netw. Appl. **14**(3), 1705–1733 (2021)
8. Deng, J., Guo, J., Xue, N., Zafeiriou, S.: ArcFace: additive angular margin loss for deep face recognition. In: Proceedings of the IEEE/CVF Conference on Computer Vision and Pattern Recognition, pp. 4690–4699 (2019)
9. Ehrlich, M., Davis, L., Lim, S.N., Shrivastava, A.: Quantization guided jpeg artifact correction. In: Proceedings of the European Conference on Computer Vision (2020)
10. Frank, J., Eisenhofer, T., Schönherr, L., Fischer, A., Kolossa, D., Holz, T.: Leveraging frequency analysis for deep fake image recognition. In: International Conference on Machine Learning, pp. 3247–3258. PMLR (2020)

11. Gentry, C., Halevi, S.: Implementing gentry's fully-homomorphic encryption scheme. In: Paterson, K.G. (ed.) EUROCRYPT 2011. LNCS, vol. 6632, pp. 129–148. Springer, Heidelberg (2011). https://doi.org/10.1007/978-3-642-20465-4_9

12. Gilad-Bachrach, R., Dowlin, N., Laine, K., Lauter, K., Naehrig, M., Wernsing, J.: CryptoNets: applying neural networks to encrypted data with high throughput and accuracy. In: International Conference on Machine Learning, pp. 201–210. PMLR (2016)

13. Gueguen, L., Sergeev, A., Kadlec, B., Liu, R., Yosinski, J.: Faster neural networks straight from jpeg. Adv. Neural. Inf. Process. Syst. **31**, 3933–3944 (2018)

14. He, K., Zhang, X., Ren, S., Sun, J.: Deep residual learning for image recognition. In: Proceedings of the IEEE Conference on Computer Vision and Pattern Recognition, pp. 770–778 (2016)

15. Huang, G.B., Mattar, M., Berg, T., Learned-Miller, E.: Labeled faces in the wild: a database forstudying face recognition in unconstrained environments. In: Workshop on Faces in 'Real-Life' Images: Detection, Alignment, and Recognition (2008)

16. Huang, J., Guan, D., Xiao, A., Lu, S.: FSDR: frequency space domain randomization for domain generalization. In: Proceedings of the IEEE/CVF Conference on Computer Vision and Pattern Recognition, pp. 6891–6902 (2021)

17. Huang, Y., Song, Z., Li, K., Arora, S.: InstaHide: instance-hiding schemes for private distributed learning. In: International Conference on Machine Learning, pp. 4507–4518. PMLR (2020)

18. Huang, Y., et al.: CurricularFace: adaptive curriculum learning loss for deep face recognition. In: Proceedings of the IEEE/CVF Conference on Computer Vision and Pattern Recognition, pp. 5901–5910 (2020)

19. Juvekar, C., Vaikuntanathan, V., Chandrakasan, A.: GAZELLE: a low latency framework for secure neural network inference. In: 27th USENIX Security Symposium (USENIX Security 2018), pp. 1651–1669. USENIX Association, Baltimore, August 2018. https://www.usenix.org/conference/usenixsecurity18/presentation/juvekar

20. Kagawade, V.C., Angadi, S.A.: Fusion of frequency domain features of face and iris traits for person identification. J. Inst. Eng. (India): Ser. B **102**, 987–996 (2021)

21. Liu, J., Juuti, M., Lu, Y., Asokan, N.: Oblivious neural network predictions via MiniONN transformations. In: Proceedings of the 2017 ACM SIGSAC Conference on Computer and Communications Security, pp. 619–631 (2017)

22. Liu, W., Wen, Y., Yu, Z., Li, M., Raj, B., Song, L.: SphereFace: deep hypersphere embedding for face recognition. In: Proceedings of the IEEE Conference on Computer Vision and Pattern Recognition, pp. 212–220 (2017)

23. Maze, B., et al.: IARPA Janus benchmark-C: face dataset and protocol. In: 2018 International Conference on Biometrics (ICB), pp. 158–165. IEEE (2018)

24. McSherry, F., Talwar, K.: Mechanism design via differential privacy. In: 48th Annual IEEE Symposium on Foundations of Computer Science (FOCS 2007), pp. 94–103. IEEE (2007)

25. Mireshghallah, F., Taram, M., Jalali, A., Elthakeb, A.T.T., Tullsen, D., Esmaeilzadeh, H.: Not all features are equal: Discovering essential features for preserving prediction privacy. In: Proceedings of the Web Conference 2021, pp. 669–680 (2021)

26. Moschoglou, S., Papaioannou, A., Sagonas, C., Deng, J., Kotsia, I., Zafeiriou, S.: AgeDB: the first manually collected, in-the-wild age database. In: Proceedings of the IEEE Conference on Computer Vision and Pattern Recognition Workshops, pp. 51–59 (2017)

27. Newton, E.M., Sweeney, L., Malin, B.: Preserving privacy by de-identifying face images. IEEE Trans. Knowl. Data Eng. **17**(2), 232–243 (2005)
28. Ronneberger, O., Fischer, P., Brox, T.: U-Net: convolutional networks for biomedical image segmentation. In: Navab, N., Hornegger, J., Wells, W.M., Frangi, A.F. (eds.) MICCAI 2015. LNCS, vol. 9351, pp. 234–241. Springer, Cham (2015). https://doi.org/10.1007/978-3-319-24574-4_28
29. Schroff, F., Kalenichenko, D., Philbin, J.: FaceNet: a unified embedding for face recognition and clustering. In: Proceedings of the IEEE Conference on Computer Vision and Pattern Recognition, pp. 815–823 (2015)
30. Sengupta, S., Chen, J.C., Castillo, C., Patel, V.M., Chellappa, R., Jacobs, D.W.: Frontal to profile face verification in the wild. In: 2016 IEEE Winter Conference on Applications of Computer Vision (WACV), pp. 1–9. IEEE (2016)
31. Wang, H., et al.: CosFace: large margin cosine loss for deep face recognition. In: Proceedings of the IEEE Conference on Computer Vision and Pattern Recognition, pp. 5265–5274 (2018)
32. Wang, H., Wu, X., Huang, Z., Xing, E.P.: High-frequency component helps explain the generalization of convolutional neural networks. In: 2020 IEEE/CVF Conference on Computer Vision and Pattern Recognition (CVPR), pp. 8681–8691. IEEE (2020)
33. Wang, X., Zhang, S., Wang, S., Fu, T., Shi, H., Mei, T.: Mis-classified vector guided softmax loss for face recognition. In: Proceedings of the AAAI Conference on Artificial Intelligence, vol. 34, pp. 12241–12248 (2020)
34. Wang, Y., Liu, J., Luo, M., Yang, L., Wang, L.: Privacy-preserving face recognition in the frequency domain (2022)
35. Whitelam, C., et al.: IARPA Janus benchmark-B face dataset. In: Proceedings of the IEEE Conference on Computer Vision and Pattern Recognition Workshops, pp. 90–98 (2017)
36. Xu, K., Qin, M., Sun, F., Wang, Y., Chen, Y.K., Ren, F.: Learning in the frequency domain. In: IEEE Conference on Computer Vision and Pattern Recognition (2020)
37. Zhang, H., Cisse, M., Dauphin, Y.N., Lopez-Paz, D.: mixup: beyond empirical risk minimization. In: International Conference on Learning Representations (2018)
38. Zheng, T., Deng, W., Hu, J.: Cross-age LFW: a database for studying cross-age face recognition in unconstrained environments. arXiv preprint arXiv:1708.08197 (2017)

Contrast-Phys: Unsupervised Video-Based Remote Physiological Measurement via Spatiotemporal Contrast

Zhaodong Sun[ID] and Xiaobai Li[(✉)][ID]

Center for Machine Vision and Signal Analysis, University of Oulu, Oulu, Finland
{zhaodong.sun,xiaobai.li}@oulu.fi

Abstract. Video-based remote physiological measurement utilizes face videos to measure the blood volume change signal, which is also called remote photoplethysmography (rPPG). Supervised methods for rPPG measurements achieve state-of-the-art performance. However, supervised rPPG methods require face videos and ground truth physiological signals for model training. In this paper, we propose an unsupervised rPPG measurement method that does not require ground truth signals for training. We use a 3DCNN model to generate multiple rPPG signals from each video in different spatiotemporal locations and train the model with a contrastive loss where rPPG signals from the same video are pulled together while those from different videos are pushed away. We test on five public datasets, including RGB videos and NIR videos. The results show that our method outperforms the previous unsupervised baseline and achieves accuracies very close to the current best supervised rPPG methods on all five datasets. Furthermore, we also demonstrate that our approach can run at a much faster speed and is more robust to noises than the previous unsupervised baseline. Our code is available at https://github.com/zhaodongsun/contrast-phys.

Keywords: Remote photoplethysmography · Face video · Unsupervised learning · Contrastive learning

1 Introduction

Traditional physiological measurement requires skin-contact sensors to measure physiological signals such as contact photoplethysmography (PPG) and electrocardiography (ECG). Some physiological parameters like heart rate (HR), respiration frequency (RF), and heart rate variability (HRV) can be derived from

Supplementary Information The online version contains supplementary material available at https://doi.org/10.1007/978-3-031-19775-8_29.

PPG signals for healthcare [42,55] and emotion analysis [29,40,57]. However, the skin-contact physiological measurement requires specific biomedical equipment like pulse oximeters, and contact sensors may cause discomfort and skin irritation. Remote physiological measurement uses a camera to record face videos for measuring remote photoplethysmography (rPPG). The weak color change in faces can be captured by cameras to obtain the rPPG signal from which several physiological parameters such as HR, RF, and HRV [38] can be measured. Video-based physiological measurement only requires off-the-shelf cameras rather than professional biomedical sensors, and is not constrained by physical distance, which has a great potential for remote healthcare [42,55] and emotion analysis applications [29,40,57].

In earlier rPPG studies [8,38,49,52], researchers proposed handcrafted features to extract rPPG signals. Later, some deep learning (DL)-based methods [7,18,23,25,32,33,35,45,56,58] were proposed, which employed supervised approaches with various network architectures for measuring rPPG signals. On one side, under certain circumstances, e.g., when head motions are involved or the videos are heterogeneous, DL-based methods could be more robust than the traditional handcrafted approaches. On the other side, DL-based rPPG methods require a large-scale dataset including face videos and ground truth physiological signals. Although face videos are comparatively easy to obtain in large amount, it is expensive to get the ground truth physiological signals which are measured by contact sensors, and synchronized with the face videos.

Can we only use face videos without ground truth physiological signals to train models for rPPG measurement? Gideon and Stent [10] proposed a self-supervised method to train rPPG measurement models without labels. They first downsample a video to get the downsampled rPPG (negative sample) and upsample the downsampled rPPG to get the reconstructed rPPG (positive sample). They also used the original video to get the anchor rPPG. Then a triplet loss is used to pull together the positive and anchor samples, and push away negative and anchor samples. However, their method has three problems. 1) They have to forward one video into their backbone model twice, which causes an extra computation burden. 2) There is still a significant gap between the performance of their unsupervised method and the state-of-the-art supervised rPPG methods [34,45,56]. 3) They showed in their paper [10] that their method is easily impacted by external periodic noise.

We propose a new unsupervised method (Contrast-Phys) to tackle the problems above. Our method was built on four observations about rPPG. 1) **rPPG spatial similarity**: rPPG signals measured from different facial areas have similar power spectrum densities (PSDs) 2) **rPPG temporal similarity**: In a short time, two rPPG signals (e.g., two consecutive 5s clips) usually present similar PSDs as the HR tends to take smooth transits in most cases. 3) **Cross-video rPPG dissimilarity**: The PSDs of rPPG signals from different videos are different. 4) **HR range constraint**: The HR should fall between 40 and 250 beats per minute (bpm), so we only care about the PSD in the frequency interval between 0.66 Hz and 4.16 Hz.

We propose to use a 3D convolutional neural network (3DCNN) to process an input video to get a spatiotemporal rPPG (ST-rPPG) block. The ST-rPPG block contains multiple rPPG signals along three dimensions of height, width, and time. According to the rPPG spatiotemporal similarity, we can randomly sample rPPG signals from the same video in different spatiotemporal locations and pull them together. According to the cross-video rPPG dissimilarity, the sampled rPPG signals from different videos are pushed away. The whole procedures are shown in Fig. 4 and Fig. 5.

The contributions of this work are 1) We propose a novel rPPG representation called spatiotemporal rPPG (ST-rPPG) block to obtain rPPG signals in spatiotemporal dimensions. 2) Based on four observations about rPPG, including rPPG spatiotemporal similarity and cross-video rPPG dissimilarity, we propose an unsupervised method based on contrastive learning. 3) We conduct experiments on five rPPG datasets (PURE [46], UBFC-rPPG [4], OBF [19], MR-NIRP [26], and MMSE-HR [59]) including RGB and NIR videos under various scenarios. Our method outperforms the previous unsupervised baseline [10] and achieves very close performance to supervised rPPG methods. Contrast-Phys also shows significant advantages with fast running speed and noise robustness compared to the previous unsupervised baseline.

2   Related Work

Video-Based Remote Physiological Measurement. Verkruysse et al. [49] first proposed that rPPG can be measured from face videos from the green channel. Several traditional handcraft methods [8,9,17,38,48,52,53] were proposed to further improve rPPG signal quality. Most rPPG methods proposed in earlier years used handcrafted procedures and did not need datasets for training, which are referred to as traditional methods. Deep learning (DL) methods for rPPG measurement are rapidly emerging. Several studies [7,23,35,45] used a 2D convolutional neural network (2DCNN) with two consecutive video frames as the input for rPPG measurement. Another type of DL-based methods [25,32,33] used a spatial-temporal signal map extracted from different facial areas as the input to feed into a 2DCNN model. Recently, 3DCNN-based methods [10,56,58] were proposed to achieved good performance on compressed videos [58]. The DL-based methods require both face videos and ground truth physiological signals, so we refer to them as supervised methods. Recently, Gideon and Stent [10] proposed an unsupervised method to train a DL model without ground truth physiological signals. However, their method falls behind some supervised methods and is not robust to external noise. In addition, the running speed is not satisfactory.

Contrastive Learning. Contrastive learning is a self-supervised learning method widely used in video and image feature embedding, which facilitates downstream task training and small dataset fine-tuning [6, 11–13, 30, 36, 39, 41, 47]. A DL model working as a feature extractor maps a high-dimensional image/video into a low dimensional feature vector. To train this DL feature extractor, features from different views of the same sample (positive pairs) are pulled together, while features from views of different samples (negative pairs) are pushed away. Data augmentations (such as cropping, blurring [6], and temporal sampling [39]) are used to obtain different views of the same sample so that the learned features are invariant to some augmentations. Previous works mentioned above use contrastive learning to let a DL model produce abstract features for downstream tasks such as image classification [6], video classification [39], face recognition [41]. On the other hand, our work uses contrastive learning to directly let a DL model produce rPPG signals, enabling unsupervised learning without ground truth physiological signals.

3   Observations About rPPG

This section describes four observations about rPPG, which are the precondition to design our method and enable unsupervised learning.

rPPG Spatial Similarity. rPPG signals from different facial areas have similar waveforms, and their PSDs are also similar. Several works [16,17,21,22,48,51,53] also exploited rPPG spatial similarity to design their methods. There might be small phase and amplitude differences between two rPPG signals from two different body skin areas [14,15]. However, when rPPG waveforms are transformed to PSDs, the phase information is erased, and the amplitude can be normalized to cancel the amplitude difference. In Fig. 1, the rPPG waveforms from four spatial areas are similar, and they have the same peaks in PSDs.

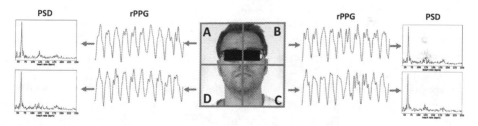

Fig. 1. Illustration of rPPG spatial similarity. The rPPG signals from four facial areas (A, B, C, D) have similar waveforms and power spectrum densities (PSDs).

rPPG Temporal Similarity. The HR does not change rapidly in a short term [10]. Stricker et al. [46] also found that the HR varies slightly in a short time interval in their dataset. Since the HR has a dominant peak in PSD, the PSD does not change rapidly, either. If we randomly sample several small windows from a short rPPG clip (e.g., 10 s), the PSDs of these windows should be similar. In Fig. 2, we sample two 5 s windows from a short 10 s rPPG signal and get the PSDs of these two windows. The two PSDs are similar and have sharp peaks at the same frequency. Since this observation is only valid under the condition of short-term rPPG signals, in the following sensitivity analysis part, we will discuss the influence of the signal length on our model performance. Overall, we can use the equation $\mathrm{PSD}\{G(v(t_1 \to t_1 + \Delta t, \mathcal{H}_1, \mathcal{W}_1))\} \approx \mathrm{PSD}\{G(v(t_2 \to t_2 + \Delta t, \mathcal{H}_2, \mathcal{W}_2))\}$ to describe spatiotemporal rPPG similarity. $v \in \mathbb{R}^{T \times H \times W \times 3}$ is a facial video, G is an rPPG measurement algorithm. We can choose one facial area with a set of height $\mathcal{H}_1$ and width $\mathcal{W}_1$, and a time interval $t_1 \to t_1 + \Delta t$ from video v to achieve one rPPG signal. We can achieve another rPPG signal similarly from the same video with $\mathcal{H}_2$, $\mathcal{W}_2$, and $t_2 \to t_2 + \Delta t$. $|t_1 - t_2|$ should be small to satisfy the condition of short-term rPPG signals.

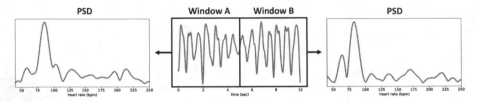

Fig. 2. Illustration of rPPG temporal similarity. The rPPG signals from two temporal windows (A, B) have similar PSDs.

Cross-video rPPG Dissimilarity. We assume rPPG signals from different face videos have different PSDs. Each video is recorded with different people and different physiological states (such as exercises and emotion status), so the HRs across different videos are likely different [1]. Even though the HRs might be similar between two videos, the PSDs might

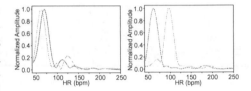

Fig. 3. The most similar (left) and most different (right) cross-video PSD pairs in the OBF dataset.

still be different since PSD also contains other physiological factors such as respiration rate [5] and HRV [37] which are unlikely to be all the same between two videos. To further validate the observation, we calculate the mean squared error for all cross-video PSD pairs in the OBF dataset [19] and show the most similar and most different cross-video PSD pairs in Fig. 3. It can be observed that the main cross-video PSD difference is the heart rate peak. The following equation describes cross-video rPPG dissimilarity. $\mathrm{PSD}\{G(v(t_1 \to t_1 + \Delta t, \mathcal{H}_1, \mathcal{W}_1))\} \neq \mathrm{PSD}\{G(v'(t_2 \to t_2 + \Delta t, \mathcal{H}_2, \mathcal{W}_2))\}$ where v and v' are two different videos. We can choose facial areas and time intervals from these two videos. The PSDs of the two rPPG signals should be different.

HR Range Constraint. The HR range for most people is between 40 and 250 bpm [2]. Most works [20,38] use this HR range for rPPG signal filtering and find the highest peak to estimate the HR. Therefore, our method will focus on PSD between 0.66 Hz and 4.16 Hz.

4   Method

The overview of Contrast-Phys is shown in Fig. 4 and Fig. 5. We describe the procedures of Contrast-Phys in this section.

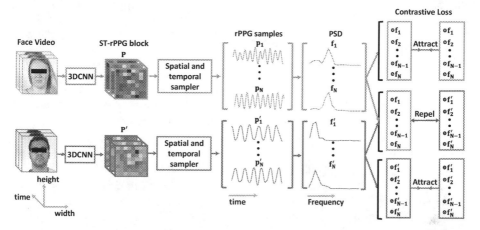

Fig. 4. Contrast-Phys Diagram. A pair of videos are fed into the same 3DCNN to generate a pair of ST-rPPG blocks. Multiple rPPG samples are sampled from the ST-rPPG blocks (The spatiotemporal sampler is illustrated in Fig. 5) and converted to PSDs. The PSDs from the same video are attracted while the PSDs from different videos are repelled.

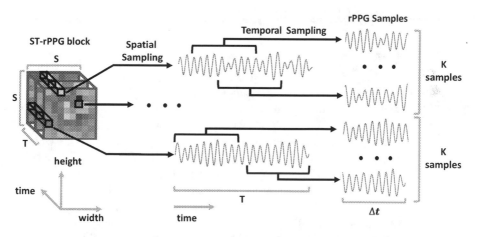

Fig. 5. Spatiotemporal sampler.

4.1   Preprocessing

The original videos are firstly preprocessed to crop the face to get the face video shown in Fig. 4 left. Facial landmarks are generated using OpenFace [3]. We first get the minimum and maximum horizontal and vertical coordinates of the landmarks to locate the central facial point for each frame. The bounding box size is 1.2 times the vertical coordinate range of landmarks from the first frame and is fixed for the following frames. After getting the central facial point of each frame and the size of the bounding box, we crop the face from each frame. The cropped faces are resized to 128×128, which are ready to be fed into our model.

4.2   Spatiotemporal rPPG (ST-rPPG) Block Representation

We modify 3DCNN-based PhysNet [56] to get the ST-rPPG block representation. The modified model has an input RGB video with the shape of $T \times 128 \times 128 \times 3$ where T is the number of frames. In the last stage of our model, we use adaptive average pooling to downsample along spatial dimensions, which can control the output spatial dimension length. This modification allows our model to output a spatiotemporal rPPG block with the shape of $T \times S \times S$ where S is spatial dimension length as shown in Fig. 5. More details about the 3DCNN model are described in the supplementary material.

The ST-rPPG block is a collection of rPPG signals in spatiotemporal dimensions. We use $P \in \mathbb{R}^{T \times S \times S}$ to denote the ST-rPPG block. Suppose we choose a spatial location (h, w) in the ST-rPPG block. In that case, the corresponding rPPG signal in this position is $P(\cdot, h, w)$ which is extracted from the receptive field of this spatial position in the original video. We can deduce that when the spatial dimension length S is small, each spatial position in the ST-rPPG block has a larger receptive field. The receptive field of each spatial position in the ST-rPPG block can cover part of the facial region, which means all spatial positions in the ST-rPPG block can include rPPG information.

4.3   rPPG Spatiotemporal Sampling

Several rPPG signals are sampled from the ST-rPPG block as illustrated in Fig. 5. For spatial sampling, we can get the rPPG signal $P(\cdot, h, w)$ at one spatial position. For temporal sampling, we can sample a short time interval from $P(\cdot, h, w)$, and the final spatiotemporal sample is $P(t \rightarrow t + \Delta t, h, w)$ where h and w are the spatial position, t is the starting time, and Δt is the time interval length. For one ST-rPPG block, we will loop over all spatial positions and sample K rPPG clips with a randomly chosen starting time t for each spatial position. Therefore, we can get $S \cdot S \cdot K$ rPPG clips from the ST-rPPG block. The more detailed sampling procedures are in our supplementary material. The sampling procedures above are used during model training. After our model is trained and used for testing, we can directly average ST-rPPG over spatial dimensions to get the rPPG signal.

4.4 Contrastive Loss Function

As illustrated in Fig. 4, we have two different videos randomly chosen from a dataset as the input. For one video, we can get one ST-rPPG block P, a set of rPPG samples $[p_1, \ldots, p_N]$ and the corresponding PSDs $[f_1, \ldots, f_N]$. For another video, we can get one ST-rPPG block P', one set of rPPG samples $[p'_1, \ldots, p'_N]$ and the corresponding PSDs $[f'_1, \ldots, f'_N]$ in the same way. As shown in Fig. 4 right, the principle of our contrastive loss is to pull together PSDs from the same video and push away PSDs from different videos. It is noted that we only use the PSD between 0.66 Hz and 4.16 Hz according to the HR range constraint in Sect. 3.

Positive Loss Term. According to rPPG spatiotemporal similarity, we can conclude that the PSDs from the spatiotemporal sampling of the same ST-rPPG block should be similar. We can use the following equations to describe this property for the two input videos. For one video, $\mathrm{PSD}\{P(t_1 \to t_1 + \Delta t, h_1, w_1)\} \approx \mathrm{PSD}\{P(t_2 \to t_2 + \Delta t, h_2, w_2)\} \implies f_i \approx f_j, i \neq j$. For another video, $\mathrm{PSD}\{P'(t_1 \to t_1 + \Delta t, h_1, w_1)\} \approx \mathrm{PSD}\{P'(t_2 \to t_2 + \Delta t, h_2, w_2)\} \implies f'_i \approx f'_j, i \neq j$.

We can use the mean squared error as the loss function to pull together PSDs (positive pairs) from the same video. The positive loss term L_p is shown below, which is normalized with the total number of positive pairs.

$$L_p = \sum_{\substack{i=1 \\ }}^{N} \sum_{\substack{j=1 \\ j \neq i}}^{N} \left(\| f_i - f_j \|^2 + \| f'_i - f'_j \|^2 \right) / (2N(N-1)) \tag{1}$$

Negative Loss Term. According to the cross-video rPPG dissimilarity, we can conclude that the PSDs from the spatiotemporal sampling of the two different ST-rPPG blocks should be different. We can use the following equation to describe this property for the two input videos. $\mathrm{PSD}\{P(t_1 \to t_1 + \Delta t, h_1, w_1)\} \neq \mathrm{PSD}\{P'(t_2 \to t_2 + \Delta t, h_2, w_2)\} \implies f_i \neq f'_j$

We use the negative mean squared error as the loss function to push away PSDs (negative pairs) from two different videos. The negative loss term L_n is shown below, which is normalized with the total number of negative pairs.

$$L_n = -\sum_{i=1}^{N} \sum_{j=1}^{N} \| f_i - f'_j \|^2 / N^2 \tag{2}$$

The overall loss function is $L = L_p + L_n$, which is the sum of the positive and negative loss terms.

Why Our Method Works. Our four rPPG observations are constraints to make the model learn to keep rPPG and exclude noises since noises do not satisfy the observations. Noises that appear in a small local region such as periodical eye blinking are excluded since the noises violate rPPG spatial similarity. Noises such as head motions/facial expressions that do not have a temporal constant frequency are excluded since they violate rPPG temporal similarity. Noises such as light flickering that exceed the heart rate range are also excluded due to the heart rate range constraint. Cross-video dissimilarity in the loss can make two videos' PSDs discriminative and show clear heart rate peaks since heart rate peaks are one of the discriminative clues between two videos' PSDs.

5   Experiments

5.1   Experimental Setup and Metrics

Datasets. We test five commonly used rPPG datasets covering RGB and NIR videos recorded under various scenarios. PURE [46], UBFC-rPPG [4], OBF [19] and MR-NIRP [26] are used for the intra-dataset testing. MMSE-HR [59] is used for cross-dataset testing. **PURE** has ten subjects' face videos recorded in six different setups, including steady and motion tasks. We use the same experimental protocol as in [25,45] to divide the training and test set. **UBFC-rPPG** includes facial videos from 42 subjects who were playing a mathematical game to increase their HRs. We use the same protocol as in [25] for the train-test split and evaluation. **OBF** has 100 healthy subjects' videos recorded before and after exercises. We use subject-independent ten-fold cross-validation as used in [33,56,58] to make fair comparison with previous results. **MR-NIRP** has NIR videos from eight subjects sitting still or doing motion tasks. The dataset is challenging due to its small scale, and weak rPPG signals in NIR [28,50]. We will use a leave-one-subject-out cross-validation protocol for our experiments. **MMSE-HR** has 102 videos from 40 subjects recorded in emotion elicitation experiments. This dataset is also challenging since spontaneous facial expressions, and head motions are involved. More details about these datasets can be found in the supplementary material.

Experimental Setup. This part shows our experimental setup during training and testing. For the spatiotemporal sampler, we evaluate different spatial resolutions and time lengths of ST-rPPG blocks in the sensitivity analysis part. According to the results, we fix the parameters in the other experiments as follows. We set $K = 4$, which means, for each spatial position in the ST-rPPG block, four rPPG samples are randomly chosen. We set the spatial resolution of the ST-rPPG block as 2×2, and the time length of the ST-rPPG block as 10 s. The time interval Δt of each rPPG sample is half of the time length of the ST-rPPG block. We use AdamW optimizer [24] to train our model with a learning rate of 10^{-5} for 30 epochs on one NVIDIA Tesla V100 GPU. For each training iteration, the inputs are two 10 s clips from two different videos, respectively. During testing, we broke each test video into non-overlapping 30 s clips and computed rPPG for each clip. We locate the highest peak in the PSD of an rPPG signal to calculate the HR. We use Neurokit2 [27] to calculate HRV metrics for the reported HRV results.

Evaluation Metrics. Following previous work [20,33,58], we use mean absolute error (MAE), root mean squared error (RMSE), and Pearson correlation coefficient (R) to evaluate the accuracy of HR measurement. Following [25], we also use standard deviation (STD), RMSE, and R to evaluate the accuracy of HRV features, including respiration frequency (RF), low-frequency power (LF) in normalized units (n.u.), high-frequency power (HF) in normalized units (n.u.), and the ratio of LF and HF power (LF/HF). For MAE, RMSE, and STD, smaller values mean lower errors, while for R, larger values close to one mean lower errors. Please check our supplementary material for more details about evaluation metrics.

5.2   Intra-dataset Testing

HR Estimation. We perform intra-dataset testing for HR estimation on PURE [46], UBFC-rPPG [4], OBF [19], and MR-NIRP [26]. Table 1 shows HR estimation results, including traditional methods, supervised methods, and unsupervised methods. The proposed Contrast-Phys largely surpasses the previous unsupervised baseline [10] and almost approaches the best supervised methods. The superior performance of Contrast-Phys is consistent on all four datasets, including the MR-NIRP [26] dataset, which contains NIR videos. The results indicate that the unsupervised Contrast-Phys can

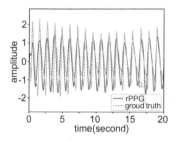

Fig. 6. rPPG waveform and ground truth PPG signal.

achieve reliable HR estimation on both RGB and NIR videos without requiring any ground truth physiological signals for training. Figure 6 also shows that the rPPG waveform from our unsupervised method is very similar to the ground truth PPG signal.

HRV Estimation. We also perform intra-dataset testing for HRV on UBFC-rPPG [4] as shown in Table 2. HRV results need to locate each systolic peak, which requires high-quality rPPG signals. Our method significantly outperforms traditional methods and the previous unsupervised baseline [10] on HRV results. There is a marginal difference in HRV results between ours and the supervised methods [25,33]. The results indicate that Contrast-Phys can achieve high-quality rPPG signals with accurate systolic peaks to calculate HRV features, which makes it feasible to be used for emotion understanding [29,40,57] and healthcare applications [42,55].

Table 1. Intra-dataset HR results. The best results are in bold, and the second-best results are underlined.

Method types	Methods	UBFC-rPPG MAE (bpm)	RMSE (bpm)	R	PURE MAE (bpm)	RMSE (bpm)	R	OBF MAE (bpm)	RMSE (bpm)	R	MR-NIRP (NIR) MAE (bpm)	RMSE (bpm)	R
Traditional	GREEN [49]	7.50	14.41	0.62	-	-	-	-	2.162	0.99	-	-	-
	ICA [38]	5.17	11.76	0.65	-	-	-	-	-	-	-	-	-
	CHROM [8]	2.37	4.91	0.89	2.07	9.92	**0.99**	-	2.733	0.98	-	-	-
	2SR [54]	-	-	-	2.44	3.06	<u>0.98</u>	-	-	-	-	-	-
	POS [52]	4.05	8.75	0.78	-	-	-	-	1.906	0.991	-	-	-
Supervised	CAN [7]	-	-	-	-	-	-	-	-	-	7.78	16.8	-0.03
	HR-CNN [45]	-	-	-	1.84	2.37	<u>0.98</u>	-	-	-	-	-	-
	SynRhythm [31]	5.59	6.82	0.72	-	-	-	-	-	-	-	-	-
	PhysNet [56]	-	-	-	2.1	2.6	**0.99**	-	1.812	0.992	3.07	7.55	<u>0.655</u>
	rPPGNet [58]	-	-	-	-	-	-	-	1.8	0.992	-	-	-
	CVD [33]	-	-	-	-	-	-	-	**1.26**	**0.996**	-	-	-
	PulseGAN [44]	1.19	2.10	0.98	-	-	-	-	-	-	-	-	-
	Dual-GAN [25]	**0.44**	**0.67**	**0.99**	**0.82**	1.31	**0.99**	-	-	-	-	-	-
	Nowara2021 [35]	-	-	-	-	-	-	-	-	-	**2.34**	**4.46**	**0.85**
Unsupervised	Gideon2021 [10]	1.85	4.28	0.93	2.3	2.9	0.99	2.83	7.88	0.825	4.75	9.14	0.61
	Ours	<u>0.64</u>	<u>1.00</u>	**0.99**	<u>1.00</u>	<u>1.40</u>	**0.99**	**0.51**	<u>1.39</u>	<u>0.994</u>	<u>2.68</u>	<u>4.77</u>	**0.85**

Table 2. HRV results on UBFC-rPPG. The best results are in bold, and the second-best results are underlined.

Method Types	Methods	LF (n.u.)			HF (n.u.)			LF/HF			RF(Hz)		
		STD	RMSE	R	STD	RMSE	R	STD	RMSE	R	STD	RMSE	R
Traditional	GREEN [49]	0.186	0.186	0.280	0.186	0.186	0.280	0.361	0.365	0.492	0.087	0.086	0.111
	ICA [38]	0.243	0.240	0.159	0.243	0.240	0.159	0.655	0.645	0.226	0.086	0.089	0.102
	POS [52]	0.171	0.169	0.479	0.171	0.169	0.479	0.405	0.399	0.518	0.109	0.107	0.087
Supervised	CVD [33]	0.053	_0.065_	0.740	0.053	_0.065_	0.740	_0.169_	_0.168_	_0.812_	_0.017_	_0.018_	0.252
	Dual-GAN [25]	**0.034**	**0.035**	**0.891**	**0.034**	**0.035**	**0.891**	**0.131**	**0.136**	**0.881**	**0.010**	**0.010**	**0.395**
Unsupervised	Gideon2021 [10]	0.091	0.139	0.694	0.091	0.139	0.694	0.525	0.691	0.684	0.061	0.098	0.103
	Ours	_0.050_	0.098	_0.798_	_0.050_	0.098	_0.798_	0.205	0.395	0.782	0.055	0.083	_0.347_

5.3   Cross-dataset Testing

We conduct cross-dataset testing on MMSE-HR [59] to test the generalization ability of our method. We train Contrast-Phys and Gideon2021 on UBFC-rPPG and test on MMSE-HR. We also show MMSE-HR cross-dataset results reported in the papers of some supervised methods [23,32,33,35,56] with different training sets. Table 3 shows the cross-dataset test results. Our method still outperforms Gideon2021 [10] and is close to supervised methods. The cross-dataset testing results demonstrate that Contrast-Phys can be trained on one dataset without ground truth physiological signals, and then generalize well to a new dataset.

5.4   Running Speed

We test the running speed of the proposed Contrast-Phys and compare it with Gideon2021 [10]. During training, the speed of our method is **802.45** frames per second (fps), while that of Gideon2021 is **387.87** fps, which is about half of our method's speed. The large difference is due to different method designs. For Gideon2021, one input video has to be fed into the model twice, i.e., firstly as the

Table 3. Cross-dataset HR estimation on MMSE-HR. The best results are in bold, and the second-best results are underlined.

Method types	Methods	MAE (bpm)	RMSE (bpm)	R
Traditional	Li2014 [20]	-	19.95	0.38
	CHROM [8]	-	13.97	0.55
	SAMC [48]	-	11.37	0.71
Supervised	RhythmNet [32]	-	7.33	0.78
	PhysNet [56]	-	13.25	0.44
	CVD [33]	-	_6.04_	0.84
	TS-CAN [23]	3.41	7.82	0.84
	Nowara2021 [35]	**2.27**	**4.90**	**0.94**
Unsupervised	Gideon2021 [10]	4.10	11.55	0.70
	Ours	_2.43_	7.34	_0.86_

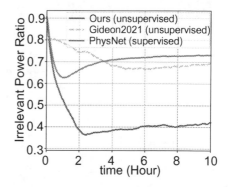

Fig. 7. Irrelevant power ratio change during training time for unsupervised and supervised methods.

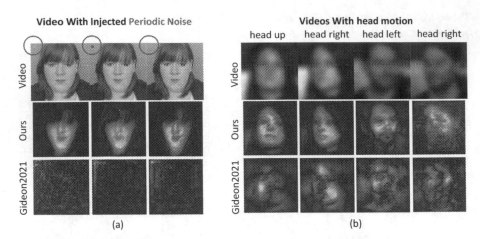

Fig. 8. Saliency maps for our method and Gideon2021 [10] (a) We add a random flashing block with the HR range between 40–250 bpm in the top left corner to all UBFC-rPPG videos. The models for our method and Gideon2021 [10] are trained on these videos with the noise. Our saliency maps have a high response on facial regions, while Gideon2021 focuses on this injected periodic noise. Table 4 also shows that our method is robust to the noise. (b) We choose the head motion moments in PURE dataset (Video frames shown here are blurred due to privacy issues.) and show the saliency maps for our method and Gideon2021 [10].

original video and then as a temporal resampled video for a second time, which causes double computation. For our method, one video is fed into the model once, which can substantially decrease computational cost compared with Gideon2021.

Furthermore, we compare the convergence speed of the two unsupervised methods and one supervised method (PhysNet [56]) using the metric of irrelevant power ratio (IPR). IPR is used in [10] to evaluate the signal quality during training, and lower IPR means higher signal quality. (more details about IPR are in the supplementary materials.) Figure 7 shows IPR with respect to time during training on the OBF dataset. Our method converges to the lowest IPR point for about 2.5 h, while Gideon2021 [10] converges to the lowest point for about 5 h. In addition, the lowest IPR for our method is about 0.36 while that for Gideon2021 [10] is about 0.66 which is higher than ours. The above evidence demonstrates that our method converges faster to a lower IPR than Gideon2021 [10]. In addition, our method also achieves lower IPR than the supervised method (PhysNet [56]).

5.5 Saliency Maps

We calculate saliency maps to illustrate the interpretability of our method. The saliency maps are obtained using a gradient-based method proposed in [43]. We fix the weights of the trained model and get the gradient of Pearson correlation with respect to the input video (More details are in the supplementary materials.). Saliency maps can highlight spatial regions from which the model estimates the rPPG signals, so the saliency map of a good rPPG model should have a large response on skin regions, as demonstrated in [7,10,35,56,58].

Figure 8 shows saliency maps under two cases, 1) periodic noise is manually injected, and 2) head motion is involved. When a periodic noise patch is injected to the left-upper corner of videos, our method is not distracted by the noise and still focuses on skin areas, while Gideon2021 is completely distracted by the noise block. We also calculate the two methods' performance on UBFC-rPPG videos with the injected noise, and the results are listed in Table 4. The results are consistent with the saliency map analysis, that the periodic noise does not impact Contrast-Phys, but fails Gideon2021 completely. Our method is robust to the noise because the noise only exists in one region, which violates rPPG spatial similarity. Figure 8(b) shows the saliency maps when head motion is involved. The saliency maps for our method focus and activate most skin areas, while saliency maps for Gideon2021 [10] show messy patterns and only partially cover facial areas during head motions.

Table 4. HR results trained on UBFC-rPPG with/without injected periodic noise shown in Fig. 8(a).

Methods	Injected Periodic Noise	MAE (bpm)	RMSE (bpm)	R
Gideon2021 [10]	w/o	1.85	4.28	0.939
	w/	22.47	25.41	0.244
Ours	w/o	0.64	1.00	0.995
	w/	0.74	1.34	0.991

5.6 Sensitivity Analysis

We perform sensitivity analysis on two variables: 1) the spatial length S of the ST-rPPG block, and 2) the temporal length T of the ST-rPPG block.

Table 5(a) shows the HR results on UBFC-rPPG, when the ST-rPPG spatial resolution is set in four levels of 1×1, 2×2, 4×4 or 8×8. Note that 1×1 means that rPPG spatial similarity is not used. The results of 1×1 are worse than other results with spatial information, which means rPPG spatial similarity improves performance. In addition, 2×2 is enough to perform well since larger resolutions do not significantly improve the HR estimation. Although 8×8 or 4×4 provides more rPPG samples, it has a smaller receptive field and thus produces noisier rPPG samples than a 2×2 block.

Table 5(b) shows the HR results on UBFC-rPPG, when the ST-rPPG temporal length is set on three levels of 5 s, 10 s and 30 s. The results indicate that 10 s is the best choice. A shorter time length (5 s) causes coarse PSD estimation, while a long time length (30 s) causes a slight violation of short-term signal condition in rPPG temporal similarity. Therefore, in these two cases (5 s and 30 s), the performance is lower than that of 10 s.

Table 5. Sensitivity Analysis: (a) HR results on UBFC-rPPG with different ST-rPPG block spatial resolutions. (b) HR results on UBFC-rPPG with different ST-rPPG block time lengths (The best results are in bold.).

<table>
<tr><td colspan="4" align="center">(a)</td><td colspan="4" align="center">(b)</td></tr>
<tr><td>Spatial Resolution</td><td>MAE (bpm)</td><td>RMSE (bpm)</td><td>R</td><td rowspan="2">Time Length</td><td rowspan="2">MAE (bpm)</td><td rowspan="2">RMSE (bpm)</td><td rowspan="2">R</td></tr>
<tr><td>1 × 1</td><td>3.14</td><td>4.06</td><td>0.963</td></tr>
<tr><td>2 × 2</td><td>**0.64**</td><td>**1.00**</td><td>**0.995**</td><td>5s</td><td>0.68</td><td>1.36</td><td>0.990</td></tr>
<tr><td>4 × 4</td><td>0.55</td><td>1.06</td><td>0.994</td><td>10s</td><td>**0.64**</td><td>**1.00**</td><td>**0.995**</td></tr>
<tr><td>8 × 8</td><td>0.60</td><td>1.09</td><td>0.993</td><td>30s</td><td>1.97</td><td>3.58</td><td>0.942</td></tr>
</table>

6   Conclusion

We propose Contrast-Phys which can be trained without ground truth physiological signals and achieve accurate rPPG measurement. Our method is based on four observations about rPPG and utilizes spatiotemporal contrast to enable unsupervised learning. Contrast-Phys significantly outperforms the previous unsupervised baseline [10] and is on par with the state-of-the-art supervised rPPG methods. In the future work, we would like to combine the supervised and the proposed unsupervised rPPG methods to further improve performance.

Acknowledgment. The study was supported by Academy of Finland (Project 323287 and 345948) and the Finnish Work Environment Fund (Project 200414). The authors also acknowledge CSC-IT Center for Science, Finland, for providing computational resources.

References

1. All about heart rate (pulse), www.heart.org/en/health-topics/high-blood-pressure/the-facts-about-high-blood-pressure/all-about-heart-rate-pulse
2. Target heart rates chart. www.heart.org/en/healthy-living/fitness/fitness-basics/target-heart-rates
3. Baltrusaitis, T., Zadeh, A., Lim, Y.C., Morency, L.P.: Openface 2.0: facial behavior analysis toolkit. In: 2018 13th IEEE international Conference on Automatic Face & Gesture Recognition (FG 2018), pp. 59–66. IEEE (2018)
4. Bobbia, S., Macwan, R., Benezeth, Y., Mansouri, A., Dubois, J.: Unsupervised skin tissue segmentation for remote photoplethysmography. Pattern Recogn. Lett. **124**, 82–90 (2019)
5. Chen, M., Zhu, Q., Wu, M., Wang, Q.: Modulation model of the photoplethysmography signal for vital sign extraction. IEEE J. Biomed. Health Inform. **25**(4), 969–977 (2020)
6. Chen, T., Kornblith, S., Norouzi, M., Hinton, G.: A simple framework for contrastive learning of visual representations. In: International Conference on Machine Learning, pp. 1597–1607. PMLR (2020)

7. Chen, W., McDuff, D.: DeepPhys video-based physiological measurement using convolutional attention networks. In: Proceedings of the European Conference on Computer Vision (ECCV), pp. 349–365 (2018)

8. De Haan, G., Jeanne, V.: Robust pulse rate from chrominance-based RPPG. IEEE Trans. Biomed. Eng. **60**(10), 2878–2886 (2013)

9. De Haan, G., Van Leest, A.: Improved motion robustness of remote-PPG by using the blood volume pulse signature. Physiol. Meas. **35**(9), 1913 (2014)

10. Gideon, J., Stent, S.: The way to my heart is through contrastive learning: remote photoplethysmography from unlabelled video. In: Proceedings of the IEEE/CVF International Conference on Computer Vision, pp. 3995–4004 (2021)

11. Grill, J.B., et al.: Bootstrap your own latent-a new approach to self-supervised learning. Adv. Neural. Inf. Process. Syst. **33**, 21271–21284 (2020)

12. Hadsell, R., Chopra, S., LeCun, Y.: Dimensionality reduction by learning an invariant mapping. In: 2006 IEEE Computer Society Conference on Computer Vision and Pattern Recognition (CVPR 2006), vol. 2, pp. 1735–1742. IEEE (2006)

13. He, K., Fan, H., Wu, Y., Xie, S., Girshick, R.: Momentum contrast for unsupervised visual representation learning. In: Proceedings of the IEEE/CVF Conference on Computer Vision and Pattern Recognition, pp. 9729–9738 (2020)

14. Kamshilin, A.A., Miridonov, S., Teplov, V., Saarenheimo, R., Nippolainen, E.: Photoplethysmographic imaging of high spatial resolution. Biomed. Opt. Exp. **2**(4), 996–1006 (2011)

15. Kamshilin, A.A., Teplov, V., Nippolainen, E., Miridonov, S., Giniatullin, R.: Variability of microcirculation detected by blood pulsation imaging. PLoS ONE **8**(2), e57117 (2013)

16. Kumar, M., Veeraraghavan, A., Sabharwal, A.: DistancePPG: robust non-contact vital signs monitoring using a camera. Biomed. Opt. Express **6**(5), 1565–1588 (2015)

17. Lam, A., Kuno, Y.: Robust heart rate measurement from video using select random patches. In: Proceedings of the IEEE International Conference on Computer Vision, pp. 3640–3648 (2015)

18. Lee, E., Chen, E., Lee, C.-Y.: Meta-rPPG: remote heart rate estimation using a Transductive meta-learner. In: Vedaldi, A., Bischof, H., Brox, T., Frahm, J.-M. (eds.) ECCV 2020. LNCS, vol. 12372, pp. 392–409. Springer, Cham (2020). https://doi.org/10.1007/978-3-030-58583-9_24

19. Li, X., et al.: The OBF database: a large face video database for remote physiological signal measurement and atrial fibrillation detection. In: 2018 13th IEEE International Conference on Automatic Face & Gesture Recognition (FG 2018), pp. 242–249. IEEE (2018)

20. Li, X., Chen, J., Zhao, G., Pietikainen, M.: Remote heart rate measurement from face videos under realistic situations. In: Proceedings of the IEEE Conference on Computer Vision and Pattern Recognition, pp. 4264–4271 (2014)

21. Liu, S.Q., Lan, X., Yuen, P.C.: Remote photoplethysmography correspondence feature for 3D mask face presentation attack detection. In: Proceedings of the European Conference on Computer Vision (ECCV), pp. 558–573 (2018)

22. Liu, S., Yuen, P.C., Zhang, S., Zhao, G.: 3D mask face anti-spoofing with remote Photoplethysmography. In: Leibe, B., Matas, J., Sebe, N., Welling, M. (eds.) ECCV 2016. LNCS, vol. 9911, pp. 85–100. Springer, Cham (2016). https://doi.org/10.1007/978-3-319-46478-7_6

23. Liu, X., Fromm, J., Patel, S., McDuff, D.: Multi-task temporal shift attention networks for on-device contactless vitals measurement. In: Larochelle, H., Ranzato, M., Hadsell, R., Balcan, M.F., Lin, H. (eds.) Advances in Neural Information Processing Systems, vol. 33, pp. 19400–19411 (2020)

24. Loshchilov, I., Hutter, F.: Decoupled weight decay regularization. In: International Conference on Learning Representations (2019). https://openreview.net/forum?id=Bkg6RiCqY7

25. Lu, H., Han, H., Zhou, S.K.: Dual-GAN: joint BVP and noise modeling for remote physiological measurement. In: Proceedings of the IEEE/CVF Conference on Computer Vision and Pattern Recognition, pp. 12404–12413 (2021)

26. Magdalena Nowara, E., Marks, T.K., Mansour, H., Veeraraghavan, A.: SparsePPG: towards driver monitoring using camera-based vital signs estimation in near-infrared. In: Proceedings of the IEEE Conference on Computer Vision and Pattern Recognition Workshops, pp. 1272–1281 (2018)

27. Makowski, D., et al.: NeuroKit2: a python toolbox for neurophysiological signal processing. Behav. Res. Methods **53**(4), 1689–1696 (2021)

28. Martinez, L.F.C., Paez, G., Strojnik, M.: Optimal wavelength selection for non-contact reflection photoplethysmography. In: 22nd Congress of the International Commission for Optics: Light for the Development of the World, vol. 8011, pp. 801191. International Society for Optics and Photonics (2011)

29. McDuff, D., Gontarek, S., Picard, R.: Remote measurement of cognitive stress via heart rate variability. In: 2014 36th Annual International Conference of the IEEE Engineering in Medicine and Biology Society, pp. 2957–2960. IEEE (2014)

30. Misra, I., Maaten, L.v.d.: Self-supervised learning of pretext-invariant representations. In: Proceedings of the IEEE/CVF Conference on Computer Vision and Pattern Recognition, pp. 6707–6717 (2020)

31. Niu, X., Han, H., Shan, S., Chen, X.: SynRhythm: learning a deep heart rate estimator from general to specific. In: 2018 24th International Conference on Pattern Recognition (ICPR), pp. 3580–3585. IEEE (2018)

32. Niu, X., Shan, S., Han, H., Chen, X.: RhythmNet: end-to-end heart rate estimation from face via spatial-temporal representation. IEEE Trans. Image Process. **29**, 2409–2423 (2019)

33. Niu, X., Yu, Z., Han, H., Li, X., Shan, S., Zhao, G.: Video-based remote physiological measurement via cross-verified feature disentangling. In: Vedaldi, A., Bischof, H., Brox, T., Frahm, J.-M. (eds.) ECCV 2020. LNCS, vol. 12347, pp. 295–310. Springer, Cham (2020). https://doi.org/10.1007/978-3-030-58536-5_18

34. Nowara, E.M., Marks, T.K., Mansour, H., Veeraraghavan, A.: Near-infrared imaging photoplethysmography during driving. IEEE Trans. Intell. Transp. Syst. **23**, 3589–3600 (2020)

35. Nowara, E.M., McDuff, D., Veeraraghavan, A.: The benefit of distraction: denoising camera-based physiological measurements using inverse attention. In: Proceedings of the IEEE/CVF International Conference on Computer Vision, pp. 4955–4964 (2021)

36. Van den Oord, A., Li, Y., Vinyals, O.: Representation learning with contrastive predictive coding. arXiv e-prints pp. arXiv-1807 (2018)

37. Pai, A., Veeraraghavan, A., Sabharwal, A.: HRVCam: robust camera-based measurement of heart rate variability. J. Biomed. Opt. **26**(2), 022707 (2021)

38. Poh, M.Z., McDuff, D.J., Picard, R.W.: Advancements in noncontact, multiparameter physiological measurements using a webcam. IEEE Trans. Biomed. Eng. **58**(1), 7–11 (2010)

39. Qian, R., et al.: Spatiotemporal contrastive video representation learning. In: Proceedings of the IEEE/CVF Conference on Computer Vision and Pattern Recognition, pp. 6964–6974 (2021)
40. Sabour, R.M., Benezeth, Y., De Oliveira, P., Chappe, J., Yang, F.: UBFC-PHYS: a multimodal database for psychophysiological studies of social stress. IEEE Trans. Affect. Comput. (Early Access 2021)
41. Schroff, F., Kalenichenko, D., Philbin, J.: FaceNet: a unified embedding for face recognition and clustering. In: Proceedings of the IEEE Conference on Computer Vision and Pattern Recognition, pp. 815–823 (2015)
42. Shi, J., Alikhani, I., Li, X., Yu, Z., Seppänen, T., Zhao, G.: Atrial fibrillation detection from face videos by fusing subtle variations. IEEE Trans. Circuits Syst. Video Technol. **30**(8), 2781–2795 (2019)
43. Simonyan, K., Vedaldi, A., Zisserman, A.: Deep inside convolutional networks: visualising image classification models and saliency maps. arXiv preprint arXiv:1312.6034 (2013)
44. Song, R., Chen, H., Cheng, J., Li, C., Liu, Y., Chen, X.: PulseGAN: learning to generate realistic pulse waveforms in remote photoplethysmography. IEEE J. Biomed. Health Inform. **25**(5), 1373–1384 (2021)
45. Špetlík, R., Franc, V., Matas, J.: Visual heart rate estimation with convolutional neural network. In: Proceedings of the British machine vision conference, Newcastle, UK, pp. 3–6 (2018)
46. Stricker, R., Müller, S., Gross, H.M.: Non-contact video-based pulse rate measurement on a mobile service robot. In: The 23rd IEEE International Symposium on Robot and Human Interactive Communication, pp. 1056–1062. IEEE (2014)
47. Tian, Y., Krishnan, D., Isola, P.: Contrastive multiview coding. In: Vedaldi, A., Bischof, H., Brox, T., Frahm, J.-M. (eds.) ECCV 2020. LNCS, vol. 12356, pp. 776–794. Springer, Cham (2020). https://doi.org/10.1007/978-3-030-58621-8_45
48. Tulyakov, S., Alameda-Pineda, X., Ricci, E., Yin, L., Cohn, J.F., Sebe, N.: Self-adaptive matrix completion for heart rate estimation from face videos under realistic conditions. In: Proceedings of the IEEE Conference on Computer Vision and Pattern Recognition, pp. 2396–2404 (2016)
49. Verkruysse, W., Svaasand, L.O., Nelson, J.S.: Remote plethysmographic imaging using ambient light. Opt. Express **16**(26), 21434–21445 (2008)
50. Vizbara, V.: Comparison of green, blue and infrared light in wrist and forehead photoplethysmography. Biomed. Eng. **17**(1) (2013)
51. Wang, W., den Brinker, A.C., De Haan, G.: Discriminative signatures for remote-PPG. IEEE Trans. Biomed. Eng. **67**(5), 1462–1473 (2019)
52. Wang, W., den Brinker, A.C., Stuijk, S., De Haan, G.: Algorithmic principles of remote PPG. IEEE Trans. Biomed. Eng. **64**(7), 1479–1491 (2016)
53. Wang, W., Stuijk, S., De Haan, G.: Exploiting spatial redundancy of image sensor for motion robust rPPG. IEEE Trans. Biomed. Eng. **62**(2), 415–425 (2014)
54. Wang, W., Stuijk, S., De Haan, G.: A novel algorithm for remote photoplethysmography: spatial subspace rotation. IEEE Trans. Biomed. Eng. **63**(9), 1974–1984 (2015)
55. Yan, B.P., et al.: Contact-free screening of atrial fibrillation by a smartphone using facial pulsatile photoplethysmographic signals. J. Am. Heart Assoc. **7**(8), e008585 (2018)
56. Yu, Z., Li, X., Zhao, G.: Remote photoplethysmograph signal measurement from facial videos using spatio-temporal networks. In: 30th British Machine Vision Conference 2019, p. 277. BMVA Press (2019)

57. Yu, Z., Li, X., Zhao, G.: Facial-video-based physiological signal measurement: recent advances and affective applications. IEEE Signal Process. Mag. **38**(6), 50–58 (2021)
58. Yu, Z., Peng, W., Li, X., Hong, X., Zhao, G.: Remote heart rate measurement from highly compressed facial videos: an end-to-end deep learning solution with video enhancement. In: Proceedings of the IEEE/CVF International Conference on Computer Vision, pp. 151–160 (2019)
59. Zhang, Z., et al.: Multimodal spontaneous emotion corpus for human behavior analysis. In: Proceedings of the IEEE conference on Computer Vision and Pattern Recognition, pp. 3438–3446 (2016)

Source-Free Domain Adaptation with Contrastive Domain Alignment and Self-supervised Exploration for Face Anti-spoofing

Yuchen Liu[1], Yabo Chen[2], Wenrui Dai[2(✉)], Mengran Gou[3], Chun-Ting Huang[3], and Hongkai Xiong[1]

[1] Department of Electronic Engineering, Shanghai Jiao Tong University, Shanghai, China
{liuyuchen6666,xionghongkai}@sjtu.edu.cn
[2] Department of Computer Science and Engineering, Shanghai Jiao Tong University, Shanghai, China
{chenyabo,daiwenrui}@sjtu.edu.cn
[3] Qualcomm AI Research, Shanghai, China
{mgou,chunting}@qti.qualcomm.com

Abstract. Despite promising success in intra-dataset tests, existing face anti-spoofing (FAS) methods suffer from poor generalization ability under domain shift. This problem can be solved by aligning source and target data. However, due to privacy and security concerns of human faces, source data are usually inaccessible during adaptation for practical deployment, where only a pre-trained source model and unlabeled target data are available. In this paper, we propose a novel Source-free Domain Adaptation framework for Face Anti-Spoofing, namely SDA-FAS, that addresses the problems of source knowledge adaptation and target data exploration under the source-free setting. For source knowledge adaptation, we present novel strategies to realize self-training and domain alignment. We develop a contrastive domain alignment module to align conditional distribution across different domains by aggregating the features of fake and real faces separately. We demonstrate in theory that the pre-trained source model is equivalent to the source data as source prototypes for supervised contrastive learning in domain alignment. The source-oriented regularization is also introduced into self-training to alleviate the self-biasing problem. For target data exploration, self-supervised learning is employed with specified patch shuffle data augmentation to explore intrinsic spoofing features for unseen attack types. To our best knowledge, SDA-FAS is the first attempt that jointly

Y. Liu, Y. Chen—Equal contribution. Qualcomm AI Research is an initiative of Qualcomm Technologies, Inc. Datasets were downloaded and evaluated by Shanghai Jiao Tong University researchers.

Supplementary Information The online version contains supplementary material available at https://doi.org/10.1007/978-3-031-19775-8_30.

optimizes the source-adapted knowledge and target self-supervised exploration for FAS. Extensive experiments on thirteen cross-dataset testing scenarios show that the proposed framework outperforms the state-of-the-art methods by a large margin.

Keywords: Face anti-spoofing · Source-free domain adaptation

1 Introduction

Face recognition (FR) systems are widely employed for human-computer interaction in our daily life. Face anti-spoofing (FAS) is crucial to protect FR systems from presentation attacks, e.g., print attack, video attack and 3D mask attack. Traditional FAS methods extract texture patterns with hand-crafted descriptors [8, 15, 24].

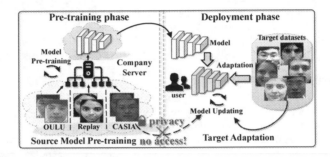

Fig. 1. A practical application scenario for face anti-spoofing. In the pre-training phase, the company builds a model based on the collected face data. When deployed on the user side, few collected unlabeled data can improve the performance through adaptation, but has distribution discrepancies with source knowledge. Moreover, due to privacy and security concerns of face data, users have no access to any source data of the company but the trained model.

With the rise of deep learning, convolutional neural networks (CNNs) have been adopted to extract deep semantic features [40, 45, 46]. Despite promising success in intra-dataset tests, these methods are dramatically degraded in cross-dataset tests where training data are from the source domain and test data are from the target domain with different distributions. The distribution discrepancies in illumination, background and resolution undermine the performance and an adaptation process is required to mitigate domain shift.

Domain adaptation (DA) based methods leverage maximum mean discrepancy (MMD) loss [16, 32] and adversarial training [12, 35, 36] to align the source and target domains, which need to access source data. Unfortunately, they might be infeasible for sensitive facial images due to the restriction by institutional policies, legal issues and privacy concerns. For example, according to the General Data Protection Regulation (GDPR) [30], institutions in the European Union

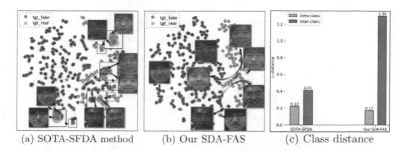

| (a) SOTA-SFDA method | (b) Our SDA-FAS | (c) Class distance |

Fig. 2. The t-SNE visualization of extracted features and corresponding faces under O & M & I→C. Same border color for faces with the same identity. **(a)** SOTA-SFDA method SHOT [17] achieves marginal distribution alignment, which is prone to map the features of real and fake faces together. **(b)** Our method with conditional distribution alignment separates them well and increases the discrimination ability. **(c)** Intra-class and inter-class distance of extracted features for SHOT and our SDA-FAS.

are regulated to protect the privacy of their data. Figure 1 illustrates a practical application scenario of *source-free domain adaptation* for FAS. A model is first pre-trained based on the (large-scale) source data and is released for deployment. In the deployment phase, the source data cannot be shared for adapting the pre-trained model to the target data, as they contain sensitive biometrics information. Besides, face images acquired under different illumination, background, resolution or using cameras with different parameters will lead to distribution discrepancies between source and target data. These distribution discrepancies have to be overcome using only the pre-trained source model and unlabeled target data. Domain generalization (DG) methods [11,26,27] learn a robust source model without exploiting the target data and achieve limited performance in practice. Consequently, Source-Free Domain Adaptation (SFDA) for face anti-spoofing is an important yet challenging problem remained to be solved.

Recently, SFDA has been considered to tackle a similar issue on image classification [1,17,41,42]. In image classification, label consistency among data with high local affinity is encouraged [41,42] or marginal distribution of source and target domains is implicitly aligned [1,17] to harmonize the clustered features in the feature space. Different from image classification, in FAS, fake faces of the same identity have similar facial features, whereas real faces of different identities differ. The intra-class distance between real faces of different identities probably exceeds the inter-class distance between real and fake faces of the same identity [11,27]. Clusters of features do not exist in FAS and SFDA models for image classification inevitably lead to degraded performance. Table 1 and Fig. 2 provide empirical results as supporting evidence, where SHOT [17], the state-of-the-art SFDA method, tends to cluster the features of real and fake faces together and obscures the discrimination ability. These problems urge a SFDA method designed specifically for FAS to achieve promising performance.

Lv et al. [20] accommodate to source-free setting for FAS by directly applying self-training but lack specific design for sufficiently exploring FAS tasks. The performance gain by adaptation is trivial (i.e., 1.9% HTER reduction on average), as shown in Table 1. To summarize, challenges to *source-free domain adaptation* for FAS include source knowledge adaptation and target data exploration.

- **Source knowledge adaptation.** Existing self-training and marginal domain alignment cannot adapt source knowledge well in FAS, especially when source data are unavailable. The target pseudo labels generated by the source model are noisy, especially under domain shift, leading to the accumulated error of self-training. Marginal distribution alignment is prone to cluster the features of real and fake faces and greatly degrades the discrimination ability for FAS.
- **Target data exploration.** Unseen attack types in the target data lead to enormous domain discrepancies where source knowledge is inapplicable and biased. It is indispensable to explore target data by itself to boost generalization ability. However, target data exploration is ignored in existing methods.

To address these issues, we propose a novel Source-free Domain Adaptation for Face Anti-Spoofing, namely SDA-FAS. Regarding source knowledge adaptation, we design novel strategies for self-training and domain alignment. We develop a contrastive domain alignment module for mitigating feature distribution discrepancies under a source-free setting. The pre-trained classifier weight is employed as the source prototypes with a theoretical guarantee of equivalence in training. We also introduce the source-oriented regularization into self-training to alleviate a self-biasing problem. For target data exploration, self-supervised learning is implemented with specified patch shuffle data augmentation to mine the intrinsic spoofing features of the target data, which also mitigates the reliance on pseudo-labels and boosts the tolerance to interfering knowledge transferred from the source domain. Contributions of this paper are summarized as below:

- We propose a novel contrastive domain alignment module to align the features of target data with the source prototypes of the same category for mitigating distribution discrepancies with theoretical support.
- We implement self-supervised learning with specified patch shuffle data augmentation to explore the target data for robust features in the case where unseen attack types emerge and source knowledge is unreliable.
- Our method is evaluated extensively on thirteen cross-dataset testing benchmarks and outperforms the state-of-the-art methods by a large margin.

To our best knowledge, SDA-FAS is the first attempt that unifies the transfer of pre-trained source knowledge and the self-exploration of unlabeled target data for FAS under a practical yet challenging source-free setting.

2   Related Work

Face Anti-spoofing. Existing face anti-spoofing (FAS) methods can be classified into three categories, i.e., handcrafted, deep learning, and DG/DA methods. Handcrafted methods extract the frame-level features using handcrafted descriptors such as LBP [8], HOG [15] and SIFT [24]. Deep learning methods boost the discrimination ability of extracted features. Yang et al. [40] first introduce CNNs into FAS, and Xu et al. [39] design a CNN-LSTM architecture to extract temporal features. Intrinsic spoofing patterns are further explored with pixel-wise supervision [44], e.g., depth maps [18], reflection maps [43] and binary masks [19]. These methods achieve remarkable performance in intra-dataset tests but degrade significantly in cross-dataset tests due to distribution discrepancies.

DG and DA have been leveraged to mitigate domain shift in cross-dataset tests. DG methods focus on extracting domain invariant features without target data. MADDG [27] learns a shared feature space with multi-adversarial learning. SSDG [11] develops a single-side DG framework by only aggregating real faces from different source domains. DA methods achieve the domain alignment using source data and unlabeled target data. Maximum mean discrepancy (MMD) loss [16,32] and adversarial training [12,35,36] are leveraged to align the feature space between the source and target domains. Quan et al. [25] present a transfer learning framework to progressively make use of unlabeled target data with reliable pseudo labels for training. However, these methods fail to work or suffer from poor performance in a practical yet challenging source-free setting, which considers the privacy and security issues of sensitive face images.

Source-Free Domain Adaptation. Domain adaptation aims at transferring knowledge from source domain to target domain. Recently, source-free domain adaptation (SFDA) has been considered to address privacy issues. PrDA [14] progressively updates the model in a self-learning manner with filtered pseudo labels. Based on the source hypothesis, SHOT [17] aligns the marginal distribution of source and target domains via information maximization. DECISION [1] further extends SHOT to a multi-source setting. TENT [34] adapts batch normalization's affine parameters with an entropy penalty. NRC [41] exploits the intrinsic cluster structure to encourage label consistency among data with high local affinity. However, existing works cannot be easily employed in FAS due to the different nature of tasks. Recently, Lv et al. [20] realize SFDA for FAS by directly using the pseudo labels for self-training, but suffer from trivial performance gain after adaptation due to the accumulated training error brought by noisy pseudo labels, especially under domain shift.

Contrastive Learning. Contrastive learning is popular for self-supervised representation learning. To obtain the best feature representations, the InfoNCE loss [22] is introduced to pull together an anchor and one positive sample (constructed by augmenting the anchor), and push apart the anchor from many negative samples. Besides, self-supervised features can be learned by only matching the similarity between the anchor and the positive sample [3,5]. Contrastive learning is also introduced into image classification in a supervised manner [13], where categorical labels are used to build positive and negative samples.

3 Proposed Method

3.1 Overview

We consider the practical source-free domain adaptation setting for face anti-spoofing, in which only a trained source model and unlabeled target domain

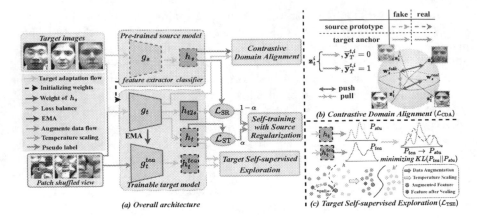

Fig. 3. (a) The overall architecture contains a pre-trained source model (in blue) and a trainable target model with three modules (in orange). For self-training with source regularization, pseudo labels $\overline{\mathbf{y}}_T^t$ and $\overline{\mathbf{y}}_T^s$ generated by target and source model supervise the outputs $\tilde{\mathbf{y}}_t$ and $\tilde{\mathbf{y}}_{t2s}$, respectively. (b) Contrastive domain alignment. The features of target data are pulled with the source prototypes of the same category (in green arrow) and pushed away from different categories (in red arrow) for conditional domain alignment. (c) Target self-supervised exploration. The original image and its patch shuffled view are sent to the student and teacher network. The output distributions are matched by minimizing the KL divergence, i.e., augmented features are pulled with features after scaling, which facilitates the learning of a compact feature space

data are available for adaptation. To recover the knowledge in the pre-trained source model, we leverage a self-training way to generate pseudo labels for target supervision. To alleviate the self-biasing problem caused by vanilla self-training, we introduce the source-oriented pseudo labels as regularization in Sect. 3.2. Considering that general SFDA methods align the marginal distribution, they could fail in adapting the source knowledge and mitigating domain shift in FAS where intra-class distances are prone to being larger than inter-class distances. Therefore, we propose a novel contrastive domain alignment module tailored for FAS that aligns target features to source prototypes for conditional distribution alignment with theoretical insights in Sect. 3.3. For unseen attack types not covered by the source knowledge, we introduce a target self-supervised exploration module with patch shuffle data augmentation to get rid of the facial structure and mine the intrinsic spoofing features in Sect. 3.4.

Figure 3 illustrates the overall architecture of our proposed framework that consists of a pre-trained source model and a trainable target model. The pre-trained source model consists of a feature extractor and a one-layer linear classifier, the parameters of which are fixed during adaptation. The feature extractor consists of a transformer encoder for feature encoding and a convolution layer for feature embedding. The target model consists of a student network and a teacher network. The student network consists of a feature extractor with multi-branch classifiers. The parameters of each target module are initialized by the parameters of the pre-trained source model.

3.2 Self-training with Source Regularization

Self-training Baseline (ST). Given the target domain data $\mathcal{D}_T = \{\mathbf{x}_T\}$ and the student network of the target model $f_t = h_t \circ g_t$ (initialized by f_s), the network output is $\tilde{\mathbf{y}}_t = h_t(g_t(\mathbf{x}_T))$ and the self-training loss is

$$\mathcal{L}_{\text{ST}} = \mathbb{1}\left(\max\left(\mathbf{c}_T^t\right) \geq \gamma\right) \mathcal{L}_{ce}(\tilde{\mathbf{y}}_t, \overline{\mathbf{y}}_T^t), \tag{1}$$

where $\mathbf{c}_T^t = \sigma(h_t(g_t(\mathbf{x}_T)))$ is the prediction confidence, $\overline{\mathbf{y}}_T^t = \text{argmax}(h_t(g_t(\mathbf{x}_T)))$ is the generated pseudo label, and $\mathbb{1} \in \{0,1\}$ is an indicator function that values 1 only when the input condition holds. γ is the confidence threshold to select out more reliable pseudo-labels.

Though self-training is effective in exploring unlabeled data [29], due to domain shift, it leads to the accumulated error and results in a self-biasing problem caused by noisy pseudo labels. As shown in Fig. 5, the accuracy of pseudo labels for ST gradually drops to about 50%, which is no better than a random guess for binary classification. Therefore, we introduce the regularization of source-oriented knowledge to alleviate the self-biasing problem.

Source-Oriented Regularization (SR). The target data $\mathcal{D}_T = \{\mathbf{x}_T\}$ are fed into the fixed pre-trained source model $f_s = h_s \circ g_s$ to obtain the source-oriented pseudo labels $\overline{\mathbf{y}}_T^s = \text{argmax}(h_s(g_s(\mathbf{x}_T)))$ and prediction confidence $\mathbf{c}_T^s = \sigma(h_s(g_s(\mathbf{x}_T)))$. The cross-entropy loss for SR compares the output $\tilde{\mathbf{y}}_{t2s} = h_{t2s}(g_t(\mathbf{x}_T))$ of h_{t2s} with $\overline{\mathbf{y}}_T^s$ as

$$\mathcal{L}_{\text{SR}} = \mathbb{1}\left(\max\left(\mathbf{c}_T^s\right) \geq \gamma\right) \mathcal{L}_{ce}(\tilde{\mathbf{y}}_{t2s}, \overline{\mathbf{y}}_T^s) \tag{2}$$

Then, ST and SR are dynamically adjusted during training. Due to domain shift, the target model produces many noisy pseudo labels in the early stage of training and generates more reliable pseudo labels as the training proceeds. Thus, we assign higher importance to SR at first and gradually increase the importance of ST. The overall loss is formulated as

$$\mathcal{L}_{\text{SSR}} = \alpha \cdot \mathcal{L}_{\text{ST}} + (1 - \alpha) \cdot \mathcal{L}_{\text{SR}}. \tag{3}$$

Here, the hyperparameter α gradually increases from 0 to 1.

3.3 Contrastive Domain Alignment

As discussed in Sect. 1, in real applications, faces are captured by various cameras under different environments, leading to distribution discrepancies in illumination, background and resolution. To mitigate the distribution discrepancies between source and target domains, DA methods employ MMD loss or adversarial learning, which requires full access to the source data. In the source-free setting, based on the source hypothesis, existing SFDA methods [1,17] align the marginal distributions of the source and target domains, i.e., $P(g_t(\mathbf{x}_S)) = P(g_t(\mathbf{x}_T))$.

However, such a marginal distribution alignment regardless of the categories suffers degraded performance in FAS. Since the intra-class distance tends to exceed the inter-class distance in FAS, features of different categories exhibit close proximity. For example, given a real subject, the corresponding fake faces with the same identity have similar facial features, while the real faces with different identities have different facial features. As shown in Fig. 2, such a marginal distribution alignment [17] may align the features of real faces with those of fake ones, which implies the different conditional distribution $P(g_t(\mathbf{x}_S)|\mathbf{y}_S) \neq P(g_t(\mathbf{x}_T)|\mathbf{y}_T)$ and affects the discrimination ability.

Thus, as shown in Fig. 3(b), we propose a contrastive domain alignment module to align the conditional distribution between the source and target domains. Due to the inaccessibility of source data, we propose to use the weights of pre-trained classifier h_s as the feature embeddings of the source prototype to compute the supervised contrastive loss.

Proposition 1. *Given a trained model $f_s = h_s \circ g_s$, where g_s is the feature extractor and h_s is the one-layer linear classifier, the ℓ_2-normalized weight vectors $\{\mathbf{w}_s^{real}, \mathbf{w}_s^{fake}\}$ of the classifier are the equivalent representation of the feature embeddings $\{\mathbf{z}_s^{real}, \mathbf{z}_s^{fake}\}$ of the source prototypes for calculating the supervised contrastive loss.*

Proof. Please refer to the supplementary material.

With the generated pseudo labels denoting the category of the feature embeddings of the target data anchor, we have the supervised contrastive loss as

$$\mathcal{L}_{\text{CDA}} = -\sum_{i=1}^{N_t} \sum_{m=1}^{M} \left[\mathbb{1}(\max(\mathbf{c}_T^{t,i}) \geq \gamma, \overline{\mathbf{y}}_T^{t,i} = m) \cdot \log \frac{\exp(\langle \mathbf{z}_t^i, \mathbf{w}_s^m \rangle / \tau)}{\sum_{j=1}^{M} \exp(\langle \mathbf{z}_t^i, \mathbf{w}_s^j \rangle / \tau)} \right], \quad (4)$$

where $\mathbf{c}_T^{t,i} = \sigma(h_t(g_t(\mathbf{x}_T^i)))$, $\overline{\mathbf{y}}_T^{t,i} = \operatorname{argmax}(h_t(g_t(\mathbf{x}_T^i)))$, $\mathbf{z}_t^i = g_t(\mathbf{x}_T^i)$, $\langle \cdot, \cdot \rangle$ denotes the inner product, τ is the temperature parameter, and M is the number of total categories. The contrastive domain alignment module has two properties: (1) pull together the feature embeddings of real (fake) faces in the target domain and those of the same category in the source domain to align the conditional distribution (green arrow in Fig. 3 (b)); (2) push apart the feature embeddings of real (fake) faces in the target domain from those of different categories in the source domain to enhance the discrimination ability (red arrow in Fig. 3 (b)).

3.4   Target Self-supervised Exploration

For FAS applications, novel fake faces are continuously evolved and it is likely to encounter diverse attack types or collecting ways unseen in the source data. For example, spoofing features of 2D attacks and 3D mask attacks are quite different. For the cases where distribution discrepancies are enormous and source knowledge fails to apply, the generalization ability will decrease. Thus, we introduce a target self-supervised exploration (TSE) module to mine the valuable information from the target domain. However, traditional data augmentation fails to fit with the spirit of FAS to capture detailed features. Taking the whole image as input will inevitably introduce global facial information. Thus, to suppress facial structure information as the biased source knowledge that leads to larger intra-class distances than inter-class distances, patch shuffle [47] is leveraged as a data augmentation strategy to destroy the face structure and learn a more compact feature space. Moreover, TSE is naturally independent of pseudo labels and can boost the tolerance to the wrongly transferred source supervision. The difference between our method and self-supervised methods [3,5] lies in the fact that we utilize the patch shuffle augmentation specifically for FAS and the target model is initialized by the pre-trained source model.

Specifically, a Siamese-like architecture is implemented to maximize the similarity of two views from one image [4], which consists of a student network (i.e., $g_t^{stu} \triangleq g_t, h_t^{stu} \triangleq h_t, f_t^{stu} \triangleq f_t$) and a teacher network $f_t^{tea} = g_t^{tea} \circ h_t^{tea}$. The student network is optimized by gradient descent, whereas the teacher network is updated with an exponential moving average (EMA). Given a target data $\mathbf{x}_T$, a patch-disordered view $\mathbf{x}_{T'}$ is obtained by splitting and splicing. We firstly divide the image into several patches and then randomly permute the image patches to form a new image as a jigsaw. The original view $\mathbf{x}_T$ and patch-permuted view $\mathbf{x}_T'$ are alternatively fed into the student and teacher networks to obtain two pairs of output probability distributions $\{P_{stu}, P'_{tea}\}$ and $\{P'_{stu}, P_{tea}\}$. Since the two views contain the same detailed real/fake features, the output should be consistent, which is matched by minimizing the Kullback-Leibler (KL) divergence.

$$\mathcal{L}_{\text{TSE}} = D_{KL}(P'_{tea}\|P_{stu}) + D_{KL}(P_{tea}\|P'_{stu}) \tag{5}$$

After updating θ_t with Eq. (5) by gradient descent, the parameters θ_t^{tea} of the teacher network are updated with an EMA as $\theta_t^{tea} \leftarrow l\theta_t^{tea} + (1-l)\theta_t$. l is the rate parameter.

The proposed framework for FAS is trained in an end-to-end manner as

$$\mathcal{L} = \mathcal{L}_{\text{SSR}} + \lambda_1 \cdot \mathcal{L}_{\text{CDA}} + \lambda_2 \cdot \mathcal{L}_{\text{TSE}}, \tag{6}$$

where λ_1 and λ_2 are hyper-parameters to balance the losses.

4   Experiments

4.1   Experimental Settings

Datasets. Evaluations are made on five public datasets: Idiap Replay-Attack [7] (denoted as I), OULU-NPU [2] (denoted as O), CASIA-MFSD [50] (denoted as C), MSU-MFSD [38] (denoted as M) and CelebA-Spoof [49] (denoted as CA). CA is significantly largest with huge diversity.

Testing Scenarios. Following [27], one dataset is treated as one domain. For simplicity, we use A & B→C for the scenario that trains on the source domains A and B, and tests on the target domain C. There are thirteen scenarios in total:

– **Multi-source Domains Cross-dataset Test:** O & C & I→M, O & M & I→C, O & C & M→I, and I & C & M→O.

Table 1. HTER and AUC for multi-source domains cross-dataset test. From top to bottom, compared methods are state-of-the-art deep learning FAS (DL-FAS), DG based FAS (DG-FAS), DA based FAS (DA-FAS), SFDA based FAS (SFDA-FAS) and state-of-the-art general SFDA methods (SOTA-SFDA). SourceOnly is our pre-trained source model and (best) is the target model after adaptation. Our average result is based on 3 independent runs with different seeds to report the mean value with standard deviation. Lv et al.(base) is the pre-trained source model and (SE) is the target model after adaptation. † indicates our reproduced results with the released code.

	Methods	O & C & I →M		O & M & I →C		O & C & M→I		I & C & M→O	
		HTER(%)↓	AUC(%)↑	HTER(%)↓	AUC(%)↑	HTER(%)↓	AUC(%)↑	HTER(%)↓	AUC(%)↑
DL-FAS	Binary CNN [40]	29.25	82.87	34.88	71.94	34.47	65.88	29.61	77.54
	Auxiliary [18]	22.72	85.88	33.52	73.15	29.14	71.69	30.17	77.61
DG-FAS	RFM [28]	17.30	90.48	13.89	93.98	20.27	88.16	16.45	91.16
	SSDG-R [11]	7.38	97.17	10.44	95.94	11.71	96.59	15.61	91.54
	D²AM [6]	15.43	91.22	12.70	95.66	20.98	85.58	15.27	90.87
DA-FAS	SDA [37]	15.4	91.8	24.5	84.4	15.6	90.1	23.1	84.3
	ADA [35]	16.9	-	24.2	-	23.1	-	25.6	-
	Wang et al. [36]	16.1	-	22.2	-	22.7	-	24.7	-
	Quan et al. [25]	7.82±1.21	97.67±1.09	4.01±0.81	98.96±0.77	10.36±1.86	97.16±1.04	14.23±0.98	93.66±0.75
SFDA-FAS	Lv et al. [20](base)	19.28	-	27.77	-	23.58	-	18.22	-
	Lv et al. [20](SE)	18.17	-	25.51	-	20.04	-	17.5	-
SOTA-SFDA	TENT† [34]	9.58	96.18	16.67	93.12	11.25	95.63	14.13	93.20
	SHOT† [17]	8.33	95.45	17.96	91.67	9.75	96.64	13.33	93.77
Ours	SourceOnly	12.50	93.71	20.00	90.53	16.25	90.99	17.26	91.80
	SDA-FAS (best)	**5.00**	**97.96**	**2.40**	**99.72**	**2.62**	**99.48**	**5.07**	**99.01**
	SDA-FAS (avg.)	**5.97±1.19**	97.38±0.54	**3.08±0.24**	99.54±0.19	3.54±0.46	99.11±0.41	6.52±1.26	98.37±0.25

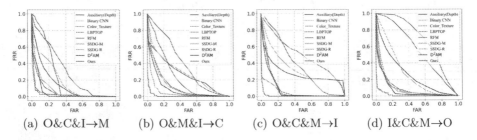

(a) O&C&I→M   (b) O&M&I→C   (c) O&C&M→I   (d) I&C&M→O

Fig. 4. ROC curves for multi-source domains cross-dataset test on O, C, I and M.

- **Limited Source Domains Cross-dataset Test:** M & I→C and M & I→O.
- **Cross-dataset Test on Large-scale CA:** M & C & O→CA.
- **Single Source Domain Cross-dataset Test:** C→I, C→M, I→C, I→M, M→C, and M→I.

Evaluation Metrics. Following [11,27], Half Total Error Rate (HTER) (half of the summation of false acceptance rate and false rejection rate) and the Area Under the Curve (AUC) are used as the evaluation metrics.

Implementation Details. Following [11], MTCNN [48] is adopted for face detection. The detected faces are normalized to $256 \times 256 \times 3$ as inputs. DeiT-S [31] pre-trained on ImageNet is used as the transformer encoder. For pre-training on the source data, we randomly specify a 0.9/0.1 train-validation split and get the optimal model based on the HTER of the validation split. For adaptation, the model is finetuned on the train set of target data and test on the test set, ensuring the test set is unseen in the whole procedure. The source code is released at https://github.com/YuchenLiu98/ECCV2022-SDA-FAS.

4.2   Experimental Results

Multi-source Domains Cross-dataset Test. Table 1 shows our SDA-FAS improves conventional deep learning FAS methods a lot by mitigating distribution discrepancies across different datasets. Besides, SDA-FAS performs better than DG based methods by exploiting unlabeled target data, as shown in Fig. 4. Moreover, SDA-FAS even outperforms the state-of-the-art DA method Quan et al. under a more challenging source-free setting, i.e., 7.71% HTER reduction and 4.71% AUC gain (lower HTER and higher AUC for better performance) for I & C & M→O that tests on the largest O dataset (among I, C, M and O). Furthermore, compared with SFDA based FAS method Lv et al. (SE), we greatly improve the performance, i.e., 3.77% vs. 20.30% HTER on average. Based on the pre-trained source model, our SDA-FAS achieves a large performance gain after adaptation with 12.7% HTER reduction on average, while Lv et al. only achieve 1.9%, validating the effectiveness of our adaptation framework. Finally, our SDA-

Table 2. HTER and AUC for test on O and C with limited source domain datasets.

Methods	M & I→C		M & I→O	
	HTER(%)	AUC(%)	HTER(%)	AUC(%)
LBPTOP [9]	45.27	54.88	47.26	50.21
SSDG-M [11]	31.89	71.29	36.01	66.88
RFM [28]	36.34	67.52	29.12	72.61
D^{2}AM [6]	32.65	72.04	27.70	75.36
SourceOnly	31.11	77.10	35.14	70.73
SDA-FAS	**15.37**	**91.35**	**22.53**	**83.54**

Table 3. HTER and AUC for test on large-scale CA.

Methods	M & C & O→CA	
	HTER(%)	AUC(%)
GRL Layer [10]	29.1	76.4
Domain-confusion [33]	33.7	70.3
Saha et al. [26]	27.1	79.2
Panwar et al. [23]	26.1	80.0
SourceOnly	29.7	77.5
SDA-FAS	**18.9**	**90.9**

Table 4. HTER(%) for single source domain cross-dataset test on C, I, and M datasets.

Methods	C→I	C→M	I→C	I→M	M→C	M→I	avg
Auxiliary [18]	27.6	-	28.4	-	-	-	-
Li et al. [16]	39.2	14.3	26.3	33.2	10.1	33.3	26.1
ADA [35]	17.5	9.3	41.6	30.5	17.7	5.1	20.3
Wang et al. [36]	15.6	**9.0**	34.2	29.0	16.8	**3.0**	17.9
USDAN-Un [12]	16.0	9.2	30.2	25.8	13.3	3.4	16.3
Lv et al. [20] (base)	21.1	-	34.4	-	-	-	-
Lv et al. [20] (SE)	18.9	-	30.1	-	-	-	-
SourceOnly	37.1	27.1	34.6	27.5	27.6	17.9	28.6
SDA-FAS	**11.5**	10.4	**19.6**	**24.1**	**10.0**	3.7	**13.2**

FAS outperforms the state-of-the-art general SFDA methods by proposing an adaptation framework specifically designed for FAS.

Limited Source Domains Cross-dataset Test. Compared with state-of-the-art DG method D^{2}AM, SDA-FAS improves the performance a lot by effectively using available unlabeled target data, i.e., 17.28% HTER reduction and 19.31% AUC gain for M & I→C, as shown in Table 2.

Cross-dataset Test on Large-Scale CA. For the most challenging test M & C & O→CA, where CA is much larger with unseen spoofing types (3D mask attacks), our SDA-FAS reduces HTER by 7.2% and increases AUC by 10.9% in comparison to the state-of-the-art DA method Panwar et al., as shown in Table 3. The promising results under a more practical source-free setting demonstrate that our method is effective and trustworthy for complex real-world scenarios.

Single Source Domain Cross-dataset Test. Table 4 shows that under a more difficult source-free setting, SDA-FAS outperforms all DA methods under four of the six tests and achieves the best average result (13.2% HTER). Besides, compared with the SFDA method Lv et al., SDA-FAS achieves a much larger performance gain after adaptation, 15.0% vs. 4.3% HTER reduction for I→C.

Table 5. Ablation studies on different components of our proposed SDA-FAS.

ST	SR	CDA	TSE	O & C & I→M		O & M & I→C		O & C & M→I		I & C & M→O	
				HTER (%)	AUC (%)	HTER (%)	AUC (%)	HTER (%)	AUC (%)	HTER (%)	AUC (%)
✓	✗	✗	✗	8.33	95.02	8.89	97.12	8.50	96.33	14.68	93.13
✓	✓	✗	✗	7.08	96.42	6.67	97.97	6.25	98.49	9.44	96.76
✓	✓	✓	✗	5.42	97.35	4.44	98.85	4.37	98.96	7.50	97.72
✓	✓	✓	✓	**5.00**	**97.96**	**2.40**	**99.72**	**2.62**	**99.48**	**5.07**	**99.01**

Table 6. HTER and AUC for unseen 3D mask attack type test on part of CA.

Methods	M & C & O→CA(3D mask)	
	HTER(%)	AUC(%)
Ours w/o TSE	20.52	89.91
Ours	**11.27**	**97.06**

Table 7. AUC(%) of the cross attack type test on C, I and M. Two attack types of unlabeled target data are used for training and tested on unseen attack type.

Methods	CASIA-MFSD (C)			Replay-Attack (I)			MSU (M)		
	Video	Cut	Warped	Video	Digital	Printed	Printed	HR	Mobile
DTN [19]	90.0	97.3	97.5	**99.9**	**99.9**	**99.6**	81.6	**99.9**	97.5
Ours	**98.3**	**97.7**	**97.6**	**99.9**	99.5	99.3	**86.3**	99.6	**97.8**

4.3   Ablation Studies

Each Component of the Network. The proposed framework and its variants are evaluated on multi-source domains cross-dataset test. Table 5 shows that, based on ST, SR improves the performance by introducing source-oriented regularization to alleviate the self-biasing problem. Besides, the performance improves with CDA added, demonstrating the effectiveness of conditional domain alignment to mitigate distribution discrepancies and enhance the discrimination ability. Moreover, TSE can further improve the performance, especially on the large test dataset (e.g., I & C&M→O), reflecting its power in self-exploring valuable information in large target data.

Portion of Target Data Used. Firstly, we randomly sample 10% and 50% of live and spoof faces in the training set for adaptation. Table 8 shows, even with 10% training samples, SDA-FAS improves the performance a lot, manifesting the validity for real scenarios with few data. For example, SDA-FAS reduces HTER by 9.44% after adaptation using only 24 unlabeled samples in C. Secondly, for extreme cases in FAS where live faces are much larger than spoof faces, we randomly sample 5%, 10% and 50% of spoof faces in the training set. With only 5% spoof faces (i.e., 9 samples), SDA-FAS reduces HTER by 8.71% after adaptation to C, demonstrating the effectiveness for more challenging scenarios.

Unseen Attack Types. To further evaluate TSE in self-exploring the target data, we reconstitute CA test set with all real faces and only 3D mask attack faces (unseen in the source data where only 2D attack types exist), and conduct experiments under M & C & O→CA (3D mask). As shown in Table 6, TSE significantly improves the performance, i.e., 9.25% HTER reduction and 7.15% AUC gain, demonstrating its effectiveness in self-exploring novel attack types in the case where the source knowledge fail to apply. Corresponding qualitative analysis is conducted in the supplementary material by visualizing a few hard 3D mask faces. Moreover, following protocols in [19], only partial attack types with unlabeled target data are tested. Table 7 shows our method outperforms DTN [19] that is fully supervised with labeled data. By adapting source knowledge, our method achieves better performance in an unsupervised manner.

Statistics of Pseudo Labels. As shown in Fig. 5, self-training (ST) results in a self-biasing problem and the accuracy of pseudo labels gradually drops to less than 50%. Self-training with source-oriented regularization (SSR) can alleviate the self-biasing problem, and the accuracy achieves a steady improvement to 70%. Moreover, with CDA mitigating domain discrepancies and TSE self-exploring target data, SDA-FAS achieves the highest accuracy exceeding 90%.

Table 8. Experiments on different portion of target train data and spoof faces. L denotes live faces and S denotes spoof faces, respectively.

Protocols	O & C & I→M		O & M & I→C		O & C & M→I		I & C & M→O	
	HTER(%)	AUC(%)	HTER(%)	AUC(%)	HTER(%)	AUC(%)	HTER(%)	AUC(%)
0% (L+S)	12.50	93.71	20.00	90.53	16.25	90.99	17.26	91.80
10% (L+S)	10.00	96.10	10.56	95.86	8.50	98.21	10.07	96.60
50% (L+S)	7.14	96.45	5.37	99.04	4.87	98.93	7.08	98.30
100%L+5%S	10.00	95.74	11.29	95.24	9.28	95.70	10.76	96.24
100%L+10%S	8.57	96.04	8.33	97.53	7.50	97.57	9.65	96.38
100%L+50%S	5.71	98.52	3.52	99.37	3.75	99.28	5.83	98.70

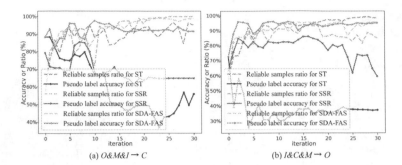

(a) $O\&M\&I \rightarrow C$ (b) $I\&C\&M \rightarrow O$

Fig. 5. Reliable samples ratio (dashed line) and pseudo labels accuracy (solid line) with respect to the updating iteration.

4.4   Visualizations

Attention Map. Figure 6 shows that, for real faces in rows 1 and 3, our method exhibits dense attention maps to effectively capture the physical structure of human faces. For the cut attack in row 2, the cut area of eyes is precisely specified, whereas the finger hint holding the paper is detected for the print attack in row 4. The attention maps suggest that SDA-FAS can model the features of live faces well and also precisely capture the intrinsic and detailed spoofing cues. Therefore, it can generalize well to the target domain.

Feature Space. We select all samples of target data for t-SNE visualizations. As shown in Fig. 7, after adaptation, the features of fake faces and real faces are better separated on the target domain compared to those before adaptation.

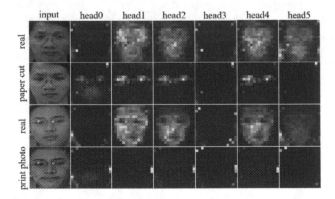

Fig. 6. Attention maps [3] from the last layer of the transformer encoder under O & M & I→C. Column 1: cropped input image. Columns 2–7: six heads of the transformer encoder. Rows 1–2: attention maps for subject 1's real face and paper-cut attack. Rows 3–4: attention maps for subject 2's real face and print photo attack.

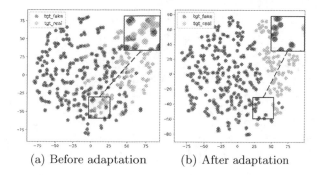

(a) Before adaptation   (b) After adaptation

Fig. 7. The t-SNE [21] visualization of the extracted features by our model with adaptation (right) and without adaptation (left) under O & M & I→C.

5 Conclusion

In this paper, we propose a novel adaptation framework for face anti-spoofing under a practical yet challenging source-free setting, which protects the security and privacy of human faces. Specifically, source-oriented regularization is introduced to alleviate the self-biasing problem of self-training. Besides, we propose a novel contrastive domain alignment module to align the conditional distribution across domains for mitigating the discrepancies. Moreover, self-supervised learning is adopted to self-explore the target data for robust features under enormous domain discrepancies where source knowledge is inapplicable. Extensive experiments validate the effectiveness of our method statistically and visually.

Acknowledgment. This work was supported in part by the National Natural Science Foundation of China under Grants 61932022, 61931023, 61971285, 62120106007, and in part by the Program of Shanghai Science and Technology Innovation Project under Grant 20511100100.

References

1. Ahmed, S.M., Raychaudhuri, D.S., Paul, S., Oymak, S., Roy-Chowdhury, A.K.: Unsupervised multi-source domain adaptation without access to source data. In: CVPR, pp. 10103–10112. IEEE (2021)
2. Boulkenafet, Z., Komulainen, J., Li, L., Feng, X., Hadid, A.: OULU-NPU: a mobile face presentation attack database with real-world variations. In: FG, pp. 612–618. IEEE (2017)
3. Caron, M., Touvron, H., Misra, I., Jégou, H., Mairal, J., Bojanowski, P., Joulin, A.: Emerging properties in self-supervised vision transformers. In: ICCV, pp. 9630–9640. IEEE (2021)
4. Chen, T., Kornblith, S., Norouzi, M., Hinton, G.: A simple framework for contrastive learning of visual representations. In: ICML, pp. 1597–1607. PMLR (2020)
5. Chen, X., He, K.: Exploring simple Siamese representation learning. In: CVPR, pp. 15750–15758. IEEE (2021)

6. Chen, Z., et al.: Generalizable representation learning for mixture domain face anti-spoofing. In: AAAI, pp. 1132–1139. AAAI Press (2021)
7. Chingovska, I., Anjos, A., Marcel, S.: On the effectiveness of local binary patterns in face anti-spoofing. In: BIOSIG, pp. 1–7 (2012)
8. de Freitas Pereira, T., Anjos, A., De Martino, J.M., Marcel, S.: *LBP*TOP based countermeasure against face spoofing attacks. In: Park, J.-I., Kim, J. (eds.) ACCV 2012. LNCS, vol. 7728, pp. 121–132. Springer, Heidelberg (2013). https://doi.org/10.1007/978-3-642-37410-4_11
9. Freitas Pereira, T., et al.: Face liveness detection using dynamic texture. EURASIP J. Image Video Process. **2014**(1), 1–15 (2014). https://doi.org/10.1186/1687-5281-2014-2
10. Ganin, Y., Lempitsky, V.: Unsupervised domain adaptation by backpropagation. In: ICML, pp. 1180–1189. PMLR (2015)
11. Jia, Y., Zhang, J., Shan, S., Chen, X.: Single-side domain generalization for face anti-spoofing. In: CVPR, pp. 8484–8493. IEEE (2020)
12. Jia, Y., Zhang, J., Shan, S., Chen, X.: Unified unsupervised and semi-supervised domain adaptation network for cross-scenario face anti-spoofing. Pattern Recogn. **115**, 107888 (2021)
13. Khosla, P., et al.: Supervised contrastive learning. In: NeurIPS, pp. 18661–18673. Curran Associates, Inc. (2020)
14. Kim, Y., Hong, S., Cho, D., Park, H., Panda, P.: Domain adaptation without source data. IEEE Trans. Artif. Intell. **2**(6), 508–518 (2020)
15. Komulainen, J., Hadid, A., Pietikäinen, M.: Context based face anti-spoofing. In: BTAS, IEEE (2013)
16. Li, H., Li, W., Cao, H., Wang, S., Huang, F., Kot, A.C.: Unsupervised domain adaptation for face anti-spoofing. IEEE Trans. Inf. Forensics Secur. **13**(7), 1794–1809 (2018)
17. Liang, J., Hu, D., Feng, J.: Do we really need to access the source data? source hypothesis transfer for unsupervised domain adaptation. In: ICML, pp. 6028–6039. PMLR (2020)
18. Liu, Y., Jourabloo, A., Liu, X.: Learning deep models for face anti-spoofing: Binary or auxiliary supervision. In: CVPR, pp. 389–398. IEEE (2018)
19. Liu, Y., Stehouwer, J., Jourabloo, A., Liu, X.: Deep tree learning for zero-shot face anti-spoofing. In: CVPR, pp. 4680–4689. IEEE (2019)
20. Lv, L., et al.: Combining dynamic image and prediction ensemble for cross-domain face anti-spoofing. In: ICASSP, pp. 2550–2554 (2021)
21. van der Maaten, L., Hinton, G.: Visualizing data using t-SNE. J. Mach. Learn. Res. **9**(86), 2579–2605 (2008)
22. van den Oord, A., Li, Y., Vinyals, O.: Representation learning with contrastive predictive coding. arXiv preprint arXiv:1807.03748 (2018)
23. Panwar, A., Singh, P., Saha, S., Paudel, D.P., Van Gool, L.: Unsupervised compound domain adaptation for face anti-spoofing. In: FG, IEEE (2021)
24. Patel, K., Han, H., Jain, A.K.: Secure face unlock: spoof detection on smartphones. IEEE Trans. Inf. Forensics Secur. **11**(10), 2268–2283 (2016)
25. Quan, R., Wu, Y., Yu, X., Yang, Y.: Progressive transfer learning for face anti-spoofing. IEEE Trans. Image Process. **30**(3), 3946–3955 (2021)
26. Saha, S., et al.: Domain agnostic feature learning for image and video based face anti-spoofing. In: CVPR Workshops, pp. 802–803. IEEE (2020)
27. Shao, R., Lan, X., Li, J., Yuen, P.C.: Multi-adversarial discriminative deep domain generalization for face presentation attack detection. In: CVPR, pp. 10023–10031. IEEE (2019)

28. Shao, R., Lan, X., Yuen, P.C.: Regularized fine-grained meta face anti-spoofing. In: AAAI, pp. 11974–11981. AAAI Press (2020)
29. Sohn, K., et al.: FixMatch: simplifying semi-supervised learning with consistency and confidence. In: NeurIPS, pp. 596–608. Curran Associates, Inc. (2020)
30. The European Parliament and The Council of the European Union: Regulation (EU) 2016/679 of the European Parliament and of the Council of 27 April 2016 on the protection of natural persons with regard to the processing of personal data and on the free movement of such data, and repealing Directive 95/46/EC (General Data Protection Regulation). Official Journal of European Union (OJ) 59(L119), pp. 1–88 (2016)
31. Touvron, H., Cord, M., Douze, M., Massa, F., Sablayrolles, A., Jégou, H.: Training data-efficient image transformers & distillation through attention. In: ICML, pp. 10347–10357. PMLR (2021)
32. Tu, X., Zhang, H., Xie, M., Luo, Y., Zhang, Y., Ma, Z.: Deep transfer across domains for face antispoofing. J. Electron. Imaging 28(4), 043001 (2019)
33. Tzeng, E., Hoffman, J., Saenko, K., Darrell, T.: Adversarial discriminative domain adaptation. In: CVPR, pp. 7167–7176. IEEE (2017)
34. Wang, D., Shelhamer, E., Liu, S., Olshausen, B., Darrell, T.: Tent: Fully test-time adaptation by entropy minimization. In: ICLR (2021)
35. Wang, G., Han, H., Shan, S., Chen, X.: Improving cross-database face presentation attack detection via adversarial domain adaptation. In: ICB, IEEE (2019)
36. Wang, G., Han, H., Shan, S., Chen, X.: Unsupervised adversarial domain adaptation for cross-domain face presentation attack detection. IEEE Trans. Inf. Forensics Secur. 16, 56–69 (2021)
37. Wang, J., Zhang, J., Bian, Y., Cai, Y., Wang, C., Pu, S.: Self-domain adaptation for face anti-spoofing. In: AAAI, pp. 2746–2754. AAAI Press (2021)
38. Wen, D., Han, H., Jain, A.K.: Face spoof detection with image distortion analysis. IEEE Trans. Inf. Forensics Secur. 10(4), 746–761 (2015)
39. Xu, Z., Li, S., Deng, W.: Learning temporal features using LSTM-CNN architecture for face anti-spoofing. In: ACPR, pp. 141–145. IEEE (2015)
40. Yang, J., Lei, Z., Li, S.Z.: Learn convolutional neural network for face anti-spoofing. arXiv preprint arXiv:1408.5601 (2014)
41. Yang, S., Wang, Y., van de Weijer, J., Herranz, L., Jui, S.: Exploiting the intrinsic neighborhood structure for source-free domain adaptation. In: NeurIPS, pp. 29393–29405. Curran Associates, Inc. (2021)
42. Yang, S., Wang, Y., van de Weijer, J., Herranz, L., Jui, S.: Generalized source-free domain adaptation. In: ICCV, pp. 8978–8987. IEEE (2021)
43. Yu, Z., Li, X., Niu, X., Shi, J., Zhao, G.: Face anti-spoofing with human material perception. In: Vedaldi, A., Bischof, H., Brox, T., Frahm, J.-M. (eds.) ECCV 2020. LNCS, vol. 12352, pp. 557–575. Springer, Cham (2020). https://doi.org/10.1007/978-3-030-58571-6_33
44. Yu, Z., Li, X., Shi, J., Xia, Z., Zhao, G.: Revisiting pixel-wise supervision for face anti-spoofing. IEEE Trans. Biomet. Behav. Ident. Sci. 3(3), 285–295 (2021)
45. Yu, Z., Wan, J., Qin, Y., Li, X., Li, S.Z., Zhao, G.: NAS-FAS: static-dynamic central difference network search for face anti-spoofing. IEEE Trans. Pattern Anal. Mach. Intell. 43(9), 3005–3023 (2021)
46. Yu, Z.,et al.: Searching central difference convolutional networks for face anti-spoofing. In: CVPR, pp. 5295–5305. IEEE (2020)
47. Zhang, K.Y., et al.: Structure destruction and content combination for face anti-spoofing. In: IJCB, IEEE (2021)

48. Zhang, K., Zhang, Z., Li, Z., Qiao, Y.: Joint face detection and alignment using multitask cascaded convolutional networks. IEEE Signal Process. Lett. **23**(10), 1499–1503 (2016)
49. Zhang, Y., et al.: CelebA-spoof: large-scale face anti-spoofing dataset with rich annotations. In: Vedaldi, A., Bischof, H., Brox, T., Frahm, J.-M. (eds.) ECCV 2020. LNCS, vol. 12357, pp. 70–85. Springer, Cham (2020). https://doi.org/10.1007/978-3-030-58610-2_5
50. Zhang, Z., Yan, J., Liu, S., Lei, Z., Yi, D., Li, S.Z.: A face antispoofing database with diverse attacks. In: ICB, pp. 26–31. IEEE (2012)

On Mitigating Hard Clusters for Face Clustering

Yingjie Chen[1,2] , Huasong Zhong[2] , Chong Chen[2(✉)] , Chen Shen[2] ,
Jianqiang Huang[2] , Tao Wang[1] , Yun Liang[1] , and Qianru Sun[3]

[1] Peking University, Beijing, China
{chenyingjie,wangtao,ericlyun}@pku.edu.cn
[2] DAMO Academy, Alibaba Group, Hangzhou, China
cheung.cc@alibaba-inc.com
[3] Singapore Management University, Singapore, Singapore
qianrusun@smu.edu.sg

Abstract. Face clustering is a promising way to scale up face recognition systems using large-scale unlabeled face images. It remains challenging to identify small or sparse face image clusters that we call hard clusters, which is caused by the heterogeneity, $i.e.$ et@tokeneonedot, high variations in size and sparsity, of the clusters. Consequently, the conventional way of using a uniform threshold (to identify clusters) often leads to a terrible misclassification for the samples that should belong to hard clusters. We tackle this problem by leveraging the neighborhood information of samples and inferring the cluster memberships (of samples) in a probabilistic way. We introduce two novel modules, Neighborhood-Diffusion-based Density (NDDe) and Transition-Probability-based Distance (TPDi), based on which we can simply apply the standard Density Peak Clustering algorithm with a uniform threshold. Our experiments on multiple benchmarks show that each module contributes to the final performance of our method, and by incorporating them into other advanced face clustering methods, these two modules can boost the performance of these methods to a new state-of-the-art. Code is available at: https://github.com/echoanran/On-Mitigating-Hard-Clusters.

Keywords: Face clustering · Diffusion density · Density peak clustering

1 Introduction

Face recognition is a classical computer vision task [13,21,34] that aims to infer person identities from face images. Scaling it up relies on more annotated data if using deeper models. Face clustering is a popular and efficient solution to reducing the annotation costs [5,15,16,27].

Y. Chen and H. Zhong—Equal contribution.

Supplementary Information The online version contains supplementary material available at https://doi.org/10.1007/978-3-031-19775-8_31.

S. Avidan et al. (Eds.): ECCV 2022, LNCS 13672, pp. 529–544, 2022.
https://doi.org/10.1007/978-3-031-19775-8_31

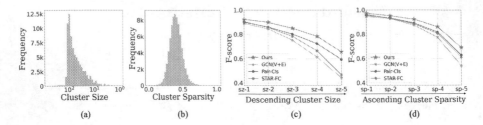

(a)   (b)   (c)   (d)

Fig. 1. (a) and (b) show the ground-truth distribution of the face identity clusters on MS1M 5.21M dataset [8]. (a) is for cluster size, $i.e.$let@tokeneonedot, the number of samples in a cluster. (b) is for cluster sparsity, which is defined as the average cosine distance of all pair of samples in a cluster, $e.g.$let@tokeneonedot, for cluster C we have Sparsity$(C) = 1 - \frac{\sum_{ij\in C, i\neq j} \text{cosine}<x_i,x_j>}{|C|(|C|-1)}$ where $|C|$ denotes the cluster size. (c) and (d) show the performances (Pairwise F-score) of three top-performing methods, GCN(V+E) [31], Pair-Cls [14], and STAR-FC [23], compared to ours, on five cluster subsets with descending size (from sz-1 to sz-5) and ascending sparsity (from sp-1 to sp-5), respectively.

Problems. Face clustering is challenging due to that 1) recognizing person identities is a fine-grained task; 2) the number of identities is always large, $e.g.$let@tokeneonedot, $77k$ on MS1M 5.21M dataset [8]; and 3) the derived face clusters are often of high variations in both size and sparsity, and small or sparse clusters—we call **hard clusters**—are hard to identify. Figure 1(a) and (b) show the distributions of ground-truth clusters on MS1M 5.21M dataset. For Fig. 1(c) and (d), we first group these clusters into five subsets based on a fixed ranking of size and sparsity, respectively, and then evaluate three top-performing methods and ours on each subset. It is clear that the performance drops significantly for hard clusters, $e.g.$let@tokeneonedot, in subsets sz-5 and sp-5, particularly on metric Recall (see Fig. 2). We think the reason is two-fold: 1) small clusters are overtaken by large ones; 2) samples of sparse clusters are wrongly taken as "on" low-density regions, $i.e.$let@tokeneonedot, the boundaries between dense clusters.

We elaborate these based on Density Peaking Clustering (DPC) [22] which has shown the impressive effectiveness in state-of-the-art face clustering works [14,31]. DPC requires point-wise density and pair-wise distance to derive clustering results. The density is usually defined as the number of neighbor points covered by an ϵ-ball around each point [4], and the distance is standard cosine distance. We find that

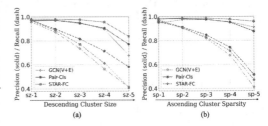

Fig. 2. Pairwise precision and recall (of the three baselines) that elaborates the results in Fig. 1(c) and (d). The recall of hard cluster subsets shows a significant drop.

both density and distance are highly influenced by the size and sparsity of latent clusters in face data. For example, 1) smaller clusters tend to have lower density as shown in Fig. 3(a), so they could be misclassified as big ones by DPC, and 2) to identify positive pairs, higher-sparsity (lower-sparsity) clusters prefer a higher (lower) distance threshold, as indicated in Fig. 3(b), so it is hard to determine a uniform threshold for DPC.

Our Solution. Our clustering framework is based on DPC, and we aim to solve the above issues by introducing new definitions of point-wise density and pair-wise distance. We propose a probabilistic method to derive a size-invariant density called Neighborhood-Diffusion-based Density (NDDe), and a sparsity-aware distance called Transition-Probability-based Distance (TPDi). Applying DPC with NDDe and TPDi can mitigate hard clusters and yield efficient face clustering with a simple and uniform threshold.

We first build a transition matrix where each row contains the normalized similarities (predicted by a pretrained model as in related works [14,23,29,31]) between a point and its K-nearest neighbors, and each column is the transition probability vector from a point to the others. Then, for NDDe, we specify a diffusion process on the matrix by 1) initializing a uniform density for each point,

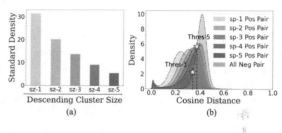

Fig. 3. (a) The average standard density of clusters on each subset. (b) The probability density function on each subset with respect to the positive pairs. "Pos" indicates "Positive", and "Neg" for "Negative".

and 2) distributing the density to its K-nearest neighbors, where the distribution strength is proportional to the transition probability, until converge. The derived NDDe is invariant to the cluster size and thus free from the issue of small clusters. We provide the theoretical justification and empirical validation in Sect. 4.2. For TPDi, we define a relative closeness that equals the inner product between two points' transition probability vectors (corresponding to two columns on the transition matrix). We assume two points are close if they have similar transition probabilities to their common neighbors. Our TPDi can yield more uniform sparsity (in clusters) than conventional distances such as cosine or Euclidean, and thus free from the issue of sparse clusters. Our justification and validation are in Sect. 4.3.

Our main contributions are threefold. 1) We inspect face clustering problem and find existing methods failed to identify hard clusters—yielding significantly low recall for small or sparse clusters. 2) To mitigate the issue of small clusters, we introduce NDDe based on the diffusion of neighborhood densities. 3) To mitigate the issue of sparse clusters, we propose the relative distance TPDi that can facilitate a uniform sparsity in different clusters. In experiments, we evaluate NDDe and TPDi on large-scale benchmarks and incorporate them into multiple baselines to show their efficiency.

2   Related Work

Face clustering has been extensively studied as an important task in the field of machine learning. Existing methods can be briefly divided into traditional methods and learning-based methods.

Traditional Methods. Traditional methods include K-means [18], HAC [24], DBSCAN [6] and ARO [20]. These methods directly perform clustering on the extracted features without any supervision, and thus they usually seem simple but have obvious defects. K-means [18] assumes the cluster shape is convex and DBSCAN [6] assumes that the compactness of different clusters is homogeneous. The performances of these two methods are limited since both assumptions are impractical for face data. The scale of unlabeled face images is usually large, and from this perspective, traditional methods are low-efficient and thus not suitable for face clustering task. The computational efficiency of HAC [24] is not acceptable when handling millions of samples. To achieve better scalability, Otto *et allet@tokeneonedot* [20] proposed ARO that uses an approximate rank-order similarity metric for clustering, but its performance is still far from satisfactory.

Learning-Based Methods. To improve the clustering performance, recent works [7,14,23,28,29,31–33] adopt a learning-based paradigm. Specifically, they first train a clustering model using a small part of data in a supervised manner and then test its performance on the rest of the data. CDP [33] proposed to aggregate the features extracted by different models, but the ensemble strategy results in a much higher computational cost. L-GCN [29] first uses Graph Convolutional Networks (GCNs) [12] to predict the linkage in an instance pivot subgraph, and then extracts the connected components as clusters. LTC [32] and GCN(V+E) [31] both adopt two-stage GCNs for clustering with the whole K-NN graph. Specifically, LTC generates a series of subgraphs as proposals and detects face clusters thereon, and GCN(V+E) learns both confidence and connectivity via GCNs. To address the low-efficiency issue of GCNs, STAR-FC [23] proposed a local graph learning strategy to simultaneously tackle the challenges of large-scale training and efficient inference. To address the noisy connections in the K-NN graph constructed in feature space, Ada-NETS [28] proposed an adaptive neighbor discovery strategy to make clean graphs for GCNs. Although GCN-based methods have achieved significant improvements, they only use shallow GCNs resulting in a lack of high-order connection information, and in addition, their efficiency remains a problem. Pair-Cls [14] proposed to use pairwise classification instead of GCNs to reduce memory consumption and inference time. Clusformer [19] proposed an automatic visual clustering method based on Transformer [25].

In general, existing learning-based methods have achieved significant improvements by focusing on developing deep models to learn better representation or pair-wise similarities, but they failed to identify and address the aforementioned hard cluster issues. In this paper, we explore face clustering task from a new perspective. Based on DPC [22], we propose a size-invariant point-wise density NDDe and a sparsity-aware pair-wise distance TPDi, which can be incorporated into multiple existing methods for better clustering performance, especially on hard clusters.

3 Preliminaries

Problem Formulation. Given N unlabelled face images with numerical feature points $X = \{x_1, x_2, \cdots, x_N\} \subset \mathbb{R}^{N \times D}$, which are extracted by deep face recognition models, face clustering aims to separate these points into disjoint groups as $X = X_1 \bigcup X_2 ... \bigcup X_m$, such that points with the same identity tend to be in the same group, while points with different identities tend to be in different groups.

Data Preprocessing. Following the general process of learning-based face clustering paradigm, the dataset X is split into a training set and a test set, $X = X_{\text{train}} \bigcup X_{\text{test}}$. For a specific learning-based face clustering method, a clustering model is first trained on X_{train} in a supervised manner, and then the clustering performance is tested on X_{test}. Without loss of generality, we always denote the features and labels as $X = \{x_1, x_2, \cdots, x_N\}$ and $l = \{l_1, l_2, \cdots, l_N\}$, respectively, for both training stage and test stage.

Density Peak Clustering (DPC). DPC [22] identifies implicit cluster centers and assigns the remaining points to these clusters by connecting each of them to the higher density point nearby, which is adopted by several state-of-the-art face clustering methods [14,31]. In this paper, we also adopt DPC as the clustering algorithm. Given point-wise density $\rho = \{\rho_1, \rho_2, \cdots, \rho_N\}$ and pair-wise distance $(d_{ij})_{N \times N}$, for each point i, DPC first finds its nearest neighbor whose density is higher than itself, $i.e.$,

$$\hat{j} = \text{argmin}_{\{j|\rho_j > \rho_i\}} d_{ij},$$

If $\hat{j}$ exists and $d_{i\hat{j}} < \tau$, then it connects i to $\hat{j}$, where τ is a connecting threshold. In this way, these connected points form many separated trees, and each tree corresponds to a final cluster. Note that τ is uniform for all clusters, so consistent point-wise density ρ and pair-wise distance $(d_{ij})_{N \times N}$ are essential for the success of DPC. To solve hard cluster issues, we propose a size-invariant density called Neighborhood-Diffusion-based Density (NDDe) and a sparsity-aware distance called Transition-Probability-based Distance (TPDi) for better ρ and $(d_{ij})_{N \times N}$.

4 Method

Figure 4 shows the overall framework consisting of four steps. First, we construct a transition matrix by learning the refined similarities between each point and its K-nearest neighbors using a model consisting of a feature encoder $\mathcal{F}$ and a Multi-Layer Perceptron (MLP). The second step uses our first novel module: computing Neighborhood-Diffusion-based Density (NDDe) by diffusing point-wise density on the neighboring transition matrix, which is invariant to cluster size. The

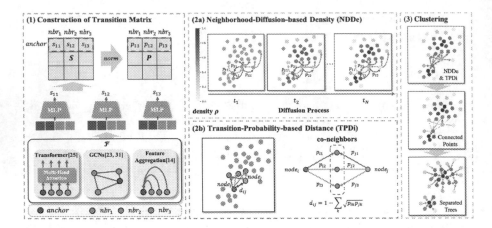

Fig. 4. Overview. Our method consists of four steps: (1) Constructing transition matrix P. Feature encoder $\mathcal{F}$ is for feature refinement, which can be Transformer, GCNs, or a feature aggregation module, and after that, an MLP is used to predict the similarity between each anchor point and its neighbors. (2a) Computing NDDe for each point through a diffusion process. (2b) Computing TPDi to measure the distance between points. (3) Applying DPC with NDDe and TPDi to obtain final clustering result.

third step is our second novel module: computing Transition-Probability-based Distance (TPDi) by introducing a relative closeness, which is aware of cluster sparsity. Fourth, we directly apply DPC with NDDe and TPDi to derive the final clustering result.

4.1 Constructing Transition Matrix

The standard way of construction the transition matrix is to compute the similarity between the deep features of pair-wise samples. The "similarity" can be the conventional cosine similarity or the learned similarity in more recent works such as [14,23,31]. To reduce the memory consumption of using GCNs [23,31] for similarity learning, Pair-Cls [14] simply learns the similarity via pair-wise classification by deciding whether two points share the same identity. However, in Pair-Cls, all pairs are completely independent during training. We argue that the similarity between a point and one of its neighbors usually depends on the similarities between the point and its other neighbors. Therefore, in our work, we adopt the same pair-wise classification, *i.e.*, using an MLP to predict the similarity, and besides that, we leverage a collaborative prediction manner by considering the similarities between each point (as an anchor) and its neighbors as a whole to improve the robustness of the prediction, similar to [14,19].

Here, we elaborate a general formulation. For a sample point i, we first find its K-nearest neighbors denoted as $\mathrm{nbr}_i = \{i_1, \cdots, i_K\}$, and then generate the following token sequence:

$$\tilde{x}_i = [x_i, x_{i_1}, x_{i_2}, \cdots, x_{i_K}].$$

Our similarity prediction model first takes $\tilde{x}_i$ as input, and outputs $K + 1$ features after feature encoder $\mathcal{F}$:

$$\{t_i, t_{i_1}, \cdots, t_{i_K}\} = \mathcal{F}(\tilde{x}_i),$$

where $\mathcal{F}$ can be Transformer [25], GCNs [23,31] or a simple feature aggregation module [14] (aggregate features of neighbors and concatenate to the feature of anchor). Then, for each neighbor i_j, $j = 1, ..., K$, t_i are concatenated with t_{i_j} and fed into an MLP with Sigmoid function to estimate the probability of that i and i_j share the same identity:

$$p_{ij} = \text{MLP}([t_i, t_{i_j}]).$$

Assuming l_{ij} is the ground-truth label, $l_{ij} = 1$ if $l_i = l_{i_j}$ and $l_{ij} = 0$ vice versa. The total loss function is formulated as:

$$\mathcal{L} = -\sum_{i=1}^{N}\sum_{j=1}^{K}(l_{ij}\log p_{ij} + (1 - l_{ij})\log(1 - p_{ij})). \tag{1}$$

Once the model converges, its predicted similarity takes the anchor's feature as well as its respective neighborhoods' features into consideration. Then, we can derive the similarity matrix $\hat{S}_{N \times N}$ by applying this model on the test set.

Finally, we assume $d_i = \sum_{j=1}^{N} \hat{s}_{ij}$ as the measure of the volume around point i, and generate the probability transition matrix P with each element as $p_{ij} = \hat{s}_{ij}/d_i$. The size of P is $N \times N$. Please note that $\hat{S}$ is a sparse matrix where each row contains $K + 1$ non-zero elements (itself and its top-K nearest neighbors). Therefore, P is also sparse. **We highlight** that the above approach is not the only way to construct the transition matrix P, and we show the results of using other approaches to obtain P in the experiment section.

4.2 Neighborhood-Diffusion-Based Density

In this section, we propose a new definition of the point-wise density, called NDDe, to alleviate the issue of small-size clusters. In the transition matrix P, each element P_{ij} denotes the probability from one point i to its specific neighbor j. It satisfies the conservation property, i.e. $\sum_j P_{ij} = 1$, which induces a Markov chain on X. Denoting $L = I - P$ as the normalized graph Laplacian, where I is the identity matrix. We can specify a diffusion process as follows,

$$\begin{cases} \frac{\partial}{\partial t}\rho_i(t) = -L\rho_i(t), \\ \rho_i(0) = 1. \end{cases} \tag{2}$$

where $\rho_i(t)$ is the density of point i at t-th step. Starting from a uniformly initialized density, the diffusion process keeps distributing the density of each point to its K-nearest neighbors, following the corresponding transition probabilities in P, until converged to a stationary distribution. The diffusion density thus can be induced as:

$$\rho_i = \lim_{t\to\infty} \rho_i(t). \tag{3}$$

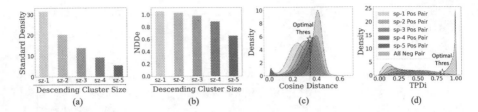

Fig. 5. (a) and (b) show the average values of the standard density and our NDDe on five cluster subsets from MS1M 5.21M dataset, respectively. NDDe is shown to be more uniform, i.e., small clusters are alleviated. (c) and (d) show the probability density functions of using the conventional cosine distance and TPDi on five cluster subsets from MS1M 5.21M dataset, respectively. Using TPDi makes it easier to decide a more uniform threshold to separate positive and negative pairs, in all subsets including the sparsest one "sp-5".

Justification of Local Properties of Diffusion Density. The diffusion process is local because each point transits its density to K-nearest neighbors and itself (based on the transition matrix $\boldsymbol{P}$). If considering the ideal situation when $\boldsymbol{P}$ is closed, which means $p_{ij} > 0$ if and only if x_i and x_j share the same identity, we have the following theorem.

Theorem 1. *Assume the dataset $\boldsymbol{X}$ can be split into m disjoint clusters: i.elet@tokeneonedot, $\boldsymbol{X} = \boldsymbol{X}_1 \bigcup ... \bigcup \boldsymbol{X}_m$. Define $\bar{\rho}_i = \frac{\sum_{j \in \boldsymbol{X}_i} \rho(j)}{|\boldsymbol{X}_i|}$ is the average density of $\boldsymbol{X}_i$, and we have $\bar{\rho}_1 = \cdots = \bar{\rho}_m = 1$ where $|\boldsymbol{X}_i|$ is the number of points in $\boldsymbol{X}_i$.*

Theorem 1 demonstrates that the average diffusion densities in all clusters are the same regardless of cluster sizes. In a dynamic sense, the diffusion process can elevate the density of latent small clusters, and thus enable DPC algorithm to identify density peaks in such clusters. To further demonstrate our claim, we divide clusters in MS1M 5.21M dataset into five subsets according to cluster sizes and calculate the average diffusion density for each subset. As shown in Fig. 5(a)(b), compared with the standard density, the average NDDe for different subsets are much more comparable.

4.3   Transition-Probability-Based Distance

In this section, we introduces our new definition of the pair-wise distance, called TPDi, to solve the issue of varying sparsity in latent face clusters. TPDi depicts the similarity between two points based on their respective transition probabilities (in $\boldsymbol{P}$) to the common neighbors. Assuming $C_{ij} = \text{nbr}_i \cap \text{nbr}_j$ contains the common neighbors in the K-nearest neighbors of both point i and j. TPDi between them is defined as:

$$d_{ij} = 1 - \sum_{c \in C_{ij}} \sqrt{p_{ic} p_{jc}}. \tag{4}$$

Algorithm 1: Pseudocode for our method

Input: Face dataset $\boldsymbol{X} = \{x_1, \ldots, x_N\}$, number of nearest neighbors K, pre-trained similarity prediction model Φ, convergence threshold ϵ, connecting threshold τ.

Output: clusters $\mathcal{C}$.

1 **procedure** CLUSTERING
2   **for** each point i:
3   Find its K-nearest neighbors $\mathrm{nbr}_i = \{i_k\}_{k=1}^K$ and construct $x_i^\star$;
4   Inference the similarities $\{s_{i,j}\}_{j=1}^K$ between them via $\Phi(x_i^\star)$;
5   Obtain the pair-wise similarity matrix $\hat{\boldsymbol{S}}_{N \times N}$, and compute $\boldsymbol{P}_{N \times N}$;
6   Compute the point-wise density $\rho_{N \times 1}$ via NDDe($\boldsymbol{P}$);
7   Compute the pair-wise distance $(d_{ij})_{N \times N}$ via TPDi($\boldsymbol{P}$);
8   Obtain clusters $\mathcal{C}$ via DPC(ρ, $(d_{ij})_{N \times N}$);
9 **end procedure**
10 **function** NDDe($\boldsymbol{P}$)
11   Initialize $\rho_{\mathrm{pre}} = \{1\}_{N \times 1}$
12   **while** $\|\rho - \rho_{\mathrm{pre}}\|_2 > \epsilon$:
13         $\rho = \rho_{\mathrm{pre}}$; $\rho_{\mathrm{pre}} = \boldsymbol{P} \times \rho$;
14   **return** ρ
15 **end function**
16 **function** TPDi($\boldsymbol{P}$)
17   **for** each pair of points i, j:
18   Compute d_{ij} as shown in Eq. 4;
19   **return** $(d_{ij})_{N \times N}$
20 **end function**

We highlight that TPDi has three impressive properties: (1) By Cauchy-Schwarz inequality, we have $\left(\sum_{c \in C_{ij}} \sqrt{p_{ic} p_{jc}}\right)^2 \leq (\sum_{c \in C_{ij}} p_{ic})(\sum_{c \in C_{ij}} p_{jc}) \leq 1$, so it is easy to check $0 \leq d_{ij} \leq 1$, which implies that d_{ij} can be a valid metric. (2) $d_{ij} = 0$ if and only if $p_{ic} = p_{jc}$ for all $c = 1, \ldots, N$, which implies that d_{ij} is small when i and j share as many as common neighbors. It is consistent with the motivation of TPDi. (3) Compared with cosine distance, TPDi of negative pairs and positive pairs are better separated, regardless of cluster sparsity (Fig. 5(c)(d)). So it is easier to choose a uniform threshold for TPDi.

Remark 1. If considering a simple case when each point transits to its neighbors with equal transition probability $\frac{1}{K}$, we have $d_{ij} = 1 - \frac{2\mathrm{Jaccard}(i,j)}{(1+\mathrm{Jaccard}(i,j))}$, where Jaccard($i, j$) is the Jaccard similarity [9]. This implies that the TPDi is a generalization of Jaccard distance, which also demonstrate the feasibility of TPDi.

4.4   Overall Algorithm

The overall clustering procedure is summarized in Algorithm 1. In our implementation, we use an iterative method as an approximation of Eq. 3.

5   Experiments

5.1   Experimental Settings

Datasets. We evaluate the proposed method on two public face clustering benchmark datasets, MS1M [8] and DeepFashion [17]. MS1M contains 5.8M images from 86K identities and the image representations are extracted by Arc-Face [5], which is a widely used face recognition model. MS1M is split into 10 almost equal parts officially. Following the same experimental protocol as in [14,23,31], we train our model on one labeled part and choose parts 1, 3, 5, 7, and 9 as unlabeled test data, resulting in five test subsets with sizes of 584K, 1.74M, 2.89M, 4.05M, and 5.21M images respectively. For DeepFashion dataset, following [31], we randomly sample 25,752 images from 3,997 categories for training and use the other 26,960 images with 3,984 categories for testing.

Metrics. The performances of face clustering methods are evaluated using two commonly used clustering metrics, Pairwise F-score (F_P) [3] and BCubed F-score (F_B) [1]. Both metrics are reflections of precision and recall.

Implementation Details. Our similarity prediction model consists of one transformer encoder layer [26] as $\mathcal{F}$ and an MLP. The input feature dimension, feedforward dimension, number of heads for $\mathcal{F}$ are set to 256, 2048, 8, respectively. LayerNorm [2] is applied before Multi-head Attention module and Feed Forward module in $\mathcal{F}$, according to [30]. Dropout is set to 0.2. The MLP consists of three linear layers ($512 \rightarrow 256, 256 \rightarrow 128, 128 \rightarrow 1$) with ReLU as the activation function for the first two layers and Sigmoid for the last layer. Adam [11] is used for optimization. For the computation of NDDe, we set the number of top nearest neighbors K to 80 for MS1M and 10 for DeepFashion (the same as previous works [14,31]). Convergence threshold ϵ is set to 0.05. Connecting threshold τ is searched within the range of $[0.5, 0.9]$ with a step of 0.05 on MS1M 584K dataset, and is fixed to 0.7 for all experiments.

5.2   Method Comparison

We compare the proposed method with a series of clustering baselines, including both traditional methods and learning-based methods. Traditional methods include K-means [18], HAC [24],DBSCAN [6], and ARO [20]. Learning-based methods include CDP [33], L-GCN [29], LTC [32], GCN (V+E) [31], Clusformer [19], Pair-Cls [14], STAR-FC [23], and Ada-NETS [28]. Since NDDe and TPDi can be incorporated into existing face clustering methods for better performance, we also incorporate them into GCN (V+E), Pair-Cls, and STAR-FC by using the three methods to obtain the transition matrix $\boldsymbol{P}$, which are denoted as GCN(V+E)++, Pair-Cls++, and STAR-FC++, respectively.

Table 1. Comparison on MS1M when training with 0.5M labeled face images and testing on five test subsets with different numbers of unlabeled face images. F_P, F_B are reported. GCN(V+E)++, Pair-Cls++ and STAR-FC++ denote incorporating NDDe and TPDi into the corresponding methods. The best results are highlighted with **bold**.

#Images	584K		1.74M		2.89M		4.05M		5.21M	
Method/metrics	F_P	F_B	F_P	F_B	F_P	F_B	F_P	F_B	F_P	F_B
K-means [18]	79.21	81.23	73.04	75.20	69.83	72.34	67.90	70.57	66.47	69.42
HAC [24]	70.63	70.46	54.40	69.53	11.08	68.62	1.40	67.69	0.37	66.96
DBSCAN [6]	67.93	67.17	63.41	66.53	52.50	66.26	45.24	44.87	44.94	44.74
ARO [20]	13.60	17.00	8.78	12.42	7.30	10.96	6.86	10.50	6.35	10.01
CDP [33]	75.02	78.70	70.75	75.82	69.51	74.58	68.62	73.62	68.06	72.92
L-GCN [29]	78.68	84.37	75.83	81.61	74.29	80.11	73.70	79.33	72.99	78.60
LTC [32]	85.66	85.52	82.41	83.01	80.32	81.10	78.98	79.84	77.87	78.86
GCN(V+E) [31]	87.93	86.09	84.04	82.84	82.10	81.24	80.45	80.09	79.30	79.25
Clusformer [19]	88.20	87.17	84.60	84.05	82.79	82.30	81.03	80.51	79.91	79.95
Pair-Cls [14]	90.67	89.54	86.91	86.25	85.06	84.55	83.51	83.49	82.41	82.40
STAR-FC [23]	91.97	90.21	88.28	86.26	86.17	84.13	84.70	82.63	83.46	81.47
Ada-NETS [28]	92.79	91.40	89.33	87.98	87.50	86.03	85.40	84.48	83.99	83.28
GCN(V+E)++	90.72	89.28	86.06	84.36	85.97	84.24	84.76	83.10	83.69	82.26
Pair-Cls++	91.70	89.94	88.17	86.50	86.49	84.76	85.25	83.50	83.74	82.61
STAR-FC++	92.35	90.50	89.03	86.94	86.70	85.16	85.38	83.93	83.94	82.95
Ours	**93.22**	**92.18**	**90.51**	**89.43**	**89.09**	**88.00**	**87.93**	**86.92**	**86.94**	**86.06**

Results on MS1M. Experimental results on MS1M dataset are shown in Table 1, which contains both F_P and F_B on five test subsets with different scales. We can observe that 1) Our method consistently outperforms the other methods in terms of both metrics, especially for large-scale subsets, e.g.let@tokeneonedot, the improvements of our method on 4.05M and 5.21M subsets are more than 2.5%. 2) By incorporating NDDe and TPDi into GCN (V+E), Pair-Cls and STAR-FC, their ++ versions achieve better clustering performance than the original versions, e.g.let@tokeneonedot, compared to GCN (V+E), the performance gains brought by GCN(V+E)++ are more than 3% on large-scale test subsets, which demonstrates that NDDe and TPDi can raise the performance of other methods to a new state-of-the-art.

Results on Hard Clusters. To demonstrate that our method is capable of tackling the issues of small clusters and sparse clusters, we conduct experiments by adding NDDe and TPDi one by one to our baseline model, i.elet@tokeneonedot, the model with the same transition matrix but the density and distance computed in the standard way. As shown in the last three rows in Table 2 and Table 3, both NDDe and TPDi have raised the performance of the baseline model to a new level, especially on hard clusters.

We also reproduce GCN(V+E), Pair-Cls and STAR-FC for comparison, all of which employ a clustering algorithm just as or similar to DPC, as shown in the first two rows in Table 2 and Table 3. It is worth noticing that the improvements brought by our method over the three top-performing methods keep increasing

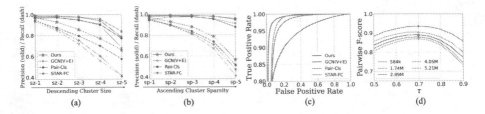

Fig. 6. (a) and (b) show Pairwise precision and recall of three baselines and our method. Significant improvements of our method in terms of recall can be observed. (c) ROC curves of the three baselines and our method. (d) Optimal threshold τ for the five test subsets of MS1M dataset.

on five cluster subsets with descending size or ascending sparsity. As shown in Fig. 6(a)(b), the improvements of our method in terms of Pairwise recall are more significant than Pairwise precision. All the experimental results show the success of our method in mitigating hard clusters, owing to NDDe and TPDi.

The Superiority of TPDi. Figure 6(c) shows the receiver operating characteristic (ROC) curves of three top-performing methods and ours, which are obtained by computing true/false positive rate at various distance threshold settings. Our method achieves the highest Area Under Curve (AUC), which illustrates that TPDi endows our method with a good measure of separability. To show that

Table 2. The effectiveness of NDDe and TPDi. F_P and F_B of five cluster subsets from MS1M 5.21M with descending size (from sz-1 to sz-5) are reported.

	sz-1		sz-2		sz-3		sz-4		sz-5		total	
	F_P	F_B	F_P	F_B	F_P	F_B	F_P	F_B	F_P	F_B	F_P	F_B
GCN(V+E)	89.06	90.52	84.52	84.81	75.17	75.84	61.28	63.03	44.15	52.49	78.77	79.08
Pair-Cls	90.02	90.65	86.03	86.20	80.21	80.80	72.37	73.72	59.28	65.30	82.19	81.63
STAR-FC	90.47	91.13	85.75	86.11	78.35	78.78	66.49	67.44	46.65	51.21	83.74	82.00
Baseline	57.54	63.45	52.39	55.89	43.84	47.62	37.84	41.77	34.67	42.23	41.49	50.76
+NDDe	83.67	86.06	78.38	78.95	69.63	70.28	60.19	61.49	49.85	54.53	72.47	74.39
+TPDi(Ours)	92.35	93.18	89.88	89.91	85.08	85.28	78.35	79.19	65.56	71.33	86.94	86.06

Table 3. The effectiveness of NDDe and TPDi. F_P and F_B of five cluster subsets from MS1M 5.21M with ascending sparsity (from sp-1 to sp-5) are reported.

	sp-1		sp-2		sp-3		sp-4		sp-5		total	
	F_P	F_B	F_P	F_B	F_P	F_B	F_P	F_B	F_P	F_B	F_P	F_B
GCN(V+E)	94.63	94.66	92.73	91.52	87.47	85.13	77.44	73.09	53.99	45.95	78.77	79.08
Pair-Cls	95.52	95.24	93.22	92.46	89.24	87.66	81.84	78.84	62.73	57.51	82.19	81.63
STAR-FC	96.18	95.27	92.92	91.50	88.50	85.96	80.78	76.54	60.81	53.56	83.74	82.00
Baseline	63.16	63.84	62.23	62.95	57.32	58.09	49.19	50.20	32.98	35.48	41.49	50.76
+NDDe	92.30	91.00	87.47	85.70	82.11	79.30	72.69	69.97	52.53	51.17	72.47	74.39
+TPDi(Ours)	97.25	96.96	95.10	94.59	92.24	91.08	86.23	84.23	69.08	64.83	86.94	86.06

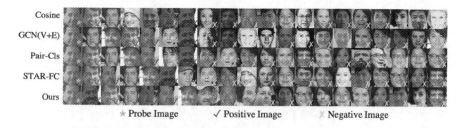

Cosine
GCN(V+E)
Pair-Cls
STAR-FC
Ours

⁕ Probe Image   ✓ Positive Image   ✗ Negative Image

Fig. 7. Top-20 images ranked by distance, using an image in hard clusters as probe.

by using TPDi, our method can yield efficient face clustering with a uniform connecting threshold τ, we conduct experiments using different τ (from 0.5 to 0.9, with a step of 0.05) on all the test subsets of MS1M dataset, as shown in Fig. 6(d). It can be observed that the best τ is the same for test subsets with varying scales. To be specific, given $\tau = 0.7$, our method consistently achieves the highest F_P on all test subsets. Figure 7 shows the discovery results of several methods with the image in the first column as a probe, and the images are ranked in ascending order of distance. We can observe that the discovery result of our method contains the most number of positive images.

Results on DeepFashion. For Deep-Fashion dataset, clustering task is much harder since it is an open set problem. It can be observed that our method also uniformly outperforms the other methods in terms of both F_P and F_B with comparable computing time, as shown in Table 4.

Table 4. Comparison on DeepFashion. #Clusters, F_P, F_B and computing time are reported.

Method	#Clusters	F_P	F_B	Time
K-means [18]	3991	32.86	53.77	573 s
HAC [24]	17410	22.54	48.7	112 s
DBSCAN [6]	14350	25.07	53.23	2.2 s
ARO [20]	10504	26.03	53.01	6.7 s
CDP [33]	6622	28.28	57.83	1.3 s
L-GCN [29]	10137	28.85	58.91	23.3 s
LTC [32]	9246	29.14	59.11	13.1 s
GCN(V+E) [31]	6079	38.47	60.06	18.5 s
Pair-Cls [14]	6018	37.67	62.17	0.6 s
STAR-FC [23]	–	37.07	60.60	–
Ada-NETS [28]	–	39.30	61.05	–
Ours	8484	40.91	63.61	4.2 s

5.3   Ablation Study

To demonstrate the effectiveness of NDDe and TPDi, we conduct an ablation study on MS1M 5.21M dataset, as shown in Table 5. All these four methods use the same transition matrix as described in Sect. 4.1. M_1 is our baseline model, which uses the standard density and cosine distance. M_2 is obtained by replacing the cosine distance in M_1 with TPDi, M_3 is obtained by replacing the standard density in M_1 with NDDe, and M_4 is the proposed method using both NDDe and TPDi as the density ρ and distance $(d_{ij})_{N \times N}$ required by DPC. Table 5 shows that both NDDe and TPDi contribute to the final clustering performance. And the improvement brought by NDDe is more significant, which illustrates that NDDe is essential for the success of our method.

5.4   Face Recognition

Table 5. Ablation study of NDDe and TPDi on MS1M. F_P and F_B are reported.

	NDDe	TPDi	584K		1.74M		2.89M		4.05M		5.21M	
			F_P	F_B	F_P	F_B	F_P	F_B	F_P	F_B	F_P	F_B
M_1			53.03	56.75	47.80	53.84	45.07	52.41	43.29	51.56	41.49	50.76
M_2		✓	61.07	59.81	59.29	58.26	58.66	57.40	58.37	57.00	57.88	56.48
M_3	✓		82.98	80.33	78.79	77.87	76.32	76.42	74.08	75.28	72.47	74.39
M_4	✓	✓	**93.22**	**92.18**	**90.51**	**89.43**	**89.09**	**88.00**	**87.93**	**86.92**	**86.94**	**86.06**

To further show the potential of our method in scaling up face recognition systems using large-scale unlabeled face images, we use our method to generate pseudo-labels for unlabeled face images and use them to train face recognition models. For a fair comparison, we adopt the same experimental setting as in [23, 31,32]. We use a fixed number of labeled data and different ratios of unlabeled data with pseudo-labels to train face recognition models and test their performance on MegaFace

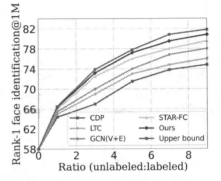

Fig. 8. Rank-1 face identification accuracy on MegaFace with 1M distractors.

benchmark [10] taking the rank-1 face identification accuracy with 1M distractors as metric. In Fig. 8, the upper bound is trained by assuming all unlabeled data have ground-truth labels, and the other five curves illustrate that all the methods benefit from an increase of the unlabeled data with pseudo-labels. And it can be observed that our method consistently achieves the highest performance given any ratio of unlabeled data, and improves the performance of the face recognition model from 58.20% to 80.80%, which is the closest to the upper bound.

6   Conclusion

In this paper, we point out a key issue in face clustering task—the low recall of hard clusters, $i.elet@tokeneonedot$, small clusters and sparse clusters. We find the reasons behind this are 1) smaller clusters tend to have a lower density, and 2) it is hard to set a uniform (distance) threshold to identify the clusters of varying sparsity. We tackle the problems by proposing two novel modules, NDDe and TPDi, which yield the size-invariant density and the sparsity-aware distance, respectively. Our extensive ablation study shows that each of them contributes to improving the recall on hard clusters, consistently on multiple face clustering benchmarks.

Acknowledgments. This work is supported by the National Key R&D Program of China under Grant 2020AAA0103901, Alibaba Group through Alibaba Research Intern Program, and Alibaba Innovative Research (AIR) programme.

References

1. Amigó, E., Gonzalo, J., Artiles, J., Verdejo, F.: A comparison of extrinsic clustering evaluation metrics based on formal constraints. Inf. Retr. **12**(4), 461–486 (2009)
2. Ba, J.L., Kiros, J.R., Hinton, G.E.: Layer normalization. arXiv preprint arXiv:1607.06450 (2016)
3. Banerjee, A., Krumpelman, C., Ghosh, J., Basu, S., Mooney, R.J.: Model-based overlapping clustering. In: Proceedings of the Eleventh ACM SIGKDD International Conference on Knowledge Discovery in Data Mining, pp. 532–537 (2005)
4. Breiman, L., Meisel, W., Purcell, E.: Variable kernel estimates of multivariate densities. Technometrics **19**(2), 135–144 (1977)
5. Deng, J., Guo, J., Xue, N., Zafeiriou, S.: ArcFace: additive angular margin loss for deep face recognition. In: CVPR (2019)
6. Ester, M., Kriegel, H.P., Sander, J., Xu, X., et al.: A density-based algorithm for discovering clusters in large spatial databases with noise. In: SIGKDD (1996)
7. Guo, S., Xu, J., Chen, D., Zhang, C., Wang, X., Zhao, R.: Density-aware feature embedding for face clustering. In: CVPR (2020)
8. Guo, Y., Zhang, L., Hu, Y., He, X., Gao, J.: MS-Celeb-1M: a dataset and benchmark for large-scale face recognition. In: Leibe, B., Matas, J., Sebe, N., Welling, M. (eds.) ECCV 2016. LNCS, vol. 9907, pp. 87–102. Springer, Cham (2016). https://doi.org/10.1007/978-3-319-46487-9_6
9. Ivchenko, G., Honov, S.: On the Jaccard similarity test. J. Math. Sci. **88**(6), 789–794 (1998)
10. Kemelmacher-Shlizerman, I., Seitz, S.M., Miller, D., Brossard, E.: The MegaFace Benchmark: 1 million faces for recognition at scale. In: CVPR (2016)
11. Kingma, D.P., Ba, J.: Adam: a method for stochastic optimization. arXiv preprint arXiv:1412.6980 (2014)
12. Kipf, T.N., Welling, M.: Semi-supervised classification with graph convolutional networks. arXiv preprint arXiv:1609.02907 (2016)
13. Kortli, Y., Jridi, M., Al Falou, A., Atri, M.: Face recognition systems: a survey. Sensors **20**(2), 342 (2020)
14. Liu, J., Qiu, D., Yan, P., Wei, X.: Learn to cluster faces via pairwise classification. In: Proceedings of the IEEE/CVF International Conference on Computer Vision, pp. 3845–3853 (2021)
15. Liu, W., Wen, Y., Yu, Z., Li, M., Raj, B., Song, L.: SphereFace: deep hypersphere embedding for face recognition. In: CVPR (2017)
16. Liu, W., Wen, Y., Yu, Z., Yang, M.: Large-margin Softmax loss for convolutional neural networks. In: ICML (2016)
17. Liu, Z., Luo, P., Qiu, S., Wang, X., Tang, X.: DeepFashion: powering robust clothes recognition and retrieval with rich annotations. In: Proceedings of the IEEE Conference on Computer Vision and Pattern Recognition, pp. 1096–1104 (2016)
18. Lloyd, S.: Least squares quantization in PCM. TIP **28**, 129–137 (1982)
19. Nguyen, X.B., Bui, D.T., Duong, C.N., Bui, T.D., Luu, K.: Clusformer: a transformer based clustering approach to unsupervised large-scale face and visual landmark recognition. In: Proceedings of the IEEE/CVF Conference on Computer Vision and Pattern Recognition, pp. 10847–10856 (2021)

20. Otto, C., Wang, D., Jain, A.K.: Clustering millions of faces by identity. TPAMI **40**, 289–303 (2017)
21. Parkhi, O.M., Vedaldi, A., Zisserman, A.: Deep face recognition (2015)
22. Rodriguez, A., Laio, A.: Clustering by fast search and find of density peaks. Science **344**(6191), 1492–1496 (2014)
23. Shen, S., et al.: Structure-aware face clustering on a large-scale graph with 107 nodes. In: Proceedings of the IEEE/CVF Conference on Computer Vision and Pattern Recognition, pp. 9085–9094 (2021)
24. Sibson, R.: Slink: an optimally efficient algorithm for the single-link cluster method. Comput. J. **16**, 30–34 (1973)
25. Vaswani, A., et al.: Attention is all you need. In: Advances in Neural Information Processing Systems 30 (2017)
26. Vaswani, A., et al.: Attention is all you need. In: NIPS (2017)
27. Wang, H., et al.: CosFace: large margin cosine loss for deep face recognition. In: CVPR (2018)
28. Wang, Y., et al.: Ada-NETS: face clustering via adaptive neighbour discovery in the structure space. arXiv preprint arXiv:2202.03800 (2022)
29. Wang, Z., Zheng, L., Li, Y., Wang, S.: Linkage based face clustering via graph convolution network. In: CVPR (2019)
30. Xiong, R., et al.: On layer normalization in the transformer architecture. In: ICML, pp. 10524–10533. PMLR (2020)
31. Yang, L., Chen, D., Zhan, X., Zhao, R., Loy, C.C., Lin, D.: Learning to cluster faces via confidence and connectivity estimation. In: CVPR (2020)
32. Yang, L., Zhan, X., Chen, D., Yan, J., Loy, C.C., Lin, D.: Learning to cluster faces on an affinity graph. In: CVPR (2019)
33. Zhan, X., Liu, Z., Yan, J., Lin, D., Loy, C.C.: Consensus-driven propagation in massive unlabeled data for face recognition. In: Ferrari, V., Hebert, M., Sminchisescu, C., Weiss, Y. (eds.) ECCV 2018. LNCS, vol. 11213, pp. 576–592. Springer, Cham (2018). https://doi.org/10.1007/978-3-030-01240-3_35
34. Zhao, W., Chellappa, R., Phillips, P.J., Rosenfeld, A.: Face recognition: a literature survey. ACM Comput. Surv. (CSUR) **35**(4), 399–458 (2003)

OneFace: One Threshold for All

Jiaheng Liu[1], Zhipeng Yu[2]([✉]), Haoyu Qin[3], Yichao Wu[3], Ding Liang[3], Gangming Zhao[4], and Ke Xu[1]

[1] State Key Lab of Software Development Environment,
Beihang University, Beijing, China
liujiaheng@buaa.edu.cn
[2] University of Chinese Academy of Sciences, Beijing, China
yuzhipeng21@mails.ucas.ac.cn
[3] SenseTime Group Limited, Science Park, Hong Kong
{qinhaoyu1,wuyichao,liangding}@sensetime.com
[4] The University of Hong Kong, Pok Fu Lam, Hong Kong

Abstract. Face recognition (FR) has witnessed remarkable progress with the surge of deep learning. Current FR evaluation protocols usually adopt different thresholds to calculate the True Accept Rate (TAR) under a pre-defined False Accept Rate (FAR) for different datasets. However, in practice, when the FR model is deployed on industry systems (e.g., hardware devices), only one fixed threshold is adopted for all scenarios to distinguish whether a face image pair belongs to the same identity. Therefore, current evaluation protocols using different thresholds for different datasets are not fully compatible with the practical evaluation scenarios with one fixed threshold, and it is critical to measure the performance of FR models by using one threshold for all datasets. In this paper, we rethink the limitations of existing evaluation protocols for FR and propose to evaluate the performance of FR models from a new perspective. Specifically, in our OneFace, we first propose the One-Threshold-for-All (OTA) evaluation protocol for FR, which utilizes one fixed threshold called as Calibration Threshold to measure the performance on different datasets. Then, to improve the performance of FR models under the OTA protocol, we propose the Threshold Consistency Penalty (TCP) to improve the consistency of the thresholds among multiple domains, which includes Implicit Domain Division (IDD) as well as Calibration and Domain Thresholds Estimation (CDTE). Extensive experimental results demonstrate the effectiveness of our method for FR.

Keywords: Face recognition · Loss function · Fairness

1 Introduction

Face recognition (FR) based on deep learning has been well investigated for many years [4–6,32,33,39,40]. Most of the progress depends on large-scale training data [10,16,46], deep neural network architectures [12,13,36], and effective loss function designs [3,5,6,29,30,37,39,49]. Recently, with the increasing deployment of FR systems, fairness in FR has attracted broad interest from research

S. Avidan et al. (Eds.): ECCV 2022, LNCS 13672, pp. 545–561, 2022.
https://doi.org/10.1007/978-3-031-19775-8_32

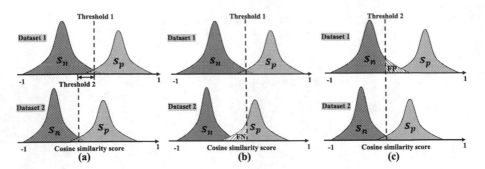

Fig. 1. Similarity distributions of two datasets. The green and red histograms mean the distributions of positive and negative pairs, respectively, where s_p and s_n denote similarities of positive and negative pairs, respectively. (a). Based on existing evaluation protocol, threshold 1 differs from threshold 2 a lot under the same FAR. (b). When we use threshold 1 of dataset 1 to evaluate the dataset 2, a large number of false negative (FN) samples from dataset 2 are produced. (c). When we use threshold 2 to evaluate the dataset 1, many false positive (FP) samples from dataset 1 are generated. (Color figure online)

communities. For example, as reported in the 2019 NIST Face Recognition Vendor Test [9], all participating FR algorithms exhibit different levels of biased performances across various demographic groups (e.g., race, gender, age). However, existing evaluation metrics cannot measure the degree of fairness on threshold across multiple datasets for FR well. Specifically, current practical FR systems usually calculate the True Accept Rate (TAR) under a pre-defined False Accept Rate (FAR) (e.g., 1e−4). As shown in Fig. 1(a), we visualize the distributions of the similarity scores from two datasets, and observe that dataset 1 and dataset 2 have different thresholds under the same FAR. Which means that current evaluation protocols adopt different thresholds for performance evaluation for different datasets. We call such phenomenon as Threshold Imbalance on different datasets. Besides, FR models seem to perform well on these two datasets under current evaluation protocols in Fig. 1(a), but in Fig. 1(b) and Fig. 1(c), the performance results of FR models are very sensitive to the changes of the thresholds. Moreover, when the FR models are deployed for industry, only one fixed threshold is adopted for all scenarios, which indicates that the current evaluation protocol with different thresholds for different datasets is not fully compatible with the practical FR. In addition, most existing FR methods are mainly evaluated under the current evaluation protocol in Fig. 1(a), which have not considered the problem of threshold imbalance.

Motivated by the above analysis, in our OneFace, we propose a new One-Threshold-for-All (OTA) evaluation protocol to better exploit the overall performance and fairness on threshold of FR models on multiple datasets, and then introduce an effective Threshold Consistency Penalty (TCP) scheme to tackle the threshold imbalance problem in the training process.

In OTA evaluation protocol, we directly measure the performance of the different datasets using only one fixed threshold, which is more consistent with the practical FR and can be easily combined with existing evaluation protocols Specifically, we call the fixed threshold for deployment as calibration threshold t_c. Given G datasets and the overall FAR (e.g., 1e−4), we introduce two types of calibration threshold estimation methods. The first type is to combine all datasets into one dataset directly and estimate t_c under the overall FAR based on the negative pairs constructed from the dataset, which is straight-forward and feasible. However, as the computation cost is proportional to the number of negative pairs, it will be unaffordable when the number of negative pairs of the whole dataset is large. Thus, we also propose another calibration threshold estimation method by using an extra dataset called as calibration dataset, where the number of negative pairs in the calibration dataset is relatively acceptable. Besides, the calibration dataset is supposed to cover images from as many domains as possible. Once the calibration dataset is prepared, we can directly adopt the threshold of this dataset under the overall FAR as t_c. After obtaining the calibration threshold, under the OTA evaluation protocol, we calculate the TAR and FAR results for these G datasets based on t_c, respectively. Meanwhile, we propose a new fairness metric γ to denote the degree of the threshold imbalance across G datasets. Specifically, we calculate the thresholds $\{t_d^g\}_{g=1}^G$ within each dataset, where t_d^g denotes the domain threshold for the g-th dataset under the overall FAR. Then, we can obtain the γ based on t_c and $\{t_d^g\}_{g=1}^G$.

To improve the performance under the OTA evaluation protocol, our TCP aims to mitigate the threshold imbalance among different domains in the training process. Specifically, first, as domain labels of the training dataset are usually not available and it is necessary to obtain the domain label for computing the domain threshold, we propose to adopt an Implicit Domain Division (IDD) module to assign the domain labels for samples implicitly following the Group-Face [18], which aims to divide the samples of each mini-batch into M domains. M is a pre-defined hyperparameter on the number of implicit domains. Then, we need to construct negative pairs to calculate the calibration and domain thresholds. Existing work [45] uses the weights of the last FC layer and the features of the current batch to construct the negative pairs. However, as discussed in VPL [6], the weights of the last FC layer update very slowly and the similarity distributions of the sample-to-prototype comparisons used in [45] are also different from distributions of the sample-to-sample comparisons used in evaluation process. Thus, this method [45] may lead to inaccurate threshold estimation for FR. To generate accurate thresholds, inspired by MoCo [11], we propose to build a feature queue to maintain the features of the previous iterations. Then, in training, we can construct the negative pairs based on the features of the current batch and the features of the feature queue. After that, we calculate the similarities of these negative pairs to obtain the calibration threshold t_c. Besides, we generate the domain threshold t_d^m for the m-th domain using the features from the m-th domain of the current batch and the features of the feature queue, where the domain labels in each mini-batch are predicted by IDD module. Finally, we adopt the ratio of t_d^m and t_c as the loss weight for the samples of m-th domain to re-weighting the samples with high domain thresholds, which makes the thresholds across domains more consistent and reduce the degree of threshold imbalance.

The contributions of our proposed OneFace are summarized as follows:

- We first investigate the limitations of the existing evaluation protocols and propose a new One-Threshold-for-All (OTA) evaluation protocol to measure the performance of different datasets under one fixed calibration threshold, which is more consistent with the industry FR scenarios.
- Our proposed Threshold Consistency Penalty (TCP) scheme can improve the fairness on threshold across the domains by penalizing the domains with high thresholds, where we introduce the Implicit Domain Division (IDD) as well as the Calibration and Domain Thresholds Estimation (CDTE).
- Extensive experiments on multiple face recognition benchmarks demonstrate the effectiveness of our proposed method.

2   Related Work

Face Recognition. Face recognition (FR) is a key technique for biometric authentication in many applications (e.g., electronic payment, video surveillance). The success of deep FR can be credited to the following three important reasons: large-scale datasets [2,16,46,50], powerful deep neural networks [25,31,32,35,36] and effective loss functions [1,4–6,17,19–24,26,28,34,37,38,48]. The mainstream of recent studies is to introduce a new objective function to maximize inter-class discriminability and intra-class compactness. For example, Triplet loss [30] enlarges the distances of negative pairs and reduce the distances of positive pairs in the Euclidean space. Recently, the angular constraint is introduced into the cross-entropy loss function to improve the discriminative ability of the learned representation [26,27]. For example, CosFace [39] and ArcFace [5] utilize a margin item for better discriminative capability of the feature representation. Besides, some mining-based loss functions (e.g., CurricularFace [14] and MV-Arc-Softmax [43]) further consider the difficulty degree of samples to emphasize the mis-classified samples and achieve better results. In addition, GroupFace [18] aggregates implicit group-aware representations to improve the discriminative ability of feature representations by using self-distributed labeling trick. Moreover, VPL [6] additionally introduces the sample-to-sample comparisons into the training process for reducing the gap between the training and evaluation processes for FR. Overall, existing methods mainly aim to improve the generalization and discriminative abilities of the learned feature representation, but they have not considered the limitations of existing evaluation protocols, where different datasets use different thresholds. In contrast, OneFace investigates the gap between existing evaluation protocols and the practical deployment scenarios from a new perspective, and we propose the OTA evaluation protocol to evaluate the performance of different datasets under the same calibration threshold.

Fairness. Recently, more and more attention has been attracted to the fairness for FR models. A straightforward way to tackle the fairness issue is to build large-scale training datasets (e.g., MS-Celeb-1M [10], Glint360k [2] and

WebFace260M [50]). Unfortunately, the time-consuming collected datasets often include unbalanced distributions of different attributes (e.g., race, gender, age), which also introduce inherent bias on different attributes. Recently, Wang et al. [41] introduce the BUPT-balanced as a balanced dataset on race, and BUPT-Globalface to reveal the real distribution of the world's population for fairness study. However, it is still difficult to collect the FR datasets with balanced distribution on different attributes and it is also unclear if the FR models trained on attribute-balanced datasets can eliminate the fairness bias completely. Therefore, some works have proposed to design effective algorithms instead of collecting datasets. For example, Wang et al. [42] propose a deep information maximization adaptation network to transfer the knowledge from Caucasians to other races, and propose another reinforcement learning-based method [41] to learn the optimal margins for different racial groups. Gong et al. [7,8] further utilize a debiasing adversarial network with four specific classifiers, where one classifier is designed for identification and the others are designed for demographic attributes. Xu et al. [45] propose to promote the consistency of instance FPRs to improve fairness across different races. In general, existing methods usually consider improving the accuracy and fairness on all races, but different thresholds are still used for different domains (e.g., races). In contrast to these methods, our proposed OneFace focuses on improving fairness of different domains with the same calibration threshold, which is more compatible with the real-world scenarios and provides new insight for FR.

3   Preliminary

In this section, we take the widely used 1:1 face verification evaluation protocol as an example to show the evaluation process of FR. False Accept Rate (FAR) and True Accept Rate (TAR) are used in face verification. Given N_p positive pairs, the TAR α is computed as follows:

$$\alpha = \frac{1}{N_p} \sum_{j=1}^{N_p} \mathbb{1}(s_p^j > t),\qquad(1)$$

where t is the chosen similarity score threshold and s_p^j is the similarity score of the j-th positive pair. $\mathbb{1}(x)$ is the indicator function, which returns 1 when x is true and returns 0 when x is false. Similarly, given N_n negative pairs, the FAR β is defined as follows:

$$\beta = \frac{1}{N_n} \sum_{i=1}^{N_n} \mathbb{1}(s_n^i > t),\qquad(2)$$

where s_n^i is the similarity score of the i-th negative pair.

In the testing process, we fix a FAR (e.g., 1e−4) and calculate the corresponding TAR to represent the performance of the FR models. For each dataset, the threshold t under the specific FAR β in Eq. 2 can be generated by the quantile of the similarity scores of all negative pairs. Then, based on the similarities of all positive pairs and the threshold t, we calculate the TAR α.

4 OneFace

In this section, we describe our OneFace framework, which includes the newly proposed One-Threshold-for-All (OTA) evaluation protocol and Threshold Consistency Penalty (TCP) scheme.

4.1 One-Threshold-for-All Evaluation Protocol

We first discuss the necessity of One-Threshold-for-All (OTA) evaluation protocol, and then introduce the details of the OTA evaluation protocol for FR, where we first estimate the calibration threshold under the overall FAR (e.g., 1e−4) and then calculate the performance results for different datasets.

Necessity of the OTA Evaluation Protocol. The above-mentioned evaluation protocol (TAR@FAR) in Sect. 3 has been adopted in many works. Nevertheless, we argue that the current evaluation protocols are not compatible with the practical FR, as different testing datasets use different thresholds even if the model and the pre-defined FAR are the same. Moreover, we can observe the Threshold Imbalance phenomenon, where these thresholds vary a lot when these datasets are from different domains. As shown in Table 1, we report the TAR and threshold results of different races from the RFW [42] dataset under the FAR of 1e−4 and 1e−5, where the thresholds are estimated within each race under the same FAR. It can be easily observed that current evaluation protocol adopts different thresholds for different races and the thresholds differ a lot in some races. For example, the threshold of African is much higher than the threshold of Caucasian under the same FAR. In contrast, when FR model is deployed in practice, only one fixed threshold score (i.e., calibration threshold) is applied for all scenarios, which means that the current evaluation protocol is not fully consistent with the practical FR applications. Therefore, it is critical to evaluate the performance of all domains using one fixed threshold for FR.

Here, we describe the evaluation process of OTA for G testing datasets. Given the overall FAR (e.g., 1e−4), we first generate the fixed calibration threshold. Then, based on the calibration threshold, we calculate the TAR and FAR results for different datasets. Finally, we also define the fairness metric to represent the degree of the threshold imbalance for these datasets.

Table 1. The threshold and TAR results based on classical 1:1 verification evaluation protocol when FAR is 1e−4 and 1e−5 on different races from the RFW dataset [42].

Results	FAR = 1e−4				FAR = 1e−5			
	African	Asian	Caucasian	Indian	African	Asian	Caucasian	Indian
Threshold	0.455	0.403	0.384	0.419	0.511	0.465	0.443	0.478
TAR	92.29	92.61	95.58	94.47	86.25	85.90	91.05	89.54

Calibration Threshold Estimation. In OTA, given G testing datasets, we describe two types of calibration threshold estimation methods. In the first type, we directly combine these G datasets into one whole dataset and extract the features of all negative pairs to calculate the similarities of these pairs. Then, we obtain the threshold score under the overall FAR (e.g., 1e−4) as the calibration threshold t_c. The first type is to estimate the threshold under the overall FAR based on the negative pairs constructed from all datasets, which is feasible and effective when the number of negative pairs is relatively acceptable. However, when the number of negative pairs increases, large computation costs for threshold estimation are needed. Therefore, we propose another calibration threshold estimation method by adopting an extra dataset (e.g., FairFace dataset [15]), which is called as calibration dataset. Specifically, we suppose that the calibration dataset covers images from as many domains as possible, which aims to make the calibration threshold t_c general and accurate. Besides, the size of the calibration dataset should be acceptable, which leads to affordable computation costs. Similarly, we calculate the similarities of all negative pairs from the calibration dataset, and obtain the calibration threshold t_c under the overall FAR.

Performance Under the Calibration Threshold. After obtaining the calibration threshold t_c, for g-th dataset, where $g \in \{1, ..., G\}$, we can easily produce the TAR α_g and FAR β_g within g-th dataset under the threshold t_c. Then, we directly adopt the mean $\mu_\alpha = \frac{1}{G} \sum_{g=1}^{G} \alpha_g$ and the variance $\sigma_\alpha^2 = \frac{1}{G} \sum_{g=1}^{G} (\alpha_g - \mu_\alpha)^2$ of all datasets to represent the overall TAR performance for these datasets. As the FAR value is usually small (e.g., 1e−4) and the magnitude of the FARs under the same threshold among different domains varies greatly, we adopt the $\log_{10}$ operation to maintain the monotonicity and simplify the calculation. Additionally, without this operation, statistics values are dominated by the large FAR value. For example, the mean of $\{1e−3, 1e−4, 1e−5\}$ is dominated by 1e−3, where 1e−5 is ignored. Thus, for the results of FAR, we utilize the mean $\mu_\beta = \frac{1}{G} \sum_{g=1}^{G} -\log_{10}\beta_g$ and the variance $\sigma_\beta^2 = \frac{1}{G} \sum_{g=1}^{G} (-\log_{10}\beta_g - \mu_\beta)^2$ of all datasets. Furthermore, we propose a fairness metric denoted as γ to represent the degree of the threshold imbalance across the G datasets, Specifically, we also compute the threshold t_d^g within g-th dataset under the overall FAR, where we call t_d^g as the domain threshold for the g-th dataset. Meanwhile, we define the deviations between G domain thresholds (i.e., $\{t_d^g\}_{g=1}^{G}$) and the calibration threshold t_c as the fairness metric γ, which is illustrated as follows:

$$\gamma = \sqrt{\frac{1}{G} \sum_{g=1}^{G} (t_d^g - t_c)^2}. \tag{3}$$

In Eq. 3, γ is larger when the degree of threshold imbalance is more serious.

4.2 Threshold Consistency Penalty

To improve the performance under the OTA evaluation protocol, we propose the Threshold Consistency Penalty (TCP) scheme to mitigate the threshold

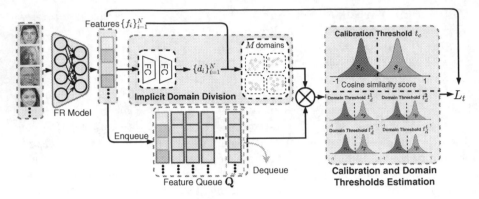

Fig. 2. The framework of our Threshold Consistency Penalty (TCP) scheme, which includes the Implicit Domain Division (IDE) as well as the Calibration and Domain Thresholds Estimation (CDTE). In each iteration, we first use the FR model to extract the features $\{f_i\}_{i=1}^N$ of each mini-batch and update the feature queue $\mathbf{Q}$, where N is the number of samples in each mini-batch. Then, we use the IDD to divide the $\{f_i\}_{i=1}^N$ into M domains implicitly, where the domain loss L_d is used. After that, we calculate the calibration threshold t_c and domain thresholds $\{t_d^m\}_{m=1}^M$. Finally, based on the t_c, $\{t_d^m\}_{m=1}^M$, $\{f_i\}_{i=1}^N$ and ground-truth identity labels, we calculate the TCP loss L_t.

imbalance among different domains in training as shown in Fig. 2, where we define a domain as a set of samples that share any common visual-or-non-visual properties for FR.

Specifically, TCP includes Implicit Domain Division (IDD) as well as Calibration and Domain Thresholds Estimation (CDTE). In IDD, we propose to divide the images of each mini-batch into several domains without additional annotations. In CDTE, we first build sufficient negative pairs using the features of the current batch and the features in our proposed feature queue. Then, we compute the similarities of these negative pairs and estimate the calibration and domain thresholds. Finally, based on the calibration and domain thresholds, we adaptively adjust the loss weights of samples from each domain.

Implicit Domain Division. As the domain labels are usually unavailable in the training dataset, our IDD is trained in a self-supervised manner, which predicts the domain label for each sample without any explicit ground-truth information. Specifically, inspired by GroupFace [18], our IDD is implemented by two fully-connected layers and a softmax layer, which takes the feature representation f_i of the i-th sample as input and predicts the domain probabilities as follows:

$$\{p_i^m\}_{m=1}^M = \mathcal{H}(f_i). \tag{4}$$

$\mathcal{H}$ denotes the neural network of IDD. M is a pre-defined hyperparameter on the number of implicit domains, which is not related to the number of evaluation datasets (i.e., G in OTA). p_i^m is the domain probability for m-th domain. Additionally, $\mathcal{H}$ is trained by the self-distributed labeling strategy. Specifically,

$\{p_i^m\}_{m=1}^M$ is the initial predictive domain probabilities for i-th sample. Following [18], to generate uniformly-distributed domain labels, we use modified probability regulated by a prior probability, where an expectation-normalized strategy is used. The updated domain probability $\tilde{p}_i^m$ for m-th domain is as follows:

$$\tilde{p}_i^m = \frac{1}{M}(p_i^m - \frac{1}{T}\sum_{i=1}^T p_i^m) + \frac{1}{M}, \tag{5}$$

where T is the number of samples to calculate the expectation value. We directly set T as the number of samples in each mini-batch. Thus, the expectation of the expectation-normalized probability $\frac{1}{T}\sum_{i=1}^T \tilde{p}_i^m = \frac{1}{M}$. The domain label $d_i \in \{1, ..., M\}$ for i-th sample is obtained as $d_i = \arg\max_m \tilde{p}_i^m$. Meanwhile, to reduce the divergence between the prediction probabilities and the generated domain label, a domain loss L_d based on cross-entropy loss is defined as follows:

$$L_d = -\frac{1}{N}\sum_{i=1}^N \log(\frac{e^{p_i^{d_i}}}{\sum_{m=1}^M e^{p_i^m}}), \tag{6}$$

where N is the number of samples in each iteration.

Calibration and Domain Thresholds Estimation. To estimate the calibration and domain thresholds in the training process, we first need to construct sufficient negative pairs with high qualities. Inspired by MoCo [11], for unsupervised learning, which adopts a memory bank from the previous mini-batches to obtain sufficient negative samples, we propose to build a feature queue $\mathbf{Q} \in \mathbb{R}^{K \times N \times d}$ to construct sufficient negative pairs, as shown in Fig. 2, where K is the number of iterations, and d represents the dimension of the feature representation extracted by the neural network for each face image. Meanwhile, as discussed in VPL [6], features drift slowly for FR models, which represents that features extracted previously can be considered as an approximation of the output of the current network within a certain number of training steps. Thus, we could set K as a relatively large value ($K = 1000$ in our work) to generate sufficient negative pairs with high qualities. Furthermore, we establish an auxiliary label queue, $\mathbf{Q}' \in \mathbb{R}^{K \times N}$ to store the identity labels for the features in $\mathbf{Q}$. In each iteration, we first extract the features $\{f_i\}_{i=1}^N$ of the current batch, where y_i is the corresponding label of f_i. Then, the features $\{f_i\}_{i=1}^N$ and the labels $\{y_i\}_{i=1}^N$ are enqueued into feature queue $\mathbf{Q}$ and label queue $\mathbf{Q}'$, respectively. After that, the features and labels from the oldest batch in $\mathbf{Q}$ and $\mathbf{Q}'$ are also dequeued. Finally, we can construct the negative pairs based on $\{f_i\}_{i=1}^N$ and $\mathbf{Q}$.

For Calibration Threshold Estimation, we calculate the similarities of all negative pairs, and generate the calibration threshold t_c in training under the overall FAR (e.g., 1e−4). For Domain Threshold Estimation, we first generate the domain labels for the samples of the current mini-batch based on IDD. Then, to accurately estimate the threshold distribution for each domain m, only the features of samples with the same domain label (i.e., m) are selected to construct domain-specific negative pairs with the features from $\mathbf{Q}$. By calculating the similarities of such domain-specific negative pairs, we can obtain the domain threshold t_d^m under the same FAR value (e.g., 1e−4) for the m-th domain. Finally, the calibration threshold t_c and domain thresholds $\{t_d^m\}_{m=1}^M$ are obtained.

Loss Formulation. After generating the calibration and domain thresholds, we define the TCP loss L_t from the domain level as follows:

$$L_t = \frac{1}{N} \sum_{m=1}^M \sum_{i \in \mathbf{T}_m} \left(\frac{t_d^m}{t_c} \cdot L_i\right). \tag{7}$$

$\mathbf{T}_m$ is an index set, which contains the indices of samples with domain label m in each mini-batch. N is the number of samples in each mini-match, and L_i denotes the classification loss for i-th sample. In our work, we utilize the widely-used ArcFace [5] loss as L_i. To this end, our TCP loss will enforce the neural network to pay more attention for these samples from domains with $t_d^m > t_c$ as there are more false positive pairs with higher scores than t_c, and we can automatically down-weight the contribution of these samples from domains with $t_d^m < t_c$ during training. In other words, our TCP loss aims to reduce the degree of threshold imbalance across multiple domains by dynamically adjusting the loss weights for M domains. It should be mentioned that L_i can be replaced with many existing loss functions [14,39]. Finally, the overall loss function of our proposed method is defined as follows:

$$L = L_t + \lambda L_d, \tag{8}$$

where λ is the loss weight for the domain loss L_d of in our IDD.

5   Experiments

In this section, we first report the results of different methods on multiple cross-domain settings under our proposed OTA evaluation protocol, where one fixed calibration threshold for different datasets. Then, we perform detailed analysis and discussion to further show the effectiveness of our method.

5.1   Implementation Details

Dataset. Our experimental settings include two settings (i.e., cross-race and cross-gender settings) as follows. For cross-race setting, we follow [45] to employ the BUPT-Balancedface [41] as the training dataset, and use the RFW dataset [42] as the testing dataset with four race groups (i.e., African, Asian, Caucasian, and Indian), where we directly use the whole RFW dataset to estimate the calibration threshold under the overall FAR for OTA evaluation protocol. For the cross-gender setting, we follow many existing works [5,14] to use the refined version of MS-1M [10] dataset as the training dataset. For the testing dataset of cross-gender setting, we split the IJB-B dataset [44] into two datasets (i.e., IJB-B(F), IJB-B(M)) manually based on the gender attribute (i.e., female or male), where we also use the whole IJB-B dataset to estimate the calibration threshold under the overall FAR for OTA evaluation protocol.

Experimental Setting. For the pre-processing of the training data and testing data, we follow [5,6,14] to generate the normalized face crops (112×112) with five landmarks detected by MTCNN [47]. For the backbone network for cross-race and cross-dataset settings, we follow the state-of-the-art method [45] to use the ResNet34 [12] to produce 512-dim feature representation. For the backbone network for cross-gender setting, we use the ResNet100 [12] for all methods. For the training process on BUPT-Balancedface [41], the initial learning rate is 0.1 and divided by 10 at the 55k, 88k, 99k iterations, where the total iteration is set as 110k. For the training process on the refined MS-1M [10], the initial learning rate is 0.1 and divided by 10 at the 110k, 190k, 220k iterations, where the total iteration is set as 240k. The batchsize is set as 512 for all experiments. For the feature queue $\mathbf{Q}$, d is set as 512, the number of iterations (i.e., K) in the feature queue and label queue is set as 1000. In training, the number of implicit domains (i.e., M) in IDD is set as 8. The loss weight (i.e., λ) of the domain loss L_d is set as 0.05. Under the OTA evaluation protocol, G is set as 4, 2 for cross-race and cross-gender settings, respectively, and we report the results of our method and recent widely-used loss functions [5,14,18,39].

5.2 Experimental Results Under the OTA Evaluation Protocol

Results on Cross-Race Setting. As shown in Table 2, for cross-race setting, we report the results of different methods among four demographic groups in the RFW dataset [42] under the OTA evaluation protocol, where we use the same calibration threshold for these groups. In Table 2, for the TAR results, when compared with methods, we observe that our method achieves higher TAR with lower variance. For the $-\log_{10}$FAR, our method also achieves lower variance, which shows the effectiveness of our method.

Results on Cross-Gender Setting. As shown in Table 3, we report the results of different methods under the OTA evaluation protocol for cross-gender setting. Specifically, we divide the original IJB-B dataset [44] into two datasets (i.e., IJB-B(F) and IJB-B(M)) based on the gender attribute, where IJB-B(F) and IJB-B(M) represents the female and male datasets, respectively. Then, we use the whole IJB-B dataset to estimate the calibration threshold under the overall FAR of 1e−4. In Table 3, we have the following observations: (1) The performance of IJB-B(F) is lower than IJB-B(M) a lot, which indicates the gender attribute influences the FR performance greatly. (2) Our method achieves better average performance results and lower variances when compared with other baseline methods on both TAR and FAR metrics.

5.3 Analysis

Analysis on the Classical Evaluation Results on the RFW Dataset. As shown in Table 4, we also report the results of different methods based on the

Table 2. The performance of different methods under the OTA evaluation protocol when overall FAR is 1e−4 for cross-race setting.

Method	TAR						$-\log_{10}$FAR					
	African	Asian	Caucasian	Indian	Avg.↑	Std.↓	African	Asian	Caucasian	Indian	Avg.↑	Std.↓
ArcFace [5]	96.71	93.80	95.33	96.22	95.51	1.107	3.030	3.791	4.052	3.519	3.598	0.378
CosFace [39]	96.44	91.83	94.36	94.99	94.41	1.667	3.024	3.810	4.053	3.477	3.591	0.386
CurricularFace [14]	96.79	93.80	95.41	96.14	95.54	1.114	2.999	3.811	4.045	3.554	**3.602**	0.389
GroupFace [18]	96.76	94.00	95.62	96.27	95.66	1.041	3.010	3.778	4.028	3.552	3.592	0.376
Ours	96.96	95.02	95.92	96.43	**96.08**	**0.715**	3.057	3.604	3.980	3.641	3.571	**0.331**

Table 3. The performance of different methods under the OTA evaluation protocol when overall FAR is 1e−4 for cross-gender setting.

Method	TAR				$-\log_{10}$FAR			
	IJB-B(F)	IJB-B(M)	Avg.↑	Std.↓	IJB-B(F)	IJB-B(M)	Avg.↑	Std.↓
ArcFace [5]	91.85	96.96	94.41	2.555	3.535	4.079	3.807	0.272
CosFace [39]	91.46	96.69	94.07	2.615	3.550	4.037	3.794	0.244
CurricularFace [14]	91.71	96.98	94.35	2.635	3.559	4.037	3.798	0.239
GroupFace [18]	91.93	97.09	94.51	2.580	3.548	4.037	3.793	0.245
Ours	92.29	97.26	**95.03**	**2.335**	3.591	4.062	**3.827**	**0.236**

classical 1:1 verification results on the RFW dataset, where different races use different thresholds. In Table 4, we observe that our method also achieves better results when compared with other methods, which further shows the effectiveness of our method. Moreover, as shown in Table 5, we also report the verification accuracy results of different methods on the RFW dataset, where different races use different thresholds. Note that the results of other methods are directly quoted from [45]. In Table 5, we observe that our method also achieves better results on most cases when compared with other methods, which further shows the effectiveness of our method.

Analysis on the Fairness Metric. As shown in Table 6, we provide the fairness results of different methods under the OTA evaluation protocol when overall FAR is 1e−4 for cross-race setting, and we observe the fairness metric in Eq. 3 of our method is also lower than other methods, which shows that our method can mitigate the threshold imbalance greatly for cross-race setting.

Analysis on the Computation Costs. No extra costs (e.g., GPU memory usage, time) are required at inference. Besides, in training, when compared with ArcFace baseline method, the training time and GPU memory usage of our method are 1.183 times and 1.005 times, respectively, which is acceptable.

Analysis on the Effectiveness of TCP. In Fig. 3(a) and Fig. 3(b), we visualize the distributions of similarity scores on the African and Caucasian from the RFW dataset of different methods (i.e., ArcFace and Ours) and the red vertical line denotes the domain threshold within each group under the FAR of 1e−4. When compared with the ArcFace, the difference of the domain thresholds between

Table 4. 1:1 verification TAR results on the RFW dataset.

Method	TAR@FAR = 1e−4			
	African	Asian	Caucasian	Indian
ArcFace [5]	92.29	92.61	95.58	94.47
CosFace [39]	91.24	90.66	94.59	92.57
CurricularFace [14]	92.41	92.88	95.56	94.73
GroupFace [18]	92.48	92.82	95.73	94.69
Xu et al. [45]	93.31	93.05	95.71	92.89
Ours	**93.43**	**93.39**	**95.93**	**95.38**

Table 5. Verification accuracy (%) on the RFW dataset.

Methods	African	Asian	Caucasian	Indian	Avg.↑	Std.↓
ArcFace [5] (R34)	93.98	93.72	96.18	94.67	94.64	1.11
CosFace [39] (R34)	92.93	92.98	95.12	93.93	93.74	1.03
RL-RBN [41] (R34)	95.00	94.82	96.27	94.68	95.19	0.93
Xu et al. [45] (R34)	95.95	95.17	96.78	96.38	96.07	**0.69**
ArcFace [5] (R100)	96.43	94.98	97.37	96.17	96.24	0.98
Xu et al. [45] (R100)	97.03	95.65	97.60	**96.82**	96.78	0.82
Ours (R100)	**97.33**	**95.95**	**98.10**	96.55	**97.01**	0.79

African and Caucasian is smaller in our method, which demonstrates that our method can mitigate the threshold imbalance among different domains.

5.4   Discussion

Discussion on the OTA Evaluation Protocol. In our work, for FR, we propose the OTA protocol to measure the fairness problem by evaluating the results of different datasets under one fixed calibration threshold, and we are not to search for one threshold given fixed mixture distributions. Specifically, when an FR model is deployed on FR systems (e.g., hardware devices), only one fixed threshold is used and it is infeasible to select a threshold for each face image. The reasons are as follows: (1) FR systems usually focus on unconstrained (in the wild) scenarios (e.g., environment), which indicates that it is difficult to define or distinguish the specific scenarios for different face images (e.g., probe/gallery). In other words, FR models are supposed to work well for different scenarios (e.g. airport/train station, sunny day/rainy day). (2) Even if the scenario is constrained, it is still difficult to define the number of domains (e.g., facial appearance), as there are many different aspects to describe the property of each domain. For example, age (old, youth, child), gender (male, female), glasses (w, w/o) and many other implicit domains that cannot be observed. (3) If we select a threshold for each image, extra costs (e.g., domain prediction model)

Table 6. The fairness of different methods under the OTA evaluation protocol when overall FAR is 1e−4 for cross-race setting.

Models	ArcFace [5]	CosFace [39]	CurricularFace [14]	GroupFace [18]	Ours
Fairness (γ)	0.038	0.043	0.037	0.038	**0.030**

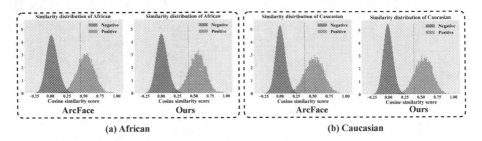

(a) African   (b) Caucasian

Fig. 3. (a) The similarity distributions of African. (b) The similarity distributions of Caucasian. The red vertical lines in (a) and (b) denote the thresholds under the FAR of 1e−4 within each race dataset. (Color figure online)

are needed, and accumulation errors will be brought by domain prediction and face verification tasks. (4) Setting different thresholds for some domains (e.g., gender, race) may also bring ethical risks. Overall, when compared with existing evaluation protocol, our OTA protocol is more consistent with real-world scenarios, and our proposed TCP method aims to align the similarity distributions of different domains and not to search for optimal thresholds.

Discussion on the Calibration Threshold. In the industry scenarios, we cannot obtain the similarity distributions of all datasets. Thus, we use the calibration threshold generated by the similarity distribution from the well-constructed calibration dataset to distinguish whether a face image pair belongs to the same identity. Specifically, to improve the robustness of the calibration threshold for practical FR, when the model is deployed, we can build the calibration dataset to generate the fixed calibration threshold, and the calibration threshold will be more suitable when the distribution of the calibration dataset is closer to the distribution of the real-world scenarios.

6   Conclusion

In our OneFace, we first investigate the limitations of the existing evaluation protocols for FR and propose the One-Threshold-for-All (OTA) evaluation protocol, which is more consistent with the deployment phase. Besides, we also propose the Threshold Consistency Penalty (TCP) scheme to improve the performance of FR models under the OTA protocol. Extensive experiments on multiple FR benchmark datasets demonstrate the effectiveness of our proposed method. Moreover, we hope our method can motivate other researchers to investigate the fairness problem on practical FR systems (e.g., more reliable fairness metric), and explore more research areas on the fairness in the future work.

Acknowledgments. This research was supported by National Natural Science Foundation of China under Grant 61932002.

References

1. An, X., et al.: Killing two birds with one stone: efficient and robust training of face recognition CNNs by partial FC. In: Proceedings of the IEEE/CVF Conference on Computer Vision and Pattern Recognition, pp. 4042–4051 (2022)
2. An, X., et al.: Partial FC: training 10 million identities on a single machine. In: Proceedings of the IEEE/CVF International Conference on Computer Vision (ICCV) Workshops, pp. 1445–1449, October 2021
3. Chopra, S., Hadsell, R., LeCun, Y.: Learning a similarity metric discriminatively, with application to face verification. In: 2005 IEEE Computer Society Conference on Computer Vision and Pattern Recognition (CVPR 2005), vol. 1, pp. 539–546. IEEE (2005)
4. Deng, J., Guo, J., Liu, T., Gong, M., Zafeiriou, S.: Sub-center arcface: boosting face recognition by large-scale noisy web faces. In: Proceedings of the IEEE Conference on European Conference on Computer Vision (2020)
5. Deng, J., Guo, J., Xue, N., Zafeiriou, S.: ArcFace: additive angular margin loss for deep face recognition. In: Proceedings of the IEEE Conference on Computer Vision and Pattern Recognition, pp. 4690–4699 (2019)
6. Deng, J., Guo, J., Yang, J., Lattas, A., Zafeiriou, S.: Variational prototype learning for deep face recognition. In: Proceedings of the IEEE/CVF Conference on Computer Vision and Pattern Recognition (CVPR), pp. 11906–11915, June 2021
7. Gong, S., Liu, X., Jain, A.K.: Jointly de-biasing face recognition and demographic attribute estimation. In: Vedaldi, A., Bischof, H., Brox, T., Frahm, J.-M. (eds.) ECCV 2020. LNCS, vol. 12374, pp. 330–347. Springer, Cham (2020). https://doi.org/10.1007/978-3-030-58526-6_20
8. Gong, S., Liu, X., Jain, A.K.: Mitigating face recognition bias via group adaptive classifier. In: Proceedings of the IEEE/CVF Conference on Computer Vision and Pattern Recognition, pp. 3414–3424 (2021)
9. Grother, P.J., Ngan, M.L., Hanaoka, K.K., et al.: Ongoing face recognition vendor test (FRVT) part 3: demographic effects. NIST Interagency/Internal Report (NISTIR), National Institute of Standards and Technology, Gaithersburg (2019)
10. Guo, Y., Zhang, L., Hu, Y., He, X., Gao, J.: MS-Celeb-1M: a dataset and benchmark for large-scale face recognition. In: Leibe, B., Matas, J., Sebe, N., Welling, M. (eds.) ECCV 2016. LNCS, vol. 9907, pp. 87–102. Springer, Cham (2016). https://doi.org/10.1007/978-3-319-46487-9_6
11. He, K., Fan, H., Wu, Y., Xie, S., Girshick, R.: Momentum contrast for unsupervised visual representation learning. In: Proceedings of the IEEE/CVF Conference on Computer Vision and Pattern Recognition, pp. 9729–9738 (2020)
12. He, K., Zhang, X., Ren, S., Sun, J.: Deep residual learning for image recognition. In: Proceedings of the IEEE Conference on Computer Vision and Pattern Recognition, pp. 770–778 (2016)
13. Hu, J., Shen, L., Sun, G.: Squeeze-and-excitation networks. In: Proceedings of the IEEE Conference on Computer Vision and Pattern Recognition, pp. 7132–7141 (2018)

14. Huang, Y., et al.: CurricularFace: adaptive curriculum learning loss for deep face recognition. In: Proceedings of the IEEE/CVF Conference on Computer Vision and Pattern Recognition, pp. 5901–5910 (2020)
15. Karkkainen, K., Joo, J.: FairFace: face attribute dataset for balanced race, gender, and age for bias measurement and mitigation. In: Proceedings of the IEEE/CVF Winter Conference on Applications of Computer Vision, pp. 1548–1558 (2021)
16. Kemelmacher-Shlizerman, I., Seitz, S.M., Miller, D., Brossard, E.: The megaface benchmark: 1 million faces for recognition at scale. In: Proceedings of the IEEE Conference on Computer Vision and Pattern Recognition, pp. 4873–4882 (2016)
17. Kim, M., Jain, A.K., Liu, X.: AdaFace: quality adaptive margin for face recognition. In: Proceedings of the IEEE/CVF Conference on Computer Vision and Pattern Recognition, pp. 18750–18759 (2022)
18. Kim, Y., Park, W., Roh, M.C., Shin, J.: GroupFace: learning latent groups and constructing group-based representations for face recognition. In: Proceedings of the IEEE/CVF Conference on Computer Vision and Pattern Recognition, pp. 5621–5630 (2020)
19. Kim, Y., Park, W., Shin, J.: BroadFace: looking at tens of thousands of people at once for face recognition. In: Vedaldi, A., Bischof, H., Brox, T., Frahm, J.-M. (eds.) ECCV 2020. LNCS, vol. 12354, pp. 536–552. Springer, Cham (2020). https://doi.org/10.1007/978-3-030-58545-7_31
20. Li, Z., et al.: Learning to auto weight: entirely data-driven and highly efficient weighting framework. In: Proceedings of the AAAI Conference on Artificial Intelligence, pp. 4788–4795 (2020)
21. Liu, C., et al.: Learning to learn across diverse data biases in deep face recognition. In: Proceedings of the IEEE/CVF Conference on Computer Vision and Pattern Recognition, pp. 4072–4082 (2022)
22. Liu, J., Qin, H., Wu, Y., Guo, J., Liang, D., Xu, K.: CoupleFace: relation matters for face recognition distillation. In: Proceedings of the European Conference on Computer Vision (2022)
23. Liu, J., Qin, H., Wu, Y., Liang, D.: AnchorFace: boosting tar@ far for practical face recognition. In: Proceedings of the AAAI Conference on Artificial Intelligence (2022)
24. Liu, J., et al.: DAM: discrepancy alignment metric for face recognition. In: Proceedings of the IEEE/CVF International Conference on Computer Vision, pp. 3814–3823 (2021)
25. Liu, J., Zhou, S., Wu, Y., Chen, K., Ouyang, W., Xu, D.: Block proposal neural architecture search. IEEE Trans. Image Process. 30, 15–25 (2020)
26. Liu, W., Wen, Y., Yu, Z., Li, M., Raj, B., Song, L.: SphereFace: deep hypersphere embedding for face recognition. In: Proceedings of the IEEE Conference on Computer Vision and Pattern Recognition, pp. 212–220 (2017)
27. Liu, W., Wen, Y., Yu, Z., Yang, M.: Large-margin softmax loss for convolutional neural networks. In: ICML, vol. 2, p. 7 (2016)
28. Meng, Q., Zhao, S., Huang, Z., Zhou, F.: MagFace: a universal representation for face recognition and quality assessment. In: Proceedings of the IEEE/CVF Conference on Computer Vision and Pattern Recognition, pp. 14225–14234 (2021)
29. Ranjan, R., Castillo, C.D., Chellappa, R.: L2-constrained softmax loss for discriminative face verification. arXiv preprint arXiv:1703.09507 (2017)
30. Schroff, F., Kalenichenko, D., Philbin, J.: FaceNet: a unified embedding for face recognition and clustering. In: CVPR, pp. 815–823 (2015)
31. Simonyan, K., Zisserman, A.: Very deep convolutional networks for large-scale image recognition. arXiv preprint arXiv:1409.1556 (2014)

32. Sun, Y., Chen, Y., Wang, X., Tang, X.: Deep learning face representation by joint identification-verification. In: Advances in Neural Information Processing Systems, pp. 1988–1996 (2014)
33. Sun, Y., Wang, X., Tang, X.: Deeply learned face representations are sparse, selective, and robust. In: Proceedings of the IEEE Conference on Computer Vision and Pattern Recognition, pp. 2892–2900 (2015)
34. Sun, Y., et al.: Circle loss: a unified perspective of pair similarity optimization. In: Proceedings of the IEEE/CVF Conference on Computer Vision and Pattern Recognition, pp. 6398–6407 (2020)
35. Szegedy, C., et al.: Going deeper with convolutions. In: Proceedings of the IEEE Conference on Computer Vision and Pattern Recognition (CVPR), June 2015
36. Taigman, Y., Yang, M., Ranzato, M., Wolf, L.: DeepFace: closing the gap to human-level performance in face verification. In: Proceedings of the IEEE Conference on Computer Vision and Pattern Recognition, pp. 1701–1708 (2014)
37. Wang, F., Cheng, J., Liu, W., Liu, H.: Additive margin softmax for face verification. IEEE Signal Process. Lett. **25**(7), 926–930 (2018)
38. Wang, F., Xiang, X., Cheng, J., Yuille, A.L.: NormFace: L2 hypersphere embedding for face verification. In: Proceedings of the 25th ACM International Conference on Multimedia, pp. 1041–1049 (2017)
39. Wang, H., et al.: CosFace: large margin cosine loss for deep face recognition. In: Proceedings of the IEEE Conference on Computer Vision and Pattern Recognition, pp. 5265–5274 (2018)
40. Wang, M., Deng, W.: Deep face recognition: a survey. arXiv 2018. arXiv preprint arXiv:1804.06655 (2018)
41. Wang, M., Deng, W.: Mitigating bias in face recognition using skewness-aware reinforcement learning. In: Proceedings of the IEEE/CVF Conference on Computer Vision and Pattern Recognition, pp. 9322–9331 (2020)
42. Wang, M., Deng, W., Hu, J., Tao, X., Huang, Y.: Racial faces in the wild: reducing racial bias by information maximization adaptation network. In: Proceedings of the IEEE/CVF International Conference on Computer Vision, pp. 692–702 (2019)
43. Wang, X., Zhang, S., Wang, S., Fu, T., Shi, H., Mei, T.: Mis-classified vector guided softmax loss for face recognition. In: Proceedings of the AAAI Conference on Artificial Intelligence, vol. 34, pp. 12241–12248 (2020)
44. Whitelam, C., et al.: IARPA Janus Benchmark-B face dataset. In: Proceedings of the IEEE Conference on Computer Vision and Pattern Recognition Workshops, pp. 90–98 (2017)
45. Xu, X., et al.: Consistent instance false positive improves fairness in face recognition. In: CVPR, pp. 578–586 (2021)
46. Yi, D., Lei, Z., Liao, S., Li, S.Z.: Learning face representation from scratch. arXiv preprint arXiv:1411.7923 (2014)
47. Zhang, K., Zhang, Z., Li, Z., Qiao, Y.: Joint face detection and alignment using multitask cascaded convolutional networks. IEEE Signal Process. Lett. **23**(10), 1499–1503 (2016)
48. Zhang, X., Fang, Z., Wen, Y., Li, Z., Qiao, Y.: Range loss for deep face recognition with long-tailed training data. In: Proceedings of the IEEE International Conference on Computer Vision, pp. 5409–5418 (2017)
49. Zhang, X., Zhao, R., Qiao, Y., Wang, X., Li, H.: AdaCos: adaptively scaling cosine logits for effectively learning deep face representations. In: Proceedings of the IEEE Conference on Computer Vision and Pattern Recognition, pp. 10823–10832 (2019)
50. Zhu, Z., et al.: WebFace260M: a benchmark unveiling the power of million-scale deep face recognition. In: CVPR, pp. 10492–10502, June 2021

Label2Label: A Language Modeling Framework for Multi-attribute Learning

Wanhua Li[1,2], Zhexuan Cao[1,2], Jianjiang Feng[1,2], Jie Zhou[1,2], and Jiwen Lu[1,2(✉)]

[1] Department of Automation, Tsinghua University, Beijing, China
{jfeng,jzhou,lujiwen}@tsinghua.edu.cn
[2] Beijing National Research Center for Information Science and Technology, Beijing, China

Abstract. Objects are usually associated with multiple attributes, and these attributes often exhibit high correlations. Modeling complex relationships between attributes poses a great challenge for multi-attribute learning. This paper proposes a simple yet generic framework named Label2Label to exploit the complex attribute correlations. Label2Label is the first attempt for multi-attribute prediction from the perspective of language modeling. Specifically, it treats each attribute label as a "word" describing the sample. As each sample is annotated with multiple attribute labels, these "words" will naturally form an unordered but meaningful "sentence", which depicts the semantic information of the corresponding sample. Inspired by the remarkable success of pre-training language models in NLP, Label2Label introduces an image-conditioned masked language model, which randomly masks some of the "word" tokens from the label "sentence" and aims to recover them based on the masked "sentence" and the context conveyed by image features. Our intuition is that the instance-wise attribute relations are well grasped if the neural net can infer the missing attributes based on the context and the remaining attribute hints. Label2Label is conceptually simple and empirically powerful. Without incorporating task-specific prior knowledge and highly specialized network designs, our approach achieves state-of-the-art results on three different multi-attribute learning tasks, compared to highly customized domain-specific methods. Code is available at https://github.com/Li-Wanhua/Label2Label.

Keywords: Multi-attribute · Language modeling · Attribute relations

1 Introduction

Attributes are mid-level semantic properties for objects which are shared across categories [14–16,31]. We can describe objects with a wide variety of attributes. For example, human beings easily perceive gender, hairstyle, expression, and so

Supplementary Information The online version contains supplementary material available at https://doi.org/10.1007/978-3-031-19775-8_33.

S. Avidan et al. (Eds.): ECCV 2022, LNCS 13672, pp. 562–579, 2022.
https://doi.org/10.1007/978-3-031-19775-8_33

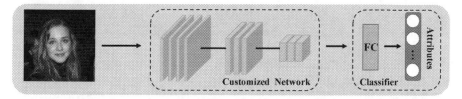

(a) Existing Multi-task Learning Framework

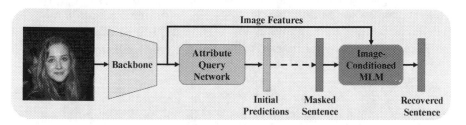

(b) Our Language Modeling Framework

Fig. 1. Comparisons of the existing multi-task learning framework and our proposed language modeling framework.

on from a facial image [30,32]. Multi-attribute learning, which aims to predict the attributes of an object accurately, is essentially a multi-label classification task [49]. As multi-attribute learning involves many important tasks, including facial attribute recognition [5,24,38], pedestrian attribute recognition [17,23,51], and cloth attribute prediction [37,59], it plays a central role in a wide range of applications, such as face identification [5], scene understanding [48], person retrieval [28], and fashion search [2].

For a given sample, many of its attributes are correlated. For example, if we observe that a person has blond hair and heavy makeup, the probability of that person being attractive is high. Another example is that the attributes of beard and woman are almost impossible to appear on a person at the same time. Modeling complex inter-attribute associations is an important challenge for multi-attribute learning. To address this challenge, most existing approaches [5,23,45,51] adopt a multi-task learning framework, which formulates multi-attribute recognition as a multi-label classification task and simultaneously learns multiple binary classifiers. To boost the performance, many methods further incorporate domain-specific prior knowledge. For example, PS-MCNN [5] divides all attributes into four groups and presents highly customized network architectures to learn shared and group-specific representations for face attributes. In addition, some methods attempt to introduce additional domain-specific guidance [24] or annotations [37]. However, these methods struggle to model sample-wise attribute relationships with a simple multi-task learning framework.

Recent years have witnessed great progress in the large-scale pre-training language models [4,11,43]. As a representative work, BERT [11] utilizes a masked language model (MLM) [52] to capture the word co-occurrence and language structure. Inspired by these methods, we propose a language modeling framework named Label2Label to model the complex instance-wise attribute relations. Specifically, we regard an attribute label as a "word", which describes the current state of the sample from a certain point of view. For example, we treat the labels "attractive" and "no eyeglasses" as two "words", which give us a sketch of the sample from different perspectives. As multiple attribute labels of each sample are used to depict the same object, these "words" can be organized as an unordered yet meaningful "sentence". For example, we can describe the human face in Fig. 1 with the sentence "attractive, not bald, brown hair, no eyeglasses, not male, wearing lipstick, ...". Although this "sentence" has no grammatical structure, it can convey some contextual semantic information. By treating multiple attribute labels as a "sentence", we exploit the correlation between attributes with a language modeling framework.

Our proposed Label2Label consists of an attribute query network (AQN) and an image-conditioned masked language model (IC-MLM). The attribute query network first generates the initial attribute predictions. Then these predictions are treated as pseudo label "sentences" and sent to the IC-MLM. Instead of simply adopting the masked language modeling framework, our IC-MLM randomly masks some "word" tokens from the pseudo label "sentence" and predicts the masked "words" conditioned on the masked "sentence" and image features. The proposed image-conditioned masked language model provides partial attribute prompts during the precise mapping from images to attribute categories, thereby facilitating the model to learn complex sample-level attribute correlations. We take facial attribute recognition as an example and show the key differences between our method and existing methods in Fig. 1.

We summarize the contributions of this paper as follows:

- We propose Label2Label to model the complex attribute relations from the perspective of language modeling. As far as we know, Label2Label is the first language modeling framework for multi-attribute learning.
- Our Label2Label proposes an image-conditioned masked language model to learn complex sample-level attribute correlations, which recovers a "sentence" from the masked one conditioned on image features.
- As a simple and generic framework, Label2Label achieves very competitive results across three multi-attribute learning tasks, compared to highly tailored task-specific approaches.

2   Related Work

Multi-attribute Recognition: Multi-attribute learning has attracted increasing interest due to its broad applications [2,5,28]. It involves many different visual tasks [18,23,37] according to the object of interest. Many works focus on domain-specific network architectures. Cao *et al.* [5] proposed a partially shared multi-task convolutional neural network (PS-MCNN) for face attribute recognition. The PS-MCNN consists of four task-specific networks and one shared network to learn shared and task-specific representations. Zhang *et al.* [59] proposed Two-Stream Networks for clothing classification and attribute recognition. Since some attributes are located in the local area of the image, many methods [17,46,51] resort to the attention mechanism. Guo *et al.* [17] presented a two-branch network and constrained the consistency between two attention heatmaps. A multi-scale visual attention and aggregation method was introduced in [46], which extracted visual attention masks with only attribute-level supervision. Tang *et al.* [51] proposed a flexible attribute localization module to learn attribute-specific regional features. Some other methods [24,37] further attempt to use additional domain-specific guidance. Semantic segmentation was employed in [24] to guide the attention of the attribute prediction. Liu *et al.* [37] learned clothing attributes with additional landmark labels. There are also some methods [50,60] to study multi-attribute recognition with insufficient data, but this is beyond the scope of this paper.

Language Modeling: Pre-training language models is a foundational problem for NLP. ELMo [43] was proposed to learn deep contextualized word representations. It was trained with a bidirectional language model objective, which combined both a forward and backward language model. ELMo representations significantly improve the performance across six NLP tasks. GPT [44] employed a standard language model objective to pre-train a language model on large unlabeled text corpora. The Transformer was used as the model architecture. The pre-trained model was fine-tuned on downstream tasks and achieved excellent results in 9 of 12 tasks. BERT [11] used a masked language model pre-training objective, which enabled BERT to learn bidirectional representations conditioned on the left and right context. BERT employed a multi-layer bidirectional Transformer encoder and advanced the state-of-the-art performance. Our work is inspired by the recent success of these methods and is the first attempt to model multi-attribute learning from the perspective of language modeling.

Transformer for Computer Vision: Transformer [53] was first proposed for sequence modeling in NLP. Recently, Transformer-based methods have been deployed in many computer vision tasks [3,19,36,42,54,56,57]. ViT [13] demonstrated that a pure transformer architecture achieved very competitive results on image classification tasks. DETR [6] formulated the object detection as a set prediction problem and employed a transformer encoder-decoder architecture. Pix2Seq [8] regarded object detection as a language modeling task and obtained competitive results. Zheng *et al.* [61] replaced the encoder of FCN with a pure transformer for semantic segmentation. Liu *et al.* [34] utilized the Transformer decoder architecture for multi-label classification. Temporal query networks were introduced in [57] for fine-grained video understanding with a query-response mechanism. There are also some efforts [9,25,41] to apply Transformer to the task of multi-label image classification. Note that the main contribution of this paper is not the use of Transformer, but modeling multi-attribute recognition from the perspective of language modeling.

3   Approach

In this section, we first give an overview of our framework. Then we present the details of the proposed attribute query network and image-conditioned masked language model. Lastly, we introduce the training objective function and inference process of our method.

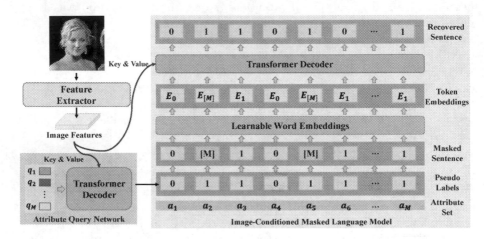

Fig. 2. The pipeline of our framework. We recover the entire label "sentence" with a Transformer decoder module, which is conditioned on the token embeddings and image features. Although there are some wrong "words" in the pseudo labels, which are shown in orange, we can treat them as another form of masks. Here E_1 or E_0 indicates the presence or absence of an attribute.

3.1   Overview

Given a sample x from a dataset $\mathcal{D}$ with M attribute types, we aim to predict the multiple attributes y to the image x. We let $\mathcal{A} = \{a_1, a_2, ..., a_M\}$ denote the attribute set, where $a_j (1 \leq j \leq M)$ represents the j-th attribute type. For simplicity, we assume that the values of all attribute types are binary. In other words, the value of a_j is 0 or 1, where 1 means that the sample has this attribute and 0 means not. However, our method can be easily extended to the case where each attribute type is multi-valued. With this assumption, we have $y \in \{0, 1\}^M$. Existing methods [5,23] usually employ a multi-tasking learning framework, which uses M binary classifiers to predict M attributes respectively. Binary cross-entropy loss is used as the objective.

This paper proposes a language modeling framework. We show the pipeline of our framework in Fig. 2. The key idea of this paper is to treat attribute labels as unordered "sentences" and use an image-conditioned masked language model to exploit the relationships between attributes. Although we can directly use the real attribute labels as the input of the IC-MLM during training, we cannot access these labels for inference. To address this issue, our Label2Label introduces an attribute query network to generate the initial attribute predictions. These predictions are then treated as pseudo-labels and used as input to the IC-MLM in the training and testing phases.

3.2 Attribute Query Network

Given an input image $x \in \mathbb{R}^{H_0 \times W_0 \times 3}$ and its corresponding label $y = \{y_j | 1 \leq j \leq M\}$, we send the image to a feature extractor to obtain the image features, where H_0 and W_0 denote the height and width of the input image respectively, y_j denotes the value of j-th attribute a_j for the sample x. As our framework is agnostic to the feature extractor, we can use any popular backbones such as ResNet-50 [20] and ViT [13]. A naive way to generate initial attribute predictions is to directly feed the extracted image features to a linear layer and learn M binary classifiers. As recent progress [12,25,34,57] shows the superiority of Transformer, we consider using the Transformer decoder to implement our attribute query network to generate initial predictions with higher quality.

Our attribute query network learns a set of permutation-invariant query vectors $Q = \{q_1, q_2, ..., q_M\}$, where each query q_j corresponds to an attribute type a_j. Then each query vector q_j pools the attribute-related features from the image features with Transformer decoder layers and generates the corresponding response vector r_j. Finally, we learn a binary classifier for each response vector to generate the initial attribute predictions.

Since many attributes are only located in some local areas of the image, using global image features is not an excellent choice. Therefore, we preserve the spatial dimensions of image features following [34]. For ResNet-50, we simply abandon the global pooling layer and employ the output of the last convolution block as the extracted features. We denote the extracted features as $X \in \mathbb{R}^{H \times W \times d}$, where H, W, and d represent the height, width, and channel of the image features respectively. To fit with the Transformer decoder, we reshape the feature to be $X' \in \mathbb{R}^{HW \times d}$. Following common practices [6,13], we add 2D-aware position embeddings $X_{pos} \in \mathbb{R}^{HW \times d}$ to the feature vectors X' to retain positional information. In this way, we obtain the visual feature vectors $\widetilde{X} = X' + X_{pos}$.

With the local visual contexts $\widetilde{X}$, the query features $Q = \{q_j \in \mathbb{R}^d | 1 \leq j \leq M\}$ are updated using multi-layer Transformer decoders. Formally, we update the query features Q_{i-1} in the i-th Transformer decoder layer as follows:

$$
\begin{aligned}
Q_{i-1}^{sa} &= \text{MultiHead}(Q_{i-1}, Q_{i-1}, Q_{i-1}), \\
Q_{i-1}^{ca} &= \text{MultiHead}(Q_{i-1}^{sa}, \widetilde{X}, X'), \\
Q_i &= \text{FFN}(Q_{i-1}^{ca}),
\end{aligned}
\tag{1}
$$

where the MultiHead() and FFN() denote the multi-head attention layer and feed-forward layer respectively. Here we set Q as Q_0. The design philosophy is that for each attribute query vector, it can give high attention scores to the interested local visual features to produce attribute-related features. This design is compatible with the locality of some attributes. Assuming that the attribute query network consists of L layers of Transformer decoders, then we denote $\boldsymbol{Q}_L$ as $\boldsymbol{R} = \{\boldsymbol{r}_1, \boldsymbol{r}_2, ..., \boldsymbol{r}_M\}$, where each response vector $\boldsymbol{r}_j \in \mathbb{R}^d$ corresponds to a query vector $\boldsymbol{q}_j$. With the response vectors, we use M independent binary classifiers to predict the attribute values $l_j = \sigma(\boldsymbol{W}_j^T \boldsymbol{r}_j + b_j)$, where $\boldsymbol{W}_j \in \mathbb{R}^d$ and $b_j \in \mathbb{R}^1$ are learnable parameters of the j-th attribute classifier, $\sigma(\cdot)$ is the sigmoid function and l_j is the predicted probability for attribute $\boldsymbol{a}_j$ of image $\boldsymbol{x}$. In the end, we read out the pseudo label "sentence" $\boldsymbol{s} = \{s_1, s_2, ..., s_M\}$ from the predictions $\{l_j\}$ with $\boldsymbol{s}_j = \mathbb{I}(l_j > 0.5)$, where $\mathbb{I}(\cdot)$ is an indicator function.

It is worth noting that the predictions from the attribute query network are not 100% correct, resulting in some wrong "words" in the generated label "sentence". However, we can treat the wrong "words" as another form of masks, because the wrong predictions account for only a small proportion. In fact, the masking strategy of the wrong word is artificially performed in some language models, such as BERT [11].

3.3   Image-Conditioned Masked Language Model

In existing multi-attribute databases, images are annotated with a variety of attribute labels. This paper is dedicated to modeling sample-wise complex attribute correlations. Instead of treating attribute labels as numbers, we regard them as "words". Since different attribute labels describe the object in an image from different perspectives, we can group them as a sequence of "words". Although the sequence is essentially an unordered "sentence" without any grammatical structure, it still conveys meaningful contextual information. In this way, we treat $\boldsymbol{y}$ as an unordered yet meaningful "sentence", where y_j is a "word".

By treating the labels as sentences, we resort to language modeling methods to mine the instance-level attribute relations effectively. In recent years, pre-training large-scale task-agnostic language models have substantially advanced the development of NLP, among which representative works include ELMo [43], GPT-3 [4], BERT [11], and so on. Inspired by the success of these methods, we consider a masked language model to learn the relationship between "words". We mask some percentage of the attribute label "sentence" $\boldsymbol{y}$ at random, and then reconstruct the entire label "sentence". Specifically, for a binary label sequence, we replace those masked "words" with a special work token [mask] to obtain the masked sentence. Then we input the masked sentence to a masked language model, which aims to recover the entire label sequence. While the MLM has proven to be an effective tool in NLP, directly using it for multi-attribute learning is not feasible. Therefore, we propose several important improvements.

Instance-Wise Attribute Relations: MLM essentially constructs the task $P(y_1, y_2, ..., y_M | \mathcal{M}(y_1), \mathcal{M}(y_2), ..., \mathcal{M}(y_M))$ to capture the "word" co-occurrence and learn the joint probability of "word" sequences $P(y_1, y_2, ..., y_M)$, where $\mathcal{M}()$

denotes the random masking operation. Such a naive approach leads to two problems. The first problem is that MLM only captures statistical attribute correlations. A diverse dataset means that the mapping $\{\mathcal{M}(y_1), \mathcal{M}(y_2), ..., \mathcal{M}(y_M)\} \mapsto \{y_1, y_2, ..., y_M\}$ is a one-to-many mapping. Therefore MLM only learns how different attributes are statistically related to each other. Meanwhile, our experiments find that this prior can be easily modeled by the attribute query network $P(y_1, y_2, ..., y_M|\boldsymbol{x})$. The second problem is that MLM and attribute query network cannot be jointly trained. Since MLM uses only the hard prediction of the attribute query network, the gradient from MLM cannot influence the training of the attribute query network. In this way, the method becomes a two-stage label refinement process, which significantly reduces the optimization efficiency.

To address these issues, we propose an image-conditioned masked language model to learn instance-wise attribute relations. Our IC-MLM captures the relations by constructing a task $P(y_1, y_2, ..., y_M|\boldsymbol{x}, \mathcal{M}(y_1), \mathcal{M}(y_2), ..., \mathcal{M}(y_M))$. Introducing an extra image condition is not trivial, as this fundamentally changes the behavior of MLM. With the conditions of image $\boldsymbol{x}$, the transformation $\{\boldsymbol{x}, \mathcal{M}(y_1), \mathcal{M}(y_2), ..., \mathcal{M}(y_M)\} \mapsto \{y_1, y_2, ..., y_M\}$ is an accurate one-to-one mapping. Our IC-MLM infers other attribute values by combining some attribute label prompts and image contexts in the precise image-to-label mapping, which facilitates the model to learn sample-level attribute relations. In addition, IC-MLM and the attribute query network can use shared image features, which enables them to be jointly optimized with a one-stage framework.

Word Embeddings: It is known that the word id is not a good word representation in NLP. Therefore, we need to map the word id to a token embedding. Instead of utilizing existing word embeddings with a large token vocabulary like BERT [11], we directly learn attribute-related word embeddings $\boldsymbol{E}$ from scratch. We use the word embedding module to map the "word" in the masked sentence to the corresponding token embedding. Since all attributes are binary, we need to build a token vocabulary with a size of $2M$ to model all possible attribute words. Also, we need to include the token embedding for the special word [mask]. This paper considers three different strategies for the [mask] token embedding. The first strategy believes the [mask] words for different attributes have different meanings, so M attribute-specific learnable token embeddings are learned, where one [mask] token embedding corresponds to one attribute. The second strategy treats the [mask] words for different attributes as the same word. Only one attribute-agnostic learnable token embedding is learned and shared by all attributes. The third strategy is based on the second strategy, which simply replaces the learnable token embedding with a fixed $\boldsymbol{0}$ vector. Our experiments find all three strategies work well while the first strategy performs best.

As mentioned earlier, we use pseudo labels $\boldsymbol{s} = \{s_1, s_2, ..., s_M\}$ as input to IC-MLM, so we actually construct $P(y_1, y_2, ..., y_M|\boldsymbol{x}, \mathcal{M}(\boldsymbol{s}_1), \mathcal{M}(\boldsymbol{s}_2), ..., \mathcal{M}(\boldsymbol{s}_M))$ as the task. We randomly mask out some "words" in the pseudo-label sequence with a probability of α to generate masked label "sentences". The "word" $\mathcal{M}(\boldsymbol{s}_j)$ in the masked label "sentences" may have three values: 0, 1, and [mask]. We use the word embedding module to map the masked labels "sentences" to a sequence of token embeddings $E = \{E_1, E_2, ..., E_M\}$ according to the word value, where $E_j \in \mathbb{R}^d$ denotes the embedding for "word" $\mathcal{M}(s_j)$.

Positional Embeddings: In BERT, the positional embedding of each word is added to its corresponding token embeddings to obtain the position information. Since our "sentences" are unordered, there is no need to introduce positional embeddings to "word" representations. We conducted experiments with positional embeddings by randomly defining some word order and found no improvement. Therefore we do not use positional embeddings for "word" representations and the learned model is permutation invariant for "words".

Architecture: In NLP, Transformer encoder layers are usually used to implement MLM, while we use multi-layer Transformer decoders to implement IC-MLM due to additional image input conditions. Following the design philosophy similar to the attribute query network, token embeddings E pool features from the local visual features X' with a cross-attention mechanism. We update the token features E_{i-1} in the i-th Transformer decoder layer as follows:

$$E_{i-1}^{sa} = \text{MultiHead}(E_{i-1}, E_{i-1}, E_{i-1}),$$
$$E_{i-1}^{ca} = \text{MultiHead}(E_{i-1}^{sa}, \widetilde{X}, X'), \tag{2}$$
$$E_i = \text{FFN}(E_{i-1}^{ca}).$$

We set E to E_0 and the number of Transformer decoder layers in IC-MLM to D. Then we denote E_D as $R' = \{r_1', r_2', ..., r_M'\}$, where r_j' corresponds to the updated feature of token E_j. In the end, we perform the final multi-attribute classification with linear projection layers. Formally, we have:

$$p_j = \sigma(W_j'^T r_j' + b_j'), 1 \leq j \leq M, \tag{3}$$

where $W_j' \in \mathbb{R}^d$ and $b_j' \in \mathbb{R}^1$ are the learnable parameters of the j-th attribute classifier, and p_j is the final predicted probability for attribute a_j of image x. Note that we are committed to recovering the entire label "sentence" and not just the masked part. In this reconstruction process, we expect our model to grasp the instance-level attribute relations.

3.4   Objective and Inference

As commonly used in most existing methods [23,38,46], we adopt the binary cross-entropy loss to train the IC-MLM. On the other hand, since most of the datasets for multi-attribute recognition are highly imbalanced, different tasks usually use different weighting strategies. The loss function for the IC-MLM is formulated as $\mathcal{L}_{mlm}(x) = \sum_{j=1}^{M} w_j(y_j \log(p_j) + (1 - y_j)\log(1 - p_j))$, where w_j is the weighting coefficient. According to different tasks, we choose different weighting strategies and always follow the most commonly used strategy for a

Table 1. Results with different Transformer decoder layers D for IC-MLM. We fix L as 1.

D	1	2	3	4
Error(%)	12.58	**12.49**	12.54	12.52

Table 2. Results with different Transformer decoder layers L for attribute query network. We fix D as 2.

L	1	2	3	4
Error(%)	**12.49**	12.52	12.50	12.58

Table 3. Results on the LFWA dataset with different mask ratios α.

α	0	0.1	0.15	0.2	0.3
Error(%)	12.55	**12.49**	12.55	12.54	12.57

Table 4. Results on the LFWA dataset with different coefficients λ.

λ	0.5	0.8	1	1.2	1.5
Error(%)	12.64	12.56	**12.49**	12.60	12.63

fair comparison. Meanwhile, to ensure the quality of the generated pseudo label sequences, we also supervise the attribute query network with the same loss function $\mathcal{L}_{aqn}(\boldsymbol{x}) = \sum_{j=1}^{M} w_j(y_j \log(l_j) + (1-y_j)\log(1-l_j))$. The final loss function $\mathcal{L}_{total}$ is a combination of the two loss functions above:

$$\mathcal{L}_{total}(\boldsymbol{x}) = \mathcal{L}_{aqn}(\boldsymbol{x}) + \lambda \mathcal{L}_{mlm}(\boldsymbol{x}), \tag{4}$$

where λ is used to balance these two losses. At inference time, we ignore the masking step and directly input the pseudo label "sentence" to the IC-MLM. Then the output of the IC-MLM is used as the final attribute prediction.

4 Experiments

In this section, we conducted extensive experiments on three multi-attribute learning tasks to validate the effectiveness of the proposed framework.

4.1 Facial Attribute Recognition

Dataset: LFWA [38] is a popular unconstrained facial attribute dataset, which consists of 13,143 facial images of 5,749 identities. Each facial image has 40 attribute annotations. Following the same evaluation protocol in [5,18,38], we partition the LFWA dataset into two sets, with 6,263 images for training and 6,880 for testing. All images are pre-cropped to a size of 250×250. We adopt the classification error for evaluation following [5,50].

Experimental Settings: We trained our model for 57 epochs with a batch size of 16. For optimization, we used an SGD optimizer with a base learning rate of 0.01 and cosine learning rate decay. The weight decay was set to 0.001. To augment the dataset, Rand-Augment [10] and Random horizontal flipping were performed. We also adopted Mixup [58] for regularization.

Table 5. Ablation experiments with different backbones.

Backbone	ResNet-50		ResNet-101		ViT-B	
Metric	Error(%)	MACs(G)	Error(%)	MACs(G)	Error(%)	MACs(G)
FC head	13.63 ± 0.02	5.30	13.05 ± 0.03	10.15	13.73 ± 0.02	16.85
AQN	13.36 ± 0.04	5.63	12.70 ± 0.02	10.48	13.32 ± 0.04	16.97
Label2Label	$\mathbf{12.49 \pm 0.02}$	6.30	$\mathbf{12.44 \pm 0.04}$	11.16	$\mathbf{12.79 \pm 0.01}$	17.23

Table 6. Results of different strategies for [Mask] embeddings.

Strategy	Error(%)
0 Vector	12.60
Attribute-agnostic	12.57
Attribute-specific	**12.49**

Table 7. Comparisons of MLM and IC-MLM.

Method	Architecture	Co-training with AQN	Error (%)
MLM	MLP	✗	13.34
	TransEncoder	✗	13.32
IC-MLM	TransDecoder	✗	13.01
	TransDecoder	✓	**12.49**

Parameters Analysis: We first analyze the influence of the number of Transformer decoder layers in the attribute query network and IC-MLM. The results are shown in Tables 1 and 2. We see that the best performance is achieved when $L = 1$ and $D = 2$. We further conduct experiments with different mask ratios α and list the results in Table 3. As we mentioned above, the wrong "words" in the pseudo label sequences also provide some form of masks. Therefore, our method performs well when $\alpha = 0$. We observe that our method attains the best performance when $\alpha = 0.1$. Table 4 shows the results with different λ, and we see that $\lambda = 1$ gives the best trade-off in (4). We consider three different strategies for [MASK] token embedding and list the results in Table 6. We see that the attribute-specific strategy achieves the best performance among them, as it better models the differences between the attributes. Unless explicitly mentioned, we adopt these optimal parameters in all subsequent experiments.

Ablation Study: To validate the effectiveness of our Label2Label, we also conduct experiments on the LFWA dataset with two baseline methods. We first consider the Attribute Query Network (AQN) method, which ignores the IC-MLM and treats the outputs of AQN in Fig. 2 as final predictions. FC Head method further replaces the Transformer decoder layers in AQN with a linear classification layer. To further verify the generalization of our method, we use different feature extraction backbone networks for ablation experiments. To better demonstrate the significance of the results, we also report the standard deviation. The results are presented in Table 5. In addition, we report the computation cost (MACs) of each method in Table 5. We observe that our method significantly outperforms FC Head and AQN across various backbones with marginal computational overhead, which illustrates the effectiveness of our method.

We then conducted experiments to show how image-conditioned MLM improves performance. The results are listed in Table 7. As we analyzed above, MLM leads to a two-stage label refinement process. We consider two network architectures to implement MLM: Transformer encoder and multilayer perceptron (MLP). The results show that none of them improve the performance of AQN (13.36%). The reason is that MLM only learns statistical attribute relations, and this prior is easily captured by AQN. Meanwhile, our IC-MLM learns instance-wise attribute relations. To see the

Table 8. Performance comparison with state-of-the-art methods on the LFWA dataset. We report the average classification error results. * indicates that additional labels are used for training, such as identity labels or segment annotations.

Method	Error(%)	Year
SSP + SSG [24]*	12.87	2017
He *et al.* [21]	14.72	2018
AFFAIR [29]	13.87	2018
GNAS [22]	13.63	2018
PS-MCNN [5]*	12.64	2018
DMM-CNN [39]	13.44	2020
SSPL [50]	13.47	2021
Label2Label	**12.49** ± 0.02	–

benefits of the additional image conditions, we still adopt the two-stage label refinement process, and train Transformer decoder layers with fixed image features. We see that performance is boosted to 13.01%, which demonstrates the effectiveness of modeling instance-wise attribute relations. We further jointly train the IC-MLM and attribute query network, which achieves significant performance improvement. These results illustrate the superiority of the proposed IC-MLM.

Comparison with State-of-the-Art Methods: Following [50], we employ ResNet50 as the backbone. We present the performance comparison on the LFWA dataset in Table 8. We observe that our method attains the best performance with a simple framework compared to highly tailored domain-specific methods. Label2Label even exceeds the methods [5,24] of using additional annotations, which further illustrates the effectiveness of our framework.

Visualization: As the Transformer decoder architecture is used to model the instance-level relations, our method can give better interpretable predictions. We visualize the attention scores in the IC-MLM with DODRIO [55]. As shown in Fig. 3, we see that related attributes tend to have higher attention scores.

4.2 Pedestrian Attribute Prediction

Dataset: The PA-100K [35] dataset is the largest pedestrian attribute dataset so far [51]. It contains 100,000 pedestrian images from 598 scenes, which are collected from real outdoor surveillance videos. All pedestrians in each image are annotated with 26 attributes including gender, handbag, and upper clothing. The dataset is randomly split into three subsets: 80% for training, 10% for validation, and 10% for testing. Following SSC [23], we merge the training set and the validation set for model training. We use five metrics: one label-based

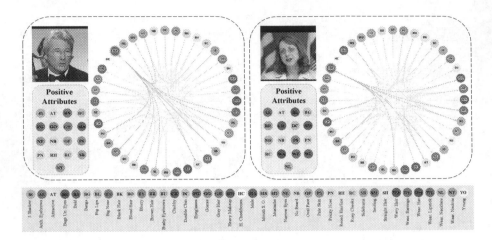

Fig. 3. Visualization of attention scores among attributes in the self-attention layer. We show the attention of the first head at layer 1 with two samples. The positive attributes of each sample are listed in the corresponding bottom-left corner.

and four instance-based. For the label-based metric, we adopt the mean accuracy (mA) metric. For instance-based metrics, we employ accuracy, precision, recall, and F1 score. As mentioned in [51], mA and F1 score are more appropriate and convincing criteria for class-imbalanced pedestrian attribute datasets.

Experimental Settings: Following the state-of-the-art methods [17,23], we adopted ResNet50 as the backbone network to extract image features. We first resize all images into 256×192 pixels. Then random flipping and random cropping were used for data augmentation. SGD optimizer was utilized with the weight decay of 0.0005. We set the initial learning rate of the backbone to 0.01. For fast convergence, we set the initial learning rate of the attribute query network and IC-MLM to 0.1. The batch size was equal to 64. We trained our model for 25 epochs using a plateau learning rate scheduler. We reduced the learning rate by a factor of 10 once learning stagnates and the patience was 4.

Results and Analysis: We report the results in Table 9. We observe that Label2Label achieves the best performance in mA, Accuracy, and F1 score. Compared to the previous state-of-the-art method SSC [23], which designs complex SPAC and SEMC modules to extract discriminative semantic features, our method achieves 0.37% performance improvements in mA. In addition, we report the re-implemented results of the MsVAA, VAC, and ALM methods in the same setting as did in [23]. Our method consistently outperforms these methods. We further show the results of the FC Head and Attribute Query Network. We see that the performance is improved by replacing the FC head with Transformer decoder layers, which shows the superiority of our attribute query network. Our

Table 9. Comparisons on the PA100K dataset. * represents the reimplementation performance using the same setting. We also report the standard deviations.

Method	mA	Accuracy	Precision	Recall	F1
DeepMAR [26]	72.70	70.39	82.24	80.42	81.32
HPNet [35]	74.21	72.19	82.97	82.09	82.53
VeSPA [47]	76.32	73.00	84.99	81.49	83.20
LGNet [33]	76.96	75.55	86.99	83.17	85.04
PGDM [27]	74.95	73.08	84.36	82.24	83.29
MsVAA [46]*	80.10	76.98	86.26	85.62	85.50
VAC [17]*	79.04	78.95	**88.41**	86.07	86.83
ALM [51]*	79.26	78.64	87.33	86.73	86.64
SSC [23]	81.87	78.89	85.98	**89.10**	86.87
FC Head	77.96 ± 0.06	75.86 ± 0.79	86.27 ± 0.13	84.16 ± 1.02	84.72 ± 0.55
AQN	80.89 ± 0.08	78.51 ± 0.08	86.15 ± 0.40	87.85 ± 0.43	86.58 ± 0.03
Label2Label	**82.24 ± 0.13**	**79.23 ± 0.13**	86.39 ± 0.32	88.57 ± 0.20	**87.08 ± 0.08**

Label2Label outperforms the attribute query network method by 1.35% for mA, which shows the effectiveness of the language modeling framework.

4.3   Clothing Attribute Recognition

Dataset: Clothing Attributes Dataset [7] consists of 1,856 images that contain clothed people. Each image is annotated with 26 clothing attributes, such as colors and patterns. We use 1,500 images for training and the rest for testing. For a fair comparison, we only use 23 binary attributes and ignore the remaining three multi-class value attributes as in [1,40]. We adopt accuracy as the metric and also report the accuracy of four clothing attribute groups following [1,40].

Experimental Settings: For a fair comparison, we utilized AlexNet to extract image features following [1,40]. We trained our model for 22 epochs using a cosine decay learning rate scheduler. We utilized an SGD optimizer with an initial learning rate of 0.05. The batch size was set to 32. For the attribute query network, we employed a 2-layer Transformer decoder ($L = 2$).

Results and Analysis: Table 10 shows the results. We observe that our Label2Label attains a total accuracy of 92.87%, which outperforms other methods with a simple framework. MG-CNN learns one CNN for each attribute, resulting in more training parameters and longer training time. Compared with the attribute query network method, our method achieves better performance on all attribute groups, which illustrates the superiority of our framework.

Table 10. The comparisons between our method and other state-of-the-art methods on the Clothing Attributes Dataset. We report accuracy and standard deviation.

Method	Colors	Patterns	Parts	Appearance	Total
S-CNN [1]	90.50	92.90	87.00	89.57	90.43
M-CNN [1]	91.72	94.26	87.96	91.51	91.70
MG-CNN [1]	93.12	95.37	88.65	91.93	92.82
Meng *et al.* [40]	91.64	96.81	89.25	89.53	92.39
FC Head	91.39 ± 0.23	96.07 ± 0.05	87.00 ± 0.27	88.21 ± 0.36	91.57 ± 0.12
AQN	91.98 ± 0.25	96.37 ± 0.23	88.19 ± 0.47	89.89 ± 0.33	92.29 ± 0.05
Label2Label	92.73 ± 0.07	96.82 ± 0.02	88.20 ± 0.09	90.88 ± 0.18	$\mathbf{92.87 \pm 0.03}$

5 Conclusions

In this paper, we have presented Label2Label, which is a simple and generic framework for multi-attribute learning. Different from the existing multi-task learning framework, we proposed a language modeling framework, which regards each attribute label as a "word". Our model learns instance-level attribute relations by the proposed image-conditioned masked language model, which randomly masks some "words" and restores them based on the remaining "sentence" and image context. Compared to well-optimized domain-specific methods, Label2Label attains competitive results on three multi-attribute learning tasks.

Acknowledgments. This work was supported in part by the National Key Research and Development Program of China under Grant 2017YFA0700802, in part by the National Natural Science Foundation of China under Grant 62125603 and Grant U1813218, in part by a grant from the Beijing Academy of Artificial Intelligence (BAAI). The authors would sincerely thank Yongming Rao and Zhiheng Li for their generous helps.

References

1. Abdulnabi, A.H., Wang, G., Lu, J., Jia, K.: Multi-task cnn model for attribute prediction. TMM **17**(11), 1949–1959 (2015)
2. Ak, K.E., Kassim, A.A., Lim, J.H., Tham, J.Y.: Learning attribute representations with localization for flexible fashion search. In: CVPR, pp. 7708–7717 (2018)
3. Bao, H., Dong, L., Wei, F.: Beit: bert pre-training of image transformers. arXiv preprint arXiv:2106.08254 (2021)
4. Brown, T.B., et al.: Language models are few-shot learners. In: NeurIPS (2020)
5. Cao, J., Li, Y., Zhang, Z.: Partially shared multi-task convolutional neural network with local constraint for face attribute learning. In: CVPR, pp. 4290–4299 (2018)
6. Carion, N., Massa, F., Synnaeve, G., Usunier, N., Kirillov, A., Zagoruyko, S.: End-to-end object detection with transformers. In: Vedaldi, A., Bischof, H., Brox, T., Frahm, J.-M. (eds.) ECCV 2020. LNCS, vol. 12346, pp. 213–229. Springer, Cham (2020). https://doi.org/10.1007/978-3-030-58452-8_13

7. Chen, H., Gallagher, A., Girod, B.: Describing clothing by semantic attributes. In: Fitzgibbon, A., Lazebnik, S., Perona, P., Sato, Y., Schmid, C. (eds.) ECCV 2012. LNCS, vol. 7574, pp. 609–623. Springer, Heidelberg (2012). https://doi.org/10.1007/978-3-642-33712-3_44

8. Chen, T., Saxena, S., Li, L., Fleet, D.J., Hinton, G.: Pix2seq: a language modeling framework for object detection. arXiv preprint arXiv:2109.10852 (2021)

9. Cheng, X., et al.: Mltr: multi-label classification with transformer. arXiv preprint arXiv:2106.06195 (2021)

10. Cubuk, E.D., Zoph, B., Shlens, J., Le, Q.: Randaugment: practical automated data augmentation with a reduced search space. In: NeurIPS, pp. 18613–18624 (2020)

11. Devlin, J., Chang, M.W., Lee, K., Toutanova, K.: Bert: pre-training of deep bidirectional transformers for language understanding. In: NAACL (2019)

12. Doersch, C., Gupta, A., Zisserman, A.: Crosstransformers: spatially-aware few-shot transfer. In: NeurIPS (2020)

13. Dosovitskiy, A., et al.: An image is worth 16×16 words: transformers for image recognition at scale. In: ICLR (2021)

14. Duan, K., Parikh, D., Crandall, D., Grauman, K.: Discovering localized attributes for fine-grained recognition. In: CVPR, pp. 3474–3481 (2012)

15. Farhadi, A., Endres, I., Hoiem, D., Forsyth, D.: Describing objects by their attributes. In: CVPR, pp. 1778–1785 (2009)

16. Feris, R.S., Lampert, C., Parikh, D.: Visual Attributes. Springer, Cham (2017). https://doi.org/10.1007/978-3-319-50077-5

17. Guo, H., Zheng, K., Fan, X., Yu, H., Wang, S.: Visual attention consistency under image transforms for multi-label image classification. In: CVPR, pp. 729–739 (2019)

18. Hand, E.M., Chellappa, R.: Attributes for improved attributes: a multi-task network utilizing implicit and explicit relationships for facial attribute classification. In: AAAI (2017)

19. He, K., Xinlei, C., Xie, S., Li, Y., Dollár, P., Girshick, R.: Masked autoencoders are scalable vision learners. arXiv preprint arXiv:2106.08254 (2021)

20. He, K., Zhang, X., Ren, S., Sun, J.: Deep residual learning for image recognition. In: CVPR, pp. 770–778 (2016)

21. He, K., et al.: Harnessing synthesized abstraction images to improve facial attribute recognition. In: IJCAI, pp. 733–740 (2018)

22. Huang, S., Li, X., Cheng, Z.Q., Zhang, Z., Hauptmann, A.: GNAS: a greedy neural architecture search method for multi-attribute learning. In: ACM MM, pp. 2049–2057 (2018)

23. Jia, J., Chen, X., Huang, K.: Spatial and semantic consistency regularizations for pedestrian attribute recognition. In: ICCV, pp. 962–971 (2021)

24. Kalayeh, M.M., Gong, B., Shah, M.: Improving facial attribute prediction using semantic segmentation. In: CVPR, pp. 6942–6950 (2017)

25. Lanchantin, J., Wang, T., Ordonez, V., Qi, Y.: General multi-label image classification with transformers. In: CVPR, pp. 16478–16488 (2021)

26. Li, D., Chen, X., Huang, K.: Multi-attribute learning for pedestrian attribute recognition in surveillance scenarios. In: ACPR, pp. 111–115 (2015)

27. Li, D., Chen, X., Zhang, Z., Huang, K.: Pose guided deep model for pedestrian attribute recognition in surveillance scenarios. In: ICME, pp. 1–6 (2018)

28. Li, D., Zhang, Z., Chen, X., Huang, K.: A richly annotated pedestrian dataset for person retrieval in real surveillance scenarios. TIP **28**(4), 1575–1590 (2018)

29. Li, J., Zhao, F., Feng, J., Roy, S., Yan, S., Sim, T.: Landmark free face attribute prediction. TIP **27**(9), 4651–4662 (2018)

30. Li, W., Duan, Y., Lu, J., Feng, J., Zhou, J.: Graph-based social relation reasoning. In: Vedaldi, A., Bischof, H., Brox, T., Frahm, J.-M. (eds.) ECCV 2020. LNCS, vol. 12360, pp. 18–34. Springer, Cham (2020). https://doi.org/10.1007/978-3-030-58555-6_2

31. Li, W., Huang, X., Lu, J., Feng, J., Zhou, J.: Learning probabilistic ordinal embeddings for uncertainty-aware regression. In: CVPR, pp. 13896–13905 (2021)

32. Li, W., Lu, J., Feng, J., Xu, C., Zhou, J., Tian, Q.: Bridgenet: a continuity-aware probabilistic network for age estimation. In: CVPR, pp. 1145–1154 (2019)

33. Liu, P., Liu, X., Yan, J., Shao, J.: Localization guided learning for pedestrian attribute recognition. In: BMVC (2018)

34. Liu, S., Zhang, L., Yang, X., Su, H., Zhu, J.: Query2label: a simple transformer way to multi-label classification. arXiv preprint arXiv:2107.10834 (2021)

35. Liu, X., et al.: Hydraplus-net: attentive deep features for pedestrian analysis. In: ICCV, pp. 350–359 (2017)

36. Liu, Z., et al.: Swin transformer: hierarchical vision transformer using shifted windows. In: ICCV (2021)

37. Liu, Z., Luo, P., Qiu, S., Wang, X., Tang, X.: Deepfashion: powering robust clothes recognition and retrieval with rich annotations. In: CVPR, pp. 1096–1104 (2016)

38. Liu, Z., Luo, P., Wang, X., Tang, X.: Deep learning face attributes in the wild. In: ICCV, pp. 3730–3738 (2015)

39. Mao, L., Yan, Y., Xue, J.H., Wang, H.: Deep multi-task multi-label CNN for effective facial attribute classification. TAC (2020)

40. Meng, Z., Adluru, N., Kim, H.J., Fung, G., Singh, V.: Efficient relative attribute learning using graph neural networks. In: ECCV, pp. 552–567 (2018)

41. Nguyen, H.D., Vu, X.S., Le, D.T.: Modular graph transformer networks for multi-label image classification. In: AAAI, pp. 9092–9100 (2021)

42. Perrett, T., Masullo, A., Burghardt, T., Mirmehdi, M., Damen, D.: Temporal-relational crosstransformers for few-shot action recognition. In: CVPR, pp. 475–484 (2021)

43. Peters, M.E., et al.: Deep contextualized word representations. In: NAACL (2018)

44. Radford, A., Narasimhan, K., Salimans, T., Sutskever, I.: Improving language understanding by generative pre-training (2018)

45. Rudd, E.M., Günther, M., Boult, T.E.: MOON: a mixed objective optimization network for the recognition of facial attributes. In: Leibe, B., Matas, J., Sebe, N., Welling, M. (eds.) ECCV 2016. LNCS, vol. 9909, pp. 19–35. Springer, Cham (2016). https://doi.org/10.1007/978-3-319-46454-1_2

46. Sarafianos, N., Xu, X., Kakadiaris, I.A.: Deep imbalanced attribute classification using visual attention aggregation. In: ECCV, pp. 680–697 (2018)

47. Sarfraz, M.S., Schumann, A., Wang, Y., Stiefelhagen, R.: Deep view-sensitive pedestrian attribute inference in an end-to-end model. In: BMVC (2017)

48. Shao, J., Kang, K., Loy, C.C., Wang, X.: Deeply learned attributes for crowded scene understanding. In: CVPR, pp. 4657–4666 (2015)

49. Shin, M.: Semi-supervised learning with a teacher-student network for generalized attribute prediction. In: Vedaldi, A., Bischof, H., Brox, T., Frahm, J.-M. (eds.) ECCV 2020. LNCS, vol. 12356, pp. 509–525. Springer, Cham (2020). https://doi.org/10.1007/978-3-030-58621-8_30

50. Shu, Y., Yan, Y., Chen, S., Xue, J.H., Shen, C., Wang, H.: Learning spatial-semantic relationship for facial attribute recognition with limited labeled data. In: CVPR, pp. 11916–11925 (2021)

51. Tang, C., Sheng, L., Zhang, Z., Hu, X.: Improving pedestrian attribute recognition with weakly-supervised multi-scale attribute-specific localization. In: ICCV, pp. 4997–5006 (2019)
52. Taylor, W.L.: "cloze procedure": a new tool for measuring readability. Journalism Q. **30**(4), 415–433 (1953)
53. Vaswani, A., et al.: Attention is all you need. In: NeurIPS, pp. 5998–6008 (2017)
54. Wang, Y., et al.: End-to-end video instance segmentation with transformers. In: CVPR, pp. 8741–8750 (2021)
55. Wang, Z.J., Turko, R., Chau, D.H.: Dodrio: exploring transformer models with interactive visualization. In: ACL (2021)
56. Yu, B., Li, W., Li, X., Lu, J., Zhou, J.: Frequency-aware spatiotemporal transformers for video inpainting detection. In: ICCV, pp. 8188–8197 (2021)
57. Zhang, C., Gupta, A., Zisserman, A.: Temporal query networks for fine-grained video understanding. In: CVPR, pp. 4486–4496 (2021)
58. Zhang, H., Cisse, M., Dauphin, Y.N., Lopez-Paz, D.: mixup: beyond empirical risk minimization. In: ICLR (2018)
59. Zhang, Y., Zhang, P., Yuan, C., Wang, Z.: Texture and shape biased two-stream networks for clothing classification and attribute recognition. In: CVPR, pp. 13538–13547 (2020)
60. Zhao, X., et al.: Recognizing part attributes with insufficient data. In: ICCV, pp. 350–360 (2019)
61. Zheng, S., et al.: Rethinking semantic segmentation from a sequence-to-sequence perspective with transformers. In: CVPR, pp. 6881–6890 (2021)

AgeTransGAN for Facial Age Transformation with Rectified Performance Metrics

Gee-Sern Hsu[✉], Rui-Cang Xie, Zhi-Ting Chen, and Yu-Hong Lin

National Taiwan University of Science and Technology, Taipei, Taiwan
{jison,m10703430,m10803432,m10903430}@mail.ntust.edu.tw

Abstract. We propose the AgeTransGAN for facial age transformation and the improvements to the metrics for performance evaluation. The AgeTransGAN is composed of an encoder-decoder generator and a conditional multitask discriminator with an age classifier embedded. The generator considers cycle-generation consistency, age classification and cross-age identity consistency to disentangle the identity and age characteristics during training. The discriminator fuses age features with the target age group label and collaborates with the embedded age classifier to warrant the desired target age generation. As many previous work use the Face++ APIs as the metrics for performance evaluation, we reveal via experiments the inappropriateness of using the Face++ as the metrics for the face verification and age estimation of juniors. To rectify the Face++ metrics, we made the Cross-Age Face (CAF) dataset which contains 4000 face images of 520 individuals taken from their childhood to seniorhood. The CAF is one of the very few datasets that offer far more images of the same individuals across large age gaps than the popular FG-Net. We use the CAF to rectify the face verification thresholds of the Face++ APIs across different age gaps. We also use the CAF and FFHQ-Aging datasets to compare the age estimation performance of the Face++ APIs and an age estimator that we made, and propose rectified metrics for performance evaluation. We compare the AgeTransGAN with state-of-the-art approaches by using the existing and rectified metrics.

1 Introduction

Facial age transformation refers to the generation of a new face image for an input face such that the generated face is the same identity as the input but at the age specified by the user. Facial age transformation is an active research topic in the fields of computer vision [1,5,13,17,23]. It is a challenging task due to the intrinsic complexity of the facial appearance variation caused by the physical aging process, which can be affected by physical condition, gender, race and other factors [4]. It has received increasing attention in recent years because

---

Supplementary Information The online version contains supplementary material available at https://doi.org/10.1007/978-3-031-19775-8_34.

of the effectiveness of the GANs [1,5,11,13,17], the availability of large facial age datasets and application potentials. It can be applied in the entertainment and cinema industry, where an actor's face often needs to appear in a younger or older age. It can also be applied to find missing juniors/seniors as the pictures for reference can be years apart.

Many approaches have been proposed in recent years [1,5,11,13,17,23]. However, many issues are yet to be addressed. The performance of the approaches is usually evaluated by target age generation and identity (ID) preservation. The former measures if the generated facial age reaches the target age, and the latter measures if the identity is kept well after the transformation. The best compromise between target age generation and ID preservation is hard to define because facial appearance does not change much across a small age gap, but it can change dramatically across a large age gap. For ID preservation, many approaches use a pretrained face model to make the generated face look similar to the input source [1,23,24], which constrains the age trait generation on the target face. Another big issue is the metrics for a fair performance evaluation. Many handle this issue by using a commercial software, e.g., the Face++ APIs [8], which is a popular choice [12,13,24]. Some turn to manual evaluation via a crowd-sourcing platform [17]. Some use proprietary age estimation models [6]. We address most of the above issues in this paper.

We propose the AgeTransGAN to handle bidirectional facial age transformation, i.e., progression and regression. The AgeTransGAN is composed of an encoder-decoder generator and a conditional multitask discriminator with an age classifier embedded. The generator explores cycle-generation consistency, age classification and cross-age identity consistency to disentangle the identity and age characteristics during training. The discriminator fuses age features with the target age group label and collaborates with the embedded age classifier to warrant the desired age traits made on the generated images.

To address the flaw of using Face++ APIs for performance evaluation and the constraint of using a pretrained face model for ID preservation, we made the Cross-Age Face (CAF) dataset which contains 4000 face images of 520 individuals taken from their childhood to seniorhood. Each face in the CAF has a ground-truth age label, and each individual has images across large age gaps. To the best of our knowledge, the CAF is one of the very few datasets that offer face images of the same individuals across large age gaps, and it contains more individuals and images than the popular FG-Net. We use the CAF to rectify the face verification thresholds of Face++ APIs which perform poorly when verifying junior and children faces. The cross-age face verification rate is an indicator for ID preservation. We also use the CAF and the FFHQ-Aging datasets to compare the age estimation performance of Face++ APIs and a tailor-made age estimator which will be released with this paper. We summarize the contributions of this paper as follows:

1. A novel framework, the AgeTransGAN, is proposed and verified effective for identity-preserving facial age transformation. The novelties include the network architecture and loss functions designed to disentangle the identity and age characteristics so that the AgeTransGAN can handle transformation across large age gaps.

2. The Conditional Multilayer Projection (CMP) discriminator is proposed to extract the multilayer age features and fuse these features with age class labels for better target age generation.
3. A novel database, the Cross-Age Face (CAF), is released with this paper. It is one of the very few databases that offer 4000 images of 520 individuals with large age gaps from early childhood to seniorhood. It can be used to rectify face verification across large age gaps and verify age estimators.
4. Experiments show that the AgeTransGAN demonstrates better performance than state-of-the-art approaches by using both the conventional evaluation metrics and the new metrics proposed in this paper. To facilitate related research, we release the trained models with this paper, https://github.com/AvLab-CV/AgeTransGAN.

In the following, we first review the related work in Sect. 2, followed by the details of the proposed approach in Sect. 3. Section 4 presents the experiments for performance evaluation, and a conclusion is given in Sect. 5.

2   Related Work

As our approach is related to high-resolution image generation, the conditional GAN and the facial age transformation, this review covers all these topics.

Motivated by the effectiveness of the adaptive instance normalization (AdaIN) [7], the StyleGAN [9] defines a new architecture for high-resolution image generation with attribute separation and stochastic variation. The generator is composed of a mapping network, a constant input, a noise addition, the AdaIN and the mixing regularization. The mapping network transfers the common latent space into an intermediate but less entangled latent space. Instead of using a common random vector as input, the StyleGAN uses the intermediate latent vector made by the mapping network. Given the intermediate latent vector, the learned affine transformations produce the styles that manipulate the layers of the generator via the AdaIN operation. To extract the styles from different layers, the generator processes the bias and noise broadcast within each style block, making the relative impact inversely proportional to the style magnitudes. The modified version, the StyleGAN2 [10], moves the bias and noise broadcast operations outside of the style block, applies a revised AdaIN to remove the artifacts made by StyleGAN, and uses the perceptual path length (PPL) as a quality metric to improve the image quality. The architectures for the generator and discriminator are modified by referring to other work for improvements.

The conditional GAN (cGAN) is considered a promising tool for handling class-conditional image generation [16]. Unlike typical GANs, the discriminators in the cGANs discriminate between the generation distribution and the target distribution given the pairs of generated data x and the conditional variable y. Most cGANs feed the conditional variable y into the discriminator by concatenating y to the input or to some feature vectors [14,18,21,25]. However, the cGAN with projection discriminator (PD) [15] considers a different perspective of incorporating the conditional information into the discriminator. It explores a projection scheme to merge the conditional requirement in the model. The discriminator takes an inner product of the embedded conditional vector y with the feature vector, leading to a significant improvement to the image generation on the ILSVRC-2012 dataset.

A significant progress has been made in facial age transformation recently. The Identity-Preserved Conditional GAN (IPC-GAN) [22] explores a cGAN module for age transfer and an identity-preserved module to preserve identity with an age classifier to enhance target age generation. The Pyramid Architecture of GAN (PA-GAN) [23] separately models the constraints for subject-specific characteristics and age-specific appearance changes, making the generated faces present the desired aging effects while keeping the personalized properties. The Global and Local Consistent Age GAN (GLCA-GAN) [11] consists of a global network and a local network. The former learns the whole face structure and simulates the aging trend, and the latter imitates the subtle local changes. The wavelet-based GAN (WL-GAN) [13] addresses the matching ambiguity between young and aged faces inherent to the unpaired training data. The Continuous Pyramid Architecture of GAN (CPA-GAN) implements adversarial learning to train a single generator and multiple parallel discriminators, resulting in smooth and continuous face aging sequences.

Different from previous work that focuses on adult faces and considers the datasets such as MORPH [19] and CACD [2], the Lifespan Age Transformation Synthesis (LATS) [17] redefines the age transformation by considering a lifespan dataset, the FFHQ-aging, in which 10 age groups are manually labeled for ages between 0 and 70+. Built on the StyleGAN [9], the LATS considers the 10 age groups as 10 domains, and applies multi-domain translation to disentangle age and identity. The Disentangled Lifespan Face Synthesis (DLFS) [5] proposes two transformation modules to disentangle the age-related shape and texture and age-insensitive identity. The disentangled latent codes are fed into a Style-GAN2 generator [10] for target face generation. Considering aging as a continuous regression process, the Style-based Age Manipulation (SAM) [1] integrates four pretrained models for age transformation: a StyleGAN-based encoder for image encoding, the ArcFace [3] for identity classification, a VGG-based model [20] for age regression and the StyleGAN2 for image generation. Different from the above approaches that either directly use the pretrained StyleGAN or made minor modifications, the proposed AgeTransGAN makes substantial modifications to the overall StyleGAN2 architecture so that the generator can better disentangle age and identity characteristics, and the discriminator can better criticize the age traits made on the generated images.

3   Proposed Approach

The proposed AgeTransGAN consists of an encoder-decoder generator $G = [G_{en}, G_{de}]$, where G_{en} is the encoder and G_{de} is the decoder, and a conditional multitask discriminator D_p. The details are presented below.

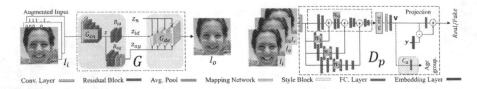

Fig. 1. [Left] The generator $G = [G_{en}, G_{de}]$ with networks B_{id} and B_{ag} for age and identity disentanglement. [Right] The Conditional Multitask Projection (CMP) discriminator D_p with four subnets $[\mathbf{n}_k]_{k=1}^{4}$ for multilayer feature extraction and an age classifier C_a. See supplementary document for details on network settings.

3.1   Encoder and Decoder

The configuration of the generator $G = [G_{en}, G_{de}]$ is shown in Fig. 1. The multitask encoder G_{en} takes the source image I_i and the target age group label y_t as input, and generates the identity latent code z_{id} and the age latent code z_{ag}. z_n is a Gaussian noise that enters the decoder G_{de} after each convolution layer. The encoder G_{en} is developed based on a modification of the StyleGAN2 discriminator, which consists of an input layer, a convolution layer, 8 downsampling residual blocks and a scalar output layer.

We first fuse the 3-layer (RGB) I_i with the target age group label y_t by augmenting I_i with the one-hot array that represents y_t using N_a layers of 0's and 1's, where N_a is the number of all target age groups. The augmented input $\hat{I}_i$ is entered into the common layers of G_{en} to generate a facial representation z. The common layers include the input layer, the convolution layer and the downsampling residual blocks. z is further processed by two independent component networks, B_{id} and B_{ag}, which we propose to disentangle the identity and age characteristics by jointly minimizing a set of specifically designed loss functions. The component network B_{id}, composed of two residual blocks, transfers z to the identity latent code z_{id} that will enter the decoder G_{de}. The other component network B_{ag}, composed of two residual blocks, a convolution layer and a mapping network, transfers z to the age latent code z_{ag}. The training with the loss functions presented below makes z_{id} capture the identity characteristics and z_{ag} capture the age characteristics. We may write z_{id} and z_{ag} as $z_{id}(I, y)$ and $z_{ag}(I, y)$ to indicate their dependence on the input I and label y. The identity latent code z_{id} and the age latent code z_{ag} will be abbreviated as the ID code and the age code, respectively, in the rest of the paper for simplicity.

The decoder G_{de} is modified from the StyleGAN2 generator with three modifications: 1) Six additional loss functions considered at training, 2) The original constant input replaced by the ID code z_{id}, and 3) The multi-stream style signals that enter the upsampling style blocks via the AdaIN are replaces by the age code z_{ag}. The generator takes I_i and the target age group label y_o as input to generate the target age output I_o. As we consider the cycle consistency between input and output during training, we also enter the generated output I_o and the input age group label y_i to the generator to reconstruct the source input I_r during training.

The configurations of G_{en} and G_{de} are shown in Fig. 1. The details of network settings are given in the supplementary document.

The loss functions considered for training the generator G include the adversarial loss, the identity loss, the cycle-consistency loss, the age class loss, the pixel-wise attribute loss and the perceptual path length regularization. The following adversarial loss $\mathcal{L}_G^{adv}$ warrants the desired properties of the generated faces.

$$\mathcal{L}_G^{adv} = \mathbb{E}_{I_i \sim p(I_i)} \log \left[D_p \left(G(I_i, y_t), y_t \right) \right] + \mathbb{E}_{I_o \sim p(I_o)} \log \left[D_p \left(G(I_o, y_i), y_i \right) \right] \quad (1)$$

The following identity (ID) loss $\mathcal{L}_{id}$ ensures the ID preservation at the output I_o by forcing the ID code z_{id} of the source I_i close to that of I_o.

$$\mathcal{L}_{id} = \| z_{id}(I_i, y_t) - z_{id}(I_o, y_i) \|_1 \quad (2)$$

The triplet loss $\mathcal{L}_t$, defined on the ID code z_{id} as shown in (3) below, moves the reconstructed ID code $z_{id}(I_r, y_i)$ closer to the source ID code $z_{id}(I_i, y_i)$ while moving the output ID code $z_{id}(I_o, y_t)$ further away from the source ID code $z_{id}(I_i, y_i)$.

$$\mathcal{L}_t = \| z_{id}(I_i, y_i) - z_{id}(I_r, y_i) \|_2^2 - \| z_{id}(I_i, y_i) - z_{id}(I_o, y_t) \|_2^2 + m_t \quad (3)$$

where m_t is the margin determined empirically.

Note that both the ID loss $\mathcal{L}_{id}$ in (2) and the triplet loss $\mathcal{L}_t$ in (3) are defined on the ID code z_{id}; but with the following differences: 1) $\mathcal{L}_{id}$ verifies the ID preservation for the transformation across all age groups/classes, i.e., $z_{id}(I_i, y_t), \forall y_t$ and $z_{id}(I_o, y_i), \forall y_i$; however, $\mathcal{L}_t$ only considers the within-class transformation, i.e., $z_{id}(I_i, y_i)$ and $z_{id}(I_o, y_t)$. 2) $\mathcal{L}_{id}$ aims to preserve the identity only between the source input and the generated output; while $\mathcal{L}_t$ aims to enhance the ID preservation between the source and the reconstructed source, and simultaneously penalize the ID preservation across age transformation.

The cycle-consistency loss $\mathcal{L}_{cyc}$ makes the age progression and regression mutually reversible, i.e., the input I_i can be reconstructed from the target I_t in the same way as the target I_t is generated from the input I_i. It is computed by the following L_1 distance between I_i and the reconstructed input $I_r = G(I_o, y_i)$.

$$\mathcal{L}_{cyc} = \| I_i - G(I_o, y_i) \|_1 \quad (4)$$

The following age class loss $\mathcal{L}_a$, which is the cross-entropy loss computed by using the age classifier C_a in the discriminator D_p, is considered when training G (and also when training D_p).

$$\mathcal{L}_a^{(g)} = \mathbb{E}_{I \sim p(I)} [- \log C_a(\mathbf{v}(I), y)] \quad (5)$$

where $\mathbf{v}(I)$ is a latent code generated within the discriminator D_p for image I, and more details are given in Sect. 3.2. The following pixel-wise attribute loss $\mathcal{L}_{px}$ is need to maintain the perceptual attribute of I_i at the output I_o.

$$\mathcal{L}_{px} = \mathbb{E}_{I_i \sim p(I_i)} \frac{1}{w \times h \times c} \| I_o - I_i \|_2^2 \quad (6)$$

where w, h, and c are the image dimension. $\mathcal{L}_{px}$ is good at keeping the background, illumination and color conditions of I_i at the generated I_o. Similar losses are used in [1,11,13,24]. Without this loss, as the settings for LATS [17] and DLFS [5], we have to crop each input face during preprocessing.

To encourage that a constant variation in the style signal results in a constant scaled change in the image, the StyleGAN2 employs the following perceptual path length regularization $\mathcal{L}_{pl}$ to make the generator smoother. We apply the same regularization on the age code z_{ag}.

$$\mathcal{L}_{pl} = \mathbb{E}_{z_{ag}} \mathbb{E}_{I_o} \left(\left\| \mathbf{J}_{z_{ag}}^T I_o \right\|_2 - a_p \right)^2 \tag{7}$$

where $\mathbf{J}_{z_{ag}} = \partial G(I_i, y_t)/\partial z_{ag}$ is the Jacobian, and a_p is a constant. $\mathbf{J}_{z_{ag}}^T I_o$ can be written as $\nabla_{z_{ag}}(G(I_i, y_t) \cdot I_o)$ for a better implementation of the needed back propagation.

The 7 loss functions in (1)–(7) are combined by the following weighted sum to train G.

$$\mathcal{L}_G = \mathcal{L}_G^{adv} + \lambda_{id}\mathcal{L}_{id} + \lambda_t\mathcal{L}_t + \lambda_{cyc}\mathcal{L}_{cyc} + \lambda_a^{(g)}\mathcal{L}_a^{(g)} + \lambda_{px}\mathcal{L}_{px} + \lambda_{pl}\mathcal{L}_{pl} \tag{8}$$

where $\lambda_{id}, \lambda_t, \lambda_{cyc}, \lambda_a^{(g)}, \lambda_{px}$ and λ_{pl} are the weights determined empirically.

3.2   CMP Discriminator

The Conditional Multitask Projection (CMP) discriminator D_p is proposed to not just distinguish the generated images from the real ones, but also force the facial traits on the generated faces close to the real facial traits shown in the training set. To attain these objectives, we make a substantial revision to the StyleGAN2 discriminator with three major modifications: 1) Embedding of a multilayer age feature extractor S_a, 2) Integration with a label projection module to make the age-dependent latent code conditional on the target age group label, and 3) Embedding of an age classifier C_a for supervising the target age generation. The configuration of D_p is shown in Fig. 1.

Multilayer Age Feature Extractor: We keep the same input layer, the convolution layer and the mini-batch standard deviation as in the StyleGAN2 discriminator, but modify the 8 downsampling residual blocks (res-blocks) for multilayer feature extraction. The 8 res-blocks is used as the base subnet $\mathbf{n}_0$ to make other subnets for extracting multilayer features. We remove the smallest two res-blocks in $\mathbf{n}_0$ to make the 6-res-block subnet $\mathbf{n}_1$. Repeating the same on $\mathbf{n}_1$ makes the 4-res-block subnet $\mathbf{n}_2$, and repeating on $\mathbf{n}_2$ makes the 2-res-block subnet $\mathbf{n}_3$. The output features from $\mathbf{n}_1$, $\mathbf{n}_2$ and $\mathbf{n}_3$ are added back to the same dimension features in the corresponding layers in $\mathbf{n}_0$, as shown in Fig. 1. Therefore, the feature output from $\mathbf{n}_0$ integrates the features from all subnets. The feature output is further processed by the mini-batch standard deviation, followed by a convolution layer and a fully-connected layer to generate an intermediate latent code $\mathbf{v}$. $\mathbf{v}$ can be written as $\mathbf{v}(I)$ as the input image I can be the generator's input I_i and the generated I_o, which are both given to D_p during training.

Label Projection Module: We design this module to make the age-dependent latent code conditional on the target age group label. It has two processing paths. One path converts $\mathbf{v}(I)$ to a scalar $v(I)$ by a fully-connected layer. The other computes the label projection, which is the projection of the age group label $y \in \mathbf{I}^{N_a}$ onto the latent code $\mathbf{v}$, as shown in Fig. 1. The computation takes the inner product of the embedded y and $\mathbf{v}(I)$, i.e., $(y^T E_m) \cdot \mathbf{v}(I)$, where E_m denotes the embedding matrix. The operation for the discriminator $D_p(I, y)$ can be written as follows:

$$D_p(I, y) = y^T E_v \cdot \mathbf{v}(I) + v(I) \tag{9}$$

where $y = y_t$ when $I = I_i$, and $y = y_i$ when $I = I_o$. The argument I in $\mathbf{v}(I)$ and $v(I)$ shows that both can be considered as the networks with I as input, i.e., $\mathbf{v}(\cdot)$ is the forward-pass of D_p without the last fully-connected layer, and $v(\cdot)$ is the forward-pass of D_p.

Age Classifier Embedding: The age classifier C_a in Fig. 1 supervises the target age generation by imposing the requirement of age classification on the latent code $\mathbf{v}$. It is made by connecting $\mathbf{v}$ to an output layer made of a softmax function. The age class loss $\mathcal{L}_a^{(d)}$ is computed on C_a in the same way as given in (5), but revised for D_p.

We consider the adversarial loss, the age class loss and the R1 regularization when training D_p. The adversarial loss $\mathcal{L}_{D_p}^{adv}$ can be computed as follows.

$$\mathcal{L}_{D_p}^{adv} = \mathbb{E}_{I_i \sim p(I_i)} \log \left[D_p \left(I_i, y_i \right) \right] + \mathbb{E}_{I_i \sim p(I_i)} \log \left[1 - D_p \left(G(I_i, y_t), y_t \right) \right]$$
$$+ \mathbb{E}_{I_o \sim p(I_o)} \log \left[1 - D_p \left(G(I_o, y_i), y_i \right) \right] \tag{10}$$

The following R1 regularization $\mathcal{L}_{r1}$ is recommended by the StyleGAN [9] as it leads to a better FID score.

$$\mathcal{L}_{r1} = \mathbb{E}_{I_i \sim p(I_i)} \left[\left\| \nabla_{I_i} D_p \left(I_i, y_i \right) \right\|^2 \right] \tag{11}$$

The overall loss for training D_p can be written as follows.

$$\mathcal{L}_{D_p} = \mathcal{L}_{D_p}^{adv} + \lambda_{r1} \mathcal{L}_{r1} + \lambda_a^{(d)} \mathcal{L}_a^{(d)} \tag{12}$$

where λ_{r1} and $\lambda_a^{(d)}$ are determined in the experiments.

4   Experiments

We first introduce the database and experimental settings in Sect. 4.1. As the Face++ APIs [8] are used as the performance metrics in many previous work [12,13,24], we follow this convention for comparison purpose but reveal the inappropriateness by experiments. We address this issue in Sect. 4.1 with proposed schemes to handle. Section 4.2 reports an ablation study that covers a comprehensive comparison across different settings on the generator and discriminator. The comparison with state-of-the-art approaches is presented in Sect. 4.3.

Table 1. Age estimation on FFHQ-aging and CAF by using Face++ API and our age estimator, better one in each category shown in boldface

Age group	0-2	3-6	7-9	10-14	15-19	20-29	30-39	40-49	50-69	70+
EAM (Estimated Age Mean) on whole FFHQ-aging dataset										
Face++	10.19	20.31	24.64	25.92	26.10	29.64	39.93	54.34	67.81	**76.96**
Our estimator	**1.50**	**5.08**	**8.96**	**13.17**	**18.94**	**24.27**	**32.08**	**42.57**	**57.57**	68.28
EAM/MAE (Mean Absolute Error) on whole CAF dataset										
Real	1.17	4.53	7.93	12.00	17.13	24.04	33.76	43.82	56.61	72.35
Face++	19/17.29	27.68/22.15	29.67/20.93	28.70/16.23	27.39/10.09	29.31/6.52	36.20/7.27	45.17/7.35	56.92/7.33	70.65/7.56
Our estimator	**1.36/1.28**	**5.65/2.16**	**8.78/3.23**	**14.03/5.68**	**17.98/5.43**	**25.37/3.46**	**34.35/3.50**	**46.37/5.57**	**53.91/5.74**	**67.46/6.71**

4.1 Databases and Experimental Settings

Due to page limit, we report our experiments on the FFHQ-Aging [17] in the main paper, and the experiments on the MORPH [19] and CACD [2] in the supplementary document. The FFHQ-Aging dataset [17] is made of ~70k images from the FFHQ dataset [9]. Each image is labeled with an age group which is not based on ground-truth but on manual annotation via crowd-sourcing [17]. 10 age groups are formed: 0–2, 3–6, 7–9, 10–14, 15–19, 20–29, 30–39, 40–49, 50–69 and ≥70 years, labeled as $G_{10}0$, $G_{10}1$, ..., $G_{10}9$, respectively. We follow the same data split as in [9] that takes the first 60k images for training and the remaining 10k for testing. $G_{10}5$ (20 ~ 29) is taken as the source set and the other nine groups as the target sets.

The weights in (8) are experimentally determined as $\lambda_{px}=10$, $\lambda_{pl}=2$, $\lambda_{cyc}=10$, $\lambda_t=0.1$, $\lambda_{id}=1$ and $\lambda_a^g=1$; and those in (12) are $\lambda_{r1}=10$ and $\lambda_a^{(d)}=1$. We chose the Adam optimizer to train G and D_p at learning rate $2e^{-4}$ on an Nvidia RTX Titan GPU. See supplementary document for more details about data preprocessing, other training and testing settings.

Metrics for Performance Evaluation

Similar to the previous work [12,13,24], we also use the public Face++ APIs [8] as the metrics for evaluating the performance, but reveal via experiments the inappropriateness of using the Face++ APIs for the face verification and age estimation of subjects younger than 20. The Face++ APIs can estimate the age of a face and allow different thresholds for face verification. We first followed the same 1:1 face verification setup as in [12,13,24], where the generated face was verified against the input face with similarity threshold 76.5 for FAR 10^{-5}.

Using the same evaluation metrics allows a fair comparison with the previous work. However, using the same similarity threshold for face verification across the entire lifespan can be inappropriate, because the facial appearance does not change much across a small age gap, but it can change dramatically across a large age gap. The dramatic change would affect the ID preservation. This fact explains that we can sometimes be surprised to see someone's face has changed so much that we cannot recognize after tens of years of separation. Besides, we also found that the Face++ APIs reported large errors when estimating the ages of infants and young children, although it performed relatively well estimating the ages of adults older than 20 years.

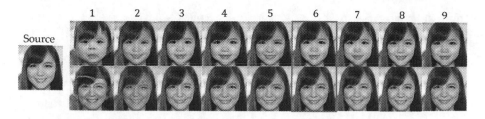

Fig. 2. Generated images with different settings: 1. B/L; 2. $B/L+\mathcal{L}_{id}^{pre}$; 3. $B/L+\mathcal{L}_{cyc}$; 4. $B/L+\mathcal{L}_{cyc}+\mathcal{L}_t$; 5. $B/L+\mathcal{L}_{cyc}+\mathcal{L}_t+\mathcal{L}_{id}^{pre}$; 6. (Best) $B/L+\mathcal{L}_{cyc}+\mathcal{L}_t+\mathcal{L}_{id}$ with D_p+C_a; 7.D_0+C_a; 8. D_pw/oC_a; and 9.$D_p+C_a(s)$

To better define the metrics needed for face verification across large age gaps and the age estimation for infants and children, we propose a rectification to the Face++ APIs usage and an age estimator that we made. For the rectification, we need a dataset with face images across large age gaps from early childhood for a sufficient number of individuals. The existing datasets cannot meet this requirement. The largest age gap for the same individual in the MORPH is less than 10 years, and it is 11 years in CACD. The FG-Net only has 1,002 images of 89 subjects although some are across large age gaps. To meet the requirement, we make the Cross-Age Face (CAF) dataset, which contains 4000 face images of 520 individuals. Each face has a ground-truth age, and each individual has images in at least 5 age groups across $G_{10}0 \sim G_{10}9$ (0 $\sim>$ 70 years). The numbers of subjects in $G_{10}0 \sim G_{10}9$ are 341, 364, 312, 399, 469, 515, 435, 296, 195, and 67, respectively. More detail of the dataset is in supplementary document.

We use the CAF dataset for the following tasks: 1) Rectify the similarity thresholds given by the Face++ APIs for face verification across various age gaps. We define the thresholds by forming intra and inter pairs for each age gap, and selecting the threshold for an allowable FAR, e.g., 10^{-4}. See Supplementary Materials for more information about the CAF and MIVIA datasets.

The comparison of face verification with and without the proposed rectification is given in the following sections. Table 1 shows the age estimation on the FFHQ-Aging and CAF datasets by using the Face++ APIs and our age estimator. Note that each image in the FFHQ-Aging does not have a ground-truth age, and only has an age-group label by crowd-sourcing annotation, so we can only compute the Estimated Age Mean (EAM) for each age group. The EAM refers to the mean of the estimated ages of all images. But each image in the CAF has a ground-truth age so we compare in terms of both the EAM and MAE (Mean Absolute Error). Table 1 reveals that Face++ APIs consistently make large errors estimating the ages of the faces younger than 20 on both datasets. Our age estimator instead presents more reliable estimated ages.

4.2   Ablation Study

To compare the effectiveness of the losses considered in (8), we first define a baseline, denoted as B/L in Table 2, which only includes the adversarial loss $\mathcal{L}_G^{adv}$,

Table 2. Performance on FFHQ-Aging for transferring $G_{10}5$ (20−29) to other 9 groups with different settings on the loss function (Top) and discriminator D_p (Bottom). Both the rectified and common thresholds used for face verification, and used for our age estimator. Best one in each category shown in **boldface**, those in red show good verification rates but poor target age generation.

Age group	0–2	3–6	7–9	10–14	15–19	30–39	40–49	50–69	70+
Face verification rate (%), rectified threshold (common threshold)									
Threshold	61.8(76.5)	68.9(76.5)	72.7(76.5)	74.2(76.5)	76.6(76.5)	76.3(76.5)	71.7(76.5)	65.2(76.5)	65.2(76.5)
B/L	61.5(3.5)	82.9(25.9)	86.45(75.9)	88.5(77.2)	79.46(80.3)	72.7(71.6)	81.1(68.5)	75.5(31.7)	70.9(27.8)
$+\,L_{id}^{pre}$	91.7(85.7)	99.2(91.3)	**100(100)**	**100(100)**	**100(100)**	**100(100)**	100(98.5)	98.7(97.4)	92.8(80.2)
$+\,L_{id}$	63.1(5.7)	82.9(25.9)	89.4(77.7)	87.9(75.8)	80.88(81.4)	82.1(81.6)	80.9(68.2)	89.4(71.5)	78.3(29.9)
$+\,L_{cyc}$	74.37(20.47)	93.1(81.7)	93.8(83.2)	93.3(83.3)	93.1(94.1)	95.5(94.3)	95.1(89.7)	93.9(80.3)	91.9(77.4)
$+\,L_{cyc}+L_t$	77.4(23.7)	95.3(85.1)	95.5(85.1)	95.4(86.6)	93.7(94.6)	95.7(93.6)	96.9(93.2)	96.2(83.7)	95.9(82.7)
$+\,L_{cyc}+L_t+L_{id}^{pre}$	**98.3(95.4)**	**100(98.4)**	100(98.6)	**100(100)**	**100(100)**	**100(100)**	100(98.6)	**100(97.4)**	**100(95.7)**
$+\,L_{cyc}+L_t+L_{id}$	80.3(10.3)	96.5(86.3)	95.9(85.4)	95.8(86.7)	**100(100)**	**100(100)**	100(97.7)	97.6(85.7)	96.8(84.7)
D_0+C_a	74.7(20.6)	92.5(80.8)	90.3(82.1)	92.4(83.5)	96.5(97.2)	**100(100)**	96.6(92.7)	94.3(81.5)	84.5(67.1)
D_p w/o C_a	75.2(22.3)	94.4(85.6)	91.2(82.4)	92.4(83.3)	95.8(96.5)	**100(100)**	95.6(90.3)	93.4(80.2)	83.8(66.6)
D_p+C_a	80.3(10.3)	**96.5(86.3)**	**95.2(85.4)**	**95.8(86.7)**	**100(100)**	**100(100)**	**100(97.7)**	**97.6(85.7)**	**96.8(84.7)**
$D_p+C_a(single)$	77.7(23.9)	94.8(85.7)	93.6(83.7)	94.3(85.3)	98.6(100)	**100(100)**	**100(96.6)**	95.2(82.9)	94.4(82.3)
EAM, ours/mean error									
Raw data (training set)	1.5/–	4.9/–	8.6/–	12.8/–	18.9/–	31.9/–	43.9/–	57.2/–	68.9/–
B/L	1.2/0.3	**4.5/0.4**	12.0/3.4	17.2/4.4	20.8/1.9	28.0/3.9	32.2/11.7	41.7/15.5	53.8/15.1
$+\,L_{id}^{pre}$	7.9/6.4	6.6/1.7	12.4/3.8	17.4/4.6	21.5/2.6	26.1/5.8	39.8/13.1	40.6/16.6	53.2(61.1)/15.7
$+\,L_{id}$	1.1/0.4	2.5/2.0	6.7/1.9	13.8/1.0	**18.6/0.3**	32.5/0.6	40.4/3.5	51.3/5.9	67.2/1.7
$+\,L_{cyc}$	**1.4/0.1**	3.3/1.6	7.0/1.6	13.2/0.4	17.2/1.2	32.8/0.9	41.4/2.5	52.6/4.6	67.5/1.4
$+\,L_{cyc}+L_t$	**1.4/0.1**	3.3/1.6	7.2/1.4	**12.6/0.2**	17.6/1.3	32.6/0.7	41.7/2.2	54.6/2.6	**69.0/0.1**
$+\,L_{cyc}+L_t+L_{id}^{pre}$	8.9/7.4	7.9/3.8	11.9/1.5	15.5/3.4	20.2/1.3	29.9/2.3	37.9/6.9	41.0/16.2	48.7/24.2
$+\,L_{cyc}+L_t+L_{id}$	1.1/0.4	**4.5/0.4**	**8.8/0.2**	13.5/0.7	18.7/0.2	**32.3/0.4**	**41.7/2.2**	**55.5/1.7**	68.4/0.5
D_0+C_a	2.6/1.1	6.2/1.3	10.0/1.4	14.3/1.5	21.2/1.3	29.1/2.8	39.9/4.0	52.3/4.9	62.2/6.7
D_p w/o C_a	3.6/2.1	7.1/2.2	10.7/2.1	15.7/2.9	20.1/1.2	28.4/3.5	36.6/7.3	50.6/7.2	61.8/7.1
D_p+C_a	**1.1/0.4**	**4.5/0.4**	**8.8/0.2**	**13.5/0.7**	**18.7/0.2**	**32.3/0.4**	**41.7/2.2**	**55.5/1.7**	**68.4/0.5**
$D_p+C_a(single)$	2.3/0.8	5.8/0.9	9.3/0.7	14.0/1.2	19.3/0.4	30.6/1.3	38.2/3.7	53.5/3.7	64.4/4.5

the age class loss $\mathcal{L}_a^{(g)}$, the pixel-wise attribute loss $\mathcal{L}_{px}$, and the perceptual path length regularization $\mathcal{L}_{pl}$. $\mathcal{L}_G^{adv}$ is needed to warrant the quality of the generated images; $\mathcal{L}_a^{(g)}$ is needed for age classification; $\mathcal{L}_{px}$ is needed to preserve the source image attribute; and $\mathcal{L}_{pl}$ is needed for image quality improvement (A comparison of the baselines with and without these losses is given in the supplementary document). We compare the performance when combining the baseline with the identity loss $\mathcal{L}_{id}$, the triplet loss $\mathcal{L}_t$ and the cycle-consistency loss $\mathcal{L}_{cyc}$. We also compare with a general way to compute the identity loss by using an off-the-shelf pretrained face encoder [22,24], and we choose the pretrained ArcFace [3].

Table 2 shows the comparisons on the FFHQ-Aging by using the common and rectified Face++ thresholds for face verification and our age estimator, where $\mathcal{L}_{id}^{pre}$ denotes the identity loss computed using the pretrained ArcFace to replace L_{id}. The performance measures in parentheses are for the common threshold 76.5 and Face++ APIs, and those out of parentheses are for rectified thresholds and our age estimator. The results can be summarized as follows.

– When $\mathcal{L}_{id}^{pre}$ is included, the ID preservation is substantially upgraded, on the cost of much deteriorating target age generation, as shown by $B/L + \mathcal{L}_{id}^{pre}$ and $B/L + \mathcal{L}_{cyc} + \mathcal{L}_t + \mathcal{L}_{id}^{pre}$. The large errors in the estimated mean ages are shown in red. Clearly $\mathcal{L}_{id}^{pre}$ can well preserve identity, but badly damage the target age generation. Figure 2 shows the generated images.

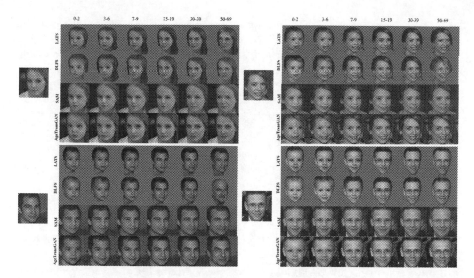

Fig. 3. Qualitative comparison with state-of-the-art methods for age transformation with the source faces on the left side.

- The triplet loss $\mathcal{L}_t$, which can only be computed with the cycle-consistency loss $\mathcal{L}_{cyc}$, demonstrates a balanced performance for ID preservation and age transformation with $B/L + \mathcal{L}_{cyc} + \mathcal{L}_t$. The performance is further enhanced when $\mathcal{L}_{id}$ is added in, resulting in the final selected settings.
- With the selected $B/L + \mathcal{L}_{cyc} + \mathcal{L}_t + \mathcal{L}_{id}$, the face verification rates with rectified thresholds show more plausible results than the common constant threshold 76.5.

To better determine the settings for the CMP discriminator D_p, we compare the performance with and without the age classifier C_a, and the condition without the label projection. We also compare the performance of using multilayer and single-layer features in D_p. The bottom part of Table 2 shows the comparison of 1) C_a with D_0, where D_0 is the discriminator D_p without the label projection; 2) D_p without C_a; 3) D_p with C_a; 4) D_p with C_a but using single-layer features, i.e., only with the subnet $\mathbf{n}_0$ in Fig. 1. Figure 2 shows the samples made by the four settings. The results can be summarized as follows:

- The performances of $D_0 + C_a$ and D_p w/o C_a are similar for both tasks of ID preservation and target age generation, although the former is slightly better for generating the children's ages.
- $D_p + C_a(single)$ with single-layer feature slightly outperforms $D_0 + C_a$ and D_p w/o C_a for both tasks.
- $D_p + C_a$ with multilayer feature outperforms $D_p + C_a(single)$ for both tasks with clear margins, especially on the youngest groups, i.e., $G_{10}0$ and $G_{10}1$.

The above comparisons have verified the settings with $D_p + C_a$, which is used for the comparison with other approaches. The margin m_t in (3) is experimentally determined as 0.5 out of a study reported in the supplementary document.

Table 3. Performance on FFHQ-Aging for transferring $G_{10}5$ $(20-29)$ to other 9 groups (only 6 groups available by using LATS and DLFS), using both common and rectified thresholds for Face++ face verification (Top), and Face++ and our age estimator for age estimation (Bottom).

Age group	0–2	3–6	7–9	10–14	15–19	30–39	40–49	50–69	70+
Verification rate (%) (common threshold)									
Threshold	61.8(76.5)	68.9(76.5)	72.7(76.5)	74.2(76.5)	76.6(76.5)	76.3(76.5)	71.7(76.5)	65.2(76.5)	65.2(76.5)
LATS [17]	51.5(5.2)	62.9(12.5)	82.7(78.9)	–	92.7(92.7)	92.7(91.5)	–	88.9(71.1)	–
DLFS [5]	52.8(12.4)	67.7(15.3)	81.9(75.2)	–	97.9(97.9)	97.5(96.8)	–	88.4(72.1)	–
SAM [1]	**93.7(54.8)**	88.3(67.8)	85.9(74.8)	88.8(82.9)	89.8(90.0)	90.8(90.5)	87.7(76.7)	83.1(46.2)	68.9(23.6)
AgeTransGAN	80.3(10.3)	**96.5(86.3)**	**95.2(85.4)**	**95.8(86.7)**	**100(100)**	**100(100)**	**100(97.7)**	**97.6(85.7)**	**96.8(84.7)**
EAM. ours/mean error									
Raw data	1.5	4.9	8.6	12.8	18.9	31.9	43.9	57.2	68.9
LATS [17]	4.6/3.1	5.4/0.5	7.6/1.0	–/–	20.4/1.5	31.6/0.3	–/–	52.1/5.1	–/–
DLFS [5]	2.0/0.5	4.1/0.8	10.6/2.7	–/–	21.6/4.5	30.2/3.6	–/–	49.5/7.1	–/–
SAM [1]	5.4/3.9	7.3/2.4	10.4/1.8	13.7/0.9	20.3/1.4	32.3/0.4	43.2/0.7	58.7/1.6	70.7/1.8
AgeTransGAN	**1.1/0.4**	**4.5/0.4**	**8.8/0.2**	**13.5/0.7**	**18.7/0.2**	32.3/0.4	41.7/2.2	55.5/1.7	68.4/0.5

Table 4. Performance on CAF for transferring $G_{10}5$ $(20-29)$ to other 9 groups (only 6 groups available by using LATS and DLFS), using rectified thresholds for Face++ face verification, and our age estimator for target age estimation.

Age group	0–2	3–6	7–9	10–14	15–19	30–39	40–49	50–69	70+
CAF									
Verification rate (%)									
LATS [17]	66.5	72.9	73.7	–	98.1	82.7	–	83.2	–
DLFS [5]	54.7	69.4	83.2	–	100	100	–	85.3	
SAM [1]	**95.2**	88.3	85.8	88.7	89.6	90.9	87.9	83.5	69.4
AgeTransGAN	88.6	**97.9**	**99.7**	**100**	**100**	**100**	**100**	**100**	**100**
EAM									
LATS [17]	4.5	5.8	10.6	–	21.8	32.2	–	44.4	–
DLFS [5]	2.0	**4.2**	10.3	–	22.4	31.2	–	51.3	
SAM [1]	6.2	7.5	10.4	14.6	21.1	**33.4**	44.8	**55.2**	68.8
AgeTransGAN	**1.8**	5.4	**8.8**	**13.8**	**16.1**	32.1	43.5	54.0	**69.3**

4.3   Comparison with SOTA Methods

Table 3 shows the comparison with LATS [17], DLFS [5], and SAM [1], which all offer pretrained models in their GitHub sites. As revealed in Table 1, the Face++ APIs performs poorly for the estimation of younger ages and our age estimator performs well, we only use the latter for the comparison. The AgeTransGAN shows the best balanced performance for both ID preservation and target age generation for transforming to most age groups. Although the SAM performs best for ID preservation on $G_{10}0$, the corresponding target age generation is the worst with mean error 3.9 years. SAM also performs best for target age generation on $G_{10}7$ and $G_{10}8$, the corresponding verification rates for ID preservation are incomparable to those of the AgeTransGAN. Figure 3 shows a qualitative comparison. The AgeTransGAN demonstrates better age traits generated on faces of different age groups while maintaining plausible levels of similarities to

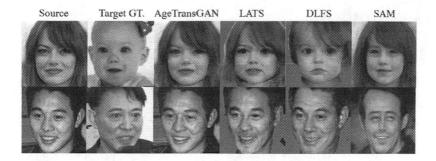

| Source | Target GT. | AgeTransGAN | LATS | DLFS | SAM |

Fig. 4. Qualitative comparison with state-of-the-art methods for age regression (female) and progression (male) on CAF.

the input (source) images. Note that the LATS and DLFS lack the attribute loss $\mathcal{L}_{px}$, all faces must be cropped during preprocessing, but the AgeTransGAN can process images with backgrounds.

Table 4 shows the comparison on the CAF dataset with faces of real ages. The AgeTransGAN outperforms other approaches for ID preservation on 8 age groups, and for target age generation on 5 age groups, showing the best overall balanced performance. The performance difference for age generation decreases considerably for the groups older than $G_{10}6$, showing that all approaches perform similarly well for generating adult faces. SAM performs best for ID preservation on $G_{10}0$, but is the worst for target age generation. Figure 4 shows the CAF samples with images generated for age progression (male) and regression (female) by all approaches. Although LATS makes beard, it does not generate sufficient wrinkles. The faces generated by SAM do not preserve some required levels of identity similarities to the source faces. The faces made by the AgeTransGAN show better qualities on identity similarity and target age traits.

5   Conclusion

We propose the AgeTransGAN for identity-preserving facial age transformation, and a rectification scheme for improving the usage of the popular metrics, Face++ APIs. The AgeTransGAN merges cycle-generation consistency, age classification and cross-age identity consistency to disentangle the identity and age characteristics, and is verified effective for balancing the performance for age transformation and identity preservation. The rectification scheme is offered with a new dataset, the CAF (Cross-Age Face), and an age estimator. We follow the conventional way to compare with other approaches, and highlight the issues with the existing metrics on the new FFHQ-Aging and CAF benchmarks. We address those issues through the rectification scheme and experiments, and verify the AgeTransGAN, the CAF dataset and our age estimator.

References

1. Alaluf, Y., Patashnik, O., Cohen-Or, D.: Only a matter of style: age transformation using a style-based regression model. ACM Trans. Graph. (TOG) **40**(4), 1–12 (2021)
2. Chen, B.C., Chen, C.S., Hsu, W.H.: Face recognition and retrieval using cross-age reference coding with cross-age celebrity dataset. TMM **17**, 804–815 (2015)
3. Deng, J., Guo, J., Xue, N., Zafeiriou, S.: ArcFace: additive angular margin loss for deep face recognition. In: CVPR (2019)
4. Fu, Y., Guo, G., Huang, T.S.: Age synthesis and estimation via faces: a survey. TPAMI **32**, 1955–1976 (2010)
5. He, S., Liao, W., Yang, M.Y., Song, Y.Z., Rosenhahn, B., Xiang, T.: Disentangled lifespan face synthesis. In: ICCV (2021)
6. He, Z., Kan, M., Shan, S., Chen, X.: S2GAN: share aging factors across ages and share aging trends among individuals. In: ICCV (2019)
7. Huang, X., Belongie, S.: Arbitrary style transfer in real-time with adaptive instance normalization. In: ICCV (2017)
8. Megvii Inc.: Face++ research toolkit. http://www.faceplusplus.com
9. Karras, T., Laine, S., Aila, T.: A style-based generator architecture for generative adversarial networks. In: CVPR (2019)
10. Karras, T., Laine, S., Aittala, M., Hellsten, J., Lehtinen, J., Aila, T.: Analyzing and improving the image quality of StyleGAN. In: CVPR (2020)
11. Li, P., Hu, Y., Li, Q., He, R., Sun, Z.: Global and local consistent age generative adversarial networks. In: ICPR (2018)
12. Li, Z., Jiang, R., Aarabi, P.: Continuous face aging via self-estimated residual age embedding. In: CVPR (2021)
13. Liu, Y., Li, Q., Sun, Z.: Attribute-aware face aging with wavelet-based generative adversarial networks. In: CVPR (2019)
14. Mirza, M., Osindero, S.: Conditional generative adversarial nets. arXiv preprint arXiv:1411.1784 (2014)
15. Miyato, T., Koyama, M.: cGANs with projection discriminator. arXiv preprint arXiv:1802.05637 (2018)
16. Odena, A., Olah, C., Shlens, J.: Conditional image synthesis with auxiliary classifier GANs. In: ICML (2017)
17. Or-El, R., Sengupta, S., Fried, O., Shechtman, E., Kemelmacher-Shlizerman, I.: Lifespan age transformation synthesis. In: Vedaldi, A., Bischof, H., Brox, T., Frahm, J.-M. (eds.) ECCV 2020. LNCS, vol. 12351, pp. 739–755. Springer, Cham (2020). https://doi.org/10.1007/978-3-030-58539-6_44
18. Reed, S., Akata, Z., Yan, X., Logeswaran, L., Schiele, B., Lee, H.: Generative adversarial text to image synthesis. arXiv preprint arXiv:1605.05396 (2016)
19. Ricanek, K., Tesafaye, T.: Morph: a longitudinal image database of normal adult age-progression. In: FG (2006)
20. Rothe, R., Timofte, R., Van Gool, L.: DEX: deep expectation of apparent age from a single image. In: ICCV (2015)
21. Sricharan, K., Bala, R., Shreve, M., Ding, H., Saketh, K., Sun, J.: Semi-supervised conditional GANs. arXiv preprint arXiv:1708.05789 (2017)
22. Wang, Z., Tang, X., Luo, W., Gao, S.: Face aging with identity-preserved conditional generative adversarial networks. In: CVPR (2018)

23. Yang, H., Huang, D., Wang, Y., Jain, A.K.: Learning face age progression: a pyramid architecture of GANs. In: CVPR (2018)
24. Yang, H., Huang, D., Wang, Y., Jain, A.K.: Learning continuous face age progression: a pyramid of GANs. TPAMI **43**, 499–515 (2019)
25. Zhu, J.Y., Park, T., Isola, P., Efros, A.A.: Unpaired image-to-image translation using cycle-consistent adversarial networks. In: ICCV (2017)

Hierarchical Contrastive Inconsistency Learning for Deepfake Video Detection

Zhihao Gu[1,2], Taiping Yao[2], Yang Chen[2], Shouhong Ding[2(✉)], and Lizhuang Ma[1,3(✉)]

[1] DMCV Lab, Shanghai Jiao Tong University, Shanghai, China
ellery-holmes@sjtu.edu.cn
[2] Youtu Lab, Tencent, Shanghai, China
{taipingyao,wizyangchen,ericshding}@tencent.com
[3] MoE Key Lab of Artificial Intelligence, Shanghai Jiao Tong University,
Shanghai, China
ma-lz@cs.sjtu.edu.cn

Abstract. With the rapid development of Deepfake techniques, the capacity of generating hyper-realistic faces has aroused public concerns in recent years. The temporal inconsistency which derives from the contrast of facial movements between pristine and forged videos can serve as an efficient cue in identifying Deepfakes. However, most existing approaches tend to impose binary supervision to model it, which restricts them to only focusing on the category-level discrepancies. In this paper, we propose a novel Hierarchical Contrastive Inconsistency Learning framework (HCIL) with a two-level contrastive paradigm. Specially, sampling multiply snippets to form the input, HCIL performs contrastive learning from both local and global perspectives to capture more general and intrinsical temporal inconsistency between real and fake videos. Moreover, we also incorporate a region-adaptive module for intra-snippet inconsistency mining and an inter-snippet fusion module for cross-snippet information fusion, which further facilitates the inconsistency learning. Extensive experiments and visualizations demonstrate the effectiveness of our method against SOTA competitors on four Deepfake video datasets, *i.e.*, FaceForensics++, Celeb-DF, DFDC, and Wild-Deepfake.

Keywords: Deepfake video detection · Inconsistency learning

1 Introduction

As the rapid development of deep learning [19–23,43,44,58,59], the resulting privacy and security concerns [7,24,56] have received numerous attention in recent years. Face manipulation technique, known as Deepfakes, is one of the most emerging threats. Since the generated faces in videos are too realistic to

---

Z. Gu and T. Yao—Equal contributions.
This work was done when Zhihao Gu was an intern at Youtu Lab.

be identified by humans, they can easily be abused and trigger severe societal or political threats. Thus it is urgent to design effective detectors for Deepfakes.

Recently, the Deepfake detection technique has achieved significant progress and developed into image-based and video-based methods. Image-based methods [2,5,6,11,25,38,52] aim to exploit various priors for mining discriminative frame-level features, including face blending boundaries [25], forgery signals on frequency spectrum [5,38] and contrast between augmented pairs [6,52]. However, the development of manipulation techniques promotes the generation of highly realistic faces, the subtle forgery cues can hardly be identified in the static images and it calls for attention to the temporal information in this area. Therefore, some researches tend to develop video-based approaches. Early works treat it as a video-level classification task and directly adopt the off-the-shelf networks like LSTM [17] and I3D [3] to deal with it, which results in inferior performance and high computational cost. Recent works [12,14,26] focus on designing efficient paradigms for modeling inconsistent facial movements between pristine and forged videos, known as temporal inconsistency. S-MIL [26] treats faces and videos as instances and bags for modeling the inconsistency in adjacent frames. LipForensics [14] extracts the high-level semantic irregularities in mouth movements to deal with the temporal inconsistency. STIL [12] exploits the difference over adjacent frames in two directions. Although the overall performance is improved, there are still some limitations. First of all, they heavily rely on the video-level binary supervision without exploring the more general and intrinsical inconsistency. Second, they either exploit the sparse sampling strategy, failing to model the local inconsistency contained in subtle motions or apply dense sampling over consecutive frames, ignoring the long-range inconsistency.

(a) Local Inconsistency   (b) Global Inconsistency

Fig. 1. Illustration of local and global inconsistency. The former one refers to the irregular facial movements within several consecutive frames. The latter one stands for the cases of partial manipulation in the video. Both of them are essential for identifying Deepfakes. Therefore, we construct the hierarchical contrastive learning from both local and global perspectives.

As illustrated in Fig. 1, the temporal inconsistency can be divided from local and global perspectives. Moreover, since it reveals the inconsistent facial movements between real and fake videos, we argue that it should be excavated through comparison, which is significantly neglected by existing works. To address this issue, we aim to introduce contrast into temporal inconsistency learning, and a hierarchical contrastive inconsistency learning framework is proposed. However, there are still several challenges: how to 1) conduct local and global contrast, and 2) extract finer local and global representations for it. To solve the former one, we sample a few snippets from each video and build the local contrastive inconsistency learning on snippet-level representations. Snippets from pristine videos are viewed as positive samples while ones from fake videos are negative samples (no matter whether the videos are partially manipulated). Then the similarity between an anchor snippet and the one from a fake video is used as the regularizer to adaptively decide whether to repel or not, leading to the weighted NCE loss. And the global contrast is directly established on video-level representations. To solve the latter one, a region-aware inconsistency module and an inter-snippet fusion module are respectively proposed to generate discriminative intra-snippet inconsistency for local contrast and promote the interaction between snippets for the global one. Overall, the proposed framework can capture essential temporal inconsistency for identifying Deepfakes and outperforms the SOTA competitors under both full and partial manipulation settings. Besides, extensive ablations and visualizations further validate its effectiveness. In summary, our main contributions can be summarized as follows:

1. We propose a novel Hierarchical Contrastive Inconsistency Learning (HCIL) framework for Deepfake Video Detection, which performs contrastive learning from both local and global perspectives to capture more general and intrinsical temporal inconsistency between real and fake videos.
2. Considering the partial forgery videos, the weighted NCE loss is specially designed to enable the snippet-level inconsistency contrast. Besides, the region-aware inconsistency module and the inter-snippet fusion module are further proposed to facilitate the inconsistency learning.
3. Extensive experiments and analysis further illustrate the effectiveness of the proposed HCIL against its competitors on several popular benchmarks.

2   Related Work

Deepfake Detection. Deepfake detection has obtained more and more attention in recent years. Early researches mainly focus on designing hand-crafted features for identification, such as face warping artifacts [28,49], eye blinking [27] and inconsistent head poses [54]. With the development of deep learning, some image-based methods are proposed to extract discriminative frame-level features for detection. [39] evaluates several well-known 2D neural networks to detect Deepfakes. X-ray [25] identifies Deepfakes by revealing whether the input image can be decomposed into the blending of two images from different sources. F^3-Net [38] exploits the frequency information as a complementary viewpoint for forgery pattern mining. All these approaches perform well in image-level detection. However, with the development of Deepfake techniques, the forgery trace can be hardly found. Recent works [13,14,26,33,37] tend to consider the temporal inconsistency as the key to distinguishing Deepfakes and propose various methods to model it. Two-branch [33] designs a two-branch network to amplify artifacts and suppress high-level face contents. S-MIL [26] introduces a multi-instance learning framework, treating faces and videos as instances and bags, for modeling the inconsistency in adjacent frames. DeepRhythm [37] conjectures that heartbeat rhythms in fake videos are entirely broken and uses CNNs to monitor them for detection. LipForensics [14] proposes to capture the semantic irregularities of mouth movements in generated videos for classification. Different from them capturing the temporal inconsistency via category-level discrepancy, we conduct two-level contrast to formulate the more general and intrinsical temporal inconsistency.

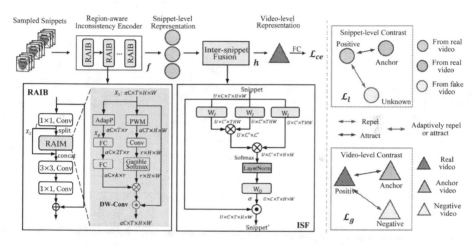

Fig. 2. Overview of the hierarchical contrastive inconsistency learning. The framework is constructed on both snippet and video-level representations. To enable the local contrast, a weighted NCE loss $\mathcal{L}_l$ is designed to adaptively decide whether snippets from fake videos need to be repelled. DW-Conv, $\otimes$, $\oplus$ and $\odot$ stand for depth-wise convolution, matrix multiplication element-wise addition and multiplication, respectively.

Contrastive Learning. The main idea behind contrastive learning is to learn visual representations via attracting similar instances while repelling dissimilar ones [6,15,52]. Recently, some works [1,10,45] attempt to introduce the contrastive learning to detect Deepfakes. DCL [42] specially designs augmentations to generate paired data and performs contrastive learning at different granularities for better generalization. SupCon [53] uses the contrast in the representation space to learn a generalizable detector. Our hierarchical contrastive framework differs from these methods in three aspects. 1) We focus on the temporal inconsistency, specially design the sampling unit called snippet, and establish the local and global contrast paradigm. 2) Not only constructing the contrastive pair is essential, so is extracting the inconsistency. Therefore, we elaborately develop the RAIM and ISF to extract region-aware local temporal inconsistency and refine the global one. 3) Considering the partial forgery in fake videos (snippet-level label unavailable), a novel weighted NCE loss is proposed to enable local contrastive learning.

Video-Analysis. Video-related tasks highly rely on the temporal modeling. Early efforts say I3D [3], exploit the 3D CNNs to capture temporal dependencies. Since they are computationally expensive, various efficient temporal modeling paradigms [30,32,47] are then proposed. TSM [30] shifts part of the channels along the temporal dimension to enable information exchange among adjacent frames. TAM [32] learns video-specific temporal kernels for capturing diverse motion patterns. Although they can be directly applied to detect Deepfakes, considering no task-specific knowledge largely impacts their performance.

3 Proposed Method

In this section, we elaborate on how to generate positive and negative pairs and conduct contrastive inconsistency learning from both local and global perspectives. In Sect. 3.1, we first give the overview of the proposed framework. Then local contrastive inconsistency learning is introduced in Sect. 3.2. Finally, we describe the global contrastive inconsistency learning in Sect. 3.3.

3.1 Overview

As mentioned in Sect. 1, compared to pristine videos, the temporal inconsistency in fake videos can be captured from both local and global perspectives. Therefore, we aim to explicitly model it via simultaneously conducting local and global contrast. Given a real video $V^+ = [S_1^+, \ldots, S_U^+]$ with U sampled snippets of shape $T \times 3 \times H \times W$ from the set $\mathcal{N}^+$ of real videos (T, H and W denote its spatiotemporal dimensions), its anchor videos $V^a = [S_1^a, \ldots, S_U^a] \in \mathcal{N}^a$ are defined as other real videos and the corresponding negative ones are the fake videos $V^- = [S_1^-, \ldots, S_U^-] \in \mathcal{N}^-$, where $\mathcal{N}^a = \mathcal{N}^+ \setminus V^+$ and $\mathcal{N}^-$ represents the set of anchor and fake videos, respectively. Then for a positive snippet $S_i^+ \in V^+$, its anchor snippet, and negative snippets are defined as $S_j^a \in V^a$ and $S_k^- \in V^-$. To enable the local contrast, we mine dynamic snippet-level representations by a region-aware inconsistency encoder and optimize a novel weighted NCE loss [36] on them. It attracts real snippets and adaptively decides whether snippets from fake videos contribute to the loss via measuring their similarity with the anchor snippets. For global contrast, an inter-snippet fusion module is proposed to fuse the cross-snippet information and the InfoNCE loss is optimized based on the video-level features. The overall framework is illustrated in Fig. 2.

3.2 Local Contrastive Inconsistency Learning

The local contrastive inconsistency learning is shown in Fig. 2. The core of it are the region-aware inconsistency module for rich local inconsistency representations learning, and the weighted NCE loss for the contrast between snippets from real and fake videos.

Region-Aware Inconsistency Module. Inspired by DRConv [4] that assigns generated spatial filters to corresponding spatial regions, we design the region-aware inconsistency module (RAIM) to mine comprehensive temporal inconsistency features based on different facial regions. As shown in Fig. 2, on the one hand, it adaptively divides the face into r regions according to the motion information by the right branch (PWM-Conv-gamble-softmax). On the other hand, r region-agnostic temporal filters are learned via the left branch (AdapP-FC-FC). Based on these branches, each region is assigned with its unique temporal filter and the corresponding temporal inconsistency is thus captured through the convolution between each region and its corresponding temporal filter.

Formally, given the input $I \in R^{C \times T \times H \times W}$, we split it along channel dimensional into two parts with the rate α. Then one part X_1 is exploited to extract the region-aware inconsistency while keeping the other part X_2 unprocessed, which is found both effective and efficient [51]. In the left branch, $X_1 \in R^{\alpha C \times T \times H \times W}$ is first spatially pooled by an adaptive average pooling (AdaP) operation, resulting in $X_p \in R^{\alpha C \times T \times r}$. Then two full connected layers FC_1 and FC_2 further deal with the temporal dimension to produce r temporal filters with kernel size k (Fig. 3):

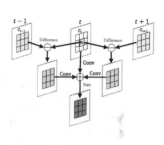

Fig. 3. Illustration of PWM.

$$[W_1, \cdots .W_r] = FC_2(ReLU(FC_1(AdaP(X_1)))), \tag{1}$$

where $W_i \in R^{\alpha C \times k}$ denotes the learned temporal kernels. We exploit the pixelwise motion (PWM) as the guidance for adaptive face division:

$$
\begin{aligned}
PWM(X_{p_0}) = &\sum_{p \in \mathcal{C}_{t-1}} w_p(X(p_0 + p) - X(p_0)) \\
+ &\sum_{p \in \mathcal{C}_{t+1}} w_p(X(p_0 + p) - X(p_0)) + \sum_{p \in \mathcal{C}_t} w_p X(p),
\end{aligned}
\tag{2}
$$

where $\mathcal{C}_{t-1}, \mathcal{C}_t$ and $\mathcal{C}_{t+1}$ stands for the 3×3 region with center position p_0. w_p represents the weight at potion $p_0 + p$. Based on the representations, a 1×1 convolution and a gamble softmax operation are conducted on each spatial location to generate a r-dimensional one-hot vector, which is used to select their temporal filters. Positions with the identical one-hot form are viewed to belong to the same facial region. Finally, X_1 is depth-wise convoluted with the temporal filters to give the region-aware inconsistency within each snippet. The RAIM is inserted right before the second convolution in each resnet block, leading to the encoder f. And snippets go through it to form the snippet-level representations.

Local (Snippet-Level) Contrast. We treat snippets from two real videos as positive pairs. However, fake videos may be partially manipulated and we can't simply treat sampled snippets from them as negative ones. To alleviate this issue, we use the normalized similarity between $S_j^a \in V^a$ and $S_k^- \in V^-$ to adjust the impact of S_k^-. A weighted NCE loss is thus proposed and formulated as:

$$\mathcal{L}_{\mathrm{NCE}}^w(q_i, p_j, \{g_l(f(S_w^-))\}_k) = -\log \frac{e^{\phi(q_i, p_j)/\tau}}{e^{\phi(q_i, p_j)/\tau} + \sum_k \left(\frac{1-\phi(p_j, n_k)}{2}\right)^\beta \cdot e^{\phi(q_i, n_k)/\tau}}, \tag{3}$$

where $g_l(\cdot) : R^C \to R^{128}$ is a projection head, $q_i = g_l(f(S_i^+))$, $p_j = g_l(f(S_j^a))$ and $n_k = g_l(f(S_w^-))_k$. $\phi(x, y)$ denotes the cosine similarity between two l_2-normalized vectors. τ refers to the temperature scalar and β is a tunable factor. The term $(\cdot)^\beta$ dynamically decides whether the snippet from fake videos contributes to the contrastive loss based on its similarity with the anchor. That is, if the snippet is pristine/forged, the term approximates $0/1$ and thus it suppresses/activates the contrast with real snippets. Then the local contrastive loss is given by:

$$\mathcal{L}_l = \frac{1}{N} \sum_{i=1}^{N} \sum_{j=1}^{N-U} \mathcal{L}_{\mathrm{NCE}}^w(g_l(f(S_i^+)), g_l(f(S_j^a)), \{g_l(f(S_w^-))\}_k), \tag{4}$$

where $N = |\mathcal{N}^+| \cdot U$, and $|\mathcal{N}^+|$ denotes the number of videos in $\mathcal{N}^+$. Note that q_i has multiple anchor sample p_j, we average their separate contrastive loss.

Analysis of the Weighted NCE Loss. We now analyze the relationship between the proposed loss and the InfoNCE loss [36]. First of all, the NCE loss can be viewed as a special case $(\beta = 0)$ of the proposed weighted NCE loss. Furthermore, we derive the relation between their derivatives:

$$\frac{\partial \mathcal{L}_{\mathrm{NCE}}^w}{\partial w} = \frac{\partial (\frac{b^w}{b})'}{\partial w} \cdot (e^{\mathcal{L}_{\mathrm{NCE}}} - 1)e^{-\mathcal{L}_{NCE}^w} + \frac{b^w}{b} \cdot \frac{e^{\mathcal{L}_{\mathrm{NCE}}}}{e^{\mathcal{L}_{\mathrm{NCE}}^w}} \cdot \frac{\partial \mathcal{L}_{\mathrm{NCE}}}{\partial w}. \tag{5}$$

where $b = \sum_k e^{\phi(q, n_k)/\tau}$ and $b^w = \sum_k (\frac{1-\phi(p, n_k)}{2})^2 e^{\phi(q, n_k)/\tau}$. If all the n_k from fake videos are forged, we then have $\frac{\partial \mathcal{L}_{NCE}^w}{\partial w} \to \frac{\partial \mathcal{L}_{NCE}}{\partial w}$. On the contrary, assume there exists a snippet $n_j \in \mathcal{N}^-$ is in-manipulated, then the learning process is less affected by n_j. Surprisingly, our solution performs on part with or even better than the InfoNCE loss as shown in Table 6.

3.3 Global Contrastive Inconsistency Learning

The global contrastive inconsistency learning is illustrated in Fig. 2. The main components are the inter-snippet fusion module (ISF) for forming the video-level representation and the NCE loss for the contrast between real and fake videos.

Inter-snippet Fusion. A common way to generate video-level representation is averaging snippet-level features along the U dimension. Inspired by [50] using the modified non-local [48] to enhance the short-term features, we instead propose to enhance the channels that reveal the intrinsical inconsistency in a similar way. To achieve this, we design an inter-snippet fusion module upon the encoder f to promote information fusion between $f(S_i) \in R^{U \times C' \times T \times H' \times W'}$ in $f(V)$, where $V = [S_1, \ldots, S_U]$. Specially, the cross-snippet interaction is defined as the self-attention operation between snippets:

$$\text{Atten} = \text{softmax}(\frac{(f(V)W_I)(W_I^T f(V)^T)}{\sqrt{C'}})f(V)W_I, \tag{6}$$

where W_I is learnable parameter of projection for dimension reduction. Then Atten is used to re-weight channels of $f(V)$ by:

$$\text{ISF}(f(V)) = \sigma(\text{Norm}(\text{Atten})W_O) \odot f(V), \tag{7}$$

where W_O is the learnable parameters of projection for dimension retrieval. $\text{Norm}(\cdot)$ is the layer-norm and σ refers to the sigmoid function.

Global (Video-Level) Contrast. We build the video-level contrast on the outputs of the inter-snippet fusion module. Different from the snippet-level contrast, labels of videos are provided and the InfoNCE loss [36] can be directly exploited:

$$\mathcal{L}_{\text{NCE}}(u_i, v_j, \{g_g(f(V_w^-))\}_k) = -\log\frac{e^{\phi(u_i, v_j)/\tau}}{e^{\phi(u_i, v_j)/\tau} + \sum_k e^{\phi(u_i, m_k)/\tau}}, \tag{8}$$

where $g_g(\cdot) : R^C \to R^{128}$ is the projection head, $u_i = g_g(f(V_i^+))$, $v_j = g_g(f(V_j^a))$, $n_k \in \mathcal{N}^-$ and $m_k = g_g(f(V_w^-))_k$. The video-level contrast can be written as:

$$\mathcal{L}_g = \frac{1}{|\mathcal{N}^+|}\sum_{i=1}^{|\mathcal{N}^+|}\mathcal{L}_{\text{NCE}}(g_g(h(f(V_i^+))), g_g(h(f(\mathcal{N}^a))), g_g(h(f(\mathcal{N}^-)))), \tag{9}$$

where $h(f(V_i^+)) = h(f(S_1^+), \ldots, f(S_U^+))$, $h(f(\mathcal{N}^a)) = \{h(f(V_i^a))\}_j$ and $h(f(\mathcal{N}^-)) = \{h(f(V_w^-))\}_k$. Note that one video u_i usually contains multiple anchor videos, we simply compute the contrastive loss separately and take their average.

3.4   Loss Function

Apart from using the contrastive loss mentioned above, we also adopt the binary cross-entropy loss $\mathcal{L}_{ce}$ to supervise the category-level discrepancies. The final loss function for training is formulated as the weighted sum of them:

$$\mathcal{L} = \mathcal{L}_{ce} + \lambda_1\mathcal{L}_l + \lambda_2\mathcal{L}_g, \tag{10}$$

where λ_1 and λ_2 are two balance factors for balancing different terms. All the projection heads for contrastive learning are discarded during inference.

4   Experiments

4.1   Datasets

We conduct all the experiments based on four popular datasets, *i.e.*, FaceForensics++ (FF++) [39], Celeb-DF [29], DFDC-Preview [9] and WildDeepfake [60]. **FaceForensics++** is comprised of 1000 original and 4000 forged videos with several visual quality, *i.e.*, high quality (HQ) and low quality (LQ). Four manipulation methods are exploited for forgery, that is DeepFakes (DF), Face2Face (F2F), FaceSwap (FS), and NeuralTextures (NT). It provides only video-level labels and nearly every frame is manipulated.
Celeb-DF contains 590 real and 5639 fake videos and all the videos are postprocessed for better visual quality. Only video-level labels are provided in it and nearly every frame is manipulated.
WildDeepfake owns 7314 face sequences of different duration from 707 Deepfake videos. Video-level labels are provided only and whether videos in it are partially forged is not mentioned. Therefore, it is relatively more difficult.
DFDC-Preview is the preview version of the DFDC dataset and consists of around 5000 videos. These videos are *partially manipulated* videos by two unknown manipulations, making it more challenging. And no frame-level labels are provided.

4.2   Experimental Settings

Implemental Details. For each Deepfake video dataset, we perform face detection with a similar strategy in [26]. λ_1, λ_2, β and τ are empirically set as 0.1, 0.01, 2 and 5. For simplification, we select the split ratio $\alpha = 0.5$ and the region number $r = 8$. The squeeze ratio is $\frac{1}{8}$ by default. ResNet-50 [16] pre-trained on the ImageNet [8] is exploited as our backbone network. During training, we equally divide a video into $U = 4$ segments and randomly sample consecutive $T = 4$ frames within them to form a snippet. Each frame is then resized into 224×224 as input. We use Adam as the optimizer and set the batch size to 12. The network is trained for 60 epochs and the initial learning rate is 10^{-4}, which is divided by 10 for every 20 epochs. Only random horizontal flip is employed as data augmentation. During testing, we centrally sample $U = 8$ snippets with $T = 4$ and also resize them into the shape of 224×224. The projection heads $g_l(\cdot)$ and $g_g(\cdot)$ are implemented as two MLP layers, transforming features to a 128-dimension space for computing similarity.

Baselines. In order to demonstrate the superiority of our HCIL, we select several video-based detectors for comparison, including Xception [39], VA-LogReg [34], D-FWA [28], FaceNetLSTM [41], Capsule [35], Co-motion [46], DoubleRNN [33], S-MIL [26], DeepRhythm [37], ADDNet-3d [60], STIL [12], DIANet [18] and TD-3DCNN [55]. Besides, for more comprehensive validation, we also re-implement some representative works in video analysis to detect Deepfakes, *i.e.*, LSTM [17], I3D [3], TEI [31], TDN [47], V4D [57], TAM [32] and DSANet [51].

4.3   State-of-the-Art Comparisons

Following [12,37], we conduct both intra-dataset evaluation and cross-dataset generalization to demonstrate the effectiveness of the proposed framework. The accuracy and the Area Under Curve (AUC) metrics are reported, respectively.

Comparison on FF++ Dataset. We first conduct the comprehensive experiments on four subsets of the FF++ dataset under both HQ and LQ settings. Table 1 illustrates the corresponding results, from which we have several observations. Firstly, since no temporal information is considered, the frame-based detectors achieve inferior accuracy compared to video analysis methods. Besides, V4D and DSANet can also achieve a comparable result to the SOTA method STIL. However, they usually employ the sparse sampling strategy and no local motion information is involved, which is also important for this task. Secondly, The proposed HCIL outperforms nearly all the competitors in all settings. Specifically, in challenging LQ NT setting, HCIL owns 94.64% accuracy, exceeding the best action recognition model V4D and the SOTA Deepfake video detector STIL by 2.14% and 2.86%, respectively. All these improvements validate that the well-designed contrastive framework prompts learning general while intrinsical inconsistency between pristine and forged videos.

Table 1. Comparisons on FF++ dataset under both HQ and LQ settings. All subsets are measured and accuracy is reported. † indicates re-implementation.

Methods	FaceForensics++ HQ				FaceForensics++ LQ			
	DF	F2F	FS	NT	DF	F2F	FS	NT
ResNet-50†	0.9893	0.9857	0.9964	0.9500	0.9536	0.8893	0.9464	0.8750
Xception	0.9893	0.9893	0.9964	0.9500	0.9678	0.9107	0.9464	0.8714
LSTM	0.9964	0.9929	0.9821	0.9393	0.9643	0.8821	0.9429	0.8821
I3D†	0.9286	0.9286	0.9643	0.9036	0.9107	0.8643	0.9143	0.7857
TEI†	0.9786	0.9714	0.9750	0.9429	0.9500	0.9107	0.9464	0.9036
TAM†	0.9929	0.9857	0.9964	0.9536	0.9714	0.9214	0.9571	0.9286
DSANet†	0.9929	0.9929	0.9964	0.9571	0.9679	0.9321	0.9536	0.9178
V4D†	0.9964	0.9929	0.9964	0.9607	0.9786	0.9357	0.9536	0.9250
TDN†	0.9821	0.9714	0.9857	0.9464	0.9571	0.9178	0.9500	0.9107
FaceNetLSTM	0.8900	0.8700	0.9000	–	–	–	–	–
Co-motion-70	0.9910	0.9325	0.9830	0.9045	–	–	–	–
DeepRhythm	0.9870	0.9890	0.9780	–	–	–	–	–
ADDNet-3d†	0.9214	0.8393	0.9250	0.7821	0.9036	0.7821	0.8000	0.6929
S-MIL	0.9857	0.9929	0.9929	0.9571	0.9679	0.9143	0.9464	0.8857
S-MIL-T	0.9964	**0.9964**	1.0	0.9429	0.9714	0.9107	0.9607	0.8679
STIL	0.9964	0.9928	1.0	0.9536	0.9821	0.9214	0.9714	0.9178
HCIL	**1.0**	0.9928	**1.0**	**0.9676**	**0.9928**	**0.9571**	**0.9750**	**0.9464**

Table 2. Comparison on Celeb-DF, DFDC, and WildDeepfake datasets. Accuracy is reported.

Methods	Celeb-DF	Wild-DF	DFDC
Xception	0.9944	0.8325	0.8458
I3D[†]	0.9923	0.6269	0.8082
TEI[†]	0.9912	0.8164	0.8697
TAM[†]	0.9923	0.8251	0.8932
V4D[†]	0.9942	0.8375	0.8739
DSANet[†]	0.9942	0.8474	0.8867
D-FWA	0.9858	–	0.8511
DIANet	–	–	0.8583
ADDNet-3D[†]	0.9516	0.6550	0.7966
S-MIL	0.9923	–	0.8378
S-MIL-T	0.9884	–	0.8511
STIL[†]	0.9961	0.8462	0.8980
HCIL	**0.9981**	**0.8586**	**0.9511**

Table 3. Cross-dataset generalization in terms of AUC. † implies re-implementation.

Methods	FF++ DF	Celeb-DF	DFDC
Xception	0.9550	0.6550	0.5939
I3D[†]	0.9541	0.7411	0.6687
TEI[†]	0.9654	0.7466	0.6742
TAM[†]	0.9704	0.6796	0.6714
V4D[†]	0.9674	0.7008	0.6734
DSANet[†]	0.9688	0.7371	0.6808
Capsule	0.9660	0.5750	–
DIANet	0.9040	0.7040	–
TD-3DCNN	–	0.5732	0.5502
DoubleRNN	0.9318	0.7341	–
ADDNet-3D[†]	0.9622	0.6085	0.6589
STIL[†]	0.9712	0.7558	0.6788
HCIL	**0.9832**	**0.7900**	**0.6921**

Comparison on Other Datasets. We also evaluate the proposed method on Celeb-DF, WildDeepfake, and DFDC datasets, as listed in Table 2. On Celeb-DF and WildDeepfake datasets, our framework consistently performs better than SOTAs (0.2% ↑ on Celeb-DF and 1.24% ↑ on Wild-DF). This is mainly because we extract the rich local inconsistency features and generate the global ones, leading to more comprehensive representations. The overall performance on Wild-DF is still low since the duration of videos varies a lot, making them difficult to deal with. On the more challenging DFDC dataset containing vast partially forged videos, HCIL still outperforms the compared works. A large performance margin can be observed (up to 6.0%). Several reasons may account for this. First, different from compared works exploiting either sparse or dense sampling strategy, we sample several snippets from each video to form the input. This strategy enables to capture partially forged parts and both local and global temporal information are covered. Second, the constructed local contrast allows the network to perform fine-grained learning. Therefore, the intrinsical inconsistencies are obtained from contrast, not only from the category-level differences.

Cross-Dataset Generalization. Following previous work [37], we first train the network on the FF++ dataset to discern the pristine videos against four manipulation types under the LQ setting. Then evaluating on FF++ DF, Celeb-DF, and DFDC datasets to measure its generalization capacity, as studied in Table 3. We achieve 98.32% AUC on FF++ DF and 69.21% on the DFDC dataset, improving the state-of-the-art competitors by about 1% on average. Larger performance gains of 3.42% are obtained on Celeb-DF. Since FF++ DF and Celeb-DF datasets are manipulated through similar face forgery techniques frame-by-frame, the detector presents relatively better generalization compared to the DFDC dataset. Note that videos in DFDC are partially forged and contain varied lighting conditions, head poses, and background. Therefore, it is harder to generalize. However, benefiting from the mechanism that learning representations from not only label-level discrepancies but also the local and global contrast, the network owns robustness to a certain degree and consequently exceeds the previous SOTA STIL by 1.3%.

4.4   Ablation Study

We conduct comprehensive ablation studies to further explore the effectiveness of the proposed modules and contrastive framework from Table 4, 5 and 6.

Study on Key Components. The core components of HCIL include the RAI module and ISF module and the specially designed hierarchical contrastive learning paradigm. We perform the ablation under both the intra and inter-dataset settings. And the corresponding results are shown in Table 4 and 5, respectively. In the FF++ dataset, only extracting the frame-level features, the baseline model has the lowest accuracy and AUC. Surprisingly, based on these frame-level representations, directly constructing the contrastive inconsistency learning from multi-hierarchy improves the performance a lot (88.93% $\rightarrow$ 91.43% on F2F and 87.50% $\rightarrow$ 90.36% on NT). Besides equipping the baseline with RAI to extract diverse region-related inconsistency within snippets, large performance gains are observed (95.36% $\rightarrow$ 98.21% on DF, 88.93% $\rightarrow$ 93.21% on F2F, 94.64% $\rightarrow$ 96.06% on FS and 87.50% $\rightarrow$ 92.14% on NT). This is reasonable that without mining the temporal relation within snippets, the general inconsistency can not be captured well via contrast. Similarly, the ISF module also improves the baseline but achieves an inferior result to the RAI module, which implies the importance of local temporal information. If constructing contrastive framework with the corresponding representations, i.e., $\mathcal{L}_s$+ RAI and $\mathcal{L}_v$+ CSF, the accuracy are further boosted (0.36% $\uparrow$ on DF, 0.72% $\uparrow$ on F2F, 0.71% $\uparrow$ on FS and 0.36% $\uparrow$ on NT). Combining RAI and ISF modules is better than contrasting based on one of them, indicating both local and global temporal features are essential for the task. No doubt, combining them all obtains the best results. Similar conclusions can be observed in cross-dataset generalization settings.

Table 4. Ablation study on key components under intra-dataset evaluation. ResNet-50 is used as baseline.

(a) RAI and CSI on FF++.					(b) Contrastive framework on FF++.			
model	DF	F2F	FS	NT	model	DF	F2F	FS	NT
Baseline	0.9536	0.8893	0.9464	0.8750	Baseline	0.9536	0.8893	0.9464	0.8750
+ RAI	0.9821	0.9321	0.9607	0.9214	+ $\mathcal{L}_l + \mathcal{L}_g$	0.9750	0.9143	0.9571	0.9036
+ ISF	0.9785	0.9286	0.9607	0.9178	+ RAI + $\mathcal{L}_l$	0.9857	0.9393	0.9678	0.9250
+ RAI + ISF	0.9857	0.9428	0.9643	0.9250	+ ISF + $\mathcal{L}_g$	0.9857	0.9325	0.9642	0.9214
+ All	**0.9928**	**0.9571**	**0.9750**	**0.9464**	+ All	**0.9928**	**0.9571**	**0.9750**	**0.9464**

Table 5. Ablation study on key components under cross-dataset generalization. ResNet-50 is used as baseline.

(c) Generalization for RAI and ISF.				(d) Generalization for contrastive framework.			
model	FF++ DF	Celeb-DF	DFDC	model	FF++ DF	Celeb-DF	DFDC
Baseline	0.9232	0.6956	0.6323	Baseline	0.9232	0.6956	0.6323
+ RAI	0.9673	0.7478	0.6575	+ $\mathcal{L}_l$ + $\mathcal{L}_g$	0.9572	0.7323	0.6518
+ ISF	0.9603	0.7392	0.6601	+ RAI + $\mathcal{L}_l$	0.9764	0.7569	0.6898
+ RAI + ISF	0.9743	0.7578	0.6665	+ ISF + $\mathcal{L}_g$	0.9711	0.7668	0.6792
+ All	**0.9832**	**0.7900**	**0.6921**	+ All	**0.9832**	**0.7900**	**0.6921**

Study on Parameter β. Equation 3 plays an important role in our contrastive inconsistency framework. β in it indeed adaptively adjusts the importance extent of each negative pair based on the similarity between anchor and snippets from fake videos. Table 6 studies the impacts of it under intra-dataset evaluation settings. In full manipulation settings, i.e., FF++, Celeb-DF and Wild-DF datasets, compared to the baseline, constructing the snippet-level contrast boosts the accuracy (0.35% ↑ on DF, 1.07% ↑ on F2F, 0.72% ↑ on FS and 1.39% ↑ on NT). And NCE loss with adaptive weights presents on part with or even slightly better performance than the vanilla NCE loss. This implies that adjusting the weights in full manipulation settings is helpful. However, the gains from it are limited (0.% ↑ on DF, 0.35% ↑ on F2F, 0.36% ↑ on FS and 0.16% ↑ on NT). In the DFDC dataset, using the NCE loss leads to slight accuracy gains (0.64%). On the contrary, our weighted NCE loss surprisingly gives a larger improvement (from 0.9331% to 0.9511%).

Table 6. Ablation study on β under intra-dataset evaluation settings. The proposed detector without $\mathcal{L}_s$ is set as baseline.

(e) FF++ dataset.					(f) Other three datasets.			
β	DF	F2F	FS	NT	β	Celeb-DF	Wild-DF	DFDC
Baseline	0.9893	0.9464	0.9678	0.9321	Baseline	0.9942	0.8472	0.9331
$\mathcal{L}_{NCE}$	**0.9928**	0.9536	**0.9750**	0.9428	$\mathcal{L}_{NCE}$	**0.9981**	0.8573	0.9395
$\mathcal{L}_{NCE}^w$	**0.9928**	**0.9571**	**0.9750**	**0.9464**	$\mathcal{L}_{NCE}^w$	**0.9981**	**0.8586**	**0.9511**

4.5   Visualization Analysis

In this section, we visualize the region-of-interest via Grad-CAM [40], as shown in Fig. 4. Similar analyses of the RAI module and the weighted NCE loss are presented in Fig. 5.

Class Activation Maps. In Fig. 4, we visualize the class activation maps against four manipulations to verify which regions the model focuses on. The forgery masks derive from the difference between the manipulated videos and the correspondingly pristine videos. The activation maps almost cover the whole faces for Deepfake that are generated from deep learning tools. Similarly, the detector also notices the swapped facial region for FaceSwap. On more challenging Face2Face and NeuralTextures whose transferred expression and mouth regions are difficult to identify, our model still successfully locates the forged areas.

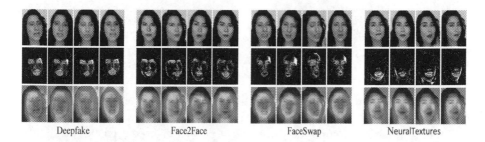

| Deepfake | Face2Face | FaceSwap | NeuralTextures |

Fig. 4. Visualization of activation maps with CAM. First row: RGB images, Second row: forgery masks, Third row: activation maps.

Impacts of RAI Module. The RAI module aims to adaptively extract dynamic local representations for snippet-level contrastive inconsistency learning. We compare it with the vanilla temporal convolution and visualize the corresponding cam maps in the third and second row of Fig. 5 (a). With the content-agnostic weights, the vanilla temporal convolution treats all the facial regions equally and is easy to focus on incomplete forgery regions (second row of left part in Fig. 5 (a)) or wrong areas (second row of right part in Fig. 5 (a)). On the contrary, the RAI module generates the region-specific temporal kernels to extract dynamic temporal information and, therefore, more complete temporal inconsistency representations can be captured for identification.

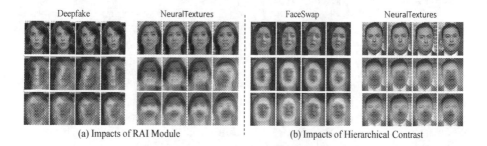

(a) Impacts of RAI Module   (b) Impacts of Hierarchical Contrast

Fig. 5. Visualization on impacts of (a) snippet-level feature extraction (second row: without RAIM, third row: with RAIM) and (b) hierarchical contrast (second row: without contrast, third row: with contrast).

Impacts of Hierarchical Contrast. The hierarchical contrast attracts positive samples while repelling negative ones from both snippet and video levels. To intuitively illustrate impacts of it, we directly visualize the activation in Fig. 5 (b). From the figure, we can observe that for the manipulation performed on small facial areas, the hierarchical contrastive loss acts as a regularizer, regularizing the attention on more accurate regions. Besides, for the manipulation performed on large facial regions, the loss instead guides the detector to focus on more comprehensive locations.

5   Conclusions

In this paper, we introduce the hierarchical contrastive inconsistency learning framework for Deepfake video detection from local and global perspectives. For local contrast, we design a region-aware inconsistency module for dynamic snippet-level representations and a novel weighted NCE loss to enable the snippet-level contrast. For global contrast, an inter-snippet fusion module is introduced for fusing cross-snippet information. The proposed framework presents superior performance and generalization on several benchmarks, and extensive visualizations also illustrate its effectiveness.

Acknowledgements. This research is supported in part by the National Key Research and Development Program of China (No. 2019YFC1521104), National Natural Science Foundation of China (No. 61972157 and No. 72192821), Shanghai Municipal Science and Technology Major Project (2021SHZDZX0102), Shanghai Science and Technology Commission (21511101200 and 21511101200) and Art major project of National Social Science Fund (I8ZD22). We also thank Shen Chen for the proof-read of our manuscript.

References

1. Beuve, N., Hamidouche, W., Deforges, O.: DmyT: dummy triplet loss for deepfake detection. In: WSMMADGD (2021)
2. Cao, J., Ma, C., Yao, T., Chen, S., Ding, S., Yang, X.: End-to-end reconstruction-classification learning for face forgery detection. In: CVPR (2022)
3. Carreira, J., Zisserman, A.: Quo vadis, action recognition? A new model and the kinetics dataset. In: CVPR (2017)
4. Chen, J., Wang, X., Guo, Z., Zhang, X., Sun, J.: Dynamic region-aware convolution. In: CVPR (2021)
5. Chen, S., Yao, T., Chen, Y., Ding, S., Li, J., Ji, R.: Local relation learning for face forgery detection. In: AAAI (2021)
6. Chen, T., Kornblith, S., Norouzi, M., Hinton, G.: A simple framework for contrastive learning of visual representations. In: ICML (2020)
7. Chen, Z., Li, B., Xu, J., Wu, S., Ding, S., Zhang, W.: Towards practical certifiable patch defense with vision transformer. In: CVPR (2022)
8. Deng, J., Dong, W., Socher, R., Li, L.J., Li, K., Fei-Fei, L.: ImageNet: a large-scale hierarchical image database. In: CVPR (2009)

9. Dolhansky, B., Howes, R., Pflaum, B., Baram, N., Ferrer, C.C.: The deepfake detection challenge (DFDC) preview dataset. In: arXiv (2019)
10. Fung, S., Lu, X., Zhang, C., Li, C.T.: Deepfakeucl: deepfake detection via unsupervised contrastive learning. In: IJCNN (2021)
11. Gu, Q., Chen, S., Yao, T., Chen, Y., Ding, S., Yi, R.: Exploiting fine-grained face forgery clues via progressive enhancement learning. In: AAAI (2021)
12. Gu, Z., Chen, Y., Yao, T., Ding, S., Li, J., Huang, F., Ma, L.: Spatiotemporal inconsistency learning for deepfake video detection. In: ACM MM (2021)
13. Gu, Z., Chen, Y., Yao, T., Ding, S., Li, J., Ma, L.: Delving into the local: dynamic inconsistency learning for deepfake video detection. In: AAAI (2022)
14. Haliassos, A., Vougioukas, K., Petridis, S., Pantic, M.: Lips don't lie: a generalisable and robust approach to face forgery detection. In: CVPR (2021)
15. He, K., Fan, H., Wu, Y., Xie, S., Girshick, R.: Momentum contrast for unsupervised visual representation learning. In: CVPR (2020)
16. He, K., Zhang, X., Ren, S., Sun, J.: Deep residual learning for image recognition. In: CVPR (2016)
17. Hochreiter, S., Schmidhuber, J.: Long short-term memory. In: NC (1997)
18. Hu, Z., Xie, H., Wang, Y., Li, J., Wang, Z., Zhang, Y.: Dynamic inconsistency-aware deepfake video detection. In: IJCAI (2021)
19. Li, B., Sun, Z., Guo, Y.: Supervae: superpixelwise variational autoencoder for salient object detection. In: AAAI (2019)
20. Li, B., Sun, Z., Li, Q., Wu, Y., Hu, A.: Group-wise deep object co-segmentation with co-attention recurrent neural network. In: ICCV (2019)
21. Li, B., Sun, Z., Tang, L., Hu, A.: Two-B-real net: two-branch network for real-time salient object detection. In: ICASSP (2019)
22. Li, B., Sun, Z., Tang, L., Sun, Y., Shi, J.: Detecting robust co-saliency with recurrent co-attention neural network. In: IJCAI (2019)
23. Li, B., Sun, Z., Wang, Q., Li, Q.: Co-saliency detection based on hierarchical consistency. In: ACM MM (2019)
24. Li, B., Xu, J., Wu, S., Ding, S., Li, J., Huang, F.: Detecting adversarial patch attacks through global-local consistency. CoRR (2021)
25. Li, L., et al.: Face X-ray for more general face forgery detection. In: CVPR (2020)
26. Li, X., et al.: Sharp multiple instance learning for deepfake video detection. In: ACM MM (2020)
27. Li, Y., Chang, M.C., Lyu, S.: In ictu oculi: Exposing AI generated fake face videos by detecting eye blinking. arXiv (2018)
28. Li, Y., Lyu, S.: Exposing deepfake videos by detecting face warping artifacts. arXiv (2018)
29. Li, Y., Yang, X., Sun, P., Qi, H., Lyu, S.: Celeb-DF: a large-scale challenging dataset for deepfake forensics. In: CVPR (2020)
30. Lin, J., Gan, C., Han, S.: TSM: temporal shift module for efficient video understanding. In: ICCV (2019)
31. Liu, Z., Luo, D., Wang, Y., Wang, L., Tai, Y., Wang, C., Li, J., Huang, F., Lu, T.: TEINet: towards an efficient architecture for video recognition. In: AAAI (2020)
32. Liu, Z., Wang, L., Wu, W., Qian, C., Lu, T.: TAM: temporal adaptive module for video recognition. In: CVPR (2021)
33. Masi, I., Killekar, A., Mascarenhas, R.M., Gurudatt, S.P., AbdAlmageed, W.: Two-branch recurrent network for isolating deepfakes in videos. In: Vedaldi, A., Bischof, H., Brox, T., Frahm, J.-M. (eds.) ECCV 2020. LNCS, vol. 12352, pp. 667–684. Springer, Cham (2020). https://doi.org/10.1007/978-3-030-58571-6_39

34. Matern, F., Riess, C., Stamminger, M.: Exploiting visual artifacts to expose deep-fakes and face manipulations. In: CVPRW (2019)
35. Nguyen, H.H., Yamagishi, J., Echizen, I.: Capsule-forensics: using capsule networks to detect forged images and videos. In: ICASSP (2019)
36. Oord, A.v.d., Li, Y., Vinyals, O.: Representation learning with contrastive predictive coding. arXiv (2018)
37. Qi, H., et al.: DeepRhythm: exposing deepfakes with attentional visual heartbeat rhythms. In: ACM MM (2020)
38. Qian, Y., Yin, G., Sheng, L., Chen, Z., Shao, J.: Thinking in frequency: face forgery detection by mining frequency-aware clues. In: Vedaldi, A., Bischof, H., Brox, T., Frahm, J.-M. (eds.) ECCV 2020. LNCS, vol. 12357, pp. 86–103. Springer, Cham (2020). https://doi.org/10.1007/978-3-030-58610-2_6
39. Rossler, A., Cozzolino, D., Verdoliva, L., Riess, C., Thies, J., Nießner, M.: Face-Forensics++: learning to detect manipulated facial images. In: ICCV (2019)
40. Selvaraju, R.R., Cogswell, M., Das, A., Vedantam, R., Parikh, D., Batra, D.: Grad-CAM: visual explanations from deep networks via gradient-based localization. In: ICCV (2017)
41. Sohrawardi, S.J., et al.: Poster: towards robust open-world detection of deepfakes. In: ACM CCCS (2019)
42. Sun, K., Yao, T., Chen, S., Ding, S., Ji, R., et al.: Dual contrastive learning for general face forgery detection. In: AAAI (2021)
43. Tang, L., Li, B.: CLASS: cross-level attention and supervision for salient objects detection. In: Ishikawa, H., Liu, C., Pajdla, T., Shi, J. (eds.) ACCV (2020)
44. Tang, L., Li, B., Zhong, Y., Ding, S., Song, M.: Disentangled high quality salient object detection. In: ICCV (2021)
45. Wang, G., Jiang, Q., Jin, X., Li, W., Cui, X.: MC-LCR: multi-modal contrastive classification by locally correlated representations for effective face forgery detection. arXiv (2021)
46. Wang, G., Zhou, J., Wu, Y.: Exposing deep-faked videos by anomalous co-motion pattern detection. arXiv (2020)
47. Wang, L., Tong, Z., Ji, B., Wu, G.: TDN: temporal difference networks for efficient action recognition. In: CVPR (2021)
48. Wang, X., Girshick, R., Gupta, A., He, K.: Non-local neural networks. In: CVPR (2018)
49. Wang, X., Yao, T., Ding, S., Ma, L.: Face manipulation detection via auxiliary supervision. In: ICONIP (2020)
50. Wu, C.Y., Feichtenhofer, C., Fan, H., He, K., Krahenbuhl, P., Girshick, R.: Long-term feature banks for detailed video understanding. In: CVPR (2019)
51. Wu, W., et al.: DSANet: Dynamic segment aggregation network for video-level representation learning. In: ACM MM (2021)
52. Wu, Z., Xiong, Y., Yu, S.X., Lin, D.: Unsupervised feature learning via non-parametric instance discrimination. In: CVPR (2018)
53. Xu, Y., Raja, K., Pedersen, M.: Supervised contrastive learning for generalizable and explainable deepfakes detection. In: WCACV (2022)
54. Yang, X., Li, Y., Lyu, S.: Exposing deep fakes using inconsistent head poses. In: ICASSP (2019)
55. Zhang, D., Li, C., Lin, F., Zeng, D., Ge, S.: Detecting deepfake videos with temporal dropout 3DCNN. In: AAAI (2021)
56. Zhang, J., et al.: Towards efficient data free black-box adversarial attack. In: CVPR (2022)

57. Zhang, S., Guo, S., Huang, W., Scott, M.R., Wang, L.: V4D: 4D convolutional neural networks for video-level representation learning. arXiv (2020)
58. Zhong, Y., Li, B., Tang, L., Kuang, S., Wu, S., Ding, S.: Detecting camouflaged object in frequency domain. In: CVPR (2022)
59. Zhong, Y., Li, B., Tang, L., Tang, H., Ding, S.: Highly efficient natural image matting. CoRR (2021)
60. Zi, B., Chang, M., Chen, J., Ma, X., Jiang, Y.G.: Wilddeepfake: a challenging real-world dataset for deepfake detection. In: ACM MM (2020)

Rethinking Robust Representation Learning Under Fine-Grained Noisy Faces

Bingqi Ma[1], Guanglu Song[1], Boxiao Liu[1,2], and Yu Liu[1(✉)]

[1] Sensetime Research, Shanghai, China
{mabingqi,songguanglu}@sensetime.com, liuboxiao@ict.ac.cn,
liuyuisanai@gmail.com
[2] SKLP, Institute of Computing Technology, CAS, Beijing, China

Abstract. Learning robust feature representation from large-scale noisy faces stands out as one of the key challenges in high-performance face recognition. Recent attempts have been made to cope with this challenge by alleviating the intra-class conflict and inter-class conflict. However, the unconstrained noise type in each conflict still makes it difficult for these algorithms to perform well. To better understand this, we reformulate the noise type of each class in a more fine-grained manner as N-identities|K^C-clusters. Different types of noisy faces can be generated by adjusting the values of N, K, and C. Based on this unified formulation, we found that the main barrier behind the noise-robust representation learning is the flexibility of the algorithm under different N, K, and C. For this potential problem, we propose a new method, named Evolving Sub-centers Learning (ESL), to find optimal hyperplanes to accurately describe the latent space of massive noisy faces. More specifically, we initialize M sub-centers for each class and ESL encourages it to be automatically aligned to N-identities|K^C-clusters faces via producing, merging, and dropping operations. Images belonging to the same identity in noisy faces can effectively converge to the same sub-center and samples with different identities will be pushed away. We inspect its effectiveness with an elaborate ablation study on the synthetic noisy dataset with different N, K, and C. Without any bells and whistles, ESL can achieve significant performance gains over state-of-the-art methods on large-scale noisy faces.

Keywords: Fine-grained noisy faces · Evolving Sub-centers Learning

1 Introduction

Owing to the rapid development of computer vision technology [22–24], face recognition [4,15,16,26,30] has made a remarkable improvement and has been widely applied in the industrial environment. Much of this progress was sparked

---

B. Ma and G. Song—Equal contributions.

Supplementary Information The online version contains supplementary material available at https://doi.org/10.1007/978-3-031-19775-8_36.

S. Avidan et al. (Eds.): ECCV 2022, LNCS 13672, pp. 614–630, 2022.
https://doi.org/10.1007/978-3-031-19775-8_36

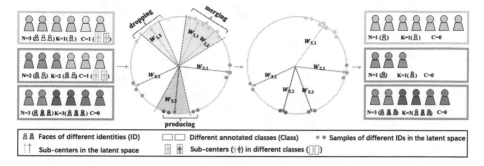

Fig. 1. Illustration of fine-grained noisy faces and ESL (Best viewed in color). For ID2 and ID3 in Class1, there is only one image for each ID, and they will be removed by the dropping operation. ID1 appears in both Class1 and Class2, so that the merging operation will merge images in Class2 with ID1 into Class1. In Class3, there are 3 IDs but only 2 sub-centers, so the producing operation will produce another valid sub-center. Our proposed ESL can be flexibly adapted to different combinations of NKC, and is more robust to unconstrained real-world noise. (Color figure online)

by the collection of large-scale web faces as well as the robust learning strategies [4,26] for representation learning. For instance, MS-Celeb-1M (MS1MV0) [8] provides more than 10 million face images with rough annotations. The growing scale of training datasets inevitably introduces unconstrained noisy faces and can easily weaken the performance of state-of-the-art methods. Learning robust feature representation from large-scale noisy faces has become an important challenge for high-performance face recognition. Conventional noisy data learning, such as recursive clustering, cleaning, and training process, suffers from high computational complexity and cumulative error. For this problem, Sub-center ArcFace [2] and SKH [13] are proposed to tackle the intra-class conflict or inter-class conflict by designing multiple sub-centers for each class. These algorithms demonstrate remarkable performance in the specific manual noise. However, they are still susceptible to the unconstrained types of real-world noisy faces. Naturally, we found that it is far from enough to just divide label noise in face recognition roughly into intra-class noise and inter-class noise. It greatly limits our understanding of the variant noise types and the exploration of noise-robust representation learning strategies.

To better understand this, we reformulate the noise data in a more fine-grained manner as N-identities|K^C-clusters faces for each class. Faces sharing *identity* (ID) means these images come from the same person. Faces annotated with the same label construct a *class*, and there may be annotation errors in the class. If there are no less than two faces for an identity, these images build a meaningful *cluster* [7]. Please refer to the Sec. 1 in the appendix for the holistic description of terms and notations. Taking the Class1 in Fig. 1 as an example, there are 3 IDs marked with ID1, ID2 and ID3, so the N in Class1 is 3. However, only ID1 contains more than 2 images, so the K in Class1 is 1. Furthermore, ID1 appears in both Class1 and Class2, which indicates one inter-class conflict, so the

Table 1. Noise type in different combination of N, K, and C. $\triangle$ represents intra-class conflict in which there are multiple clusters in the class. $\square$ represents intra-class conflict where there are outlier faces in the class. $\Diamond$ represents the inter-class conflict in which there are multiple clusters with the same identify in different classes.

	$N = K = 1$	$N = K > 1$	$N > K > 1$	$N > K = 1$	$N > K = 0$
$C = 0$	–	$\triangle$	$\triangle\ \square$	$\square$	$\square$
$C > 0$	$\Diamond$	$\triangle\ \Diamond$	$\triangle\ \square\ \Diamond$	$\square\ \Diamond$	–

C in Class1 is 1. As shown in Table 1, our proposed N-identities|K^C-clusters formulation can clearly represent different fine-grained noisy data.

However, if N and K are larger than the predefined sub-center number in Sub-center ArcFace [2] and SKH [13], images without corresponding sub-center will lead to intra-class conflict. If C exceeds the sub-center number in SKH [13], extra conflicted clusters will bring inter-class conflict. Both intra-class conflict and inter-class conflict will lead to the wrong gradient, which would dramatically impair the representation learning process.

In this paper, we constructively propose a flexible method, named Evolving Sub-centers Learning (ESL), to solve this problem caused by unconstrained N, K, and C. More specifically, we initialize M sub-centers for each class first. Images belonging to the same identity will be pushed close to the corresponding positive sub-center and away from all other negative sub-centers. Owning to elaborate designed producing, dropping, and merging operations, ESL encourages the number of sub-centers to be automatically aligned to N-identities|K^C-clusters faces. As shown in Fig. 1, our proposed ESL can be flexibly adapted to different combinations of NKC, and is more robust to unconstrained real-world noise. We inspect its effectiveness with elaborate ablation study on variant N-identities|K^C-clusters faces. Without any bells and whistles, ESL can achieve significant performance gains over current state-of-the-art methods in large-scale noisy faces. To sum up, the key contributions of this paper are as follows:

- We reformulate the noise type of faces in each class into a more fine-grained manner as N-identities|K^C-clusters. Based on this, we reveal that the key to robust representation learning strategies under real-world noise is the flexibility of the algorithm to the variation of N, K, and C.
- We introduce a general flexible method, named Evolving Sub-centers Learning (ESL), to improve the robustness of feature representation on noisy training data. The proposed ESL enjoys scalability to different combinations of N, K, and C, which is more robust to unconstrained real-world noise.
- Without relying on any annotation post-processing or iterative training, ESL can easily achieve significant performance gains over state-of-the-art methods on large-scale noisy faces.

2   Related Work

Loss function for Face Recognition. Deep face recognition models rely heavily on the loss function to learn discriminate feature representation. Previous works [4,6,12,14,20,25,26,28,31] usually leverage the margin penalty to opti-mize the intra-class distance and the inter-class distance. Facenet [20] uses the Triplet to force that faces in different classes have a large Euclidean distance than faces in the same class. However, the Triplet loss can only optimize a subset of all classes in each iteration, which would lead to an under-fitting phenomenon. It is still a challenging task to enumerate the positive pairs and negative pairs with a growing number of training data. Compared with the sample-to-sample optimization strategy, Liu et al. [14] proposes the angular softmax loss which enables convolutional neural networks to learn angularly discriminative features. Wang et al. [26] reformulates Softmax-base loss into a cosine loss and introduces a cosine margin term to further maximize the decision margin in the angular space. Deng et al. [4] directly introduces a fixed margin, maintaining the consis-tency of the margin in the angular space. Liu et al. [16] adopts the hard example mining strategy to re-weight temperature in the Softmax-base loss function for more effective representation learning.

Dataset for Face Recognition. Large-scale training data can significantly improve the performance of face recognition models. MS1MV0 [8], in which there are about 100K identities and 1M faces, is the most commonly used face recognition dataset. MS1MV3 is a cleaned version from MS1MV0 with a semi-automatic approach [5]. An et al. [1] cleans and merges existing public face recog-nition datasets, then obtains Glint360K with 17M faces and 360K IDs. Recently, Zhu et al. [35] proposes a large-scale face recognition dataset WebFace260M and a automatically cleaning pipeline. By iterative training and cleaning, they pro-posed well-cleaned subset with 42M images and 2M IDs.

Face Recognition under Noisy Data. Iterative training and cleaning is an effective data cleanup method. However, it is extremely inefficient as the face number increases. Recent works [4,10,13,27,32–34] focus on efficient noisy data cleanup methods. Zhong et al. [32] decouples head data and tail data of a long-tail distribution and designs a noise-robust loss function to learn the sample-to-center and sample-to-sample feature representation. Deng et al. [2] designs multiple centers for each class, splitting clean faces and noisy faces into different centers to deal with the inter-class noise. Liu et al. [13] leverages multiple hyper-planes with a greedy switching mechanism to alleviate both inter-class noise and intra-class noise. However, these methods are sensitive to hyper-parameter and can not tackle the complex noisy data distribution.

3   The Proposed Approach

In this section, we are committed to eliminating the unconstrained real-world noise via a flexible and scalable learning manner, named Evolving Sub-centers Learning, that can be easily plugged into any loss functions. The pipeline of ESL is as shown in Fig. 2. We will first introduce our proposed ESL and then give a deep analysis to better understand its effectiveness and flexibility under fine-grained noisy faces. Finally, we conduct a detailed comparison between ESL and the current state-of-the-art noise-robust learning strategies.

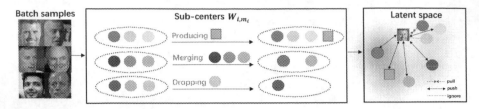

Fig. 2. The pipeline of the Evolving Sub-centers Learning. We initialize M sub-centers for each class and they will evolve adaptively to align the data distribution. It pushes the images belonging to an identity close to the specific sub-center and away from all other negative sub-centers. The sub-center with a confusing similarity to the current sample will be ignored in the latent space. This can effectively dispose of the label conflict caused by fine-grained noisy faces.

3.1   Evolving Sub-centers Learning

In face recognition tasks, the unified loss function can be formulated as:

$$\mathcal{L}(x_i) = -\log \frac{e^{\hat{f}_{i,y_i}}}{e^{\hat{f}_{i,y_i}} + \sum_{j=1,j\neq y_i}^{S} e^{f_{i,j}}}, \tag{1}$$

where i is the index of face images, y_i represents the label ID of image I_i and S indicates the total class number in the training data. Let the x_i and W_j denote the feature representation of face image I_i and the j-th class center, the logits $\hat{f}_{i,y_i}$ and $f_{i,j}$ can be computed by:

$$\hat{f}_{i,y_i} = s \cdot [m_1 \cdot \cos(\theta_{i,y_i} + m_2) - m_3], \tag{2}$$

$$f_{i,j} = s \cdot cos(\theta_{i,j}), \tag{3}$$

where s is the re-scale parameter and $\theta_{i,j}$ is the angle between the x_i and W_j normalized with $\mathcal{L}_2$ manner. For ArcFace with $m_1 = 1$ and $m_3 = 0$, we can compute the θ_{i,y_i} by:

$$\theta_{i,y_i} = arccos(\frac{W_{y_i}^T}{||W_{y_i}^T||_2} \frac{x_i}{||x_i||_2}). \tag{4}$$

As shown in SKH [13], Eq. (1) will easily get wrong loss under $N > 1$ which indicates at least two different identities exist in the images currently labeled as the same identity. In this paper, we address this problem by proposing the idea of using class-specific sub-centers for each class, which can be directly adopted by any loss functions and will significantly increase its robustness. As illustrated in Fig. 2, we init M_j sub-centers for j-th class where each center is dominated by a learnable vector $W_{j,m_j}, m_j \in [1, M_j]$. The original class weight $W_j \in \mathbb{R}^{1 \times D}$ can be replaced by all sub-centers $W_j \in \mathbb{R}^{1 \times M_j \times D}$. Based on this, Eq. (1) can be re-written as:

$$\mathcal{L}(x_i) = -\log \frac{e^{\hat{f}_{i,y_i,m}}}{e^{\hat{f}_{i,y_i,m}} + \sum_{\substack{j\in[1,C],m_j\in[1,M_j] \\ (j,m_j)\neq(y_i,m)}} (1 - \mathbb{1}\{\cos(\theta_{i,j,m_j}) > \mathcal{D}_{j,m_j}\})e^{f_{i,j,m_j}}}, \tag{5}$$

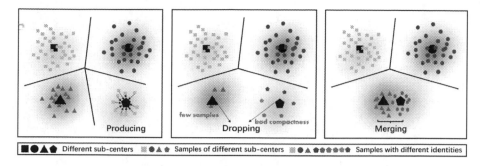

Fig. 3. Illustration of the sub-centers producing, dropping and merging. The instances with black color indicate the sub-centers. The instance belonging to each sub-center is represented by the same shape and different colors mean different identities. (Color figure online)

where indicator function $\mathbb{1}\{\cos(\theta_{i,j,m_j}) > \mathcal{D}_{j,m_j}\}$ returns 1 when $\cos(\theta_{i,j,m_j}) > \mathcal{D}_{j,m_j}$ and 0 otherwise. We calculate the mean μ_{j,m_j} and standard deviation σ_{j,m_j} of the cosine similarity between the sub-center W_{j,m_j} and the samples belonging to it. $\mathcal{D}_{j,m_j}$ can be generated by:

$$\mathcal{D}_{j,m_j} = \mu_{j,m_j} + \lambda_1 \sigma_{j,m_j} \tag{6}$$

$\hat{f}_{i,y_i,m}$ and f_{i,j,m_j} are computed by:

$$\hat{f}_{i,y_i,m} = s \cdot [m_1 \cdot \cos(\theta_{i,y_i,m} + m_2) - m_3], \tag{7}$$

$$f_{i,j,m_j} = s \cdot cos(\theta_{i,j,m_j}), \tag{8}$$

where θ_{i,j,m_j} is the angle between the feature representation x_i of the i-th face image and the m_j-th sub-center W_{j,m_j} in j-th class. We determine m for sample I_i by the nearest distance priority manner as:

$$m = \arg\max_{m_{y_i}} \cos(\theta_{i,y_i,m_{y_i}}). \ \ m_{y_i} \in [1, M_{y_i}] \tag{9}$$

Given an initial M_j for each class, Eq. (5) can capture the unconstrained distribution of the whole training data with potential label noise. It pushes the images belonging to the same identity close to a specific sub-center and away from all other negative sub-centers. Meanwhile, the sub-center with a confusing similarity to the current sample will be ignored to dispose of the label conflict. To make it more flexible with unconstrained changes in N, K, and C, we further introduce the producing, merging, and dropping operations as shown in Fig. 3.

Sub-centers Producing. Based on the aforementioned design, it can effectively alleviate the conflict caused by label noise when $M_j > N > 1$. However, the unconstrained N makes it difficult to select the appropriate M_j for each class. To make it more flexible, we introduce the sub-centers producing operation to automatically align the sub-centers and the actual identity number in each class.

Given $\mathcal{N}$ images with label y and assigned to sub-center m with Eq. 9, a new sub-center W_{y,M_y+m} can be generated by:

$$W_{y,M_y+m} = \frac{1}{\mathcal{T}} \sum_{i=1}^{\mathcal{N}} \mathbb{1}\{cos(\theta_{i,y,m}) < \mu_{y,m} - \lambda_2\sigma_{y,m}\}x_i, \quad if \ \mathcal{T} > 0, \qquad (10)$$

where $\mathcal{T} = \sum_{i=1}^{\mathcal{N}} \mathbb{1}\{cos(\theta_{i,y,m}) < \mu_{y,m} - \lambda_2\sigma_{y,m}\}$. If $\mathcal{T} = 0$, there is no new sub-center to be formed. After this, we can progressively produce new sub-center to house additional identities beyond M_j. It effectively improves intra-class compactness and reduces the conflict caused by unconstrained N and K.

Sub-centers Dropping. As demonstrated by [2,13], many state-of-the-art methods are susceptible to the outlier faces (the image number belonging to an identity is less than 2, $N > K \geq 1$). These outlier images are hard to be pushed close to any corresponding positive sub-center. During the producing process, outlier images from each sub-center will generate a new sub-center. The dropping operation should remove the sub-center from outlier images but preserve the sub-center with a valid identity. Considering the standard deviation can not reflect the density of a distribution, we just leverage μ_{i,m_i} as the metric. The condition of dropping can be formulated as:

$$\mathcal{J}(W_{i,m_i}) = \mathbb{1}\{\mu_{i,m_i} \leq \lambda_3\}. \qquad (11)$$

If the μ_{i,m_i} is less than λ_3, we will ignore these images during the training process and then erase the specific sub-center.

Sub-centers Merging. Using sub-centers for each class can dramatically improve the robustness under noise. However, the inter-class discrepancy will be inevitably affected by inter-class conflict caused by the shared identity between different sub-centers. SKH [13] sets the same fixed number of sub-centers for each class, which can not handle complex inter-class conflict with unconstrained C. Meanwhile, the sub-center strategy undermines the intra-class compactness as the samples in a clean class also converge to different sub-centers. To deal with this potential problem, we employ the sub-centers merging operation to aggregate different $W_{*,*}$. The condition of merging can be formulated as:

$$\mathcal{J}(W_{i,m_i}, W_{j,m_j}) = \mathbb{1}\{W_{i,m_i}^T W_{j,m_j} \geq \max(\mu_{j,m_j} + \lambda_4\sigma_{j,m_j}, \mu_{i,m_i} + \lambda_4\sigma_{i,m_i})\}, \qquad (12)$$

where W_{i,m_i} and W_{j,m_j} are normalized with L_2 manner. According to Eq. (12), we merge multiple sub-centers satisfying $\mathcal{J}(*,*) = 1$ into a group and combine them into a single sub-center as following:

$$W_{*,new} = \frac{1}{|G|} \sum_{(p,m_p)\in G} W_{p,m_p}, \qquad (13)$$

where G and $|G|$ indicate the merged group and its sub-center number. Furthermore, images belonging to G will be assigned to a new label (we directly select the minimum label ID in the G as the target).

3.2   Progressive Training Framework

At the training stage, we perform the sub-centers producing, dropping and merging operations progressively to effectively alleviate the label conflict cause by unconstrained N, K, and C. The training framework is summarized in Algorithm 1.

Algorithm 1. Evolving Sub-centers Learning

Input: Training data set $\mathcal{X}$, label set $\mathcal{Y}$, total training epoch E, start
epoch ε for ESL.

Initialize: Label number C, sub-centers number M_* and $W_{*,*}$ for each
class.

$e \leftarrow 0$;

while $e < E$ **do**

    sample data X, Y from $\mathcal{X}$, $\mathcal{Y}$;

    compute loss function $L(X, Y)$ by Eq. (5) and update model;

    generating μ_* and σ_* for each sub-center;

    if $e > \varepsilon$ **then**

        for $i = 1$ **to** C **do**

            for $j = 1$ **to** M_i **do**

                `// Producing`

                computing $\mathcal{T}$ via Eq. (10);

                if $\mathcal{T} > 0$ **then**

                    computing W_{i,M_i+j} via Eq. (10);

                `// Dropping`

                generating $\mathcal{J}(W_{i,j})$ via Eq. (11);

                if $\mathcal{J}(W_{i,j})$ **then**

                    dropping sub-center $W_{i,j}$;

                    generating $\mathcal{X}_i$ as images with label i;

                    foreach *image* X *in* $\mathcal{X}_i$ **do**

                        computing m via Eq.(9);

                        if $j = m$ **then**

                            dropping image X;

    `// Merging`

    generating vertex set V with each sub-center W in $W_{*,*}$;

    generating edge set E with (W_i, W_j) if $\mathcal{J}(W_i, W_j) = 1$ via Eq. (12);

    generating graph $G = (V, E)$;

    foreach *connected component* g *in* G **do**

        generating new sub-center via Eq. (13);

        relabeling images belonging to sub-centers in g;

  $e = e + 1$;

In this manner, ESL is able to capture the complex distribution of the whole training data with unconstrained label noise. It tends to automatically adjust the sub-centers to align the distribution of N, K, and C in the given datasets.

This allows it to be flexible in solving the real-world noise while preventing the network from damaging the inter-class discrepancy on clean faces.

3.3   Robustness Analysis on Fine-Grained Noisy Faces

When applied to the practical N-identities|K^C-clusters faces, the key challenge is to process different combinations of N, K, and C. In Table 1, we have analyzed the noise type in different combinations of N, K, and C. Now we investigate the robustness of ESL on fine-grained noisy faces.

To simplify the analysis, we first only consider the circumstance when $\mathbf{C} = \mathbf{0}$. (1) $\mathbf{N} = \mathbf{K} = \mathbf{1}, \mathbf{C} = \mathbf{0}$. This phenomenon indicates the training dataset is absolutely clean. In this manner, most of the feature learning strategies can perform excellent accuracy. However, introducing the sub-centers for each class will damage the intra-class compactness and degrade the performance. The sub-centers merging allows ESL to progressively aggregate the sub-centers via Eq. (12) to maintain the intra-class compactness. (2) $\mathbf{N} = \mathbf{K} > \mathbf{1}, \mathbf{C} = \mathbf{0}$. This means there are several identities existing in a specific class. $N = K$ represents the images for each identity are enough to form a valid cluster in the latent space and there are no outlier images. Under this manner, with appropriate hyper-parameter, the state-of-the-art methods Sub-center ArcFace [2] and SKH [13] can effectively cope with this label conflict. However, the unconstrained N and K still make them ineffective even with some performance gains. In ESL, the sub-centers producing via Eq. (10) adaptively produce new sub-centers to accommodate the external identities beyond the initialized sub-center number if there are fewer sub-centers than identities in the class. If the number of sub-centers is larger than the identity number, the merging strategy will merge clusters with the same identity to keep the intra-class compactness. (3) $\mathbf{N} > \mathbf{K} > \mathbf{1}, \mathbf{C} = \mathbf{0}$. Besides the conflict clusters, several identities can not converge to valid clusters in the latent space. We find that this is caused by the few-shot samples in each identity. It lacks intra-class diversity, which prevents the network from effective optimization and leads to the collapse of the feature dimension. To deal with these indiscoverable outliers, we design the sub-centers dropping operation to discard these sub-centers with few samples or slack intra-class compactness based on Eq. 11 in ESL. This is based on our observation that these sub-centers are not dominated by any one identity. Multiple outliers try to compete for the dominance, leading to bad compactness. (4) $\mathbf{N} > \mathbf{K} = \mathbf{1}, \mathbf{C} = \mathbf{0}$. It indicates that there is one valid identity and several outlier images in this class. ESL will enable the dropping strategy to remove the noise images and keep valid faces. (5) $\mathbf{N} > \mathbf{K} = \mathbf{0}, \mathbf{C} = \mathbf{0}$. It indicates each identity in the class only owns few-shot samples. We proposed dropping operation will discard all the sub-centers in this class. For $\mathbf{C} > \mathbf{0}$, there are multiple clusters with the same identity but different labels. This introduces the inter-class conflict. The state-of-the-art method SKH [13] can not perform well under unconstrained C. For this potential conflict, Eq. (12) in ESL can also accurately alleviate this by dynamically adjusting the label of images belonging to the merged sub-centers.

Table 2. Comparison with other noise-robust learning strategies under different types of noise. + indicates the method can solve the noise under the specific setting. +++ indicates the method can solve the noise problem. − indicates the method can not handle the problem.

Method	C	$N = K = 1$	$N = K > 1$	$N > K > 1$	$N > 1 \geq K$
ArcFace [4]	$C = 0$	+++	−	−	−
	$C > 0$	−	−	−	−
NT [10]	$C = 0$	+++	−	+	+
	$C > 0$	−	−	−	−
NR [32]	$C = 0$	+++	−	+	+
	$C > 0$	−	−	−	−
Sub-center [2]	$C = 0$	+	+	+	+
	$C > 0$	−	−	−	−
SKH [13]	$C = 0$	+	+	+	+
	$C > 0$	+	+	+	+
ESL	$C = 0$	+++	+++	+++	+++
	$C > 0$	+++	+++	+++	+++

3.4   Comparison with Other Noise-Robust Learning Strategies

The main difference between the proposed ESL and other methods [2,4,10,13,32] is that ESL is less affected by the unconstrained N, K, and C from the real-world noise. It's more flexible to face recognition under different types of noise while keeping extreme simplicity, only adding three sub-centers operation. To better demonstrate this, we make a detailed comparison with other methods under fine-grained noisy faces as shown in Table 2. The superiority of our method is mainly due to the flexible sub-center evolving strategy, which can handle variant intra-class noise and inter-class noise simultaneously.

4   Experiments

4.1   Experimental Settings

Datasets. MS1MV0 [8] and MS1MV3 [5] are popular academic face recognition datasets. MS1MV0 [8] is raw data that is collected from the search engine based on a name list, in which there is around 50% noise. MS1MV3 [5] is the cleaned version of MS1MV0 [8] by a semi-automatic pipeline. To further explore the effectiveness of our proposed ESL, we also elaborately construct synthetic noisy datasets. We establish intra-conflict, inter-class conflict, and mixture conflict noisy datasets, which will be detailed introduced in the supplementary material. As for the performance evaluation, we tackle the True Accept Rate (TAR) at a specific False Accept Rate (FAR) as the metric. We mainly consider the performance on IJB-B [29] dataset and IJB-C [17] dataset. Moreover, we also report the results on LFW [11], CFP-FP [21] and AgeDB-30 [18].

Table 3. Experiments of different settings on MS1MV0 and synthetic mixture noisy dataset comparing with state-of-the-art methods.

Method	Dataset	IJB-B			IJB-C		
		$1e-3$	$1e-4$	$1e-5$	$1e-3$	$1e-4$	$1e-5$
ArcFace [4]	MS1MV0	93.27	87.87	74.74	94.59	90.27	81.11
Sub-center ArcFace $M=3$ [2]	MS1MV0	94.88	91.70	85.62	95.98	93.72	90.59
Co-ming [27]	MS1MV0	94.99	91.80	85.57	95.95	93.82	90.71
NT [10]	MS1MV0	94.79	91.57	85.56	95.86	93.65	90.48
NR [32]	MS1MV0	94.77	91.58	85.53	95.88	93.60	90.41
SKH + ArcFace $M=3$ [13]	MS1MV0	95.89	93.50	89.34	96.85	95.25	93.00
ESL + ArcFace	MS1MV0	**96.61**	**94.60**	**91.15**	**97.58**	**96.23**	**94.24**
ArcFace [4]	Mixture of noises	93.17	87.54	74.02	94.99	90.03	82.40
Sub-center ArcFace $M=3$ [2]	Mixture of noises	92.83	86.80	73.11	94.20	89.32	81.43
SKH + ArcFace $M=4$ [13]	Mixture of noises	95.76	93.62	89.18	96.89	95.16	92.71
ESL + ArcFace	Mixture of noises	**96.48**	**94.51**	**90.95**	**97.62**	**96.22**	**93.60**

Implementation Details. Following ArcFace [4], we generate aligned faces with RetinaFace [3] and resize images to (112×112). We employ ResNet-50 [9] as backbone network to extract 512-D feature embedding. For the experiments in our paper, we initialize the learning rate with 0.1 and divide it by 10 at 100K, 160K, and 220K iteration. The total training iteration number is set as 240K. We adopt an SGD optimizer, then set momentum as 0.9 and weight decay as $5e-4$. The model is trained on 8 NVIDIA A100 GPUs with a total batch size of 512. The experiments are implemented with Pytorch [19] framework. For experiments on ESL, we set the initial number of sub-centers for each class as 3. The λ_1, λ_2, λ_3 and λ_4 is separately set as 2, 2, 0.25, and 3.

4.2   Comparison with State-of-the-Art

We conduct extensive experiments to investigate our proposed Evolving Sub-centers Learning. In Table 3, we compare ESL with state-of-the-art methods on both real-word noisy dataset MS1MV0 [8] and the synthetic mixture of noise dataset. Without special instructions, the noise ratio is 50%.

When training on noisy data, ArcFace has an obvious performance drop. It demonstrates that noise samples would do dramatically harm to the optimization process. ESL can easily outperform current methods by an obvious margin. To be specific, ESL can outperform Sub-center ArcFace [2] by 2.51% and SKH [13] by 0.98% on IJB-C dataset. On the synthetic mixture noisy dataset, we make a grid search of the sub-center number in Sub-center ArcFace [2] and SKH [13]. Sub-center ArcFace [2] achieves the best performance when $M=3$ and SKH [13] achieves the best performance when $M=4$. ESL can easily outperform Sub-center ArcFace [2] by 6.9% and SKH [13] by 1.06% on IJB-C dataset. Our proposed ESL can handle the fine-grained intra-class conflict and inter-class conflict under unconstrained N, K, and C, which brings significant performance improvement.

Table 4. Ablation experiments to explore the hyperparameters.

λ_1(Eq. (5))	λ_2(Eq. (10))	λ_3(Eq. (11))	λ_4(Eq. (12))	M_j(Eq. (5))	TAR@FAR=-4
2	2	0.25	3	3	**96.22**
1	2	0.25	3	3	95.73
3	2	0.25	3	3	95.88
2	1	0.25	3	3	96.11
2	3	0.25	3	3	95.89
2	2	0.2	3	3	96.05
2	2	0.3	3	3	96.18
2	2	0.25	1	3	95.07
2	2	0.25	2	3	95.82
2	2	0.25	3	1	96.03
2	2	0.25	3	2	96.14
2	2	0.25	3	4	96.20
2	2	0.25	3	5	96.19

Table 5. Ablation experiments to verify the effectiveness of proposed operations.

ArcFace	Class-specific sub-center	Merging	Producing	Dropping	Dataset	IJB-C		
						1e−3	1e−4	1e−5
✓	✗	✗	✗	✗	Mixture of noises	94.99	90.03	82.40
✓	✓	✗	✗	✗	Mixture of noises	95.35	93.76	90.88
✓	✓	✓	✗	✗	Mixture of noises	96.02	94.51	92.14
✓	✓	✓	✓	✗	Mixture of noises	97.23	95.54	92.98
✓	✓	✓	✓	✓	Mixture of noises	**97.62**	**96.22**	**93.60**

4.3 Ablation Study

Exploration on Hyperparameters. The hyperparameters in our proposed ESL contain the initial sub-center number for each class and the λ in each proposed operation. In Table 4, we investigate the impact of each hyperparameter.

Effectiveness of Proposed Operations. To demonstrate the effectiveness of our proposed ESL, we decouple each operation to ablate each of them on the mixture noisy dataset in Table 5.

We take turns adding each component to the original ArcFace [4] baseline. Due to the gradient conflict from massive fine-grained intra-class noise and inter-class noise, ArcFace [4] only achieves limited performance. Sub-center loss introduces sub-centers for each identity to deal with the intra-class conflict. Meanwhile, the ignore strategy can ease part of conflict from inter-class noise. Sub-center loss brings 3.73% performance improvement. The merging operation aims to merge images that share the same identity but belong to different sub-centers. The merging operation boosts the performance by 0.75%. The producing operation can automatically align the sub-centers and the actual identity number in each class, which improves intra-class compactness effectively. It further brings 1.03% performance improvement. The dropping operation tends to drop the outlier faces without the specific positive sub-center. These faces are hard to optimize and would harm the optimization process. It can obtain a significant performance gain by 0.68% under the fine-grained noisy dataset.

Table 6. Ablation experiments to compare ESL with posterior cleaning methods. The GPU hour is measured on NVIDIA A100 GPU. $M = n \downarrow 1$ indicates the posterior data clean strategy proposed in Sub-center Arcface [2].

Method	Dataset	Posterior clean	GPU hour	IJB-C		
				$1e-3$	$1e-4$	$1e-5$
Sub-center ArcFace $M = 3 \downarrow 1$	MS1MV0	✓	128	97.40	95.92	94.03
SKH + ArcFace $M = 3 \downarrow 1$	MS1MV0	✓	128	96.55	96.26	94.18
ESL + ArcFace	MS1MV0	✗	80	**97.58**	96.23	**94.24**
ArcFace	MS1MV3	✗	64	97.64	96.44	94.66
Sub-center ArcFace $M = 3 \downarrow 1$	Mixture of noises	✓	128	97.13	95.89	92.67
SKH + ArcFace $M = 4 \downarrow 1$	Mixture of noises	✓	128	97.46	96.14	92.87
ESL + ArcFace	Mixture of noises	✗	80	**97.62**	**96.22**	**93.60**

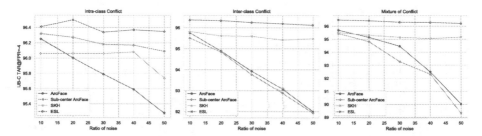

Fig. 4. Experiments of ArcFace, Sub-center ArcFace, SKH and ESL on different noise ratio. We tackle the TAR@FAR $= -4$ on IJB-C dataset as evaluation metric.

Efficiency of ESL. Deng et al. [2] and Liu et al. [13] adopt a posterior data clean strategy to filter out noise samples in an offline manner. Deng et al. [2] searches for the intra-class margin to drop the outlier samples for each domain center. Liu et al. [13] further introduces inter-class margin to merge samples belonging to different centers. For each margin setting, they should train for 20 epochs to verify its effectiveness, which is extremely time and computation resources consuming. In Table 6, we compare ESL with these posterior cleaning strategies. ESL can also achieve better performance on both MS1MV0 [8] and synthetic mixture noisy dataset. Meanwhile, there is only a slight gap between ESL and ArcFace [4] training on cleaned MS1MV3.

Robustness under Various Noise Ratio. To further investigate the effectiveness of our proposed ESL, we conduct sufficient experiments under various noise ratios. As shown in Fig. 4, we visualize the relationship between noise ratio and evaluation results. ESL can remain robustness under different noise ratio and surpass Sub-center ArcFace [2] and SKH [13] by a large margin.

In Table 7, we also compare our proposed ESL with other methods on the cleaned MS1MV3 dataset. Samples in a clean class would converge to different sub-centers so that the performance of Sub-center ArcFace [2] slightly drops. SKH [13] leads to a significant performance drop when directly training on the cleaned dataset. The restraint of SKH [13] forces each hyperplane to contain a subset of all IDs in the cleaned dataset, which does great harm to the inter-class

Table 7. Experiments on cleaned MS1MV3 dataset. For IJB-B and IJB-C dataset, we adopt the TPR@FPR $= -4$ as evaluation metric.

Method	Dataset	IJB-B	IJB-C	LFW	CFP-FP	AgeDB-30
ArcFace	MS1MV3	95.04	96.44	**99.83**	98.57	98.12
Sub-center ArcFace M $= 3$	MS1MV3	94.84	96.35	99.75	98.50	98.14
Sub-center ArcFace M $= 3 \downarrow 1$	MS1MV3	94.87	96.43	99.78	98.52	98.19
SKH + ArcFace M $= 3$	MS1MV3	93.50	95.25	99.78	98.59	98.23
SKH + ArcFace M $= 3 \downarrow 1$	MS1MV3	94.98	96.48	99.77	98.70	98.25
ESL + ArcFace	MS1MV3	**95.12**	**96.50**	99.80	**98.72**	**98.43**

Table 8. Experiments on CosFace loss function.

Method	Dataset	IJB-B			IJB-C		
		$1e-3$	$1e-4$	$1e-5$	$1e-3$	$1e-4$	$1e-5$
CosFace	Mixture of noises	93.44	86.87	74.20	95.15	90.56	83.01
Sub-center CosFace M $= 3$	Mixture of noises	91.85	84.40	69.88	94.25	89.19	80.25
SKH + CosFace M $= 4$	Mixture of noises	95.07	93.15	87.13	96.28	94.46	91.87
ESL + CosFace	Mixture of noises	**96.52**	**94.64**	**88.93**	**97.50**	**96.10**	**93.51**

representation learning. Compare with Sub-center ArcFace [2] and SKH [13], our proposed ESL can further boost the performance on the cleaned dataset, which further verifies the generalization of ESL.

Generalization on Other Loss Function. We also verify the generalization ability of proposed ESL on CosFace [26], which is another popular loss function for deep face recognition. In Table 8, we can observe that ESL can significantly outperform Sub-center [2] and SKH [13] by a large margin.

5   Conclusions

In this paper, We reformulate the noise type of faces in each class into a more fine-grained manner as N-identities|K^C-clusters. The key to robust representation learning strategies under real-world noise is the flexibility of the algorithm to the variation of N, K, and C. Furthermore, we introduce a general flexible method, named Evolving Sub-centers Learning (ESL), to improve the robustness of feature representation on noisy training data. The proposed ESL enjoys scalability to different combinations of N, K, and C, which is more robust to unconstrained real-world noise. Extensive experiments on noisy data training demonstrate the effectiveness of ESL and it provides a new state-of-the-art for noise-robust representation learning on large-scale noisy faces.

Acknowledgments. The work was supported by the National Key R&D Program of China under Grant 2021ZD0201300.

References

1. An, X., et al.: Partial FC: training 10 million identities on a single machine. In: Proceedings of the IEEE/CVF International Conference on Computer Vision, pp. 1445–1449 (2021)
2. Deng, J., Guo, J., Liu, T., Gong, M., Zafeiriou, S.: Sub-center ArcFace: boosting face recognition by large-scale noisy web faces. In: Vedaldi, A., Bischof, H., Brox, T., Frahm, J.-M. (eds.) ECCV 2020. LNCS, vol. 12356, pp. 741–757. Springer, Cham (2020). https://doi.org/10.1007/978-3-030-58621-8_43
3. Deng, J., Guo, J., Ververas, E., Kotsia, I., Zafeiriou, S.: RetinaFace: single-shot multi-level face localisation in the wild. In: Proceedings of the IEEE/CVF Conference on Computer Vision and Pattern Recognition, pp. 5203–5212 (2020)
4. Deng, J., Guo, J., Xue, N., Zafeiriou, S.: ArcFace: additive angular margin loss for deep face recognition. In: Proceedings of the IEEE/CVF Conference on Computer Vision and Pattern Recognition, pp. 4690–4699 (2019)
5. Deng, J., Guo, J., Zhang, D., Deng, Y., Lu, X., Shi, S.: Lightweight face recognition challenge. In: Proceedings of the IEEE/CVF International Conference on Computer Vision Workshops (2019)
6. Deng, J., Zhou, Y., Zafeiriou, S.: Marginal loss for deep face recognition. In: Proceedings of the IEEE Conference on Computer Vision and Pattern Recognition Workshops, pp. 60–68 (2017)
7. Du, H., Shi, H., Liu, Y., Wang, J., Lei, Z., Zeng, D., Mei, T.: Semi-Siamese training for shallow face learning. In: Vedaldi, A., Bischof, H., Brox, T., Frahm, J.-M. (eds.) ECCV 2020. LNCS, vol. 12349, pp. 36–53. Springer, Cham (2020). https://doi.org/10.1007/978-3-030-58548-8_3
8. Guo, Y., Zhang, L., Hu, Y., He, X., Gao, J.: MS-Celeb-1M: a dataset and benchmark for large-scale face recognition. In: Leibe, B., Matas, J., Sebe, N., Welling, M. (eds.) ECCV 2016. LNCS, vol. 9907, pp. 87–102. Springer, Cham (2016). https://doi.org/10.1007/978-3-319-46487-9_6
9. He, K., Zhang, X., Ren, S., Sun, J.: Deep residual learning for image recognition. In: Proceedings of the IEEE Conference on Computer Vision and Pattern Recognition, pp. 770–778 (2016)
10. Hu, W., Huang, Y., Zhang, F., Li, R.: Noise-tolerant paradigm for training face recognition CNNs. In: Proceedings of the IEEE/CVF Conference on Computer Vision and Pattern Recognition, pp. 11887–11896 (2019)
11. Huang, G.B., Mattar, M., Berg, T., Learned-Miller, E.: Labeled faces in the wild: a database for studying face recognition in unconstrained environments. In: Workshop on Faces in 'Real-Life' Images: Detection, Alignment, and Recognition (2008)
12. Huang, Y., et al.: CurricularFace: adaptive curriculum learning loss for deep face recognition. In: Proceedings of the IEEE/CVF Conference on Computer Vision and Pattern Recognition, pp. 5901–5910 (2020)
13. Liu, B., Song, G., Zhang, M., You, H., Liu, Y.: Switchable k-class hyperplanes for noise-robust representation learning. In: Proceedings of the IEEE/CVF International Conference on Computer Vision, pp. 3019–3028 (2021)
14. Liu, W., Wen, Y., Yu, Z., Li, M., Raj, B., Song, L.: SphereFace: deep hypersphere embedding for face recognition. In: Proceedings of the IEEE Conference on Computer Vision and Pattern Recognition, pp. 212–220 (2017)
15. Liu, Yu., Song, G., Shao, J., Jin, X., Wang, X.: Transductive centroid projection for semi-supervised large-scale recognition. In: Ferrari, V., Hebert, M., Sminchisescu, C., Weiss, Y. (eds.) ECCV 2018. LNCS, vol. 11209, pp. 72–89. Springer, Cham (2018). https://doi.org/10.1007/978-3-030-01228-1_5

16. Liu, Y., et al.: Towards flops-constrained face recognition. In: Proceedings of the IEEE/CVF International Conference on Computer Vision Workshops (2019)
17. Maze, B., et al.: IARPA Janus benchmark-C: face dataset and protocol. In: 2018 International Conference on Biometrics (ICB), pp. 158–165. IEEE (2018)
18. Moschoglou, S., Papaioannou, A., Sagonas, C., Deng, J., Kotsia, I., Zafeiriou, S.: AgeDB: the first manually collected, in-the-wild age database. In: Proceedings of the IEEE Conference on Computer Vision and Pattern Recognition Workshops, pp. 51–59 (2017)
19. Paszke, A., et al.: PyTorch: an imperative style, high-performance deep learning library. In: Advances in Neural Information Processing Systems 32 (2019)
20. Schroff, F., Kalenichenko, D., Philbin, J.: FaceNet: a unified embedding for face recognition and clustering. In: Proceedings of the IEEE Conference on Computer Vision and Pattern Recognition, pp. 815–823 (2015)
21. Sengupta, S., Chen, J.C., Castillo, C., Patel, V.M., Chellappa, R., Jacobs, D.W.: Frontal to profile face verification in the wild. In: 2016 IEEE Winter Conference on Applications of Computer Vision (WACV), pp. 1–9. IEEE (2016)
22. Song, G., Leng, B., Liu, Y., Hetang, C., Cai, S.: Region-based quality estimation network for large-scale person re-identification. In: Proceedings of the AAAI Conference on Artificial Intelligence, vol. 32 (2018)
23. Song, G., Liu, Y., Jiang, M., Wang, Y., Yan, J., Leng, B.: Beyond trade-off: accelerate FCN-based face detector with higher accuracy. In: Proceedings of the IEEE Conference on Computer Vision and Pattern Recognition, pp. 7756–7764 (2018)
24. Song, G., Liu, Y., Wang, X.: Revisiting the sibling head in object detector. In: Proceedings of the IEEE/CVF Conference on Computer Vision and Pattern Recognition, pp. 11563–11572 (2020)
25. Sun, Y., Wang, X., Tang, X.: Deep learning face representation from predicting 10,000 classes. In: Proceedings of the IEEE Conference on Computer Vision and Pattern Recognition, pp. 1891–1898 (2014)
26. Wang, H., et al.: CosFace: large margin cosine loss for deep face recognition. In: Proceedings of the IEEE Conference on Computer Vision and Pattern Recognition, pp. 5265–5274 (2018)
27. Wang, X., Wang, S., Wang, J., Shi, H., Mei, T.: Co-mining: deep face recognition with noisy labels. In: Proceedings of the IEEE/CVF International Conference on Computer Vision, pp. 9358–9367 (2019)
28. Wang, X., Zhang, S., Wang, S., Fu, T., Shi, H., Mei, T.: Mis-classified vector guided Softmax loss for face recognition. In: Proceedings of the AAAI Conference on Artificial Intelligence, vol. 34, pp. 12241–12248 (2020)
29. Whitelam, C., et al.: IARPA Janus benchmark-B face dataset. In: Proceedings of the IEEE Conference on Computer Vision and Pattern Recognition Workshops, pp. 90–98 (2017)
30. Zhang, M., Song, G., Zhou, H., Liu, Yu.: Discriminability distillation in group representation learning. In: Vedaldi, A., Bischof, H., Brox, T., Frahm, J.-M. (eds.) ECCV 2020. LNCS, vol. 12355, pp. 1–19. Springer, Cham (2020). https://doi.org/10.1007/978-3-030-58607-2_1
31. Zhang, X., Zhao, R., Qiao, Y., Wang, X., Li, H.: AdaCos: adaptively scaling cosine logits for effectively learning deep face representations. In: Proceedings of the IEEE/CVF Conference on Computer Vision and Pattern Recognition, pp. 10823–10832 (2019)
32. Zhong, Y., et al.: Unequal-training for deep face recognition with long-tailed noisy data. In: Proceedings of the IEEE/CVF Conference on Computer Vision and Pattern Recognition, pp. 7812–7821 (2019)

33. Zhu, M., Martínez, A.M.: Optimal subclass discovery for discriminant analysis. In: 2004 Conference on Computer Vision and Pattern Recognition Workshop. IEEE (2004)
34. Zhu, M., Martinez, A.M.: Subclass discriminant analysis. IEEE Trans. Pattern Anal. Mach. Intell. **28**(8), 1274–1286 (2006)
35. Zhu, Z., et al.: WebFace260M: a benchmark unveiling the power of million-scale deep face recognition. In: Proceedings of the IEEE/CVF Conference on Computer Vision and Pattern Recognition, pp. 10492–10502 (2021)

Teaching Where to Look: Attention Similarity Knowledge Distillation for Low Resolution Face Recognition

Sungho Shin⬤, Joosoon Lee⬤, Junseok Lee⬤, Yeonguk Yu⬤,
and Kyoobin Lee$^{(\boxtimes)}$⬤

School of Integrated Technology (SIT), Gwangju Institute of Science and Technology
(GIST), Cheomdan-gwagiro 123, Buk-gu, Gwangju 61005, Republic of Korea
{hogili89,joosoon1111,junseoklee,yeon_guk,kyoobinlee}@gist.ac.kr

Abstract. Deep learning has achieved outstanding performance for face recognition benchmarks, but performance reduces significantly for low resolution (LR) images. We propose an attention similarity knowledge distillation approach, which transfers attention maps obtained from a high resolution (HR) network as a teacher into an LR network as a student to boost LR recognition performance. Inspired by humans being able to approximate an object's region from an LR image based on prior knowledge obtained from HR images, we designed the knowledge distillation loss using the cosine similarity to make the student network's attention resemble the teacher network's attention. Experiments on various LR face related benchmarks confirmed the proposed method generally improved recognition performances on LR settings, outperforming state-of-the-art results by simply transferring well-constructed attention maps. The code and pretrained models are publicly available in the https://github.com/gist-ailab/teaching-where-to-look.

Keywords: Attention similarity knowledge distillation · Cosine similarity · Low resolution face recognition

1 Introduction

Recent face recognition model recognizes the identity of a given face image from the 1M distractors with an accuracy of 99.087% [15]. However, most face recognition benchmarks such as MegaFace [15], CASIA [33], and MS-Celeb-1M [10] contain high resolution (HR) images that differ significantly from real-world environments, typically captured by surveillance cameras. When deep learning approaches are directly applied to low resolution (LR) images after being trained on HR images, significant performance degradation occurred [1,21,30] (Fig. 1).

Supplementary Information The online version contains supplementary material available at https://doi.org/10.1007/978-3-031-19775-8_37.

S. Avidan et al. (Eds.): ECCV 2022, LNCS 13672, pp. 631–647, 2022.
https://doi.org/10.1007/978-3-031-19775-8_37

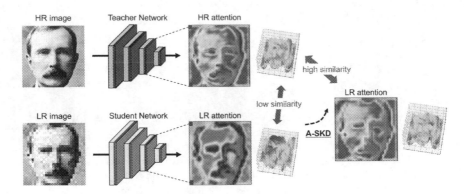

Fig. 1. Proposed attention similarity knowledge distillation (A-SKD) concept for low resolution (LR) face recognition problem. Well-constructed attention maps from the HR network are transferred to the LR network by forming high similarity between them for guiding the LR network to focus on detailed parts captured by the HR network. Face images and attention maps are from the AgeDB-30 [23].

To overcome the LR problem associated with face recognition, prior knowledge extracted from HR face images is used to compensate spatial information loss. Depending on the approach of transferring the prior knowledge to LR image domain, LR face recognition methods are categorized into two types: super-resolution and knowledge distillation based approaches. Super-resolution based approaches utilize generative models to improve LR images to HR before input to recognition networks [7,9,11,16,28,30]. Following the development of super-resolution methods, LR images can be successfully reconstructed into HR images and recognized by a network trained on HR images [5,6,19,26]. However, super-resolution models incur high computational costs for both training and inference, even larger than the costs required for recognition networks. Furthermore, generating HR from LR images is an ill-posed problem, i.e., many HR images can match with a single LR image [4]; hence the identity of a LR image can be altered.

To combat this, knowledge distillation based methods have been proposed to transfer prior knowledge from HR images to models trained on LR face images [8, 22,37]. When the resolution of face images is degraded, face recognition models cannot capture accurate features for identification due to spatial information loss. In particular, features from detailed facial parts are difficult to be captured from a few pixels on LR images, e.g. eyes, nose, and mouth [18]. Previous studies mainly focused on feature based knowledge distillation (F-KD) methods to encourage the LR network's features to mimic the HR network's features by reducing the Euclidean distance between them [8,22,37]. The original concept of F-KD was proposed as a lightweight student model to mimic features from over-parameterized teacher models [34]. Because teacher model's features would generally include more information than the student model, F-KD approaches improve the accuracy of the student model. Similarly, informative features from the HR network are distilled to the LR network in the LR face recognition problems.

This study proposes the attention similarity knowledge distillation approach to distill well-constructed attention maps from an HR network into an LR network by increasing similarity between them. The approach was motivated by the observation that humans can approximate an object's regions from LR images based on prior knowledge learned from previously viewed HR images. Kumar et al. proposed that guiding the LR face recognition network to generate facial keypoints (e.g., eyes, ears, nose, and lips) improved recognition performance by directing the network's attention to the informative regions [18]. Thus, we designed the prior knowledge as an attention map and transferred the knowledge by increasing similarity between the HR and LR networks' attention maps.

Experiments on LR face recognition, face detection, and general object classification demonstrated that the attention mechanism was the best prior knowledge obtainable from the HR networks and similarity was the best method for transferring knowledge to the LR networks. Ablation studies and attention analyses demonstrated the proposed A-SKD effectiveness.

2   Related Works

Knowledge Distillation. Hinton et al. first proposed the knowledge distillation approach to transfer knowledge from a teacher network into a smaller student network [12]. Soft logits from a teacher network were distilled into a student network by reducing the Kullback-Leibler (KL) divergence score, which quantifies the difference between the teacher and student logits distributions. Various F-KD methods were subsequently proposed to distill intermediate representations [25,27,34,35]. FitNet reduced the Euclidean distance between teacher and student network's features to boost student network training [27]. Zagoruyko et al. proposed attention transfer (AT) to reduce the distance between teacher and student network's attention maps rather than distilling entire features [35]. Since attention maps are calculated by applying channel-wise pooling to feature vectors, activation levels for each feature can be distilled efficiently. Relational knowledge distillation (RKD) recently confirmed significant performance gain by distilling structural relationships for features across teacher and student networks [25].

Feature Guided LR Face Recognition. Various approaches that distill well-constructed features from the HR face recognition network to the LR network have been proposed to improve LR face recognition performances [8,22,37]. Conventional knowledge distillation methods assume that over-parameterized teacher networks extract richer information and it can be transferred to smaller student networks. Similarly, LR face recognition studies focused on transferring knowledge from networks trained on highly informative inputs to networks trained on less informative inputs. Zhu et al. introduced knowledge distillation approach for LR object classification [37], confirming that simple logit distillation from the HR to LR network significantly improved LR classification performance, even superior to super-resolution based methods. F-KD [22] and hybrid order relational knowledge distillation (HORKD) [8], which is the variant of RKD [25], methods were subsequently applied to LR face recognition problems to transfer intermediate representations from the HR network.

Another approach is to guide the LR network by training it to generate keypoints (e.g. eyes, ears, nose, and lips) [18]. An auxiliary layer is added to generate keypoints, and hence guide the network to focus on specific facial characteristics. It is well known that facial parts such as eyes and ears are important for recognition [17,18], hence LR face recognition networks guided by keypoints achieve better performance. Inspired by this, we designed the attention distillation method that guides the LR network to focus on important regions of the HR network. However, attention distillation methods have not been previously explored for LR face recognition. We investigated the efficient attention distillation methods for LR settings and proposed the cosine similarity as the distance measure between HR and LR network's attention maps.

3   Method

3.1   Low Resolution Image Generation

We require HR and LR face image pairs to distill the HR network's knowledge to the LR network. Following the protocol for LR image generation in super-resolution studies [5,6,19,26], we applied bicubic interpolation to down-sample HR images with 2×, 4×, and 8× ratios. Gaussian blur was then added to generate realistic LR images. Finally, the downsized images were resized to the original image size using bicubic interpolation. Figure 2 presents sample LR images.

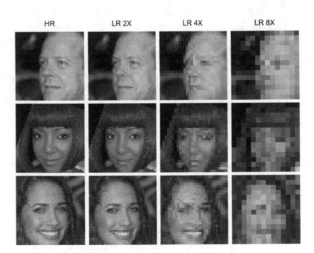

Fig. 2. The samples of HR and LR images from the training dataset (CASIA [33]) with the down-sampling ratios of 2×, 4×, and 8×.

3.2   Face Recognition with Attention Modules

Face Recognition Network. ArcFace [3] is a SOTA face recognition network comprising convolutional neural network (CNN) backbone and angular margin introduced to softmax loss. Conventional softmax loss can be expressed as

$$L_{softmax} = -\frac{1}{N} \sum_{i=1}^{N} log \frac{e^{W_{y_i}^T x_i + b_{y_i}}}{\sum_{j=1}^{n} e^{W_j^T x_i + b_j}}, \tag{1}$$

where $x_i \in \mathbb{R}^d$ is the embedded feature of the i-th sample belonging to the y_i-th class; N and n are the batch size and the number of classes, respectively; $W_j \in \mathbb{R}^d$ denotes the j-th column of the last fully connected layer's weight $W \in \mathbb{R}^{d \times n}$ and $b_j \in \mathbb{R}^n$ is the bias term for the j-th class.

For simplicity, the bias term is fixed to 0 as in [20]. Then the logit of the j-th class can be represented as $W_j^T x_i = \|W_j\|\|x_i\|cos(\theta_j)$, where θ_j denotes the angle between the W_j and x_i. Following previous approaches [20,29], ArcFace set $\|W_j\| = 1$ and $\|x_i\| = 1$ via l_2 normalisation to maximize θ_j among inter-class and minimize θ_j among intra-class samples. Further, constant linear angular margin (m) was introduced to avoid convergence difficulty. The ArcFace [3] loss can be expressed as

$$L_{arcface} = -\frac{1}{N} \sum_{i=1}^{N} log \frac{e^{s(cos(\theta_{y_i}+m))}}{e^{s(cos(\theta_{y_i}+m))}+\sum_{j=1,j \neq y_i}^{n} e^{s(cos(\theta_j))}}, \tag{2}$$

where s is the re-scale factor and m is the additive angular margin penalty between x_i and W_{y_i}.

Attention. Attention is a simple and effective method to guide feature focus on important regions for recognition. Let $\mathbf{f}_i = \mathcal{H}_i(\mathbf{x})$ be intermediate feature outputs from the i-th layer of the CNN. Attention maps about $\mathbf{f}_i$ can be represented as the $\mathcal{A}_i(\mathbf{f}_i)$, where $\mathcal{A}_i(\cdot)$ is attention module.

Many attention mechanisms have been proposed; AT [35] simply applied channel-wise pooling to features to estimate spatial attention maps. SENet [13] and CBAM [31] utilized parametric transformations, e.g. convolution layers, to represent attention maps. Estimated attention maps were multiplied with the features and passed to a successive layer. Trainable parameters in attention module are updated to improve performance during back-propagation, forming accurate attention maps. Attention mechanisms can be expressed as

$$\mathbf{f_i'} = \mathcal{A}_i^c(\mathbf{f_i}) \otimes \mathbf{f_i} \tag{3}$$

and

$$\mathbf{f_i''} = \mathcal{A}_i^s(\mathbf{f_i'}) \otimes \mathbf{f_i'}, \tag{4}$$

where $\mathcal{A}_i^c(\cdot)$ and $\mathcal{A}_i^s(\cdot)$ are attention modules for channel and spatial attention maps, respectively (Fig. 3).

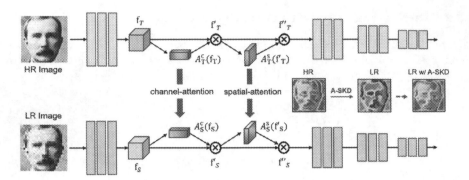

Fig. 3. Proposed A-SKD framework. The LR network formulates precise attention maps by referencing well-constructed channel and spatial attention maps obtained from the HR network, focusing on detailed facial parts which are helpful for the face recognition. We only show the attention distillation for the first block.

Features are refined twice by multiplying channel and spatial attention maps in order (3) and (4). Any parametric attention transformation could be employed for the proposed A-SKD, and we adopted the popular CBAM [31] module,

$$\mathcal{A}^c(\mathbf{f}) = \sigma(FC(AvgPool(\mathbf{f})) + FC(MaxPool(\mathbf{f}))) \tag{5}$$

and

$$\mathcal{A}^s(\mathbf{f}) = \sigma(f^{7\times7}(AvgPool(\mathbf{f}); MaxPool(\mathbf{f}))), \tag{6}$$

where $\sigma(\cdot)$ is the sigmoid function; and $FC(\cdot)$ and $f^{7\times7}(\cdot)$ are fully connected and convolution layers with 7×7 filters, respectively.

3.3   Proposed Attention Similarity Knowledge Distillation Framework

Unlike the conventional knowledge distillation, the network size of teacher and student network is same for A-SKD. Instead, the teacher network is trained on HR images whereas the student network is trained on LR images. Due to the resolution differences, features from both networks are difficult to be identical. Therefore, we propose to distill well-constructed attention maps from the HR network into the LR network instead of features.

$$
\begin{aligned}
\rho_i &= 1 - \langle \mathcal{A}_{T,i}(\mathbf{f}_{T,i}), \mathcal{A}_{S,i}(\mathbf{f}_{S,i}) \rangle \\
&= 1 - \frac{\mathcal{A}_{T,i}(\mathbf{f}_{T,i})}{\|\mathcal{A}_{T,i}(\mathbf{f}_{T,i})\|_2} \cdot \frac{\mathcal{A}_{S,i}(\mathbf{f}_{S,i})}{\|\mathcal{A}_{S,i}(\mathbf{f}_{S,i})\|_2},
\end{aligned}
\tag{7}
$$

where ρ_i is the cosine distance between attention maps from the i-th layer of the teacher and student networks; $\langle \cdot, \cdot \rangle$ denotes the cosine similarity; $\|\cdot\|_2$ denotes L2-norm; $\mathcal{A}_i(\mathbf{f}_i)$ denotes the attention maps for the i-th layer features; and T

and S denote the teacher and student network, respectively. Thus, $\mathcal{A}_{T,i}(\mathbf{f}_{T,i})$ and $\mathcal{A}_{S,i}(\mathbf{f}_{S,i})$ are attention maps estimated from the i-th layer of the teacher and student network's features, respectively. Reducing the cosine distance between HR and LR attention maps increases the similarity between them.

Distillation loss for A-SKD is calculated as

$$\mathcal{L}_{distill} = \sum_{i=1}^{N} \frac{(\rho_i^s + \rho_i^c)}{2} \tag{8}$$

which average the cosine distance for channel and spatial attention maps between the HR and LR networks, and sums them across layers ($i = 1, 2, 3, ..., N$) of the backbone. N is the number of layers utilized for the distillation.

Total loss for the LR face recognition network is the sum of target task's loss and distillation loss (8) weighted by the factor ($\lambda_{distill}$). In this work, we utilized the ArcFace loss (2) as a target task's loss.

$$\mathcal{L}_{total} = \mathcal{L}_{arcface} + \lambda_{distill} * \mathcal{L}_{distill}. \tag{9}$$

Further, our method can be utilized in conjunction with the logit distillation by simply adding the logit distillation loss [12] to our loss function (9). Since logit is the final output of the network, incorporating the logit distillation loss allows the LR network to make the same decision as the HR network based on the refined attention maps.

4   Experiments

4.1   Settings

Datasets. We employed the CASIA [33] dataset for training, which is a large face recognition benchmark comprising approximately 0.5M face images for 10K identities. Each sample in CASIA was down-sampled to construct the HR-LR paired face dataset. For the evaluation, the manually down-sampled face recognition benchmark (AgeDB-30 [23]) and the popular LR face recognition benchmark (TinyFace [1]) were employed. Since AgeDB-30 have similar resolution to CASIA, networks trained on down-sampled CASIA images were validated on AgeDB-30 down-sampled images with matching ratio. In contrast, the real-world LR benchmark (TinyFace) comprises face images with the resolution of 24×24 in average when they are aligned. Therefore, they were validated using a network trained on CASIA images down-sampled to 24×24 pixels.

Task and Metrics. Face recognition was performed for two scenarios: face verification and identification. Face verification is where the network determines whether paired images are for the same person, i.e., 1:1 comparison. To evaluate verification performance, accuracy was determined using validation sets constructed from probe and gallery set pairs following the LFW protocol [14]. Face identification is where the network recognize the identity of a probe image by measuring similarity against all gallery images, i.e., 1:N comparison. This study

employed the smaller AgeDB-30 dataset for the face verification; and larger
TinyFace dataset for the face identification.

Comparison with Other Methods. Typically, the distillation of intermediate representation is performed concurrently with the target task's loss. Previous
distillation methods in the experiments utilized the both face recognition and
distillation loss, albeit face recognition loss of varying forms. In addition, some
feature distillation approach reported their performances with the logit distillation loss. In order to conduct a fair comparison, we re-implemented the prior distillation methods with the same face recognition loss (ArcFace [3]) and without
the logit distillation loss. Further, our method requires the parametric attention
modules for the distillation. Therefore, we utilized the same backbone network
with CBAM attention modules for all methods; we combined the CBAM modules to all convolution layers, with the exception of the stem convolution layer
and the convolution layer with a kernel size of 1.

Implementation Details. We followed the ArcFace protocol for data preprocessing: detecting face regions using the MTCNN [36] face detector, cropping
around the face region, and resizing the resultant portion to 112×112 pixel
using bilinear interpolation. The backbone network was ResNet-50 with CBAM
attention module. For the main experiments, we distilled the attention maps for
every convolution layers with the exception of the stem convolution layer and
the convolution layer with a kernel size of 1. Weight factors for distillation (9)
$\lambda_{distill} = 5$. This weight factors generally achieved the superior results not only
for the face recognition benchmarks, but also for the ImageNet [2]. Learning
rate = 0.1 initially, divided by 10 at 6, 11, 15, and 17 epochs. SGD optimizer
was utilized for the training with batch size = 128. Training completed after 20
epochs. The baseline refers to the LR network that has not been subjected to any
knowledge distillation methods. For the hyperparameter search, we divided 20%
of the training set into the validation set and conducted a random search. After
the hyperparameter search, we trained the network using the whole training set
and and performed the evaluation on the AgeDB-30 and TinyFace.

4.2   Face Recognition Benchmark Results

Evaluation on AgeDB-30. Table 1 shows LR face recognition performance
on AgeDB-30 with various down-sample ratios depending on distillation methods. Except for HORKD, previous distillation methods [22,35] exhibited only
slight improvement or even reduced performance when the downsampling ratios
increase. This indicates that reducing the L2 distance between the HR and LR
network's features is ineffective. In contrast, HORKD improved LR recognition
performance by distilling the relational knowledge of the HR network's features.
When the input's resolution decrease, the intermediate features are hard to be
identical with the features from the HR network. Instead the relation among the
features of the HR network can be transferred to the LR network despite the
spatial information loss; this was the reason of HORKD's superior performances
even for the 4× and 8× settings.

Table 1. Proposed A-SKD approach compared with baseline and previous SOTA methods on AgeDB-30 with 2×, 4×, and 8× down-sampled ratios. L, F, SA, and CA indicate distillation types of logit, feature, spatial attention, and channel attention, respectively. Ver-ACC denotes the verification accuracy. Base refers to the LR network that has not been subjected to any knowledge distillation methods.

Resolution	Method	Distill type	Loss function	Ver-ACC (%) (AgeDB-30)
1×	Base	–	–	93.78
2×	Base	–	–	92.83
	F-KD [22]	F	L2	93.05
	AT [35]	SA	L2	92.93
	HORKD [8]	F	L1+Huber	93.13
	A-SKD (**Ours**)	SA+CA	Cosine	**93.35**
	A-SKD+KD (**Ours**)	SA+CA+L	Cosine+KLdiv	**93.58**
4×	Base	–	–	87.74
	F-KD [22]	F	L2	87.72
	AT [35]	SA	L2	87.75
	HORKD [8]	F	L1+Huber	88.08
	A-SKD (**Ours**)	SA+CA	Cosine	**88.58**
	A-SKD+KD (**Ours**)	SA+CA+L	Cosine+KLdiv	**89.15**
8×	Base	–	–	77.75
	F-KD [22]	F	L2	77.85
	AT [35]	SA	L2	77.40
	HORKD [8]	F	L1+Huber	78.27
	A-SKD (**Ours**)	SA+CA	Cosine	**79.00**
	A-SKD+KD (**Ours**)	SA+CA+L	Cosine+KLdiv	**79.45**

However, attention maps from the HORKD exhibit similar pattern to LR baseline network rather than the HR network in the Fig. 4. HR attention maps are highly activated in facial landmarks, such as eyes, lips, and beard, which are helpful features for face recognition [18]. In contrast, detailed facial parts are less activated for LR attention maps because those parts are represented with a few pixels. Although HORKD boosts LR recognition performance by transferring HR relational knowledge, it still failed to capture detailed facial features crucial for recognition. The proposed A-SKD method directs the LR network's attention toward detailed facial parts that are well represented by the HR network's attention maps.

Based on the refined attention maps, A-SKD outperforms the HORKD and other knowledge distillation methods for all cases. AgeDB-30 verification accuracy increased 0.6%, 1.0%, and 1.6% compared with baseline for 2×, 4×, and 8× down-resolution ratios, respectively. In addition, when A-SKD is combined with logit distillation (KD), the verification accuracy increased significantly for all settings. From the results, we confirmed that the attention knowledge from the HR network can be transferred to the LR network and led to significant improvements that were superior to the previous SOTA method.

Evaluation on TinyFace. Unlike the face verification, the identification task requires to select a target person's image from the gallery set consists of a large number of face images. Therefore, the identification performances decrease significantly when the resolution of face images are degraded. Table 2 showed the identification performances on the TinyFace benchmark. When the AT [35] was applied, the rank-1 identification accuracy decreased 13.34% compared to the baseline. However, our approach improved the rank-1 accuracy 13.56% compared to the baseline, even outperforming the HORKD method. This demonstrated that the parametric attention modules (CBAM) and cosine similarity loss are the key factors for transferring the HR network's knowledge into the LR network via attention maps. The proposed method is generalized well to real-world LR face identification task which is not manually down-sampled.

Table 2. Evaluation results on TinyFace identification benchmark depending on the distillation methods. Acc@K denotes the rank-K accuracy (%).

	ACC@1	ACC@5	ACC@10	ACC@50
Base	42.19	50.62	53.67	60.41
AT [35]	36.56	45.68	49.03	56.44
HORKD [8]	45.49	54.80	58.26	64.30
A-SKD (Ours)	**47.91**	**56.55**	**59.92**	**66.60**

5 Discussion

Attention Correlation Analysis. Figure 5 shows Pearsons correlation between attention maps from the HR and LR networks for the different distillation methods. Spatial and channel attention maps from the four blocks for models other than A-SKD have a low correlation between the HR and LR networks, with a magnitude lower than 0.5. In particular, spatial attention maps obtained from the first block of the LR baseline and HORKD network have negative correlation with the HR network ($r = -0.39$ and -0.29, respectively).

Figure 4 shows that spatial attention maps from LR baseline and HORKD networks are highly activated in skin regions, which are less influenced by resolution degradation, in contrast to the HR network. This guides the LR network to the opposite directions from the HR network. However, spatial attention maps from A-SKD exhibit strong positive correlation with those from the HR network, highlighting detailed facial attributes such as beard, hair, and eyes. Through the A-SKD, the LR network learned where to focus by generating precise attention maps similar to those for the HR network. Consequently, Pearsons correlation, i.e., the similarity measure between HR and LR attention maps, was significantly improved for all blocks, with a magnitude higher than 0.6. Thus the proposed

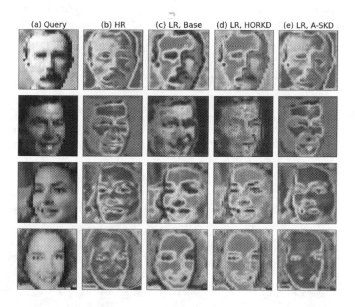

Fig. 4. Normalized spatial attention maps from the first block for different distillation methods. Red and blue regions indicate high and low attention, respectively. Face images and attention maps are from the AgeDB-30. (Color figure online)

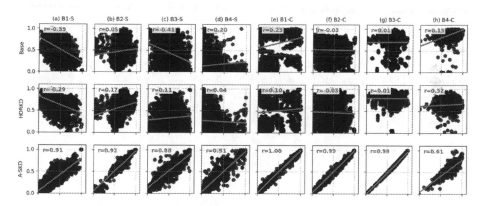

Fig. 5. Pixel level Pearsons correlation between the HR and LR network's attention maps for different distillation methods. B$\{i\}$-$\{$S, C$\}$ indicates Pearsons correlation for spatial or channel attention maps obtained from the i-th ResNet block between the HR and LR networks; and r is Pearsons correlation coefficient representing linear relationships between input variables. Base refers to the LR network that has not been subjected to any knowledge distillation methods. Pearsons correlation is measured using the AgeDB-30.

A-SKD approach achieved superior efficacy and success compared with previous feature based SOTA methods.

Comparison with Attention Transfer [35]. Primary distinctions between AT [35] and A-SKD include the cosine similarity loss, parametric attention modules, and distillation of both channel and spatial attention maps. Correlation analysis for A-SKD confirmed that the cosine similarity loss is an effective strategy for transferring attention knowledge. Distilling AT attention maps using the cosine similarity rather than the L2 loss increased AgeDB-30 verification accuracy by 0.32%p (Table 3). AT calculates attention maps using channel-wise pooling, a non-parametric layer; whereas A-SKD calculates attention maps using parametric layers comprising fully connected and convolution layers. When the input image resolution degrades, the student network's feature representation diverges from that of the teacher network. Therefore, it is difficult to match the attention maps of the student network obtained by the non-parametric module with those of the teacher network. Instead, A-SKD employs the parametric module for the attention maps extraction and the cosine similarity loss for the distillation; therefore, the attention maps from the student network can be adaptively trained to be similar to the attention maps from the teacher network despite the differences in the features. Finally, A-SKD distills both spatial and channel attention maps in contrast to AT which only considered spatial attention maps. We confirmed A-SKD with spatial and channel attention additionally improved AgeDB-30 verification accuracy by 0.34%p compared with spatial-only attention. This comparison results also confirmed that A-SKD, designed for attention distillation on LR settings, is the most effective approach for transferring attention knowledge.

Table 3. Comparing attention transfer (AT) [35] and proposed A-SKD on AgeDB-30 benchmark down-sampled with 8× ratio. AT* indicates the cosine similarity loss was utilized for attention transfer rather than the original L2 loss. SA and CA indicate spatial and channel attention maps, respectively.

Method	Type	Transformation	Loss function	Ver-ACC (%) (AgeDB-30)
AT [35]	SA	Non-parametric layer	L2	77.40
AT*	SA	Non-parametric layer	Cosine	77.72
A-SKD	SA	Parametric layer	Cosine	78.66
A-SKD	SA+CA	Parametric layer	Cosine	**79.00**

6   Extension to Other Tasks

6.1   Object Classification

We conducted experiments for object classification on LR images using the 4× down-sampled ImageNet [2]. For the backbone network, we utilized the ResNet18 with CBAM attention modules. We compared our method to other knowledge distillation methods (AT [35] and RKD [25]) which are widely utilized in the classification domains. We re-implemented those methods using its original hyperparameters. Usually, AT and RKD were utilized along with the logit distillation for the ImageNet; therefore, we performed the AT, RKD, and A-SKD in conjunction with the logit distillation in the Table 4. Training details are provided in the Supplementary Information.

Table 4 shows that A-SKD outperformed the other methods on the LR ImageNet classification task. Park et al. demonstrated that introducing the accurate attention maps led the significant improvement on classification performances [24,31]. When the attention maps were distilled from the teacher network, student network could focus on informative regions by forming precise attention maps similar with the teacher's one. Thus, our method can be generalized to general object classification task, not restricted to face related tasks.

Table 4. Proposed A-SKD performance on low resolution ImageNet classification. All distillation methods were performed in conjunction with the logit distillation.

Resolution	Method	ACC (%)
1×	Base	70.13
4×	Base	65.34
	AT [35]	65.79
	RKD [25]	65.95
	A-SKD	**66.52**

6.2 Face Detection

Face detection is a sub-task of object detection to recognize human faces in an image and estimate their location(s). We utilized TinaFace [38], a deep learning face detection model, integrated with the CBAM attention module to extend the proposed A-SKD approach to face detection. Experiments were conducted on the WIDER FACE [32] dataset (32,203 images containing 393,703 faces captured from real-world environments) with images categorized on face detection difficulty: easy, medium, and hard. LR images were generated with 16× and 32× down-resolution ratios, and further training and distillation details are provided in the Supplementary Information.

Table 5 shows that A-SKD improved the overall detection performance by distilling well-constructed attention maps, providing significant increases of mean average precision (mAP) for the easy (15.72% for 16× and 7.54% for 32×), medium (14.15% for 16× and 12.59% for 32×), and hard (33.98% for 16× and 14.43% for 32×) level detection tasks. Small faces were well detected in the LR images after distillation as illustrated in Fig. 6. Thus the proposed A-SKD approach can be successfully employed for many LR machine vision tasks.

Table 5. Proposed A-SKD performance on LR face detection. mAP is mean average precision; easy, medium, and hard are pre-assessed detection difficulty.

Resolution	Model	mAP (%)		
		Easy	Medium	Hard
1×	Base	95.56	95.07	91.45
16×	Base	54.38	52.73	35.29
	A-SKD	**62.93**	**60.19**	**47.28**
32×	Base	31.15	26.68	14.00
	A-SKD	**33.50**	**30.04**	**16.02**

Fig. 6. Qualitative results for LR face detection before and after applying A-SKD. Small faces were better detected after A-SKD. The face images are from the evaluation set of WIDERFACE.

7   Conclusion

We verified that attention maps constructed from HR images were simple and effective knowledge that can be transferred to LR recognition networks to compensate for spatial information loss. The proposed A-SKD framework enabled any student network to focus on target regions under LR circumstances and generalized well for various LR machine vision tasks by simply transferring well-constructed HR attention maps. Thus, A-SKD could replace conventional KD methods offering improved simplicity and efficiency and could be widely applicable to LR vision tasks, which have not been strongly studied previously, without being limited to face related tasks.

Acknowledgments. This work was supported by the ICT R&D program of MSIT/IITP[2020-0-00857, Development of Cloud Robot Intelligence Augmentation, Sharing and Framework Technology to Integrate and Enhance the Intelligence of Multiple Robots. And also, this work was partially supported by Korea Institute of Energy Technology Evaluation and Planning (KETEP) grant funded by the Korea government

(MOTIE) (No. 20202910100030) and supported by Electronics and Telecommunications Research Institute (ETRI) grant funded by the Korean government. [22ZR1100, A Study of Hyper-Connected Thinking Internet Technology by autonomous connecting, controlling and evolving ways].

References

1. Cheng, Z., Zhu, X., Gong, S.: Low-resolution face recognition. In: ACCV (2018)
2. Deng, J., Dong, W., Socher, R., Li, L.J., Li, K., Fei-Fei, L.: ImageNet: a large-scale hierarchical image database. In: 2009 IEEE Conference on Computer Vision and Pattern Recognition, pp. 248–255 (2009). https://doi.org/10.1109/CVPR.2009.5206848
3. Deng, J., Guo, J., Xue, N., Zafeiriou, S.: ArcFace: additive angular margin loss for deep face recognition. In: Proceedings of the IEEE Computer Society Conference on Computer Vision and Pattern Recognition, June 2019, pp. 4685–4694, January 2018. http://arxiv.org/abs/1801.07698
4. Dong, C., Loy, C.C., He, K., Tang, X.: Image super-resolution using deep convolutional networks. IEEE Trans. Pattern Anal. Mach. Intell. **38**, 295–307 (2016)
5. Dong, C., Loy, C.C., Tang, X.: Accelerating the super-resolution convolutional neural network. In: Leibe, B., Matas, J., Sebe, N., Welling, M. (eds.) ECCV 2016. LNCS, vol. 9906, pp. 391–407. Springer, Cham (2016). https://doi.org/10.1007/978-3-319-46475-6_25
6. Flusser, J., Farokhi, S., Höschl, C., Suk, T., Zitová, B., Pedone, M.: Recognition of images degraded by Gaussian blur. IEEE Trans. Image Process. **25**(2), 790–806 (2016). https://doi.org/10.1109/TIP.2015.2512108
7. Fookes, C., Lin, F., Chandran, V., Sridharan, S.: Evaluation of image resolution and super-resolution on face recognition performance. J. Vis. Commun. Image Represent. **23**(1), 75–93 (2012). https://doi.org/10.1016/j.jvcir.2011.06.004
8. Ge, S., et al.: Look one and more: distilling hybrid order relational knowledge for cross-resolution image recognition. In: Proceedings of the AAAI Conference on Artificial Intelligence, vol. 34, no. 07, pp. 10845–10852 (2020)
9. Gunturk, B.K., Batur, A.U., Altunbasak, Y., Hayes, M.H., Mersereau, R.M.: Eigenface-domain super-resolution for face recognition. IEEE Trans. Image Process. **12**(5), 597–606 (2003). https://doi.org/10.1109/TIP.2003.811513
10. Guo, Y., Zhang, L., Hu, Y., He, X., Gao, J.: MS-Celeb-1M: a dataset and benchmark for large-scale face recognition. In: Leibe, B., Matas, J., Sebe, N., Welling, M. (eds.) ECCV 2016. LNCS, vol. 9907, pp. 87–102. Springer, Cham (2016). https://doi.org/10.1007/978-3-319-46487-9_6
11. Hennings-Yeomans, P.H., Baker, S., Kumar, B.V.: Simultaneous super-resolution and feature extraction for recognition of low-resolution faces. In: 26th IEEE Conference on Computer Vision and Pattern Recognition, CVPR (2008). https://doi.org/10.1109/CVPR.2008.4587810
12. Hinton, G., Vinyals, O., Dean, J.: Distilling the knowledge in a neural network. In: NIPS Deep Learning and Representation Learning Workshop (2015). http://arxiv.org/abs/1503.02531
13. Hu, J., Shen, L., Albanie, S., Sun, G., Wu, E.: Squeeze-and-excitation networks. IEEE Trans. Pattern Anal. Mach. Intell. **42**, 2011–2023 (2020)
14. Huang, G.B., Mattar, M.A., Berg, T.L., Learned-Miller, E.: Labeled faces in the wild: A database for studying face recognition in unconstrained environments (2008)

15. Kemelmacher-Shlizerman, I., Seitz, S.M., Miller, D., Brossard, E.: The MegaFace benchmark: 1 million faces for recognition at scale. In: 2016 IEEE Conference on Computer Vision and Pattern Recognition (CVPR), pp. 4873–4882 (2016). https://doi.org/10.1109/CVPR.2016.527

16. Kong, H., Zhao, J., Tu, X., Xing, J., Shen, S., Feng, J.: Cross-resolution face recognition via prior-aided face hallucination and residual knowledge distillation. arXiv, May 2019. http://arxiv.org/abs/1905.10777

17. Köstinger, M., Wohlhart, P., Roth, P.M., Bischof, H.: Annotated facial landmarks in the wild: A large-scale, real-world database for facial landmark localization. In: 2011 IEEE International Conference on Computer Vision Workshops (ICCV Workshops), pp. 2144–2151 (2011)

18. Kumar, A., Chellappa, R.: S2ld: Semi-supervised landmark detection in low resolution images and impact on face verification. In: 2020 IEEE/CVF Conference on Computer Vision and Pattern Recognition Workshops (CVPRW), pp. 3275–3283 (2020)

19. Ledig, C., et al.: Photo-realistic single image super-resolution using a generative adversarial network. In: 2017 IEEE Conference on Computer Vision and Pattern Recognition (CVPR), pp. 105–114 (2017)

20. Liu, W., Wen, Y., Yu, Z., Li, M., Raj, B., Song, L.: SphereFace: deep hypersphere embedding for face recognition. In: 2017 IEEE Conference on Computer Vision and Pattern Recognition (CVPR), pp. 6738–6746 (2017)

21. Lui, Y.M., Bolme, D., Draper, B.A., Beveridge, J.R., Givens, G., Phillips, P.J.: A meta-analysis of face recognition covariates. In: Proceedings of the 3rd IEEE International Conference on Biometrics: Theory, Applications and Systems, BTAS 2009, pp. 139–146. IEEE Press (2009)

22. Massoli, F.V., Amato, G., Falchi, F.: Cross-resolution learning for face recognition. Image Vis. Comput. **99**, 103927 (2020). https://doi.org/10.1016/j.imavis.2020.103927

23. Moschoglou, S., Papaioannou, A., Sagonas, C., Deng, J., Kotsia, I., Zafeiriou, S.: AgeDB: the first manually collected, in-the-wild age database, pp. 1997–2005 (2017). https://doi.org/10.1109/CVPRW.2017.250

24. Park, J., Woo, S., Lee, J.Y., Kweon, I.S.: BAM: bottleneck attention module. In: BMVC (2018)

25. Park, W., Kim, D., Lu, Y., Cho, M.: Relational knowledge distillation. In: Proceedings of the IEEE Computer Society Conference on Computer Vision and Pattern Recognition, June 2019, pp. 3962–3971, January 2020. http://arxiv.org/abs/1904.05068

26. Pei, Y., Huang, Y., Zou, Q., Zhang, X., Wang, S.: Effects of image degradation and degradation removal to CNN-based image classification. IEEE Trans. Pattern Anal. Mach. Intell. **43**(4), 1239–1253 (2021). https://doi.org/10.1109/TPAMI.2019.2950923

27. Romero, A., Ballas, N., Kahou, S.E., Chassang, A., Gatta, C., Bengio, Y.: FitNets: hints for thin deep nets. CoRR abs/1412.6550 (2015)

28. Tran, L., Yin, X., Liu, X.: Disentangled representation learning GAN for pose-invariant face recognition. In: Proceedings of 30th IEEE Conference on Computer Vision and Pattern Recognition, CVPR 2017, January 2017, pp. 1283–1292. Institute of Electrical and Electronics Engineers Inc., November 2017. https://doi.org/10.1109/CVPR.2017.141

29. Wang, H., Wang, Y., Zhou, Z., Ji, X., Li, Z., Gong, D., Zhou, J., Liu, W.: CosFace: large margin cosine loss for deep face recognition. In: 2018 IEEE/CVF Conference on Computer Vision and Pattern Recognition, pp. 5265–5274 (2018)

30. Wilman, W.W.Z., Yuen, P.C.: Very low resolution face recognition problem. In: 2010 Fourth IEEE International Conference on Biometrics: Theory, Applications and Systems (BTAS), pp. 1–6 (2010). https://doi.org/10.1109/BTAS.2010.5634490

31. Woo, S., Park, J., Lee, J.-Y., Kweon, I.S.: CBAM: convolutional block attention module. In: Ferrari, V., Hebert, M., Sminchisescu, C., Weiss, Y. (eds.) ECCV 2018. LNCS, vol. 11211, pp. 3–19. Springer, Cham (2018). https://doi.org/10.1007/978-3-030-01234-2_1. http://arxiv.org/abs/1807.06521

32. Yang, S., Luo, P., Loy, C.C., Tang, X.: WIDER FACE: a face detection benchmark. In: 2016 IEEE Conference on Computer Vision and Pattern Recognition (CVPR), pp. 5525–5533 (2016). https://doi.org/10.1109/CVPR.2016.596

33. Yi, D., Lei, Z., Liao, S., Li, S.Z.: Learning face representation from Scratch, November 2014. http://arxiv.org/abs/1411.7923

34. Yim, J., Joo, D., Bae, J.H., Kim, J.: A gift from knowledge distillation: Fast optimization, network minimization and transfer learning. In: 2017 IEEE Conference on Computer Vision and Pattern Recognition (CVPR), pp. 7130–7138 (2017)

35. Zagoruyko, S., Komodakis, N.: Paying more attention to attention: improving the performance of convolutional neural networks via attention transfer. In: 5th International Conference on Learning Representations, ICLR 2017 - Conference Track Proceedings, December 2016

36. Zhang, K., Zhang, Z., Li, Z., Qiao, Y.: Joint face detection and alignment using multi-task cascaded convolutional networks. IEEE Signal Process. Lett. **23**(10), 1499–1503 (2016). https://doi.org/10.1109/LSP.2016.2603342. http://arxiv.org/abs/1604.02878

37. Zhu, M., Han, K., Zhang, C., Lin, J., Wang, Y.: Low-resolution visual recognition via deep feature distillation. In: ICASSP 2019–2019 IEEE International Conference on Acoustics, Speech and Signal Processing (ICASSP), pp. 3762–3766 (2019). https://doi.org/10.1109/ICASSP.2019.8682926

38. Zhu, Y., Cai, H., Zhang, S., Wang, C., Xiong, Y.: TinaFace: strong but simple baseline for face detection. arXiv abs/2011.13183 (2020)

Teaching with Soft Label Smoothing for Mitigating Noisy Labels in Facial Expressions

Tohar Lukov$^{(\boxtimes)}$, Na Zhao, Gim Hee Lee, and Ser-Nam Lim

Department of Computer Science, National University of Singapore,
Singapore, Singapore
tohar@u.nus.edu, {zhaona,gimhee.lee}@nus.edu.sg

Abstract. Recent studies have highlighted the problem of noisy labels in large scale in-the-wild facial expressions datasets due to the uncertainties caused by ambiguous facial expressions, low-quality facial images, and the subjectiveness of annotators. To solve the problem of noisy labels, we propose Soft Label Smoothing (SLS), which smooths out multiple high-confidence classes in the logits by assigning them a probability based on the corresponding confidence, and at the same time assigning a fixed low probability to the low-confidence classes. Specifically, we introduce what we call the Smooth Operator Framework for Teaching (SOFT), based on a mean-teacher (MT) architecture where SLS is applied over the teacher's logits. We find that the smoothed teacher's logit provides a beneficial supervision to the student via a consistency loss – at 30% noise rate, SLS leads to 15% reduction in the error rate compared with MT. Overall, SOFT beats the state of the art at mitigating noisy labels by a significant margin for both symmetric and asymmetric noise. Our code is available at https://github.com/toharl/soft.

Keywords: Noisy labels · Facial expression recognition

1 Introduction

The problem of noisy labels in facial expressions datasets can be attributed to a few factors. On one hand, facial expressions can be fairly ambiguous, which leads to subjectiveness in the annotations. On the other hand, the prevalence of low-quality facial images, especially those collected from the wild, can also degrade the quality of the labels significantly. Mitigating the effect of noisy labels has thus become an important area of research in facial expression recognition (FER), where the goal is to prevent deep learning models from overfitting to the noisy labels.

Earlier studies [60,68] have shown that representing ground truth labels with label distributions (*i.e.*, multiple classes with different intensity) instead of

_____

Supplementary Information The online version contains supplementary material available at https://doi.org/10.1007/978-3-031-19775-8_38.

one-hot label can help to mitigate the presence of noisy labels. Intuitively, facial expressions are often compound in nature, *e.g.*, an expression can appear both angry and sad at the same time, and a label distribution helps to capture the intricacy much better. To this end, researchers have looked into label distribution learning (LDL) [68], label enhancement (LE) [5,44,60], and label smoothing regularization (LSR) [36,51]. Here, while the goals of LDL, LE and LSR are similar, LSR is in an analytic form, and thus is much more efficient in comparison.

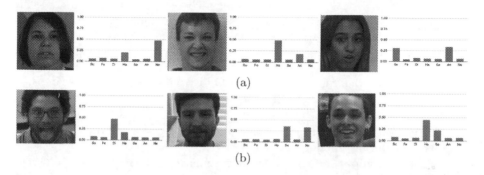

Fig. 1. Label distributions generated by applying SLS to the prediction logits of the teacher. The given one-hot ground truth label is denoted in red. Su, Fe, Di, Ha, Sa, An, Ne denote Surprise, Fear, Disgust, Happiness, Sadness, Anger, Neutral, respectively (Due to license restrictions, the images shown were not the actual images from RAF-DB from which the histograms were generated, but from the DFDC dataset [10] that have similar expressions). (a) Although the true label is ambiguous, SLS produced a label distribution that is better at describing the compound facial expression, even though the highest confidence class does not correspond to the ground truth. (b) While the highest class is predicted correctly as the ground truth class, the label distribution is better at describing the facial expression with more than one expression compared to that with only a single expression. (Color figure online)

Inspired by these work, we propose in this paper that logits smoothing can similarly help to handle noisy labels effectively. Indeed, one of the main findings in this work is that by smoothing the logits, we are able to achieve significant performance boost in FER in the face of noisy labels. We propose Soft Label Smoothing (SLS) for logits, which smooths out multiple high-confidence classes in the logits by assigning them a probability based on the corresponding confidence, and at the same time assigning a fixed low probability to the low-confidence classes. While LSR can also be utilized for smoothing logits, it differs from SLS as it only assigns a fixed high probability to the highest confidence class. We say SLS is instance-aware as the distribution it produces varies per sample while LSR is only class-aware. Our study shows that SLS has a clear advantage over LSR.

We also further consider that the logits produced by a model for SLS to smooth should also have some tolerance to noisy labels. We are motivated by the mean-teacher (MT) architecture introduced in [52], which [39] shown can be utilized to detect noisy labels by considering the discrepancies between the student and teacher's logits. Drawing from the success of MT in [39], we conjecture that a MT network by itself already has a stabilizing effect against noisy labels with the teacher "keeping the student in check" by "retaining" historical information as it is updated by the exponential moving average of the student's network. In this work, we employ a consistency loss that penalizes discrepancies between the student and teacher's logits, which are smoothed with SLS. With such a framework, which we refer to as the Smooth Operator Framework for Teaching (SOFT), we observe a significant boost in performance for FER, even with the challenging asymmetric (class-dependent) noise [4,40] where labels are switched to corresponding labels with the highest confusion instead of just random. In summary, the main contributions of our paper are as follow:

- We propose a novel framework, named SOFT, to mitigate label noise in FER. SOFT consists of a MT with SLS applied on the teacher's logits.
- Our simple approach does not require additional datasets, or label distributions annotations (such as in LDL), and does not cause additional computational cost during training.
- Our model produces state of the art performance for the FER task at different levels of noisy labels.

2   Related Work

2.1   Facial Expression Recognition

FER algorithms can be divided into two categories: handcrafted and learning-based techniques. Examples of the traditional handcrafted features based methods are SIFT [8], HOG [9], Histograms of local binary patterns [43] and Gabor wavelet coefficients [34]. Learning-based strategies [53,61] have become the majority with the development of deep learning and demonstrate high performance. Several employed two stream network to fuse face images with landmarks [67] and optical flows [50]. Some, such as [7,30,61], leverage the differences between expressive and neutral facial expressions. Recently, Ruan et al. [42] use a convolutional neural network to extract basic features which are then decomposed into a set of facial action-aware latent features that efficiently represent expression similarities across different expressions. However, all of these methods are not designed to deal with noisy labels and ambiguous facial expressions.

2.2   Learning with Noisy Labels

Deep learning with noisy labels has been extensively studied for classification tasks [47]. One line of work proposes a robust architecture by adding a noise adaptation layer [6,17,49] to the network in order to learn the label transition matrix or developing a dedicated architecture [18,22,59]. Another line of work studies regularization methods to improve the generalizability. Explicit regularization [28,58], such as dropout [48] and weight decay [31], modifies the training loss while implicit regularization [20,63] includes augmentation [45] and label smoothing [36,51] which prevents the model from assigning a full probability to samples with corrupted labels. Reducing overfitting during training increases the robustness to noisy labels. Other methods propose noise-tolerant loss functions such as absolute mean error [16], generalized cross entropy [65] or modifying the loss value by loss correction [25,40], loss reweighting [35,56], or label refurbishment [41,46]. Another key concept is sample selection [37,46] to select the clean samples from the noisy dataset and update the network only for them, which has been shown to work well when combined with other approaches [3,46].

The problem of noisy labels in facial expressions datasets is mainly caused by ambiguous facial expressions, the subjectiveness of annotators, and low-quality facial images. Label distribution learning [12–15,26,33] and label enhancement have been proposed for mitigating ambiguity in related tasks such as head pose and facial age estimation. Zhou et al. [68] is the first to address this by learning the mapping from the expression images to the emotion distributions with their emotion distribution learning (EDL) method. However, this method assumes the availability of label distributions as ground truth which is expensive to obtain. To address the unavailable issue of label distributions, Xu et al. [60] propose label enhancement (LE) mechanism utilising one-hot label. However, the proposed approach has a high time complexity due to K-NN search, limiting the size of the dataset that can be used for training.

Zeng et al. [62] address inconsistencies in annotations of FER datasets by obtaining multiple labels for each image with human annotations and predicted pseudo labels, followed by learning a model (IPA2LT) to fit the latent truth from the inconsistent pseudo labels. Wang et al. [54] suppress uncertain samples by learning an uncertainty score and utilising a relabeling mechanism in an attempt to correct the noisy labels. However, this work does not take into account of compound expression. Moreover, these methods treat the inconsistency as noise while ignoring noisy labels caused by ambiguous facial expressions. Chen et al. [5] construct nearest neighbor graphs for label distribution learning, which requires additional datasets for related auxiliary tasks. To deal with ambiguity, She et al. [44] introduce the Distribution Mining and the pairwise Uncertainty Estimation (DMUE) approach. DMUE works by constructing multiple auxiliary branches as the number of classes in order to discover the label distributions of samples.

3 Our Approach

3.1 Background and Notation

Given a FER labelled dataset (X, Y), each sample x is annotated by a one-hot label over C classes, $y \in \{0, 1\}^C$. However, for a dataset with noisy labels, the labels can be either wrong or ambiguous. This poses a challenge for training deep neural networks as they suffer from the memorization effect, fitting to the noisy labels with their large capability, and causing performance degradation.

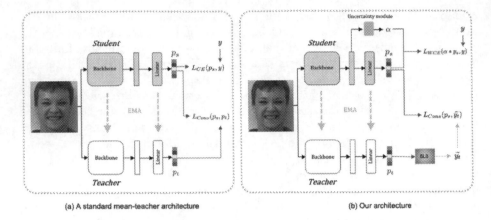

(a) A standard mean-teacher architecture   (b) Our architecture

Fig. 2. Comparison between (a) standard mean-teacher architecture and (b) our architecture to learn from noisy labels in FER. Note that the given label, y, might be noisy. Our architecture is an extension of the mean-teacher architecture. We enhance the teacher's logits by applying our SLS on them. The student consists of an uncertainty module to predict the uncertainty score. The classification loss of the student is the cross entropy (CE) loss weighted by this predicted uncertainty score, which we will refer to as the weighted CE (WCE) loss following nomenclature in the literature.

Label Smoothing [51]. Label Smoothing Regularization [51] (LSR) has been widely used for regularization which improves generalization and calibration. When applying LSR, the one-hot label is modified such that the label is mixed with a uniform mixture over all possible labels. More formally, given one-hot label $y \in \{0, 1\}^C$, LSR produces $\widetilde{y} \in \mathbb{R}^{1 \times C}$ and is formulated as:

$$\widetilde{y} = (\widetilde{y_1}, \widetilde{y_2}, ..., \widetilde{y_C}), \tag{1}$$

where

$$\widetilde{y_i} = \begin{cases} 1 - \epsilon, & i = l \\ \frac{\epsilon}{C-1}, & \text{otherwise.} \end{cases} \tag{2}$$

Here, ϵ is a hyper-parameter that is used to smooth the distribution and l is the index of the ground truth class.

Mean-Teacher. The mean-teacher architecture [52] is originally introduced for semi-supervised learning. We adapt it to deal with noisy labels in FER. Two networks (*i.e.* student and teacher) with the same architecture are forced to output consistent predictions, despite random noise introduced to the inputs or networks. The random noise can be implemented for example by applying dropout layers or random augmentations over the input for each branch. In our training, we use the latter. As can be seen in Fig. 2(a), the inputs to the student and teacher networks are the same. The output logits from the student network are supervised by the given labels via a classification loss (*i.e.* cross entropy) and the teacher's logits via a consistency loss, respectively.

3.2 Overview of Our Method

Our architecture, shown in Fig. 2(b), is based on a mean-teacher where the given ground truth label (y) might be noisy with an unknown noise rate. We enhance the teacher's logits by applying our soft label smoothing (SLS). The student consists of an uncertainty module that predicts an uncertainty score to compute a weighted cross entropy (WCE) loss. During testing, the uncertainty module is removed. The best performing branch, either the student or teacher, over the validation set can be used for inference.

3.3 Soft Label Smoothing (SLS)

Our SLS preserves the high-confidences of multiple (top-k) predictions and unifies the remaining low-confidence predictions. The number of the high-confidence classes, k, depends on each instance as a result. More formally, for a sample x, we denote the logits, $p(x) \in \mathbb{R}^{1 \times C}$. We further obtain $q = softmax(p)$ which is a distribution vector, $||q||_1 = 1$. We then define k as the number of elements above a threshold τ, which is empirically tuned:

$$k = \sum_{i=1}^{n} [q_i > \tau].$$

(3)

Here, [...] are the Iverson brackets. Our SLS is then formulated as:

$$\widetilde{y}_i = \begin{cases} \frac{q_i}{\sum_{j=1}^{C} q_j [q_j > \tau]} (1 - \epsilon), & q_i > \tau, \\ \frac{\epsilon}{C-k}, & \text{otherwise.} \end{cases}$$

(4)

Note that for samples with $k = 1$, SLS behaves like LSR. We show in our experiments later that $k > 1$ for a significant portion of the samples at the optimal τ, thus allowing SLS to play its intended role. In our framework, we apply SLS over the teacher's logits. The weights of the teacher network [52], θ', are updated only by the exponential moving average (EMA) of the weights from the student network θ:

$$\theta' \leftarrow \omega\theta' + (1 - \omega)\theta,$$

(5)

where $\omega \in [0, 1]$ denotes the decay rate.

3.4   Loss Function

Consistency Loss. The student is supervised by the consistency loss which is formulated as the kullback-leibler (KL) divergence between the students' logits, p_s, and the teachers' soft labels after applying our SLS, $\widetilde{y}_t = SLS(p_t)$:

$$L_{Cons} = D_{KL}(p_s \parallel \widetilde{y}_t). \tag{6}$$

Logit Weighted Cross-Entropy Loss. For a sample x_i, we denote the feature vector produced by a backbone network as $f(x_i)$. This is given to the uncertainty module [27,54] which we attach to the student network and predict an uncertainty score α_i. The uncertainty module consists of a linear layer followed by a sigmoid function σ. It is formulated as:

$$\alpha_i = \sigma(W_u^\top(f(x_i))), \tag{7}$$

where W_u denotes the parameters of the linear layer. The logit weighted cross-entropy loss is formulated as:

$$L_{WCE} = -\frac{1}{N} \sum_{i=1}^{N} log \frac{e^{\alpha_i W_{y_i}^\top f(x_i)}}{\sum_{j=1}^{C} e^{\alpha_i W_j^\top f(x_i)}}, \tag{8}$$

where $f(x_i)$ is the feature vector and α_i is the uncertainty score of the i-th sample labelled as the y_i-th class (*i.e.* y_i denotes the index of the annotated class). $W_j^\top f(x_i)$ is the logit of the j-th class of sample i.

Total Loss. The total loss function for training the student network is given as:

$$L_{total} = L_{WCE}(\alpha * p_s, y) + \lambda L_{Cons}(p_s, \widetilde{y}_t), \tag{9}$$

where the score α is given by the uncertainty module. λ is the weight to control the contribution of the consistency loss.

4   Experiments

In this section, we first describe the three datasets we used in our experiments. We then present the ablation studies we conducted to demonstrate the efficacy of each component in our approach. Next, we present results that demonstrate the robustness of our approach against noisy labels by injecting varying amount of "synthetic" noise into the training dataset. Finally, we provide comparative results with several state-of-the-art approaches on the three datasets.

4.1   Datasets

RAF-DB [32] includes 30,000 face images that have been tagged with basic or compound expressions by 40 experienced human annotators. Only images with seven expressions (neutral, happiness, surprise, sadness, anger, disgust, fear) are utilised in our experiment, resulting in 12,271 images for training and 3,068 images for testing. For measurement, the overall sample accuracy is adopted.

AffectNet [38] is a very large dataset with both category and Valence-Arousal classifications. It comprises almost one million images retrieved from the Internet by querying expression-related keywords in three search engines, 450,000 of which are manually annotated with eight expression labels ('contempt' is annotated in addition to the expressions in RAF-DB). We train and test either on these eight emotions or only seven emotions (without 'contempt'), which we denote by AffectNet-8 and AffectNet-7 respectively. For measurement, the mean class accuracy on the validation set is employed following [44,54,55].

FERPlus [2] is an extension to FER2013 [19], including 28,709 training images and 3,589 testing images resized to 48×48 grayscale pixels. Ten crowd-sourced annotators assign each image to one of eight categories as in Affectnet. The most popular vote category is chosen as the label for each image as in [44,54,55].

Since the noisy labels in these datasets are unknown, we follow [44,54] to inject synthetic label noise to them, in order to assess the denoising ability of our method. Specifically, we flip each original label y to label y' by a label transition matrix T, where $T_{ij} = Pr[y' = j | y = i]$. We use two types of noise, symmetric (*i.e.* uniform) [44,54] and asymmetric (*i.e.* class-dependent) [4,40]. For symmetric noise, a label is randomly switched to another label to simulate noisy labels. On the other hand, asymmetric noise switches a label to the label that it is most often confused with (which can be identified from a confusion matrix). Figure 5(b) illustrates asymmetric and symmetric noise transition matrices for 30% noise rate, respectively.

Table 1. Mean Accuracy of the different components of our model. U denotes the use of an uncertainty module and SLS denotes our smoothing function over the teacher logits in a MT. Acc denotes using a pretrained model on the facial dataset, MS-Celeb-1M. Acc_i stands for training from a pretrained model on ImageNet.

MT	U	SLS	Acc_i	Acc
×	×	×	71.80	75.12
√	×	×	79.43	83.89
×	√	×	73.82	82.75
×	×	√	80.93	85.33
√	√	×	80.31	85.13
√	×	√	82.56	86.82
√	√	√	**83.18**	**86.94**

4.2   Implementation Details

We first detect and align input face images using MTCNN [64]. We then resize them to 224×224 pixels, followed by augmenting them with random cropping and horizontal flipping. As for our backbone network, we adopt the commonly used ResNet-18 [24] in FER. For a fair comparison, the backbone network is pre-trained on MS-Celeb-1M with the standard routine [44,54,55]. The student network is trained by an Adam optimizer. We first pre-train the student with SLS for 6 epochs, and then add the uncertainty module for another 80 epochs with a learning rate of 0.0001 and a batch size of 64. The parameters of the teacher network is updated using EMA (see Eq. 5), and the weight decay is set to 0.999 following the original MT paper [52]. The loss weight λ (see Eq. 9) is set to 10. We use the better performing branch that is evaluated over the validation set (*i.e.* the student without the uncertainty module or the teacher) for inference.

4.3   Ablation Studies

SLS Effects. In the following, we examine the effects of SLS. All experiments are performed on RAF-DB with 30% injected symmetric noise unless specified otherwise. Additional experiments, including for asymmetric noise and other noise ratios, are provided in the supplementary material.

Mitigating Label Noise. We confirm empirically that SLS is beneficial for mitigating label noise. In Table 1, we isolate the effect of SLS and observe that SLS alone leads to an additional error reduction of 15% when added to a MT. To assess the effect of SLS for mitigating noisy labels, we show that SLS not only improves the predictions on clean samples, but also corrects the predictions of the noisy samples. To demonstrate that, in Table 2, following [36], we report performance with and without SLS on the noisy and clean parts of the training data. The noisy part refers to the 30% of the training labels that are randomly selected and switched in a symmetric noise setting.

Table 2. Performance (Acc) of our model trained with 30% symmetric noise on different parts of the training data. The training data is separated into clean and noisy parts (see Subsect. 4.3). By applying SLS, not only is the test and train accuracy higher, the accuracy on the noisy part is higher ("Noisy Correct") as well, with the original correct labels predicted instead of the injected noisy labels. Moreover, under "Noisy Noise", which denotes the number of samples in the noisy part predicted with the injected wrong labels, SLS was able to reduce the number significantly.

SLS	Test	Train	Noisy correct	Noisy noise
×	85.13	87.81	64.25	28.33
✓	**86.94**	**89.67**	**75.78**	**13.57**

Compared to performance without SLS, under "Noisy Correct" in the table, SLS causes a substantial number of the predictions in the noisy part to be corrected to the original correct labels, while at the same time under "Noisy Noise", SLS reduces by 52%, the noisy part from being predicted with the injected wrong labels.

Design of SLS. We now present our findings on the three main design components of SLS, namely: 1) instance-awareness, 2) non-zero low confidences, and 3) applying on the teacher's logits. To assess the contribution of each one of these components, in Table 3a, we present ablation studies by performing experiments with different smoothing methods. Table 3b compares the performance between

Table 3. The effects of different design components for SLS. In both tables the first row is a MT with uncertainty module without smoothing. Table (a) shows the three main design choices in SLS: 1) Applied over the teacher's logits, 2) Instance-aware smoothing, 3) Non zero low-confidences. Table (b) shows a comparison of the branch to which SLS is applied on (i.e., the student's logits, the teacher's logits or both). See Subsect. 4.3 for more details.

(a) Different Smooth Methods					(b) Different Targets		
Smooth method	Teacher	Ins-aware	Non-zero	Acc	Student	Teacher	Acc
×	×	×	×	85.13	×	×	85.13
LSR	×	×	×	85.43	√	×	85.23
LSR*	√	×	×	85.85	×	√	**86.94**
SLS(0)	√	√	×	85.30	√	√	85.36
SLS	√	√	√	**86.94**			

where SLS is applied on, namely, the student's logits, the teacher's logits or both of them. The first row in both tables is a MT with uncertainty module without applying smoothing at all. We now discuss our findings here in details:

1. **Instance-awareness.** SLS is an instance-aware smoothing mechanism. For each sample, it utilizes the original confidence of the multiple high-confidence predictions. In Table 3a, we isolate this effect by comparing SLS with LSR*. LSR* denotes a version of LSR, where instead of a one-hot label, the input is now the logits, which is first transformed into a one-hot label by taking the top-1 prediction. LSR* is not considered as instance-aware since it ignores the original intensities of each sample, instead it produces the same smoothed label for all the samples with the same top-1 predicted class. We note that both SLS and LSR* are applied on the teacher logits. We observe that SLS performs better than LSR* by 1% which shows the advantage of being instance-aware.

2. **Non-zero low confidences.** The instance-awareness of SLS is derived from the utilization of multiple high confidence classes in its calculations. We are also curious on the effect of the way SLS handles low confidence classes. In Table 3a, we compare SLS with SLS(0). Referring to Eq. 4, the latter is SLS with $\epsilon = 0$ which zeros out the low confidence classes. SLS outperforms the accuracy of SLS(0) by 1.6%, demonstrating the contribution of the non zero mechanism.

3. **The teacher benefit.** We explore the benefit of applying SLS over the teacher's logits in a MT vs in a vanilla network in Table 1, rows 4 (SLS) and 6 (MT+SLS). We train a vanilla network consisting of a backbone and classification layer. We apply SLS over the logits and supervise the network with a consistency loss between the produced logits and the logits after applying SLS. The results demonstrates that indeed MT increases the performance by 1.5% over the vanilla network. We also explore the benefit of applying SLS

over the teacher's logits vs the student's logits in Table 3b. In Table 3b, we can see that indeed applying SLS to the teacher's logits is more beneficial to the performance (with an increase of 1.7%) when compared to applying it on the student's logits or both. This is an interesting result since one would naturally think that applying to both the teacher's and student's logits could produce better results. Our conjecture is that this is due to the student being more prone to the noisy labels since it is the classification branch that ingests them, while the teacher is a more stable branch. As a result, applying SLS on the student could in fact accentuate the noisy labels.

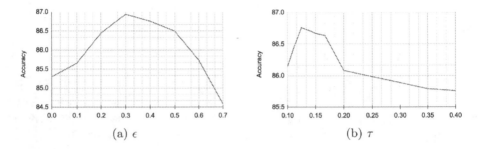

(a) ϵ (b) τ

Fig. 3. Mean Accuracy (%) on the RAF-DB dataset with 30% symmetric noise for (a) varying smoothing parameter ϵ, and (b) varying τ.

Evaluation of Varying ϵ. Figure 3(a) shows the Mean Accuracy(%) of varying smoothing parameter ϵ, on RAF-DB with 30% symmetric noise. Performance peaks at $\epsilon = 0.3$, after which performance degrades with higher values of ϵ until it is worse than the baseline at $\epsilon = 0$. For these high ϵ values, the label distribution produced by SLS is no longer meaningful.

Evaluation of Varying τ. Figure 3(b) shows the Mean Accuracy (%) of varying τ on RAF-DB with 30% symmetric noise. Referring to Eq. 3, for each sample, the value of τ and the original confidences q, influence the number of high confidences k. Here, we initialize all experiments with a MT trained for 6 epochs (without SLS). We also log the number of samples in the training data with more than one high confidence ($k > 1$) at the beginning of training with SLS. Performance peaks at $\tau = \frac{1}{8}$ for which we observe that 65% of samples have $k > 1$. Then, performance degrades with higher values of τ and lesser samples with $k > 1$ until τ reaches 0.4, by when all samples have only one high-confidence class.

Qualitative Results. After our model is trained with SLS in an MT architecture, we extract the logits of the teacher corresponding to different training images. We observe that the label distributions are consistent with how human would categorize the facial expressions. In Fig. 1(b), we present examples for label distribution where the ground truth corresponds to the highest confidence class in the logits. In Fig. 1(a), on the other hand, even though the highest confidence class is different from the ground truth, the former is actually plausible and might be even better than the ground truth label.

4.4 Comparison to the State of the Art

Our ablations, particularly Table 1 and 3a, show that SLS in MT produces the best performance. In addition, an uncertainty module added to the student branch provides an additional modest improvement. In this section onwards, we will refer to SOFT as one that includes SLS and an uncertainty module in a MT network.

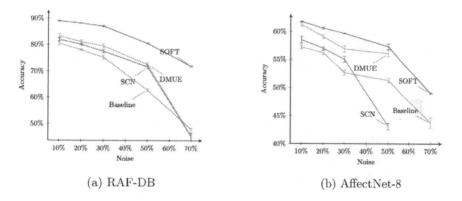

(a) RAF-DB   (b) AffectNet-8

Fig. 4. Accuracy on RAF-DB and AffectNet-8 with injected symmetric noise. SOFT consistently beats the other methods and performs well at noise rates as high as 70%. For AffectNet-8, the released code for DMUE and SCN does not work at 70% noise rate. A detailed table can be found in the Appendix.

Evaluation on Synthetic Symmetric Noise. We quantitatively evaluate the robustness of SOFT against mislabelled annotations on RAF-DB and AffectNet. We inject varying amount of symmetric noise to the training set. We compare SOFT's performance with the state-of-the-art noise-tolerant FER methods, SCN [54] and DMUE [44], as shown in Fig. 4. For a fair comparison, we follow the same experimental settings such that all methods are pre-trained on MS-Celeb-1M with ResNet-18 as the backbone. Similar to the other methods' experimental settings, the baseline network is build with the same ResNet-18 backbone and fully-connected layer for classification. After three repetitions of each experiment, the mean accuracy and standard deviation on the testing set are presented. SOFT consistently beats the vanilla baseline and the other two state-of-the-art methods, SCN and DMUE. The improvement from SOFT becomes more significant as the noise ratio increases. We outperform the best performing method, DMUE, by 5.7%, 7%, 7.5% and 26.5% on RAF-DB with 10%, 20%, 30% and 70% noise rates, respectively. Also, we outperform DMUE by 0.1%, 1.9%, 2.8% on AffectNet-8 with 10%, 20%, 30% noise rates, respectively. With the noise rate as high as 70%, we improve the accuracy by 5.2% over the vanilla baseline on AffectNet-8. Please see a detailed table in the Appendix.

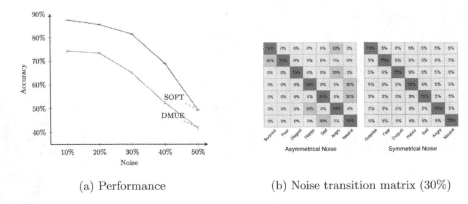

(a) Performance   (b) Noise transition matrix (30%)

Fig. 5. (a) Accuracy on RAF-DB with injected asymmetric noise. SOFT consistently and significantly beats the state-of-the-art, DMUE. (b) Illustration of asymmetric and symmetric noise transition matrix for 30% noise rate as an example. We flip each original label y to label y' by the label transition matrix T, where $T_{ij} = Pr[y' = j | y = i]$.

Evaluation on Synthetic Asymmetric Noise. In Fig. 5(a), we also show experiments on asymmetric (or class-dependent) noise [4,40], for which true labels are more likely to be flipped to a specific label. For example, 'Surprise' is most likely to be confused with 'Anger'. An example for asymmetric noise transition is given by the left matrix in Fig. 5(b). This type of noise is a better representation of real-world corruption and ambiguity, but has not been investigated by previous methods for handling noisy labels in FER. We first obtain a confusion matrix after training a vanilla network. Subsequently, we use the top-1 mis-labeled class to construct the noise transition matrix for each class. As shown in Fig. 5(a), SOFT consistently and significantly outperforms the state-of-the-art, DMUE, by 13%, 12%, 16% on RAF-DB with 10%, 20%, 30% noise rates, respectively. A detailed table reporting the comparison can be found in the Appendix. Beyond a 30% noise rate, the performance decreases significantly for both methods. Intuitively, at high asymmetric noise ratios, for example at 50% and assuming an evenly distributed training set, clean samples for a given label will likely only be present 50% of the time, with its sole wrong-label counterpart being the other 50%. This makes it very challenging for any models to learn the correct pattern. This phenomenon is not as severe in a symmetric noise setting, since the corresponding wrong labels are evenly distributed among the rest of the labels, so that the clean samples for a given class still dominate for that label, until we hit noise ratio of about 85%.

Comparison on Benchmarks. In the previous experiments, we demonstrated the noise-tolerance ability of SOFT. We now verify that this does not cause performance degradation for the original FER datasets (i.e., without any injected noise). Table 4 compares SOFT to the state-of-the-art FER methods on AffectNet (7 or 8 classes), RAF-DB and FERPlus datasets. Among all the FER

Table 4. Comparison of FER state-of-the-art accuracy (without synthetic noise). $^+$ denotes both AffectNet and RAF-DB are used as the training set. * denotes using extra label distribution instead of one-hot label. Refer to Sect. 4.4 on the preprocessing procedure that we utilized for each of these datasets.

(a) AffectNet-7

Method	Noise-tolerant	Acc
LDL-ALSG$^+$ [5]	Y	59.35
CAKE [29]	N	61.70
DDA-Loss [11]	N	62.34
EfficientFace [66]	N	63.70
DAN [57]	N	65.69
SOFT	Y	**66.13**

(b) AffectNet-8

Method	Noise-tolerant	Acc
IPA2LT$^+$ [62]	Y	55.71
RAN [55]	N	59.50
EfficientFace [66]	N	59.89
SCN [54]	Y	60.23
DAN [57]	N	62.09
DMUE [44]	Y	**62.84**
SOFT	Y	62.69

(c) RAF-DB

Method	Noise-tolerant	Acc
LDL-ALSG$^+$ [5]	Y	85.53
IPA2LT$^+$ [62]	Y	86.77
SCN [54]	Y	87.03
SCN$^+$ [54]	Y	88.14
DMUE [44]	Y	88.76
DAN [57]	N	89.70
SOFT	Y	**90.42**

(d) FERPlus

Method	Noise-tolerant	Acc
PLD* [2]	N	85.10
SeNet50* [1]	N	88.80
SCN [54]	Y	88.01
RAN [55]	N	88.55
DMUE [44]	Y	**88.64**
SOFT	Y	88.60

methods, LDL-ALSG, IPA2LT, SCN and DMUE are the only noise-tolerant FER methods, among which DMUE achieves the best results. SOFT outperforms these recent state-of-the-art methods on RAF-DB, Affectnet-7 (with 90.42% and 66.13%, respectively) and is comparable with the other methods on AffectNet-8 and FERPlus (with 62.69% and 88.6%, respectively).

5   Conclusion

We have introduced in this paper a novel solution to deal with noisy labels in FER - SOFT, which incorporates a soft label smoothing technique (SLS) into the Mean-Teacher paradigm. Through extensive ablations and benchmarks that we presented in this paper, we show that there is strong empirical support for SOFT. What is unexplored in this paper, however, is the applicability of SOFT to other use cases outside of FER, as well as the effect of SLS on other "student-teacher like" architectures, including contrastive and self-supervised learning frameworks [21,23], both of which also compare pairs of samples during training. We are hopeful that the findings about SOFT would be valuable in helping researchers pursue these directions.

References

1. Albanie, S., Nagrani, A., Vedaldi, A., Zisserman, A.: Emotion recognition in speech using cross-modal transfer in the wild. In: Proceedings of the 26th ACM International Conference on Multimedia, pp. 292–301 (2018)
2. Barsoum, E., Zhang, C., Ferrer, C.C., Zhang, Z.: Training deep networks for facial expression recognition with crowd-sourced label distribution. In: Proceedings of the 18th ACM International Conference on Multimodal Interaction, pp. 279–283 (2016)
3. Berthelot, D., Carlini, N., Goodfellow, I., Papernot, N., Oliver, A., Raffel, C.A.: Mixmatch: a holistic approach to semi-supervised learning. In: Advances in Neural Information Processing Systems, vol. 32 (2019)
4. Blanchard, G., Flaska, M., Handy, G., Pozzi, S., Scott, C.: Classification with asymmetric label noise: consistency and maximal denoising. Electron. J. Stat. $10(2)$, 2780–2824 (2016)
5. Chen, S., Wang, J., Chen, Y., Shi, Z., Geng, X., Rui, Y.: Label distribution learning on auxiliary label space graphs for facial expression recognition. In: Proceedings of the IEEE/CVF Conference on Computer Vision and Pattern Recognition, pp. 13984–13993 (2020)
6. Chen, X., Gupta, A.: Webly supervised learning of convolutional networks. In: Proceedings of the IEEE International Conference on Computer Vision, pp. 1431–1439 (2015)
7. Chen, Y., Wang, J., Chen, S., Shi, Z., Cai, J.: Facial motion prior networks for facial expression recognition. In: 2019 IEEE Visual Communications and Image Processing (VCIP), pp. 1–4. IEEE (2019)
8. Cheung, W., Hamarneh, G.: n-sift: n-dimensional scale invariant feature transform. IEEE Trans. Image Process. $18(9)$, 2012–2021 (2009)
9. Dalal, N., Triggs, B.: Histograms of oriented gradients for human detection. In: 2005 IEEE Computer Society Conference on Computer Vision and Pattern Recognition (CVPR 2005), vol. 1, pp. 886–893. IEEE (2005)
10. Dolhansky, B., et al.: The deepfake detection challenge (DFDC) dataset. arXiv preprint arXiv:2006.07397 (2020)
11. Farzaneh, A.H., Qi, X.: Discriminant distribution-agnostic loss for facial expression recognition in the wild. In: Proceedings of the IEEE/CVF Conference on Computer Vision and Pattern Recognition Workshops, pp. 406–407 (2020)
12. Gao, B.B., Xing, C., Xie, C.W., Wu, J., Geng, X.: Deep label distribution learning with label ambiguity. IEEE Trans. Image Process. $26(6)$, 2825–2838 (2017)
13. Geng, X.: Label distribution learning. IEEE Trans. Knowl. Data Eng. $28(7)$, 1734–1748 (2016)
14. Geng, X., Qian, X., Huo, Z., Zhang, Y.: Head pose estimation based on multivariate label distribution. IEEE Trans. Pattern Anal. Mach. Intell. (2020)
15. Geng, X., Yin, C., Zhou, Z.H.: Facial age estimation by learning from label distributions. IEEE Trans. Pattern Anal. Mach. Intell. $35(10)$, 2401–2412 (2013)
16. Ghosh, A., Kumar, H., Sastry, P.: Robust loss functions under label noise for deep neural networks. In: Proceedings of the AAAI Conference on Artificial Intelligence, vol. 31 (2017)
17. Goldberger, J., Ben-Reuven, E.: Training deep neural-networks using a noise adaptation layer (2016)
18. Goodfellow, I., et al.: Generative adversarial nets. In: Advances in Neural Information Processing Systems, vol. 27 (2014)

19. Goodfellow, I.J., et al.: Challenges in representation learning: a report on three machine learning contests. In: Lee, M., Hirose, A., Hou, Z.-G., Kil, R.M. (eds.) ICONIP 2013. LNCS, vol. 8228, pp. 117–124. Springer, Heidelberg (2013). https://doi.org/10.1007/978-3-642-42051-1_16

20. Goodfellow, I.J., Shlens, J., Szegedy, C.: Explaining and harnessing adversarial examples. arXiv preprint arXiv:1412.6572 (2014)

21. Grill, J.B., et al.: Bootstrap your own latent-a new approach to self-supervised learning. Adv. Neural. Inf. Process. Syst. **33**, 21271–21284 (2020)

22. Han, B., et al.: Masking: a new perspective of noisy supervision. In: Advances in Neural Information Processing Systems, vol. 31 (2018)

23. He, K., Fan, H., Wu, Y., Xie, S., Girshick, R.: Momentum contrast for unsupervised visual representation learning. In: Proceedings of the IEEE/CVF Conference on Computer Vision and Pattern Recognition, pp. 9729–9738 (2020)

24. He, K., Zhang, X., Ren, S., Sun, J.: Deep residual learning for image recognition. In: Proceedings of the IEEE Conference on Computer Vision and Pattern Recognition, pp. 770–778 (2016)

25. Hendrycks, D., Mazeika, M., Wilson, D., Gimpel, K.: Using trusted data to train deep networks on labels corrupted by severe noise. In: Advances in Neural Information Processing Systems, vol. 31 (2018)

26. Hou, P., Geng, X., Zhang, M.L.: Multi-label manifold learning. In: Proceedings of the AAAI Conference on Artificial Intelligence, vol. 30 (2016)

27. Hu, W., Huang, Y., Zhang, F., Li, R.: Noise-tolerant paradigm for training face recognition CNNs. In: Proceedings of the IEEE/CVF Conference on Computer Vision and Pattern Recognition, pp. 11887–11896 (2019)

28. Jenni, S., Favaro, P.: Deep bilevel learning. In: Proceedings of the European Conference on Computer Vision (ECCV), pp. 618–633 (2018)

29. Kervadec, C., Vielzeuf, V., Pateux, S., Lechervy, A., Jurie, F.: Cake: compact and accurate k-dimensional representation of emotion. arXiv preprint arXiv:1807.11215 (2018)

30. Kim, Y., Yoo, B., Kwak, Y., Choi, C., Kim, J.: Deep generative-contrastive networks for facial expression recognition. arXiv preprint arXiv:1703.07140 (2017)

31. Krogh, A., Hertz, J.: Ba simple weight decay can improve generalization. In: Moody, J.E., Hanson, S.J., Lippmann, R.P. (eds.) Advances in Neural Information Processing Systems, vol. 4 (1992)

32. Li, S., Deng, W., Du, J.: Reliable crowdsourcing and deep locality-preserving learning for expression recognition in the wild. In: Proceedings of the IEEE Conference on Computer Vision and Pattern Recognition, pp. 2852–2861 (2017)

33. Li, Y.K., Zhang, M.L., Geng, X.: Leveraging implicit relative labeling-importance information for effective multi-label learning. In: 2015 IEEE International Conference on Data Mining, pp. 251–260. IEEE (2015)

34. Liu, C., Wechsler, H.: Gabor feature based classification using the enhanced fisher linear discriminant model for face recognition. IEEE Trans. Image Process. **11**(4), 467–476 (2002)

35. Liu, T., Tao, D.: Classification with noisy labels by importance reweighting. IEEE Trans. Pattern Anal. Mach. Intell. **38**(3), 447–461 (2015)

36. Lukasik, M., Bhojanapalli, S., Menon, A., Kumar, S.: Does label smoothing mitigate label noise? In: International Conference on Machine Learning, pp. 6448–6458. PMLR (2020)

37. Malach, E., Shalev-Shwartz, S.: Decoupling "when to update" from "how to update". In: Advances in Neural Information Processing Systems, vol. 30 (2017)

38. Mollahosseini, A., Hasani, B., Mahoor, M.H.: Affectnet: a database for facial expression, valence, and arousal computing in the wild. IEEE Trans. Affect. Comput. **10**(1), 18–31 (2017)
39. Nguyen, D.T., Mummadi, C.K., Ngo, T.P.N., Nguyen, T.H.P., Beggel, L., Brox, T.: Self: learning to filter noisy labels with self-ensembling. arXiv preprint arXiv:1910.01842 (2019)
40. Patrini, G., Rozza, A., Krishna Menon, A., Nock, R., Qu, L.: Making deep neural networks robust to label noise: A loss correction approach. In: Proceedings of the IEEE Conference on Computer Vision and Pattern Recognition, pp. 1944–1952 (2017)
41. Reed, S., Lee, H., Anguelov, D., Szegedy, C., Erhan, D., Rabinovich, A.: Training deep neural networks on noisy labels with bootstrapping. arXiv preprint arXiv:1412.6596 (2014)
42. Ruan, D., Yan, Y., Lai, S., Chai, Z., Shen, C., Wang, H.: Feature decomposition and reconstruction learning for effective facial expression recognition. In: Proceedings of the IEEE/CVF Conference on Computer Vision and Pattern Recognition, pp. 7660–7669 (2021)
43. Shan, C., Gong, S., McOwan, P.W.: Facial expression recognition based on local binary patterns: a comprehensive study. Image Vis. Comput. **27**(6), 803–816 (2009)
44. She, J., Hu, Y., Shi, H., Wang, J., Shen, Q., Mei, T.: Dive into ambiguity: latent distribution mining and pairwise uncertainty estimation for facial expression recognition. In: Proceedings of the IEEE/CVF Conference on Computer Vision and Pattern Recognition, pp. 6248–6257 (2021)
45. Shorten, C., Khoshgoftaar, T.M.: A survey on image data augmentation for deep learning. J. Big Data **6**(1), 1–48 (2019)
46. Song, H., Kim, M., Lee, J.G.: Selfie: refurbishing unclean samples for robust deep learning. In: International Conference on Machine Learning, pp. 5907–5915. PMLR (2019)
47. Song, H., Kim, M., Park, D., Shin, Y., Lee, J.G.: Learning from noisy labels with deep neural networks: a survey. arXiv preprint arXiv:2007.08199 (2020)
48. Srivastava, N., Hinton, G., Krizhevsky, A., Sutskever, I., Salakhutdinov, R.: Dropout: a simple way to prevent neural networks from overfitting. J. Mach. Learn. Res. **15**(1), 1929–1958 (2014)
49. Sukhbaatar, S., Bruna, J., Paluri, M., Bourdev, L., Fergus, R.: Training convolutional networks with noisy labels. arXiv preprint arXiv:1406.2080 (2014)
50. Sun, N., Li, Q., Huan, R., Liu, J., Han, G.: Deep spatial-temporal feature fusion for facial expression recognition in static images. Pattern Recogn. Lett. **119**, 49–61 (2019)
51. Szegedy, C., Vanhoucke, V., Ioffe, S., Shlens, J., Wojna, Z.: Rethinking the inception architecture for computer vision. In: Proceedings of the IEEE Conference on Computer Vision and Pattern Recognition, pp. 2818–2826 (2016)
52. Tarvainen, A., Valpola, H.: Mean teachers are better role models: weight-averaged consistency targets improve semi-supervised deep learning results. arXiv preprint arXiv:1703.01780 (2017)
53. Wang, C., Wang, S., Liang, G.: Identity-and pose-robust facial expression recognition through adversarial feature learning. In: Proceedings of the 27th ACM International Conference on Multimedia, pp. 238–246 (2019)
54. Wang, K., Peng, X., Yang, J., Lu, S., Qiao, Y.: Suppressing uncertainties for large-scale facial expression recognition. In: Proceedings of the IEEE/CVF Conference on Computer Vision and Pattern Recognition, pp. 6897–6906 (2020)

55. Wang, K., Peng, X., Yang, J., Meng, D., Qiao, Y.: Region attention networks for pose and occlusion robust facial expression recognition. IEEE Trans. Image Process. **29**, 4057–4069 (2020)
56. Wang, R., Liu, T., Tao, D.: Multiclass learning with partially corrupted labels. IEEE Trans. Neural Netw. Learn. Syst. **29**(6), 2568–2580 (2017)
57. Wen, Z., Lin, W., Wang, T., Xu, G.: Distract your attention: multi-head cross attention network for facial expression recognition. arXiv preprint arXiv:2109.07270 (2021)
58. Xia, X., et al.: Robust early-learning: hindering the memorization of noisy labels. In: International Conference on Learning Representations (2020)
59. Xiao, T., Xia, T., Yang, Y., Huang, C., Wang, X.: Learning from massive noisy labeled data for image classification. In: Proceedings of the IEEE Conference on Computer Vision and Pattern Recognition, pp. 2691–2699 (2015)
60. Xu, N., Liu, Y.P., Geng, X.: Label enhancement for label distribution learning. IEEE Trans. Knowl. Data Eng. **33**(4), 1632–1643 (2019)
61. Yang, H., Ciftci, U., Yin, L.: Facial expression recognition by de-expression residue learning. In: Proceedings of the IEEE Conference on Computer Vision and Pattern Recognition, pp. 2168–2177 (2018)
62. Zeng, J., Shan, S., Chen, X.: Facial expression recognition with inconsistently annotated datasets. In: Proceedings of the European Conference on Computer Vision (ECCV), pp. 222–237 (2018)
63. Zhang, H., Cisse, M., Dauphin, Y.N., Lopez-Paz, D.: mixup: beyond empirical risk minimization. arXiv preprint arXiv:1710.09412 (2017)
64. Zhang, K., Zhang, Z., Li, Z., Qiao, Y.: Joint face detection and alignment using multitask cascaded convolutional networks. IEEE Signal Process. Lett. **23**(10), 1499–1503 (2016)
65. Zhang, Z., Sabuncu, M.: Generalized cross entropy loss for training deep neural networks with noisy labels. In: Advances in Neural Information Processing Systems, vol. 31 (2018)
66. Zhao, Z., Liu, Q., Zhou, F.: Robust lightweight facial expression recognition network with label distribution training. In: Proceedings of the AAAI Conference on Artificial Intelligence, vol. 35, pp. 3510–3519 (2021)
67. Zhong, L., Liu, Q., Yang, P., Liu, B., Huang, J., Metaxas, D.N.: Learning active facial patches for expression analysis. In: 2012 IEEE Conference on Computer Vision and Pattern Recognition, pp. 2562–2569. IEEE (2012)
68. Zhou, Y., Xue, H., Geng, X.: Emotion distribution recognition from facial expressions. In: Proceedings of the 23rd ACM International Conference on Multimedia, pp. 1247–1250 (2015)

Learning Dynamic Facial Radiance Fields for Few-Shot Talking Head Synthesis

Shuai Shen[1,2], Wanhua Li[1,2], Zheng Zhu[3], Yueqi Duan[4], Jie Zhou[1,2], and Jiwen Lu[1,2(✉)]

[1] Department of Automation, Tsinghua University, Beijing, China
shens19@mails.tsinghua.edu.cn,
{li-wh17,duanyueqi,jzhou,lujiwen}@tsinghua.org.cn
[2] Beijing National Research Center for Information Science and Technology, Beijing, China
[3] PhiGent Robotics, Beijing, China
zhengzhu@ieee.org
[4] Department of Electronic Engineering, Tsinghua University, Beijing, China

Abstract. Talking head synthesis is an emerging technology with wide applications in film dubbing, virtual avatars and online education. Recent NeRF-based methods generate more natural talking videos, as they better capture the 3D structural information of faces. However, a specific model needs to be trained for each identity with a large dataset. In this paper, we propose Dynamic Facial Radiance Fields (DFRF) for few-shot talking head synthesis, which can rapidly generalize to an unseen identity with few training data. Different from the existing NeRF-based methods which directly encode the 3D geometry and appearance of a specific person into the network, our DFRF conditions face radiance field on 2D appearance images to learn the face prior. Thus the facial radiance field can be flexibly adjusted to the new identity with few reference images. Additionally, for better modeling of the facial deformations, we propose a differentiable face warping module conditioned on audio signals to deform all reference images to the query space. Extensive experiments show that with only tens of seconds of training clip available, our proposed DFRF can synthesize natural and high-quality audio-driven talking head videos for novel identities with only 40k iterations. We highly recommend readers view our supplementary video for intuitive comparisons. Code is available in https://sstzal.github.io/DFRF/.

Keywords: Few-shot talking head synthesis · Neural radiance fields

1 Introduction

Audio-driven talking head synthesis is an ongoing research topic with a variety of applications including filmmaking, virtual avatars, video conferencing and

---

Supplementary Information The online version contains supplementary material available at https://doi.org/10.1007/978-3-031-19775-8_39.

S. Avidan et al. (Eds.): ECCV 2022, LNCS 13672, pp. 666–682, 2022.
https://doi.org/10.1007/978-3-031-19775-8_39

online education [4,17,45,51,53,55]. Existing talking head generation methods can be roughly divided into 2D-based and 3D-based ones. Conventional 2D-based methods usually depend on GAN model [6,11,16] or image-to-image translation [12,53–55]. However, due to the lack of 3D structure modeling, most of these approaches struggle in generating vivid and natural talking styles. Another genre for talking head synthesis [4,36,39,49] relies on the 3D morphable face model (3DMM) [2,40,57]. Benefit from the 3D-aware modeling, they can generate more vivid talking faces than 2D-based methods. Since the use of intermediate 3DMM parameters leads to some information loss, the audio-lip consistency of the generated videos may be affected [17].

Fig. 1. We propose Dynamic Facial Radiance Fields (DFRF), a learning framework for few-shot talking head synthesis within a small number of training iterations. Given only a 15 s video clip of Obama for 10k iterations training, our DFRF rapidly generalizes to this specific identity including the scene, and synthesizes photo-realistic talking head sequence as shown in row (c). In contrast, NeRF [27] and AD-NeRF [17] fail to produce plausible results in such a few-shot setting within limited training iterations.

More recently, the emerging Neural Radiance Fields (NeRF) based talking head methods [17,27,47] have achieved great performance improvement. They map audio features to a dynamic radiance field for talking portraits rendering without introducing extra intermediate representation. However, they directly encode the 3D geometry and appearance of a specific person into the radiance field, thereby failing to generalize to novel identities. A specific model needs to be trained for each novel identity with high computational cost. Moreover, a large training dataset is required, which cannot meet some practical scenarios where only a few data is available. As shown in Fig. 1, given only a 15 s training clip, AD-NeRF [17] renders some blurry faces after 10k training iterations.

In this paper, we study this more challenging setting, few-shot talking head synthesis, for the aforementioned practical application scenarios. For an arbitrary new identity with merely a short training video clip available, the model should generalize to this specific person within a few iterations of fine-tuning. There are three key features of the few-shot talking head synthesis *i.e.* limited training video, fast convergence, and realistic generation results. To this end, we propose a Dynamic audio-driven Facial Radiance Field (DFRF) for few-shot talking head synthesis. A reference mechanism is designed to learn the generic mapping from a few observed frames to the talking face with corresponding appearance (including the same identity, hairstyle and makeup). Specifically, with some 2D observations as references, the 3D query point can be projected back to the 2D image space of these references respectively and draw the corresponding pixel information to guide the following synthesis and rendering. A prior assumption for such projection operation is that two intersecting rays in 3D volume space should correspond to the same color [27,29]. This conception holds for static scenes, yet talking heads are deformable objects and such naive warping may lead to some mismatch. We therefore introduce a differentiable face warping module for better modeling the facial dynamics when talking. This face warping module is realized as a 3D point-wise deformation field conditioned on audio signals to warp all reference images to the query space.

Extensive experiments show that our proposed DFRF can generate realistic and natural talking head videos with few training data and training iterations. Figure 1 shows the visual comparison with NeRF [27] and AD-NeRF [17]. Given only a 15-second video clip of Obama for 10k training iterations, our proposed DFRF quickly generalizes to this specific identity and synthesizes photo-realistic talking head results. In contrast, NeRF and AD-NeRF fail to produce plausible results in such few-shot setting within limited training iterations. To summarize, we make the following contributions:

- We propose a dynamic facial radiance field conditioned on the 3D aware reference image features. The facial field can rapidly generalize to novel identities with only 15 s clip for fine-tuning.
- For better modeling the face dynamics of talking head, we learn a 3D point-wise face warping module conditioned on audio signals for each reference image to warp it to the query space.
- The proposed DFRF can generate vivid and natural talking head videos using only a handful of training data with limited iterations, which far surpasses other NeRF-based methods under the same setting. We highly recommend readers view the supplementary videos for better comparisons.

2   Related Work

2D-Based Talking-Head Synthesis. Talking-head synthesis aims to animate portraits with given audios. 2D-based methods usually employ GANs [6,11,16,31] or image-to-image translation [12,53–56] as the core technologies, and use some intermediate parameters such as 2D landmarks [5,7,11,25,55] to realize the synthesis task. There are also some works focusing on the few-shot talking head generation [12,23,26,44,51]. Zakharov *et al.* [51] propose a few-shot adversarial learning approach through pre-training high-capacity generator and discriminator via meta-learning. Wang *et al.* [44] realize one-shot talking head generation by predicting flow-based motion fields. Meshry *et al.* [26] disentangle the spatial and style information for few-shot talking head synthesis. However, since these 2D-based methods cannot grasp the 3D structure of head, the naturalness and vividness of the generated talking videos are inferior to the 3D-based methods.

3D-Based Talking-Head Synthesis. A series of 3D model-based methods [4,9,13,19–21,36,37,39] generate talking heads by utilizing 3D Morphable Models (3DMM) [2,34,40,57]. Taking advantage of 3D structure modeling, these approaches can achieve more natural talking style than 2D methods. Representative methods [37,39] have generated realistic and natural talking head videos. However, since their networks are optimized on a specific identity for idiosyncrasies learning, per-identity training on a large dataset is needed. Another common limitation is the information loss brought by the use of intermediate 3DMM parameters [2]. In contrast, our proposed method gets rid of such computationally expensive per-identity training settings while generating high-quality videos. More recently, the emerging NeRF [27] provides a new technique for 3D-aware talking head synthesis. Guo *et al.* [17] are the first to apply NeRF into the area of talking head synthesis and have achieved better visual quality. Yao *et al.* [47] further disentangle lip movements and personalized attributes. However both of them suffer in the few-shot learning setting.

Neural Radiance Fields. Neural Radiance Fields (NeRF) [27] store the information of 3D geometry and appearance in terms of voxel grids [35,38] with a fully-connected network. The invention of this technology has inspired a series of following works. pi-GAN [3] proposes a generative model with NeRF as the backbone for static face generation while our method learns a dynamic radiance field. Since the original NeRF is designed for static scenes, some works try to extend this technique to the dynamic domain [14,15,29,32,41]. Gafni *et al.* [14] encode the expression parameters into the NeRF for dynamic faces rendering. [29,32,41] encode non-rigid scenes via ray bending into a canonical space. [45] represents face as compact 3D keypoints and performs keypoint driven animation. i3DMM [48] generates faces relying on geometry latent code. However, these methods need to optimize the model to every scene independently requiring a large dataset, while our method realizes fast generalization across identities based on easily accessible 2D reference images. There are also some other works that try to improve NeRF's generalization capabilities [42,43,50], yet their research are limited to static scenes.

3   Methodology

3.1   Problem Statement

Some limitations of existing talking head technologies hinder them from practical applications. 2D-based methods struggle to generate a natural talking style [39]. Classical 3D-based approaches have information loss due to the use of 3DMM intermediate representations [17]. NeRF-based ones synthesize superior talking head videos, however the computational cost is relatively high since a specific model needs to be trained for each identity. And a large dataset is required for training. We therefore focus on a more challenging setting for the talking head synthesis task. For an arbitrary person with merely a short training video clip available, a personalized audio-driven portrait animation model with high-quality synthesis results should be constructed within only a few iterations of fine-tuning. Three core features of this setting can be summarized as: limited training data, fast convergence and excellent generation effect.

To this end, we propose a Dynamic Facial Radiance Field (DFRF) for few-shot talking head synthesis. The image features are introduced as a condition to build a fast mapping from reference images to the corresponding facial radiance field. For better modeling the facial deformations, we further design a differentiable face warping module to warp reference images to the query space. Specifically, for fast convergence, a base model is firstly trained across different identities to capture the structure information of the head and establish a generic mapping from audio to lip motions. On this basis, efficient fine-tuning is performed to quickly generalize to a new target identity. In the following, we will detail these designs.

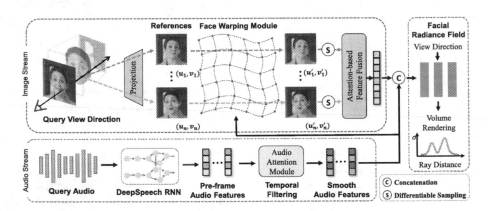

Fig. 2. Overview of the proposed Dynamic Facial Radiance Fields (DFRF).

3.2 Dynamic Facial Radiance Field

The emerging NeRFs [27] provide a powerful and elegant framework for 3D scene representation. It encodes a scene into a 3D volume space with a MLP $\mathcal{F}_\theta$. The 3D volume can then be rendered into images by integrating colors and densities along camera rays [10,28,33]. Specifically, using $\mathcal{P}$ as the collection of all 3D points in the voxel space, with a 3D query point $p = (x, y, z) \in \mathcal{P}$ and a 2D view direction $d = (\theta, \phi)$ as input, this MLP infers the corresponding RGB color c and density σ, which can be formulated as $(c, \sigma) = \mathcal{F}_\theta (p, d)$.

In this work, we employ NeRF as the backbone for 3D-aware talking head modeling. The talking head task focuses on the audio-driven face animation.

However, the original NeRF is designed for only static scenes. We therefore provide the missing deformation channel by introducing audio condition as shown in the audio stream of Fig. 2. We firstly use a pre-trained RNN-based Deep-Speech [18] module to extract the per-frame audio feature. For inter-frame consistency, a temporal filtering module [39] is further introduced to compute smooth audio features A, which can be denoted as the self-attention-based fusion of its neighbor audio features. Taking these audio feature sequences A as the condition, we can learn the audio-lip mapping. This audio-driven facial radiance field can be denoted as $(c, \sigma) = \mathcal{F}_\theta (p, d, A)$.

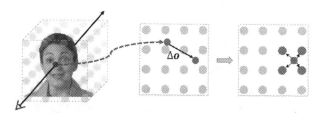

Fig. 3. Visualization of the differentiable face warping. A query 3D point (purple) is projected to the reference image space (red). Then an offset Δo is learned to warp it to the query space (green), where its feature is computed by bilinear interpolation. (Color figure online)

Since the identity information is implicitly encoded into the facial radiance field, and no explicit identity feature is provided when rendering, this facial radiance field is person specific. For each new identity, it needs to be optimized from scratch on a large dataset. This leads to expensive calculation costs and requires long training videos. To get rid of these restrictions, we design a reference mechanism to empower a well-trained base model to quickly generalize to new person categories, with only a short clip of the target person available. An overview of this reference-based architecture is shown in Fig. 2. Specifically, taken N reference images $M = \left\{ M_n \in \mathbb{R}^{H \times W} | 1 \leq n \leq N \right\}$ and their corresponding camera position $\{T_n\}$ as input, a two-layer convolutional network is used to calculate their pixel aligned image features $F = \left\{ F_n \in \mathbb{R}^{H \times W \times D} | 1 \leq n \leq N \right\}$ without down sampling. Feature dimension D is set as 128 in this work, and H, W indicates the height and width of an image respectively. The use of multiple reference images provides better multi-view informations. For a 3D query point $p = (x, y, z) \in \mathcal{P}$, we project it back to the 2D image spaces of these references using intrinsics $\{K_n\}$ and camera poses $\{R_n, T_n\}$ and get the corresponding 2D coordinate. Using $p_n^{ref} = (u_n, v_n)$ to denote the 2D coordinate in the n-th reference image, this projection can be formulated as:

$$p_n^{ref} = \mathcal{M}(p, K_n, R_n, T_n), \tag{1}$$

where $\mathcal{M}$ is the traditional mapping from world space to image space. These corresponding pixel-level features $\{F_n(u_n, v_n)\} \in \mathbb{R}^{N \times D}$ from N references are then sampled after a rounding operation and fused with an attention-based module [24] to get the final feature $\tilde{F} = Aggregation(\{F_n(u_n, v_n)\}) \in \mathbb{R}^D$. These feature grids contain rich information about identity and appearance. Using them as an additional condition for our facial radiance field makes the model possible to quickly generalize to a new face appearance from a few observed frames. This dual-driven facial radiance field can be finally formulated as:

$$(c, \sigma) = \mathcal{F}_\theta \left(p, d, A, \tilde{F} \right). \tag{2}$$

3.3   Differentiable Face Warping

In Sect. 3.2, we project the query 3D point back to the 2D image spaces of these reference images as Eq. (1) to get the conditioned pixel features. This operation bases on the prior knowledge in NeRF that intersecting rays casting from different viewpoints should correspond to the same physical location and thus yield the same color [29]. This strict spatial mapping relationship holds for rigid scenes yet the talking face is dynamic. When speaking, the lip and other facial muscles moves according to the pronunciation. Applying Eq. (1) directly on a deformable talking face may result in the key points mismatch. For example, a 3D point near the corner of the mouth in the standard volume space is mapped back to the pixel space of a reference image. If the reference face shows a different mouth shape, the mapped point may fall away from the desired real mouth corner. Such inaccurate mapping results in incorrect pixel feature conditions from reference images, which further affects the prediction of deformations of talking mouth.

To tackle this limitation, we propose an audio-conditioned and 3D point-wise face warping module $\mathcal{D}_\eta$. It regresses offsets $\Delta o = (\Delta u, \Delta v)$ for every projected point p^{ref} under the specific deformations, just as shown in the image stream of Fig. 2. Specifically, $\mathcal{D}_\eta$ is realized as a deformation field with a three-layer MLP, where η is the learnable parameters. To regress the offset Δo, dynamics differences between the query image and these reference images need to be effectively exploited. The audio information A reflects the dynamics of the query image, while the deformations of the reference images can be seen through image features $\{F_n\}$ implicitly. We therefore take these two parts together with the query 3D point coordinate p as the input for $\mathcal{D}_\eta$. The process to predict the offset with the face warping module $\mathcal{D}_\eta$ can be formulated as:

$$\Delta o_n = \mathcal{D}_\eta(p, A, F_n(u_n, v_n)). \tag{3}$$

The predicted offset o_n is then added to the p_n^{ref} as shown in Fig. 3 to get the exact corresponding coordinate $p_n^{ref'}$ for the 3D query point p,

$$p_n^{ref'} = p_n^{ref} + \Delta o_n = (u_n', v_n'), \tag{4}$$

where $u_n' = u_n + \Delta u_n$ and $v_n' = v_n + \Delta v_n$.

Since the hard index operation $F_n(u_n', v_n')$ is not differentiable, the gradient cannot be back propagated to this warpping module. We therefore introduce a soft index function to realize the differentiable warping, where the feature of each pixel is obtained through features interpolation of its surrounding points by bilinear sampling. In this way, the deformation field $\mathcal{D}_\eta$ and the facial radiance field $\mathcal{F}_\theta$ can be jointly optimized end to end. A visualization of this soft index operation is shown in Fig. 3. For the green point, its pixel feature is computed through the features of its four nearest neighbours by bilinear interpolation. To better constrain the training process of this warping module, we introduce a regularization term L_r to limit the value of predicted offsets in a reasonable range to prevent distortions,

$$L_r = \frac{1}{N \cdot |\mathcal{P}|} \sum_{p \in \mathcal{P}} \sum_{n=1}^{N} \sqrt{\Delta u_n^2 + \Delta v_n^2}, \tag{5}$$

where $\mathcal{P}$ is the collection of all 3D points in the voxel space, and N is the number of reference images. Furthermore, we argue that the points with low density are more likely to be background areas that should have low deformation offset. In these regions, stronger regularization constraints should be imposed. For more reasonable constraint, we change the above L_r as:

$$L_r{}' = (1 - \sigma) \cdot L_r, \tag{6}$$

where σ indicates the density of these points. The dynamic facial radiance field can finally be formulated as:

$$(c, \sigma) = \mathcal{F}_\theta \left(p, d, A, \tilde{F}' \right), \tag{7}$$

where $\tilde{F}' = Aggregation(\{F_n(u_n', v_n')\})$.

With this face warping module, all reference images can be transformed to the query space for better modeling the talking face deformations. The ablation study in Sect. 4.2 has proven the effectiveness of this component in producing more accurate and audio-synchronized mouth movements.

3.4 Volume Rendering

The volume rendering is used to integrate the colors c and densities σ from Eq. (7) into face images. We treat the background, torso and neck parts together as the rendering 'background' and restore it frame by frame from the original videos. We set the color of the last point of each ray as the corresponding background pixel to render a natural background including the torso part. Here we follow the setting in the original NeRF, and the accumulated color C of a camera ray r under the condition of audio signal A and image features $\tilde{F}'$ is:

$$C \left(r; \theta, \eta, R, T, A, \tilde{F}' \right) = \int_{z_{near}}^{z_{far}} \sigma(t) \cdot c(t) \cdot T(t) \, dt, \tag{8}$$

where θ and η are the learnable parameters for the facial radiance field $\mathcal{F}_\theta$ and the face warping module $\mathcal{D}_\eta$ respectively. R is the rotation matrix and

T is the translation vector. $T(t) = exp\left(-\int_{z_{near}}^{t} \sigma(r(s)) ds\right)$ is the integral transmittance along camera ray, where z_{near} and z_{far} are the near and far bound of the camera ray. We follow the NeRF to design a MSE loss as $L_{MSE} = \|C - I\|^2$, where I is the ground truth color. Coupled with the regularization term in Eq. (6), the overall loss function can be formulated as:

$$L = L_{MSE} + \lambda \cdot L_r'. \tag{9}$$

3.5   Implementation Details

We train only one base radiance field across different identities from coarse to fine. In the coarse training stage, the facial radiance field $\mathcal{F}_\theta$ as Eq. (2) is trained under the supervision of L_{MSE} to grasp the structure of the head and establish a general mapping from audio to lip motions. Then we add the face warping module into training as Eq. (7) to jointly optimize the offset regression network $\mathcal{D}_\eta$ and the $\mathcal{F}_\theta$ end to end with the loss function L in Eq. (9).

For an arbitrary unseen identity with only a short training clip available, we only need tens of seconds of his/her speaking video for fine-tuning based on the well-trained base model. After short iterations of fine-tuning, the personalized mouth pronunciation patterns can be learned, and the rendered image quality is greatly improved. Then this fine-tuned model can be used for inference.

4   Experiments

4.1   Experimental Settings

Dataset. AD-NeRF [17] collects several high-resolution videos in natural scenes to better evaluate the performance in practical application. Following this practice, we collect 12 public videos with an average length of 3 min from 11 identities from the YouTube. The protagonists of these videos are all celebrities like news anchors, entrepreneurs or presidents. We resample all videos to 25 FPS and set the resolution as 512×512. We select three videos from different races and languages (English and Chinese), and combine them into a three-minute video to train the base model. For other videos, we split each of them into three training sets of the length of 10 s, 15 s and 20 s. Then the remaining part is used as the test set. There is no overlap between the training set and the test set. All videos and the corresponding identities used in the following experiments are unseen when training the base model. These data will be released for reproduction.

Head Pose. Following the AD-NeRF, we estimate head poses based on Face2Face [40]. To get temporally smooth poses, we further apply the bundle adjustment [1] as a temporal filtering. The camera poses $\{R_n, T_n\}$ are the inverse of head poses, where R is the rotation matrix and T is the translation vector.

Metrics. We conduct performance evaluations through some quantitative metrics and visual results. Peak Signal-to-Noise Ratio (PSNR↑), Structure SIMilarity (SSIM↑) [46] and Learned Perceptual Image Patch Similarity (LPIPS↓) [52]

Table 1. Quantitative comparisons with different numbers of reference images.

Reference	1		2		4		6	
Metric	PSNR↑	LPIPS↓	PSNR↑	LPIPS↓	PSNR↑	LPIPS↓	PSNR↑	LPIPS↓
	31.03	**0.019**	31.19	**0.019**	**31.23**	**0.019**	**31.23**	0.020

Table 2. Method comparisons when using different lengths of training videos.

Method	NeRF [27]			AD-NeRF [17]			Ours		
	10 s	15 s	20 s	10 s	15 s	20 s	10 s	15 s	20 s
PSNR↑	19.83	19.77	8.02	31.21	**31.32**	30.90	30.95	30.75	30.96
SSIM↑	0.773	0.781	0.003	0.948	0.947	**0.949**	0.948	0.947	**0.949**
LPIPS↓	0.237	0.239	1.058	0.039	0.041	0.040	**0.036**	**0.036**	**0.036**
SyncNet↓↑	–	–	–	15/1.313	−14/0.654	−5/0.932	0/3.447	0/4.105	**0/4.346**

Table 3. Ablation study to investigate the contribution of the proposed differentiable face warping module. 'w' indicates the model equipped with the face warping module.

Method	Test set A				Test set B			
	PSNR↑	SSIM↑	LPIPS↓	SyncNet↓↑	PSNR↑	SSIM↑	LPIPS↓	SyncNet↓↑
GT	–	–	–	4/7.762	–	–	–	3/8.947
w/o	29.50	0.907	0.057	−1/4.152	28.98	**0.899**	0.104	−2/2.852
w	**29.66**	**0.911**	**0.053**	**0/4.822**	**29.14**	**0.899**	**0.101**	**0/4.183**

are used as image quality metrics. PSNR tends to give higher scores to blurry images [29]. We therefore recommend the more representative perceptual metrics LPIPS. We further use the SyncNet (offset↓/confidence↑) [8] to measure the audio-visual synchronization. The SyncNet offset is better with smaller absolute value. Here we use the '↓' as a brief indication.

Training Details. Our code is based on PyTorch [30]. All experiments are performed on an RTX 3090. The coefficient λ in Eq. (9) is set as 5e-8. We train the base model with an Adam solver [22] for 300k iterations and then jointly train it with the offset regression network for another 100k iterations.

4.2 Ablation Study

The Number of Reference Images. In this work, we learn a generic rendering from arbitrary reference face images to talking head with the corresponding appearance (including identity, hairstyle and makeup). Here we perform experiments to investigate the performance gains from various reference face images. We select different numbers of references and fine-tune the base model for 10k iterations on 15 s video clip respectively. Quantitative comparisons in Table 1 show that our method is robust to the number of reference images. According to results, we uniformly use four references in the following experiments.

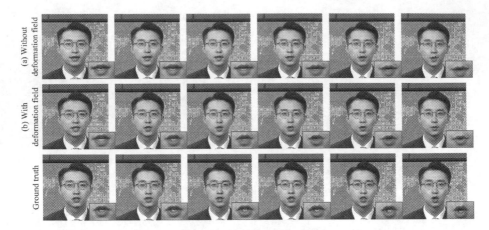

Fig. 4. Ablation study on the proposed face warping module. The ground truth sequence shows a pout-like expression. Generated results from the model equipped with the deformation field reproduce such pronunciation trend well in line (b), while results in line (a) hardly reflect such lip motions.

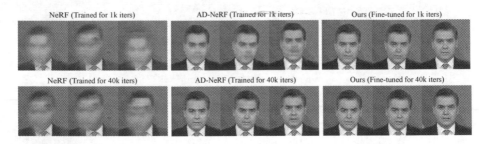

Fig. 5. Visual comparison using 15 s training clip for different training iterations.

Impact of the Length of Training Data. In this subsection, we investigate the impact of different amounts of training data. We fine-tune the proposed DFRF with 10 s, 15 s, and 20 s training videos for 50k iterations. For fair comparisons, we train NeRF and AD-NeRF with the same data and iterations. It is worth noting that we have tried to pre-train NeRF and AD-NeRF across identities following DFRF. However since they lack the ability to generalize between different identities, such per-training fails to learn the general audio-lip mapping. Experimental results in Table 2 show that tens of seconds of data are insufficient for NeRF training. PSNR tends to give higher scores to blurry images [29], so we recommend LPIPS as more representative metrics for visual quality. In comparison, our method is able to acquire more prior knowledge about the general audio-lip mapping from the base model, thus achieving better audio-visual sync with limited training data. With only a 10 s training video, the proposed DFRF can achieve superior 0.036 LPIPS and 3.447 SyncNet confidence, while AD-NeRF struggles in the lip-audio sync.

Table 4. Method comparisons on two test sets using 15 s training clip for different training iterations. More visual results can be seen in Fig. 5 and Fig. 6.

Method		Test set A				Test set B			
		PSNR↑	SSIM↑	LPIPS↓	SyncNet↓↑	PSNR↑	SSIM↑	LPIPS↓	SyncNet↓↑
Ground-truth		–	–	–	0/7.217	–	–	–	–1/7.762
NeRF	1k	16.88	0.708	0.198	–	14.69	0.397	0.442	–
	10k	13.98	0.531	0.338	–	15.24	0.396	0.427	–
	40k	15.87	0.556	0.306	–	15.91	0.405	0.394	–
AD-NeRF	1k	27.38	0.901	0.084	–15/0.136	27.61	0.863	0.115	14/0.798
	10k	29.14	0.931	0.057	–14/0.467	30.07	0.905	0.083	–2/0.964
	40k	29.45	0.936	0.039	–14/0.729	**30.72**	0.909	0.059	–2/1.017
Ours	1k	28.96	0.933	0.040	–1/2.996	29.05	0.892	0.076	0/3.157
	10k	29.33	0.935	0.043	0/4.246	29.68	0.905	0.063	0/4.038
	40k	**29.48**	**0.937**	**0.037**	1/4.431	30.44	**0.925**	**0.045**	0/4.951

Effect of Differentiable Face Warping. In DFRF, we propose an audio conditioned differentiable face warping module for better modeling the dynamics of talking face. Here we conduct an ablation study to investigate the contribution of this component. Table 3 shows the generated results with and without warping module on two test sets. All models are fine-tuned on 15 s videos for 50k iterations. Without this module, the query 3D point cannot be mapped to the exact corresponding point in the reference image, especially in some areas with rich dynamics. Therefore, the dynamics of the speaking mouth are affected to some extent, which is reflected in the audio-visual sync (SyncNet score). In contrast, the model equipped with the deformation field can significantly improve the SyncNet confidence and the visual quality also has slight improvement. Figure 4 further shows some visual results for more intuitive comparisons. In this video sequence, the ground truth shows a pout-like expression. The generated results (b) with the deformation field show such pronunciation trend well, while results in (a) hardly reflect this kind of lip motions.

4.3   Method Comparisons

Method Comparisons in the Few-Shot Setting. In this section, we perform method comparisons on two test sets using a 15 s training clip for different training iterations. Quantitative results in Table 4 show that our proposed method far surpasses NeRF and AD-NeRF in the perceptual image quality metric LPIPS. PSNR tends to give higher scores to blurry images [29] which can be proved in the visualization in Fig. 6, so we recommend LPIPS as more representative metrics for visual quality. We also achieve higher audio-lip synchronization indicated by the SyncNet score while AD-NeRF nearly fails on this indicator. Figure 5 visualizes the generated frames of the three methods. Under the same 1k training iterations, the visual quality of our method is far superior to others. When training for 40k iterations, the AD-NeRF achieves acceptable visual quality, however some face details are missing. The visual gap with our method can be seen obviously from the zoomed-in details in Fig. 6. We show two generated talking sequences driven by the same audio from our method and AD-NeRF with 15 s training clip after 40k iterations in Fig. 6. Compared with the ground truth, our method shows more accurate audio-lip synchronization than AD-NeRF. For example, in the fifth frame, the rendered face from AD-NeRF opens the mouth wrongly. We zoom in some facial details for clearer comparison. It can be seen that our method has generated more realistic details such as sharper hair texture, more obvious wrinkles, brighter pupils and more accurate mouth shape. In our supplementary video, we further add the visual comparison with AD-NeRF when it is trained to convergence (400k iterations).

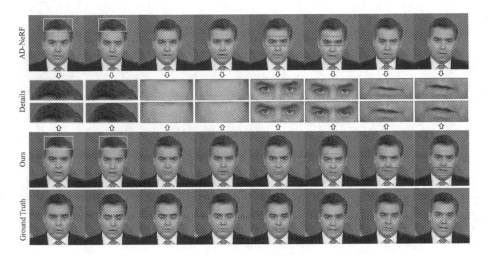

Fig. 6. Comparison with AD-NeRF using the same 15 s training clip for 40k training iterations. We zoom in on some facial details for better visual quality comparison.

Table 5. Method comparisons with two non-NeRF based methods SO [37] and NVP [39] and the AD-NeRF [17] under the setting with more training data.

Method	Test set A			Test set B			Few-shot
	PSNR↑	LPIPS↓	SyncNet↓↑	PSNR↑	LPIPS↓	SyncNet↓↑	Method
Suwajanakorn *et al.* [37]	–	–	3/4.301	–	–	–	✕
NVP [39]	–	–	–	–	–	-1/4.677	✕
AD-NeRF [17]	33.20	0.032	0/5.289	33.85	0.028	0/4.200	✕
Ours	**33.28**	**0.029**	**1/5.301**	**34.65**	**0.027**	**1/5.755**	✓

Method Comparisons with More Training Data. Our DFRF is far superior to others in the few-shot learning setting. For more comprehensive evaluations, we further compare the DFRF with some recent high-performance non-NeRF 3D-based methods [37,39] and the AD-NeRF [17] with more training data (180 s training clip). Since the source of [37,39] are not fully open, we follow the AD-NeRF to collect two test sets from the demos of [37,39] for method comparisons, and the results are shown in Table 5. Our method still surpasses others with long training clip up to 180 s, since the proposed face warping module better models the talking face dynamics. Moreover, our DFRF is the only method that works in the few-shot learning setting. In the supplementary video, we further include more comparisons with 2D-based (non-NeRF based) methods.

Cross-Language Results. We further verify the performance of our method driven by audios with different languages and genders. We select four models trained with 15 s training clips from different languages (source), then conduct inference with driven audios cross six languages and different genders (target). We also list the self-driven (source and target are from the same identity) results (the second column) for reference. SyncNet (offset/confidence)($\downarrow/\uparrow$) scores in Table 6 shows that our method produces reasonable lip-audio synchronization in such cross language setting.

Table 6. SyncNet scores under the cross language setting.

Source-target	Same identity	English (male)	Chinese (male)	Russian (male)	French (female)	Spanish (female)	German (female)
English (male)	$-3/5.042$	$-2/3.805$	$-2/4.879$	$-1/4.118$	$-2/3.019$	$-2/4.986$	$-1/4.820$
Chinese (male)	$0/4.486$	$-2/3.029$	$-2/3.534$	$-3/4.206$	$-3/3.931$	$-3/4.085$	$-2/4.494$
Russian (male)	$-1/4.431$	$-2/2.831$	$-2/4.397$	$-2/5.109$	$-3/4.307$	$-1/5.011$	$-1/5.008$
French (male)	$-2/4.132$	$-2/3.193$	$-3/3.383$	$-3/4.088$	$-2/3.339$	$-2/3.728$	$-1/3.529$

4.4 Applications and Ethical Considerations

The talking head synthesis technique can be used in a variety of practical scenarios, including correcting pronunciation, re-dubbing, virtual avatars, online education, electronic game making and providing speech comprehension for hearing impaired people. However, the talking head technology may bring some potential misuse issues. We are committed to combating these malicious behaviors and advocate more attention to the active application of this technology. We support those organizations that devote themselves to identifying fake defamatory videos, and are willing to provide them with the generated videos to expand the training set for automatic identification technology. Meanwhile, any individual or organization should obtain our permission before using our code, and it is recommended to use a watermark to indicate the generated video.

5   Conclusion

In this paper, we have proposed a dynamic facial radiance field for few-shot talking head synthesis. We employ audio signals coupled with 3D-aware image features as the condition for fast generalizing to novel identities. To better model the mouth motions of talking head, we further learn an audio-conditioned face warping module to deform all reference images to the query space. Extensive experiments show the superiority of our method in generating natural talking videos with limited training data and iterations.

Acknowledgement. This work was supported in part by the National Key Research and Development Program of China under Grant 2017YFA0700802, in part by the National Natural Science Foundation of China under Grant 62125603 and Grant U1813218, in part by a grant from the Beijing Academy of Artificial Intelligence (BAAI).

References

1. Andrew, A.M.: Multiple view geometry in computer vision. Kybernetes (2001)
2. Blanz, V., Vetter, T.: A morphable model for the synthesis of 3D faces. In: Annual Conference on Computer Graphics and Interactive Techniques (1999)
3. Chan, E.R., Monteiro, M., Kellnhofer, P., Wu, J., Wetzstein, G.: pi-GAN: periodic implicit generative adversarial networks for 3D-aware image synthesis. In: CVPR (2021)
4. Chen, L., et al.: Talking-head generation with rhythmic head motion. In: Vedaldi, A., Bischof, H., Brox, T., Frahm, J.-M. (eds.) ECCV 2020. LNCS, vol. 12354, pp. 35–51. Springer, Cham (2020). https://doi.org/10.1007/978-3-030-58545-7_3
5. Chen, L., Maddox, R.K., Duan, Z., Xu, C.: Hierarchical cross-modal talking face generation with dynamic pixel-wise loss. In: CVPR (2019)
6. Christos Doukas, M., Zafeiriou, S., Sharmanska, V.: HeadGAN: video-and-audio-driven talking head synthesis. arXiv (2020)
7. Chung, J.S., Jamaludin, A., Zisserman, A.: You said that? In: BMVC (2017)
8. Chung, J.S., Zisserman, A.: Out of time: automated lip sync in the wild. In: ACCV (2016)
9. Cudeiro, D., Bolkart, T., Laidlaw, C., Ranjan, A., Black, M.J.: Capture, learning, and synthesis of 3D speaking styles. In: CVPR (2019)
10. Curless, B., Levoy, M.: A volumetric method for building complex models from range images. In: Annual Conference on Computer Graphics and Interactive Techniques (1996)
11. Das, D., Biswas, S., Sinha, S., Bhowmick, B.: Speech-driven facial animation using cascaded GANs for learning of motion and texture. In: Vedaldi, A., Bischof, H., Brox, T., Frahm, J.-M. (eds.) ECCV 2020. LNCS, vol. 12375, pp. 408–424. Springer, Cham (2020). https://doi.org/10.1007/978-3-030-58577-8_25
12. Eskimez, S.E., Zhang, Y., Duan, Z.: Speech driven talking face generation from a single image and an emotion condition. TMM **24**, 3480–3490 (2021)
13. Fried, O., et al.: Text-based editing of talking-head video. TOG **38**, 1–14 (2019)
14. Gafni, G., Thies, J., Zollhofer, M., Nießner, M.: Dynamic neural radiance fields for monocular 4D facial avatar reconstruction. In: CVPR (2021)

15. Gao, C., Shih, Y., Lai, W.S., Liang, C.K., Huang, J.B.: Portrait neural radiance fields from a single image. arXiv (2020)
16. Gu, K., Zhou, Y., Huang, T.: FLNet: landmark driven fetching and learning network for faithful talking facial animation synthesis. In: AAAI (2020)
17. Guo, Y., Chen, K., Liang, S., Liu, Y., Bao, H., Zhang, J.: AD-NeRF: audio driven neural radiance fields for talking head synthesis. In: ECCV (2021)
18. Hannun, A., et al.: Deep speech: Scaling up end-to-end speech recognition. arXiv (2014)
19. Jaderberg, M., Simonyan, K., Zisserman, A., et al.: Spatial transformer networks. In: NeurIPS (2015)
20. Ji, X., et al.: Audio-driven emotional video portraits. In: CVPR (2021)
21. Karras, T., Aila, T., Laine, S., Herva, A., Lehtinen, J.: Audio-driven facial animation by joint end-to-end learning of pose and emotion. TOG **36**, 1–12 (2017)
22. Kingma, D.P., Ba, J.: Adam: a method for stochastic optimization. arXiv (2014)
23. Kumar, N., Goel, S., Narang, A., Hasan, M.: Robust one shot audio to video generation. In: CVPRW (2020)
24. Locatello, F., et al.: Object-centric learning with slot attention. arXiv (2020)
25. Lu, Y., Chai, J., Cao, X.: Live speech portraits: real-time photorealistic talking-head animation. TOG **40**, 1–17 (2021)
26. Meshry, M., Suri, S., Davis, L.S., Shrivastava, A.: Learned spatial representations for few-shot talking-head synthesis. arXiv (2021)
27. Mildenhall, B., Srinivasan, P.P., Tancik, M., Barron, J.T., Ramamoorthi, R., Ng, R.: NeRF: representing scenes as neural radiance fields for view synthesis. In: Vedaldi, A., Bischof, H., Brox, T., Frahm, J.-M. (eds.) ECCV 2020. LNCS, vol. 12346, pp. 405–421. Springer, Cham (2020). https://doi.org/10.1007/978-3-030-58452-8_24
28. Niemeyer, M., Mescheder, L., Oechsle, M., Geiger, A.: Differentiable volumetric rendering: learning implicit 3D representations without 3d supervision. In: CVPR (2020)
29. Park, K., et al.: Nerfies: deformable neural radiance fields. In: ICCV (2021)
30. Paszke, A., et al.: PyTorch: an imperative style, high-performance deep learning library. In: NeurIPS (2019)
31. Prajwal, K., Mukhopadhyay, R., Namboodiri, V.P., Jawahar, C.: A lip sync expert is all you need for speech to lip generation in the wild. In: ACM MM (2020)
32. Pumarola, A., Corona, E., Pons-Moll, G., Moreno-Noguer, F.: D-NeRF: neural radiance fields for dynamic scenes. In: CVPR (2021)
33. Seitz, S.M., Dyer, C.R.: Photorealistic scene reconstruction by voxel coloring. IJCV **35**, 151–173 (1999). https://doi.org/10.1023/A:1008176507526
34. Shang, J., Shen, T., Li, S., Zhou, L., Zhen, M., Fang, T., Quan, L.: Self-supervised monocular 3D face reconstruction by occlusion-aware multi-view geometry consistency. In: Vedaldi, A., Bischof, H., Brox, T., Frahm, J.-M. (eds.) ECCV 2020. LNCS, vol. 12360, pp. 53–70. Springer, Cham (2020). https://doi.org/10.1007/978-3-030-58555-6_4
35. Sitzmann, V., Zollhöfer, M., Wetzstein, G.: Scene representation networks: Continuous 3D-structure-aware neural scene representations. arXiv (2019)
36. Song, L., Wu, W., Qian, C., He, R., Loy, C.C.: Everybody's talkin': let me talk as you want. arXiv (2020)
37. Suwajanakorn, S., Seitz, S.M., Kemelmacher-Shlizerman, I.: Synthesizing Obama: learning lip sync from audio. TOG **36**, 1–13 (2017)
38. Tewari, A., et al.: State of the art on neural rendering. In: Computer Graphics Forum (2020)

39. Thies, J., Elgharib, M., Tewari, A., Theobalt, C., Nießner, M.: Neural voice puppetry: audio-driven facial reenactment. In: Vedaldi, A., Bischof, H., Brox, T., Frahm, J.-M. (eds.) ECCV 2020. LNCS, vol. 12361, pp. 716–731. Springer, Cham (2020). https://doi.org/10.1007/978-3-030-58517-4_42

40. Thies, J., Zollhofer, M., Stamminger, M., Theobalt, C., Nießner, M.: Face2Face: real-time face capture and reenactment of RGB videos. In: CVPR (2016)

41. Tretschk, E., Tewari, A., Golyanik, V., Zollhofer, M., Lassner, C., Theobalt, C.: Non-rigid neural radiance fields: reconstruction and novel view synthesis of a dynamic scene from monocular video. In: ICCV (2021)

42. Trevithick, A., Yang, B.: GRF: learning a general radiance field for 3D representation and rendering. In: ICCV (2021)

43. Wang, Q., et al.: IBRNet: learning multi-view image-based rendering. In: CVPR (2021)

44. Wang, S., Li, L., Ding, Y., Fan, C., Yu, X.: Audio2Head: audio-driven one-shot talking-head generation with natural head motion. arXiv (2021)

45. Wang, T.C., Mallya, A., Liu, M.Y.: One-shot free-view neural talking-head synthesis for video conferencing. In: CVPR (2021)

46. Wang, Z., Bovik, A.C., Sheikh, H.R., Simoncelli, E.P.: Image quality assessment: from error visibility to structural similarity. TIP **13**, 600–612 (2004)

47. Yao, S., Zhong, R., Yan, Y., Zhai, G., Yang, X.: DFA-NeRF: personalized talking head generation via disentangled face attributes neural rendering. arXiv (2022)

48. Yenamandra, T., et al.: i3DMM: deep implicit 3D morphable model of human heads. In: CVPR (2021)

49. Yi, R., Ye, Z., Zhang, J., Bao, H., Liu, Y.J.: Audio-driven talking face video generation with learning-based personalized head pose. arXiv (2020)

50. Yu, A., Ye, V., Tancik, M., Kanazawa, A.: pixelNeRF: neural radiance fields from one or few images. In: CVPR (2021)

51. Zakharov, E., Shysheya, A., Burkov, E., Lempitsky, V.: Few-shot adversarial learning of realistic neural talking head models. In: ICCV (2019)

52. Zhang, R., Isola, P., Efros, A.A., Shechtman, E., Wang, O.: The unreasonable effectiveness of deep features as a perceptual metric. In: CVPR (2018)

53. Zhang, X., Wu, X., Zhai, X., Ben, X., Tu, C.: DAVD-Net: deep audio-aided video decompression of talking heads. In: CVPR (2020)

54. Zhou, H., Liu, Y., Liu, Z., Luo, P., Wang, X.: Talking face generation by adversarially disentangled audio-visual representation. In: AAAI (2019)

55. Zhou, Y., Han, X., Shechtman, E., Echevarria, J., Kalogerakis, E., Li, D.: MakeItTalk: speaker-aware talking-head animation. TOG **39**, 1–15 (2020)

56. Zhu, H., Huang, H., Li, Y., Zheng, A., He, R.: Arbitrary talking face generation via attentional audio-visual coherence learning. In: IJCAI (2020)

57. Zollhöfer, M., et al.: State of the art on monocular 3D face reconstruction, tracking, and applications. In: Computer Graphics Forum (2018)

CoupleFace: Relation Matters for Face Recognition Distillation

Jiaheng Liu[1], Haoyu Qin[2(✉)], Yichao Wu[2], Jinyang Guo[3], Ding Liang[2], and Ke Xu[1]

[1] State Key Lab of Software Development Environment, Beihang University, Beijing, China
liujiaheng@buaa.edu.cn
[2] SenseTime Group Limited, Beijing, China
{qinhaoyu1,wuyichao,liangding}@sensetime.com
[3] SKLSDE, Institute of Artificial Intelligence, Beihang University, Beijing, China

Abstract. Knowledge distillation is an effective method to improve the performance of a lightweight neural network (i.e., student model) by transferring the knowledge of a well-performed neural network (i.e., teacher model), which has been widely applied in many computer vision tasks, including face recognition (FR). Nevertheless, the current FR distillation methods usually utilize the Feature Consistency Distillation (FCD) (e.g., L_2 distance) on the learned embeddings extracted by the teacher and student models for each sample, which is not able to fully transfer the knowledge from the teacher to the student for FR. In this work, we observe that mutual relation knowledge between samples is also important to improve the discriminative ability of the learned representation of the student model, and propose an effective FR distillation method called CoupleFace by additionally introducing the Mutual Relation Distillation (MRD) into existing distillation framework. Specifically, in MRD, we first propose to mine the informative mutual relations, and then introduce the Relation-Aware Distillation (RAD) loss to transfer the mutual relation knowledge of the teacher model to the student model. Extensive experimental results on multiple benchmark datasets demonstrate the effectiveness of our proposed CoupleFace for FR. Moreover, based on our proposed CoupleFace, we have won the first place in the ICCV21 Masked Face Recognition Challenge (MS1M track).

Keywords: Face recognition · Knowledge distillation · Loss function

1 Introduction

Face recognition (FR) has been well investigated for decades. Most of the progress is credited to large-scale training datasets [19,59], resource-intensive networks with millions of parameters [13,38] and effective loss functions [4,47].

S. Avidan et al. (Eds.): ECCV 2022, LNCS 13672, pp. 683–700, 2022.
https://doi.org/10.1007/978-3-031-19775-8_40

In practice, FR models are often deployed on mobile and embedded devices, which are incompatible with the large neural networks (e.g., ResNet-101 [13]). Besides, as shown in Fig. 1(a), the capacities of the lightweight neural networks (e.g., MobileNetV2 [35]) cannot be fully exploited when they are only supervised by existing popular FR loss functions (e.g., ArcFace [4]). Therefore, how to develop lightweight and effective FR models for real-world applications has been investigated in recent years. For example, knowledge distillation [14] is being actively discussed to produce lightweight and effective neural networks, which transfers the knowledge from an effective teacher model to a lightweight student model.

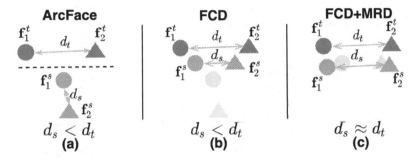

Fig. 1. The illustration of different methods for two samples from different classes. $\mathbf{f}_1^t$ and $\mathbf{f}_2^t$ are extracted by teacher model, and $\mathbf{f}_1^s$ and $\mathbf{f}_2^s$ are extracted by student model. d_t and d_s denote the distances (i.e., $1 - \cos(\mathbf{f}_1^t, \mathbf{f}_2^t)$ and $1 - \cos(\mathbf{f}_1^s, \mathbf{f}_2^s)$) of the embeddings, where $\cos(\cdot, \cdot)$ measures the cosine similarity of two features. Model is better when distance is larger. (a). The teacher and student models are both trained by ArcFace [4]. (b). The teacher model is trained by ArcFace. The student model is trained by using FCD. (c). The teacher model is trained by ArcFace. The student model is trained by using both FCD and MRD in CoupleFace.

Most existing knowledge distillation works usually aim to guide the student to mimic the behavior of the teacher by introducing probability constraints (e.g., KL divergence [14]) between the predictions of teacher and student models, which are not well-designed for FR. In contrast, as improving the discriminative ability of the feature embedding is the core problem for FR, it is important to enable the student model to share the same embedding space with the teacher model for similarity comparison. Thus, a simple and straightforward FR distillation method is to directly minimize the L_2 distance of the embeddings extracted by teacher and student models [37,49], which aims to align the embedding spaces between the teacher and student models. We call this simple method as Feature Consistency Distillation (FCD) as shown in Fig. 1(b), and FCD has been widely used in practice to improve the lightweight neural networks for FR.

In Fig. 1(b), when FCD is used, feature embeddings extracted by student model get close to the corresponding feature embeddings extracted by teacher model. However, there are some cases that the distances of embeddings (i.e., d_s) from different classes of student model are still smaller than d_t of teacher model. Similarly, there are also some cases that d_s from the same class is larger than d_t. In practice, when student model is deployed, smaller d_s for negative pair or larger d_s for positive pair usually leads to false recognition, which indicates that FCD is not sufficient to transfer the knowledge of teacher model to student model. In Fig. 1(c), we define the cosine similarity between samples as mutual relation and observe that $d_s \approx d_t$ when we additionally utilize the mutual relation information of teacher model to distill the corresponding mutual relation information of student model. Thus, we propose to introduce the Mutual Relation Distillation (MRD) into the existing FR distillation by reducing the gap between teacher and student models with respect to the mutual relation information.

Since FCD is able to align the embedding space of student model to teacher model well, the student model can distinguish most image pairs easily, which indicates that the differences of the most mutual relations between teacher and student models are relatively small. In other words, most mutual relations cannot provide valuable knowledge and affect the similarity distribution of all image pairs for FR. Therefore, how to generate sufficient informative mutual relations efficiently in MRD is a challenging issue.

Moreover, recent metric learning works [11,39,53] have shown that hard negative samples are crucial for improving the discriminative ability of feature embedding. But these works are not well-designed for mining mutual relations between samples for FR distillation. To this end, we propose to mine informative mutual relations in MRD. Moreover, as the number of positive pairs is usually small, we focus on mining the mutual relations of negative pairs.

Overall, in our work, we propose an effective FR distillation method referred to as CoupleFace, including FCD and MRD. Specifically, we first pre-train the teacher model on the large-scale training dataset. Then, in FCD, we calculate the L_2 distance of the embeddings extracted by the teacher and student models to generate the FCD loss for aligning the embedding spaces of the teacher and student models. In MRD, we first propose the Informative Mutual Relation Mining module to generate informative mutual relations efficiently in the training process, where the informative prototype set generation and memory-updating strategies are introduced to improve the mining efficiency. Then, we introduce the Relation-Aware Distillation (RAD) loss to exploit the mutual relation knowledge by using valid mutual relations and filtering out relations with subtle differences, which aims to better transfer the informative mutual relation knowledge from the teacher model to the student model.

The contributions are summarized as follows:

- In our work, we first investigate the importance of mutual relation knowledge for FR distillation, and propose an effective distillation framework called

CoupleFace, which consists of Feature Consistency Distillation (FCD) and Mutual Relation Distillation (MRD).
- In MRD, we propose to obtain informative mutual relations efficiently in our Informative Mutual Relation Mining module, where the informative prototype set generation and memory-updating strategies are used. Then, we introduce the Relation-Aware Distillation (RAD) loss to better transfer the mutual relation knowledge.
- Extensive experiments on multiple benchmark datasets demonstrate the effectiveness and generalization ability of our proposed CoupleFace method.

2   Related Work

Face Recognition. FR aims to maximize the inter-class discriminative ability and the intra-class compactness. The success of FR can be summarized into the three factors: effective loss functions [3–5,21,23,24,26,28,31,41,46,47,58], large-scale datasets [1,18,19,57,59], and powerful deep neural networks [27,38, 40,42,43]. The loss function design is the main-stream research direction for FR, which improves the generalization and discriminative abilities of the learned feature representation. For example, Triplet loss [36] is proposed to enlarge the distances of negative pairs and reduce the distances of positive pairs. Recently, the angular constraint is applied into the cross-entropy loss function in many angular-based loss functions [28,29]. Besides, CosFace [48] and ArcFace [4] further utilize a margin item for better discriminative capability of the feature representation. Moreover, some mining-based loss functions (e.g., Curricular-Face [16] and MV-Arc-Softmax [50]) take the difficulty degree of samples into consideration and achieve promising results. The recent work VPL [5] additionally introduces the sample-to-sample comparisons to reduce the gap between the training and evaluation processes for FR. In contrast, we propose to design an effective distillation loss function to improve the lightweight neural network.

Knowledge Distillation. As a representative type of model compression and acceleration methods [9,10,22], knowledge distillation aims to distill knowledge from a powerful teacher model into a lightweight student model [14], which has been applied in many computer vision tasks [2,6,7,15,17,25,32,33,44,54–56]. Many distillation methods have been proposed by utilizing different kinds of representation as knowledge for better performance. For example, FitNet [34] uses the middle-level hints from hidden layers of the teacher model to guide the training process of the student model. CRD [44] utilizes a contrastive-based objective function for transferring knowledge between deep networks. Some relation-based knowledge distillation methods (e.g., CCKD [33], RKD [32]) utilize the relation knowledge to improve the student model. Recently, knowledge distillation has also been applied to improve the performance of lightweight network (e.g., MobileNetV2 [35]) for FR. For example, EC-KD [49] proposes a position-aware exclusivity strategy to encourage diversity among different filters of the same layer to alleviate the low capability of student models. When compared with existing works, CoupleFace is well-designed for FR distillation by considering to mine informative mutual relations and transferring the relation knowledge of the teacher model using Relation-Aware Distillation loss.

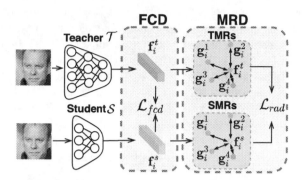

Fig. 2. The framework of CoupleFace for FR distillation. In FCD, we use the $\mathcal{L}_{fcd}$ loss between $\mathbf{f}_i^s$ and $\mathbf{f}_i^t$ to align the embedding spaces of the teacher and student models. In MRD, we first mine the informative features $\{\mathbf{g}_i^k\}_{k=1}^K$ (see Sect. 3.3), where K is the number of features, and calculate the teacher mutual relations (TMRs) and student mutual relations (SMRs) based on $\mathbf{f}_i^s$, $\mathbf{f}_i^t$ and $\{\mathbf{g}_i^k\}_{k=1}^K$. Then, we minimize the $\mathcal{L}_{rad}$ loss to transfer the mutual relation knowledge of the teacher model to student model.

3   Method

In this section, we introduce the details of our CoupleFace in Fig. 2, which contains Feature Consistency Distillation (FCD) and Mutual Relation Distillation (MRD) for FR distillation. The overall pipeline is as follows. First, we train the teacher model on a large-scale dataset. Then, in the distillation process of student model, we extract the feature embeddings based on the teacher and student models for each face image. After that, in FCD, we compute the Feature Consistency Distillation (FCD) loss $\mathcal{L}_{fcd}$ based on L_2 distance of the feature embeddings. Meanwhile, in MRD, we first build the informative mutual relations using the feature embeddings of teacher and student models with the mined informative features as shown in Fig. 3, and then calculate the Relation-Aware Distillation (RAD) loss $\mathcal{L}_{rad}$ to transfer the mutual relation knowledge.

3.1   Preliminary on Face Recognition Distillation

In this section, we define some notations in CoupleFace, and discuss the necessity of FR distillation when compared with traditional knowledge distillation.

Notations. We denote the teacher model as $\mathcal{T}$ and the student model as $\mathcal{S}$. For each sample x_i, the corresponding identity label is y_i, and the corresponding features extracted by $\mathcal{T}$ and $\mathcal{S}$ are denoted as $\mathbf{f}_i^t$ and $\mathbf{f}_i^s$, respectively.

Necessity of Face Recognition Distillation. For the traditional knowledge distillation of image classification, the existing methods usually utilize the probability consistency [14] (e.g., KL divergence) to align the prediction probabilities from $\mathcal{S}$ with the prediction probabilities from $\mathcal{T}$. However, the traditional knowledge distillation techniques are usually incompatible with FR. In practice, for FR, we can only obtain a pre-trained $\mathcal{T}$ but have no idea about how it was trained (e.g., the training datasets, loss functions). Therefore, the probability consistency loss is not available when the number of identities of the training dataset for $\mathcal{T}$ is different from the current dataset for $\mathcal{S}$ or $\mathcal{T}$ is trained by other metric learning based loss functions (e.g., triplet loss [36]). Besides, FR models are trained to generate discriminative feature embeddings for similarity comparison in the open-set setting rather than an effective classifier for the close-set classification. Thus, aligning the embedding spaces between $\mathcal{S}$ and $\mathcal{T}$ is more important for FR distillation.

3.2   Feature Consistency Distillation

In Feature Consistency Distillation (FCD), to boost the performance of $\mathcal{S}$ for FR, a simple and effective Feature Consistency Distillation (FCD) loss $\mathcal{L}_{fcd}$ is widely adopted in practice, which is defined as follows:

$$\mathcal{L}_{fcd} = \frac{1}{2N} \sum_{i=1}^{N} \left\| \frac{\mathbf{f}_i^t}{||\mathbf{f}_i^t||_2} - \frac{\mathbf{f}_i^s}{||\mathbf{f}_i^s||_2} \right\|^2, \tag{1}$$

where N is the number of face images for each mini-batch.

3.3   Mutual Relation Distillation

In this section, we describe the Mutual Relation Distillation (MRD) of CoupleFace in detail. First, we discuss the necessity of MRD for FR distillation. Then, we describe how to generate the informative mutual relations by using our informative mutual relation mining strategy. Finally, we introduce the Relation-Aware Distillation (RAD) loss to transfer the mutual relation knowledge.

Necessity of Mutual Relation Distillation. First, we define a pair of embeddings as a couple. Given a couple $(\mathbf{f}_i^g, \mathbf{f}_j^g)$, $g \in \{s,t\}$ denotes $\mathcal{S}$ or $\mathcal{T}$, the mutual relation $\mathrm{R}(\mathbf{f}_i^g, \mathbf{f}_j^g)$ of this couple is defined as follows:

$$\mathrm{R}(\mathbf{f}_i^g, \mathbf{f}_j^g) = \cos(\mathbf{f}_i^g, \mathbf{f}_j^g), \tag{2}$$

where $\cos(\cdot, \cdot)$ measures the cosine similarity between two features. Given two couples $(\mathbf{f}_i^t, \mathbf{f}_i^s)$ and $(\mathbf{f}_j^t, \mathbf{f}_j^s)$, when FCD loss is applied on $\mathcal{S}$, the mutual relations $\mathrm{R}(\mathbf{f}_i^t, \mathbf{f}_i^s)$ and $\mathrm{R}(\mathbf{f}_j^t, \mathbf{f}_j^s)$ will be maximized. However, in practice, when $\mathcal{S}$ is deployed, the mutual relation $\mathrm{R}(\mathbf{f}_i^s, \mathbf{f}_j^s)$ will be used to measure the similarity of this couple $(\mathbf{f}_i^s, \mathbf{f}_j^s)$ for FR. Thus, if we only use the FCD loss in FR distillation, the optimization on the mutual relation $\mathrm{R}(\mathbf{f}_i^s, \mathbf{f}_j^s)$ is ignored, which limits the further improvement of $\mathcal{S}$ for FR. Meanwhile, $\mathcal{T}$ with superior performance is able to provide an effective mutual relation $\mathrm{R}(\mathbf{f}_i^t, \mathbf{f}_j^t)$ as the ground-truth to distill the mutual relation $\mathrm{R}(\mathbf{f}_i^s, \mathbf{f}_j^s)$ from $\mathcal{S}$. Therefore, we introduce the MRD into the existing FR distillation framework.

However, direct optimization on mutual relation $\mathrm{R}(\mathbf{f}_i^s, \mathbf{f}_j^s)$ may not be a good choice in practice. Due to the batch size limitation and randomly sampling strategy in the training process, the mutual relations for these couples $\{(\mathbf{f}_i^s, \mathbf{f}_j^s)\}_{j=1, j \neq i}^N$ constructed across the mini-batch cannot support an effective and efficient mutual relation distillation. To this end, in CoupleFace, we propose to optimize $\mathrm{R}(\mathbf{f}_i^s, \mathbf{f}_j^t)$ instead of $\mathrm{R}(\mathbf{f}_i^s, \mathbf{f}_j^s)$ for following reasons. First, the quantity of mutual relations for these couples $\{(\mathbf{f}_i^s, \mathbf{f}_j^t)\}_{j=1, j \neq i}^L$ can be very large, where L is the number of samples of the dataset and $\mathbf{f}_j^t$ can be pre-calculated using teacher model. Second, the quality of the mutual relation for couple $(\mathbf{f}_i^s, \mathbf{f}_j^t)$ can be guaranteed as it can be mined from sufficient couples. Third, $\mathbf{f}_j^s$ is almost the same as $\mathbf{f}_j^t$ from the perspective of $\mathbf{f}_i^s$ when the FCD loss between $\mathbf{f}_j^s$ and $\mathbf{f}_j^t$ is fast converged, which represents that the mutual relation $\mathrm{R}(\mathbf{f}_i^s, \mathbf{f}_j^t)$ of couple $(\mathbf{f}_i^s, \mathbf{f}_j^t)$ is an ideal approximation of $\mathrm{R}(\mathbf{f}_i^s, \mathbf{f}_j^s)$ of couple $(\mathbf{f}_i^s, \mathbf{f}_j^s)$. Therefore, we propose to use the mutual relation $\mathrm{R}(\mathbf{f}_i^t, \mathbf{f}_j^t)$ to distill $\mathrm{R}(\mathbf{f}_i^s, \mathbf{f}_j^t)$.

Note that we call $\mathrm{R}(\mathbf{f}_i^t, \mathbf{f}_j^t)$ and $\mathrm{R}(\mathbf{f}_i^s, \mathbf{f}_j^t)$ as teacher mutual relation (**TMR**) and student mutual relation (**SMR**), respectively. To sum up, in MRD, we propose to utilize TMRs to distill the SMRs in the training process of FR.

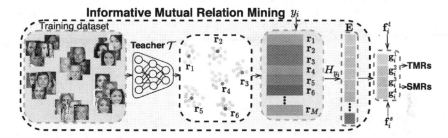

Fig. 3. We first pre-calculate identity prototypes $\{\mathbf{r}_m\}_{m=1}^M$ and generate the informative prototype set H_{y_i} for identity y_i using our informative prototype set generation strategy, where M is the number of identities across the training dataset. Then, we maintain a feature bank $\mathbf{E} \in \mathbb{R}^{M \times d}$ to store the feature embeddings of $\mathcal{T}$, which is updated in each iteration using our memory-updating strategy. Finally, we obtain K informative features $\{\mathbf{g}_i^k\}_{k=1}^K$ based on H_{y_i} and $\mathbf{E}$, and construct the SMRs and the TMRs based on $\mathbf{f}_i^s$, $\mathbf{f}_i^t$ and $\{\mathbf{g}_i^k\}_{k=1}^K$. Here, we set $K = 4$ for better illustration.

Informative Mutual Relation Mining. In this section, we describe how to generate informative mutual relations efficiently in Fig. 3 of CoupleFace.

Intuitively, for $\mathbf{f}_i^s$ and $\mathbf{f}_i^t$, a straightforward way is to construct the couples across all training samples and generate SMRs and TMRs. However, the computation cost is very large in this way, which is not applicable in practice. An alternative way is to generate SMRs and TMRs across the mini-batch. However, the batch size is relatively small and most mutual relations cannot provide valuable knowledge to improve $\mathcal{S}$, as it is easy to distinguish most image pairs and only the hard image pairs will greatly affect the performance of FR model. Meanwhile, recent metric learning works [11,39,53] show that hard negative samples are crucial for improving the discriminative ability of the embeddings. Therefore, in MRD, we propose to mine informative mutual relations among negative pairs to reduce the computation cost and improve $\mathcal{S}$ as shown in Fig. 3. Specifically, we use an informative prototype set generation strategy to find a set of most similar identities called informative prototype set H_{y_i} for identity y_i. Then, we utilize a memory-updating strategy to build the informative mutual relations based on $\mathbf{f}_i^s$, $\mathbf{f}_i^t$ and the features belonging to H_{y_i} efficiently.

Informative Prototype Set Generation. First, we extract the features of the training data by using a well-performed trained model. In our work, we directly use $\mathcal{T}$. The generated features can be denoted as $\{\mathbf{f}_i^t\}_{i=1}^L$ and the corresponding identity label is y_i for $\mathbf{f}_i^t$. For the training dataset, we denote the number of samples as L, the number of identities as M, and the identity label set as $\{m\}_{m=1}^M$. For each identity m, we calculate the identity prototype $\mathbf{r}_m$ as follows:

$$\mathbf{r}_m = \frac{1}{l_m} \sum_{i=1, y_i=m}^L \frac{\mathbf{f}_i^t}{||\mathbf{f}_i^t||_2}, \tag{3}$$

where l_m is the number of samples in the training dataset for identity m. Then, all identity prototypes of the dataset can be denoted as $\{\mathbf{r}_m\}_{m=1}^M$. After that, to find the informative prototype set H_m for $\mathbf{r}_m$, we calculate the cosine similarity between $\mathbf{r}_m$ and $\mathbf{r}_n$, where $n \in \{m\}_{m=1}^M$ and $n \neq m$. Afterwards, we select the top K (e.g., $K = 100$) identities with the largest similarities to construct H_m for identity m, where H_m contains K identity labels. Finally, the informative prototype set for identity y_i is denoted as H_{y_i}.

Memory-Updating. As the number of samples belonging to H_{y_i} is also relatively large, we further propose a memory-updating strategy to reduce the computation cost while making full use of all samples belonging to H_{y_i} inspired by MoCo [12]. Specifically, we maintain a feature bank $\mathbf{E} \in \mathbb{R}^{M \times d}$ to store feature embeddings extracted by $\mathcal{T}$, where only one embedding is preserved for each identity and d is the dimension (e.g., 512) of the embedding. At the beginning of the training process for $\mathcal{S}$, we initialize the feature bank $\mathbf{E}$ by randomly selecting one feature embedding generated by $\mathcal{T}$ for each identity. Then, in each iteration, we first obtain the feature embeddings $\{\mathbf{f}_i^t\}_{i=1}^N$ extracted by $\mathcal{T}$, where N is the size of mini-batch, and we update the feature bank $\mathbf{E}$ by setting $\mathbf{E}[y_i] = \mathbf{f}_i^t$, where $[\cdot]$ means to obtain features from $\mathbf{E}$ based on identity y_i.

Based on the informative prototype set H_{y_i} for y_i, we can obtain K informative negative features $\mathbf{G}_i = \mathbf{E}[H_{y_i}]$, where $\mathbf{G}_i \in \mathbb{R}^{K \times d}$. Meanwhile, we denote each feature in $\mathbf{G}_i$ as $\mathbf{g}_i^k$, where $k \in \{1, ..., K\}$. Finally, a set of couples $\{(\mathbf{f}_i^s, \mathbf{g}_i^k)\}_{k=1}^K$ is constructed for $\mathbf{f}_i^s$ and we can calculate the informative SMRs using these couples. Similarly, we can also generate the informative TMRs based on a set of couples $\{(\mathbf{f}_i^t, \mathbf{g}_i^k)\}_{k=1}^K$ as the ground-truth of these SMRs.

Relation-Aware Distillation Loss. Based on the mined TMRs and SMRs, the Relation-Aware Distillation (**RAD**) loss can be easily defined as follows:

$$\mathcal{L}_{rad} = \frac{1}{NK} \sum_{i=1}^{N} \sum_{k=1}^{K} |\cos(\mathbf{f}_i^s, \mathbf{g}_i^k) - \cos(\mathbf{f}_i^t, \mathbf{g}_i^k)|. \qquad (4)$$

However, the teacher model is not always better than the student model for each case in the training dataset. As illustrated in Fig. 4 of Sect. 4.4, we observe that $\cos(\mathbf{f}_i^t, \mathbf{g}_i^k) > \cos(\mathbf{f}_i^s, \mathbf{g}_i^k)$ does exist between the mined TMRs and SMRs. Therefore, if we directly use the Eq. (4) to transfer the mutual relation knowledge of $\mathcal{T}$, $\mathcal{S}$ will be misled in some cases, which may degrade the performance of $\mathcal{S}$.

To this end, we propose only to use the valid mutual relations when $\cos(\mathbf{f}_i^t, \mathbf{g}_i^k) < \cos(\mathbf{f}_i^s, \mathbf{g}_i^k)$, and we reformulate the RAD loss of Eq. (4) as follows:

$$\mathcal{L}_{rad} = \frac{1}{N'} \sum_{i=1}^{N} \sum_{k=1}^{K} \max(\cos(\mathbf{f}_i^s, \mathbf{g}_i^k) - \cos(\mathbf{f}_i^t, \mathbf{g}_i^k), 0), \qquad (5)$$

where N' is the number of valid mutual relations across the mini-batch. Thus, RAD loss of Eq. (5) will only affect the gradient when $\cos(\mathbf{f}_i^t, \mathbf{g}_i^k) < \cos(\mathbf{f}_i^s, \mathbf{g}_i^k)$ and transfer the accurate mutual relation knowledge from $\mathcal{T}$ to $\mathcal{S}$ in MRD.

We mine informative mutual relations for each identity. But there exists some identities, which can be easily distinguished from other identities, which indicates that the differences between the mined SMRs and TMRs for these identities are still subtle. Inspired by the hinge loss [8], we further propose a more effective variant of our RAD loss by introducing a margin q as follows:

$$\mathcal{L}_{rad} = \frac{1}{N'} \sum_{i=1}^{N} \sum_{k=1}^{K} \max(\cos(\mathbf{f}_i^s, \mathbf{g}_i^k) - \cos(\mathbf{f}_i^t, \mathbf{g}_i^k) - q, 0). \qquad (6)$$

Intuitively, the RAD loss in Eq. (6) further filters out the mutual relations with subtle differences between the SMRs and TMRs and pays attention to these SMRs, which are far away from their corresponding TMRs.

3.4 Loss Function of CoupleFace

The overall loss function of CoupleFace is defined as follows:

$$\mathcal{L} = \mathcal{L}_{fcd} + \alpha \cdot \mathcal{L}_{rad} + \beta \cdot \mathcal{L}_{ce}, \qquad (7)$$

where α and β are the weights of RAD loss $\mathcal{L}_{rad}$ and recognition loss $\mathcal{L}_{ce}$ (e.g., ArcFace [4]), respectively. We also provide an algorithm in Alg. 1.

4 Experiments

Datasets. For training, the mini version of Glint360K [1] named as Glint-Mini [18] is used, where Glint-Mini [18] contains 5.2M images of 91k identities. For testing, we use four datasets (i.e., IJB-B [51], IJB-C [30], and MegaFace [19]).

Algorithm 1 CoupleFace

Input: Pre-trained teacher model $\mathcal{T}$; Randomly initialized student model $\mathcal{S}$; Current
batch with N images; The dimension of feature representation d; The number of
identities M; The training dataset with L images; The feature bank $\mathbf{E} \in \mathbb{R}^{M \times d}$;

1: Extract all features $\{\mathbf{f}_i^t\}_{i=1}^L$ of the dataset using $\mathcal{T}$;
2: Generate all identity prototypes $\{\mathbf{r}_m\}_{m=1}^M$ based on $\{\mathbf{f}_i^t\}_{i=1}^L$ according to Eq. (3);
3: Based on $\{\mathbf{r}_m\}_{m=1}^M$, calculate informative prototype set $\{H_{y_i}\}_{i=1}^L$ for each sample
   according to the label y_i;
4: Initialize feature bank $\mathbf{E}$ using $\{\mathbf{f}_i^t\}_{i=1}^L$;
5: **for** each iteration in the training process **do**
6:     Get features $\{\mathbf{f}_i^t\}_{i=1}^N$ extracted by $\mathcal{T}$ from $\{\mathbf{f}_i^t\}_{i=1}^L$;
7:     Get features $\{\mathbf{f}_i^s\}_{i=1}^N$ extracted by $\mathcal{S}$;
8:     Calculate $\mathcal{L}_{fcd}$ of $\{\mathbf{f}_i^s\}_{i=1}^N$ and $\{\mathbf{f}_i^t\}_{i=1}^N$ by Eq. (1);
9:     Update feature bank $\mathbf{E}$ using $\{\mathbf{f}_i^t\}_{i=1}^N$;
10:    **for** each feature $\mathbf{f}_i^s$ in $\{\mathbf{f}_i^s\}_{i=1}^N$ **do**
11:        Get K informative negative features $\{\mathbf{g}_i^k\}_{k=1}^K$ from $\mathbf{E}$ using H_{y_i};
12:        Build TMRs and SMRs by $\{(\mathbf{f}_i^s, \mathbf{g}_i^k)\}_{k=1}^K$ and $\{(\mathbf{f}_i^t, \mathbf{g}_i^k)\}_{k=1}^K$, respectively;
13:    **end for**
14:    Calculate $\mathcal{L}_{rad}$ using TMRs and SMRs by Eq. (6);
15:    Update the parameters of $\mathcal{S}$ by minimizing the loss function
       $\mathcal{L} = \mathcal{L}_{fcd} + \alpha \cdot \mathcal{L}_{rad} + \beta \cdot \mathcal{L}_{ce}$;
16: **end for**
Output: The optimized student model $\mathcal{S}$.

Experimental Setting. For the pre-processing of the training data, we follow
the recent works [3,4,20] to generate the normalized face crops (112×112). For
teacher models, we use the widely used large neural networks (e.g., ResNet-34,
ResNet-50 and ResNet-100 [13]). For student models, we use MobileNetV2 [35]
and ResNet-18 [13]. For all models, the feature dimension is 512. For the training
process of all models based on ArcFace loss, the initial learning rate is 0.1 and
divided by 10 at the 100k, 160k, 180k iterations. The batch size and the total
iteration are set as 512 and 200k, respectively. For the distillation process, the
initial learning rate is 0.1 and divided by 10 at the 45k, 70k, 90k iterations. The
batch size and the total iteration are set as 512 and 100k, respectively. In the
informative mutual relation mining stage, we set the number of most similar
identities (i.e., K) as 100. In Eq. (6), we set the margin (i.e., q) as 0.03. The
loss weight α is set as 1, where β is set as 0 in the first 100k iterations, and is
set as 0.01 in CoupleFace+ of Table 1. In the following experiments of different
distillation methods, by default, we use the ResNet-50 (R-50), MobileNetV2
(MBNet) as $\mathcal{T}$ and $\mathcal{S}$, respectively.

4.1   Results on the IJB-B and IJB-C Datasets

As shown in Table 1, the first two rows represent the performance of models
trained by using the ArcFace loss function [4]. We compare our method with
classical KD [14], FCD, CCKD [33], SP [45], RKD [32], EC-KD [49]. For FCD,

we only use the FCD loss of Eq. (1) to align the embedding space of the student and teacher models, which is a very strong baseline to improve the performance of student model for FR. For these methods (i.e., CCKD [33], SP [45] and RKD [32]), we combine these methods with FCD loss instead of the classical KD loss to achieve better performance. For EC-KD proposed for FR, we reimplement this method. In Table 1, FCD is much better than classical KD, which indicates the importance of aligning embedding space for FR when compared with classical KD. Moreover, we observe that CoupleFace achieves significant performance improvements when compared with existing methods, which demonstrates the effectiveness of CoupleFace. For the CoupleFace+, we first pretrain student by CoupleFace, and then train student by CoupleFace with ArcFace by setting β in Eq. (7) as 0.01 for another 100k iterations, better results are obtained.

Table 1. Results (TAR@FAR) on IJB-B and IJB-C of different methods.

Models	Method	IJB-B		IJB-C	
		1e−4	1e−5	1e−4	1e−5
R-50 [13]	ArcFace [4]	93.89	89.61	95.75	93.44
MBNet [35]	ArcFace [4]	85.97	75.81	88.95	82.64
MBNet [35]	KD [14]	86.12	75.99	89.03	82.69
	FCD	90.34	81.92	92.68	87.74
	CCKD [33]	90.72	83.34	93.17	89.11
	RKD [32]	90.32	82.45	92.33	88.12
	SP [45]	90.52	82.88	92.71	88.52
	EC-KD [49]	90.59	83.54	92.85	88.32
	CoupleFace	91.18	84.63	93.18	89.57
	CoupleFace+	**91.48**	**85.12**	**93.37**	**89.85**

4.2   Results on the MegaFace Dataset

In Table 2, we also provide the results of CoupleFace on MegaFace [52], and we observe that CoupleFace is better than other methods. For example, when compared with the FCD baseline, our method improves the rank-1 accuracy by 0.62% on MegaFace under the distractor size as 10^6.

4.3   Ablation Study

The Effect of Different Variants of RAD Loss. In MRD, we propose three variants of RAD loss (i.e., Eq. (4), Eq. (5) and Eq. (6)). To analyze the effect of different variants, we also perform additional experiments based on Eq. (4) and Eq. (5) and report the results of MBNet on IJB-B and IJB-C. In Table 3,

Table 2. Rank-1 accuracy with different distractors on MegaFace.

Models	Method	Distractors		
		10^4	10^5	10^6
R-50 [13]	ArcFace [4]	99.40	98.98	98.33
MBNet [35]	ArcFace [4]	94.56	90.25	84.64
MBNet [35]	KD [14]	94.46	90.25	84.65
	FCD	97.81	96.39	93.65
	CCKD [33]	98.07	96.43	93.90
	RKD [32]	98.06	96.41	93.84
	SP [45]	98.01	96.58	93.95
	EC-KD [49]	98.00	96.41	93.85
	CoupleFace	**98.09**	**96.74**	**94.27**

for CoupleFace-A, we replace the RAD loss of Eq. (6) with Eq. (4), which means that we transfer the mutual relations by distilling all TMRs to the corresponding SMRs without any selection process. For CoupleFace-B, we replace the RAD loss of Eq. (6) with Eq. (5) without using the margin item. In Table 3, we observe that CoupleFace-B outperforms CoupleFace-A a lot, which shows the effectiveness of only using the mutual relations when $\cos(\mathbf{f}_i^t, \mathbf{g}_i^k) < \cos(\mathbf{f}_i^s, \mathbf{g}_i^k)$. Moreover, CoupleFace also outperforms CoupleFace-B, so it is necessary to emphasize the distillation on these SMRs, which are far from their corresponding TMRs.

Table 3. Results on IJB-B and IJB-C of different methods.

Methods	IJB-B		IJB-C	
	1e−4	1e−5	1e−4	1e−5
CoupleFace	91.18	84.63	93.18	89.57
CoupleFace-A	90.73	83.68	92.75	88.35
CoupleFace-B	90.88	84.23	93.02	88.99
CoupleFace-C	90.65	83.18	92.52	88.89
CoupleFace-D	90.85	83.73	92.86	89.04
CoupleFace-E	90.78	83.32	92.78	88.79

The Effect of Informative Mutual Relation Mining. To demonstrate the effect of mining the informative mutual relations, we further propose three alternative variants of CoupleFace (i.e., CoupleFace-C, CoupleFace-D, CoupleFace-E). Specifically, for CoupleFace-C, we propose to directly optimize mutual relation $R(\mathbf{f}_i^s, \mathbf{f}_j^s)$ without the process of mining mutual relations, where only mutual relations from limited couples $\{(\mathbf{f}_i^s, \mathbf{f}_j^s)\}_{j=1, j\neq i}^N$ across the mini-batch are constructed.

For CoupleFace-D, we randomly select $K = 100$ identities to construct the H_{y_i} for each identity y_i, while for CoupleFace-E, we propose to build the TMRs and SMRs across the mini-batch without mining process, and compute the RAD loss from them. As shown in Table 3, we observe that CoupleFace is much better than three alternative variants, which demonstrates that it is beneficial to mine the informative mutual relations in CoupleFace.

The Effect of Informative Prototype Set Generation When Using Different Models. In our work, we directly use $\mathcal{T}$ (i.e., R-50) to generate the informative prototype set H_{y_i} for each identity y_i. Here, we propose to use other pre-trained models (i.e., **MBNet** and ResNet-100 (**R-100**)) trained by ArcFace loss to generate H_{y_i} for y_i, and we call these alternative methods as CoupleFace-MBN and CoupleFace-RN100, respectively. In Table 4, we report the results of MBNet on IJB-B and IJB-C after using CoupleFace, CoupleFace-MBN and CoupleFace-RN100. We observe that CoupleFace achieves comparable performance with CoupleFace-RN100, and higher performance than CoupleFace-MBN. For this phenomenon, we assume that when using a more effective model, we will generate more discriminative identity prototypes $\{\mathbf{r}_m\}_{m=1}^M$, which leads to generating a more accurate informative prototype set H_{y_i}. Thus, it is beneficial to use effective models for obtaining the informative prototype set.

Table 4. Results on IJB-B and IJB-C of different methods.

Methods	IJB-B		IJB-C	
	1e−4	1e−5	1e−4	1e−5
CoupleFace	91.18	84.63	93.18	89.57
CoupleFace-MBN	91.06	83.91	92.95	88.94
CoupleFace-RN100	91.15	84.65	93.16	89.58

4.4   Further Analysis

Visualization on the Differences of SMRs and TMRs. To further analyze the effect of CoupleFace, we visualize the distributions of the differences between SMRs and TMRs for both FCD and CoupleFace in Fig. 4. Specifically, we use the models of MBNet in the 20,000th, 60,000th, 100,000th iterations for both FCD and CoupleFace. The first and the second rows show the results of FCD and CoupleFace, respectively. In Fig. 4, during training, for CoupleFace, we observe that the differences of SMRs and TMRs gradually decrease at the right side of the red line, which demonstrates the effect of minimizing the RAD loss of Eq. (6). Besides, when compared with FCD, most SMRs are less than or approximate to TMRs in CoupleFace, which indicates that CoupleFace transfers the mutual relation knowledge of the teacher model to the student model well.

Visualization on the Distributions of Similarity Scores. We visualize the distributions of similarity scores on IJB-C of **MBNet** based on different methods

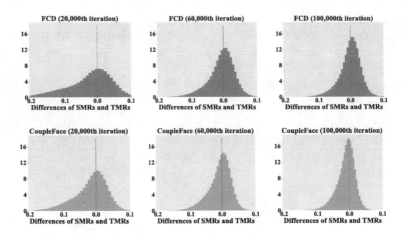

Fig. 4. The distributions of differences between the SMRs and TMRs of different iterations for FCD and CoupleFace.

in Fig. 5. Specifically, we still use the **R-50** as $\mathcal{T}$ to distill **MBNet** in FCD and CoupleFace. As shown in Fig. 5, when compared with FCD, the similarity distributions of positive pairs and negative pairs in CoupleFace are more compact and separable, which further shows the effectiveness of CoupleFace.

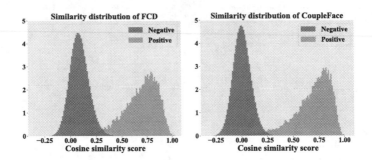

Fig. 5. Cosine similarity distributions of the positive pairs and negative pairs.

Comparison with Existing Relation-Based Knowledge Distillation Methods (RB-KDs). The differences between CoupleFace and existing RB-KDs (e.g., CCKD, RKD [32,33]) are as follows. (1) General KDs are usually incompatible with FR. Existing RB-KDs are proposed for general vision tasks (e.g., close-set classification). In contrast, it is non-trivial to transfer relation knowledge for open-set FR well and CoupleFace is well-designed for FR distillation. (2) Mining is considered. We observe that most mutual relations cannot provide valuable knowledge and affect the similarity distribution for FR, so how to produce sufficient informative mutual relations efficiently is a challenging issue. In CoupleFace, we propose to mine informative mutual relations in

MRD, while existing RB-KDs have not discussed the mining process. (3) Loss function is intrinsically different. The RAD loss in Eq. (6) aims to better exploit the mutual relation knowledge by using valid mutual relations and filtering out the mutual relations with subtle differences, which are not discussed in existing works. (4) Better performance. CoupleFace outperforms existing RB-KDs a lot.

Computation Costs. When compared with FCD, training time and GPU memory use of CoupleFace are 1.056 times and 1.002 times, respectively. which further demonstrates the efficiency of our proposed CoupleFace.

5   Conclusion

In our work, we investigate the importance of mutual relation knowledge for FR distillation and propose an effective FR distillation method named as Couple-Face. When compared with existing methods using Feature Consistency Distillation (FCD), CoupleFace further introduces the Mutual Relation Distillation (MRD), where we propose to mine the informative mutual relations and utilize the Relation-Aware Distillation (RAD) loss to transfer the mutual relation knowledge from the teacher model to the student model. Extensive experiments on multiple FR benchmark datasets demonstrate the effectiveness of Couple-Face. In our future work, we will continue to explore what kind of information is important for FR distillation and develop more effective distillation methods.

Acknowledgments. This research was supported by National Natural Science Foundation of China under Grant 61932002.

References

1. An, X., et al.: Partial FC: training 10 million identities on a single machine. In: Proceedings of the IEEE/CVF International Conference on Computer Vision (ICCV) Workshops, pp. 1445–1449, October 2021
2. David, S., Sergey, A.: MarginDistillation: distillation for face recognition neural networks with margin-based Softmax. Int. J. Comput. Inf. Eng. **15**(3), 206–210 (2021)
3. Deng, J., Guo, J., Liu, T., Gong, M., Zafeiriou, S.: Sub-center ArcFace: boosting face recognition by large-scale noisy web faces. In: Proceedings of the IEEE Conference on European Conference on Computer Vision (2020)
4. Deng, J., Guo, J., Xue, N., Zafeiriou, S.: ArcFace: additive angular margin loss for deep face recognition. In: Proceedings of the IEEE Conference on Computer Vision and Pattern Recognition, pp. 4690–4699 (2019)
5. Deng, J., Guo, J., Yang, J., Lattas, A., Zafeiriou, S.: Variational prototype learning for deep face recognition. In: Proceedings of the IEEE/CVF Conference on Computer Vision and Pattern Recognition (CVPR). pp. 11906–11915, June 2021
6. Fang, Z., Wang, J., Wang, L., Zhang, L., Yang, Y., Liu, Z.: (SEED): self-supervised distillation for visual representation. In: International Conference on Learning Representations (2021)

7. Feng, Y., Wang, H., Hu, H.R., Yu, L., Wang, W., Wang, S.: Triplet distillation for deep face recognition. In: 2020 IEEE International Conference on Image Processing (ICIP), pp. 808–812. IEEE (2020)

8. Gentile, C., Warmuth, M.K.: Linear hinge loss and average margin. In: Advances In Neural Information Processing Systems 11, pp. 225–231 (1998)

9. Guo, J., Liu, J., Xu, D.: JointPruning: pruning networks along multiple dimensions for efficient point cloud processing. IEEE Trans. Circuits Syst. Video Technol. **32**(6), 3659–3672 (2021)

10. Guo, J., Ouyang, W., Xu, D.: Multi-dimensional pruning: a unified framework for model compression. In: Proceedings of the IEEE/CVF Conference on Computer Vision and Pattern Recognition, pp. 1508–1517 (2020)

11. Harwood, B., Vijay Kumar, B.G., Carneiro, G., Reid, I., Drummond, T.: Smart mining for deep metric learning. In: Proceedings of the IEEE International Conference on Computer Vision, pp. 2821–2829 (2017)

12. He, K., Fan, H., Wu, Y., Xie, S., Girshick, R.: Momentum contrast for unsupervised visual representation learning. In: Proceedings of the IEEE/CVF Conference on Computer Vision and Pattern Recognition, pp. 9729–9738 (2020)

13. He, K., Zhang, X., Ren, S., Sun, J.: Deep residual learning for image recognition. In: Proceedings of the IEEE Conference on Computer Vision and Pattern Recognition, pp. 770–778 (2016)

14. Hinton, G., Vinyals, O., Dean, J.: Distilling the knowledge in a neural network. arXiv preprint arXiv:1503.02531 (2015)

15. Huang, Y., Shen, P., Tai, Y., Li, S., Liu, X., Li, J., Huang, F., Ji, R.: Improving face recognition from hard samples via distribution distillation loss. In: Vedaldi, A., Bischof, H., Brox, T., Frahm, J.-M. (eds.) ECCV 2020. LNCS, vol. 12375, pp. 138–154. Springer, Cham (2020). https://doi.org/10.1007/978-3-030-58577-8_9

16. Huang, Y., et al.: CurricularFace: adaptive curriculum learning loss for deep face recognition. In: Proceedings of the IEEE/CVF Conference on Computer Vision and Pattern Recognition, pp. 5901–5910 (2020)

17. Huang, Y., Wu, J., Xu, X., Ding, S.: Evaluation-oriented knowledge distillation for deep face recognition. In: Proceedings of the IEEE/CVF Conference on Computer Vision and Pattern Recognition (CVPR), pp. 18740–18749, June 2022

18. InsightFace: Glint-mini face recognition dataset (2021). https://github.com/deepinsight/insightface/tree/master/recognition/_datasets_

19. Kemelmacher-Shlizerman, I., Seitz, S.M., Miller, D., Brossard, E.: The MegaFace benchmark: 1 million faces for recognition at scale. In: Proceedings of the IEEE Conference on Computer Vision and Pattern Recognition, pp. 4873–4882 (2016)

20. Kim, Y., Park, W., Shin, J.: BroadFace: looking at tens of thousands of people at once for face recognition. In: Vedaldi, A., Bischof, H., Brox, T., Frahm, J.-M. (eds.) ECCV 2020. LNCS, vol. 12354, pp. 536–552. Springer, Cham (2020). https://doi.org/10.1007/978-3-030-58545-7_31

21. Li, Z., Wu, Y., Chen, K., Wu, Y., Zhou, S., Liu, J., Yan, J.: Learning to auto weight: entirely data-driven and highly efficient weighting framework. In: Proceedings of the AAAI Conference on Artificial Intelligence, pp. 4788–4795 (2020)

22. Liu, J., Guo, J., Xu, D.: APSNet: towards adaptive point sampling for efficient 3D action recognition. IEEE Trans. Image Process. **31**, 5287–5302 (2022)

23. Liu, J., Qin, H., Wu, Y., Liang, D.: AnchorFace: Boosting TAR@FAR for practical face recognition. In: Proceedings of the AAAI Conference on Artificial Intelligence (2022)

24. Liu, J., Wu, Y., Wu, Y., Li, C., Hu, X., Liang, D., Wang, M.: Dam: Discrepancy alignment metric for face recognition. In: Proceedings of the IEEE/CVF International Conference on Computer Vision, pp. 3814–3823 (2021)
25. Liu, J., Yu, T., Peng, H., Sun, M., Li, P.: Cross-lingual cross-modal consolidation for effective multilingual video corpus moment retrieval. In: NAACL-HLT (2022)
26. Liu, J., et al.: OneFace: one threshold for all. In: Farinella, T. (ed.) ECCV 2022. LNCS, vol. 13672, pp. 545–561. Springer, Cham (2022)
27. Liu, J., Zhou, S., Wu, Y., Chen, K., Ouyang, W., Xu, D.: Block proposal neural architecture search. IEEE Trans. Image Process. **30**, 15–25 (2020)
28. Liu, W., Wen, Y., Yu, Z., Li, M., Raj, B., Song, L.: SphereFace: deep hypersphere embedding for face recognition. In: Proceedings of the IEEE Conference on Computer Vision and Pattern Recognition, pp. 212–220 (2017)
29. Liu, W., Wen, Y., Yu, Z., Yang, M.: Large-margin Softmax loss for convolutional neural networks. In: ICML, vol. 2, p. 7 (2016)
30. Maze, B., et al.: IARPA Janus benchmark-C: face dataset and protocol. In: 2018 International Conference on Biometrics (ICB), pp. 158–165. IEEE (2018)
31. Meng, Q., Zhao, S., Huang, Z., Zhou, F.: MagFace: a universal representation for face recognition and quality assessment. In: Proceedings of the IEEE/CVF Conference on Computer Vision and Pattern Recognition, pp. 14225–14234 (2021)
32. Park, W., Kim, D., Lu, Y., Cho, M.: Relational knowledge distillation. In: Proceedings of the IEEE/CVF Conference on Computer Vision and Pattern Recognition, pp. 3967–3976 (2019)
33. Peng, B., et al.: Correlation congruence for knowledge distillation. In: ICCV, October 2019
34. Romero, A., Ballas, N., Kahou, S.E., Chassang, A., Gatta, C., Bengio, Y.: FitNets: hints for thin deep nets. arXiv preprint arXiv:1412.6550 (2014)
35. Sandler, M., Howard, A., Zhu, M., Zhmoginov, A., Chen, L.C.: MobileNetV2: inverted residuals and linear bottlenecks. In: Proceedings of the IEEE Conference on Computer Vision and Pattern Recognition, pp. 4510–4520 (2018)
36. Schroff, F., Kalenichenko, D., Philbin, J.: FaceNet: a unified embedding for face recognition and clustering. In: CVPR, pp. 815–823 (2015)
37. Shi, W., Ren, G., Chen, Y., Yan, S.: ProxylessKD: direct knowledge distillation with inherited classifier for face recognition. arXiv preprint arXiv:2011.00265 (2020)
38. Simonyan, K., Zisserman, A.: Very deep convolutional networks for large-scale image recognition. arXiv preprint arXiv:1409.1556 (2014)
39. Suh, Y., Han, B., Kim, W., Lee, K.M.: Stochastic class-based hard example mining for deep metric learning. In: Proceedings of the IEEE/CVF Conference on Computer Vision and Pattern Recognition, pp. 7251–7259 (2019)
40. Sun, Y., Chen, Y., Wang, X., Tang, X.: Deep learning face representation by joint identification-verification. In: Advances in Neural Information Processing Systems, pp. 1988–1996 (2014)
41. Sun, Y., et al.: Circle loss: a unified perspective of pair similarity optimization. In: Proceedings of the IEEE/CVF Conference on Computer Vision and Pattern Recognition, pp. 6398–6407 (2020)
42. Szegedy, C., et al.: Going deeper with convolutions. In: Proceedings of the IEEE Conference on Computer Vision and Pattern Recognition (CVPR), June 2015
43. Taigman, Y., Yang, M., Ranzato, M., Wolf, L.: DeepFace: closing the gap to human-level performance in face verification. In: Proceedings of the IEEE Conference on Computer Vision and Pattern Recognition, pp. 1701–1708 (2014)

44. Tian, Y., Krishnan, D., Isola, P.: Contrastive representation distillation. In: ICLR (2020)
45. Tung, F., Mori, G.: Similarity-preserving knowledge distillation. In: Proceedings of the IEEE/CVF International Conference on Computer Vision (ICCV), October 2019
46. Wang, F., Cheng, J., Liu, W., Liu, H.: Additive margin Softmax for face verification. IEEE Signal Proc. Lett. **25**(7), 926–930 (2018)
47. Wang, F., Xiang, X., Cheng, J., Yuille, A.L.: NormFace: L2 hypersphere embedding for face verification. In: Proceedings of the 25th ACM International Conference on Multimedia, pp. 1041–1049 (2017)
48. Wang, H., et al.: CosFace: large margin cosine loss for deep face recognition. In: Proceedings of the IEEE Conference on Computer Vision and Pattern Recognition, pp. 5265–5274 (2018)
49. Wang, X., Fu, T., Liao, S., Wang, S., Lei, Z., Mei, T.: Exclusivity-consistency regularized knowledge distillation for face recognition. In: Vedaldi, A., Bischof, H., Brox, T., Frahm, J.-M. (eds.) ECCV 2020. LNCS, vol. 12369, pp. 325–342. Springer, Cham (2020). https://doi.org/10.1007/978-3-030-58586-0_20
50. Wang, X., Zhang, S., Wang, S., Fu, T., Shi, H., Mei, T.: Mis-classified vector guided softmax loss for face recognition. In: Proceedings of the AAAI Conference on Artificial Intelligence, vol. 34, pp. 12241–12248 (2020)
51. Whitelam, C., et al.: IARPA Janus Benchmark-B face dataset. In: Proceedings of the IEEE Conference on Computer Vision and Pattern Recognition Workshops, pp. 90–98 (2017)
52. Wolf, L., Hassner, T., Maoz, I.: Face recognition in unconstrained videos with matched background similarity. IEEE (2011)
53. Wu, C.Y., Manmatha, R., Smola, A.J., Krahenbuhl, P.: Sampling matters in deep embedding learning. In: Proceedings of the IEEE International Conference on Computer Vision, pp. 2840–2848 (2017)
54. Yang, C., An, Z., Cai, L., Xu, Y.: Hierarchical self-supervised augmented knowledge distillation. In: Proceedings of the Thirtieth International Joint Conference on Artificial Intelligence, pp. 1217–1223 (2021)
55. Yang, C., An, Z., Cai, L., Xu, Y.: Knowledge distillation using hierarchical self-supervision augmented distribution. IEEE Tran. Neural Netw. Learn. Syst. (2022)
56. Yang, C., Zhou, H., An, Z., Jiang, X., Xu, Y., Zhang, Q.: Cross-image relational knowledge distillation for semantic segmentation. In: Proceedings of the IEEE/CVF Conference on Computer Vision and Pattern Recognition, pp. 12319–12328 (2022)
57. Yi, D., Lei, Z., Liao, S., Li, S.Z.: Learning face representation from scratch. arXiv preprint arXiv:1411.7923 (2014)
58. Zhang, X., Fang, Z., Wen, Y., Li, Z., Qiao, Y.: Range loss for deep face recognition with long-tailed training data. In: Proceedings of the IEEE International Conference on Computer Vision, pp. 5409–5418 (2017)
59. Zhu, Z., et al.: WebFace260M: a benchmark unveiling the power of million-scale deep face recognition. In: CVPR, pp. 10492–10502, June 2021

Controllable and Guided Face Synthesis for Unconstrained Face Recognition

Feng Liu[⊠] ⓘ, Minchul Kim ⓘ, Anil Jain ⓘ, and Xiaoming Liu ⓘ

Computer Science and Engineering, Michigan State University, East Lansing, USA
{liufeng6,kimminc2,jain,liuxm}@msu.edu

Abstract. Although significant advances have been made in face recognition (FR), FR in unconstrained environments remains challenging due to the domain gap between the semi-constrained training datasets and unconstrained testing scenarios. To address this problem, we propose a controllable face synthesis model (CFSM) that can mimic the distribution of target datasets in a style latent space. CFSM learns a linear subspace with orthogonal bases in the style latent space with precise control over the diversity and degree of synthesis. Furthermore, the pre-trained synthesis model can be guided by the FR model, making the resulting images more beneficial for FR model training. Besides, target dataset distributions are characterized by the learned orthogonal bases, which can be utilized to measure the distributional similarity among face datasets. Our approach yields significant performance gains on unconstrained benchmarks, such as IJB-B, IJB-C, TinyFace and IJB-S (+5.76% Rank1). Code is available at http://cvlab.cse.msu.edu/project-cfsm.html.

Keywords: Face synthesis · Model training · Target dataset distribution · Unconstrained face recognition

1 Introduction

Face recognition (FR) is now one of the most well-studied problems in the area of computer vision and pattern recognition. The rapid progress in face recognition accuracy can be attributed to developments in deep neural network models [25, 29,71,74], sophisticated design of loss functions [12,33,48,49,53,73,81–84,86,93], and large-scale training datasets, *e.g.*, MS-Celeb-1M [24] and WebFace260M [95].

Despite this progress, state-of-the-art (SoTA) FR models do not work well on real-world surveillance imagery (unconstrained) due to the domain shift issue, that is, the large-scale training datasets (semi-constrained) obtained via web-crawled celebrity faces lack in-the-wild variations, such as inherent sensor noise, low resolution, motion blur, turbulence effect, *etc.*. For instance, 1:1 verification accuracy reported by one of the SoTA models [68] on unconstrained IJB-S [34]

Supplementary Information The online version contains supplementary material available at https://doi.org/10.1007/978-3-031-19775-8_41.

S. Avidan et al. (Eds.): ECCV 2022, LNCS 13672, pp. 701–719, 2022.
https://doi.org/10.1007/978-3-031-19775-8_41

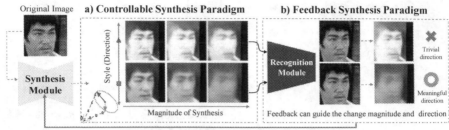

Fig. 1. (a) Given an input face image, our controllable face synthesis model (CFSM) enables precise control of the direction and magnitude of the targeted styles in the generated images. The latent style has both the direction and the magnitude, where the direction linearly combines the learned bases to control the *type* of style, while the magnitude controls the *degree* of style. (b) CFSM can incorporate the feedback provided by the FR model to generate synthetic training data that can benefit the FR model training and improve generalization to the unconstrained testing scenarios.

dataset is about 30% lower than on semi-constrained LFW [30]. A potential remedy to such a performance gap is to assemble a large-scale unconstrained face dataset. However, constructing such a training dataset with tens of thousands of subjects is prohibitively difficult with high manual labeling cost.

An alternative solution is to develop facial image generation models that can synthesize face images with desired properties. Face translation or synthesis using GANs [10,37,40,41,85,90] or 3D face reconstruction [2,27,42,63,76,77,94] has been well studied in photo-realistic image generation. However, most of these methods mainly focus on face image restoration or editing, and hence do not lead to better face recognition accuracies. A recent line of research [44,62,70,79] adopts disentangled face synthesis [13,75,78], which can provide control over explicit facial properties (pose, expression and illumination) for generating additional synthetic data for varied training data distributions. However, the hand-crafted categorization of facial properties and lack of design for cross-domain translation limits their generalizability to challenging testing scenarios. Shi *et al.* [69] propose to use an unlabeled dataset to boost unconstrained face recognition. However, all of the previous methods can be considered as performing *blind* data augmentation, *i.e.*, without the feedback of the FR model, which is required to provide critical information for improving the FR performance.

In Fig. 1, we show the difference between a blind and feedback-based face synthesis paradigm. For blind face synthesis, the FR model does not take part in the synthesis process, so there is no guidance from the FR model to avoid trivial synthesis. With feedback from the FR model, as in Fig. 1 (b), synthesized images can be more relevant to increasing the FR performance. Therefore, it is the goal of our paper to allow the FR model to guide the face synthesis towards creating synthetic datasets that can improve the FR performance.

It is not trivial to incorporate the signal from the FR model, as the direct manipulation of an input image towards decreasing the FR loss results in adversarial images that are not analogous to the real image distribution [23]. We thus propose to learn manipulation in the subspace of the *style space* of the target properties, so that the control can be accomplished 1) in low-dimensions, 2) semantically meaningful along with various quality factors.

In light of this, this paper aims to answer these three questions:

1. *Can we learn a face synthesis model that can discover the styles in the target unconstrained data, which enables us to precisely control and increase the diversity of the labeled training samples?*
2. *Can we incorporate the feedback provided by the FR model in generating synthetic training data, towards facilitating FR model training?*
3. *Additionally, as a by-product of our proposed style based synthesis, can we model the distribution of a target dataset, so that it allows us to quantify the distributional similarity among face datasets?*

Towards this end, we propose a face synthesis model that is 1) controllable in the synthesis process and 2) guided in the sense that the sample generation is aided by the signal from the FR model. Specifically, given a labeled training sample set, our controllable face synthesis model (CFSM) is trained to discover different attributes of the unconstrained target dataset in a style latent space. To learn the explicit degree and direction that control the styles in an unsupervised manner, we embed one linear subspace model with orthogonal bases into the style latent space. Within generative adversarial training, the face synthesis model seeks to capture the principal variations of the data distribution, and the style feature magnitude controls the degree of manipulation in the synthesis process.

More importantly, to extract the feedback of the FR model, we apply adversarial perturbations (FGSM) in the learned style latent space to guide the sample generation. This feedback is rendered meaningful and efficient because the manipulation is in the low dimensional style space as opposed to in the high dimensional image space. With the feedback from the FR model, the synthesized images are more beneficial to the FR performance, leading to significantly improved generalization capabilities of the FR models trained with them. It is worth noting that our pre-trained synthesis models could be a plug-in to any SoTA FR model. Unlike the conventional face synthesis models that focus on high quality realistic facial images, our face synthesis module is a conditional mapping from one image to a set of style shifted images that match the distribution of the target unconstrained dataset towards boosting its FR performance.

Additionally, the learned orthogonal bases characterize the target dataset distribution that could be utilized to quantify distribution similarity between datasets. The quantification of datasets has broad impact on various aspects. For example, knowing the dataset distribution similarity could be utilized to gauge the expected performance of FR systems in new datasets. Likewise, given a choice of various datasets to train an FR model, one can find one closest to the testing scenario of interest. Finally, when a new face dataset is captured in the future, we may also access its similarity to existing datasets in terms of styles, in addition to typical metrics such as number of subjects, demographics, etc.

In summary, the contributions of this work include:

- We show that a controllable face synthesis model with linear subspace style representation can generate facial images of the target dataset style, with precise control in the magnitude and type of style.
- We show that FR model performance can be greatly increased by synthesized images when the feedback of the FR model is used to optimize the latent style coefficient during image synthesis.
- Our learned linear subspace model can characterize the target dataset distribution for quantifying the distribution similarity between face datasets.
- Our approach yields significant performance gains on unconstrained face recognition benchmarks, such as IJB-B, IJB-C, IJB-S and TinyFace.

2   Prior Work

Controllable Face Synthesis. With the remarkable ability of GANs [22], face synthesis has seen rapid developments, such as StyleGAN [40] and its variations [38,39,41] which can generate high-fidelity face images from random noises. Lately, GANs have seen widespread use in face image translation or manipulation [6,10,13,28,47,61,72,88]. These methods typically adopt an encoder-decoder/generator-discriminator paradigm where the encoder embeds images into disentangled latent representations characterizing different face properties. Another line of works incorporates 3D prior (*i.e.*, 3DMM [5]) into GAN for 3D-controllable face synthesis [11,16,17,42,52,59,60,67]. Also, EigenGAN [26] introduces the linear subspace model into each generator layer, which enables to discover layer-wise interpretable variations. Unfortunately, these methods mainly focus on high-quality face generation or editing on pose, illumination and age, which has a well-defined semantic meaning. However, style or domain differences are hard to be factorized at the semantic level. Therefore, we utilize learned bases to cover unconstrained dataset attributes, such as resolution, noise, *etc.*.

Face Synthesis for Recognition. Early attempts exploit disentangled face synthesis to generate additional synthetic training data for either reducing the negative effects of dataset bias in FR [44,62,64,70,79] or more efficient training of pose-invariant FR models [78,91,92], resulting in increased FR accuracy. However, these models only control limited face properties, such as pose, illumination and expression, which are not adequate for bridging the domain gap between the semi-constrained and unconstrained face data. The most pertinent study to our work is [69], which proposes to generalize face representations with auxiliary unlabeled data. Our framework differs in two aspects: i) our synthesis model is precisely-controllable in the style latent space, in both magnitude and direction, and ii) our synthesis model incorporates *guidance* from the FR model, which significantly improves the generalizability to unconstrained FR.

Domain Generalization and Adaptation. Domain Generalization (DG) aims to make DNN perform well on unseen domains [18,19,46,55,56]. Conventionally, for DG, few labeled samples are provided for the target domain to

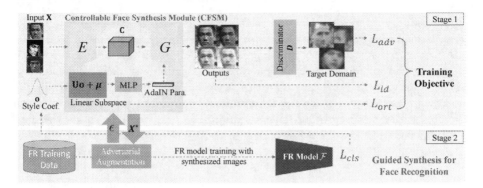

Fig. 2. Overview of the proposed method. Top (Stage 1): Pipeline for training the controllable face synthesis module that mimics the distribution of the target domain. $\mathcal{L}_{adv}$ ensures target domain similarity, $\mathcal{L}_{id}$ enforces the magnitude of **o** to control the degree of synthesis, and $\mathcal{L}_{ort}$ factorizes the target domain style with linear bases. Bottom (Stage 2): Pipeline for using the pre-trained face synthesis module for the purpose of training an FR model. The synthesis module works as an augmentation to the training data. We adversarially update **o** to maximize $\mathcal{L}_{cls}$ of a given FR model.

generalize. Popular DG methods utilize auxiliary losses such as Maximum Mean Discrepancy or domain adversarial loss to learn a shared feature space across multiple source domains [18,46,56]. In contrast, our method falls into the category of Unsupervised Domain Adaptation where adaptation is achieved by adversarial loss, contrastive loss or image translation, and learning a shared feature space that works for both the original and target domains [15,36,57,58,65,80]. Our method augments data resembling the target domain with unlabeled images.

Dataset Distances. It is important to characterize and contrast datasets in computer vision research. In recent years, various notions of dataset similarity have been proposed [1,4,50]. Alpha-distance and discrepancy distance [4,50] measures a dissimilarity that depends on a loss function and the predictor. To avoid the dependency on the label space, [1] proposes OT distance, an optimal transport distance in the feature space. However, it still depends on the ability of the predictor to create a separable feature space across domains. Moreover, a feature extractor trained on one domain may not predict the resulting features as separable in a new domain. In contrast, we propose to utilize the learned linear bases for latent style codes, which are optimized for synthesizing images in target domains, to measure the dataset distance. The style-based distance has the benefit of not being dependent on the feature space or the label space.

3 Proposed Method

3.1 Controllable Face Synthesis Model

Generally, for face recognition model training, we are given a labeled semi-constrained dataset that consists of n face images $\mathcal{X} = \{\mathbf{X}\}_{i=1}^{n}$ and the corresponding identity labels. Meanwhile, similar to the work in [69], we assume the availability of an unlabeled target face dataset with m images $\mathcal{Y} = \{\mathbf{Y}\}_{i=1}^m$, which contains a large variety of unconstrained factors. Our goal is to learn a style latent space where the face synthesis model, given an input image from the semi-constrained dataset, can generate new face images of the same subject, whose style is similar to the target dataset. Due to the lack of corresponding images, we seek an unsupervised algorithm that can learn to translate between domains without paired input-output examples. In addition, we hope this face synthesis model has explicit dimensions to control the unconstrained attributes.

Our face synthesis model is not designed to translate the intrinsic proper-ties between faces, i.e., pose, identity or expression. It is designed to focus on capturing the unconstrained imaging environment factors in unconstrained face images, such as noise, low resolution, motion blur, turbulence effects, etc. These variations are not present in large-scale labeled training data for face recognition.

Multimodal Image Translation Network. We adopt a multimodal image-to-image translation network [32, 45] to discover the underlying style distribution in the target domain. Specifically, as shown in Fig. 2, our face synthesis generator consists of an encoder E and a decoder G. Given an input image $\mathbf{X} \in \mathbb{R}^{W \times H \times 3}$, the encoder first extracts its content features $\mathbf{C} = E(\mathbf{X})$. Then, the decoder generates the output image $\hat{\mathbf{X}} \in \mathbb{R}^{W \times H \times 3}$, conditioned on both the content features and a random style latent code $\mathbf{z} \in \mathcal{Z}^d$: $\hat{\mathbf{X}} = G(\mathbf{C}, \mathbf{z})$. Here, the style code $\mathbf{z}$ is utilized to control the style of the output image.

Inspired by recent works that use affine transformation parameters in nor-malization layers to represent image styles [14, 31, 32, 40], we equip the residual blocks in the decoder D with Adaptive Instance Normalization (AdaIN) lay-ers [31], whose parameters are dynamically generated by a multilayer perceptron (MLP) from the style code $\mathbf{z}$. Formally, the decoder process can be presented as

$$\hat{\mathbf{X}} = G(\mathbf{C}, \mathrm{MLP}(\mathbf{z})). \tag{1}$$

It is worth noting that such G can model continuous distributions which enables us to generate multimodal outputs from a given input.

We employ one adversarial discriminator D to match the distribution of the synthesized images to the target data distribution; images generated by the model should be indistinguishable from real images in the target domain. The discriminator loss can be described as:

$$\mathcal{L}_D = -\mathbb{E}_{\mathbf{Y} \sim \mathcal{Y}}[\log(D(\mathbf{Y}))] - \mathbb{E}_{\mathbf{X} \sim \mathcal{X}, \mathbf{z} \sim \mathcal{Z}}[\log(1 - D(\hat{\mathbf{X}}))]. \tag{2}$$

The adversarial loss for the generator (including E and G) is then defined as:

$$\mathcal{L}_{adv} = -\mathbb{E}_{\mathbf{X} \sim \mathcal{X}, \mathbf{z} \sim \mathcal{Z}}[\log(D(\hat{\mathbf{X}}))]. \tag{3}$$

Domain-Aware Linear Subspace Model. To enable precise control of the targeted face properties, achieving flexible image generation, we propose to embed a linear subspace model with orthogonal bases into the style latent space. As illustrated in Fig. 2, a random style coefficient $\mathbf{o} \sim \mathcal{N}_q(\mathbf{0}, \mathbf{I})$ can be used to linearly combine the bases and form a new style code $\mathbf{z}$, as in

$$\mathbf{z} = \mathbf{U}\mathbf{o} + \boldsymbol{\mu}, \tag{4}$$

where $\mathbf{U} = [\mathbf{u}_1, \cdots, \mathbf{u}_q] \in \mathbb{R}^{d \times q}$ is the orthonormal basis of the subspace. $\boldsymbol{\mu} \in \mathbb{R}^d$ denotes the mean style. This equation relates a q-dimensional coefficient $\mathbf{o}$ to a corresponding d-dimensional style vector ($q << d$) by an affine transformation and translation. During training, both $\mathbf{U}$ and $\boldsymbol{\mu}$ are learnable parameters. The entire bases $\mathbf{U}$ are optimized with the orthogonality constraint [26]: $\mathcal{L}_{ort} = |\mathbf{U}^T\mathbf{U} - \mathbf{I}|_1$, where $\mathbf{I}$ is an identity matrix.

The isotropic prior distribution of $\mathbf{o}$ does not indicate which directions are useful. However, with the help of the subspace model, each basis vector in $\mathbf{U}$ identifies a latent direction that allows control over target image attributes that vary from straightforward high-level face properties. This mechanism is algorithmically simple, yet leads to effective control without requiring ad-hoc supervision. Accordingly, Eq. 1 can be updated as $\hat{\mathbf{X}} = G(\mathbf{C}, \mathrm{MLP}(\mathbf{Uo} + \boldsymbol{\mu}))$.

Magnitude of the Style Coefficient and Identity Preservation. Although the adversarial learning (Eq. 2 and 3) could encourage the face synthesis module to characterize the attributes in the target data, it cannot ensure the identity information is maintained in the output face image. Hence, the cosine similarity S_C between the face feature vectors $f(\mathbf{X})$ and $f(\hat{\mathbf{X}})$ is used to enforce identity preservation: $\mathcal{L}_{id} = 1 - S_C(f(\mathbf{X}), f(\hat{\mathbf{X}}))$, where $f(\cdot)$ represents a pre-trained feature extractor, *i.e.*, ArcFace [12] in our implementation.

Besides identifying the meaningful latent direction, we continue to explore the property of the magnitude $a = ||\mathbf{o}||$ of the style coefficient. We expect the magnitude can measure the degree of identity-preservation in the synthesized image $\hat{\mathbf{X}}$. In other words, $S_C(f(\mathbf{X}), f(\hat{\mathbf{X}}))$ monotonically increases when the magnitude a is decreased. To realize this goal, we re-formulate the identity loss:

$$\mathcal{L}_{id} = \left\| \left(1 - S_C(f(\mathbf{X}), f(\hat{\mathbf{X}}))\right) - g(a) \right\|_2^2, \tag{5}$$

where $g(a)$ is a function with respect to a. We assume the magnitude a is bounded in $[l_a, u_a]$. In our implementation, we define $g(a)$ as a linear function on $[l_a, u_a]$ with $g(l_a) = l_m$, $g(u_a) = u_m$: $g(a) = (a - l_a)\frac{u_m - l_m}{u_a - l_a} + l_m$.

By simultaneously learning the direction and magnitude of the style latent coefficients, our model becomes *precisely controllable* in capturing the variability of faces in the target domain. To our knowledge, this is the first method which is able to explore the complete set of two properties associated with the style, namely direction and magnitude, in unsupervised multimodal face translation.

Model Learning. The total loss for the generator (including encoder E, decoder G and domain-aware linear subspace model), with weights λ_i, is

$$\mathcal{L}_\mathcal{G} = \lambda_{adv}\mathcal{L}_{adv} + \lambda_{ort}\mathcal{L}_{ort} + \lambda_{id}\mathcal{L}_{id}. \tag{6}$$

3.2   Guided Face Synthesis for Face Recognition

In this section, we introduce how to incorporate the *pre-trained* face synthesis module into deep face representation learning, enhancing the generalizability to unconstrained FR. It is effectively addressing, *which synthetic images,*

*when added as an augmentation to the data, will increase the performance of the
learned FR model in the unconstrained scenarios?*

Formally, the FR model is trained to learn a mapping $\mathcal{F}$, such that $\mathcal{F}(\mathbf{X})$ is
discriminative for different subjects. If $\mathcal{F}$ is only trained on the domain defined
by semi-constrained $\mathcal{X}$, it does not generalize well to unconstrained scenarios.
However, $\mathbf{X}$ with identity label l in a training batch may be augmented with a
random style coefficient $\mathbf{o}$ to produce a synthesized image $\hat{\mathbf{X}}$ with CFSM.

However, such data synthesis with random style coefficients may generate
either extremely easy or hard samples, which may be redundant or detrimental
to the FR training. To address this issue, we introduce an adversarial regularization strategy to *guide* the data augmentation process, so that the face synthesis
module is able to generate meaningful samples for the FR model. Specifically,
for a given pre-trained CFSM, we apply adversarial perturbations in the learned
style latent space, in the direction of maximizing the FR model loss. Mathematically, given the perturbation budget ϵ, the adversary tries to find a style latent
perturbation $\boldsymbol{\delta} \in \mathbb{R}^d$ to maximize the classification loss function $\mathcal{L}_{cla}$:

$$\boldsymbol{\delta}^* = \underset{||\boldsymbol{\delta}||_\infty < \epsilon}{\arg\max}\, \mathcal{L}_{cla}\left(\mathcal{F}(\mathbf{X}^*), l\right), \text{where } \mathbf{X}^* = G(E(\mathbf{X}), \mathrm{MLP}(\mathbf{U}(\mathbf{o} + \boldsymbol{\delta}) + \boldsymbol{\mu})). \quad (7)$$

Here, $\mathbf{X}^*$ denotes the perturbed synthesized image. $\mathcal{L}_{cla}$ could be any
classification-based loss, *e.g.,* popular angular margin-based loss, ArcFace [12]
in our implementation. In this work, for efficiency, we adopt the one-step Fast
Gradient Sign Method (FGSM) [23] to obtain $\boldsymbol{\delta}^*$ and subsequently update $\mathbf{o}$:

$$\mathbf{o}^* = \mathbf{o} + \boldsymbol{\delta}^*, \quad \boldsymbol{\delta}^* = \epsilon \cdot \mathrm{sgn}\left(\nabla_{\mathbf{z}}\mathcal{L}_{cla}\left(\mathcal{F}(\mathbf{X}^*), l\right)\right), \quad (8)$$

where $\nabla\mathcal{L}_{cla}(\cdot, \cdot)$ denotes the gradi-
ent of $\mathcal{L}_{cla}(\cdot, \cdot)$ w.r.t. $\mathbf{o}$, and $\mathrm{sgn}(\cdot)$ is
the sign function.

Finally, based on the adversarial-
based augmented face images, we further optimize the face embedding
model $\mathcal{F}$ via the objective:

$$\min_\theta \mathcal{L}_{cla}([\mathbf{X}^*, \mathbf{X}], l), \quad (9)$$

where θ indicates the parameters of
FR model $\mathcal{F}$ and $[\cdot]$ refers to concatenation in the batch dimension.
In other words, it encourages to
search for the best perturbations in
the learned style latent space in the

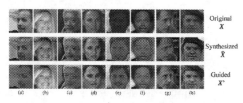

Fig. 3. Plot of mini-batch samples augmented with CFSM during training of the
FR model. Top: Original images. Middle:
Synthesized results before the feedback of
the FR model. Bottom: Synthesized results
after the feedback. The guide from the
FR model can vary the images' style for
increased difficulty (a–d), and preventing
the images from identity lost (e–h).

direction of maximal difficulty for the FR model. Examples within a mini-batch
are shown in Fig. 3.

Dataset Distribution Measure. As mentioned above, as a by-product of our
learned face synthesis model, we obtain a target-specific linear subspace model,

which can characterize the variations in the target dataset. Such learned linear subspace models allow us to quantify the distribution similarity among different datasets. For example, given two unlabeled datasets A and B, we can learn the corresponding linear subspace models $\{\mathbf{U}_A, \boldsymbol{\mu}_A\}$ and $\{\mathbf{U}_B, \boldsymbol{\mu}_B\}$. We define the distribution similarity between them as

$$\mathcal{S}(A,B) = \frac{1}{q}\left(\sum_i^q S_C(\mathbf{u}_A^i + \boldsymbol{\mu}_A, \mathbf{u}_B^i + \boldsymbol{\mu}_B)\right), \tag{10}$$

where $S_C(\cdot, \cdot)$ denotes the Cosine Similarity between the corresponding basis vectors in $\mathbf{U}_A$ and $\mathbf{U}_B$ respectively, and q is the number of the basis vectors.

Measuring the distance or similarity between datasets is a fundamental concept underlying research areas such as domain adaptation and transfer learning. However, the solution to this problem typically involves measuring feature distance with respect to a learned model, which may be susceptible to modal failure that the model may encounter in unseen domains. In this work, we provide an alternative solution via learned style bases vectors, that are directly optimized to capture the characteristics of the target dataset. We hope our method could provide new understandings and creative insights in measuring the dataset similarity. For visualizations of S among different datasets, please refer to Sect. 4.3.

3.3   Implementation Details

All face images are aligned and resized into 112×112 pixels. The network architecture of the face synthesis model is given in the supplementary (***Supp***). In the main experiments, we set $q = 10$, $d = 128$, $l_a = 0$, $u_a = 6$, $l_m = 0.05$, $u_m = 0.65$, $\lambda_{adv} = 1$, $\lambda_{ort} = 1$, $\lambda_{id} = 8$, $\epsilon = 0.314$. For more details, refer to Sect. 4 or ***Supp***.

4   Experimental Results

4.1   Comparison with SoTA FR Methods

Datasets. Following the experimental setting of [69], we use **MS-Celeb-1M** [24] as our labeled training dataset. MS-Celeb-1M is a large-scale public face dataset with web-crawled celebrity photos. For a fair comparison, we use the cleaned MS1M-V2 (3.9M images of 85.7K classes) from [69]. For our target data, **WiderFace** [89] is used. WiderFace is a dataset collected for face detection in challenging scenarios, with a diverse set of unconstrained variations. It is a suitable target dataset for training CFSM, as we aim to bridge the gap between the semi-constrained training faces and the faces in challenging testing scenarios. We follow [69] and use 70K face images from WiderFace. For evaluation, we test on four **unconstrained** face recognition benchmarks: IJB-B, IJB-C, IJB-S and TinyFace. These 4 datasets represent real-world testing scenarios where faces are significantly different from the semi-constrained training dataset.

Table 1. Comparison with state-of-the-art methods on the IJB-B benchmark. '*' denotes a subset of data selected by the authors.

Method	Train data, #labeled(+#unlabeled)	Backbone	Verification			Identification	
			1e−5	1e−4	1e−3	Rank1	Rank5
VGGFace2 [7]	VGGFace2, 3.3M	SE-ResNet-50	70.50	83.10	90.80	90.20	94.6
AFRN [35]	VGGFace2-*, 3.1M	ResNet-101	77.10	88.50	94.90	**97.30**	**97.60**
ArcFace [12]	MS1MV2, 5.8M	ResNet-50	84.28	91.66	94.81	92.95	95.60
MagFace [53]	MS1MV2, 5.8M	ResNet-50	83.87	91.47	94.67	–	–
Shi *et al.* [69]	Cleaned MS1MV2, 3.9M(+70K)	ResNet-50	88.19	92.78	95.86	95.86	96.72
ArcFace	Cleaned MS1MV2, 3.9M	ResNet-50	87.26	94.01	95.95	94.61	96.52
ArcFace+Ours	Cleaned MS1MV2, 3.9M(+70K)	ResNet-50	**90.95**	**94.61**	**96.21**	94.96	96.84

Table 2. Comparison with state-of-the-art methods on the IJB-C benchmark.

Method	Train data, #labeled(+#unlabeled)	Backbone	Verification			Identification	
			1e−6	1e−5	1e−4	Rank1	Rank5
VGGFace2 [7]	VGGFace2, 3.3M	SE-ResNet-50	–	76.80	86.20	91.40	95.10
AFRN [35]	VGGFace2-*, 3.1M	ResNet-101	–	88.30	93.00	95.70	**97.60**
PFE [68]	MS1M-*, 4.4M	ResNet-64	–	89.64	93.25	95.49	97.17
DUL [8]	MS1M-*, 3.6M	ResNet-64	–	90.23	94.20	95.70	97.60
ArcFace [12]	MS1MV2, 5.8M	ResNet-50	80.52	88.36	92.52	93.26	95.33
MagFace [53]	MS1MV2, 5.8M	ResNet-50	81.69	88.95	93.34	–	–
Shi *et al.* [69]	Cleaned MS1MV2, 3.9M(+70K)	ResNet-50	87.92	91.86	94.66	95.61	97.13
ArcFace	Cleaned MS1MV2, 3.9M	ResNet-50	87.24	93.32	95.61	95.89	97.08
ArcFace+Ours	Cleaned MS1MV2, 3.9M(+70K)	ResNet-50	**89.34**	**94.06**	**95.90**	**96.31**	97.48

- **IJB-B** [87] contains both high-quality celebrity photos collected in the wild and low-quality photos or video frames with large variations. It consists of $1,845$ subjects with 21.8K still images and 55K frames from $7,011$ videos.
- **IJB-C** [51] is an extension of IJB-B, which includes about $3,500$ subjects with a total of $31,334$ images and $117,542$ unconstrained video frames.
- **IJB-S** [34] is an extremely challenging benchmark where the images were collected in real-world surveillance environments. The dataset contains 202 subjects with an average of 12 videos per subject. Each subject also has 7 high-quality enrollment photos under different poses. We test on three protocols, Surveillance-to-Still (**V2S**), Surveillance-to-Booking (**V2B**) and Surveillance-to-Surveillance (**V2V**). The first/second notation in the protocol refers to the probe/gallery image source. 'Surveillance' (V) refers to the surveillance video, 'still' (S) refers to the frontal high-quality enrollment image and 'Booking' (B) refers to the 7 high-quality enrollment images.
- **TinyFace** [9] consists of $5,139$ labelled facial identities given by $169,403$ native low resolution face images, which is created to facilitate the investigation of unconstrained low-resolution face recognition.

Table 3. Comparison with state-of-the-art methods on three protocols of the IJB-S and TinyFace benchmark. The performance is reported in terms of rank retrieval (closed-set) and TAR@FAR (open-set). It is worth noting that MARN [21] is a multi-mode aggregation method and is fine-tuned on UMDFaceVideo [3], a video dataset.

Method	Labeled train data	Backbone	IJB-S V2S				IJB-S V2B				IJB-S V2V				TinyFace	
			Rank1	Rank5	1%	10%	Rank1	Rank5	1%	10%	Rank1	Rank5	1%	10%	Rank1	Rank5
C-FAN [20]	MS1M-*	ResNet-64	50.82	61.16	16.44	24.19	53.04	62.67	27.40	29.70	10.05	17.55	0.11	0.68	–	–
MARN [21]	MS1M-*	ResNet-64	58.14	64.11	21.47	–	59.26	65.93	32.07	–	22.25	34.16	0.19	–	–	–
PFE [68]	MS1M-*	ResNet-64	50.16	58.33	31.88	35.33	53.60	61.75	35.99	39.82	9.20	20.82	0.84	2.83	–	–
ArcFace [12]	MS1MV2	ResNet-50	50.39	60.42	32.39	42.99	52.25	61.19	34.87	43.50	–	–	–	–	–	–
Shi et al. [69]	MS1MV2-*	ResNet-50	59.29	66.91	39.92	50.49	60.58	67.70	32.39	44.32	17.35	28.34	1.16	5.37	–	–
ArcFace [12]	MS1MV2-*	ResNet-50	58.78	66.40	40.99	50.45	60.66	67.43	43.12	51.38	14.81	26.72	2.51	5.72	62.21	66.85
ArcFace+Ours*	MS1MV2-*	ResNet-50	61.69	68.33	43.99	53.34	62.20	69.50	44.38	53.49	18.14	31.34	2.09	4.51	62.39	67.36
ArcFace+Ours	MS1MV2-*	ResNet-50	**63.86**	**69.95**	**47.86**	**56.44**	**65.95**	**71.16**	**47.28**	**57.24**	**21.38**	**35.11**	**2.96**	**7.41**	**63.01**	**68.21**
AdaFace [43]	WebFace12M	IResNet-100	71.35	76.24	59.40	**66.34**	71.93	76.56	59.37	**66.68**	36.71	50.03	4.62	11.84	72.29	74.97
AdaFace+Ours	WebFace12M	IResNet-100	**72.54**	**77.59**	**60.94**	66.02	**72.65**	**78.18**	**60.26**	65.88	**39.14**	**50.91**	**5.05**	**13.17**	**73.87**	**76.77**

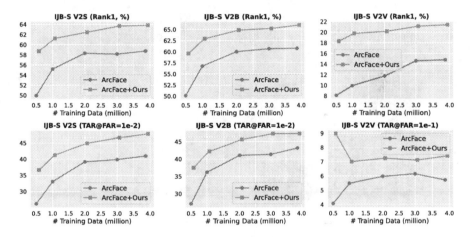

Fig. 4. Comparison on three IJB-S protocols with varied number of training data on the x-axis (maximum is 3.9M). The plots shows that using guided CFSM as an augmentation (ArcFace+Ours) can lead to higher performance in all settings. Note that CFSM trained on 70K unlabeled data is more useful than 3M original data as shown by the higher **V2V** performance of **Ours** with 0.5M than Baseline 3.9M.

Experiment Setting. We first train CFSM with ~10% of MS-Celeb-1M training data ($n = 0.4M$) as the source domain, and WiderFace as the target domain ($m = 70K$). The model is trained for $125,000$ steps with a batch size of 32. Adam optimizer is used with $\beta_1 = 0.5$ and $\beta_2 = 0.99$ at a learning rate of $1e-4$.

For the FR model training, we adopt ResNet-50 as modified in [12] as the backbone and use ArcFace loss function [12] for training. We also train a model without using CFSM (*i.e.*, replication of ArcFace) for comparison, denoted as **ArcFace**. The efficacy of our method (**ArcFace+Ours**) is validated by training an FR model with the guided face synthesis as the auxiliary data augmentation during training according to Eq. 9.

Results. Tables 1 and 2 respectively show the face verification and identi-fication results on IJB-B and IJB-C datasets. Our approach achieves SoTA performance on most of the protocols. For IJB-B, performance increase from using CFSM (**ArcFace+Ours**) is 3.69% for TAR@FAR $= 1e-5$, and 2.10% for TAR@FAR $= 1e-6$ on IJB-C. Since both IJB-B and IJB-C are a mixture of high quality images and low quality videos, the performance gains with the aug-mented data indicate that our model can generalize to both high and low quality scenarios. In Table 3, we show the comparisons on IJB-S and TinyFace. With our CFSM (**ArcFace+Ours**), ArcFace model outperforms all the baselines in both face identification and verification tasks, and achieves a new SoTA performance.

4.2 Ablation and Analysis

In this experiment, we compare the face verification and identification perfor-mance on the *most challenging* IJB-S and TinyFace datasets.

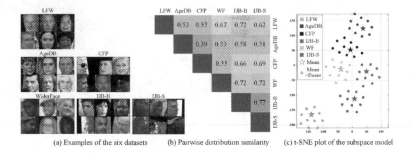

(a) Examples of the six datasets   (b) Pairwise distribution similarity   (c) t-SNE plot of the subspace model

Fig. 5. (a) Few examples from each dataset. Note differences in style. For example, AgeDB contains old grayscale photos, WiderFace has mostly low resolution faces, and IJB-S includes extreme unconstrained attributes (*i.e.,* noise, motion blur or turbulence effect). **(b)** shows the pairwise distribution similarity scores among datasets that are calculated using the learned subspace via Eq. 10. Note both IJB-B and WiderFace have high similarity scores with IJB-S. **(c)** The t-SNE plot of the learned $[\mathbf{u}_1, ..., \mathbf{u}_{10}]$ and mean style μ. The dots represent $\mathbf{u}_i + \mu$ and the stars denote μ.

Large-Scale Training Data vs. Augmentation. To further validate the applicability of our CFSM as an augmentation in different training settings, we adopt IResNet-100 [12] as the FR model backbone and utilize SoTA AdaFace loss function [43] and large-scale WebFace12M [95] dataset for training. As compared in Table 3, our model still improves unconstrained face recognition accuracies by a promising margin (Rank1: +2.43% on the IJB-S V2V protocol and +1.58% on TinyFace) on a large-scale training dataset (WebFace12M).

Effect of Guidance in CFSM. To validate the effectiveness of the proposed *controllable* and *guided* face synthesis model in face recognition, we train a FR model with CFSM as augmentation but with random style coefficients (**Ours***), and compare with guided CFSM (**Ours**). Table 3 shows that the synthesis with random coefficients does not bring significant benefit to unconstrained IJB-S dataset performance. However, when samples are generated with guided CFSM, the trained FR model performs much better. Figure 3 shows the effect of guidance in the training images. For low quality images, too much degradation leads to images with altered identity. Figure 3 shows that guided CFSM avoids synthesizing bad quality images that are unidentifiable.

Effect of the Number of Labeled Training Data. To validate the effect of the number of labeled training data, we train a series of models by adjusting the number of labeled training samples from 0.5M to 3.9M and report results both on **ArcFace** and **ArcFace+Ours** settings. Figure 4 shows the performance on various IJB-S protocols. For the full data usage setting, our model trained with guided CFSM as an augmentation outperforms the baseline by a large margin. Also note that the proposed method trained with 1/8th (0.5M) labeled data still achieves comparable performance, or even better than the baseline with 3.9M labeled data on **V2V** protocol. This is due to CFSM generating target data-specific augmentations, thus demonstrating the value of our controllable and guided face synthesis, which can significantly boost unconstrained FR performance.

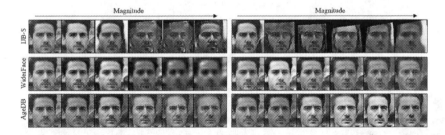

Fig. 6. Interpretable magnitude of the style coefficients. Given an input image, we randomly sample two sets of style coefficients o_1 (left) and o_2 (right) for all 3 models (respectively trained with the IJB-S, WiderFace and AgeDB datasets as the target data). We dynamically adjust the magnitude of these two coefficients by $0.5a$, a, $1.5a$, $2a$, $3a$, $4a$, where $a = \frac{o}{||o||}$. As can be seen, our model indeed realizes the goal of changing the degree of style synthesis with the coefficient magnitude.

4.3   Analysis and Visualizations of the Face Synthesis Model

In this experiment, we quantitatively evaluate the distributional similarity between datasets based on the learned linear subspace model for face synthesis. To this end, we choose 6 face datasets that are publicly available and popular for face recognition testing. These datasets are LFW [30], AgeDB-30 [54], CFP-FP [66], IJB-B [87], WiderFace (WF) [89] and IJB-S [34]. Figure 5(a) shows examples from these 6 datasets. Each dataset has its own style. For example, CFP-FP includes profile faces, WiderFace has mostly low resolution faces, and IJB-S contains extreme unconstrained attributes. During training, for each dataset, we randomly select 12K images as our target data to train the synthesis model. For the source data, we use the same subset of MS-Celeb-1M as in Sect. 4.1.

Distribution Similarity. Based on the learned dataset-specific linear subspace model, we calculate the pairwise distribution similarity score via Eq. 10. As shown in Fig. 5(b), the score reflects the style correlation between datasets. For instance, strong correlations among IJB-B, IJB-S and WiderFace (WF) are observed. We further visualize the learned basis vectors $[\mathbf{u}_1, ..., \mathbf{u}_q]$ and the mean style μ in Fig. 5(c). The basis vectors are well clustered and the discriminative grouping indicates the correlation between dataset-specific models.

Visualizations of Style Latent Spaces. Figure 6 shows face images generated by the learned CFSM. It can be seen that when the magnitude increases, the corresponding synthesized faces reveal more dataset-specific style variations. This implies the magnitude of the style code is a good indicator of image quality.

We also visualize the learned $\mathbf{U}$ of 6 models in Fig. 7. As we move along a basis vector of the learned subspace, the synthesized images change their style in dataset-specific ways. For instance, with target as WiderFace or IJB-S, synthesized images show various low quality styles such as blurring or turbulence effect. CFP dataset contains cropped images, and the "crop" style manifests in certain directions. Also, we can observe the learned $\mathbf{U}$ are different among datasets, which further verifies that our learned linear subspace model in CFSM is able to capture the variations in target datasets.

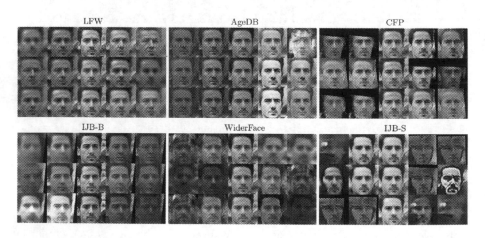

Fig. 7. Given a single input image, we visualize the synthesized images by traversing along with the learned orthonormal basis $\mathbf{U}$ in the 6 dataset-specific models. For each dataset, Rows 1−3 illustrate the first 3 basis vectors traversed. Columns 1−5 show the directions which are scaled to emphasize their effect, *i.e.*, only one element of style coefficient $\mathbf{o}$ varies from -3σ to 3σ while other $q-1$ elements remain 0.

5   Conclusions

We answer the fundamental question of *"How can image synthesis benefit the end goal of improving the recognition task?"* Our controllable face synthesis model (CFSM) with adversarial feedback of FR model shows the merit of task-oriented image manipulation, evidenced by significant performance increases in unconstrained face datasets (IJB-B, IJB-C, IJB-S and TinyFace). In a broader context, it shows that adversarial manipulation could go beyond being an attacker, and serve to increase recognition accuracies in vision tasks. Meanwhile, we define a dataset similarity metric based on the learned style bases, which capture the style differences in a label or predictor agnostic way. We believe that our research has presented the power of a controllable and guided face synthesis model for unconstrained FR and provides an understanding of dataset differences.

Acknowledgments. This research is based upon work supported in part by the Office of the Director of National Intelligence (ODNI), Intelligence Advanced Research Projects Activity (IARPA), via 2022-21102100004. The views and conclusions contained herein are those of the authors and should not be interpreted as necessarily representing the official policies, either expressed or implied, of ODNI, IARPA, or the U.S. Government. The U.S. Government is authorized to reproduce and distribute reprints for governmental purposes notwithstanding any copyright annotation therein.

References

1. Alvarez-Melis, D., Fusi, N.: Geometric dataset distances via optimal transport. In: Advances in Neural Information Processing Systems 33, pp. 21428–21439 (2020)
2. Bai, Z., Cui, Z., Liu, X., Tan, P.: Riggable 3D face reconstruction via in-network optimization. In: CVPR (2021)
3. Bansal, A., Castillo, C., Ranjan, R., Chellappa, R.: The do's and don'ts for CNN-based face verification. In: ICCVW (2017)
4. Ben-David, S., Blitzer, J., Crammer, K., Pereira, F.: Analysis of representations for domain adaptation. In: NeurIPS (2006)
5. Blanz, V., Vetter, T.: A morphable model for the synthesis of 3D faces. In: SIGGRAPH (1999)
6. Bulat, A., Yang, J., Tzimiropoulos, G.: To learn image super-resolution, use a GAN to learn how to do image degradation first. In: Ferrari, V., Hebert, M., Sminchisescu, C., Weiss, Y. (eds.) ECCV 2018. LNCS, vol. 11210, pp. 187–202. Springer, Cham (2018). https://doi.org/10.1007/978-3-030-01231-1_12
7. Cao, Q., Shen, L., Xie, W., Parkhi, O.M., Zisserman, A.: VGGface2: a dataset for recognising faces across pose and age. In: FG (2018)
8. Chang, J., Lan, Z., Cheng, C., Wei, Y.: Data uncertainty learning in face recognition. In: CVPR (2020)
9. Cheng, Z., Zhu, X., Gong, S.: Low-resolution face recognition. In: Jawahar, C.V., Li, H., Mori, G., Schindler, K. (eds.) ACCV 2018. LNCS, vol. 11363, pp. 605–621. Springer, Cham (2019). https://doi.org/10.1007/978-3-030-20893-6_38
10. Choi, Y., Choi, M., Kim, M., Ha, J.W., Kim, S., Choo, J.: StarGAN: unified generative adversarial networks for multi-domain image-to-image translation. In: CVPR (2018)

11. Deng, J., Cheng, S., Xue, N., Zhou, Y., Zafeiriou, S.: UV-GAN: adversarial facial UV map completion for pose-invariant face recognition. In: CVPR (2018)
12. Deng, J., Guo, J., Xue, N., Zafeiriou, S.: ArcFace: additive angular margin loss for deep face recognition. In: CVPR (2019)
13. Deng, Y., Yang, J., Chen, D., Wen, F., Tong, X.: Disentangled and controllable face image generation via 3D imitative-contrastive learning. In: CVPR (2020)
14. Dumoulin, V., Shlens, J., Kudlur, M.: A learned representation for artistic style. In: ICLR (2017)
15. Ganin, Y., Lempitsky, V.: Unsupervised domain adaptation by backpropagation. In: ICML (2015)
16. Gecer, B., Bhattarai, B., Kittler, J., Kim, T.-K.: Semi-supervised adversarial learning to generate photorealistic face images of new identities from 3D morphable model. In: Ferrari, V., Hebert, M., Sminchisescu, C., Weiss, Y. (eds.) ECCV 2018. LNCS, vol. 11215, pp. 230–248. Springer, Cham (2018). https://doi.org/10.1007/978-3-030-01252-6_14
17. Geng, Z., Cao, C., Tulyakov, S.: 3D guided fine-grained face manipulation. In: CVPR (2019)
18. Ghifary, M., Balduzzi, D., Kleijn, W.B., Zhang, M.: Scatter component analysis: a unified framework for domain adaptation and domain generalization. PAMI **39**, 1414–1430 (2016)
19. Ghifary, M., Kleijn, W.B., Zhang, M., Balduzzi, D.: Domain generalization for object recognition with multi-task autoencoders. In: ICCV (2015)
20. Gong, S., Shi, Y., Kalka, N.D., Jain, A.K.: Video face recognition: component-wise feature aggregation network. In: ICB (2019)
21. Gong, S., Shi, Y., Jain, A.: Low quality video face recognition: multi-mode aggregation recurrent network. In: ICCVW (2019)
22. Goodfellow, I., et al.: Generative adversarial nets. In: NeurIPS (2014)
23. Goodfellow, I.J., Shlens, J., Szegedy, C.: Explaining and harnessing adversarial examples. In: ICLR (2015)
24. Guo, Y., Zhang, L., Hu, Y., He, X., Gao, J.: MS-Celeb-1M: a dataset and benchmark for large-scale face recognition. In: Leibe, B., Matas, J., Sebe, N., Welling, M. (eds.) ECCV 2016. LNCS, vol. 9907, pp. 87–102. Springer, Cham (2016). https://doi.org/10.1007/978-3-319-46487-9_6
25. He, K., Zhang, X., Ren, S., Sun, J.: Deep residual learning for image recognition. In: CVPR (2016)
26. He, Z., Kan, M., Shan, S.: EigenGAN: layer-wise eigen-learning for GANs. In: ICCV (2021)
27. Hou, A., Sarkis, M., Bi, N., Tong, Y., Liu, X.: Face relighting with geometrically consistent shadows. In: CVPR (2022)
28. Hu, Q., Szabó, A., Portenier, T., Favaro, P., Zwicker, M.: Disentangling factors of variation by mixing them. In: CVPR (2018)
29. Huang, G., Liu, Z., Van Der Maaten, L., Weinberger, K.Q.: Densely connected convolutional networks. In: CVPR (2017)
30. Huang, G.B., Ramesh, M., Berg, T., Learned-Miller, E.: Labeled faces in the wild: a database for studying face recognition in unconstrained environments. Technical report 07-49, University of Massachusetts, Amherst (2007)
31. Huang, X., Belongie, S.: Arbitrary style transfer in real-time with adaptive instance normalization. In: ICCV (2017)

32. Huang, X., Liu, M.-Y., Belongie, S., Kautz, J.: Multimodal unsupervised image-to-image translation. In: Ferrari, V., Hebert, M., Sminchisescu, C., Weiss, Y. (eds.) ECCV 2018. LNCS, vol. 11207, pp. 179–196. Springer, Cham (2018). https://doi.org/10.1007/978-3-030-01219-9_11

33. Huang, Y., et al.: CurricularFace: adaptive curriculum learning loss for deep face recognition. In: CVPR (2020)

34. Kalka, N.D., et al.: IJB-S: IARPA Janus surveillance video benchmark. In: BTAS (2018)

35. Kang, B.N., Kim, Y., Jun, B., Kim, D.: Attentional feature-pair relation networks for accurate face recognition. In: ICCV (2019)

36. Kang, G., Jiang, L., Yang, Y., Hauptmann, A.G.: Contrastive adaptation network for unsupervised domain adaptation. In: CVPR (2019)

37. Karras, T., Aila, T., Laine, S., Lehtinen, J.: Progressive growing of GANs for improved quality, stability, and variation. In: ICLR (2018)

38. Karras, T., Aittala, M., Hellsten, J., Laine, S., Lehtinen, J., Aila, T.: Training generative adversarial networks with limited data. In: NeurIPS (2020)

39. Karras, T., et al.: Alias-free generative adversarial networks. In: NeurIPS (2021)

40. Karras, T., Laine, S., Aila, T.: A style-based generator architecture for generative adversarial networks. In: CVPR (2019)

41. Karras, T., Laine, S., Aittala, M., Hellsten, J., Lehtinen, J., Aila, T.: Analyzing and improving the image quality of StyleGAN. In: CVPR (2020)

42. Kim, H., et al.: Deep video portraits. TOG **37**, 1–14 (2018)

43. Kim, M., Jain, A.K., Liu, X.: AdaFace: quality adaptive margin for face recognition. In: CVPR (2022)

44. Kortylewski, A., Egger, B., Schneider, A., Gerig, T., Morel-Forster, A., Vetter, T.: Analyzing and reducing the damage of dataset bias to face recognition with synthetic data. In: CVPRW (2019)

45. Lee, H.-Y., Tseng, H.-Y., Huang, J.-B., Singh, M., Yang, M.-H.: Diverse image-to-image translation via disentangled representations. In: Ferrari, V., Hebert, M., Sminchisescu, C., Weiss, Y. (eds.) ECCV 2018. LNCS, vol. 11205, pp. 36–52. Springer, Cham (2018). https://doi.org/10.1007/978-3-030-01246-5_3

46. Li, H., Pan, S.J., Wang, S., Kot, A.C.: Domain generalization with adversarial feature learning. In: CVPR (2018)

47. Lin, J., Xia, Y., Qin, T., Chen, Z., Liu, T.Y.: Conditional image-to-image translation. In: CVPR (2018)

48. Liu, H., Zhu, X., Lei, Z., Li, S.Z.: AdaptiveFace: adaptive margin and sampling for face recognition. In: CVPR (2019)

49. Liu, W., Wen, Y., Yu, Z., Li, M., Raj, B., Song, L.: SphereFace: deep hypersphere embedding for face recognition. In: CVPR (2017)

50. Mansour, Y., Mohri, M., Rostamizadeh, A.: Domain adaptation: learning bounds and algorithms. In: COLT (2009)

51. Maze, B., et al.: IARPA Janus Benchmark-C: face dataset and protocol. In: ICB (2018)

52. Medin, S.C., et al.: MOST-GAN: 3D morphable StyleGAN for disentangled face image manipulation. In: AAAI (2022)

53. Meng, Q., Zhao, S., Huang, Z., Zhou, F.: MagFace: a universal representation for face recognition and quality assessment. In: CVPR (2021)

54. Moschoglou, S., Papaioannou, A., Sagonas, C., Deng, J., Kotsia, I., Zafeiriou, S.: AgeDB: the first manually collected, in-the-wild age database. In: CVPRW (2017)

55. Motiian, S., Piccirilli, M., Adjeroh, D.A., Doretto, G.: Unified deep supervised domain adaptation and generalization. In: ICCV (2017)

56. Muandet, K., Balduzzi, D., Schölkopf, B.: Domain generalization via invariant feature representation. In: ICML (2013)
57. Murez, Z., Kolouri, S., Kriegman, D., Ramamoorthi, R., Kim, K.: Image to image translation for domain adaptation. In: CVPR (2018)
58. Nam, H., Lee, H., Park, J., Yoon, W., Yoo, D.: Reducing domain gap by reducing style bias. In: CVPR (2021)
59. Nguyen-Phuoc, T., Li, C., Theis, L., Richardt, C., Yang, Y.L.: HoloGAN: unsupervised learning of 3D representations from natural images. In: ICCV (2019)
60. Piao, J., Qian, C., Li, H.: Semi-supervised monocular 3D face reconstruction with end-to-end shape-preserved domain transfer. In: ICCV (2019)
61. Pumarola, A., Agudo, A., Martinez, A.M., Sanfeliu, A., Moreno-Noguer, F.: GANimation: anatomically-aware facial animation from a single image. In: Ferrari, V., Hebert, M., Sminchisescu, C., Weiss, Y. (eds.) ECCV 2018. LNCS, vol. 11214, pp. 835–851. Springer, Cham (2018). https://doi.org/10.1007/978-3-030-01249-6_50
62. Qiu, H., Yu, B., Gong, D., Li, Z., Liu, W., Tao, D.: SynFace: face recognition with synthetic data. In: ICCV (2021)
63. Rossler, A., Cozzolino, D., Verdoliva, L., Riess, C., Thies, J., Nießner, M.: FaceForensics++: learning to detect manipulated facial images. In: ICCV (2019)
64. Ruiz, N., Theobald, B.J., Ranjan, A., Abdelaziz, A.H., Apostoloff, N.: MorphGAN: one-shot face synthesis GAN for detecting recognition bias. arXiv preprint arXiv:2012.05225 (2020)
65. Saito, K., Watanabe, K., Ushiku, Y., Harada, T.: Maximum classifier discrepancy for unsupervised domain adaptation. In: CVPR (2018)
66. Sengupta, S., Chen, J.C., Castillo, C., Patel, V.M., Chellappa, R., Jacobs, D.W.: Frontal to profile face verification in the wild. In: WACV (2016)
67. Shen, Y., Zhou, B., Luo, P., Tang, X.: FaceFeat-GAN: a two-stage approach for identity-preserving face synthesis. arXiv preprint arXiv:1812.01288 (2018)
68. Shi, Y., Jain, A.K.: Probabilistic face embeddings. In: ICCV (2019)
69. Shi, Y., Jain, A.K.: Boosting unconstrained face recognition with auxiliary unlabeled data. In: CVPRW (2021)
70. Shi, Y., Yu, X., Sohn, K., Chandraker, M., Jain, A.K.: Towards universal representation learning for deep face recognition. In: CVPR (2020)
71. Simonyan, K., Zisserman, A.: Very deep convolutional networks for large-scale image recognition. arXiv preprint arXiv:1409.1556 (2014)
72. Sun, T., et al.: Single image portrait relighting. TOG **39**, 1–13 (2019)
73. Sun, Y., et al.: Circle loss: a unified perspective of pair similarity optimization. In: CVPR (2020)
74. Tan, M., Le, Q.: EfficientNet: rethinking model scaling for convolutional neural networks. In: ICML (2019)
75. Tewari, A., et al.: StyleRig: Rigging styleGAN for 3D control over portrait images. In: CVPR (2020)
76. Thies, J., Zollhofer, M., Stamminger, M., Theobalt, C., Nießner, M.: Face2Face: real-time face capture and reenactment of RGB videos. In: CVPR (2016)
77. Tran, L., Liu, X.: Nonlinear 3D face morphable model. In: CVPR (2018)
78. Tran, L., Yin, X., Liu, X.: Disentangled representation learning GAN for pose-invariant face recognition. In: CVPR (2017)
79. Trigueros, D.S., Meng, L., Hartnett, M.: Generating photo-realistic training data to improve face recognition accuracy. Neural Netw. **134**, 86–94 (2021)
80. Tzeng, E., Hoffman, J., Saenko, K., Darrell, T.: Adversarial discriminative domain adaptation. In: CVPR (2017)

81. Wang, F., Cheng, J., Liu, W., Liu, H.: Additive margin Softmax for face verification. SPL **25**, 926–930 (2018)
82. Wang, F., Xiang, X., Cheng, J., Yuille, A.L.: NormFace: L2 hypersphere embedding for face verification. In: ACMMM (2017)
83. Wang, H., et al.: CosFace: large margin cosine loss for deep face recognition. In: CVPR (2018)
84. Wang, X., Zhang, S., Wang, S., Fu, T., Shi, H., Mei, T.: Mis-classified vector guided Softmax loss for face recognition. In: AAAI (2020)
85. Wang, X., Li, Y., Zhang, H., Shan, Y.: Towards real-world blind face restoration with generative facial prior. In: CVPR (2021)
86. Wen, Y., Zhang, K., Li, Z., Qiao, Yu.: A discriminative feature learning approach for deep face recognition. In: Leibe, B., Matas, J., Sebe, N., Welling, M. (eds.) ECCV 2016. LNCS, vol. 9911, pp. 499–515. Springer, Cham (2016). https://doi.org/10.1007/978-3-319-46478-7_31
87. Whitelam, C., et al.: IARPA Janus Benchmark-B face dataset. In: CVPRW (2017)
88. Xiao, T., Hong, J., Ma, J.: ELEGANT: exchanging latent encodings with GAN for transferring multiple face attributes. In: Ferrari, V., Hebert, M., Sminchisescu, C., Weiss, Y. (eds.) ECCV 2018. LNCS, vol. 11214, pp. 172–187. Springer, Cham (2018). https://doi.org/10.1007/978-3-030-01249-6_11
89. Yang, S., Luo, P., Loy, C.C., Tang, X.: WIDER face: a face detection benchmark. In: CVPR (2016)
90. Yang, T., Ren, P., Xie, X., Zhang, L.: GAN prior embedded network for blind face restoration in the wild. In: CVPR (2021)
91. Yin, X., Yu, X., Sohn, K., Liu, X., Chandraker, M.: Towards large-pose face frontalization in the wild. In: ICCV (2017)
92. Zhao, J., Xiong, L., Li, J., Xing, J., Yan, S., Feng, J.: 3D-aided dual-agent GANs for unconstrained face recognition. TPAMI **41**, 2380–2394 (2018)
93. Zheng, Y., Pal, D.K., Savvides, M.: Ring loss: convex feature normalization for face recognition. In: CVPR (2018)
94. Zhu, X., Lei, Z., Liu, X., Shi, H., Li, S.Z.: Face alignment across large poses: a 3D solution. In: CVPR (2016)
95. Zhu, Z., et al.: WebFace260M: a benchmark unveiling the power of million-scale deep face recognition. In: CVPR (2021)

Towards Robust Face Recognition
with Comprehensive Search

Manyuan Zhang[1,2], Guanglu Song[2], Yu Liu[2(✉)], and Hongsheng Li[1]

[1] Multimedia Laboratory, The Chinese University of Hong Kong, Shatin, Hong Kong
zhangmanyuan@link.cuhk.edu.hk, hsli@ee.cuhk.edu.hk
[2] SenseTime Research, Shatin, Hong Kong
songguanglu@sensetime.com, liuyuisanai@gmail.com

Abstract. Data cleaning, architecture, and loss function design are important factors contributing to high-performance face recognition. Previously, the research community tries to improve the performance of each single aspect but failed to present a unified solution on the joint search of the optimal designs for all three aspects. In this paper, we for the first time identify that these aspects are tightly coupled to each other. Optimizing the design of each aspect actually greatly limits the performance and biases the algorithmic design. Specifically, we find that the optimal model architecture or loss function is closely coupled with the data cleaning. To eliminate the bias of single-aspect research and provide an overall understanding of the face recognition model design, we first carefully design the search space for each aspect, then a comprehensive search method is introduced to jointly search optimal data cleaning, architecture, and loss function design. In our framework, we make the proposed comprehensive search as flexible as possible, by using an innovative reinforcement learning based approach. Extensive experiments on million-level face recognition benchmarks demonstrate the effectiveness of our newly-designed search space for each aspect and the comprehensive search. We outperform expert algorithms developed for each single research track by large margins. More importantly, we analyze the difference between our searched optimal design and the independent design of the single factors. We point out that strong models tend to optimize with more difficult training datasets and loss functions. Our empirical study can provide guidance in future research towards more robust face recognition systems.

Keywords: Face recognition · Comprehensive search

1 Introduction

Large-scale face recognition is a fundamental problem and of great practical value in computer vision. It is challenging to learn robust feature representations from

S. Avidan et al. (Eds.): ECCV 2022, LNCS 13672, pp. 720–736, 2022.
https://doi.org/10.1007/978-3-031-19775-8_42

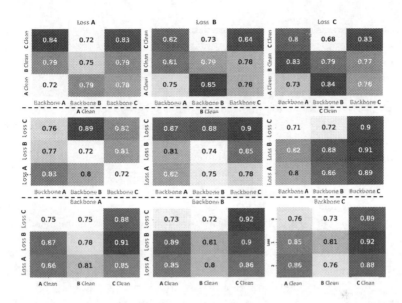

Fig. 1. Given a fixed component (referred to the title of each table), we explore the optimal combinations between the other components (horizontal and vertical axes of each table). The optimal design of each component actually changes following the modifications of the other two, which suggests that it is sub-optimal to consider them separately. A, B, C represent different designs for each component. And the value is the validation accuracy.

million-level datasets. Recently, the vision community has rapidly improved the performance of face recognition. To a large extent, these advances have been driven by three aspects: large-scale noisy data cleaning strategies [1,2,9,13,34], margin-based loss function formulations [6,17,28,31], and proper architecture designs [4,38].

Intuitively, large-scale datasets are important for face representation learning. Along with the development of various large-scale datasets [1,2,6,9,26,42], extensive works [5,26,30] tried to conduct automatic or semi-automatic data cleaning to alleviate the influence of noisy data. In addition, learning robust feature representation with margin-based loss functions [6,17,20,22,28,37] can effectively enhance the representations. The advanced architecture design also plays an import role for robust face recognition, including many common backbones [10,23,36,38] and some specially designed backbones [3,4].

However, there lacks a unified understanding of optimal designs of all the three aspects. Previous methods all delved into one aspect while ignoring the other two, therefore neglecting the coupling between these aspects. For instance, many loss functions [12,20,41] rely on hard sample mining to boost face recognition performance. However, this approach would be quite sensitive to noisy data. More specifically, many data cleaning methods [6,42] leverage the sample-to-class similarity scores to filter out outliers and merge similar classes. They pre-define a confidence threshold and samples with scores lower than the threshold are treated as the noise. Obviously, different thresholds would lead to different loss function designs.

To further evaluate this argument, we conduct extensive experiments to explore the coupling among these individual research aspects. We randomly select some combinations of different cleaning strategies, loss function designs, and backbones. As shown in Fig. 1, different combinations lead to significantly different performances. The optimal design of each component actually changes following the modifications of the other two. Separate studies greatly limit the performance and would bias each aspect's algorithmic design.

To eliminate the bias of single-aspect research and provide an overall understanding of the face recognition models, we first propose new design space for each aspect, then deliberately design a comprehensive search method to jointly search data cleaning, architectures, and loss function design. Undoubtedly, the proper design of search space for each aspect will greatly contribute to the comprehensive search result. For the data cleaning, we introduce the innovative sample discriminability-guided data cleaning search strategy to deal with the inter- and inter-class noise. We consider the relationship between a sample with its class centroid and the hardest-negative class to determine whether the sample is a noisy sample or not. It is worth mentioning that the search-based data cleaning approach can effectively alleviate the problem of tangled judgment on whether the edge samples are hard positive or noisy samples. We clean the data simply based on whether the searched cleanup result helps the training model form a more robust representation and whether it can cooperate well with the design of the other two aspects. For the loss function design formulation, we follow the successful margin-based loss function [6,27,28], but emphasize the importance of the scales of positive and negative samples, namely the scale-aware margin-based loss function design. As for the backbone design, we optimize the width and depth expansion ratio for the existing base, which have been shown to be the most important two factors in network design [25].

After carefully designing the search space for each aspect, we conduct the comprehensive search to jointly optimize them and explore their collaboration patterns. Obviously, it is hard to simply integrate data cleaning, architectures, and loss function design together. The main barrier is how to effectively design and explore such a complex and huge search space. To solve challenge, given the searchable formulation, we view the comprehensive search as a sequence prediction and generate their hyper-parameters as a sequence of tokens following the order of data cleaning → loss function design → architecture design. Every prediction is produced by a softmax classifier and then fed into the next time step as input to influence the generation of the next hyper-parameter. We adopt a Recurrent Neural Network (RNN) as the controller to explore the search space. Once the controller RNN finishes generating the data clean strategy, an architecture, and a loss function, a joint solution based on these components is built and trained. The parameters of RNN are then optimized to maximize the expected validation accuracy of the searched components. The proposed comprehensive search effectively explores the learning of the coupling between different components and enables a unified system.

Although some previous works utilize search-based methods for auto loss function [15,29] design, to our best knowledge, we are the first to jointly explore all three research fields and try to understand their entanglement. And our proposed scale-aware margin-based loss function can also surpass them substantially for the comparison in loss function design alone. Extensive experiments show the superiority of our newly-proposed search space for each aspect, then the comprehensive search will further outperform former expert algorithms developed for each single research track by large margins. In summary, the main contributions of this paper can be summarized as follows:

- We identify an ignored problem of face recognition that the data cleaning strategy, loss function design, and backbone architecture are coupled to each other. Separate studies on each factor greatly limit their performance.
- We carefully design search space for each aspect and propose the comprehensive search method to jointly explore data cleaning strategy, loss function formulation, and backbone architecture design. We introduce an innovative reinforcement learning based search framework to carry out the research and learns a unified system to achieve robust face recognition.
- We conduct extensive experiments on million-level face recognition tasks and evaluation on various benchmarks, including LFW, SLLFW, CALFW, CPLFW, IJB-B, and IJB-C. It demonstrates the superiority of our search space design for each aspect and the huge performance gained from the integration together by the comprehensive search.

2   Related Work

2.1   Deep Face Recognition

The proposal of large-scale noise-control datasets, strong backbone architectures, and well-designed loss functions have all greatly advanced the face recognition community. From the initial CASIA-Webface [34] to the recent large-scale datasets Glint360K [1] and Webface260M [42], these datasets have been proposed to greatly improve the accuracy of face recognition models. However, these datasets rely on internet search engines and therefore contain a large percentage of noise. How to deal with these noises and learn face recognition with noisy labels has drawn much attention [5,16,33,41,42]. Zhu [42] introduces the cleaning automatically by self-training (CAST) pipeline, which introduces a self-training process to auto clean inter-class and intra-class noise by an iteration manner and achieves good performance. For the model design, some architectures have been proposed to achieve efficient face recognition, such as PolyNet [38] and MobileFaceNet [4]. For sophisticated face recognition loss function design, efficient margin-based softmax approaches [6,17,18,27,28] were proposed by modifying the softmax loss function to achieve tighter intra-class compactness and more sparse inter-class separation, which achieve state-of-the-art performance.

Since separate design on each component has achieved good performance, all the above methods ignore the relationship among data cleaning strategy, loss function formulation, and training backbone. Disentangled studies greatly limit performance.

2.2   Auto-ML for Face Recognition

Auto-ML methods aim to automatically design suitable machine learning systems. Reinforcement learning guided search [24,43] automate the design process and can easily outperform manually designed components in a given search space. Some former works use the searching methods to optimize the training loss function for face recognition. Li *et al.*[15] proposes the auto loss function search method from a hyper-parameter optimization perspective. Wang *et al.* [29] develop a unified formulation for the prevalent margin-based softmax loss. Then the random and reward-guided methods are designed to search for the best candidate. The above methods only focus on the loss function search. In this work, we consider all three aspects for face recognition. We first re-design the search space for each aspect, then jointly consider them and achieve excellent performance.

3   Method

To eliminate the bias of single-aspect research and provide an overall understanding of the face recognition model design, we propose the comprehensive search method to jointly search optimal data cleaning strategy, loss function formulation, and backbone architecture design for robust face recognition. In this section, we will introduce the details of our comprehensive search. The key components of comprehensive search include three aspects, *i.e.* search space, search objective, and search algorithms. In the following, we will first introduce our newly-designed search space for each aspect, then couple them together to form the comprehensive search space. The search objective to evaluate the searched solution and the innovative reinforcement learning-based reward-guided comprehensive approach to identify the optimal combination design are introduced consequently.

3.1   Comprehensive Search Space Design

Discriminabiliy-Guided Data Cleaning Search Strategy. There are two kinds of data noise for face recognition datasets, the 'outliers' and 'label flips', corresponding to intra-class and inter-class noise respectively. An 'outlier' noise means an image does not belong to any class of the datasets. A 'label flip' noise refers to an image that is wrongly labeled with the incorrect class label. To clean the noise, we conduct inter-class filtering and intra-class merging. As for the intra-class noise cleaning, we filter the samples by their *discriminability* [35]. The *discriminability* is defined as:

$$\mathcal{D}_i = \frac{dist_{ip}}{\max\{dist_{in} \mid n \in [1, K], n \neq p\}},\tag{1}$$

where p is the positive class label for sample i and $n, n \in [1, K], n \neq p$ is the negative class label. K is the number of class. $dist$ represents the feature similarity. We use the cosine similarity as the similarity indicator here. The *discriminability* is the ratio between the feature's similarity with the centroid of its class and the similarity from the hardest-negative class. It can successfully distinguish the outliers and can be used to filter the samples whose *discriminability* is lower than a pre-defined confidence threshold τ_{intra}. As for the inter-class noise, we simply merge a pair of classes whose class center similarity is higher than the threshold τ_{inter}. The τ_{intra} and τ_{inter} are the two hyper-parameters that we need to optimize for the data cleaning strategy.

One of the most difficult problems in past manual or semi-automatic data cleaning pipelines is how to classify edge samples as hard positive or noisy samples. Many loss functions [20, 22] rely on strategies such as hard sample mining to further boost performance. However, edge samples are often hard to distinguish. By conducting the search-based data cleaning, we have eliminated this tangle. The result of data cleaning, *i.e.* whether an edge sample is filtered or not, only depends on whether the cleanup result ultimately helps the training model to form a more robust feature representation, or whether it can cooperate well with the design of the other two aspects.

Scale-Aware Margin-Based Loss Function Design. Margin-based softmax loss functions [6,17,27,28] have been proposed in recent years to enhance the feature discrimination for face recognition. In summary, they can be defined in a uniform formulation. Suppose $\boldsymbol{w}_k \in \mathbb{R}^d$ is the k-th class's weight ($k \in \{1, 2, \ldots, K\}$) and K is the number of classes. $\boldsymbol{x} \in \mathbb{R}^d$ denotes the face feature. The formulation for the margin-based loss function can be written as follows:

$$\mathcal{L} = -\log \frac{e^{s_p f(m, \theta_{w_y, x})}}{e^{s_p f(m, \theta_{w_y, x})} + \sum_{k \neq y}^{K} e^{s_n \cos(\theta_{w_k, x})}},\tag{2}$$

where $\theta_{w_k, x}$ is the angle between $\boldsymbol{w}_k$ and $\boldsymbol{x}$. Suppose $\cos(\theta_{w_k, x}) = \boldsymbol{w}_k^T \boldsymbol{x}$ is the cosine similarity of the k-th class weight and feature, $f(m, \theta_{w_y, x}) \leq \cos(\theta_{w_y, x})$ is the carefully designed margin function and can be summarized into the combined version $f(m, \theta_{w_y, x}) = \cos(m_1 \theta_{w_y, x} + m_2) - m_3$. m_1 and m_2 are the multiplicative and additive angular margin, and m_3 is the additive cosine margin. Previous work [6,27,28] has mainly focused on the margin optimization, but ignored the importance of tuning the positive sample scale s_p and the negative sample scale s_n, which control the optimization difficulty of positive and negative sample. We experimentally prove that face recognition accuracy can be further hugely improved by optimizating the s_p and s_n jointly. Overall, m_1, m_2, m_3, s_p and s_n are the hyper-parameters we need to search for the loss function.

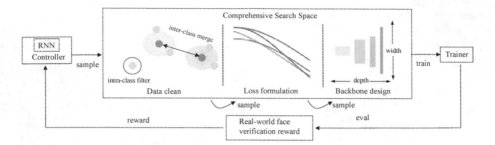

Fig. 2. The overall pipeline for our innovative reinforcement learning-based comprehensive search for robust face recognition. We design it for the sample-eval-update loop. Firstly, the RL agent samples a batch of combinations that contains the hyperparameters for data cleaning strategy, loss function formulation, and backbone architecture design. For each sampled combination, we train it on the target face recognition task to get its accuracy and reward. After that, the agent will be updated by maximizing the expected reward until converged.

Algorithm 1 . Comprehensive Search

Input: Training set $\mathcal{S}_t = \{(\boldsymbol{x}_i, y_i)\}_{i=1}^n$, validation set $\mathcal{S}_v$, total searching epochs T, agent policy π_θ and its initialized parameter θ_0.

    for $t = 1$ to T do **do**

        Sample B combinations of hyper-parameters $C_1, \ldots C_B$ for data cleaning strategy τ_{intra}, τ_{inter}, loss function formulation m_1, m_2, m_3, s_p, s_n and backbone architecture design $\mathcal{W}$ and $\mathcal{D}$ via RNN sequencely.

        Train the combinations $C_1, \ldots C_B$ for one epoch separately on the training set $\mathcal{S}_t = \{(\boldsymbol{x}_i, y_i)\}_{i=1}^n$ with the sampled hyper-parameters.

        Evaluate the trained combinations $C_1, \ldots C_B$ on the validation set $\mathcal{S}_v$ to get reward $\mathrm{R}(C_1), \ldots \mathrm{R}(C_B)$ via Eq. 3.

        Update θ_t by $\max \mathbb{E}_P \mathrm{R}(C)$

        Update $\pi_{\theta_{t+1}} = \pi_{\theta_t}$

    end for

Output: Final policy π_θ

Effective Depth and Width Search for the Backbone. Following [25], the width expansion ratio $\mathcal{W}$ and the depth expansion ratio $\mathcal{D}$ are the two most important factors for neural network architecture design, so we utilize those two factors to search backbone architecture for face recognition. The base model we selected is modified MobileNet [11], which has been proved successful for face recognition and its search cost is affordable.

3.2 Comprehensive Search Objective and Algorithm

Joint search with data cleaning strategy, loss function design, and architecture is not easy to perform due to the complex and huge search space. To enable fast and flexible comprehensive search, we view the comprehensive search as a sequence

prediction and generate their hyper-parameters as a sequence of tokens following the order of data cleaning → loss function design → architecture
design. Every prediction of each step is produced by a softmax classifier and
then fed into the next time step as input to influence the generation of the next
hyper-parameter. We adopt the Recurrent Neural Network (RNN) as the controller to generate the search parameters. Once the controller has completed the
generation of the data clean strategy, the loss function, and the architecture,
a joint solution based on these components is built and trained. The parameters of RNN are optimized to maximize the expected validation accuracy of the
searched components. The overall searching process can be optimized with the
sample-eval-update loop as shown in Algorithm 1 and Fig. 2.

To solve the real-world face recognition problem, we test all candidate joint
solutions and use the performance as the search criterion. Another object that
needs to be considered is the computational cost, since we search for the depth
expansion ratio and width expansion ratio for the backbone, we expect the new
backbone to have a similar computational budget with the original backbone.
To this end, we use a weighted product method to approximate Pareto optimal
solutions. The final reward function can be formulated as:

$$R(C) = \text{ACC}(C) \times \left[\frac{\text{COST}(C)}{\text{TAR}} \right]^{\alpha} \tag{3}$$

where C is the sampled combination of designs of the three aspects, $\text{ACC}(C)$
is the accuracy on real-world face verification task of the sampled combination,
$\text{COST}(C)$ is the computation cost (FLOPs) of the combination, TAR is the
target computation cost which we set it to the original backbone, and α is the
weight factor.

Following [24], we use Proximal Policy Optimization (PPO) [21] to optimize
the RL agent to find Pareto optimal solutions for our comprehensive search
problem. For each sampled combination in the search space, we map it to a list
of tokens, which are determined by a sequence of action $a_{1:T}$ from the RL agent
with policy π_θ. The overall objective is to maximize the expected reward:

$$\mathbb{J} = \mathbb{E}_{P(a_{1:T};m)} R(C) \tag{4}$$

For each sample C, we train it on the target task to get its accuracy $\text{ACC}(C)$
and its computation cost $\text{COST}(C)$. We then calculate the reward $R(C)$ using
Eq. 3. After that, the agent with policy π_θ will be updated by maximizing the
expected reward in Eq. 4 until the parameter θ converges. Then we will re-train
the top combinations C with the highest $R(C)$ by full train and select the best
combination as our searched results.

4　Experiment

4.1　Datasets

Training Data. To conduct the comprehensive search for robust face recognition, we use the MS1MV2 [6] as our training set, which contains 5.8 million face images from 85K identities. To prove the good generation ability of our comprehensive search, we also experiment on the recently introduced large-scale dataset Glint360K [1], which contains 17 million images of 360K individuals. Note that the MS1MV2 and Glint360K are the two largest publicly available face recognition datasets.

Test Data. During the sample-eval-update loop, we evaluate our combination on the re-organised MegaFace [14] verification benchmark. After retraining, we test the final model at most popular face benchmarks including LFW [13], CALFW [40], SLLFW [8], CPLFW [39], IJB-B [32] and IJB-C [19]. We follow the unrestricted with labelled outside data protocol [13].

4.2　Implementation Details

Data Processing. All the faces in the training images are detected by Retina Face [7]. Alignment by five landmarks is conducted and the face is cropped to 112×112. Images are normalized by subtracting 127.5 and dividing by 128. For the data cleaning strategy, we use the ResNet-50 [10] model trained on MS1MV2 to extract feature to conduct intra-class and inter-class cleaning.

Searching. During the searching process, each sampled solution is trained on the training set and evaluated on the validation set. The weighted accuracy of TAR@FAR at 10^{-3}, 10^{-4}, 10^{-5} by 0.5, 0.25 and 0.25 are used as the ACC. After that, we re-train the top reward solution. We choose the top 20 solutions under the computation budget and fully train them on the target task and then report the final performance on popular face benchmarks. A recurrent neural network (RNN) is utilized as the controller to generate the parameter combination. We use Adam optimizer with learning rate of 5×10^{-4} and momentum 0.9 to update the controller. For the sampled solution's training, we only train each solution for one epoch. The search process needs around 1,000 samples to converge and costs around 37 GPU days (NVIDIA A100, FP16 training).

Retraining. After choosing the top 20 combinations, we fully retrain them. We use SGD optimizer with weight decay 5×10^{-4} and momentum 0.9. We train models on 8 NVIDIA GPUs with batch size 1024 for 100K iterations. The initial learning rate is set to 0.1 and decays by 0.1 at iterations 40K, 60K, and 80K for MS1MV2. As for Glint360K, we train for batch size 1024 with 150k iterations totally and learning rate decay at 60k, 90k, and 13k.

Table 1. Verification performance (%) of different search space combinations on the LFW, SLLFW, CALFW, CPLFW, SLLFW, IJB-B and IJB-C benchmarks. The last row represents the baseline performance, which combines the best previous hand-crafted design for each track. The training set is MS1MV2.

Search space			LFW	SLLFW	CALFW	CPLFW	IJB-B		IJB-C	
Data	Loss	Backbone					10^{-5}	10^{-4}	10^{-5}	10^{-4}
✓	✓	✓	99.55	**98.78**	**94.75**	**84.80**	**85.66**	**91.46**	**90.45**	**93.58**
×	✓	✓	**99.58**	98.43	94.15	84.33	83.67	91.20	89.55	93.24
✓	×	×	99.43	98.15	94.00	82.48	82.35	90.72	88.47	92.84
×	✓	×	99.52	98.27	94.03	84.30	83.26	90.80	88.88	92.91
×	×	✓	99.35	98.22	94.23	82.78	80.62	90.40	87.92	92.70
×	×	×	99.43	98.12	94.03	82.30	79.71	90.44	87.29	92.61

4.3 Results

We test our comprehensive search method on widely-used face verification benchmarks including LFW, SLLFW, CALFW, CPLFW, IJB-B, and IJB-C. The results are shown in Table 1. The bold numbers in each column represent the best results. In Table 1, the last line represents the baseline results without any search-based design compo-

Table 2. The best combinations searched with our comprehensive search on MS1MV2 dataset.

Best combination								
τ_{intra}	τ_{inter}	m_1	m_2	m_3	s_p	s_n	$\mathcal{D}$	$\mathcal{W}$
0.3	0.62	1.15	0.22	0	40	48	1.47	0.84
0.22	0.76	1.0	0.32	0	40	40	1.22	0.91
0.26	0.62	1.2	0.36	0	32	32	1.47	0.84

nent. For the baseline, the training dataset is semi-automatically cleaned by [6], the loss function is finely designed margin-based loss function ArcFace [6], and the backbone architecture is modified MobileNet [11] that is specially designed for face recognition. From the results, we can see that our comprehensive search outperforms the baseline by a large margin on all test benchmarks, $i.e.$ 0.97% performance gain for IJB-C at 1e−4 and 3.16% at 1e−5. The huge improvement demonstrates the great potential of comprehensive search compared to manually design components separately. What's more, the search for each component also boosts the performance greatly, which shows the excellent and flexible design of our newly-designed search space for each component and the limitation of hand-crafted design. From the result of row 1 and rows 3–5 in Table 1, we can also observe that searching for each component separately may not achieve the best performance. If we comprehensively search for all three aspects jointly, we can further boost the performance significantly. Comparing rows 1 and 2 in Table 1, the addition of search for data cleaning strategy improves the search result dramatically, which has been ignored in previous searching methods [15,29]. We show some combinations of top verification accuracy we searched on MS1MV2 in Table 2. The results show a very different design preference of searching from the previous manual design for each component separately.

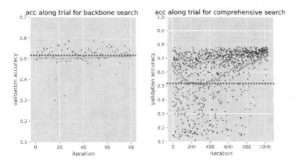

Fig. 3. Visualization of the accuracy on validation set along the search process. **Left**: The accuracy for searching backbone architecture only. **Right**: The accuracy for comprehensive search. The yellow line represents the moving average of all sampled combinations, and the red dot line indicates the baseline. (Color figure online)

4.4   Searching Process

We visualize the searching process in Fig. 3. The yellow line represents the moving average of the accuracy of the sampled solution. From the figure, we can observe that the validation accuracy has been improved gradually through the searching, which suggests that the agent has learned a robust policy that can sample high-quality hyper-parameter combinations. We also compare the searching process for comprehensive search and backbone architecture search in Fig. 3. From the figure, we can also see that by enhancing with flexible data cleaning strategy and loss function formulation searching, the validation accuracy of sampled combination has been improved greatly, which confirms the validity of our comprehensive search. In addition, all the rewards of comprehensive and backbone search outperform the baseline combination (represented by the red dot line), which shows the effectiveness of searching.

4.5   Ablation Study

Effect of *Discriminabiliy*-Guided Data Cleaning Strategy. For the data cleaning strategy, we propose the *discriminabiliy*-guided data intra-class filtering and inter-class merging according to class center similarity to determine 'outlier' and 'label flip' samples respectively. From the results of rows 3 and 6 in Table 1, we can observe that only searching for data cleaning strategy can outperform baseline whose dataset is semi-automatically cleaned hugely, and achieve performance gains of 1.18% at 10^{-5} on IJB-C and 2.64% at 10^{-5} on IJB-B benchmark. Note that the result is customized, as we have stated before, the best candidate for cleaning hyper-parameters is highly related to the loss function formulation and backbone architecture design.

Effect of Cleaning Feature Extraction Model. In the cleaning process, we rely on the feature extracted by a pre-trained model to calculate *discriminabiliy* and inter-class similarity. Intuitively, the more discriminate features would lead to better cleaning accuracy. We conduct an ablation study to study the influence. We use the ResNet-50 backbone trained on MS1MV2 and Glint360K to extract features and the results are shown in Table 3. The better features would lead to better cleaning accuracy, but the impact is quite limited.

Table 3. Results of different feature extracted model.

Train Dataset	IJB-C	
	10^{-5}	10^{-4}
MS1MV2	88.80	92.88
Glint360K	**88.84**	**92.92**

Effect of Scale-Aware Margin-Based Loss Function. For the loss function design, we formulate it as the scale-aware margin-based loss function. The results of rows 4 and 6 in Table 1 demonstrate that the search of loss function can improve the performance significantly compared to the widely used hand-crafted design loss ArcFace, *i.e.* 1.59% gain at IJB-C 10^{-5}. Furthermore, compared to the other two components data and backbone, the separate searching of loss achieves the largest improvement, which demonstrates that sophisticated loss function design is crucial for learning discriminate feature representation.

Effect of Depth and Width Search. For backbone architecture design, we only search for the width expansion ratio W and the depth expansion ratio D. We constrain the new backbone to share the same FLOPs with the base model. From Table 1 rows 5 and 6, the simple width and depth search can still boost performance, which indicates the limitation of hand-crafted backbone design.

4.6   Search on Glint360K

To show the good generalizability of our comprehensive search, we also conduct experiments on the newly introduced large-scale dataset Glint360K [1]. The results are shown in Table 4. From the table, we can see that the comprehensive search improves the performance significantly on all face benchmarks, which demonstrates good generalizability. Our solution improves 3.73% at 10^{-5}

Table 4. Verification performance (%) of different search space combinations on LFW, SLLFW, CALFW, CPLFW, SLLFW, IJB-B and IJB-C benchmarks. The last row represents the baseline performance, which combines the best previous hand-crafted design for each track. The training set is Glint360K.

Search Space			LFW	SLLFW	CALFW	CPLFW	IJB-B		IJB-C	
Data	Loss	Backbone					10^{-5}	10^{-4}	10^{-5}	10^{-4}
✓	✓	✓	**99.63**	**98.75**	**94.28**	**86.20**	**86.06**	**92.55**	**90.06**	**94.36**
×	×	✓	99.38	98.11	94.07	85.10	80.96	91.13	86.68	93.25
Y	Y	Y	99.34	98.05	93.83	84.35	79.83	90.92	86.33	93.16

Table 5. Verification performance (%) of dataset transferability experiment on IJB-B and IJB-C.

Search on	Train on	IJB-B		IJB-C	
		10^{-5}	10^{-4}	10^{-5}	10^{-4}
Glint360K	MS1MV2	**84.26**	**91.12**	**89.41**	**93.18**
MS1MV2	MS1MV2	85.66	91.46	90.45	93.58
–	MS1MV2	79.71	90.44	87.29	92.61

Table 6. Verification performance (%) of comprehensive search for ResNet50 on IJB-B and IJB-C benchmarks with MS1MV2. The last row represents the baseline performance.

Transfer Data	Loss	IJB-B		IJB-C	
		10^{-5}	10^{-4}	10^{-5}	10^{-4}
✓	✓	**90.11**	**94.55**	**94.11**	**96.04**
✗	✗	89.11	94.24	93.68	95.76

for IJB-C and 6.23% at 10^{-5} for IJB-B respectively. The main reason is that our proposed comprehensive search can effectively capture the intrinsic connections among the different aspects and find the best combinations. Moreover, the best combinations in Glint360K of the three aspects are different from the results searched on MS1MV2, which validates our observation that the three aspects are coupled to each other.

4.7   Transferability

Since the three aspects are highly related and the best combination would be changed for different training datasets, we explore the transferability of our comprehensive search in this subsection. We use the best combination searched on Glint360K to train models on MS1MV2. Results are shown in Table 5. The results show that even the transfer learning reduces performance gains, we still achieve significant performance improvements compared to the baseline. The result also suggests that it is better to search and test with the same dataset and the best candidate of one component would be changed along with the other two.

4.8   Search with Larger Backbone

In the above experiments, we have performed the search based on the modified MobileNet whose FLOPs is around 0.33G. In this section, we perform a joint search based on ResNet50 whose FLOPs is around 6G. We search for data cleaning strategy and loss function, and the results are shown in Table 6. The performance improvement verifies the generalization of our joint search.

4.9   Discussion

In order to explore the intrinsic relationship among these three aspects, we analyse the results of the search to find the best combination pattern. To facilitate the analysis, we defined the difficulty of the data and the difficulty of the loss function separately. For data cleaning, a lower τ_{intra} as well as a higher τ_{inter} implies greater training difficulty. For loss function, stricter margins and larger

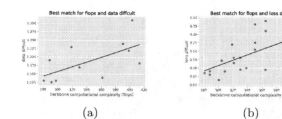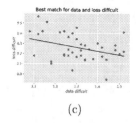

| (a) | (b) | (c) |

Fig. 4. Visualization of best matches among data difficult, loss difficult and backbone computational complexity (FLOPs).

S_n/S_p imply more difficult optimizations. They are defined as follows,

$$\text{Difficulty}_{data} = 1 - \tau_{intra} + \tau_{inter}$$
$$\text{Difficulty}_{loss} = \frac{s_n * (m_1 - 1 + m_2 + m_3)}{s_p} \tag{5}$$

We visualize the relationship between data difficulty, loss difficulty, and backbone architecture in Fig. 4. In Fig. (4a) and (4b), larger models (higher FLOPs) tend to train with greater data difficulty and loss difficulty. A larger model implies a stronger fitting ability and is, therefore, able to handle more complex optimization problems. So it can learn more from difficult samples and strict optimization objectives. For the best matching of loss function and data cleaning, as Fig. (4c) suggests, models tend to choose easier optimization loss functions under severe data. These findings provide a fresh perspective for the design of face recognition pipeline.

5   Conclusion

In this work, we explore the relationship among the data cleaning strategy, loss function formulation, and backbone architecture design for robust face recognition. Previously, people optimize them separately but fail to present a unified understanding of integrating them. We provide a fresh perspective, coupling them and optimizing jointly. We propose an innovative reinforcement learning-based comprehensive search for the best combination. Extensive experiments have proven the excellence of our method for robust face recognition.

Acknowledgement. Hongsheng Li is also a Principal Investigator of Centre for Perceptual and Interactive Intelligence Limited (CPII). This work is supported in part by CPII, in part by the General Research Fund through the Research Grants Council of Hong Kong under Grants (Nos. 14204021, 14207319), in part by CUHK Strategic Fund.

References

1. An, X., et al.: Partial FC: training 10 million identities on a single machine. In: Proceedings of the IEEE/CVF International Conference on Computer Vision, pp. 1445–1449 (2021)
2. Cao, Q., Shen, L., Xie, W., Parkhi, O.M., Zisserman, A.: VGGFace2: a dataset for recognising faces across pose and age. In: 2018 13th IEEE International Conference on Automatic Face & Gesture Recognition (FG 2018), pp. 67–74. IEEE (2018)
3. Chang, W.Y., Tsai, M.Y., Lo, S.C.: ResSaNet: a hybrid backbone of residual block and self-attention module for masked face recognition. In: Proceedings of the IEEE/CVF International Conference on Computer Vision, pp. 1468–1476 (2021)
4. Chen, S., Liu, Y., Gao, X., Han, Z.: MobileFaceNets: efficient CNNs for accurate real-time face verification on mobile devices. In: Zhou, J., Wang, Y., Sun, Z., Jia, Z., Feng, J., Shan, S., Ubul, K., Guo, Z. (eds.) CCBR 2018. LNCS, vol. 10996, pp. 428–438. Springer, Cham (2018). https://doi.org/10.1007/978-3-319-97909-0_46
5. Deng, J., Guo, J., Liu, T., Gong, M., Zafeiriou, S.: Sub-center ArcFace: boosting face recognition by large-scale noisy web faces. In: Vedaldi, A., Bischof, H., Brox, T., Frahm, J.-M. (eds.) ECCV 2020. LNCS, vol. 12356, pp. 741–757. Springer, Cham (2020). https://doi.org/10.1007/978-3-030-58621-8_43
6. Deng, J., Guo, J., Xue, N., Zafeiriou, S.: ArcFace: additive angular margin loss for deep face recognition. In: Proceedings of the IEEE/CVF Conference on Computer Vision and Pattern Recognition, pp. 4690–4699 (2019)
7. Deng, J., Guo, J., Zhou, Y., Yu, J., Kotsia, I., Zafeiriou, S.: RetinaFace: single-stage dense face localisation in the wild. arXiv preprint arXiv:1905.00641 (2019)
8. Deng, W., Hu, J., Zhang, N., Chen, B., Guo, J.: Fine-grained face verification: FGLFW database, baselines, and human-DCMN partnership. Pattern Recognit. **66**, 63–73 (2017)
9. Guo, Y., Zhang, L., Hu, Y., He, X., Gao, J.: MS-celeb-1M: a dataset and benchmark for large-scale face recognition. In: Leibe, B., Matas, J., Sebe, N., Welling, M. (eds.) ECCV 2016. LNCS, vol. 9907, pp. 87–102. Springer, Cham (2016). https://doi.org/10.1007/978-3-319-46487-9_6
10. He, K., Zhang, X., Ren, S., Sun, J.: Deep residual learning for image recognition. In: Proceedings of the IEEE Conference on Computer Vision and Pattern Recognition, pp. 770–778 (2016)
11. Howard, A.G., et al.: MobileNets: efficient convolutional neural networks for mobile vision applications. arXiv preprint arXiv:1704.04861 (2017)
12. Hu, W., Huang, Y., Zhang, F., Li, R.: Noise-tolerant paradigm for training face recognition CNNs. In: Proceedings of the IEEE/CVF Conference on Computer Vision and Pattern Recognition, pp. 11887–11896 (2019)
13. Huang, G.B., Mattar, M., Berg, T., Learned-Miller, E.: Labeled faces in the wild: a database forstudying face recognition in unconstrained environments. In: Workshop on Faces in 'Real-Life' Images: Detection, Alignment, and Recognition (2008)
14. Kemelmacher-Shlizerman, I., Seitz, S.M., Miller, D., Brossard, E.: The MegaFace benchmark: 1 million faces for recognition at scale. In: Proceedings of the IEEE Conference on Computer Vision and Pattern Recognition, pp. 4873–4882 (2016)
15. Li, C., et al.: AM-LFS: AutoML for loss function search. In: Proceedings of the IEEE/CVF International Conference on Computer Vision, pp. 8410–8419 (2019)
16. Liu, B., Song, G., Zhang, M., You, H., Liu, Y.: Switchable k-class hyperplanes for noise-robust representation learning. In: Proceedings of the IEEE/CVF International Conference on Computer Vision, pp. 3019–3028 (2021)

17. Liu, W., Wen, Y., Yu, Z., Li, M., Raj, B., Song, L.: SphereFace: deep hypersphere embedding for face recognition. In: Proceedings of the IEEE Conference on Computer Vision and Pattern Recognition, pp. 212–220 (2017)
18. Liu, Y., et al.: Towards flops-constrained face recognition. In: Proceedings of the IEEE/CVF International Conference on Computer Vision Workshops (2019)
19. Maze, B., et al.: IARPA Janus benchmark-C: face dataset and protocol. In: 2018 International Conference on Biometrics (ICB), pp. 158–165. IEEE (2018)
20. Schroff, F., Kalenichenko, D., Philbin, J.: FaceNet: a unified embedding for face recognition and clustering. In: Proceedings of the IEEE Conference on Computer Vision and Pattern Recognition, pp. 815–823 (2015)
21. Schulman, J., Wolski, F., Dhariwal, P., Radford, A., Klimov, O.: Proximal policy optimization algorithms. arXiv preprint arXiv:1707.06347 (2017)
22. Sun, Y., Wang, X., Tang, X.: Deeply learned face representations are sparse, selective, and robust. In: Proceedings of the IEEE Conference on Computer Vision and Pattern Recognition, pp. 2892–2900 (2015)
23. Szegedy, C., et al.: Going deeper with convolutions. In: Proceedings of the IEEE Conference on Computer Vision and Pattern Recognition, pp. 1–9 (2015)
24. Tan, M., et al.: MnasNet: platform-aware neural architecture search for mobile. In: Proceedings of the IEEE/CVF Conference on Computer Vision and Pattern Recognition, pp. 2820–2828 (2019)
25. Tan, M., Le, Q.: EfficientNet: rethinking model scaling for convolutional neural networks. In: International Conference on Machine Learning, pp. 6105–6114. PMLR (2019)
26. Wang, F., et al.: The devil of face recognition is in the noise. In: Ferrari, V., Hebert, M., Sminchisescu, C., Weiss, Y. (eds.) ECCV 2018. LNCS, vol. 11213, pp. 780–795. Springer, Cham (2018). https://doi.org/10.1007/978-3-030-01240-3_47
27. Wang, F., Cheng, J., Liu, W., Liu, H.: Additive margin softmax for face verification. IEEE Signal Process. Lett. **25**(7), 926–930 (2018)
28. Wang, H., et al.: CosFace: large margin cosine loss for deep face recognition. In: Proceedings of the IEEE Conference on Computer Vision and Pattern Recognition, pp. 5265–5274 (2018)
29. Wang, X., Wang, S., Chi, C., Zhang, S., Mei, T.: Loss function search for face recognition. In: International Conference on Machine Learning, pp. 10029–10038. PMLR (2020)
30. Wang, X., Wang, S., Wang, J., Shi, H., Mei, T.: Co-mining: deep face recognition with noisy labels. In: Proceedings of the IEEE/CVF International Conference on Computer Vision, pp. 9358–9367 (2019)
31. Wen, Y., Zhang, K., Li, Z., Qiao, Yu.: A discriminative feature learning approach for deep face recognition. In: Leibe, B., Matas, J., Sebe, N., Welling, M. (eds.) ECCV 2016. LNCS, vol. 9911, pp. 499–515. Springer, Cham (2016). https://doi.org/10.1007/978-3-319-46478-7_31
32. Whitelam, C., et al.: IARPA Janus benchmark-B face dataset. In: Proceedings of the IEEE Conference on Computer Vision and Pattern Recognition Workshops, pp. 90–98 (2017)
33. Wu, X., He, R., Sun, Z., Tan, T.: A light CNN for deep face representation with noisy labels. IEEE Trans. Inf. Forensics Secur. **13**(11), 2884–2896 (2018)
34. Yi, D., Lei, Z., Liao, S., Li, S.Z.: Learning face representation from scratch. arXiv preprint arXiv:1411.7923 (2014)
35. Zhang, M., Song, G., Zhou, H., Liu, Yu.: Discriminability distillation in group representation learning. In: Vedaldi, A., Bischof, H., Brox, T., Frahm, J.-M. (eds.)

ECCV 2020. LNCS, vol. 12355, pp. 1–19. Springer, Cham (2020). https://doi.org/10.1007/978-3-030-58607-2_1

36. Zhang, X., Zhou, X., Lin, M., Sun, J.: ShuffleNet: an extremely efficient convolutional neural network for mobile devices. In: Proceedings of the IEEE Conference on Computer Vision and Pattern Recognition, pp. 6848–6856 (2018)

37. Zhang, X., Zhao, R., Qiao, Y., Wang, X., Li, H.: AdaCos: adaptively scaling cosine logits for effectively learning deep face representations. In: Proceedings of the IEEE/CVF Conference on Computer Vision and Pattern Recognition, pp. 10823–10832 (2019)

38. Zhang, X., Li, Z., Change Loy, C., Lin, D.: PolyNet: a pursuit of structural diversity in very deep networks. In: Proceedings of the IEEE Conference on Computer Vision and Pattern Recognition, pp. 718–726 (2017)

39. Zheng, T., Deng, W.: Cross-pose LFW: a database for studying cross-pose face recognition in unconstrained environments. Beijing University of Posts and Telecommunications, Technical report 5, 7 (2018)

40. Zheng, T., Deng, W., Hu, J.: Cross-age LFW: a database for studying cross-age face recognition in unconstrained environments. arXiv preprint arXiv:1708.08197 (2017)

41. Zhong, Y., Deng, W., Wang, M., Hu, J., Peng, J., Tao, X., Huang, Y.: Unequal-training for deep face recognition with long-tailed noisy data. In: Proceedings of the IEEE/CVF Conference on Computer Vision and Pattern Recognition, pp. 7812–7821 (2019)

42. Zhu, Z., et al.: WebFace260M: a benchmark unveiling the power of million-scale deep face recognition. In: Proceedings of the IEEE/CVF Conference on Computer Vision and Pattern Recognition, pp. 10492–10502 (2021)

43. Zoph, B., Le, Q.V.: Neural architecture search with reinforcement learning. arXiv preprint arXiv:1611.01578 (2016)

Towards Unbiased Label Distribution Learning for Facial Pose Estimation Using Anisotropic Spherical Gaussian

Zhiwen Cao[1(✉)], Dongfang Liu[2], Qifan Wang[3], and Yingjie Chen[1]

[1] Purdue University, West Lafayette, USA
{cao270,victorchen}@purdue.edu
[2] Rochester Institute of Technology, Rochester, USA
dongfang.liu@rit.edu
[3] Meta AI, Menlo Park, USA
wqfcr@fb.com

Abstract. Facial pose estimation refers to the task of predicting face orientation from a single RGB image. It is an important research topic with a wide range of applications in computer vision. Label distribution learning (LDL) based methods have been recently proposed for facial pose estimation, which achieve promising results. However, there are two major issues in existing LDL methods. First, the expectations of label distributions are biased, leading to a *biased pose estimation*. Second, *fixed* distribution parameters are applied for all learning samples, severely limiting the model capability. In this paper, we propose an Anisotropic Spherical Gaussian (ASG)-based LDL approach for facial pose estimation. In particular, our approach adopts the spherical Gaussian distribution on a unit sphere which constantly generates *unbiased expectation*. Meanwhile, we introduce a new loss function that allows the network to learn the distribution parameter for each learning sample *flexibly*. Extensive experimental results show that our method sets new state-of-the-art records on AFLW2000 and BIWI datasets.

Keywords: Facial pose estimation · Anisotropic spherical Gaussian · Label distribution learning

1 Introduction

The task of facial pose estimation is to estimate the orientation of the face from a single RGB image. It plays an important role in many real-world applications, including driver's monitoring system [16,33], human-computer interaction [4,29] and face alignment [3,44]. With the recent advance of deep learning in computer vision [5,6,19,25,26,43], learning-based facial pose estimation has become

---

Z. Cao and D. Liu—Equal contributions.

Q. Wang—The analysis and all work described in this paper was performed by the authors at Purdue and RIT. Qifan Wang served as an advisor to the project.

S. Avidan et al. (Eds.): ECCV 2022, LNCS 13672, pp. 737–753, 2022.
https://doi.org/10.1007/978-3-031-19775-8_43

a dominant approach, achieving promising results [1,2,35,38]. However, as a general problem in deep learning, data shortage also limits the concurrent methods for facial pose estimation to achieve superior performance. How to effectively estimate the facial pose with limited data remains a challenge, which is the focus of this work.

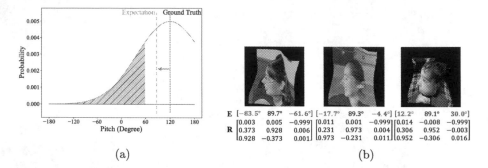

(a)   (b)

Fig. 1. (a) **Example of Biased Expectation** for 1D Gaussian label distribution. The distribution from original pitch $= 120°$ in the range $(-180°, +180°]$ gives a biased expectation of the label. The condition becomes worse as the original angle gets closer to $180°$ or $-180°$. (b) **E** and **R** denote *Euler angles* and *rotation matrix*, respectively. Euler angles have inconsistent representations for profile faces. The Red and **Blue** values show evident discrepancies when denoting similar profile poses. Applying LDL on Euler angles inevitably introduces heavily noisy supervision. In contrast to Euler angles, the corresponding elements in rotation matrices are close to each other. All samples are from the 300W-LP dataset [45]. (Color figure online)

Recently, label distribution learning (LDL) has been introduced to address the issue of insufficient training data. These LDL methods aim at reconstructing new labels of the distribution around the original ones for training, which promote the learning of facial images not only from their own labels but also the adjacent ones. LDL has shown its effectiveness in tasks such as facial age estimation [15], facial attractiveness estimation [9] and crowd counting [49]. However, the exploration of LDL application to facial pose estimation is insufficient.

To date, LDL in the task of facial pose estimation are mainly applied on Euler angles which are known as pitch, yaw and roll. A seminal work from [14] proposed to use a 2D Gaussian Distribution to describe the probability distribution between pitch and yaw in the range of $(-90°, +90°)$. Liu *et al.* [28] followed the track and converted each Euler angle label to a 1D Gaussian distribution. They also expand the task to the one of wild range. Therefore, each face image corresponds to three 1D Gaussian distributions (*i.e.*, ρ_{pitch}, ρ_{yaw} and ρ_{roll}). For instance, an original label of the pitch angle ($120°$) can be used to generate a Gaussian distribution in $(-180°, 180°]$ (see Fig. 1a). Through predicting the probability of each integer degree in set $\mathcal{S} = \{-179°, -178°, \cdots, 179°, 180°\}$ and compute the cross entropy loss, the task can be considered as the combination of both regression and classification.

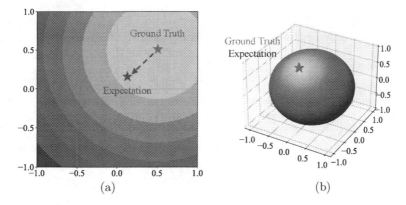

Fig. 2. (a) 2D distribution [2] generated by the first two elements of a column vector of rotation matrices (we omit the third element for visualization). The limit of range $[-1, 1]$ in two directions results in biased expectation. **(b) Spherical Gaussian Distribution** (ours) guarantees *unbiased expectation*.

Albeit being simple and effective, applying LDL on Euler angles has several obstacles: ① Euler angle is not a continuous rotation representation and LDL deteriorates the issue by contributing the learning of adjacent angles. The discontinuity is embodied in the Euler angle labels of profile faces (see Fig. 1b). Since similar profile images have very different Euler angle labels, converting the angles to distributions cannot help the learning of adjacent labels; ② Gaussian distribution labeling on Euler angles leads to biased expectations. Since the angle is limited to a certain range $(-180°, 180°]$, probabilities assigned in the shadow area make the expectation of labels incorrectly shift to left (see Fig. 1a); and ③ concurrent LDL methods utilize the variance of Gaussian distribution as a hyper-parameter which is fixed during training. This is computationally inefficient because they need to perform an exhaustive search to cherry pick the best parameter setting. Besides, using the same distribution for all the poses is not aligned with the real situation. Since faces at different poses have diverse contributions to adjacent faces, the network should learn the distribution parameters adaptively.

The first issue was studied in [2], which identifies the discontinuity issue of Euler angle and proposed a vector-based representation to train the network. In other words, they let the neural network learn the rotation matrices from facial images. Rotation matrices can form a continuous special orthogonal group $SO(3)$ and can circumvent the problem of discontinuity. However, they still failed to recognize the issue of biased expectation. Since every element of the rotation matrix stays in the range of $[-1, 1]$, they convert each element to a Gaussian distribution in range $[-1, 1]$ and let the network learn the distribution in an element-wise manner. Consequently, the issue of biased expectation is inevitably similar to Euler angles. To our best knowledge, the second and third issues remain largely under-explored in existing literature.

In light of the foregoing discussions, we are motivated to present our Anisotropic Spherical Gaussian (ASG)-based label distribution learning method for facial pose estimation. Specifically, we treat each column vector of the rotation matrix as an entity and map them to a spherical Gaussian distribution respectively. Due to the symmetric distribution of ASG, our approach guarantees for an unbiased expectation of label distributions. The difference between ASG and the method in [2] is demonstrated in Fig. 2. Armed with the spherical Gaussian distribution, we further design a new loss function for the network to learn the distribution parameters adaptively during the training stage. This enables every facial image to adjust contribution to adjacent poses based on its pose. Ablation studies show that it can transcend the cherry-picked parameter by at least 4.0% when trained on 3000W-LP and tested on AFLW2000 dataset.

Our method enjoys a few attractive qualities: ❶ it ensures the network learns the distribution with **unbiased expectation**. Since most existing methods have biased expectations (unless the original ground truth is exactly in the middle), we observe significant performance gain from our method; ❷ the capacity of learnable ASG distribution parameters allows the network to adjust the parameter for each pose, enabling a **fine-grained** prediction; ❸ all the performance achievement comes from optimization on **representation of rotation** without increasing the size of neural networks. Our approach achieves state-of-the-art performance with a very light-weighted backbone network, *i.e.*, ResNet18 [19]. Specifically, we decrease the Mean Absolute Error (MAE) by 0.27° (6.9% ↓) compared to [1] and 0.19° (5.0% ↓) compared to [38] when tested on AFLW2000 dataset [51]; and ❹, our method is the **first attempt** that adopts directional statistics in the task of pose estimation. We believe it can help invoke more thoughts for further exploration in the community. Our contributions are summarized below:

- We propose a novel ASG-LDL method which encodes each column vector of the rotation matrix as an anisotropic spherical gaussian on a unit sphere. Our method addresses the issue of biased expectation that is under-explored in previous works.
- We propose a novel training paradigm that allows the network to learn the distribution parameters adaptively. The flexibility allows the network to learn individual distribution parameters for each pose.
- We conduct extensive experiments on two benchmarks. Experimental results show the effectiveness of our method. With a light-weight ResNet-18 as the backbone, our method achieves state-of-the-art results and outperforms many strong baselines with a heavier backbone (*i.e.*, ResNet-50).

2   Related Work

This section summarizes the recent progresses in the related fields regarding facial pose estimation, label distribution learning and spherical Gaussian distribution.

Facial Pose Estimation. Recently, landmark-free learning based methods have become popular. By training an end-to-end deep neural network, it can estimate the face poses using global information and can be more robust to the environment variations. [35] puts forward a CNN with a multi-loss function that performs binned classification to regress three Euler angles. [46] proposes a fine-grained structure by learning global spatial feature importance that improves the results. [20] formulates face pose estimation using quaternion-annotated labels to avoid the ambiguity problem in Euler angle representation. [1] proposes a Faster RCNN based network to regress 6DoF pose of faces by performing pose estimation and face alignment simultaneously. [38] puts forward a multi-modal network that can perform three tasks of head pose estimation, landmark-based face alignment and localization of face simultaneously. By the combination of three tasks they achieve state-of-the-art results. All of the above methods perform the training process through direct regression. Differently, we approach the problem as a label distribution learning task.

Label Distribution Learning. Label distribution learning [30] is a learning paradigm that is first proposed for facial age estimation [15,27]. [15] finds that faces at close ages look similar. Therefore, they map each face image to a label distribution which covers a certain number of ages. Through this way one face image can contribute to not only the learning of its chronological age, but also the learning of its adjacent ages. LDL also shows it effectiveness in similar tasks such as facial attractiveness estimation [9], crowd counting [49] and movie rating prediction [13] *etc.* [8] applies a similar approach on ordinal regression such as image ranking and monocular depth estimation. [2] shows that the evaluation metric, mean absolute error of Euler angles (MAE), cannot reflect the actual performance especially for profile faces. Instead, they propose to use mean absolute error of vectors (MAEV) as a new metric. However, all the methods give biased expectation from the distribution, which severely limits the performance of neural network.

Spherical Gaussian Distribution. Spherical Gaussian (SG) distribution, also known as von Mises-Fisher distribution [11], is commonly used to simulate the properties of illumination and reflection in computer graphics. [18] uses SG to estimate multiple light sources and reflectance properties. [39] approximates the normal distribution function (NDF) by a mixture of SGs. [7] uses SG for the approximation of Bidirectional Transmittance Distribution Function (BTDF) for real-time estimation of environment lighting. However, SG only describes an isotropic distribution. [42] further proposes the ASG distribution which can describe an anisotropic distribution for rendering applications. Inspired by the above work, we successfully extend ASG to the field of head pose estimation.

6D Object Pose Estimation. 6D object pose estimation includes estimation of 3D location and 3D orientation. The latter task resembles our head pose estimation. The approaches for 6D object pose estimation can be generally classified into two categories. The first type such as [34,37,47] first capture instance information and keypoints from images to determine locations of objects, then build

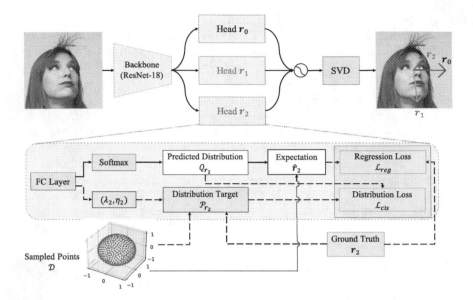

Fig. 3. The overall framework. The dashed lines are only used in the training stage. For simplicity, we only visualize one head r_2 but each head has the same working pipeline.

the correspondence between the 2D and 3D keypoints. After that, they obtain 6D pose estimation by solving the PnP problem [24]. The other category of methods such as [12,31,41] use neural networks to estimate orientation of objects directly. To our knowledge, the use of spherical Gaussian is a new attempt in the task of pose estimation.

3   Proposed Method

3.1   Overview

Our overall framework is illustrated in Fig. 3. The network learns the pose information from a cropped human facial image. To demonstrate the advantage of our approach, we choose light-weighted ResNet-18 as our backbone network. We append three heads to the ResNet-18 backbone as each head corresponds to one pose vector. They work collectively to perform the facial pose estimation. During training, the backbone first extracts the features from the input image and then feeds them to each of the heads, which is supervised by the classification and regression loss respectively. During inference, the three heads work collaboratively to predict the rotation matrix through singular value decomposition (SVD). We elaborate our method in the following sections.

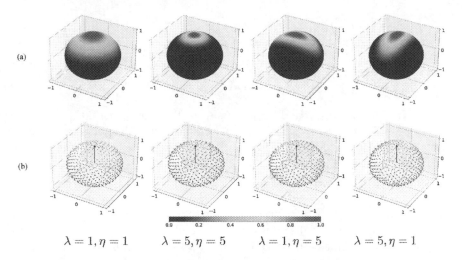

(a)

(b)

$\lambda = 1, \eta = 1$ $\lambda = 5, \eta = 5$ $\lambda = 1, \eta = 5$ $\lambda = 5, \eta = 1$

Fig. 4. Visualization of ASG distributions of different λ and η when $r = [0, 0, 1]^T$. a) ASG distribution on a unit sphere; b) visualization of sampled points with probabilities when $M = 600$.

3.2 Motivation

When we use a rotation matrix $\mathbf{R}_{3\times3} = [\mathbf{r}_0, \mathbf{r}_1, \mathbf{r}_2]$ to describe facial poses, the three column vectors $\mathbf{r}_0$, $\mathbf{r}_1$, $\mathbf{r}_2$ are equivalent to three pose vectors in Fig. 3, $i.e.$ left (blue), down (green) and front (red) vectors respectively [2]. Therefore, for a ground truth pose vector $\mathbf{r}_i, i = \{0, 1, 2\}$, any direction $\mathbf{v}$ surrounding it can also be regarded as an alternative legitimate label. The smaller the angle difference is, the more likely that the vector $\mathbf{v}$ is a valid label. Therefore, all the probabilities of a direction $\mathbf{v}$, that can be considered as a legitimate label, constitute a probability distribution on a unit sphere. Intuitively, the probability distribution can be represented using an isotropic spherical Gaussian (SG) model, since the probability is only related to the angle between $\mathbf{v}$ and $\mathbf{r}_i$. However, human faces change at different rates when rotating along different axes. For example, rolling a face with 45° does not change the observed area of the face, while nodding or raising the face for 45° makes large portion of facial area self-occluded. Based on this observation, we propose to use ASG distribution which is able to capture the anisotropic features along different axes.

3.3 Label Distribution Construction

All three pose vectors constitute an orthogonal coordinate system. For each ground truth pose vector $\mathbf{r}_i$, we can calculate the portion G^i that a direction $\mathbf{v}$ accounts for a full class description of the sample:

$$G^i(\boldsymbol{v}; \boldsymbol{R}, [\lambda, \eta]) = c \cdot \mathrm{S}(\boldsymbol{v}; \boldsymbol{r}_i) \cdot e^{-\lambda(\boldsymbol{v} \cdot \boldsymbol{r}_j)^2 - \eta(\boldsymbol{v} \cdot \boldsymbol{r}_k)^2}$$
$$\text{where } i = \{0, 1, 2\}$$
$$j = (i + 1) \bmod 3$$
$$k = (i + 2) \bmod 3. \tag{1}$$

Here, $\boldsymbol{R} = [\boldsymbol{r_0}, \boldsymbol{r_1}, \boldsymbol{r_2}]$. λ and η are the parameters that control the decreasing speed of possibility along $\boldsymbol{r}_j$ and $\boldsymbol{r}_k$. Figure 4a illustrates the spherical Gaussian distribution of different λ and η. c is the normalization term that ensures the sum of probability distribution to be 1. $S(\boldsymbol{v}; \boldsymbol{r}_i) = \max(\boldsymbol{v} \cdot \boldsymbol{r}_i, 0)$ is the smooth term. Since the exponential part $B(\boldsymbol{v}) = e^{-\lambda(\boldsymbol{v} \cdot \boldsymbol{r}_j)^2 - \mu(\boldsymbol{v} \cdot \boldsymbol{r}_k)^2}$, also known as Bingham distribution [23], is antipodally symmetric and has two peaks at $\boldsymbol{v} = \pm\boldsymbol{r}_i$. We keep only the peak of $\boldsymbol{v} = \boldsymbol{r}_i$ with the smooth term $S(\boldsymbol{v}; \boldsymbol{r}_i)$.

To convert a vector to a distribution, we first adopt spherical Fibonacci lattice algorithm [17] to sample M near-equidistant points from an unit sphere, denoted by $\mathcal{D} = \{\boldsymbol{d}_1, \boldsymbol{d}_2, \cdots, \boldsymbol{d}_M\}$ where $\boldsymbol{d}_i \in \mathbb{R}^3$ (see Fig. 3). Note that we only perform the sampling once, thus all pose vectors share a same set of sampled points. During the training stage, for any ground truth vector label $\boldsymbol{r}_i$, the network first predicts parameters λ and η and then use them to calculate the probabilities for all the sampled points $\mathcal{P}_{\boldsymbol{r}_i} = \{p_1^i, p_2^i, \cdots, p_M^i\}$. The probability of point k can be obtained by the normalization:

$$p_k^i = \frac{\exp\{\mathrm{G}^i(\boldsymbol{v}_k; \boldsymbol{R}, [\lambda, \eta]\}}{\sum_{j=1}^{M} \exp\{\mathrm{G}^i(\boldsymbol{v}_j; \boldsymbol{R}, [\lambda, \eta]\}}. \tag{2}$$

The process of label distribution generation is applied on all three column vectors $\boldsymbol{r}_0, \boldsymbol{r}_1$ and $\boldsymbol{r}_2$. Therefore, we can obtain three sets of probability distribution $\mathcal{P}_{\boldsymbol{r}_0}, \mathcal{P}_{\boldsymbol{r}_1}$ and $\mathcal{P}_{\boldsymbol{r}_2}$ with the same size of M. The probability distribution on sampled points are visualized in Fig. 4b.

3.4   Working Pipeline

Training. In the training stage, the backbone-encoded features are first fed into three heads separately (See Fig. 3). Each head has one fully connected (FC) layer, which outputs a vector with size of $M+2$. The first M elements denote the ASG probabilities of sampled points for the corresponding pose vector, which is normalized by a softmax layer to generate the probability distribution $\mathcal{Q}_{\boldsymbol{r}_i} = \{q_1^i, q_2^i, \cdots, q_M^i\}$. Therefore the expectation of the distribution is given by:

$$\hat{\boldsymbol{r}}_i = \mathbb{E}_{\mathcal{Q}_{\boldsymbol{r}_i}}[\mathcal{D}] = \sum_{k=1}^{M} q_k^i \boldsymbol{d}_k. \tag{3}$$

The last two elements of the output vector from the FC layer correspond to the parameters (λ_i, η_i). In conjunction with the sampled point set $\mathcal{D}$ and the ground truth vector r_i, the network is able to generate the distribution target $\mathcal{P}_{r_i}$ using Eq. 1 and 2.

Loss Function. To supervise our method, our training loss consists of two terms: classification loss $\mathcal{L}_{cls}$ and regression loss $\mathcal{L}_{reg}$. The overall loss $\mathcal{L}$ is given by:

$$\mathcal{L} = \mathcal{L}_{cls} + \alpha L_{reg}. \tag{4}$$

More concretely, we adopt mean square error (MSE) loss function for regression $\mathcal{L}_{reg} = \text{MSE}\,(r_i, \hat{r}_i)$ and Kullback-Liebler (KL) divergence for classification $\mathcal{L}_{cls} = \text{D}_{\text{KL}}(\mathcal{P}_{r_i}\|\mathcal{Q}_{r_i})$. The value of the trade-off parameter α is in the range of $[0, 1]$. We find its optimal value in our experiments.

Inference. In the inference stage, we first concatenate the three pose vectors $\hat{r}_0, \hat{r}_1, \hat{r}_2$ generated by the three heads from the learned network to obtain matrix $\hat{R} = [\hat{r}_0, \hat{r}_1, \hat{r}_2]$. We then obtain its closest rotation matrix through singular value decomposition (SVD). Given a matrix $\hat{R} = U\Sigma V^T$, its closest rotation matrix is obtained by $R = UV^T$.

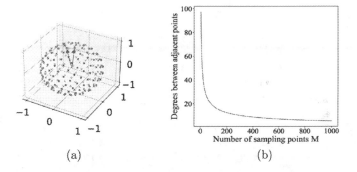

(a)   (b)

Fig. 5. (a) Visualization of the angle between two adjacent points. **(b)** Relationship between the number of sampling points M and the angle between two adjacent points.

4 Experiments

4.1 Datasets and Metrics

We conduct an extensive set of experiments to evaluate our approach on three benchmarks: 300W-LP [51], AFLW2000 [52] and BIWI [10]. **300W-LP** is a synthesized dataset which contains 122,450 images with large varieties in facial poses and identities. Image samples in 300W-LP are synthesized from 300W dataset [36] which includes around 4,000 images. **AFLW2000** contains the first 2,000 images of the popular AFLW [32] dataset with diverse facial poses in the wild. The dataset is commonly used as the test set to evaluate model performances. **BIWI** is collected in an indoor environment with an RGB-D camera. It provides accurate ground truth labels. This dataset is also widely used for depth-based facial pose estimation. Since bounding boxes of human heads are not provided in BIWI, we use MTCNN [48] to detect and crop the face areas.

To ensure a fair comparison with different methods, we follow the same experiment scenarios applied in [20,35,46] and discard the test samples with Euler angles beyond the range of $[-99°, 99°]$. **Scenario 1:** We train our network on 300W-LP and test on both AFLW2000 and BIWI datasets. **Scenario 2:** We perform the 3-fold cross validation on the BIWI dataset. We randomly split the BIWI dataset into 3 groups. Each group contains 8 videos and the videos of the same person appear only in one group. We use mean absolute error of Euler angles (MAE) as our metric.

Table 1. MAE and MAEV results of **different representations** of rotation under scenario 1 and 2. All use the ResNet-18 as backbone. We highlight the **best** results.

Train	Test	Representation	Euler Angle Errors				Vector Errors			
			Pitch	Yaw	Roll	MAE	Left	down	front	MAEV
300W-LP	AFLW 2000	Euler angles	6.36	4.64	4.84	5.28	6.71	5.97	7.62	6.76
		Lie algebra	5.62	3.92	4.04	4.52	5.84	5.13	6.52	5.83
		Quaternion	5.77	4.01	4.20	4.66	5.63	5.62	6.57	5.94
		Rotation matrix	**5.46**	**3.71**	**3.77**	**4.31**	**5.52**	**4.97**	**5.92**	**5.47**
	BIWI (all)	Euler angles	6.43	4.22	4.08	4.91	6.08	5.72	6.13	5.98
		Lie algebra	5.87	**3.39**	3.73	4.33	5.82	5.66	5.42	5.63
		Quaternion	6.11	3.54	**3.61**	4.42	5.79	5.88	5.61	5.76
		Rotation matrix	**5.43**	3.52	3.63	**4.19**	**5.74**	**5.10**	**5.12**	**5.32**
BIWI (70%)	BIWI (30%)	Euler angles	4.07	3.76	3.73	3.85	4.52	4.89	4.57	4.66
		Lie algebra	3.46	3.21	3.11	3.26	4.31	4.22	4.18	4.24
		Quaternion	3.52	3.35	3.24	3.37	4.51	4.32	4.20	4.34
		Rotation matrix	**3.08**	**3.16**	**3.01**	**3.08**	**4.12**	**4.16**	**4.02**	**4.10**

4.2   Implementation Detail

There are two hyper-parameters in our approach. One is the coefficient α for the regression loss term $\mathcal{L}_{reg}$. Another one is the number of sampled points M. Figure 5 shows the relationship between number of sampled points M and angle between adjacent points. We set $\alpha = 0.2$ and $M = 600$ in our experiments.

We implement our proposed approach based on PyTorch and adopt ResNet-18 [19] as the backbone. In training, we adopt Adam optimizer with the initial learning rate of 0.0001. The total training epoch is set to be 50 with the decay rate of 0.95 for every epoch. Batch size is set to be 64 and every image is resized to 224×224. All the experiments are conducted on a RTX 2080 Ti GPU.

We augment training images with random crop, noise and random zoom with scale from 0.8 to 1.2.

Table 2. Comparison with state-of-the-art methods on the AFLW2000 and BIWI datasets. All methods are trained on 300W-LP. We highlight the **best** results and **our** results.

Method	Backbone	AFLW2000				BIWI (full)			
		Pitch	Yaw	Roll	MAE	Pitch	Yaw	Roll	MAE
3DDFA [51]	Two-stream	27.09	4.71	28.43	20.08	41.90	5.50	13.22	20.21
Dlib [22]	-	11.25	8.49	22.83	14.19	13.00	11.86	19.56	14.81
HPE [21]	ResNet-50	6.18	4.87	4.80	5.28	5.18	4.57	3.12	4.29
Hopenet [35]	ResNet-50	7.12	5.31	6.13	6.19	5.89	6.01	3.72	5.20
Quatnet [20]	GoogLeNet	5.62	3.97	3.92	4.50	5.49	4.01	2.94	4.15
Liu *et al.* [28]	ResNet-50	5.06	**3.03**	3.68	3.93	5.61	4.12	3.15	4.29
FSA-Net [46]	SSR-Net	6.34	4.96	4.78	5.36	5.21	4.56	3.07	4.28
TriNet [2]	ResNet-50	5.77	4.20	4.04	4.67	4.75	**3.05**	4.11	3.97
MNN [38]	Encoder-Decoder	**4.69**	3.34	3.48	3.83	4.61	3.98	**2.39**	3.66
img2pose [1]	ResNet-18	5.03	3.43	3.28	3.91	3.55	4.57	3.24	3.79
Ours	ResNet-18	**4.74**	3.08	**3.11**	**3.64**	**3.52**	4.21	**3.10**	**3.61**

4.3 Analysis of Rotation Representations

Even though there are multiple ways to describe a rotation and the most commonly used ones include Euler angles, quaternion, Lie algebra and rotation matrices, it remains under-studied that which representation is the best option for the task of facial pose estimation. [2] briefly discussed Euler Angle and quaternions. However, they omitted Lie algebra and did not provide any experimental support. We implement a thorough comparison between the performances of different representations using the same backbone of ResNet-18 (see Table 1). Since MAE is not an accurate measure for profile faces, we also adopt mean absolute error of vectors (MAEV) to make a comprehensive comparison. Experiments show that rotation matrices achieve the best result among all representations under both scenarios.

The experimental results accord with the continuity properties of each representation. As shown by the work [40,50], it needs at least 5 dimensions to describe the rotation continuously, otherwise it incurs discontinuity issue similar to Euler angles. Both Euler angle and Lie algebra $\in \mathbb{R}^3$ and quaternion $\in \mathbb{R}^4$. Therefore, none of them can describe the rotation continuously. Here we include some cases when the phenomenons of discontinuity occur. For a unit quaternion $q = w + xi + yj + zk$, where $w^2 + x^2 + y^2 + z^2 = 1$. Then $(1, 0, 0, 0)$ and $(-1, 0, 0, 0)$ represents the same rotation. For Lie algebra $\mathfrak{so}(3)$ which is denoted by an anti-symmetric matrix $\phi^\wedge$ where $\phi = \theta a$ and $a \in \mathbb{R}^3$, $||a||_2 = 1, \theta \in [-\pi, +\pi]$. For any a, faces have similar appearances when θ approaches π and $-\pi$. Therefore, the rotation matrix is the best representation in terms of the performance and continuity property.

Table 3. Comparison between **direct regression** and **distribution learning**. Results are obtained on the AFLW2000 and BIWI benchmarks.

Training set	300W-LP		BIWI (70%)
Testing set	ALFW2000	BIWI (full)	BIWI (30%)
Direct regression	4.31	4.19	3.08
SG learning	3.79	3.71	2.93
ASG learning	**3.64**	**3.61**	**2.77**

4.4   Comparison with State-of-the-Arts

We compare the performance of our method with other state-of-the-art methods (see Table 2) under scenario 1. Since the training/test set division in scenario 2 is arbitrary and thus is not adopted by methods [1,38], we choose only scenario 1 for comparison. The results of the compared methods are directly cited from their original papers. Liu *et al.* [28] is the first work that follows the distribution learning paradigm for wild pose estimation. Different from our work, they convert the Euler angles to 3 Gaussian distributions with the same variance. Even though they use the ResNet-50 as the backbone which is deeper than our ResNet-18, our ASG-based distribution learning surpasses their performance. FSA-Net [46] and TriNet [2] take advantage of the combination of attention module and capsule network and append them to the backbone network to improve the learning ability of the network. Even though both have more complex structures and more parameters than our network, their performance is inferior to ours.

It is worth mentioning that some of the methods such as MNN [38] and img2pose [1] also use face landmarks in a weakly supervised manner to help improve the performance of network. To highlight the effectiveness of our distri-

Fig. 6. Qualitative comparison of different methods. Trained on 300W-LP and tested on AFLW2000.

bution learning strategy, we make the network learn the pose estimation without the landmark labels. Experiments show that even though less information is provided, our network still achieves better performance. The qualitative results are demonstrated in Fig. 6. It can be seen that our approach makes more accurate predictions when faces are partially occluded.

Table 4. Comparison between adaptive **parameters learning** and **fixed** ASG parameters. Results are obtained on the AFLW2000 and BIWI benchmarks.

Training set	300W-LP		BIWI (70%)
Testing set	AFLW2000	BIWI (full)	BIWI (30%)
$\lambda = 1, \eta = 1$	3.79	3.84	2.92
$\lambda = 5, \eta = 5$	3.86	3.97	3.03
$\lambda = 1, \eta = 5$	3.92	4.03	2.97
Adaptive parameters	**3.64**	**3.61**	**2.77**

Table 5. Comparison of the effects of **different loss terms**. Results are obtained on the AFLW2000 and BIWI benchmarks.

Training set	300W-LP		BIWI (70%)
Testing set	ALFW2000	BIWI (full)	BIWI (30%)
$\mathcal{L}_{cls}$	3.67	3.68	2.81
$\mathcal{L}_{reg}$	4.31	4.19	3.08
$\mathcal{L}_{cls} + \mathcal{L}_{reg}$	**3.64**	**3.61**	**2.77**

4.5 Ablation Study

In this section, we investigate the effectiveness of our method by carrying out ablation experiments on the adaptive ASG label distribution learning and different loss components.

Distribution Learning vs. Regression. We examine the advantages of the ASG distribution to isotropic SG distribution and use the direct regression of the rotation matrix as baseline. Results are shown in Table 3. Our ASG distribution can effectively improve the performance compared with other two baseline methods.

Adaptive Parameters vs. Fixed Parameters. We conduct experiments to compare the performance of methods with adaptive parameters and fixed parameters (see Table 4). While achieving superior performance over the fixed parameters, our adaptive parameter learning is computationally efficient as it avoids the exhaustive search of parameters. All the ASG parameters λ and η learned by the samples in the 300W-LP dataset are demonstrated in Fig. 7. We

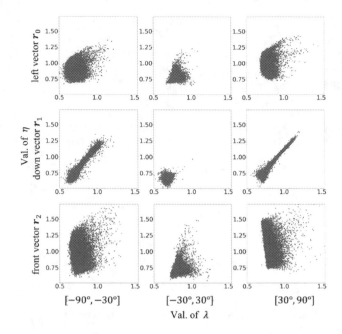

Fig. 7. Visualization of η and λ distribution for three pose vectors r_0, r_1 and r_2 of all the samples in different ranges of Yaw. All trained on the 300W-LP dataset.

divide the angle of yaw into three equal ranges. It is worth noting that our learning behavior follows a clear pattern. For instance, the parameter distributions in the first and third columns resemble each other. Because turning faces to left and right results in symmetric images, the parameters of ASG should be similar. This is reflected by the distribution of parameters.

Loss Functions. We examine the effectiveness of each loss term (see Table 5). Notice when only $\mathcal{L}_{reg}$ is applied, the network is supervised by the arbitrary distribution with expectation of the same as the ground truth. The above results confirm that the classification term and regression term can work collaboratively to operate effective label distribution learning for the facial pose estimation.

5   Conclusion

In this paper, we introduce a novel ASG-based Label Distribution Learning method for estimating facial pose. This is the first attempt to include directional statistics in the estimation of pose. We anticipate that this work will illustrate potential future directions for the community to investigate.

References

1. Albiero, V., Chen, X., Yin, X., Pang, G., Hassner, T.: img2pose: face alignment and detection via 6DoF, face pose estimation. In: CVPR (2021)

2. Cao, Z., Chu, Z., Liu, D., Chen, Y.: A vector-based representation to enhance head pose estimation. In: WACV (2021)
3. Chang, F.J., Tuan Tran, A., Hassner, T., Masi, I., Nevatia, R., Medioni, G.: Face-poseNet: making a case for landmark-free face alignment. In: ICCV Workshops (2017)
4. Chen, Z., Liu, Z., Hu, H., Bai, J., Lian, S., Shi, F., Wang, K.: A realistic face-to-face conversation system based on deep neural networks. In: ICCV (2019)
5. Cheng, Z., et al.: Physical attack on monocular depth estimation with optimal adversarial patches. In: ECCV (2022)
6. Cui, Y., Yan, L., Cao, Z., Liu, D.: TF-blender: temporal feature blender for video object detection. In: ICCV (2021)
7. De Rousiers, C., Bousseau, A., Subr, K., Holzschuch, N., Ramamoorthi, R.: Real-time rough refraction. In: Symposium on Interactive 3D Graphics and Games, pp. 111–118 (2011)
8. Diaz, R., Marathe, A.: Soft labels for ordinal regression. In: CVPR (2019)
9. Fan, Y.Y., et al.: Label distribution-based facial attractiveness computation by deep residual learning. IEEE Trans. Multimedia $20(8)$, 2196–2208 (2017)
10. Fanelli, G., Dantone, M., Gall, J., Fossati, A., Van Gool, L.: Random forests for real time 3D face analysis. Int. J. Comput. Vis. $101(3)$, 437–458 (2013)
11. Fisher, R.A.: Dispersion on a sphere. Proc. R. Soc. London Ser. A Math. Phys. Sci. $217(1130)$, 295–305 (1953)
12. Gao, G., Lauri, M., Zhang, J., Frintrop, S.: Occlusion resistant object rotation regression from point cloud segments. In: Leal-Taixé, L., Roth, S. (eds.) ECCV 2018. LNCS, vol. 11129, pp. 716–729. Springer, Cham (2019). https://doi.org/10.1007/978-3-030-11009-3_44
13. Geng, X., Hou, P.: Pre-release prediction of crowd opinion on movies by label distribution learning. In: IJCAI (2015)
14. Geng, X., Xia, Y.: Head pose estimation based on multivariate label distribution. In: CVPR (2014)
15. Geng, X., Yin, C., Zhou, Z.H.: Facial age estimation by learning from label distributions. IEEE Trans. Pattern Anal. Mach. Intell. $35(10)$, 2401–2412 (2013). https://doi.org/10.1109/TPAMI.2013.51
16. Geronimo, D., Lopez, A.M., Sappa, A.D., Graf, T.: Survey of pedestrian detection for advanced driver assistance systems. IEEE Trans. Pattern Anal. Mach. Intell. $32(7)$, 1239–1258 (2009)
17. González, Á.: Measurement of areas on a sphere using Fibonacci and latitude-longitude lattices. Math. Geosci. $42(1)$, 49–64 (2010)
18. Hara, K., Nishino, K., Ikeuchi, K.: Multiple light sources and reflectance property estimation based on a mixture of spherical distributions. In: ICCV (2005)
19. He, K., Zhang, X., Ren, S., Sun, J.: Deep residual learning for image recognition. In: CVPR (2016)
20. Hsu, H.W., Wu, T.Y., Wan, S., Wong, W.H., Lee, C.Y.: QuatNet: quaternion-based head pose estimation with multiregression loss. IEEE Trans. Multimedia $21(4)$, 1035–1046 (2018)
21. Huang, B., Chen, R., Xu, W., Zhou, Q.: Improving head pose estimation using two-stage ensembles with top-k regression. Image Vis. Comput. 93, 103827 (2020)
22. Kazemi, V., Sullivan, J.: One millisecond face alignment with an ensemble of regression trees. In: CVPR (2014)
23. Kent, J.T.: The Fisher-Bingham distribution on the sphere. J. R. Stat. Soc. Ser. B (Methodol.) $44(1)$, 71–80 (1982)

24. Lepetit, V., Moreno-Noguer, F., Fua, P.: EPnP: an accurate O(n) solution to the PnP problem. Int. J. Comput. Vis. **81**(2), 155 (2009)
25. Liu, D., Cui, Y., Tan, W., Chen, Y.: SG-Net: spatial granularity network for one-stage video instance segmentation. In: CVPR (2021)
26. Liu, et al..: DenserNet: weakly supervised visual localization using multi-scale feature aggregation. In: AAAI (2021)
27. Liu, X., et al.: AgeNet: deeply learned regressor and classifier for robust apparent age estimation. In: ICCVW (2015)
28. Liu, Z., Chen, Z., Bai, J., Li, S., Lian, S.: Facial pose estimation by deep learning from label distributions. In: CVPR Workshops (2019)
29. Liu, Z., Hu, H., Wang, Z., Wang, K., Bai, J., Lian, S.: Video synthesis of human upper body with realistic face. In: 2019 IEEE International Symposium on Mixed and Augmented Reality Adjunct (ISMAR-Adjunct), pp. 200–202. IEEE (2019)
30. Liu, Z., et al.: Unveiling the power of mixup for stronger classifiers. arXiv preprint arXiv:2103.13027 (2021)
31. Mahendran, S., Ali, H., Vidal, R.: 3D pose regression using convolutional neural networks. In: ICCV Workshops (2017)
32. Koestinger, M., Wohlhart, P., Roth, P.M., Bischof, H.: Annotated facial landmarks in the wild: a large-scale, real-world database for facial landmark localization. In: Proceedings of the First IEEE International Workshop on Benchmarking Facial Image Analysis Technologies (2011)
33. Murphy-Chutorian, E., Doshi, A., Trivedi, M.M.: Head pose estimation for driver assistance systems: a robust algorithm and experimental evaluation. In: 2007 IEEE Intelligent Transportation Systems Conference, pp. 709–714. IEEE (2007)
34. Peng, S., Liu, Y., Huang, Q., Zhou, X., Bao, H.: PVNet: pixel-wise voting network for 6DoF pose estimation. In: CVPR (2019)
35. Ruiz, N., Chong, E., Rehg, J.M.: Fine-grained head pose estimation without key-points. In: CVPR Workshops (2018)
36. Sagonas, C., Tzimiropoulos, G., Zafeiriou, S., Pantic, M.: 300 faces in-the-wild challenge: the first facial landmark localization challenge. In: ICCV Workshops (2013)
37. Song, C., Song, J., Huang, Q.: HybridPose: 6D object pose estimation under hybrid representations. In: CVPR (2020)
38. Valle, R., Buenaposada, J.M., Baumela, L.: Multi-task head pose estimation in-the-wild. IEEE Trans. Pattern Anal. Mach. Intell. **43**, 2874–2881 (2020)
39. Wang, J., Ren, P., Gong, M., Snyder, J., Guo, B.: All-frequency rendering of dynamic, spatially-varying reflectance. In: ACM SIGGRAPH Asia 2009 papers, pp. 1–10 (2009)
40. Xiang, S.: Eliminating topological errors in neural network rotation estimation using self-selecting ensembles. ACM Trans. Graph. (TOG) **40**(4), 1–21 (2021)
41. Xiang, Y., Schmidt, T., Narayanan, V., Fox, D.: PoseCNN: a convolutional neural network for 6D object pose estimation in cluttered scenes. arXiv preprint arXiv:1711.00199 (2017)
42. Xu, K., Sun, W.L., Dong, Z., Zhao, D.Y., Wu, R.D., Hu, S.M.: Anisotropic spherical gaussians. ACM Trans. Graph. (TOG) **32**(6), 1–11 (2013)
43. Yan, L., et al.: GL-RG: global-local representation granularity for video captioning. In: IJCAI (2022)
44. Yang, H., Mou, W., Zhang, Y., Patras, I., Gunes, H., Robinson, P.: Face alignment assisted by head pose estimation. arXiv preprint arXiv:1507.03148 (2015)
45. Yang, S., Luo, P., Loy, C.C., Tang, X.: Wider face: a face detection benchmark. In: CVPR (2016)

46. Yang, T.Y., Chen, Y.T., Lin, Y.Y., Chuang, Y.Y.: FSA-net: learning fine-grained structure aggregation for head pose estimation from a single image. In: CVPR (2019)
47. Zakharov, S., Shugurov, I., Ilic, S.: DPOD: 6D pose object detector and refiner. In: ICCV (2019)
48. Zhang, K., Zhang, Z., Li, Z., Qiao, Y.: Joint face detection and alignment using multitask cascaded convolutional networks. IEEE Signal Process. Lett. **23**(10), 1499–1503 (2016)
49. Zhang, Z., Wang, M., Geng, X.: Crowd counting in public video surveillance by label distribution learning. Neurocomputing **166**, 151–163 (2015)
50. Zhou, Y., Barnes, C., Lu, J., Yang, J., Li, H.: On the continuity of rotation representations in neural networks. In: CVPR (2019)
51. Zhu, X., Lei, Z., Liu, X., Shi, H., Li, S.Z.: Face alignment across large poses: a 3D solution. In: CVPR (2016)
52. Zhu, X., Lei, Z., Yan, J., Yi, D., Li, S.Z.: High-fidelity pose and expression normalization for face recognition in the wild. In: CVPR (2015)

Author Index

Printed in the United States
by Baker & Taylor Publisher Services